AF788117

Petroleum and Basin Evolution

Springer
Berlin
Heidelberg
New York
Barcelona
Budapest
Hong Kong
London
Milan
Paris
Santa Clara
Singapore
Tokyo

D.H. Welte   B. Horsfield   D.R. Baker   (Eds.)

Petroleum and Basin Evolution

Insights from Petroleum Geochemistry, Geology and Basin Modeling

With 214 Figures and 38 Tables

Springer

Prof. Dr. Dr. h.c. Dietrich H. Welte
Institut für Erdöl und Organische Geochemie (ICG-4)
Forschungszentrum Jülich GmbH
52425 Jülich
Germany

Dr. Brian Horsfield
Institut für Erdöl und Organische Geochemie (ICG-4)
Forschungszentrum Jülich GmbH
52425 Jülich
Germany

Prof. Donald R. Baker
Department of Geology and Geophysics
The Wiess School of Natural Sciences
Rice University
Houston, TX 77251
USA

ISBN-13: 978-3-642-64400-9 Springer-Verlag Berlin Heidelberg New York

Library of Congress Cataloging-in-Publication Data. Petroleum and basin evolution/D.H. Welte, B. Horsfield, D.R. Baker, editors. p. cm. Includes bibliographical references and index.

ISBN-13: 978-3-642-64400-9 e-ISBN-13: 978-3-642-60423-2

DOI: 10.1007/978-3-642-60423-2

1. Petroleum–Geology.
2. Sedimentary basins–Mathematical models. 3. Geochemistry. I. Welte, D.H. (Dietrich H.), 1935- . II. Horsfield, B. (Brian), 1951- . III. Baker. Donald R., 1927- . TN870.5.P4747 1997 553.2′8–dc20 97-23073

This work is subject to copyright. All rights are reserved, whether the whole or part of the material is concerned, specifically the rights of translation, reprinting, reuse of illustrations, recitation, broadcasting, reproduction on microfilm or in any other ways, and storage in data banks. Duplication of this publication or parts thereof is permitted only under the provisions of the German Copyright Law of September 9, 1965, in its current version, and permission for use must always be obtained from Springer-Verlag. Violations are liable for prosecution under the German Copyright Law.

© Springer-Verlag Berlin Heidelberg 1997

Softcover reprint of the hardcover 1st edition 1997

The use of general descriptive names, registered names, trademarks, etc. in this publication does not imply, even in the absence of a specific statement, that such names are exempt from the relevant protective laws and regulations and therefore free for general use.

Cover design: Springer-Verlag, E. Kirchner

Typesetting: Scientific Publishing Services (P) Ltd, Madras

SPIN: 10009836 32/3136/SPS – 5 4 3 2 1 0 – Printed on acid-free paper

Preface

This book has been prepared by the collaborative effort of two somewhat separate technical groups: the researchers at the Institute for Petroleum and Organic Geochemistry, Forschungszentrum Jülich (KFA), and the technical staff of Integrated Exploration Systems (IES). One of us, Donald R. Baker, from Rice University, Houston, has spent so much time at KFA as a guest scientist and researcher that it is most appropriate for him to contribute to the book. During its more than 20-year history the KFA group has made numerous and significant contributions to the understanding of petroleum evolution. The KFA researchers have emphasized both the field and laboratory approaches to such important problems as source rock recognition and evaluation, oil and gas generation, maturation of organic matter, expulsion and migration of hydrocarbons, and crude oil composition and alteration. IES Jülich has been a leader in the development and application of numerical simulation (basin modeling) procedures. The cooperation between the two groups has resulted in a very fruitful synergy effect both in the development of modeling software and in its application.

The purpose of the present volume developed out of the 1994 publication by the American Association of Petroleum Geologists of a collection of individually authored papers entitled *The Petroleum System – From Source to Trap*, edited by L.B. Magoon and W.G. Dow. The idea was to describe the "petroleum systems approach" and thereby to "provide a mechanism for evaluating migration from active source rock to the trap". In 1984 when my colleague and friend Bernard Tissot and I published the second edition of our book *Petroleum Formation and Occurrence* (Springer-Verlag), we wrote in the preface that in the geosciences "computer modeling is here to stay." As an intellectual result we expected a "better chance for a much-needed synthesis of the principal fields in geosciences such as geology, geophysics, and geochemistry." Both of these volumes aspired to serve the same purpose: to understand interrelated processes of petroleum generation, migration, and accumulation and to put these in the perspective of the geological framework of basin evolution.

Although substantial progress has been made along these lines, there is still room for further advances and for more synergy. The present volume strives to generate more synergy among the various fields of the geosciences and to improve the methodology for a holistic approach in petroleum exploration and production. It is clear that this book cannot be the "last word" about what is a very complex subject. Instead we treat only a few of the major topics and questions in detail. The book is thus considered as a mosaic of important building blocks bringing together the evolution of petroleum and the geological framework of the dynamics of sedimentary basins. Basin modeling occupies an important place in this context.

We foresee that *earth scientists in general* will be attracted by the philosophy and approach presented in this book. It paves the way for combining the insight and views into the subsurface provided by modern seismic, advanced interpretation and visualization, with the understanding and results from an online interactive basin modeling capability. This combination certainly helps to unravel complex geological processes during basin evolution and permits a more rigorous and quantified assessment of a basin as a whole. For the same reasons we also expect *petroleum explorationists* to benefit substantially from the book.

In addition, we hope that *organic geochemists* will gain considerable new insight from our treatment of the current set of geochemical data coupled to the dynamic aspects of basin history. Geochemical data and interpretation results, such as those on source rock maturity and petroleum generation and expulsion, can now be extended in their application from single boreholes to entire regions by means of basin modeling. The direct coupling of basin models with seismic data bases will bring a new and exciting dimension to geochemistry in the not-too-distant future.

The authors want to take this opportunity to express their grateful thanks to the managerial, technical and secretarial staffs of the Institute for Petroleum and Organic Geochemistry at KFA and IES for their assistance and support throughout the research programs described here and in the preparation of the book. The book is a testimony to their professionalism and technical excellence. Finally, I wish to extend my personal thanks to Mr. Klaus Hebben (Monaco) for his long-term support and encouragement.

Jülich, Germany
August 1996

D.H. Welte

Contents

List of Contributors

Prof. D.R. Baker, Department of Geology and Geophysics, The Wiess School of Natural Sciences, Rice University, Houston, TX 77251, USA

Dr. T. Hantschel, IES Gesellschaft für Integrierte Explorationssysteme mbH, Bastionstr. 11–19, 52428 Jülich, Germany

Dr. B. Horsfield, Institut für Erdöl und Organische Geochemie (ICG-4), Forschungszentrum Jülich GmbH, 52425 Jülich, Germany

Dr. B. Krooss, Institut für Erdöl und Organische Geochemie (ICG-4), Forschungszentrum Jülich GmbH, 52425 Jülich, Germany

Prof. Dr. D. Leythaeuser, Geologisches Institut, Universität zu Köln, Zülpicher Str. 49a, 50674 Köln, Germany

Priv.-Doz. Dr. R. Littke, Institut für Erdöl und Organische Geochemie (ICG-4), Forschungszentrum Jülich GmbH, 52425 Jülich, Germany

Dr. U. Mann, Institut für Erdöl und Organische Geochemie (ICG-4), Forschungszentrum Jülich GmbH, 52425 Jülich, Germany

Dr. H.S. Poelchau, Institut für Erdöl und Organische Geochemie (ICG-4), Forschungszentrum Jülich GmbH, 52425 Jülich, Germany

Dr. M. Radke, Institut für Erdöl und Organische Geochemie (ICG-4), Forschungszentrum Jülich GmbH, 52425 Jülich, Germany

Prof. Dr. J. Rullkötter, Institut für Chemie und Biologie des Meeres (ICBM), Universität Oldenburg, Carl-von-Ossietzky-Str. 9–11, 26111 Oldenburg, Germany

Univ.-Doz. Dr. R.F. Sachsenhofer, Institut für Geowissenschaften, Montan-Universität Leoben, 8700 Leoben, Austria

Dr. R.G. Schaefer, Institut für Erdöl und Organische Geochemie (ICG-4), Forschungszentrum Jülich GmbH, 52425 Jülich, Germany

Dr. H.J. Schenk, Institut für Erdöl und Organische Geochemie (ICG-4), Forschungszentrum Jülich GmbH, 52425 Jülich, Germany

Prof. Dr. K. Schwochau, Institut für Erdöl und Organische Geochemie (ICG-4), Forschungszentrum Jülich GmbH, 52425 Jülich, Germany

Prof. Dr. Dr. h.c. D.H. Welte, Institut für Erdöl und Organische Geochemie (ICG-4), Forschungszentrum Jülich GmbH, 52425 Jülich, Germany

Dr. B. Wygrala, IES Gesellschaft für Integrierte Explorationssysteme mbH, Bastionstr. 11–19, 52428 Jülich, Germany

Prof. Dr. M.N. Yalçın, Department of Earth Sciences, TÜBITAK, Marmara Research Centre, P.O. Box 21, 41470 Gebze-Kocaeli, Turkey

Personal Profiles of the Contributors and Editors

Donald R. Baker

- BS in Geology, 1950, California Institute of Technology
- PhD Geology, 1955, Princeton University, USA

Present Affiliation and Position
Professor Emeritus, Department of Geology and Geophysics, The Wiess School of Natural Sciences, Rice University, Houston, Texas, USA.

Scientific Interests
Petroleum geology and geochemistry, organic geochemistry and sedimentologic controls on the nature and accumulation of organic matter in sediments.

Thomas Hantschel

- Diplom (equivalent of a master's degree) in Physics, 1985, University Halle-Wittenberg, Germany
- Dr. rer. nat. (in physics), 1987, University Halle-Wittenberg, Germany

Present Affiliation and Position
Senior physicist at Integrated Exploration Systems (IES) GmbH, Jülich, Germany.

Scientific Interests
Theoretical physics, numerical analysis of basin evolution and petroleum systems.

Brian Horsfield

- BSc (Hons) in Geology, 1973, University of Durham, England
- PhD (in organic geochemistry), 1978, University of Newcastle Upon Tyne, England

Present Affiliation and Position
Research group leader of the Institute for Petroleum and Organic Geochemistry, Department for Chemistry and Dynamics of the Geosphere, at the Forschungszentrum Jülich GmbH (KFA) in Jülich, Germany.

Scientific Interests
Origin and behaviour of petroleum and its precursors in geological systems.

Bernhard M. Krooss

- Diplom (equivalent of a master's degree) in Chemistry, 1979, at the Technical University (RWTH) of Aachen, Germany
- Dr. rer. nat. (in physical chemistry), 1985, Technical University (RWTH) of Aachen

Present Affiliation and Position
Member of the basin modeling group of the Institute for Petroleum and Organic Geochemistry, Department for Chemistry and Dynamics of the Geosphere, at the Forschungszentrum Jülich GmbH (KFA) in Jülich, Germany.

Scientific Interests
Transport processes in sedimentary rocks, natural gas generation, dynamics of geologic systems.

Detlev Leythaeuser

- Diplom (equivalent of a master's degree) in Geology, 1965, University of Würzburg, Germany
- Dr. rer. nat. (in geology and chemistry), 1968, University of Würzburg, Germany
- External professor (professorship II), University of Oslo, 1984–1993

Present Affiliation and Position
Professor of Geology (Lehrstuhl für Allgemeine Geologie), Geologisches Institut, Universität zu Köln, Germany.

Scientific Interests
Petroleum geology and geochemistry, organic geochemistry.

Ralf Littke

- Diplom (equivalent of a master's degree) in Geology, 1981, University of Bochum, Germany
- Dr. rer. nat., 1985, University of Bochum, Germany
- Dr. rer. nat. habil., 1993, University of Bochum, Germany

Present Affiliation and Position
Vice director and leader of the "basin modeling" group at the Institute for Petroleum and Organic Geochemistry, Department for Chemistry and Dynamics of the Geosphere, at the Forschungszentrum Jülich GmbH (KFA) in Jülich, Germany.
and
Professor of Organic Sedimentology and Organic Geochemistry at the Ruhr-Universität Bochum, Germany.

Scientific Interests
Petroleum geology and geochemistry, organic matter sedimentation processes, basin modeling and the origin of organic and inorganic gases in sedimentary basins.

Ulrich Mann

- Diplom (equivalent of a master's degree) in Mineralogy, 1977, University of Heidelberg, Germany
- Dr. rer. nat. (in mineralogy and geology), 1980, University of Heidelberg, Germany

Present Affiliation and Position
Research scientist and co-ordinator of Petroleum Geochemistry - Analytical Services at the Institute of Petroleum and Organic Geochemistry, Department for Chemistry and Dynamics of the Geosphere at the Forschungszentrum Jülich GmbH (KFA) in Jülich, Germany.
and
Lecturer in Petroleum Geology, Petroleum Geochemistry and Basin Modeling at the Friedrich Alexander University Erlangen-Nürnberg, Germany.

Scientific Interests
Interdisciplinary approach to petroleum migration pathways based on sedimentology, petrophysics, organic geochemistry and numerical simulation.

Harald S. Poelchau

- MS Geology, 1963, University of Colorado, Boulder, USA
- PhD Earth Sciences, 1974, Scripps Institution of Oceanography, University of California, San Diego, USA

Present Affiliation and Position
Project geologist in the basin modeling group of the Institute for Petroleum and Organic Geochemistry, Department for Chemistry and Dynamics of the Geosphere, at the Forschungszentrum Jülich GmbH (KFA) in Jülich, Germany.

Scientific Interests
Quantitative geology and computer application, sedimentology, geothermics, reservoir geology, and silicoflagellates.

Matthias Radke

- First academic degree in Chemistry (Dipl.-Chem.), 1969, University of Hamburg, Germany
- Dr. rer. nat. (dissertation in chemistry under the guidance of Prof. Dr. rer. nat. Wolfgang Walter), 1972, University of Hamburg, Germany

Present Affiliation and Position
Research associate at the Institute for Petroleum and Organic Geochemistry, Department for Chemistry and Dynamics of the Geosphere, at the Forschungszentrum Jülich GmbH (KFA) in Jülich, Germany.

Scientific Interests
Organic geochemistry, analytical chemistry, petroleum geology and geochemistry.

Jürgen Rullkötter

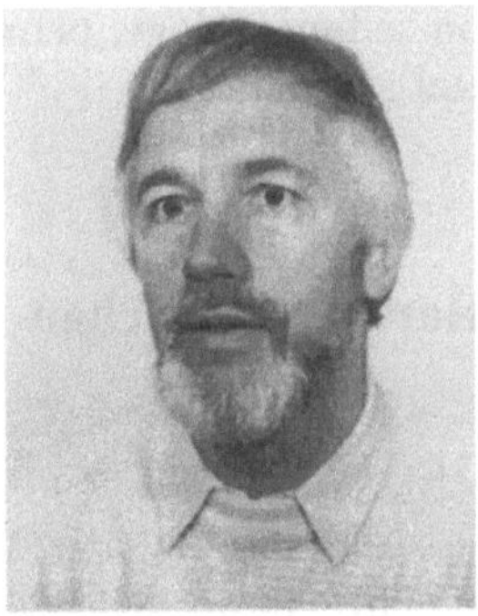

- Diplom (equivalent of a master's degree) in Chemistry, 1971, Technical University of Braunschweig, Germany
- Dr. rer. nat. (in chemistry), 1974, University of Cologne, Germany

Present Affiliation and Position
Professor of Organic Geochemistry at the Carl von Ossietzky University Oldenburg, Germany; Head of Organic Geochemistry Group at the Institute of Chemistry and Biology of the Marine Environment (ICBM).

Scientific Interests
Structural and isotopic composition of molecular sedimentary organic constituents; application to ecology and climate developments in coastal areas and the deep sea; global organic carbon cycle.

Reinhard F. Sachsenhofer

- Diplom (equivalent of a master's degree) in Economic Geology, 1985, University of Leoben, Austria
- Dr. mont., 1987, University of Leoben, Austria
- Habilitation (postdoctoral degree), 1994, University of Leoben, Austria

Present Affiliation and Position
Associate professor at the Department of Geosciences, University of Leoben, Austria.

Scientific Interests
Basin analysis including basin modeling, petroleum and coal geology, organic petrology and geochemistry.

Rainer G. Schaefer

- First academic degree in Chemistry (Dipl.-Chem.), 1967, University of Berlin (FU), Germany
- Dr. rer. nat. (dissertation in chemistry under the guidance of Prof. Dr. Günther Wilke), 1970, Max-Planck-Institut für Kohlenforschung, Mülheim/Ruhr, University of Bochum, Germany

Present Affiliation and Position
Research associate at the Institute for Petroleum and Organic Geochemistry, Department for Chemistry and Dynamics of the Geosphere, at the Forschungszentrum Jülich GmbH (KFA) in Jülich, Germany.

Scientific Interests
Organic geochemistry, petroleum geochemistry, low-molecular-weight hydrocarbons, reaction kinetics, gas chromatography and related analytical methods.

Hans J. Schenk

- Diplom (equivalent of a master's degree) in Chemistry, 1968, University of Frankfurt, Germany
- Dr. rer. nat. (in theoretical chemistry), 1971, University of Bonn, Germany

Present Affiliation and Position
Research scientist at the Institute for Petroleum and Organic Geochemistry, Department for Chemistry and Dynamics of the Geosphere, at the Forschungszentrum Jülich GmbH (KFA) in Jülich, Germany.

Scientific Interests
Organic geochemistry, spectroscopy, kinetics and quantification of organic geochemical processes.

Klaus Schwochau

- Diplom (equivalent of a master's degree) in Chemistry, 1959, University of Mainz, Germany
- Dr. rer. nat. (in chemistry), 1961, University of Cologne, Germany

Present Affiliation and Position
Guest scientist at the Institute for Petroleum and Organic Geochemistry, Department for Chemistry and Dynamics of the Geosphere, at the Forschungszentrum Jülich GmbH (KFA) in Jülich, Germany,
and
Professor of Inorganic Chemistry at the Rheinisch-Westfälische Technische Hochschule Aachen (RWTH-Aachen), Germany.

Scientific Interests
Inorganic chemistry, chemistry of radioactive elements, coordination chemistry of technetium, kinetics of petroleum formation.

Dietrich H. Welte

- Diplom (equivalent of a master's degree) in Geology, 1957, University of Würzburg, Germany
- Dr. rer. nat. (in geology and chemistry), 1959, University of Würzburg, Germany
- Dr. rer. nat. h.c., 1995, University of Bochum, Germany

Present Affiliation and Position
Managing director of the Department for Chemistry and Dynamics of the Geosphere, at the Forschungszentrum Jülich GmbH (KFA) in Jülich, Germany,
and
Professor of Geology and Geochemistry of Petroleum and Coal Deposits at the Rheinisch-Westfälische Technische Hochschule Aachen (RWTH-Aachen), Germany.

Scientific Interests
Organic geochemistry, petroleum geology and geochemistry, basin modeling and quantification of geological processes.

Björn P. Wygrala

- Diplom (equivalent of a master's degree) in Geology, 1979, University of Cologne, Germany
- Dr. rer. nat. (in geology), 1989, University of Cologne, Germany

Present Affiliation and Position
Senior geologist at Integrated Exploration Systems (IES) GmbH, Jülich, Germany.

Scientific Interests
Applied basin modeling, quantification and sensitivity analysis of geologic parameters and processes in petroleum exploration.

M. Namık Yalçın

- Diplom (equivalent of a master's degree) in Geology, 1972, University of Istanbul, Turkey
- Dr. rer. nat. (in geology and petroleum geology), 1977, University of Istanbul, Turkey

Present Affiliation and Position
Professor of Petroleum Geology and Geochemistry at the University of Istanbul, Turkey,
and
Head of the Department of Earth Sciences at the Marmara Research Centre of Scientific and Technical Research Council of Turkey (TÜBITAK) in Gebze, Turkey.

Scientific Interests
Geology and organic geochemistry of fossil fuels, quantitative basin analysis and basin modeling.

Introduction

Sedimentary basins are the sites where petroleum is formed by chemical reactions from sedimentary biogenic precursor material, where it is redistributed by migration via permeable pathways, and where it is dumped and stored in reservoir rocks or dissipated and destroyed by chemical or biochemical reactions. As in a chemical reactor, the type of product formed, the processes responsible for the formation, and the separation of the product from the precursor or reactant mixture are constrained by a few critical parameters and the prevailing physical and chemical conditions. The counterpart of the reactor is the sedimentary basin, and the counterpart of the original feedstock is sedimentary organic matter. The chemical reaction is controlled in nature and in the laboratory principally by the course of temperature and pressure changes. In the laboratory the removal of the product is normally achieved by a special plumbing system connecting the site of reaction with an appropriate storage vessel. In nature the product removal occurs by migration out of the source rocks and via permeable migration pathways through carrier rocks, to the place of storage in a suitable reservoir rock.

However, there is a profound difference between nature and laboratory conditions. Under natural conditions there are continuously changing ambient conditions during the evolution of a sedimentary basin. As a consequence of this the essential parameters, such as the chemical composition of the precursor material, temperature, and pressure are also a function of basin evolution and hence variables in space and time. By contrast, in the laboratory it is possible to keep the ambient conditions of the reaction vessel and of its peripheral system stable and to exert a rigorous control on the decisive parameters influencing the reaction.

In essence the natural system, i.e., the evolving sedimentary basin and relevant chemical reactions and transport processes, is much more complex than any laboratory scenario. Moreover, the interrelated chain of processes leading to the end product "petroleum" is highly nonlinear. Therefore any predictive interpretation, or, better, "hypothesis testing," concerning the formation and occurrence of petroleum makes it absolutely necessary to retrace the changing ambient conditions and to reconstruct the most important parameter fields. The best possible choice to achieve this is "basin modeling" or "basin simulation" as it has evolved over the past decade.

Chapter 1 discusses sedimentary basins and geological processes occurring during their evolution from the point of view of basin modeling. It is obvious that in this context the documentation of a rigorous time sequence in terms of geological events, such as deposition of sediments, nondeposition, uplift (tectonics), and erosion, are of utmost importance. In short, this chapter should prepare the nonspecialist in the wide field of "classical basin analysis" with the most important concepts and facts needed to understand and apply basin modeling. Chapter 2 can be considered as an expansion of Chapter 1. It focuses on the thermal history of sedimentary basins and thus on mechanisms of thermal transport and essential parameters influencing heat transfer and temperature distribution. Chapter 2 presents six case studies of the thermal history of individual basins.

The principal chemical reactions of interest for explorationists and for organic geochemists concern the maturation of kerogen and the generation of petroleum. These are treated in Chapter 3, with special emphasis on tools that can be used in assessing the timing and intensity of hydrocarbon generation. This Chapter also provides valuable information about temperature-induced structural chemical changes of organic material on the molecular level and prepares the reader for the considerations of reaction kinetics as a means to predict and quantify petroleum generation.

Modern kinetic models describe the restructuring of kerogen in marine source rocks and the formation of hydrocarbons, and permit the extrapolation from experimental to natural geological heating conditions in sedimentary basins. The necessary background information about these kinetic models, the differences between laboratory and field data, and their applicability to the real world of sedimentary basins is treated in Chapter 4.

Chapter 5 considers the deposition of source rocks, and Chapter 6 presents a new approach to product prediction of first-formed petroleum in source rocks. Chapter 5 focuses on such depositional settings as marine silled basins, anoxic continental shelves, progradational submarine fans, fluvio-deltaic basins, and anoxic lakes. Source rock facies, as described in these depositional settings, strongly influences the composition and structural make-up of kerogen. An approach to the predictive appraisal of the most probable petroleum product expected, upon maturation of the respective kerogen, is presented in Chapter 6. This discusses the application of experimental pyrolysis and the rigorous comparison of pyrolysis products with extractable organic matter from various types of source rock facies at different stages of maturity.

An integrated approach to petroleum migration is presented in Chapter 7. This is a crucial chapter for the book, as well as for the understanding of petroleum systems and for the petroleum industry. The chapter discusses migration mechanisms, pathways and efficiency, and most importantly migration modeling. Whereas in the past the emphasis in basin modeling was mainly on the reconstruction of temperature histories and on the timing and location of petroleum generation, here the emphasis is on the investigation of postgeneration processes in a basin. There is a rather detailed discussion of the physical and geological conditions of migration, including, for instance, the multidimensional effects of changing complex geometries and pressures. At the end of Chapter 7 two case histories are presented which apply migration modeling for oil and gas.

Chapter 1
Basin Simulation and the Design of the Conceptual Basin Model

Chapter 1: Overview and Insights

A sedimentary basin can be compared to a chemical reactor in which petroleum is formed. Unlike in a reactor, however, where reaction conditions such as composition of feedstock, temperature, pressure, and duration of reaction can be technically controlled, this is not so in sedimentary basins. Here the circumstantial conditions leading to the formation of petroleum must be laboriously reconstructed in retrospect. This is best accomplished with computer-aided basin modeling. This first chapter discusses important aspects of the evolution of sedimentary basins and crucial processes of their postdepositional history. The understanding of these is the prerequisite to reconstruct successfully the generation, migration, and accumulation of petroleum.

The first step in basin modeling is the acquisition of data from general and regional geological knowledge, from wells, various logging techniques, and most of all from seismic. Hence the use of such important aspects as stratigraphic and lithological sequences of the sedimentary fill are described in this chapter along with the proper knowledge about material properties of rocks relevant for the transfer and storage of thermal energy and fluids. The proper analysis of the sequential geometrical arrangement of rock strata and their chronostratigraphy form the basis for the establishment of a time sequence for the most important geological events during basin evolution. Such events are the primary processes of deposition, nondeposition, uplift and erosion. The definition of these events, together with a rigorous time sequence based on absolute geological ages, forms the backbone of the conceptual basin model. The identification of source rocks, carrier and reservoir rocks, along with the definition of important material parameters (e.g., amount and type of kerogen, type of lithology, porosity, permeability) are part of a conceptual basin model. Similar to many other models, the successful numerical simulation of basin development depends on a well-defined conceptual model.

Once the conceptual model of the basin filling framework is established, the postdepositional processes acting on the basin fill must be considered. Of interest are the changes in material properties such as permeability and porosity, thermal properties and the maturation of organic matter. All of these are modeled during basin simulation and crucial for the development of the temperature history of the basin, petroleum generation, and the transport of heat and fluids. The results of all these calculations must be evaluated against measured calibration values and against the available geological knowledge in order to test the various hypotheses of basin evolution.

Basin Simulation and the Design of the Conceptual Basin Model

H.S. Poelchau[1], D.R. Baker[2], Th. Hantschel[3], B. Horsfield[1], and B. Wygrala[3]

1.1 Introduction: Integrated Basin Analysis

A sedimentary geologic basin is the result of the sum of all geologic, geophysical, and geochemical processes having acted on its component parts during its entire geologic history. For the geoscientist faced with the interpretation of a basin this presents an enormous puzzle since the evidence of multiple historical events and processes is often fragmentary and has been altered, modified, or destroyed by events of more recent age. For the scientist trying quantitatively to model the history of the basin this means that only forward models can be designed and applied: basin evolution consists mostly of irreversible processes which do not lend themselves to inverse modeling (Cross and Harbaugh 1990).

Understanding the geologic facts and data in terms of the basin history and evolution requires background knowledge of the many processes having acted on a basin, as well as their interactions. This can be achieved only when data from all possible sources are integrated to arrive at a unified and consistent picture. Data and measurements obtained from wells, exploration seismic, remote sensing, and outcrops are used primarily to construct a model of present-day conditions, i.e., the final result of all processes having acted on the basin. In a second step they can be interpreted to shed light on past conditions and events. Gaps can often be filled through integration of available data and their geologic interpretation and interpolation.

The space-time continuum of generation, migration, and accumulation of hydrocarbons involves many complex, dynamic, and multivariable processes which occur during the long basin history and are part of "integrated basin analysis" (Welte and Yalçın 1988). These processes are induced and steadily driven by progressive changes in physical conditions (pVTX) that occur during basin evolution. Sediments deposited at the surface under ambient conditions, saturated with connate water, and with diverse compositions are in pro-

[1]Institut für Erdöl und Organische Geochemie (ICG-4), Forschungszentrum Jülich GmbH, 52425 Jülich, Germany
[2]Department of Geology and Geophysics, The Wiess School of Natural Sciences Rice University, Houston, TX 77251, USA
[3]IES Gesellschaft für Integrierte Explorationssysteme mbH, Bastionstr. 11–19, 52428 Jülich, Germany

Welte et al. (eds)
Petroleum and Basin Evolution
© Springer-Verlag Berlin Heidelberg 1997

nounced disequilibrium with the conditions imposed during basin subsidence and progressive burial. Consequently, in attempting to reestablish equilibrium they respond in a variety of ways to the differing and changing conditions. Thus, the formation of a petroleum deposit results from the direct response of the basin fill to many geologic events and processes and hence is intimately linked to sedimentary basin evolution.

Integrated basin analysis requires a broad utilization of geologic, geophysical, and geochemical data to develop a comprehensive and quantitative expression of basin history. Specifically, integrated quantitative basin analysis, through numerical modeling, shows the changes in the controls and the products which occur during the long geologic history of a sedimentary basin. Dynamic controls and products of an evolving sedimentary basin which are of special interest and importance to petroleum formation and occurrence include:

- Rate(s) of subsidence, uplift and deformation of basin fill
- Paleogeography, paleobathymetry, paleoclimate
- Depositional conditions and products (i.e., sedimentation rates, environments, facies, organic matter accumulation)
- Hydrodynamics (fluid pressure distribution and patterns)
- Rock properties, such as porosity, permeability, density, thermal conductivity, heat capacity, compressibility
- Fluid properties (water, oil, and gas), such as composition, density, viscosity, thermal properties
- Heat transfer (conduction and convection), i.e., thermal history
- Organic matter transformation (hydrocarbon generation), kinetic processes
- Redistribution (transport) of fluids, especially primary and secondary migration of oil and gas
- Trap formation and competence (stratigraphic, structural, seals)
- Accumulation, alteration, and loss of petroleum

Basin modeling is the temporal reconstruction of basin history and specifically refers to the procedure of establishing the sequential record of changes in controls and products which have occurred during the long geologic history of a basin. Petroleum accumulations are one of the many products of great economic interest. Computer modeling of controlling basin-evolution processes, when reiterated to simulate geologic time, provides temporal reconstructions of the various elements of basin history. Thus, basin modeling can provide a view of petroleum formation in terms of whole basin history.

From the perspective of geologic processes, there are four types of models commonly discussed in the literature that are applied to basin studies:

1. Basin-fill models simulate the sedimentation and stratigraphic aspects of assembling the sediment layers which fill and bury a basin. In most cases these are stochastic forward models that can produce any number of sediment stacks or fills within the given starting and boundary conditions. They range from large-scale, basin-wide models (Lawrence et al. 1990) to models of individual depositional system suites (Bitzer and Pflug 1990). Inverse

stratigraphic models are still the exception; this approach has recently been tried by Lessenger and Cross (1995).

2. Tectonic deformation models range from simulating folding and faulting on a smaller scale to crustal deformation on the basin scale (Jordan 1981; Steckler and Watts 1978). These are usually deterministic forward models (e.g., Jones and Linsser 1986) with the goal of reconstructing present-day structural cross sections (cross section balancing) or basin profiles, but some programs also use inverse modeling techniques (De Paor and Bradley 1988).
3. Ground water fluid flow models, especially for hydrogeology and hydrodynamics, are usually applied to movement of water in shallow porous media over "human" time scales and, with few exceptions (Bethke 1989), consider no burial and compaction or changing geometry (e.g., McDonald and Harbaugh 1988)
4. The term "basin modeling" has acquired a meaning referring mostly to simulating the thermal history of a basin for a given geologic and depositional history and, associated with it, the timing (and volume) of hydrocarbon generation as well as migration and accumulation (Welte 1988; Ungerer et al. 1990). These are generally deterministic forward models.

These types of models have somewhat different objectives and are rarely integrated into one simulation system, mainly because coupling would be become too difficult and CPU intensive. Some attempts have been made to integrate ground water (type 3) models with chemical reaction models (Ortoleva et al. 1987; Bethke et al. 1988). Another interesting suggestion (Stephenson 1990) has been to combine features of basin-fill and tectonic deformation models to control synthetic stratigraphy directly with input from crustal evolution models.

Only the "basin model" (type 4) is addressed extensively in this chapter. Similar to many other models, the basin model depends on a well-defined conceptual model which is based on the integrated sum of all available geologic data. The simulation is then performed on the discretized, numerical representation of the conceptual model. Output of the simulation is compared with the input and with independent calibration parameters, and if necessary the conceptual model is adjusted or modified to lead to a better match between simulation and calibration data.

This chapter discusses the principal fundamental processes involved in basin evolution which affect primarily basin simulation. The main emphasis is on the type of data needed to build the conceptual model, the common data sources, and the considerations involved in defining and adjusting the conceptual model. It follows the steps which the model builder must take from describing the architecture or geometry of the basin, defining the physical stratigraphy, chronology, and the physical properties of the basin-fill materials, identifying the postdepositional processes, their timing, and their kinetics, to finding the various temperature and maturation measures needed for calibrating the model and the final output of temperature history and hydrocarbon generation. The treatment of these topics is to a large degree based on ideas and experience gained from working with the basin simulation programs PDI-PC and PetroMod developed by IES Jülich.

1.2 The Conceptual Basin Model

The evolutionary history of a geologic sedimentary basin consists of a sequence of events in time during which a large number of physical processes have acted on the basin materials in three-dimensional space. The computer simulation requires the quantification of all defining parameters, and the conceptual model provides the temporal framework which is needed to structure the input data (Wygrala 1988). The conceptual model is a condensed description of the geologic evolution of the basin (Welte and Yalçin 1987).

The conceptual model is related to and includes the concept of the "petroleum system" (Magoon 1988; Magoon and Dow 1994) within the basin model. It contains the definition and history of source rocks, carrier rocks, reservoir rocks, and seals. Used in a two- or three-dimensional simulation program, it can be applied to testing the petroleum system ideas and modifying them where necessary.

For all types of basin simulation, i.e., modeling a single well in one dimension, a geologic profile in two dimensions, or an area of a basin in three dimensions, the design of a conceptual model of basin history is the first and most critical step. Briefly, the conceptual model is a formulation, suitable for numerical treatment (discretization) of the principal elements of a basin history. It must be based on the interpretation of conventional geologic, geophysical, and geochemical observational data placed in a temporal framework. Stratigraphic analysis provides the most crucial input to the conceptual model. From such analysis, basin history is subdivided into an uninterrupted sequence of *events*, of specified age and duration, which takes place during the evolution of the basin. Each stratigraphic event represents a time span during which one of three basic geologic processes prevailed, i.e., accumulation of a layer (deposition), nondeposition (hiatus), or uplift and erosion (unconformity). Similarly, structural and tectonic events, for example, folding, faulting, fracturing, salt piercement, etc., may be included in the conceptual model when reasonable geologic age limits and durations can be assigned. The resulting conceptual model of the geologic basin history provides the numerical "pattern," both physical and temporal, which provides the essential input for the basin simulation program.

Clearly, stratigraphic history modeled as a sequence of events is the most basic and common requirement for the design of all basin models. However, the conceptual model has certain shortcomings when used in more advanced basin modeling systems where a multitude of postdepositional processes are considered which can change or add parts of the stratigraphic section. This concerns, for instance, diagenetic alteration of only parts of selected layers such that a new physical stratigraphic entity is created long after deposition. Other examples are interstratal hydrothermal deposits or intrusive sills. A second problem is the rigid time frame of a limited number of events extending over the entire cross section. Advanced finite element modeling systems are capable of much higher and varying time (and space) resolution that can be adjusted to local differences in geologic details. Therefore, the "conceptual model" idea as used up to now must be refined and adjusted to the new requirements, especially once three-dimensional modeling becomes a generally used reality.

1.3 Definition and Classification of Basins and Their Thermal Regimes

For the purpose of modeling thermal and geologic history, basins are commonly classified in terms of their tectonic origin or their plate-tectonic position (Bally and Snelson 1980; Helwig 1985). Crustal behavior is the prime cause of basin subsidence and uplift, and determines much of the heat flow budget and temperature history.

Using the first classification parameter, geodynamic or tectonic causes, the two most common crustal basin models are the extensional and the flexural basin. The extensional model first proposed by McKenzie (1978) and elaborated on by Beaumont et al. (1982) is driven by the stretching and thinning of the lithosphere with an associated heating and cooling cycle. The flexural basin is best typified by the compressive foreland model of Jordan (1981) and Beaumont (1981) which assumes loading of the lithosphere by overthrusting, causing a depression in front of the thrust and an associated fore-bulge. A variety of models also exist for intracratonic basins (Quinlan 1987). All these models are significant mainly because they are process oriented and attempt to predict the geometry of a sedimentary basin, the history of subsidence and uplift, the associated stratigraphic patterns and types of lithologies (Stephenson 1990), and most importantly the changes in heat flow from the crust (McKenzie 1981).

Other subsidence mechanisms for basin formation listed by Dickinson (1993) are lithospheric cooling, asthenospheric flow, and crustal densification by phase change. Dickinson (1993) cautions, however, against becoming stuck on a quasistatic basin taxonomy in which processes are inferred from basin type. Individual basins are the result of a complex interaction of mixed processes and are often to some extent unique. Thus, "basin type should be seen as a function rather than a predictor of process."

Although crustal evolution and behavior is certainly a fundamental aspect of basin development, it is often difficult to obtain adequate data to understand just what the crustal history has been. In fact, many inferences about the lower lithosphere and its development are taken from the sedimentary record of the basins created in the crust. It is therefore tempting to propose using the temperature distribution and history in the shallow lithosphere, established with basin modeling and calibrated with sedimentary maturity parameters, for testing the simulation results of deep crustal evolution and magmatic activity (e.g., Neugebauer and Walzebuck (1987); McKenzie (1981), etc.). In some instances this may be possible and lead to better constrained crustal models. In many cases, however, the geologic architecture and history of deep crust and mantle is even less well definable (constrained) than that of shallow basin models. As Wilson (1993) points out, "there is a wide range of magmatic responses to rifting of the continental lithosphere," reflecting regional variations in the thermal state of the asthenosphere, the amount of lithospheric extension, and preexisting crustal structures and their tectono-magmatic history.

Perhaps most interesting is the recent theoretical and modeling work on vertical heat transport through the lower crust to the base of the sedimentary section. As Neugebauer (1994) has shown with his simulations, dynamic-thermal magma transport through the subcrustal lithosphere can take place within 10^6 years (for viscosities of 10^{22} Pa/s) and be emplaced in the crust-mantle transition to correspond with seismic evidence. This plutonic transport can bring an excess heat of 50°–200° C to the lower crust transition zone. By contrast, porous flow velocities through the crust are estimated at 10^{-9} m/s for a minimum crustal permeability of 10^{-16} m^2. Under these conditions, flow takes 5–10 m.y. through 15 km crust and shorter along fault zones. This compares with about 30 m.y. for purely conductive transport. These effects can cause 50–100 mW/m^2 excess heat flow at the base of a sedimentary basin.

Nunn (1994) uses similar arguments to propose the possibility of free thermal convection through microfracture permeability in the upper crust, beneath sedimentary basins. This rapid convective heat transport to the base of the sediments cools the crust, causing additional, rapid, anomalous subsidence. The convective system shuts down when the microfractures get cemented or healed, starting a period of slow crustal warming by conduction. This process could be cyclical. Thermal "plumes" reaching the sedimentary basin fill could increase heat flow at the surface by 100–220 mW/m^2. This concept is attractive because there is evidence from fluid inclusions and other thermal indicators pointing to short pulses of heating (Leischner 1994; Robinson and Gluyas 1992; Poelchau and Zwach 1994).

The second parameter used in classification and characterization of basins is the position in the plate tectonic system (Fig. 1.1). Since the theory of the "new global tectonics" was established (Isacks et al. 1968) much research has

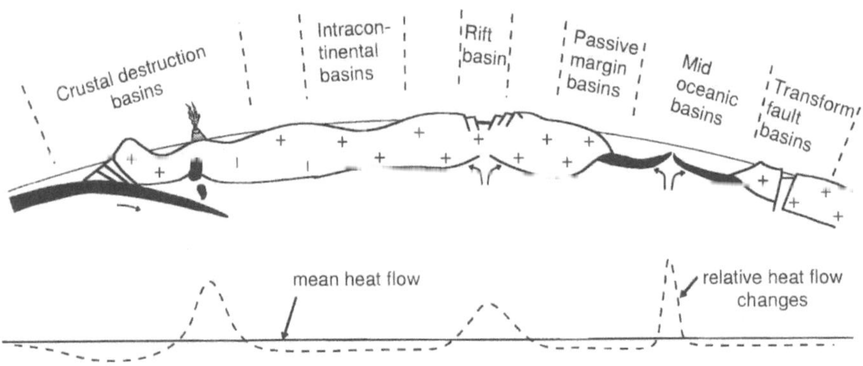

Fig. 1.1. Plate tectonic style of different basin types and typical changes in heat flow with evolution of a basin through plate tectonic stages. (In part modified after Stoneley 1981)

Table 1.1. Typical heat flow values for sedimentary basins (in mW m^{-2}). (after Allen and Allen 1990)

	Mean	Range
Extensional basins		
Active ocean ridges and volcanoes	120	120–205
Active (syn-rift) back-arc basins	85	67–120
Active (syn-rift) rift or passive margin	80	65–110
Thermally subsiding (postrift) rift or passive margin	50	40–65
Compressional basins		
Collisional foldbelt	70	40–97
Ocean trench foreland basin (foothills→margin)	40	40–80
Fore-arc basin unrelated to arc magmatism	35	20–45
Strike-slip basins		
Active strike-slip, deep lithosphere involvement	100	80–120
Active strike-slip, shallow thin-skinned (crustal) extension only	60	50–69
Basement		
Precambrian shield	40	30–55
Oceanic crust (>200 Ma)	35	30–40
Approx. global average heat flow	65	60–70

accumulated on the crustal evolution at plate boundaries and within the plates. Basins can be situated on passive or divergent margins (usually related to the extensional type of basin), on compressive or convergent margins, or on transform or strike-slip faults. Each of these positions show characteristic thermal behavior (Fig. 1.1) which can be used to estimate heat flow history associated with basin evolution (Condie 1976; McKenzie 1981). Allen and Allen (1990) have succinctly summarized typical basin heat flow ranges in various tectonic situations (Table 1.1).

Subsidence and uplift history can be derived from crustal models such as that of McKenzie (1978) or Beaumont (1981). Several newer approaches and calculations are found in the book by Beaumont and Tankard (1987). However, these models require knowledge of crustal thickness and, in the case of extensional basins, the stretching factor (β). This means that in many cases a thorough analysis of the stratigraphic column and sediment succession probably gives a more accurate subsidence history than that predicted by the theoretical models. The theoretical approach is useful, however, where erosional hiatuses are difficult to estimate, and a limit on maximum burial needs to be established. This has been shown to be applicable especially in basin-fill modeling including lithospheric stresses and temporal stress changes (Cloetingh 1992; Kooi and Cloetingh 1989). Some of these stress models can simulate observed seismic stratigraphic sequences and erosional hiatuses on basin flanks, and fore-bulges similar to those observed in nature.

1.3.1 Temperature and Heat Flow History

The crux in basin modeling is the reconstruction of paleotemperature both over geologic time and its spatial variation in the basin. There are several approaches used for paleotemperature estimation and for paleoheat flow. The topic is treated in detail by Yalçin et al. (Chap. 2, this Vol.) and is summarized briefly here.

Temperature, the most important variable in basin modeling, is calculated by various simulation programs in one of two ways. Either the temperature at depth (z) is computed from a given (or assumed) geothermal gradient and surface temperature:

$$T = z \cdot \text{grad } T + T_{surface}$$

or from heat flow (q) at the base of the section, thermal conductivity (λ) and surface temperature:

$$T = q \cdot z / \lambda + T_{surface}.$$

These two approaches or philosophies are quite different, both in the way the models are constructed and in the way the results behave. In both cases, however, the problem consists in making numerical estimates of past conditions needed as input for the modeling.

An example of the first method is presented by Tissot and Espitalié (1975) who outlined a method for the calculation of ancient geothermal gradients. The procedure is essentially empirical and dependent upon the availability of a vitrinite reflectance-depth (time) profile through the basin sequence. Using the kinetic approach developed for the simulation of petroleum generation, they designed a similar model for the thermal maturation of vitrinite using an iterative technique for a best fit. The second method, starting with heat flow, is described by Yükler et al. (1978) who took a more rigorous and process oriented approach for the reconstruction of paleotemperature.

Input for the heat flow driven basin models consists of heat flow values specified for each geologic event and each grid point. Recent values can often be read off regional heat flow maps based on well data or surface measurements. For past heat flow values the only reasonable approach consists in using analogies of the basin to be modeled with known plate tectonic situations. For this reason it is important to classify the basin within a plate tectonic framework and use crustal evolution models to help in estimating heat flow history.

Defining and tracking the heat source can be accomplished in several ways. One method includes, and tries to define, the lower crust as part of the heat generation and transfer system using present observations of thermal regimes of typical tectonic settings, and includes radiogenic heat sources as well. The crustal evolution model is used, and defined quantitatively, to calculate the heat flow system. The other method defines the heat flow density at an arbitrary depth, usually at the top of the "economic" basement or the base of the sedimentary section of interest, and modifies the heat flow history (perhaps using a crustal evolution model as a qualitative guide) as one of the parameters to fit the simulation to the calibration observations. The first method is geologically more attractive because it tries to include all relevant thermal processes, but it suffers from our general ignorance of the specifics in deep crustal structures. The second approach is more pragmatic and simple-minded in that it does not make many assumptions about deep crustal heat sources.

1.4 The Filling of the Sedimentary Basin: Stratigraphy and Lithofacies

Starting with the initial stages of a basin as a crustal depression there are transport, sedimentation, and infilling with products of erosion and reworking (clastic sediments) and biogenic materials or chemical precipitates (mainly carbonates and evaporites). The process of infilling of the basin is controlled by the crustal evolution (i.e., the tectonic basin type), orogenic development of the hinterland supplying the sediments, and the prevailing climate. Definition, classification, or even prediction of the types of sedimentation and, where possible, the details of sedimentary environments is one of the tasks of basin analysis. The primary purpose of this analysis is of course to assess the occurrence within the basin architecture of the three types of rocks of greatest interest for the petroleum system: hydrocarbon source rocks, porous carrier, and reservoir rocks, and rocks of low permeability serving as seals for hydrocarbon accumulations or aquitards for lateral fluid movement. Of further importance is the ability to predict lateral facies relationships as well as to foresee the typical future development of each sedimentary facet of the basin in terms of diagenesis, porosity development, reservoir or source rock potential. The processes of basin filling are the essential agents that determine the final character of the conceptual basin model used in basin simulation.

1.4.1 Chronostratigraphy: Definition of Events

Basin modeling requires first of all a well-designed geologic time framework. The conceptual model constructed at key well locations must define the time-stratigraphic sequence of events, i.e., time spans during which either deposition, erosion, or nondeposition has occurred. This means that the physical sequence of layers (as depth or thickness) from wells or seismic sections must be converted into an uninterrupted sequence of events with absolute ages for each event boundary.

The methods of establishing a time framework for the conceptual model depend very much on the exploration maturity of the basin, i.e., how much work is available from previous studies. If a general stratigraphic column with basin wide correlations exists – as it does in many basins of the world (e.g., the Correlation of Stratigraphic Units of North America Project) – the formations are usually assigned to global geologic periods, stages, etc. These assignments may be based on biostratigraphic determinations, on lithostratigraphic correlations or, in rare cases, on radiometric age dates. Sometimes magnetic reversal stratigraphy can also be applied, or there may be fission track ages available. Normally, absolute ages can be assigned only approximately for formation or event boundaries, based on how these boundaries correlate with chronostratigraphic stage boundaries.

An additional complication arises due to the application of different geologic time scales. Scales that have been commonly used are those of Harland et al. (1982), van Eysinga (1975), and Palmer (1983). Recently the time scale of

Haq et al. (1987) has gained widespread acceptance as a compilation and interpretation of many available radiometric dates together with biostratigraphic data, magnetostratigraphy, and depositional sequences. These dates have been issued in a new chart (Haq and van Eysinga 1987) and are somewhat modified from the original paper and extended beyond the Triassic. The latest additions to the spectrum of time scales are the new, revised scale of Harland et al. (1990) and that of Gradstein et al. (1994) for the Mesozoic.

Comparison of the time scales (Table 1.2) shows clearly that considerable differences exist between different authors. These discrepancies are due in part to conflicting radiometric databases. K/Ar dates from glauconites tend to be younger than those derived from tuffs (Menning 1989). Besides the analytic precision of the radiometric ages there is the question of the biostratigraphic precision: how closely the dated samples are bracketed by paleontologic determinations. Statistical methods can be used to determine the best estimates from the available data. A study by Bayer (1987) based on the same data as the Harland et al. (1982) time scale resulted in quite different dates (Table 1.2) and showed that the statistical error for some stage boundaries can be very high, sometimes considerably exceeding the duration of a stage.

Ages from these composite time scales have the advantage that potentially erroneous dates are averaged out. On the other hand, as the high error margins in some periods indicate, some dates might be skewed in the wrong direction. As new dates become available, stage boundaries for specific basin studies may be adjusted by local data. This can have a strong effect on duration of events and sometimes on calculated sedimentation rates (see Sect. 1.4.3); for instance, Ar/Ar dates in a recent paper by Hess and Lippolt (1986) moved the upper Carboniferous boundary back by 10 m.y. and considerably shortened the time available for deposition of Stephanian age rocks.

Adequate knowledge of duration and sequence of events can be crucial for the results of simulation. For instance, a depositional event followed by first erosion and then nondeposition results in different physical properties than one followed by nondeposition and then erosion, although the hiatus may look the same and represent the same amount of missing time. The difference affects chiefly the compaction processes of the sediments which influence porosity and permeability development, and in turn heat capacity and conductivity, and therefore ultimately the thermal history of the sediment column. Also, the time it takes to deposit a layer, i.e., the accumulation rate, can make a difference in the compaction behavior of the rock and possibly lead to overpressured formations in the case of rapid deposition.

The normal procedure to establish event ages and apply these to all points in the basin model is as follows: (1) use (micro)paleontology to assign the rocks to a biostratigraphic zone or stage; (2) correlate from the location of biostratigraphic information over the basin using log characteristics, seismic sections, cuttings, cores or additional paleontologic data; (3) find the appropriate absolute age for the stage from one of the time scales.

In many basins or study areas, formation tops are already assigned, either marked on a log, entered on a scout ticket, or available from electronic databases. However, a large effort is usually needed to verify tops and correlations,

Table 1.2. Geologic time scales. Age at Base (Begin) [Ma] (Compiled by H.S. Poelchau)

System (Period)	Series (Epoch)		Stage (Age)	Gradstein et al. (1994)	Harland 89 (1990)	Haq et al. (1987)	Menning (1989)	Bayer (1987)	Computed min. error	COSUNA[a] (1983)	Palmer (1983)	Harland 82 (Cambridge)
Quaternary	Holocene					0.01						
	Pleistocene		Milazzian			0.40						
			Sicilian			0.70						
			Emilian			1.00						
			Calabrian		1.64	1.65	1.6			1.7	1.6	2.0
Tertiary	Pilocene		Piacenzian		3.4	3.5				4.6	3.4	
			Zanclian		5.2	5.2	5.2			5.3	5.3	5.1
	Miocene	Upper	Messinian		6.7	6.3				6.7	6.7	
			Tortonian		10.4	10.2				10.8	11.2	11.3
		Middle	Serravallian		14.2	15.2				15.4	15.1	
			Langhian		16.3	16.2				17	16.6	
		Lower	Burdigalian		21.5	20.0				23	21.8	
			Aquitanian		23.3	25.2	24			25	23.7	24.6
	Oligocene	Upper	Chattian		29.3	30.0				33	30.0	32.8
		Lower	Rupelian		35.4	36.0	36			38	36.6	38.0
	Eocene	Upper	Priabonian		38.6	39.4				41	40.0	42.0
		Middle	Bartonian		42.1	42.0				45	43.6	
			Lutetian		50.0	49.0				50	52.0	50.5
		Lower	Ypresian		56.5	54.0	55			55	57.8	54.9
	Paleocene	Upper	Thanetian		60.5	60.2				62	60.6	60.2
		Lower	Danian	65.0 ± 0.1	65.0	66.5	66	64.8	+1	67	66.4	65.0
Cretaceous	Upper	Senonian	Maastrichtian	71.3 ± 0.5	74.0	74.0	72	71.5	+2	72	74.5	73.0
		"	Campanian	83.5 ± 0.5	84.0	84.0	83	81.1	−2 +2	80	84.0	83.0
		"	Santonian	85.8 ± 0.5	86.5	88.0	87	85.4	−2 +3	85	87.5	87.5
		"	Coniacian	89.0 ± 0.5	88.5	89.0	89	87.9	+1	90	88.5	88.5
			Turonian	93.5 ± 0.2	90.5	92.0	92	90.8	−1 +2	92	91.0	91.0
			Cenomanian	98.9 ± 0.6	97.0	96.0	97	97.6	+1	100	97.5	97.5
	Lower		Albian	112.2 ± 1.1	112.0	108.0	110			108	113	113

Table 1.2 (*Contd.*)

System (Period)	Series (Epoch)		Stage (Age)	Gradstein et al. (1994)	Harland 89 (1990)	Haq et al. (1987)	Menning (1989)	Bayer (1987)	Computed min. error	COSUNA[a] (1983)	Palmer (1983)	Harland 82 (Cambridge)
			Aptian	121.0±1.4	124.5	113.0	117	113.7	-3 + 3	115	119	119
			Barremian	127.0±1.6	132.0	116.5	122	122.2	-3 + 3	125	124	125
		Neocomian	Hauterivian	132.0±1.9	135.0	121.0	128	125.0	-1 + 3	130	131	131
		"	Valanginian	137.0±2.2	140.5	128.0	135			135	138	138
		"	Berriasian	144.2±2.6	145.6	131.0	140	134.7	-1 + 2	140	144	144
Jurassic	Upper (Malm)	Tithonian	Portlandian	150.7±3.0	152.1	136.0	148		-5 + 2	145	152	150
			Kimmeridgian	154.1±3.3	154.7	146.0	154	151.0	-3 + 3	155	156	156
			Oxfordian	159.4±3.6	157.1	152.0	161			160	163	163
			Callovian	164.4±3.8	161.3	157.0	167	156.4	-3 +13	165	169	169
		Middle (Dogger)	Barthonian	169.2±4.0	166.1	165.0	174	161.9	-3 +34	170	176	175
			Bajocian	176.5±4.0	173.5	171.0	182			175	183	181
			Aalenian	180.1±4.0	178.0	179.0	186			180	187	188
			Toarcian	189.6±4.0	187.0	186.0	192	171.6	-2 +28	185	193	194
		Lower (Lias)	Pilensbachian	195.3±3.9	194.5	194.0	198	198.1	-25 +12	190	198	200
			Sinemurian	201.9±3.9	203.5	201.0	204	209.3	-3 +10	195	204	206
			Hettangian	205.7±4.0	208.0	210.0	208	212.4	-2 + 6	200	208	213
Triassic	Upper (Keuper)		Rhaetian	209.6±4.1	210.0	215.0		215.7	-6 +18	215		219
			Norian	220.7±4.4	223.0	223.0	222	218.4	-6 +15	220	225	225
			Carnian	227.4±4.5	235.0	231.0	229			230	230	231
	Middle (Muschelkalk)		Ladinian	234.3±4.7		236.0	234	235.7	+ 2	240	235	238
			Anisian	241.7±4.8	241.0	240.0	240			245	240	243
	Lower (Bunts.)		Scythian	248.2±4.8	245.0	250.0	247			250	245	248
Permian	Upper (Zechstein)		Tatarian			255.0	262	239.1	-3 + 8	255	253	253
			Kazanian		256.0		268	240.4	-4 +22		258	258
			Kungurian		260.0			257.1	+14	270	263	263

	Lower	Artinskian	269.0	270.1	+13	275	268	268
		Sakmarian	282.0			285		
		Asselian	290.0	285.8	−2 + 2	290	286	286
Carbon-iferous	U. (Pennsylvanian)	Stephanian	303.0	299.2	−3 + 3	310	296	296
	Middle	Westphalian		321.9	−3 + 9	315	315	315
		U. Namurian	323.0			330	320	320
	L. (Mississippian)	L. Namurian	333.0	333.2	+ 8	340	333	333
		Visean	350.0	349.2	−4 + 5	355	352	352
		Tournaisian	363.0	353.4	−3 + 5	365	360	360
Devonian	Upper	Fammennian	367.0	362.8	−3 + 4	380	367	367
		Frasnian	377.0	372.5	−3 +25	385	374	374
	Middle	Givetian	381.0	386.6	−11 +14		380	380
		Eifelian	386.0			390	387	387
	Lower	Emsian	390.0			395	394	394
		Siegenian	396.0	397.0	+12	400	401	401
		Gedinnian	409.0			405	408	408
Silurian	Upper	Pridolian	411.0	420.9	−8 + 6			
		Ludlovian	424.0	424.3	−3 + 3	415	421	421
	Lower	Wenlockian	430.0	427.0	+11	420	428	428
		Llandoverian	439.0			425	438	438
Ordovician	Upper	Ashgillian	443.0	434.7	−3 +16		448	448
		Caradocian	464.0			460	458	458
	Middle	Llandeilian	469.0	464.4	−1 + 2	475	468	468
		Llanvirnian	476.0	470.7	−5 + 6	485	478	478
	Lower	Arenigian	493.0	488.4	−4 + 5	490	488	488
		Tremadocian	510.0		−12 +10	500	505	505
Cambrian	Upper	Merionethian	517.0	516.5	−12 +19	515	523	523
	Middle	StDavidian	536.0	543.2	+18	540	540	540
	Lower	Caerfaian	570.0	594.3		570	570	590

[a]See Salvador (1995)

since most databases contain errors and tops are rarely picked consistently. The tools available to cull out erroneous data include structural and isopach mapping of horizons to eliminate "bulls-eyes," cross sections to check on correlations, and integration with depth converted seismic profiles. Modern commercial mapping packages can be an excellent aid in automating some of these processes such as generating various maps and producing cross sections hung on a variety of datums.

1.4.2
Physical Stratigraphy: Definition of Layers

Basin development is subdivided into an uninterrupted sequence of events (geochronologic units) which take place during the evolution of the basin. Each event represents the time span during which either deposition, erosion, or nondeposition occurred (Wygrala 1988). Physical sedimentary units which result from geologic processes during a single event are called layers. If a layer is not removed subsequently during an erosional event it forms part of the present stratigraphic column of layers. Events must be defined in terms of their duration as part of the conceptual model, whereas the corresponding layers are defined in terms of their local (present) thickness as well as their lithologic and physical properties. Physical stratigraphy is therefore the most important control of the conceptual model and basin simulation.

Thicknesses of stratigraphic units, usually for some key wells, can be derived from well logs, sample logs, and cores. Of special importance for lateral correlation and for interpolation of thicknesses and facies changes is the integration of seismic sections. Whenever seismic is available, depth-converted sections should be used to establish the first framework of the stratigraphic geometry as well as the initial interpretation of seismic stratigraphy. There are now several commercial programs available to make this task easier and interactive; they also allow the integration of well data with seismic sections to produce a fine-tuned stratigraphic cross section.

The total stratigraphic column must be subdivided into a reasonable number of physically uniform units which can be separated by, but should not contain, any major hiatus or unconformity. Dividing or grouping of units is often a difficult decision. The rationale is to provide a framework which can be described numerically and on which the mathematical simulation can be performed. This requires that subdivisions are chosen such that one set of representative physical properties (e.g., compressibility, heat conductivity) can be assigned to each unit. This can be done on a purely lithologic basis, for example, from logs, or based on stratigraphic or facies packages from cross sections or seismic profiles using the concepts of sequence stratigraphy.

Sequence stratigraphy (Vail et al. 1977) is one of the tools available from seismic stratigraphy. Together with the more recent concepts of depositional suites or facies systems tracts (Brown and Fisher 1977), it is possible to gain a much better understanding of lateral and vertical changes in the layers making up the basin fill. Although available for some time, these concepts have recently gained increased acceptance and use (e.g., Posamentier and Vail 1988; Van

Wagoner et al. 1990) due to the improvements in seismic processing and data quality. The fine details that can be worked out with these methods are normally not needed for basin modeling. However, the facies sequence concept implied is useful for estimating lateral changes in lithology and physical properties of the layers, i.e., for interpolation between data points (wells) and, especially, for interpretation of seismic sections.

The basic elements of sequence stratigraphy are *depositional sequences* – coherent, genetically related packets of sediments with chronostratigraphic meaning, bounded by unconformities or paraconformities. These boundaries are of critical significance both for the definition of the depositional sequence itself and its relationship to the under- and overlying sequences, and also for the geologic time missing in the sedimentary record. The geometric relationships of the boundaries with the internal stratification (onlap, downlap, etc.) reflect the evolution of the basin and the relative importance of the competing forces of tectonic basin subsidence and global (eustatic) sea level fluctuations. The tectonic movements of the crust can be caused either by isostatic adjustment to loading of sediments and overthrust plates or by thermal disturbances such as expansion or contraction. Eustatic sea level changes are due to changes of total volume of ocean water relative to total volume of ocean basin accommodation space. The resulting sea level fluctuations usually occur on a time scale shorter than those of isostatic adjustments and combine with crustal subsidence and sediment filling to produce a *relative* sea level history.

Each depositional sequence contains a succession of *systems tracts* which are identified genetically as lowstand, highstand, transgressive or shelf margin (Posamentier and Vail 1988). Each systems tract (Brown and Fisher 1977) is in turn a sequence of contemporaneous, linked *depositional systems* – an assemblage of depositional environments typical for each of these four types. Lowstand tracts consist of basin floor deposits, slope fans, and continental slope wedges. Highstand tracts include subaerial (alluvial, fluvial, deltaic) and shelf deposits. Condensed sections are typical for transgressive tracts.

Depositional environments can be classified in terms of energy of the transporting medium. This is quite useful because the physical properties of the resulting sediments reflect the energy level of deposition. Transport energy primarily influences the carrying capacity of the medium, hence the size of particles carried, and secondly the sorting of particles. Thus, high-energy environments such as wave dominated shore lines or alluvial fans tend to produce well sorted or coarse grained deposits with better porosity and permeability and low organic matter concentration. If porosity is preserved, these rocks serve as carrier beds or reservoir rocks. Conversely, environments in slowly moving or still waters are sites for accumulation of fine grained sediments having lower permeabilities and decreased oxygenation. These tend to have higher concentrations and better preservation of organic matter and may become petroleum source rocks or reservoir seals.

Depositional environments have typical lithofacies associations. Such facies models (e.g., Walker 1984) can facilitate estimation of initial (at time of deposition) properties such as porosity, permeability, and associated parameters as thermal conductivity and heat capacity. In addition, two other parameters important for thermal history simulation, water depth and sediment-water interface temperature, can also be constrained with the aid of the facies concept. Unfortunately, very little quantitative material has been published (e.g., Weber 1982), although many data on distribution of these measurements for various depositional environments have been accumulated by oil company research laboratories. Nevertheless, compilations of facies models (Walker 1984; Galloway and Hobday 1983) and listings of dimensions and lithologies (Potter 1967; Visher 1984) can give at least some idea what range of values to expect in certain environments, how far various facies might extend laterally, and thus help to fill in the typical gaps of information between existing wells and outcrops.

In basin modeling it is important to decide on which level of stratigraphic detail the definition of events and layers should be based. As a first approach, depositional sequences are probably best suited for events in terms of boundaries and chronostratigraphic definition. With more detailed knowledge, these events can be split into systems tracts to allow for more detailed definition of physical properties of lithostratigraphic bundles.

For basins where subsurface data are sparse it may eventually be possible to fill the data gaps in physical stratigraphy with basin-fill simulation. Perlmutter and Matthews (1990), for instance, have simulated basin-fill and facies sequences with input of climatic zones and translated the results into synthetic seismic sections. These could be combined with the few available subsurface data in an immature exploration province to form the basis for a basin simulation framework.

1.4.3 Accumulation Rates and Subsidence: The Burial History

Sediment deposition, as with many geologic processes, tends to be episodic. Especially in high-energy environments many depositional events are wiped out by one erosional event, leaving only a small percentage of sediment preserved in the geologic record. Hence one needs to distinguish between short-term and long-term sedimentation rates. The latter is usually recorded and calculated when the accumulated thickness of a layer or a formation is divided by the time between its lower and upper boundary. The older literature uses the term sedimentation rate for short-term (ephemeral) rates and accumulation rates for the sediments preserved over the long run. This usage is still preferred by Füchtbauer (1988) but is slowly fading.

Einsele (1992) follows the usage established in the Deep Sea Drilling Project reports and differentiates between sedimentation rates (S) which refer to the thickness (z) of the sediment (given in $m/10^6a$ or Bubnoff units; see Fischer 1969) and accumulation rates (A) in terms of mass (m) sedimented per area (a) and time (t):

$$S = \frac{\partial z}{\partial t}, \text{and } A = \frac{m}{\partial t \cdot a} = S \cdot \rho_{bulk}$$

where ρ_{bulk} is the bulk sediment density.

Accumulation rates can be calculated from sedimentation rates when porosity and grain density, or dry bulk density, of the sediment are known. Since this is rarely the case, sedimentation rates are usually the only measure reported. However, since layer thickness and porosity decreases with depth because of compaction, sedimentation rates are only an approximate parameter, especially when comparing compacted, lithified rocks with shallow, poorly consolidated sediments.

Sedimentation rates are of interest because they can give an indication of the depositional environment (Fig. 1.2), for example, where sediments were

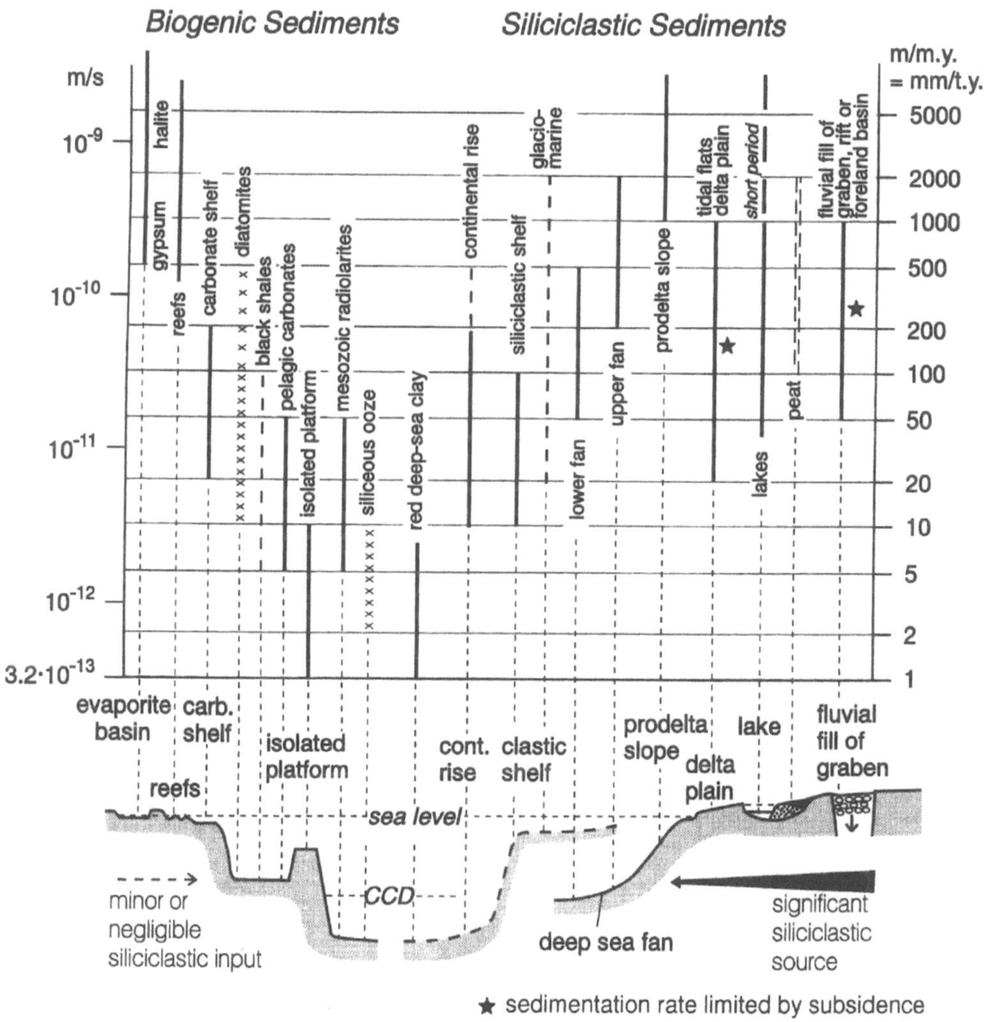

Fig. 1.2. Summary diagram of typical sedimentation rates in various depositional environments. Note that the scale is logarithmic. (Einsele 1992)

supplied in large quantities and laid down rapidly or where parts of basins were starved and only fine grained particles accumulated very slowly. Primarily, sedimentation rates are important because they reflect, albeit imperfectly, the subsidence rates of the basin. In addition, they influence the preservation of organic matter and quality of source rocks as discussed in detail by Littke et al. (Chap. 5, this Vol.).

Accuracy of age dating and the use of different time scales can strongly influence the appearance of sedimentation rate diagrams (Fig. 1.3). In the illustrated example, assignment of new dates from the Haq et al. (1987) time scale, to replace those of the original report by Scholle (1977), resulted in much higher sedimentation rates in the upper Cretaceous and reduced the very high sedimentation rates at beginning of the Cretaceous. By comparison, a burial history diagram which plots cumulative thickness against time is less strongly affected by errors or changes in geologic age dates. Thus, sedimentation rate diagrams are a useful and sensitive tool to point out unreasonable age assignments which show up as high (or too low) apparent rates of deposition, and which should be corrected before basin modeling.

Subsidence, that is the depth of burial changing with geologic time, can be displayed in the form of a burial history diagram plotted as depth against age

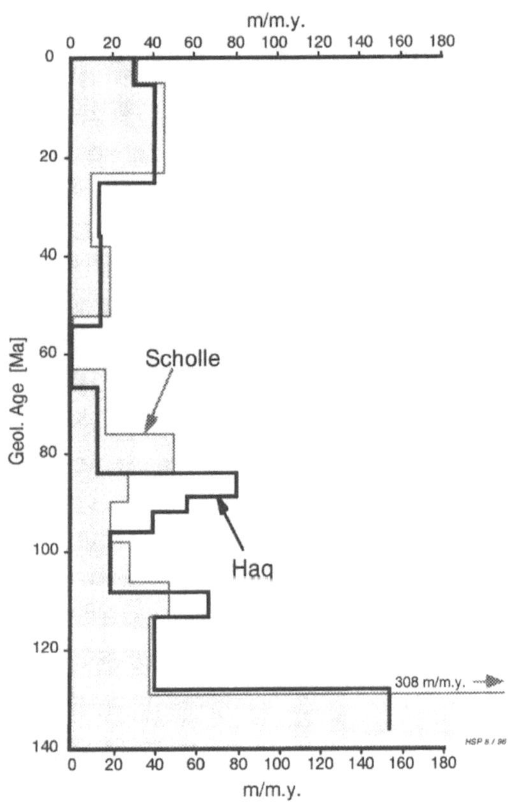

Fig. 1.3. Sedimentation rates of formations in the COST B-2 well calculated for Scholle (1977) age dates and using the Haq et al. (1987) geologic time scale. Note the large differences caused by the two time scales in the upper Cretaceous and at the base of the section

for a particular well or area. Van Hinte (1978) has popularized this concept under the name of geohistory diagram. Burial history diagrams are useful as a background for showing temperature history and for estimating timing of hydrocarbon generation when vitrinite reflectance or other maturity parameters are overlaid on the subsidence curves. Plotting cumulative thickness of layers against age results in a set of curves such as Fig. 1.4 which show the subsidence and sediment accumulation through time for a particular well.

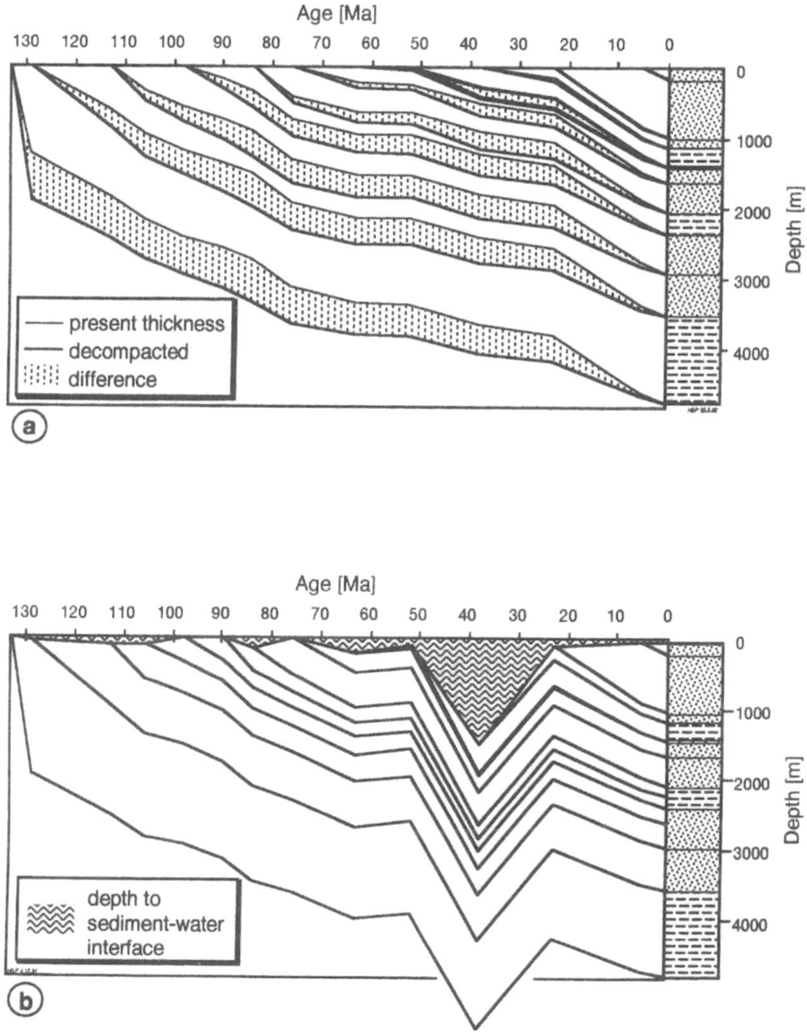

Fig. 1.4a,b. Burial history diagrams (COST B-2 well). **a** Note the large discrepancy between present (compacted) and decompacted thicknesses. The decompaction was calculated using the Sclater and Christie (1980) backstripping technique as implemented in a program by Gildner (1990). **b** Note the change in geometry when water depth is considered. Unit 4 (uppermost shale) is interpreted as lower slope environment in 1500 m water depth

Since present thicknesses, as measured in a well, are of compacted sediments, this initial plot (thin lines in Fig. 1.4a) must be corrected to reflect the actual porosity at the geologic time plotted; the layers must be decompacted. This can be achieved by the "backstripping" method or by iterative forward modeling using assumptions about original porosities of various lithologies. Both methods require data on compaction coefficients and matrix density of the sediments. Figure 1.4a shows geohistory curves corrected with the backstripping algorithm and parameters of Sclater and Christie (1980).

An additional correction of burial history may be necessary due to varying paleobathymetry and sea level. Depending on the magnitude of water depth variations this correction can make quite a difference (Bertram and Milton 1988). Figure 1.4b shows the rather extreme effect of assigning a lower slope environment of 1500 m water depth to the uppermost shale layer (unit 4) during its time of deposition between 51 and 38 Ma. Since the base line of the diagram is correlative with the crustal subsidence of the basin, excursions such as the one shown on Fig. 1.4b should give cause to reconsidering water depth estimates unless the tectonic history of the area agrees with it. The effects of water depth on the basin simulation are only indirect. Water depth does not affect effective stress and compaction but must be considered in calculating pore pressure and sediment-surface temperature.

1.4.4 Paleogeography and Paleotemperature

Paleogeography, that is, the topography and bathymetry of the geologic basin and its sediment source areas through geologic history, as well as paleoclimate, are important boundary conditions of basin simulation. Although often neglected for lack of available data, the importance of including paleogeography cannot be overstated. Some of the effects on the basin history are:

- Differences in compaction history due to varying amounts of overburden erosion
- Heat flow and temperature history affected by sediment-water interface temperatures
- Sedimentation rates as a function of elevation and erosion of hinterland source areas
- Sediment types and grain size as a function of sediment supply and depositional gradients
- Recharge with meteoric waters affecting fluid flow directions and diagenetic processes

The first two points enter directly into the basin simulation as input while most of the other effects should be considered, at least indirectly.

It is often very difficult to obtain quantitative data on the distribution of elevations of the paleosurfaces and on paleotemperatures. Paleontology is probably the most common source of paleobathymetric information. The study of microfossils has become especially useful and in some cases almost routine for this kind of information; for example, the presence of arenaceous forams as evidence of nearshore shelf deposits, certain kinds of diatoms or charophytes as fresh water fossils, or planktonic forams for open ocean. Benthonic forams have been fairly well zoned as to typical depth of occurrence and are routinely used to indicate environments from shelf to lower slope, as well as brackish and marshy facies (Haq and Boersma 1978).

Micropaleontology can also give some indication of water temperature. Statistical evaluation of foram populations were used extensively in the CLIMAP program of the North Atlantic (McIntyre 1976) and various species of silicoflagellates have shown different temperature preferences (Poelchau 1974; Jendrzejewski and Zarillo 1972).

A further source of water temperature have been the $\delta^{18}O$ stable isotope data from fossil shells, especially those from benthonic and planktonic forams studied by, for example, Emiliani (1955) and Shackleton and Opdyke (1973). Although in part influenced by the ice volume effect, these data have been used successfully to establish time-temperature curves in various parts of the world ocean for sediments of at least Pleistocene and Tertiary ages.

Surface water temperature is a function of latitude, oceanic currents and long-term climatic change as well as seasonal events. Temperature at the sediment-water interface depends in part on water depth, and is on the shelf influenced by long-term changes in surface water temperature, i.e., by climatic change. This means that long-term average surface temperature should be factored into estimating bottom temperature as well as water depth derived from depositional environment or benthonic fossils. Wygrala (1989) has synthesized surface temperature trends from Frakes (1979) as a time-latitude diagram (Fig. 1.5) which can be quite useful for estimating values for shallow water sediments within global climatic belts. To improve on this approach one might consider the equatorial asymmetry of temperature distributions seen in today's oceans and try to adjust paleo-asymmetries and ocean currents based on landmass distribution in plate tectonic reconstructions.

In greater water depths beyond the shelf edge, mean temperature profiles typical for present oceans or bottom temperature measurement transects (Fig. 1.6) give a starting point for estimates. However, water temperatures are also affected by different oceanic water masses overlying each other with characteristic temperatures and salinities which distinguish and separate each mass due to density differences. These physical properties are mostly determined by the conditions prevailing in the area of origin for each water mass, for example, circumpolar areas for the cold, deep bottom water mass. For ancient oceans the types and distributions of water masses are rarely known and can only be estimated in analogy to present oceanographic conditions.

Paleotopography is probably the most difficult type of information to reconstruct. It becomes important for two reasons: meteoric water recharge and large-scale seasonal temperature changes. Hydrology of the basin, especially for the shallower formations, is largely affected by the climate (precipitation and evaporation) and by the topography of recharge areas. Elevation determines the hydraulic head, which influences the amount and rate of water movement into the basin. This can also affect the migration of petroleum fluids. Hydraulic head is also an important parameter for estimating (or

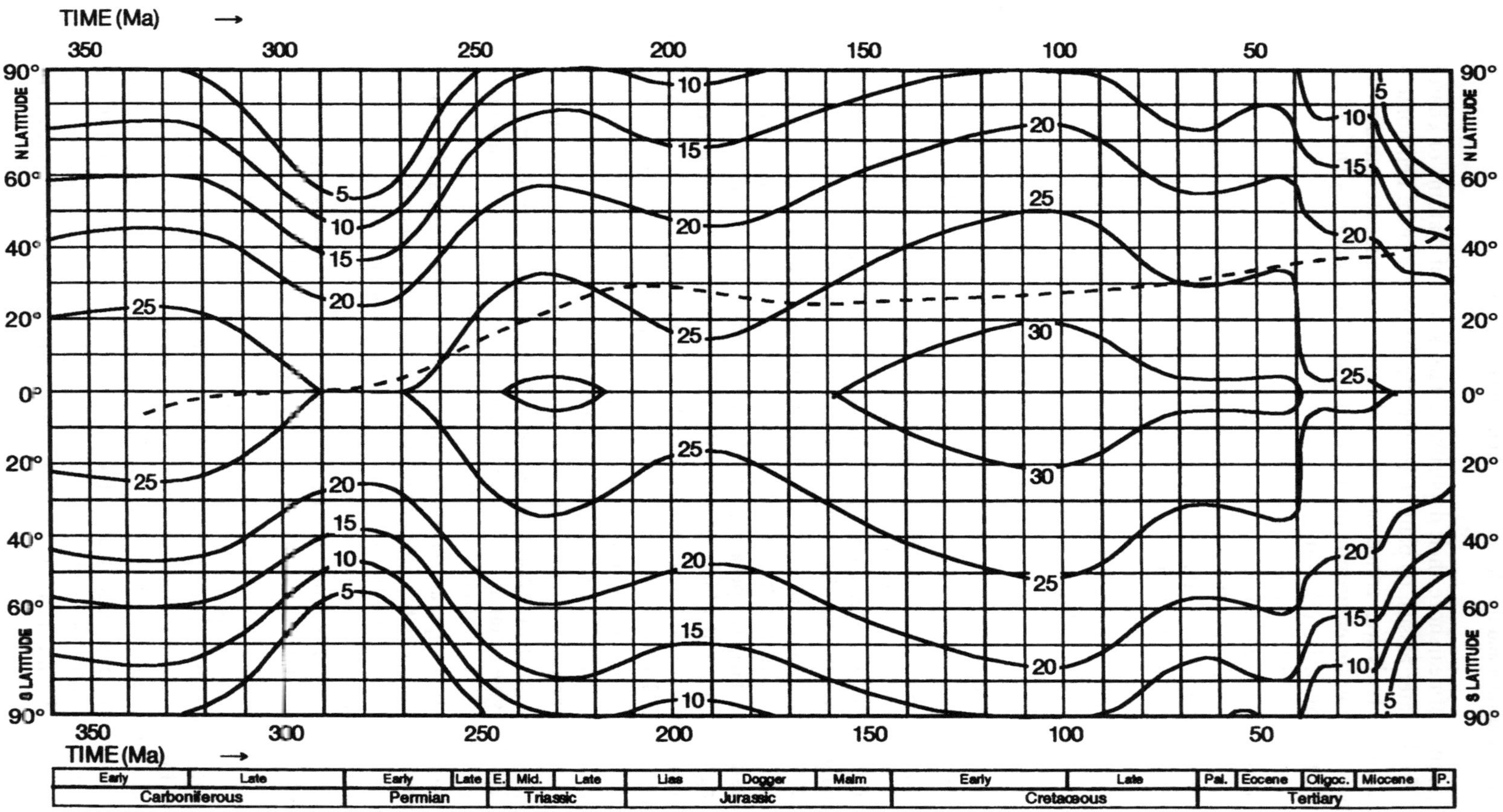

TIME (Ma)
350
300
250
200
150
100
50
90°
60°
40°
20°
0°
20°
40°
60°
90°
N LATITUDE
S LATITUDE
5
10
15
20
25
30
Early
Late
Carboniferous
Early
Late
Permian
E.
Mid.
Late
Triassic
Lias
Dogger
Malm
Jurassic
Early
Late
Cretaceous
Pal.
Eocene
Oligoc.
Miocene
P.
Tertiary

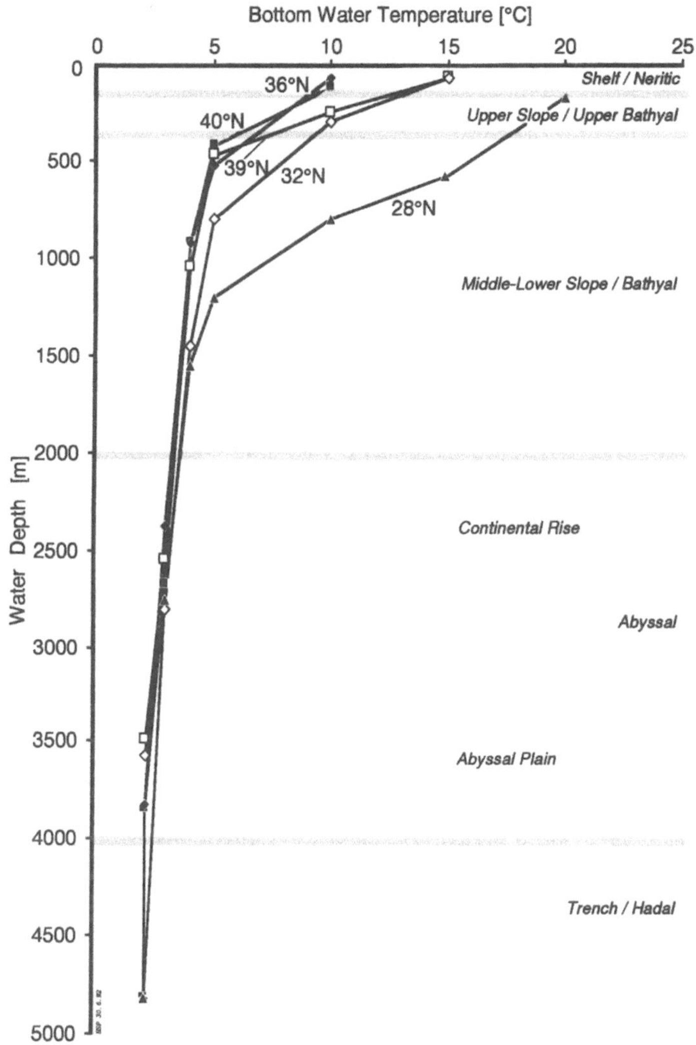

Fig. 1.6. Depth profiles of bottom water temperatures for several transects in the Northwest Atlantic Ocean margin. (Data from Emery and Uchupi 1972; faunal zones after Berggren in Haq and Boersma 1978)

simulating) diagenetic processes that depend on transport of certain ions (see Sect. 1.5.1). Surface temperature, one of the input variables for basin simulation, is influenced by surface elevation, and the average depends on large-scale seasonal differences.

Fig. 1.5. Geologic age–latitude plot of average ocean surface temperatures. (From Wygrala 1989, based on data from Frakes 1979)

Using techniques for estimating the amount of eroded overburden (discussed below in section 1.5.2) one can get order of magnitude estimates of past topography. The difficulty here is that uplift and erosion often go hand in hand due to isostatic adjustment. This means that erosion estimates yield maximum uplift values that do not necessarily reflect the actual topographic elevation at any one time.

1.5
Postdepositional Processes

1.5.1
Compaction and the Evolution of Rock Physical Properties

Compaction is the largely irreversible process of sediment volume reduction due to overburden loading, grain rearrangement (packing), grain solution, etc., and can be described primarily as a function of pore space (porosity) reduction. The reduction of the solids volume plays only a minor role. Bulk compressibility describes compaction as a function of stress. As the pore space is filled with fluids which must be expelled during the compaction process, the ability of the sediments to transmit fluids, or permeability, is the additional controlling parameter in compaction processes. In order to model compaction the general path of porosity, permeability and compressibility decrease with depth must be known.

Porosity

Initial porosities of loosely packed (i.e., uncompacted and uncemented) sediments depend on sorting, grain shape, grain size, and sedimentation rates and can range from approx. 25–55% for sandstones (Houseknecht 1987), 50–90% for shales (= mudstone = clay = argillaceous mud, as used here), 40–95% for limestones (e.g., 44% and 55% for grainstones and packstones, respectively (Enos and Sawatsky 1981), and 70–95% for deep sea calcareous ooze). The extreme range for carbonate rocks is due mainly to the wider possible range of grain sizes and to the shape of organic debris. The sediment to water contact is often transitional and is not always well defined: for example, argillaceous sediments can show a gradual upward transition from a sediment with a porosity exceeding 80%, to a fluid with a gradually decreasing percentage of particles carried in suspension. Porosity values can vary widely within the uppermost sediments (e.g., see Haenel 1979; their Fig. 4: 40–70% porosity in the uppermost 1 m). A definite boundary and therefore a definite initial porosity value cannot be defined (Füchtbauer and Müller 1970; Larsen and Chilingar 1979; Chilingarian 1983). However, initial porosities are one of several parameters which are required to simulate compaction processes; a value must therefore be defined which is a representative average.

General porosity values for sediments are mostly given as porosity-depth curves for a specific lithotype. However, porosity-depth curves describe only the present condition and do not necessarily indicate the reduction of porosity versus time and/or depth, i.e., the compaction history (Chapman 1981). Decreasing porosities with depth can be a function of overburden thickness, time, lithology, depositional environment, pressure development (including overpressuring), diagenesis, and tectonic stress (Chilingarian 1983). A wide range of porosity-depth trends is possible, and Figs. 1.7–1.9 show examples of porosity-depth trends for the basic lithotypes sand, shale, and limestone.

A multitude of factors interactively influence porosity development and compaction (e.g., see Chilingarian 1983, p. 86ff and Table 3 for sandstones, p. 69ff for shales), and interpretations using individual factors can be misleading. The following examples are given to stress this point.

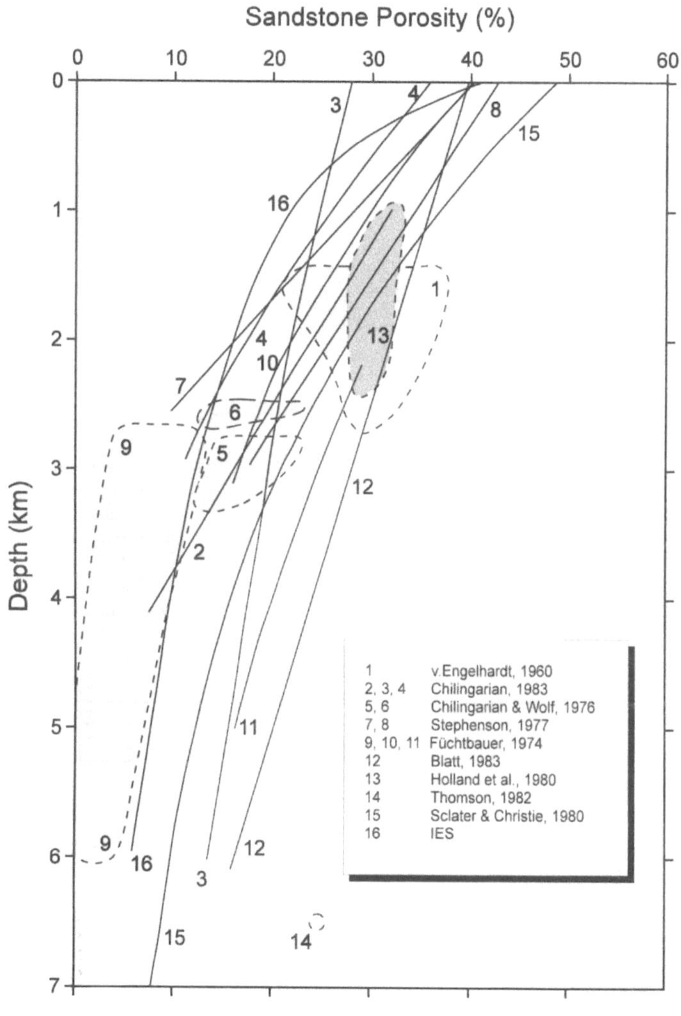

Fig. 1.7. Porosity-depth curves for sandstones from various published sources. (Wygrala 1989)

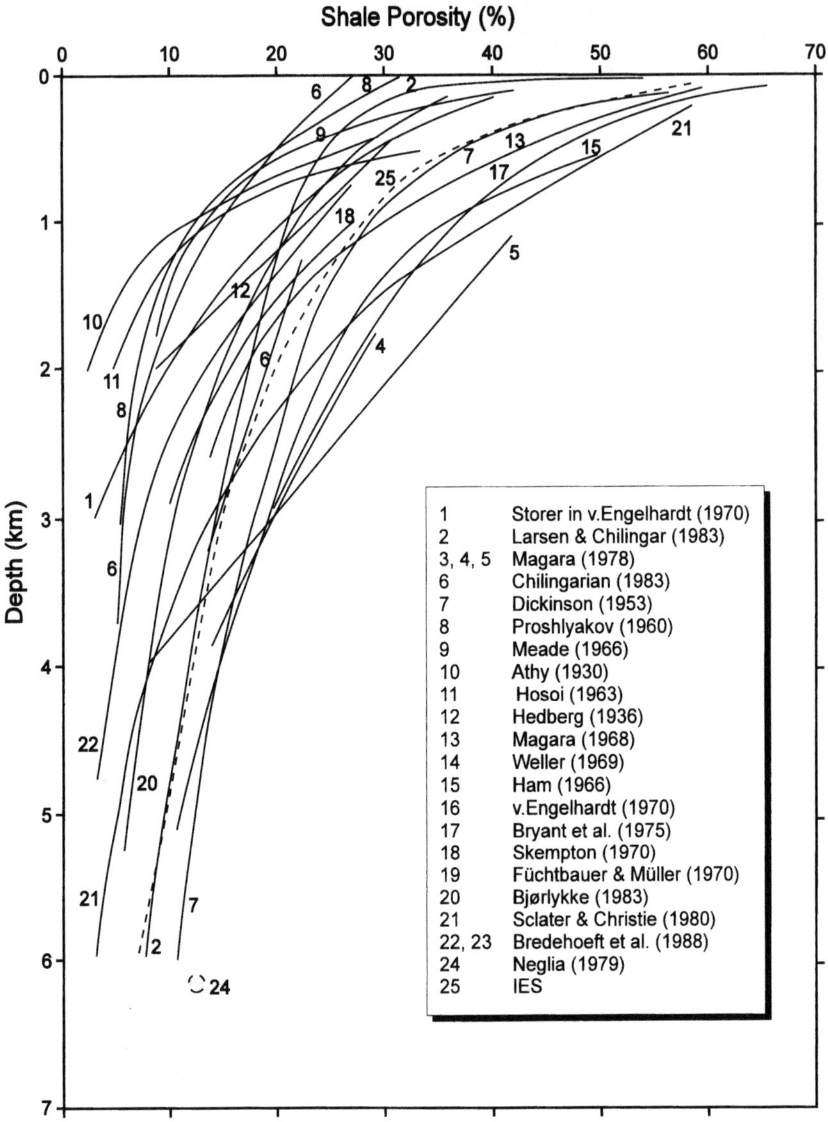

Fig. 1.8. Porosity-depth curves for shales from various published sources. (Wygrala 1989)

Overpressures develop when pore fluids cannot be expelled from low-permeability sediments which are exposed to rapid overburden pressure increases due to high sedimentation rates; compaction is retarded, and porosities remain higher than normal as the overburden pressure is partially balanced by the increased pore pressures (Chapman 1981).

Temperature affects fluid viscosities and therefore also controls fluid expulsion rates. Higher temperatures should theoretically lead to more rapid compaction and reduced porosities. However, higher temperatures also in-

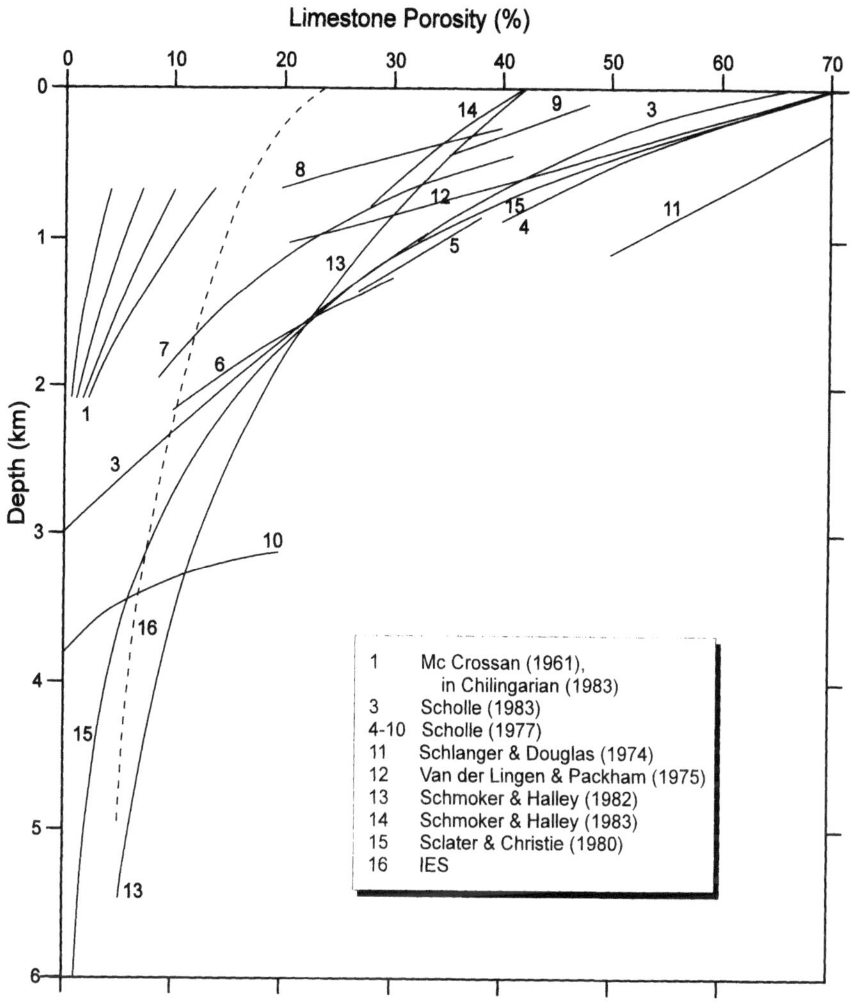

Fig. 1.9. Porosity-depth curves for limestones from various published sources. (Wygrala 1989)

fluence solution and cementation processes which can negate the direct effect of viscosities (Gregory 1977; Stephenson 1977). The general conclusion is, however, that the effect of temperatures on porosity is negligible when compared to other factors (Schopper 1982).

Diagenetic processes can either reduce or enhance porosity. Minerals in contact with pore water are subject to chemical alteration or dissolution, depending on the ion balance in rock and fluid and the temperature (and to some degree pressure). The pore fluid may precipitate certain minerals as cement, depending on the relative saturation of ions needed to form these cements. The critical saturation is of course temperature dependent.

Most basin modeling systems simulate porosity change as physical compaction with burial but do not deal with the fact that a large amount of porosity is lost (and some also gained) through chemical diagenesis. The choices are then to ignore the effects of diagenesis or hope that they cancel out, to adjust the compaction or porosity/depth functions in such a way as to average in the porosity lost due to diagenesis, or to attempt modeling of chemical reactions coupled with fluid flow and diffusion transport as part of, or parallel to, basin modeling.

Of the many possible diagenetic mineral reactions there are a limited number which are volumetrically important and common enough to be encountered in most sandstones. They concern mostly silica or quartz, carbonate cements, feldspar dissolution, and clay mineral reactions. Some of these are listed in Table 1.3. Reactions can be grouped as to typical depth or temperature: shallow reactions occurring between surface and about 75° C and deeper reactions at greater than 100° C.

Shallow diagenetic events such as kaolinite formation are often a response to fresh (meteoric) water flowing through the sediments (Bjørlykke et al. 1989). Early events can strongly influence the later diagenetic evolution of a sandstone. Early carbonate filling of pores creates a potential for later secondary solution porosity (McBride 1984), and chlorite rims in volcanic rich sediments (Tillman and Almon 1979) appear to have prevented later quartz overgrowth and porosity occlusion. In general, original differences in texture, i.e., heterogeneities of permeability, tend to be enhanced by diagenesis and determine the final porosity distribution.

The type of diagenetic reactions possible and the manner of porosity destruction or enhancement are largely determined by the original composition and texture of the sediment. Volcanogenic sandstones tend to develop zeolite or iron-rich clay cements as well as early silica precipitation. Marine sands are usually rich in biogenic carbonate particles and tend toward carbonate cementation. Arkosic (feldspar-rich) sediments give rise to pores filled with kaolinite but may also develop good secondary porosity.

Among the deeper diagenetic events, pressure solution of quartz grains around 100° C and the accompanying loss of porosity through pore-filling overgrowth and additional compaction appears to be dominant and very common, especially in the North Sea (Bjørlykke et al. 1986) and many other

Table 1.3. List of common diagenetic reactions and their temperature ranges in sandstones. (Bjørlykke et al. 1989)

Source mineral(s)	Pore water	Reaction Temp (° C)	Product mineral(s)
Aragonite and high-Mg calcite		20–50	Low-Mg calcite (sparry)
Feldspar	$-Na^+$, $-K^+$	20–100	Kaolinite
Fine-grained low-Mg calcite		60–70	Low-Mg calcite (poikilotopic)
Amorphous silica and opal CT		60–80	Quartz
Smectite	$+K^+$	50–100	Illite+quartz
K-feldspar	$+Na^+$, $-K^+$	>65	Albite
Quartz (pressure solution)		100–150	Quartz
K-feldspar+kaolinite		120–130	Illite+quartz

areas, such as the Alberta Deep Basin (Zwach and Poelchau 1993). Albitization of feldspars also has been observed in many cases (Boles 1982).

In the case of carbonate rocks, cementation starts almost immediately after deposition (Friedman 1975), but loss of porosity continues with increasing burial depth (Schmoker and Halley 1982). Therefore the compaction behavior of carbonates tends to be quite different from that of siliciclastic sediments.

The conclusion here is that no single factor is sufficient to determine the porosity-depth values and no generally valid porosity/depth functions exist (Bjørlykke et al. 1989); a quantitative multiparameter approach is required. Internal factors which control porosities, for example, temperatures, fluid viscosities, and permeabilities, are taken into account by the simulation program but external factors such as cementation or tectonic effects can be handled only indirectly.

Permeability and Porosity-Permeability Relationships

Permeability describes the ability of a sediment to transmit fluids and/or gases, which in a completely saturated unit volume is a function of the hydraulic head, flow rate (v), and viscosity (μ) of the fluid or gas:

$$k = v \cdot \frac{\mu}{\rho \cdot g \cdot \frac{dh}{dl}}$$

Permeability is measured in millidarcy (mD) or m^2 (1 darcy = 9.87×10^{-13} m^2). In a sediment it is controlled by the pore size distribution or the number and type of pore interconnections, by the effective porosity, and by the nature of the permeating fluid.

Measurable permeability values in sedimentary rocks range over at least ten orders of magnitude, with maximum values of several darcies for loose, unconsolidated sands. Sandstone permeabilities range over seven orders of magnitude from 10^5 mD for clean sands to 10^{-2} mD for consolidated, tight sandstones. Shales exhibit a similar wide range but are clearly separated from sandstones with values from 10^{-1}–10^{-8} mD. Typical values are 200–1000 mD in consolidated sands, 1–10 mD in tight sands, and 10^{-1}–10^{-4} mD or less in consolidated shales. Limestones show the widest range of permeability variations due to cementation processes and to the higher likelihood of fracturing in more competent, lithified rocks. Unlike porosity data, permeability trends vs. depth are difficult to establish due to the influence of such local factors as packing, sorting, and cementation (Engelhardt 1960).

Anisotropy of permeabilities is especially pronounced in shales, i.e., in sediments composed of platy minerals or fragments, and the ratio of horizontal to vertical permeability can exceed 10 to 1; permeability anisotropies in sandstones can reach 4 to 1 (Zoback and Byerlee 1975). However, thickness and permeability contrasts of adjacent layers should also be taken into account as they can considerably alter the *bulk* permeability anisotropy of sedimentary units when compared to measurements in individual samples (Rose 1983). Permeability anisotropy could appear to have a controlling effect on flow direction; however, the overriding factor in fluid transport is the *fluid potential gradient*, which forces flow in a generally vertical direction during burial in spite of permeability anisotropies (Chapman 1981). Flow is usually upwards, but occasionally downwards, for example, from compacting shales into underlying, more porous units. Permeability anisotropies must be taken into account during the simulation and fluids moved according to the pressure gradients; this means that fluid movements from an overpressured shale unit into underlying sands are simulated correctly.

Permeability generally decreases as a function of porosity with burial depth. Certain sediment types, such as soils, do not adhere to this general rule (Brace 1975) and investigations of specific sites such as a reservoir or sedimentary unit at a certain depth may not show a discernible relationship between porosity and permeability (Bjørlykke 1983). However, the relationship does exist as basic, subparallel trends in all types of sedimentary rocks during their burial from the surface throughout the depth range of hydrocarbon generation and accumulation, i.e., to at least 8000 m.

The porosity-permeability relationship is controlled by primary depositional factors such as grain size, distribution, shape, orientation, packing, and surface area (Rieke and Chilingarian 1974), as well as by later secondary factors such as fracturing, cementation, and other mineralogic processes. Functions such as the Kozeny-Carman equation have been developed to define the relationship, but the number of variables is too large to allow these generalizations to cover all types of lithologies and burial histories. It is, however, possible to define porosity-permeability relationships for specific lithologies which represent an acceptable mean from the range of possible values (Fig. 1.10). As most of the data is from reservoirs, however, it is difficult to assess whether the available data is representative for sediments in general.

Cementation can reduce porosities in sandstones considerably without significantly reducing permeability (Bjørlykke 1983); however, the resulting general trend of porosity vs. permeability in sandstones is quite well defined, with permeabilities mostly ranging over only approximately two orders of magnitude for any selected porosity value.

Data on shale porosity-permeability relationships are not widely available due to the absence of perceived, direct economic importance (i.e., reservoir analysis), due to inherent problems in measuring core samples disturbed by drilling and to difficulties in indirect estimates in wells caused by the low permeability values (Magara 1978). Based on studies reported in the literature (Bryant et al. 1975; Neglia 1979) shale permeability values mostly range over 3–4 orders of magnitude at a specific level of porosity. Permeabilities are between 1 and 10^{-3} mD at 70% porosity and decrease to 10^{-2} to less than 10^{-6} mD at porosity values of 5%.

Porosity and permeability in limestones are less closely related than in the other lithotypes due to the influence of cementation and fracturing. Complete lack of permeability in porous limestones is a common occurrence. Poorly defined porosity-permeability relationships in carbonates in general necessitate adaptations to specific, local conditions and burial histories. An extensive selection of permeability/porosity data is available in Schopper (1982, p. 289ff).

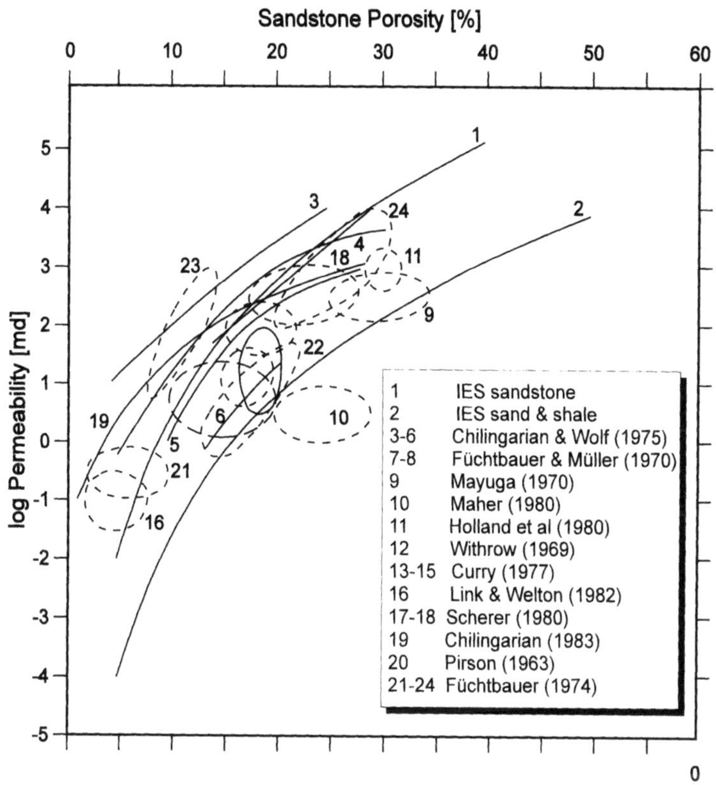

Fig. 1.10. Sandstone porosity-permeability relationships from published sources. The small circles indicate average values whereas the ellipsoids show ranges within a reservoir or sedimentary unit; the curves represent means for value ranges. Values affected by cementation are also included. (Wygrala 1989)

A general porosity-permeability trend for specific lithotypes based on bulk porosity and permeability information from well logs and flow tests appears to be a reasonable, pragmatic approach towards defining generally applicable values for numerical simulation purposes, as long as the option of adapting and calibrating these relationships to specific situations is still available.

Compaction Model

The exact determination of the rock porosities and their changes are very important wherever dynamic aspects play a significant part in basin evolution. The geometric description of basin layers and elements is directly connected with the porosity changes of the rocks. In addition, most of the rock properties, such as conductivity, heat capacity, density, and elastic or plastic modulus, strongly depend on rock porosity.

Empirical or statistical average porosity-depth curves, fitted as exponential or polynomial functions, are generally used to describe porosity changes. In these kinematic descriptions the causes of processes (forces, potentials, loads, stresses) are not taken into account. These approximations are suitable if steady state conditions occur, i.e., if fluids are expelled at the same rate as load is applied. Nonsteady behavior is caused during phases of high sedimentation rates or by material inhomogeneities (salt intrusions) or highly dynamic tectonic events. This usually means that overpressure builds up because fluids are not able to escape fast enough.

The most important assumption in these models is that the depositional overburden load is the cause of the compaction of the lower subsiding layers. The overburden load is responsible for the stress field in the basin. Smith (1971) proposed to use the vertical component of the effective stress potential as the measure for the compaction strength. The effective stress tensor σ is the difference of the single-axis stress tensor of the external overburden load (total stress) τ and the sphere tensor of the fluid pressure (pore pressure) p acting against the total stress. In other words, compaction is caused by that part of the vertical overburden stress that is not balanced by the pore pressure. If the pore pressure is close to the overburden stress, as in overpressured sediments, little or no compaction takes place. The vertical components are identical to the main scalar values of both p and τ. The compaction law used in the simulation can be formulated as follows:

$$\frac{\partial \phi}{\partial t} = -(1-\phi) \cdot C \cdot \frac{\partial u^{\sigma}}{\partial t} \quad \text{and} \quad u^{\sigma} = u^{\tau} - u^{p}$$

where ϕ is porosity, C is compaction modulus of the rock matrix, u^{i} is the hydraulic potential of the effective stress, pore pressure, and total stress, respectively. The compaction modulus depends strongly on porosity. With decreasing porosities it must tend to very low values in order to stop the compaction process.

Thermal Properties of Rocks

Thermal conductivity and heat capacity of rocks determine to a large degree the distribution of heat in a basin and the maturation of organic matter. Bulk rock properties used in basin modeling are a function of relative proportion of mineral grains (or nonporous matrix) and pore-filling fluids and their re-

Table 1.4. Thermal conductivities of some geologic materials

	$Wm^{-1} K^{-1}$	Source
Earth's crust	2.0–2.5	Mean value range, Kappelmeyer and Haenel (1974)
Rocks	1.2–5.9	Sass et al. (1971)
Sandstones	2.5	Clark (1966)
Shales	1.1–2.1	Clark (1966), Blackwell and Steele (1989)
Limestones	2.5–3	Clark (1966), Robertson (1967)
Water	0.6	At 20° C
Oil	0.15	At 20° C
Ice	2.1	Gretener (1981)
Air	0.025	CRC (1974) Handbook
Methane	0.033	CRC (1974) Handbook

spective physical properties. Published values of thermal conductivities range widely for various types of rocks (Table 1.4). This is due to the wide range of porosities involved. Unfortunately, in many cases the porosity measurements are omitted so that many literature values are practically useless for comparison or modeling.

Thermal conductivities of sediments are more a function of their water content than of their mineral constituents (Ratcliffe 1960). Wet samples have higher thermal conductivities than dry samples, as water has a higher thermal conductivity than air. This means for thermal conductivity λ that always: $\lambda_{matrix} > \lambda_{wet\ rock} > \lambda_{dry\ rock}$. Therefore, conductivity becomes a function of porosity, or the state of compaction, as well as the mineralogy and the pore-filling fluid.

All pore-filling fluids have lower λ values than rocks (Fig. 1.11). This causes bulk thermal conductivities to decrease with increasing porosity. In other words, bulk thermal conductivities should increase with progressing compaction. Porosity and the type of pore fluid are the most important controlling factors, as the thermal conductivity of a water saturated sediment can increase by a factor of 2–5 during the course of compaction from its initial porosity value to its final value (Woodside and Messmer 1961). Water has considerably higher conductivity (0.6 $Wm^{-1}\ K^{-1}$) than gas (0.03 $Wm^{-1}\ K^{-1}$), air and even

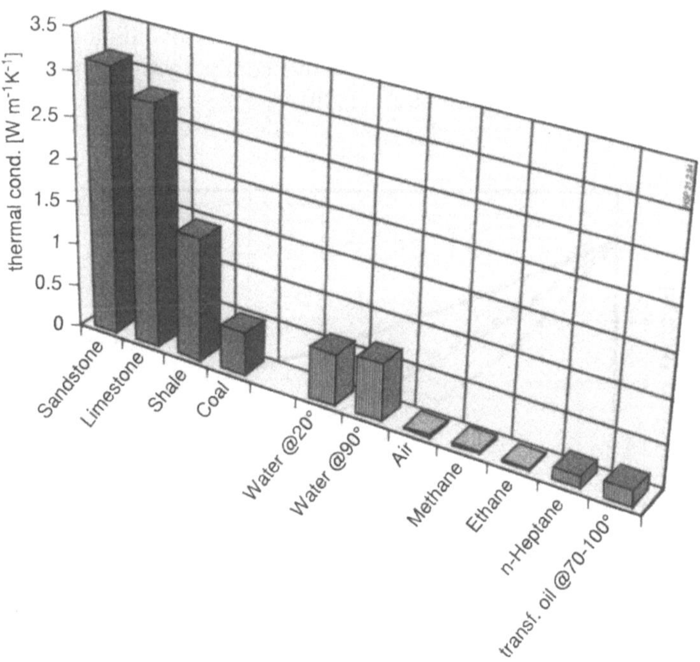

Fig. 1.11. Thermal conductivities of selected rocks and fluids. The values for rocks are matrix conductivities for 0% porosity at 20° C. (Data from IES 1993, Blackwell and Steele 1989; Clark 1966; Weast 1974)

oil. Therefore we should see differences in temperature distribution depending on the degree of hydrocarbon saturation of reservoir and source rocks. Examples for the effect of porosity and different pore fluids on bulk thermal conductivity are plotted in Fig. 1.12.

Most sedimentary rocks are anisotropic with higher horizontal than vertical thermal conductivity values (Gretener 1981). However, as heat flow follows temperature gradients, which are usually more or less vertical, increased horizontal λ values need not necessarily have an important effect. Kappelmeyer and Haenel (1974) give anisotropy ratios (horizontal to vertical) of 1.04 for quartzitic sandstone, up to 1.28 for sandstone, and up to 2.5 for shales.

The effect of temperature is to decrease conductivity of well-conducting lithotypes (e.g., sandstone), while values for shales of less than 25% porosity are virtually independent of temperature, which leads to lower thermal conductivity contrasts at higher temperatures. This means that corrections for the effect of temperature are usually only required for lithologies with thermal conductivities of more than 2.5 $Wm^{-1} K^{-1}$ (Gretener 1981; Sekiguchi 1984).

To calculate the effective thermal conductivity of a rock for use in modeling from the component mixture of minerals and pore-filling fluids, it is necessary to use the appropriate end member values of matrix thermal conductivities. The mixing formula used to calculate bulk thermal conductivity λ (and heat capacity) is based on the geometric mean:

$$\lambda = \lambda_1^{f_1} \cdot \lambda_2^{f_2} \cdot \lambda_3^{f_3} \ldots$$

where $f_{1\ldots3}$ are the fractions of the components of lithology and pore fluids ($\Sigma f_i = 1.0$), and $\lambda_{1\ldots3}$ are their respective conductivities. This is similar to the approach used by Brigaud et al. (1990).

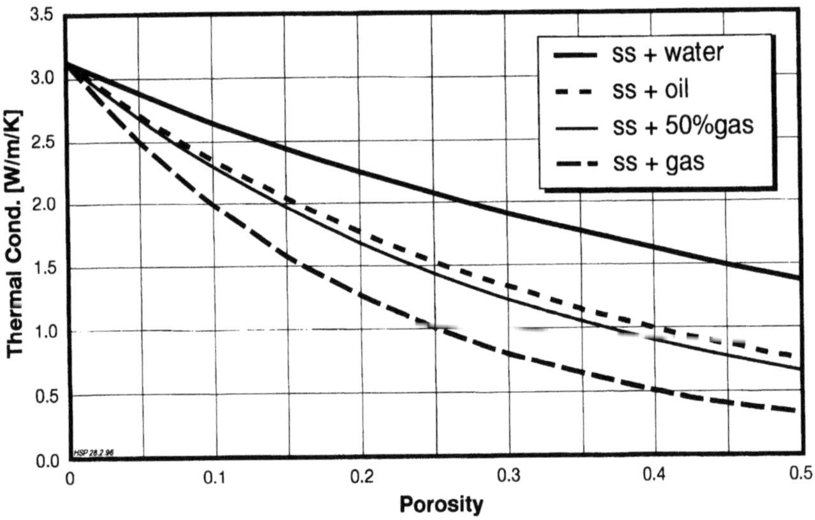

Fig. 1.12. Thermal conductivity of sandstone as function of porosity and pore fluid (at ambient T and p)

Thermal conductivities used for calculation of bulk conductivities are usually derived from laboratory measurements of individual minerals or rocks with measured porosities (Somerton 1992). These values do not always agree with thermal conductivities derived from in situ field measurements, for example, thermal logs (Blackwell and Steele 1989), or values deemed appropriate through modeling experience. The discrepancy is important to keep in mind when choosing λ values for modeling since the effect on modeling results can be significant. Laboratory measurements of individual minerals reported in the literature are often higher than what should be the thermal conductivity of the corresponding rock in situ. For instance, Brigaud et al. (1990) list 7.8 $Wm^{-1}K^{-1}$ for quartz and 7.01 $Wm^{-1} K^{-1}$ for sandstone while IES uses an empirical value of 3.12 $Wm^{-1} K^{-1}$ for sandstone based on modeling experience.

Similar discrepancies exist for shale. Blackwell and Steele (1989) point out that their shale λ values calculated for in situ temperature and temperature gradient logs (assuming constant heat flow) are often ca. 50% lower than most literature values. Their range of λ_{shale} is 1.1 to 1.3 $Wm^{-1} K^{-1}$ (unspecified porosity, but probably between 10% and 30%), while many laboratory measurements exceed 2.0. To obtain a λ of 1.1–1.3 for porosities of 30% to 10% one needs to reduce the λ_{matrix} of shale from the commonly used 1.95 to 1.45 $Wm^{-1} K^{-1}$. Robertson (1967) has compiled numerous thermal conductivity data from the literature and plotted them for each lithology against solidity (the complement of porosity). His 0% quartz line (presumably pure shale) intersects the 1.0 solidity or 0% porosity line at 1.5 $Wm^{-1} K^{-1}$. Extrapolating a line through the lowest shale λ values on his Fig. 10 gives an intercept of 1.3 at 0% porosity. These data are for water filled pores at 27° C and 5 MPa.

While the difference between laboratory measurements and in situ values could be blamed on unrealistic laboratory setup or sample disturbance, one might also speculate that, especially in shales and coals, the effective in situ conductivity is lowered by absorbed gas or pore-filling gas that has escaped from samples at surface conditions before laboratory measurements.

Under geologic conditions bulk thermal conductivity of rocks is affected by many factors simultaneously. Figure 1.13 shows the effect various overlapping controls occurring with increasing burial and temperature on a sandstone. Specifically note: (1) mineral (rock matrix) conductivity decreases with depth as a function of temperature; (2) pore fluid conductivity increases with temperature; (3) the proportion of pore fluid relative to the solid mineral matter decreases with compaction because of decreasing porosity. Pressure has only minor direct influence on conductivity within the normal geologic environment. The effective thermal conductivity of a formation therefore evolves in different ways depending on the parameters temperature, pore fluid composition and porosity. The initial trend is towards increasing conductivity due to rapidly declining porosity and increasing proportion of the conductive mineral matrix. With increasing burial, as temperature rises, the effect of declining mineral conductivity takes over and the bulk conductivity shows a reversed trend toward reduced values. However, when the aqueous pore fluid is replaced by gas, the conductivity continues to increase and the reversal point of the curve can move to much greater depth (or temperature). This is because gas

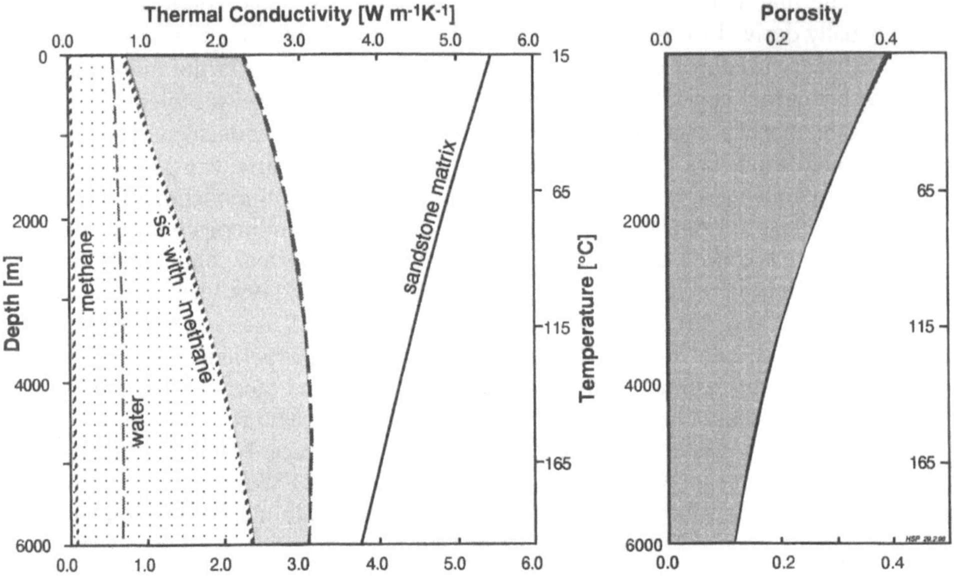

Fig. 1.13. Influence of porosity, temperature, and pore-filling fluids on the effective thermal conductivity of a sandstone. Note the large difference between sandstone filled with methane and sandstone filled with water. Note also that thermal conductivity of gas filled sandstones keeps increasing to much greater depth than that of water filled sandstones. (after Zwach et al. 1994)

conductivity rises almost linearly with temperature while pore water conductivity levels out (Fig. 1.13). On the other hand, when the gas saturation increases slowly, the conductivity curve lies between the water and gas curves and eventually joins the gas curve with an overall effect of decreasing conductivity with depth.

The gas effect on thermal conductivity and temperature history was tested with a modified code of the program package PetroMod (IES 1993) taking into account hydrocarbon saturation changing with geologic time and differing in space. Figure 1.14 shows the thermal insulation effect of a "gas bubble" in just one part of the section, representing a spatially and temporally limited gas field. The change in isotherms and maturation of the source rock is clearly visible.

Gas fields usually have a limited extent in a basin but can still have a marked local effect on the temperature history and distribution. This effect should not be ignored. By contrast, some sedimentary basins, such as the Alberta Deep Basin and the San Juan Basin, have deeply buried sections that are completely gas saturated beneath a water-saturated zone (Berry 1959; Silver 1968; Masters 1979, 1984). Clearly, gas saturation must be considered for the calibration of the basin thermal history with the consequence that input parameters such as heat flow would have to be lowered in order to match measured vitrinite reflectance. Studies in the Alberta Deep Basin (Zwach et al. 1994) have shown

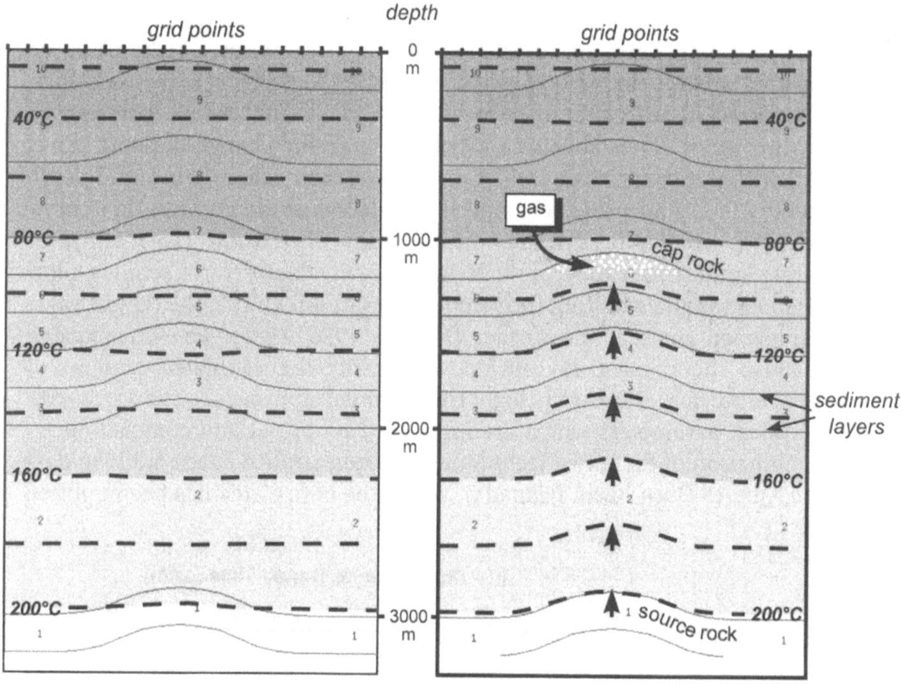

Fig. 1.14. Simulation of a schematic hydrocarbon trap comparing the effects of water and gas saturation on temperature. *Left*, All sediments water filled; *right*, gas filling in structurally high position of layer 6. *Thin lines*, layer boundaries in a sand-shale sequence; *dashed lines*, isotherms showing the simulated present-day temperature field. Gas was generated in layer 1 and has migrated upward into a sandstone reservoir (layer 6) beneath a shale cap rock (layer 7). Gas saturation in the trap was assumed to be present for the past 20 m.y. (Zwach et al. 1994)

how different the reconstructed heat flow history can be when gas is partially or completely substituted for water as pore-filling fluid. Conversely, comparing water and gas filled models for the same heat flow history demonstrates greatly increased vitrinite reflectance and much higher petroleum generation.

1.5.2 Erosion of Overburden and the Estimation of Maximum Burial

Stratigraphers have commonly estimated the amount of erosion by comparison with sections that appear complete or by projecting cross sections from less eroded parts of a basin to reconstruct pre-erosion thicknesses and geometries (e.g., Kalkreuth and McMechan 1989). This method has its limitations where nearby sections are incomplete or unavailable, and it also must assume that no irregular thickness changes were present.

A second group of methods is based on using the irreversible effects of burial and compaction on the physical properties (density, porosity, velocity and resistivity changes, or maturation, etc.) of shales to restore and estimate the maximum depth of burial. One of the first to apply this thinking was Athy (1930) who estimated erosion of 4000–5400 ft in northeastern Oklahoma based on measured shale densities. Physical properties logs, especially sonic logs, have been used in several studies to estimate maximum burial. This can be done in two ways: (1) establishing a typical curve for change of sonic travel time (Δt) with depth for shales which are still at their maximum burial depth (Jankowsky 1962) and projecting uplifted shale values down onto this curve; the difference in depth indicates the amount of erosion (Fig. 1.15); or (2) plotting sonic travel time of shale intervals vs. depth (on a semilog plot) and extrapolating the resulting (hopefully) straight line to Δt_0, the typical value of uncompacted shale at the surface (Magara 1976, 1986). The extrapolated height above the present-day surface represents the estimated amount of missing overburden or erosion (Fig. 1.16). Similar approaches can work with other physical parameters which are influenced by burial and compaction.

A combination of these two techniques has been applied successfully to data from the Alberta Deep Basin (Canada), where the entire area has been uplifted

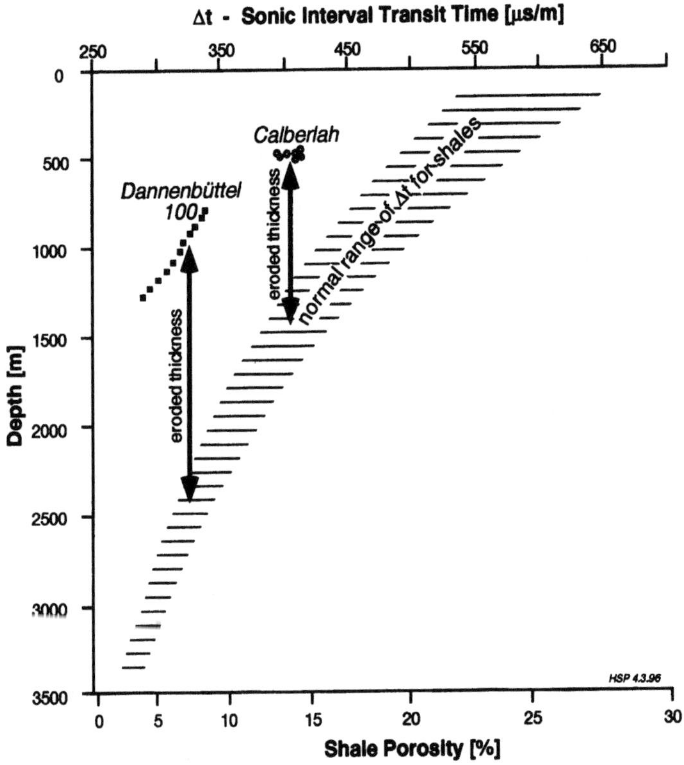

Fig. 1.15. Typical curve for sonic travel time (Δt) and porosity vs depth for Lias epsilon (Jurassic) shales at maximum burial depth; this figure illustrates method (1) for estimating thickness of eroded overburden. Data from Gifhorn Trough, northwestern Germany. (Jankowsky 1962)

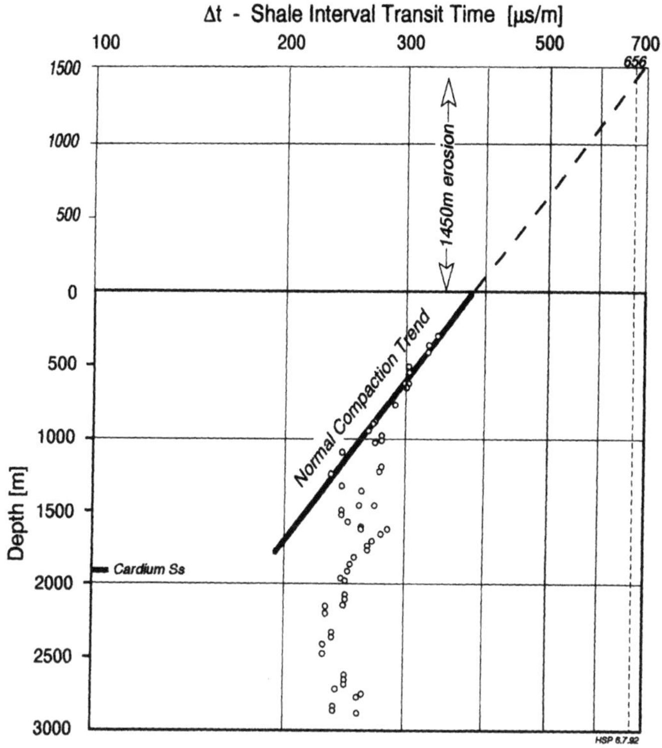

Fig. 1.16. Estimate of erosion of overburden in a well based on extrapolation of sonic interval transit times vs depth to an uncompacted shale value of 656 μs/m; example for method (2) decribed in text. Data from Western Canadian Sedimentary Basin (Alberta). (Reprinted from Magara K., Geological models of petroleum entrapment, 1986, pp 1–328, with kind permission from Elsevier Science - NL, Sara Burgerhartstraat 25, 1055 KV Amsterdam, The Netherlands)

and no data exist, for construction of a standard burial curve (Poelchau and Zwach 1994). First, the amount of erosion was estimated with the extrapolation technique (2) for several wells widely distributed over the study area. These points were used to calculate maximum depth of burial for one of the younger shale formations for which average Δt values could be determined from the sonic logs, and then fitted to Magara's (1986) model:

$$\ln(\Delta t - \Delta t_m) = -c\, z + \ln(\Delta t_0 - \Delta t_m)$$

where Δt is sonic interval travel time of a particular shale formation, c is a compaction factor, z is maximum depth of burial, and o and m the indices of uncompacted shale and clay matrix, respectively. The resulting regression equation is:

$$z = A \cdot \ln(\Delta t - \Delta t_m) + B$$

This is equivalent to the standard burial curve (1) and allows calculation of maximum depth of burial for this shale layer in all wells having sonic logs in the study area and, by extension, permits mapping of the thickness of overburden removed by erosion.

Maturation of organic matter, i.e., vitrinite reflectance (VR), has been used in analogous ways. Connolly (1989) has estimated erosion by extrapolating log VR trends to a surface value of 0.2%. This approach is based on earlier attempts by Hacquebard (1977) who used coal rank and moisture to determine the maximum burial of Mannville coals in Alberta. Weiss (1985) has used a more refined approach (Fig. 1.17). Matching measured vitrinite reflectance to vitrinite reflectance calculated via the time-temperature index (TTI; Waples 1980) for various overburden and temperature gradients, he established groups of curves for several shales and estimated the most likely temperature gradients and overburden for several wells in Alberta.

Use of vitrinite reflection for estimation of overburden or erosion is relatively easy where data are available but has two distinct disadvantages. The extrapolation approach assumes that only the temperature associated with the maximum burial depth has influenced the maturation. Clearly this is not always the case since points at different depth may have resided at these temperatures for varying lengths of time. Secondly, since vitrinite reflectance is commonly used for calibrating the basin simulation temperature history, it should be reserved for that purpose and not, even indirectly, enter the input data for the simulation.

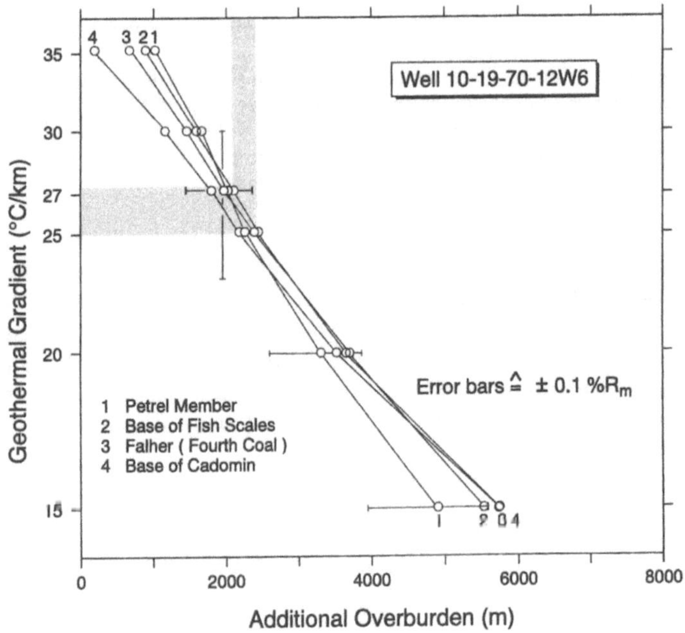

Fig. 1.17. Calculated thickness of eroded overburden in several formations for various assumed geothermal gradients. The values plotted are the result of iteratively calculating the vitrinite reflectance from TTI with varying overburden until they match the measured vitrinite reflectance. (Weiss 1985)

1.5.3 Methods of Predicting Diagenesis

Prediction of porosity (and permeability) before drilling has always been in great demand. Because of the many highly varying parameters affecting porosity development, most attempts in the literature have been empirical (e.g., Bloch 1991; Bjørlykke et al. 1989; Scherer 1987). In all of these attempts the problem has been that no global scheme exists for prediction, that each empirically derived equation or curve is fairly specific for one basin or formation with its attending geologic idiosyncrasies.

Timing and sequence of diagenetic mineral reactions in a formation are usually worked out with standard petrographic observations in thin sections. Modern tools are now available to aid in this interpretation. For instance, cathodoluminescence microscopy (Sippel 1968), and especially the new technique with a hot cathode (Neuser et al. 1989), is capable of resolving several cement generations in the pore space. Application of stable isotope analysis of cements and fluid inclusion homogenization temperatures has helped to bracket cementation temperatures and pore fluid composition (e.g., Tilley 1988). Fission track analysis of apatites can supply some dates and temperatures in the diagenetic history (e.g., Meyer et al. 1989; Leischner 1994). Although these methods can supply an approximate history of diagenetic events, their timing is still mostly subject to speculation (except perhaps for fission track ages). Also, despite the advances in measuring paleotemperature, detailed distinctions among various cements are rarely possible because of postevent exchanges in open systems, and because microanalytical techniques are not sufficiently refined.

An alternative approach to estimating temperature ranges of diagenetic reactions is available using thermodynamic equations and constants (Helgeson 1968; Bowers et al. 1984; Bjørkum and Gjelsvik 1988). Programs such as SUPCRT (Flowers 1986) and SOLMINEQ (Kharaka et al. 1988) can compute equilibrium constants of many reactions for wide range of temperatures and pressures and supply the data for construction of phase diagrams which can be used to determine at what temperatures and pressures (ranges) and pore water compositions certain minerals precipitate, dissolve, or are transformed. Together with the data gained from petrography this can be used to constrain temperatures and timing during basin development.

The most sophisticated approach has been to model the causal processes that modify porosity and permeability. This involves the integrated simulation of transport of pore water carrying dissolved chemical species and the reactions between pore water and solid mineral grains at the local temperature and pressure. Given the proper boundary conditions, transfer and reaction coefficients, and the ability to solve the algorithms quickly and efficiently on a supercomputer, this approach could model porosity evolution in sufficient detail to show at least the trends to be expected. Unfortunately, none of these requirements can be fully met at this time. The early models attempted to simply couple existing transport models with chemical reaction model (e.g., Crowe 1988; Vogt 1990) using two-step iterations. More recent models are integrating the two types of processes with more or less simultaneous solutions. Several studies have made inroads to model at least a simplified system, either by limiting the number of reactions (Ondrak 1992), solving only for one dimension and constant temperature (Lichtner 1988) or working with only a few cells or limited time steps (Ortoleva et al. 1987). These models can, for fairly specific and simplified conditions, provide possible scenarios which can be checked against observations. Thus certain hypotheses may be ruled out or accepted.

Zwach and Poelchau (1993) provided an example of the way in which petrography, basin modeling, and diagenesis modeling can be integrated. In a study of the Albian Cadotte Sandstone of the Alberta Deep Basin in Canada, basin modeling provided temperature, pressure, and porosity data as well as the structural geometry of a cross section of the sandstone at the time of deepest subsidence (48 Ma). These data were used as input to a coupled flow and chemical reaction model developed by Ondrak (1992). Simulating the amount of quartz overgrowth due to regional flow of cooling pore water which might occlude the remaining porosity produced results that essentially repudiated the hypothesis of regional flow cementation. Instead, Zwach and Poelchau (1993) postulated from petrographic observations that pressure solution of chert grains provided most of the silica cement found on quartz grains. This supported the alternative hypothesis that small scale, closed system processes had affected the porosity much more than regional flow.

In general, whether a system is open or closed and the scale on which pore fluid transport affects diagenesis is still a major question. Bjørlykke has repeatedly published evidence (e.g., Bjørlykke et al. 1988) that most deep diagenesis is an isochemical process in a closed system with little long range transport. The exception is meteoric water advective transport which normally takes place early in shallow rocks (Nedkvitne and Bjørlykke 1992). In many cases, such as quartz pressure solution and overgrowth, the main transport process appears to be very small scale intergrain diffusion, requiring no advective transport at all but still accomplishing major porosity changes (Zwach 1995).

1.5.4
Structural Deformation History

Structural or tectonic deformation in sedimentary rocks can be defined as the result of processes which directly affect the rocks and their geometrical relationships within the sedimentary sequence, as opposed to sedimentation and erosion processes which affect only the surficial layers of a sedimentary basin. Tectonic processes can be indicated for example by faults, folds, or the movements of less competent materials such as salt layers. The two extreme types of regional stress systems which control tectonic processes in sedimentary basins can be described as extensional, where the depositional area is stretched, and the previously deposited units must accommodate a larger area, or as compressional, where forces act to shorten crustal sections.

The structural deformation history is a crucial controlling factor on both conductive and convective heat transport. Conductive heat transport is a function of the changing bulk thermal conductivities of each lithologic unit through time, and tectonic processes can modify the geometry, i.e., the relative positions of lithologic units, and also affect their physical properties. The structural deformation history can also influence convective heat transport, for example by creating faults which can either open or close off fluid flow pathways. The timing and location of the generation and expulsion of hydrocarbons is directly controlled by the resulting thermal histories, while the subsequent migration and accumulation of hydrocarbons and mineral-bearing fluids are significantly influenced by the structural history.

It is important to avoid the attractions of anomalous local or small-scale geologic features when interpreting their effect on bulk geologic processes. A typical example is the tendency to overemphasize the importance of faults (i.e., permeability anomalies) as possible pathways for fluid movements. These movements are a function of the fluid potential gradients and not just of permeabilities. Even if horizontal permeabilities in shales are significantly higher than the vertical values, the fluid quantities can be much larger moving in a vertical direction than horizontally as the total surface area through which movement occurs is obviously much larger for vertical than for horizontal movements (Chapman 1981). The relative importance of the possible effects must be considered. For example, do the volume of fluids and the associated convective heat moving through a fault zone during a certain time unit really have a noticeable effect when compared to the total amount of heat being transported through the system by conduction? The answer to this question can be given only by a rigorous, quantitative analysis of the volumes and velocities of transported fluids with an integrated conduction/convection model.

In the following sections, examples for the direct and indirect controls exerted on the hydrocarbon generation and accumulation process by the structural history are provided from both extensional and compressional deformation systems. The conclusion in each case is that generalized interpretations of the role of structural deformation processes in sedimentary basin histories virtually always represent potentially dangerous oversimplifications.

Growth Faulting

Although not necessarily restricted to extensional tectonic environments, growth (or syndepositional) faults and their effects provide valuable insights into the tectonic factors which control the movements of both water and hydrocarbons. Hooper (1991) provides a concise review of the often contradictory arguments concerning fault properties and their effects on fluid flow. Direct physical evidence can be given which, for most indicators, supports both fault sealing and fault opening properties, often along the same fault. This evidence includes in situ permeability measurements, where permeabilities are reduced due to mineral breakdown or increased due to dilatation effects prior to fault movements. In addition, shale smears on fault planes are indicative of permeability reductions, and mineralizations give proof of at least temporary increases. Fluid properties such as compositions (hydrocarbons and/or water), temperatures and salinities, but also potential gradients and flow directions can provide clear evidence for faults acting either as seals or as conduits on a local as well as on a regional scale. Indications are that faults can be both seals and conduits, and that flow can occur periodically along faults which are otherwise sealed, especially during phases of increased fault activity (seismic pumping).

Many examples of growth fault systems with well-documented measured data have been published, especially in the Gulf of Mexico (e.g., Flemings et al. 1992), which conclusively prove the existence of fault-induced perturbations of pressure and temperature fields in an abnormally pressured environment. However, resolution generally does not suffice to determine whether the fault or fault zone is open, providing a path for fluid movement and the concomitant perturbations, or whether it is acting as a seal and therefore forcing fluids to ascend in front of the fault zone.

Overthrusting

Overthrusting represents an extreme case of a compressional tectonic environment and provides a particular challenge for explorationists. Displacement distances of thrust sheets can exceed 100 km and thicknesses are commonly in the range of 4–6 km (Gretener 1972). Consequently, during overthrusting, the volume and velocity of mass movements can be more extreme than under any other type of geologic process, and their combined effects can result in drastic changes of geologic parameters such as temperatures and pressures. The dimensions of thrust sheets and the general lack of pronounced frictional effects indicate that friction is reduced during the thrusting process. The classic papers of Hubbert and Rubey (1959; 1961), and Rubey and Hubbert (1959) showed that overpressured pore fluids can reduce effective stress and, in effect, float the rock sequence overlying the overpressured zone. The moving thrust sheet can even create its own overpressured basal zone due to the extreme loading rates (Gretener 1969). More refined models (e.g., De Bremaecker 1987) have since been developed, but the general conclusions on the role of overpressures and the negligible effects of frictional heat during thrust sheet movements remain valid.

In order to determine the possible thermal effects of thrust sheet movements, Oxburgh and Turcotte (1974) presented a concept on which virtually all subsequent work has been directly based. Overthrusting is modeled with an "instantaneous emplacement" approach: the duration of the overthrusting process is "brief," a sawtooth-shaped temperature profile is created (i.e., an extreme temperature inversion), and heat flows down from the overlying thrust sheet until temperatures in the entire sequence requilibrate. Angevine and Turcotte (1983) applied the concept to problems of hydrocarbon generation below thrust sheets.

Wygrala et al. (1990) argued that instantaneous emplacement is not the proper concept and applied a two-dimensional model to test the sensitivity of various parameters. The model makes it possible to test such complex factors as the gradual emplacement of the total thickness of the thrust sheet, thrust plane angles, thrust sheet movement rates, and erosion during overthrusting as well as the role of thermal conductivity and heat capacity of the lithologies. The study showed that one-dimensional instantaneous emplacement concepts oversimplify the actual processes as equilibration occurs mostly during the time of thrust sheet emplacement, and that disequilibrium, or temperature inversion, occurs only when thick sheets (approx. 4+ km) are emplaced with high movement rates (approx. 5 cm/year; Fig. 1.18). The sawtooth-shaped temperature profile of Oxburgh and Turcotte (1974) cannot be generated under realistic geologic conditions. The total amount of heat transported in the thrust sheet itself (by convective or mass transport) is negligible compared to the total amount of heat flowing through the sequence from the basement during the overthrusting process.

The implications for hydrocarbon generation are that there are no simple solutions due to the number of the variables which must be taken into account or at least subjected to a sensitivity analysis. One-dimensional "short-cuts," for example by simply "depositing" a thrust sheet with high sedimentation rates, can be misleading. One-dimensional simulations, though mathematically correct, only permit an approximation of temperature histories in thrustbelts, even if the results may be sufficiently accurate in some cases to provide a qualitative calibration and interpretation of thermal histories.

Salt Movement

The thermal effects of salt movements are another case where generalizations are common, but where the problems are often oversimplified, leading to inaccurate or even erroneous conclusions. The most widespread generalization on the thermal effects of salt bodies is that they act as conductive "pipes" due to their relatively higher thermal conductivities and lower porosities. Therefore, temperatures in a salt dome are higher than in the adjacent rocks at any given depth level. However, Lewis and Rose (1970), using a simple model, show clearly that thermal conditions in and around highly conductive bodies can vary depending on the type and geometry of the structure as well as on the depth level relative to the structure at which the temperature or the heat flow values are being compared. Both temperature and heat flow distributions are perturbed. Near the base of an isolated highly conductive body temperatures are lower than in the adjacent rocks, while near the middle level of the body there are no lateral temperature differences, and near the top of the body temperatures are higher within the dome than in the adjacent rocks. These conclusions are made for a completely buried dome and additional complexities ("cold" vs. "hot" domes) arise when a dome has penetrated the entire overlying sedimentary sequence to reach the surface.

Simple models with uniform heat flow values at a sufficient depth below the base of the highly conductive salt dome, show that heat flow values within the dome (i.e., from below the base to above the top) are higher than in the adjacent rocks, but in the immediate lateral vicinity of the dome there is a zone where heat flow is even lower than in the surrounding rocks at greater dis-

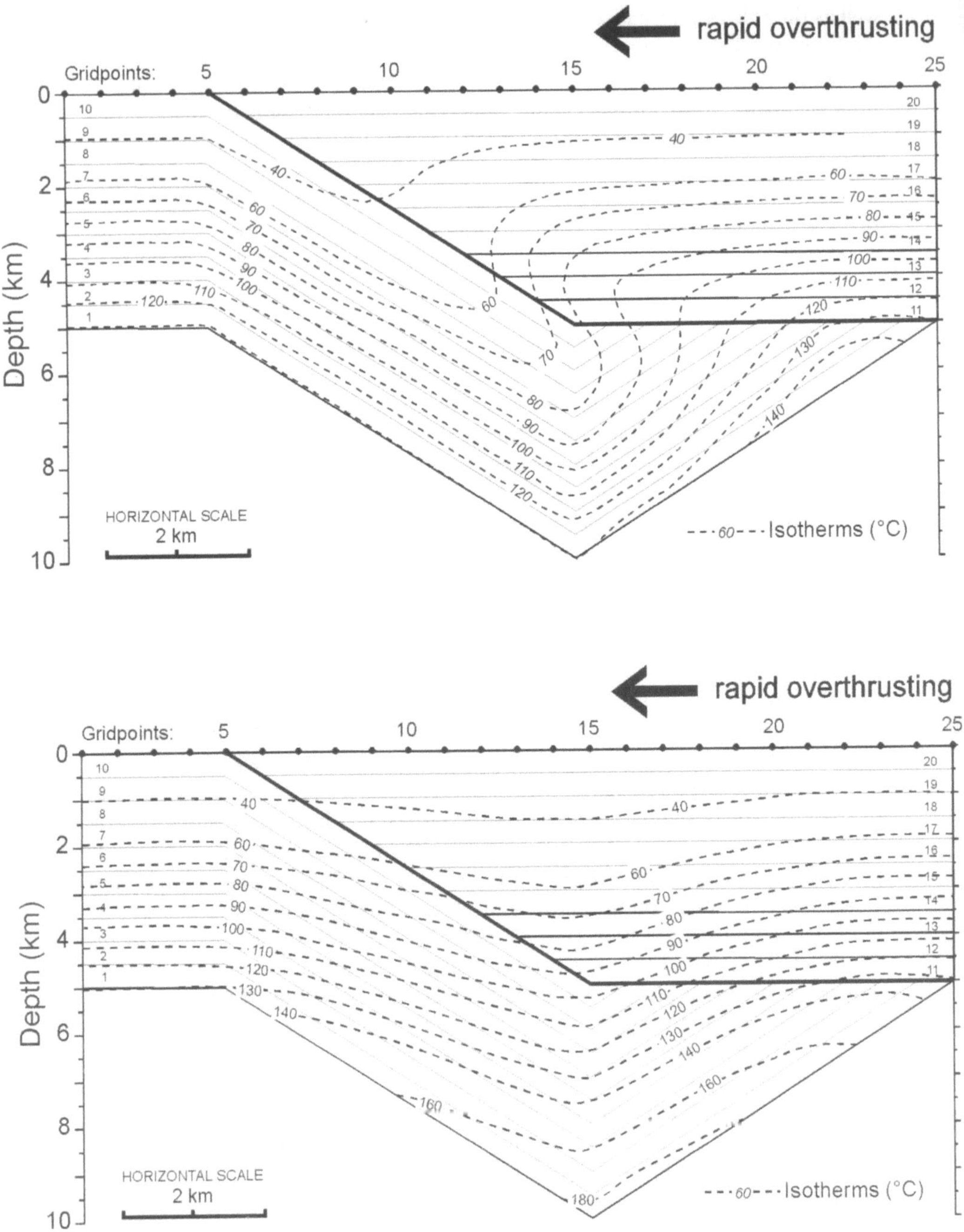

Fig. 1.18. Simulation of heat distribution compared for rapid overthrusting (*left*) and slow overthrusting (*right*). *Upper panels*, temperature distribution at the end of thrust movement (rapid emplacement is accomplished in 100,000 years, slow in 1 m.y.). *Lower panels*, progress of temperature equilibration 1 m.y. later. (Wygrala et al. 1990)

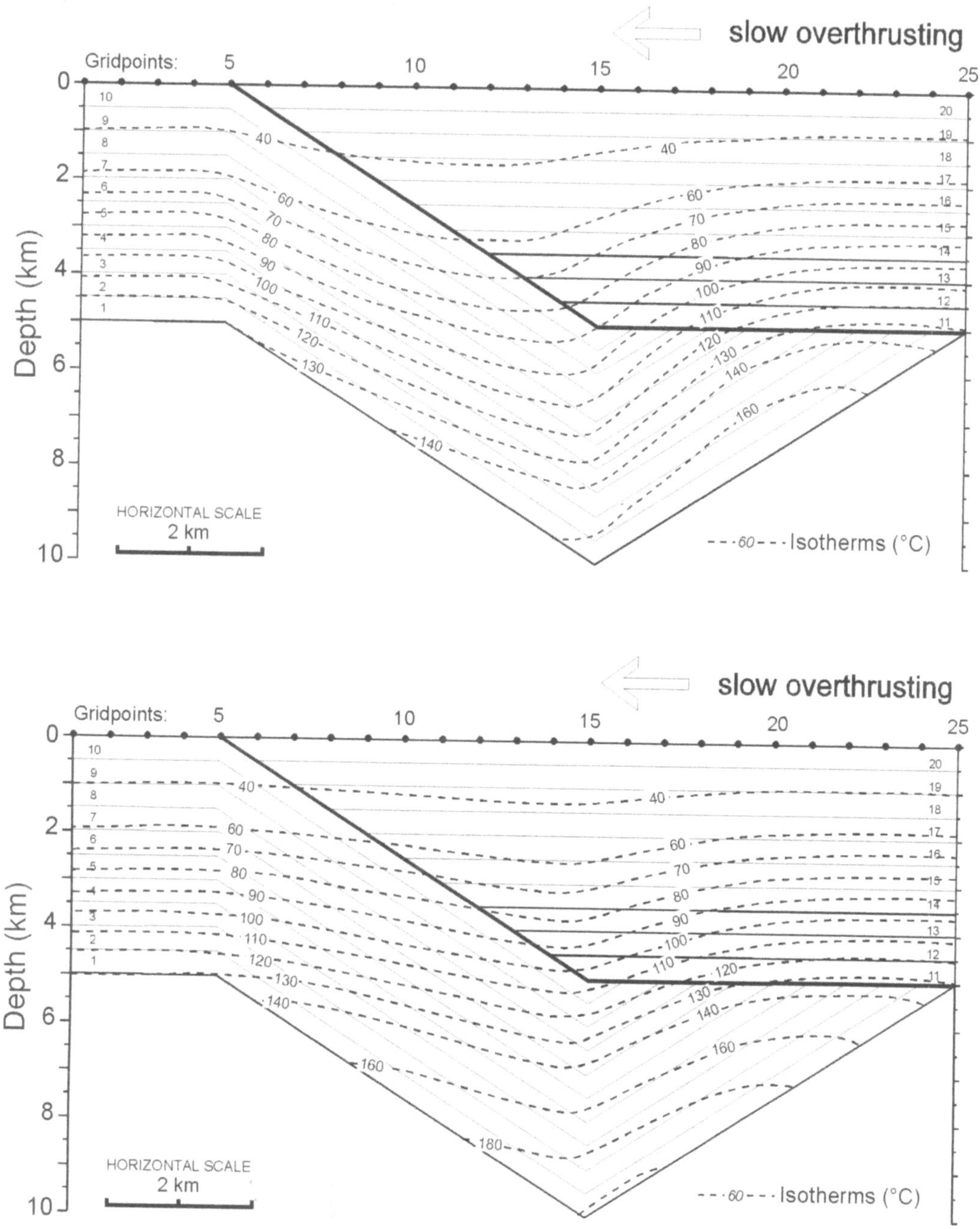

Fig. 1.18 (*Contd.*)

tances. Then, at a sufficient distance above the top of the submerged, highly conductive body, heat flow values can once again be uniform.

The implications of these temperature and heat flow perturbation patterns can be considerable, for example, when thermal and hydrocarbon generation histories are being simulated in the vicinity of salt domes. One-dimensional models are frequently corrected only by arbitrarily increasing heat flow values (or "temperature gradients"), and the "calibrated" values are then used for hydrocarbon generation modeling. The simple examples above show, however, that heat flow values are often reduced near salt domes. Factors to be taken into account include the timing and the changing geometry of salt movement as well as the distance of the simulated well from the salt body. Two-dimensional models are a definite improvement if they can take lateral effects into account, and if the modeled section cuts across a salt wall, but the typical dome-shaped geometry of salt bodies necessitates three-dimensional modeling for more accurate thermal history analysis.

1.5.5
Petroleum Generation and Estimation of Petroleum Yield

One of the ultimate goals in basin modeling is to assess the amount of petroleum that could have been generated in a petroleum system from source rocks and accumulated in reservoirs. The principal limit to petroleum abundance in every petroleum system is the ability of the source rock to provide a sufficient charge to associated carrier beds. Without adequate generation, there can be no accumulation of petroleum no matter how favorable reservoir quality, sealing and trapping are.

Source rocks situated within the diagenesis zone of maturation (vitrinite reflectance <0.5%; see Radke et al., this Vol.) are usually immature. The subsequent breakdown of kerogen in the zone of catagenesis (vitrinite reflectance = 0.5–2.0%) to yield petroleum can be discerned by a significant and accelerated increase in the relative concentrations of extractable hydrocarbon relative to total organic carbon. This phenomenon has been observed in the Los Angeles and Ventura Basins (Philippi 1965), the Paris Basin (Tissot and Pelet 1971), the Uinta Basin (Tissot et al. 1987b), in German coals of increasing rank (Leythaeuser and Welte 1969) and many other locations. The oil generation process results in a 4- to 12-fold increase in hydrocarbon/organic carbon ratios (Philippi 1965; Magoon and Claypool 1983; and many other studies).

Generated petroleum-like compounds are thought to result from a multitude of quasi-irreversible, assumed first-order thermal cracking reactions (Huck and Karweil 1955; Hanbaba and Jüntgen 1969; Tissot 1969) whose overall rate is mainly governed by kerogen structure and the extent of thermal stress over geologic time (see also Schenk et al., Chap. 4, this Vol.). The progressive loss of components from marine kerogens proceeds according to bond strengths with weaker bonds breaking before stronger ones, leaving behind a hydrogen-depleted residue. Kerogen at an optimum stage of liquid hydrocarbon generation is said to be mature. The early and middle part of catagenesis is dominated by oil generation from kerogen and accumulation within the pore and fracture system of the source rock.

Temperature is considered to be of overriding importance in generating petroleum from organic matter enclosed in source rocks (Philippi 1965; Welte 1965; Vassoyevich et al. 1969). Insufficient heating gives insignificant conversion, whereas overheating results in the breakdown of primary products to lighter liquids and gases plus an aromatic residue.

Kerogen type exerts a major influence on yield, petroleum type and timing of generation. Absolute yields generated in nature appear to decrease in the order of kerogen types I, II, and III, which is in line with predictions from kerogen pyrolysis (Espitalié et al. 1977).

Kinetic models (see also Schenk et al., this Vol.) can be used in basin simulation to determine the extent of petroleum generation using experimental data on immature samples (Ungerer and Pelet 1987; Burnham et al. 1987; Schaefer et al. 1990). Kerogens are heated at various heating rates differing by two to three orders of magnitudes, and the experimental pyrolysis yield curves are modeled using varying fitting techniques (Burnham et al. 1988). Conceptually, the models consider an infinite set of parallel cracking reactions obeying first-order kinetics whose rate constants can be determined by the Arrhenius equation. The resulting continuum of activation energies is arbitrarily discretized, for example, in intervals of 2 kcal/mol between 40 and 80 kcal/mol (Ungerer and Pelet 1987), a range which was found to be sufficient to encompass all kinds of sedimentary organic matter types. The frequency factor (A) initially is kept constant for each reaction but is used for optimization in the fitting procedure.

With these types of kinetic models the influence of time and temperature on the formation of oil and gas from kerogen can be determined in some detail. Most importantly, it was found that reaction rates do not double for each temperature increase of 10° C as in the Lopatin/Waples TTI model (Waples 1980), but rather for a 3–5° C increase (Sweeney et al. 1987; Tissot et al. 1987a).

All of the foregoing discussion refers only to the yields and types of petroleum generated within source rocks. This is only the first step of petroleum formation. Losses during passage through migration pathways and fractionations associated with either primary, secondary, or tertiary migration can play an overriding influence in controlling both yields and compositions of petroleums in reservoirs (Thompson 1988; England et al. 1987; Leythaeuser et al. 1987; Leythaeuser and Poelchau 1991). Any assessment of in-place petroleum reserves must clearly consider these influences sequentially.

1.6 Optimization and Calibration: Testing and Evaluation of the Model

Simulation of the basin evolution based on the discretized, numerical, conceptual model results in output of a large number of data for each layer at each event time at each grid point. Thicknesses of compacting layers and porosities, temperatures and maturity parameters, thermal conductivities and permeabilities, pressure and overpressure all are recalculated at each time step and can be graphically displayed against time or depth.

Graphic comparison of calculated values with the measured observations quickly indicates whether the conceptual model matches the geologic reality. If the match is good, the geologic assumptions made for the conceptual model can be said to be reasonable although not necessarily true. If the match is poor it can be concluded that the hypothesis of the conceptual model is false, i.e., that certain assumptions need to be changed. Assumptions to be changed are usually those that cannot be measured or calculated directly or accurately. They may include past water depth and sediment surface temperature, heat flow, thermal conductivities, or other physical properties of average lithologies, amount and duration of erosion, or length of an hiatus.

The systematic testing and evaluation of results, i.e., solutions and outputs, is a crucial and distinctive step of all well designed modeling studies. Although there is some confusion in terminology, the procedures are commonly referred to as optimization, calibration, checking, and sensitivity analysis (Yükler 1979; Wygrala 1989; Hermanrud 1993).

Checking, a term applied by Yükler (1987), refers simply to the direct comparison of modeled solutions to comparable observational data. Differences between calculated and observed values provide a basis for estimation of probable errors in simulated outputs. When calculated outputs are within "acceptable" error limits (which of course must be specified) of observed values, the solution is judged as "successful" (see Welte and Yükler 1981).

If results are beyond acceptable error limits, the model is deemed unsuccessful, and optimization procedures must be applied. This is a most critical and significant part of the testing and evaluation of numerical modeling procedures. Specifically, optimization involves the changing or adjustment of input data, pertaining both to the geologic conceptual model and the simulation equations, followed by recalculation of new solutions which are again checked against observational data. This is a "trial and error" procedure, typically involving successive trials of different inputs, until acceptable solutions are obtained, i.e., outputs of calculated results agree within the acceptable error limits with observed values. Thus, optimization of the model searches for a set of inputs which yield results that best satisfy or coincide with available observational data. Analogous to riflery, optimization has logically also been called "calibration," because the procedure involves making corrections (e.g., adjusting sights for range and elevation inputs) so that the target (measured observational data) is directly "hit" by the shot (i.e., output values)!

Calibration is always limited by the availability of measured observational data. Because of its substantial influence on most aspects of basin modeling, calibration of thermal history outputs is always essential. This is usually accomplished by the comparison of calculated present-time temperature profiles and calculated vitrinite reflectance, compared to present-day measured borehole temperatures and observed vitrinite reflectance-values.

1.6.1 Temperature Calibration

Temperature data are the most easily available data for calibration, albeit usually restricted to present time. The most commonly used temperature parameters are present-day bottom hole temperatures (BHT) and borehole temperature logs.

BHT temperatures should be corrected for circulation of cooler drilling or completion fluids. This is carried out with the Horner plot extrapolation when two or more temperatures measured during different logging runs are available from the same depth at different times after circulation has stopped (example, Fig. 1.19). Because of the difficulty of correcting BHTs to steady-state temperature they tend to be lower than actual formation temperature but are rarely too high.

A similar kind of data are temperatures from drill stem tests (DST) or formation tester (FT). They usually have been corrected already by the Horner plot technique and often are available together with corrected formation pressures.

Temperature logs can be a source of good temperature data if they have been run after the well has been shut in for a sufficiently long time (weeks or months) to reach a state near equilibrium. In some countries (e.g., Russia) this seems to be done routinely, while in other areas (e.g., North America) this is usually not the case. Such temperature data are usually close to real formation temperatures but tend to be lower than DST extrapolated temperatures in the lower part of the well and higher in the upper part.

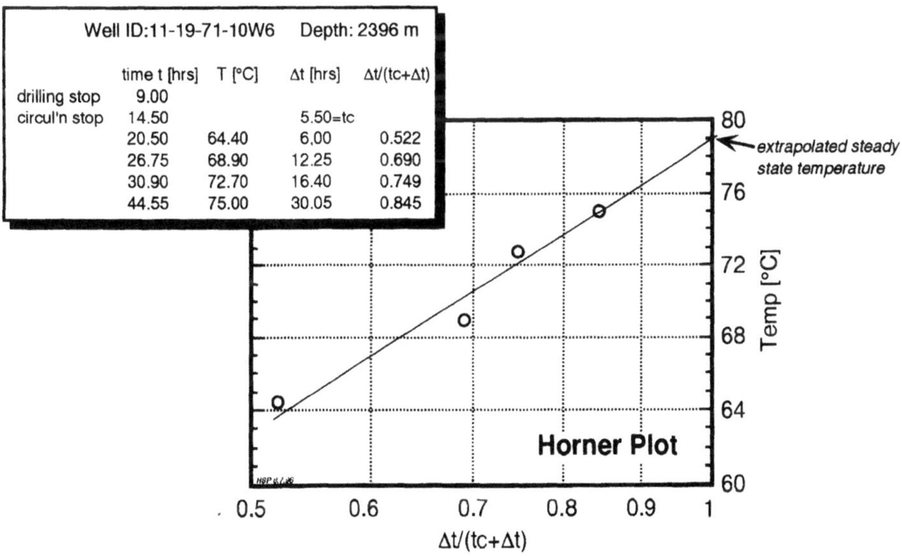

Fig. 1.19. Example of Horner Plot extrapolation method to correct BHT for cooling during drilling and circulation. The corrected steady state temperature for this formation is 79° C

Another occasional source of data are gas production temperatures which should be slightly lower than formation temperature but often seem to fit well with BHT temperature trends.

Simulated present-day temperature is affected by recent heat flow, surface temperature, and by thermal conductivity of the underlying and overlying layers. It is not affected by the heat flow history (except for the most recent part). Therefore, present-day temperature is fairly simple to match, but the match has little significance for the thermal history of the basin. Needed are some indicators for actual paleo-temperatures for various geologic events.

Inorganic paleo-temperature indicators provide such additional constraints on the model. Certain geothermometers from water analyses (Kharaka and Mariner 1989) and from diagenetic studies (fluid inclusion and fission track temperatures) can furnish maximum fluid temperatures or fix points on the time-temperature scale and are useful for supplementary checking. Leischner (1994) has demonstrated how useful and important such a comparison and integration of different inorganic and organic temperature indicators is for the final calibration of the model (see also a detailed report on his study in Yalçin, this Vol.). Petrographic observations often suggest a history of short-term hydrothermal events, but these events tend not to affect organic maturity to a great degree.

1.6.2
Vitrinite Reflectance Kinetics and Other Organic Calibration Parameters

Vitrinite reflectance is the most widely used parameter to indicate the maturity of the source rock. Since it is almost routinely measured and abundantly available, it is also the primary calibration parameter for modeling the temperature history.

For calibration, the increase in vitrinite reflectance is simulated in the basin modeling program in accordance with the conceptual model and the temperature history. Several "kinetic" models for vitrinite reflectance evolution have been designed and used extensively. The oldest and most widely used was Lopatin's model (Lopatin 1971), popularized as TTI index by Waples (1980). This rather simple model is based on the premise of "dark room" kinetics that the reaction rate doubles with each 10° C increase in temperature. The method is simple and easy to apply but does not give proper results for basins with very low or high heating rates (Quigley et al. 1988).

Reaction kinetic models using the Arrhenius equation (see Schenk et al., this Vol.) and a distribution of activation energies have been more successful. The two models used predominantly have been proposed by Larter (1988); Larter (1989) and by Burnham and Sweeney (1989) and Sweeney and Burnham (1990). The Larter model is based on the observed reduction of phenolic moieties during the maturation of vitrinite and shows good matches in the reflectance range of 0.45 to 1.5%. The Burnham/Sweeney model describes the chemical changes in vitrinite with four overlapping reactions: the successive release of water, CO_2, higher hydrocarbons and methane. The decrease in these components results in a reduction of the H/C ratio in vitrinite which can be correlated with the increase in reflectance (Leischner 1994). The four reactions are each described by an activation energy distribution in the spectrum from 38 to 74 kcal/mol and a constant frequency factor of 10^{13} s^{-1}. A simplified, but nevertheless equally effective model, the so called EASY%Ro model (Sweeney and Burnham 1990), combines the four reactions into one spectrum of activation energies. This model has been most successful for calibration purposes because it is applicable for maturation values as high as 4.6% vitrinite reflectance.

Although vitrinite reflectance is the most commonly available and routinely measured maturation parameter used for calibration, it is not always the best. In some situations vitrinite reflectance may be suppressed by weathering, bitumen retention, or microbial alteration. In oxic environments (e.g., carbonates) vitrinite reflectance can be too high. Also, there are some lithologies in which vitrinites are rare or absent. In these cases other organic parameters can be used instead (see Radke et al., this Vol., for details). Leischner (1994) has studied and compared a large number of these parameters and also constrained them with inorganic, diagenetic measures.

Pyrolysis (RockEval) T_{max} values are an alternative maturation parameter that has been correlated with vitrinite reflectance, especially for type III kerogen (Espitalié et al. 1984). Provided that the vitrinite reflectance values based on T_{max} fall within the established trend, they can be used as an additional constraint. Sweeney (1990) has published a program for simulating T_{max} evolution with temperature history as result of petroleum generation, that can also be used for further calibration.

Another group of organic maturation parameters is based on the isomerization and aromatization of certain organic compounds in the kerogen. One such commonly applied parameter is the Methylphenantrene Index or MPI (Radke and Welte 1983) which has been correlated with vitrinite reflectance.

Hopane or sterane isomerization is sometimes applied as additional maturation calibration parameter. The kinetic reactions have been worked out by Rullkötter and Marzi (1988) and Mackenzie and McKenzie (1983) and can be used directly in the simulation eliminating the need for recalculation into vitrinite reflectance.

1.6.3 Clay Kinetics as Temperature History Indicator

The transformation of smectite to illite (S/I) has been known to take place over a range of depths suggesting a kinetic process involving temperature as well as time (Hower et al. 1976; Burst 1969, etc). Both experimental and subsurface data involving different heating rates have been interpreted to reflect a first order kinetic reaction, and kinetic models have been fitted to these observations (Eberl and Hower 1976; Pytte and Reynolds 1989).

Vasseur and Velde (1993) have recently developed a computer model to devolve S/I transformation curves and have suggested a two step model involving two sequential processes and sets of kinetic parameters. This theory has been based on data from numerous different basins and settings. Each of these settings resulted in different numerical values for activation energy (Ea) and frequency factor (A). However, when plotted on an A vs. Ea diagram all values fall almost on a straight line. This means that mathematically the magnitude of the two parameters is less important than their ratio, although chemically some of these values may be meaningless.

In practice then it is possible to find parameters for a kinetic equation that models clay transformation satisfactorily to a given temperature history. This model can then be used to provide an additional calibration of the temperature history of a basin. S/I transformation can be especially useful because the depths of most rapid change are usually somewhat shallower than those of the vitrinite transformation. For deeper sediments the process of crystallinity increase of illite with temperature can be used as an additional calibration technique.

1.6.4 Compaction or Porosity Optimization

Forward modeling of compaction, in contrast to back-stripping models, makes assumptions about initial porosity as well as compressibility of each lithology. This means that the final compacted thickness may not match the thickness measured in the bore hole. Therefore, an adjustment of the assumed initial values is often necessary before a match with present thicknesses can be achieved. In some systems (as in the PDI and PetroMod programs) the balance between thickness and porosity is optimized automatically. The resulting present-day porosity distribution can be checked against measured core porosity or, better, against average formation porosity calculated from porosity logs, such as sonic or neutron logs.

Using sonic logs, the porosities can be calculated from digitized (or hand picked) log values using the Wyllie equation:

$$\phi = \frac{\Delta t_{log} - \Delta t_m}{\Delta t_f - \Delta t_m}$$

where Δt_{log} is the transit time read off the sonic log curve (in μs/ft), Δt_m is the matrix transit time of the rock, and Δt_f is the transit time of the pore fluid. Constants commonly used are $\Delta t_m = 55$ for sandstone and 60 for shale, and $\Delta t_f = 185$ for normal saline pore waters (see Asquith 1982, p. 66). Shale and sandstone porosities should be calculated separately using the gamma ray curve with a shale-line cutoff to distinguish the two lithologies. Then an average for each lithology can be computed and a weighted average formation porosity compared with the simulation.

1.6.5 Sensitivity Analysis

Both the optimization and subsequent calibration procedures raise questions concerning the influence that different input parameters have on simulation results. These effects are evaluated through the procedure of sensitivity analysis. This method simply involves making finite changes of input parameters (ordinarily one at a time) for the previously optimized "successful" model followed by recomputation of results. Sensitivity is defined as the rate of change of the resulting output with respect to change of the input parameter (Zwach 1995). For example, the influence of changes in the magnitude of an assumed basement heat flow on the calculated temperature history and on successively simulated compaction and hydrocarbon generation outputs can be evaluated by systematically changing heat flow inputs and doing iterative computations. Further, the effect of variations in kinetic parameters (e.g., activation energies and prexponential factors) on hydrocarbon generation calculations is commonly evaluated. In principle, the relative influence and importance of any input parameter in governing the nature of a particular output result for the model can be evaluated.

As one might expect, sensitivity analysis also plays an important role during the optimization procedures. Specifically, a knowledge of the sensitivity of outputs to variations in the magnitude of input parameters guides the modeler in making appropriate modifications to the model. In particular, when coupled with information on the probable natural range and inherent errors, i.e., uncertainty in the magnitude of input parameters, optimization of the model can be accomplished in a more realistic and quantitative fashion. Similarly, once a "successful" model has been designated, again using expected ranges and errors of input data, sensitivity analysis is used to estimate probable error, i.e., uncertainty, in the simulation results. An excellent example of the use of sensitivity analysis is given by Zwach (1995) who tested the effect of change of a whole series of input parameters (event resolution, heat flow variation and timing, length of erosional hiatus, thickness of eroded overburden, surface temperature, and gas saturation) on simulated vitrinite reflectance.

1.6.6
The End Result

The end result of the optimization and calibration procedure is a finalized "successful" or acceptable basin model. The conditions, controls, parameters, etc. (i.e., collective inputs) of preliminary models are, through optimization, redefined, modified, and revised to yield the inclusive specifications of the "successful" model. Where possible, additional outputs of the successful model should be checked against comparable observational data. However, as pointed out by Yükler (1987, p. 393), "checking parameters should be chosen in such a way that they are independent of the optimization parameters."

Availability and quality of calibration data pose serious limitations on the recognition and selection of a "successful" model. In particular, most observational, measured data available for calibration pertain only to the present time, for example, present-day borehole temperature, vitrinite reflectance, hydrocarbon yield, and porosity. Measured calibration controls representative at a particular time (age) in the geologic past are very difficult to establish. For example, although some use of geothermometers, such as fluid inclusion homogenization temperatures, apatite fission tracks, oxygen isotopes, and mineral reactions have been utilized, these observations are infrequently available and often are inherently difficult to apply. As a result, "present-day" calibration is the rule. This means that the pathway of a temporal solution remains in most cases unconstrained. That is, alternative pathways leading to the same present-day results exist and may be modeled. Without calibration controls from the geologic past along the geologic temporal path, the actual pathway of the basin cannot be established unequivocally. This problem was recently clarified by Leischner and Welte (1993), who illustrated two alternative thermal history models for a well in the Lower Saxony Basin of northwestern Germany. When constrained by geologically dated fluid-inclusion homogenization temperatures, the thermal history (temperature temporal pathway) of the section differed significantly from that obtained by the standard procedure of calibration with only present-day temperature and vitrinite reflectance. Similarly, there is a clear need to develop other paleo-calibration control data, not only for temperature modeling but also for fluid expulsion, hydrocarbon generation, and migration models in order to provide acceptable constraints on the temporal pathways of these important simulation histories.

1.7 Conclusion: A Note of Caution and Outlook

Basin simulation programs or systems are constantly improving and changing at a rapid pace to incorporate ever more geologic processes and features demanded by the user. The programs are expanded from one- to two-dimensional, and now the first integrated three-dimensional systems are almost on the market. The user interfaces are more and more advanced to make it easier for the user, especially the first time user, to get started and apply the method without unnecessary effort. The resulting programs are capable of modeling geologic processes and situations that far exceed what the initial simple one-dimensional systems could do, but with the problem that it is much more difficult for the user to know and understand what is in fact taking place in the program underneath the shiny interface.

Three areas of caution should be kept in mind when applying a modern basin simulation system: the default input data, especially for boundary conditions and physical parameters; the often unforeseen effect of interaction of the many processes being modeled; and finally the improvements and changes incorporated in every new version.

Much of the input is user supplied and explicitly visible to the user. However, to facilitate the input for the user, default values are offered that are geologically reasonable or probable. The temptation, for the routine user as much as for the first time user with little time to spare, is to accept default values without checking if they are appropriate for the particular situation. In many cases, this has no detrimental effect on the results since the default values are chosen to be geologically reasonable and applicable in a majority of basins. Yet, there are studies with unusual or different conditions when suddenly the defaults are wrong and produce a faulty simulation.

A similar problem is the assignment of physical parameters to the cells of the individual layers (formations). This is usually done by assigning a lithology (label) that refers to a set of parameters in a library table that is on the average typical for that lithology. In this context it is easy to overlook that these parameter assignments are just a first approximation. The label "shaly sand" points to a collection of parameter values (density, thermal conductivity, heat capacity, compressibility, etc.) which may be typical for the average "shaly sand." The particular layer may have actual values far from that average, and perhaps closer to another lithology with a different label, say "carbonaceous sand," even though "shaly sand" is correct as lithology. What counts are the values of the physical properties that the program uses for the calculations, not the lithology label. Experimentation with different lithologies will often show a better match of the results than the initially assigned lithology.

A geologic system is complex because of the many interactions of the multitude of geologic, physical, and chemical processes taking place. The same complexity exists in a good basin simulation system that tries to imitate the geologic reality. As a result, it is difficult for the user to keep control of what is actually going on in the simulation of his study. On the one hand, the actual calculation processes may be much too complicated for the average user to follow from the programming and numerical mathematics standpoint (let alone the problem of black box proprietary software). On the other hand, some of the interactions are so complex that the results are hardly predictable. A good study therefore requires careful tracing of the behavior of various variables using sensitivity analysis such as shown in Zwach (1995). Only then can the results be used with reasonable confidence.

The third area of caution concerns the rapid evolution and succession of new, improved versions of simulation programs. New versions can have radically new concepts, approaches, and numerical solutions, or they may provide incremental improvements, new features or geologic process considerations, or "bug fixes." Experience has shown that studies done on an old version will usually not repeat identical results when run on a new version, for reasons which are not always clear. They may relate to slightly different approaches in handling data, equations or grid elements. There may be different defaults set up, or different limiting parameters in the calculations, or the input may be interpolated with slightly different methods. All, of course, are improvements over the old versions. In most cases, however, studies carried out on different versions of the same basin modeling system are not exactly comparable. Indeed, once the old version of a program is replaced on the workstation by a new one, the results of older studies may be essentially irreproducible.

New developments in basin modeling appear to head in three directions. One is the increasing integration with seismic interpretation for model building. This takes the form of front end modules with interactive on-screen interpretation of digital seismic in terms of seismic stratigraphy and sequence stratigraphy. Such programs should allow the user to construct a cross section with chronostratigraphic interpretation including age assignments of erosion and nondeposition, facilitate the input and comparison with well data, permit the use of systems tract and seismic facies analysis, and, finally, build an input file for the actual basin modeling. Some versions of such input modules exist already but need to be improved to make model building easier and faster.

Another new development is the inclusion in the simulation of processes that are not due to the normal mechanical compaction. For instance, specific diagenetic processes influencing porosity development, or igneous intrusions affecting the temperature history and maturation can be used as options to make the simulation fit the geologic reality. Similarly, the inclusion of gas saturation to calculate lowered thermal conductivity has recently been made possible. Other, more specialized processes will certainly be included in the future.

The main trend now is toward real three-dimensional modeling systems. These will have to be coupled with those other types of models mentioned in the beginning: basin-fill simulation to help build the three-dimensional stratigraphy, and tectonic and structural deformation models to construct the various stages of deformation of the basin with geologic time. This is essential to trace the changing paths of fluid movement and heat distribution with time. The main difficulty to overcome, however, will not be the programming, the numerical techniques or even the high computer run time and storage requirements, but the acquisition of input data to fill the elements of the basin matrix with geologic and physical properties that should reflect the geologic reality. But once this is achieved, this integrated, interactive kind of basin modeling can bring the petroleum systems approach to life.

References

Allen PA, Allen JR (1990) Basin analysis: principles and applications. Blackwell, Oxford, pp 1–451

Angevine CL, Turcotte DL (1983) Oil generation in overthrust belts. AAPG Bull 67: 235–241

Asquith GB (1982) Basic well log analysis for geologists. AAPG Methods in Exploration. Am Assoc Petrol Geol, Tulsa, pp 1–216

Athy LF (1930) Density, porosity and compaction of sedimentary rocks. AAPG Bull 14: 1–24

Bally AW, Snelson S (1980) Realms of subsidence. facts and principles of world oil occurrence. Mem Can Soc Petrol Geol, pp 9–94

Bayer U (1987) Chronometric calibration of a comparative time scale for the Mesozoic and Paleozoic. Geol Rundsch 76: 485–503

Beaumont C (1981) Foreland basins. Geophys J R Astron Soc 65: 291–329

Beaumont C, Tankard AJ (1987) Sedimentary basins and basin-forming mechanisms. Mem Can Soc Petrol Geol 12, Calgary

Beaumont C, Keen CE, Boutilier R (1982) On the evolution of rifted continental margins: comparison of models and observations for the Nova Scotian margin. Geophys J R Astron Soc 70: 667–715

Berry FAF (1959) Hydrodynamics and geochemistry of the Jurassic and Cretaceous systems in the San Juan basin, northwestern New Mexico and southwestern Colorado. Diss, Stanford University

Bertram GT, Milton NJ (1988) Reconstructing basin evolution from sedimentary thickness; the importance of paleobathymetric control, with reference to the North Sea. Basin Res 1: 247–257

Bethke CM (1989) Modeling subsurface flow in sedimentary basins. In: Poelchau HS, Mann U(eds) Geologic modeling – aspects of integrated basin analysis and numerical simulation. Geol Rundsch 78: 129–154

Bethke CM, Harrison WJ, Upson C, Altaner SP (1988) Supercomputer analysis of sedimentary basins. Science 239: 233–324

Bitzer K, Pflug R (1990) DEPO3D: A three-dimensional model for simulating clastic sedimentation and isostatic compensation in sedimentary basins. In: Cross TA (ed) Quantitative dynamic stratigraphy. Prentice Hall, Englewood Cliffs, pp 335–348

Bjørkum PA, Gjelsvik N (1988) An isochemical model for formation of authigenic kaolinite, K-feldspar and illite in sediments. J Sed Petrol 58: 506–511

Bjørlykke K (1983) Diagenetic reactions in sandstones. In: Parker A, Sellwood BW (eds) Sediment diagenesis. Reidel, Dordrecht, pp 169–213

Bjørlykke K, Aagaard P, Dypvik H, Hastings DS, Harper AS (1986) Diagenesis and reservoir properties of Jurassic sandstones from the Haltenbanken area, offshore mid Norway. Habitat of hydrocarbons on the Norwegian Continental Shelf. Proc Norwegian Petrol Soc, Symp Stavanger, Oct 1985, pp 275–286

Bjørlykke K, Mo A, Palm E (1988) Modelling of thermal convection in sedimentary basins and its relevance to diagenetic reactions. Mar Petrol Geol 5: 338–351

Bjørlykke K, Ramm M, Saigal GC (1989) Sandstone diagenesis and porosity modification during basin evolution. In: Poelchau HS, Mann U (eds) Geologic modeling – aspects of integrated basin analysis and numerical simulation. Geol Rundsch 78: 243–268

Blackwell DD, Steele JL (1989) Thermal conductivity of sedimentary rocks: measurement and significance. In: Naeser ND, McCulloh TH (eds) Thermal history of sedimentary basins – methods and case histories. Springer, Berlin Heidelberg New York, pp 14–36

Bloch S (1991) Empirical prediction of porosity and permeability in sandstones. AAPG Bull 75: 1145–1160

Boles JR (1982) Active albitization of plagioclase, Gulf Coast Tertiary. Am J Sci 282: 165–180

Bowers TS, Jackson KJ, Helgeson HC (1984) Equilibrium activity diagrams for coexisting minerals and aqueous solutions at pressures and temperatures to 5 kb and 600 °C. Springer, Berlin Heidelberg New York, pp 1–397

Brace JS (1975) Abnormal formation pressure. AAPG Bull 59: 957–973

Brigaud F, Chapman DS, Le Douaran S (1990) Estimating thermal conductivity in sedimentary basins using lithologic data and geophysical well logs. AAPG Bull 74: 1459–1477

Brown LF, Fisher WL (1977) Seismic stratigraphic interpretation of depositional systems. In: Payton CE (ed) Seismic stratigraphy – applications to hydrocarbon exploration. AAPG Mem 26: 213–248

Bryant WR, Hottman W, Trabant P (1975) Permeability of unconsolidated and consolidated marine sediments, Gulf of Mexico. Mar Geotech 1: 1–14

Burnham AK, Sweeney JJ (1989) A chemical kinetic model of vitrinite maturation and reflectance. Geochim Cosmochim Acta 53: 2649–2657

Burnham AK, Braun RL, Gregg HR, Samoun AL (1987) Comparison of methods for measuring kerogen pyrolysis rates and fitting kinetic parameters. Lawrence Livermore Natl Labs preprint, Livermore, California, pp 1–27

Burnham AK, Braun RL, Samoun AM (1988) Further comparison of methods for measuring kerogen pyrolysis rates and fitting kinetic parameters. In: Mattavelli L, Novelli L (eds) Advances in organic geochemistry. Org Geochem 13: 839–845

Burst JF (1969) Diagenesis of Gulf Coast clayey sediments and its possible relation to petroleum migration. AAPG Bull 53: 73–93

Chapman RE (1981) Geology and water. Developments in applied earth sciences 1 Nijhoff-Junk, The Hague, pp 1–228

Chilingarian GV (1983) Compactional diagenesis. In: Parker A, Sellwood BW (eds) Sediment diagenesis. Reidel, pp 57–168

Clark SP Jr (1966) Handbook of physical constants. Geol Soc Am Mem 97: 587

Cloetingh S (1992) Intraplate stress and sedimentary basin evolution. In: Brown GC, Hawkesworth CJ, Wilson RCL (eds) Understanding the earth – a new synthesis. Cambridge Univ Press, Cambridge, pp 415–434

Condie KC (1976) Plate tectonics and crustal evolution. Pergamon Press, New York, pp 1–288

Connolly CA (1989) Thermal history and diagenesis of the Wilrich Member shale, Spirit River Formation, northwest Alberta. Bull Can Petrol Geol 37: 182–197

Cross TA, Harbaugh JW (1990) Quantitative dynamic stratigraphy: a workshop, a philosophy, a methodology. In: Cross TA (ed) Quantitative dynamic stratigraphy. Prentice Hall, Englewood Cliffs, pp 3–20

Crowe AS (1988) A numerical model for simulating mass transport and reactions during diagenesis in clastic sedimentary basins. PhD Diss, Alberta Univ
De Bremaecker JC (1987) Thrust sheet motion and earth quake mechanisms. Earth Planet Sci Lett 83: 159–166
De Paor DG, Bradley DC (1988) Balanced sections in thrust belts. Part II: Computerized line and area balancing. Geobyte: 33–37
Dickinson WR (1993) Basin geodynamics. Basin Res 5: 196–197
Eberl DD, Hower J (1976) Kinetics of illite formation. GSA Bull 87: 1326–1330
Einsele G (1992) Sedimentary basins – evolution, facies, and sediment budget. Springer, Berlin Heidelberg New York
Emery KO, Uchupi E (1972) Western North Atlantic Ocean: topography, rocks, structure, water, life, and sediments. AAPG Mem 17: 1–532
Emiliani C (1955) Pleistocene temperatures. J Geol 63: 538–578
Engelhardt WV (1960) Der Porenraum der Sedimente. Springer, Berlin Heidelberg New York
England WA, Mackenzie AS, Mann DM, Quigley TM (1987) The movement and entrapment of petroleum fluids in the subsurface. J Geol Soc 144: 327–347
Enos P, Sawatsky LH (1981) Pore space in Holocene carbonate sediments. AAPG Bull 63: 445 (Abstr)
Espitalié J, LaPorte JL, Madec M, Marquis F, Leplat P, Paulet J, Boutefeu A (1977) Méthode rapide de caractérisation des roches mères de leur potentiel pétrolier et de leur degré d'evolution. Rev Inst Fr Pét 32: 23–42
Espitalié J, Marquis F, Barsony I (1984) Geochemical logging. In: Voorhees KJ (ed) Analytical pyrolysis techniques and applications. Butterworths, London, pp 276–304
Fischer AG (1969) Geological time-distance rates: the Bubnoff unit. Geol Soc Am Bull 80: 549–552
Flemings PB, Alexander LL, Spiegelman M, Anderson RN (1992) Fluid expulsion from abnormally pressured strata along growth faults in the Eugene Island Block 331 area, Gulf of Mexico. AAPG Annu Meet, Abstr, Calgary, p 42
Flowers GC (1986) Computation of the thermodynamic properties of reactions involving minerals and aqueous solutions with the aid of the personal computer. Comput Geosci 12: 361–379
Frakes LA (1979) Climates throughout geological time. Elsevier, Amsterdam, pp 1–310
Friedman GM (1975) The making and unmaking of limestones or the downs and ups of porosity. J Sed Petrol 45: 379–398
Füchtbauer H (1988) Sedimente und Sedimentgesteine. Schweizerbart'sche Verlagsbuchhandlung, Stuttgart, 1141 pp
Füchtbauer H, Müller G (1970) Sedimente und Sedimentgesteine. Schweizerbart'sche Verlagsbuchhandlung, Stuttgart
Galloway WE, Hobday DK (1983) Terrigenous clastic depositional systems. Springer, Berlin Heidelberg New York, pp 1–423
Gildner RF (1990) STRIPPER: an interactive backstripping, decompaction and geohistory program. In: Cross TA (ed) Quantitative dynamic stratigraphy. Prentice Hall, Englewood Cliffs, pp 165–180
Gradstein FM, Agterberg FP, Ogg JG, Hardenbol J, van Veen P, Thierry J, Huang Z (1994) A Mesozoic time scale. J Geophys Res 99: 24051–24074
Gregory AR (1977) Aspects of rock physics from laboratory and log data that are important to seismic interpretation. Seismic stratigraphy – applications to hydrocarbon exploration 26. AAPG Mem, Tulsa, pp 15–46
Gretener PE (1969) Fluid pressure in porous media – its importance in geology: a review. Bull Can Petrol Geol 17: 255–295
Gretener PE (1972) Thoughts on overthrust faulting in a layered sequence. Bull Can Petrol Geol 20: 583–607
Gretener PE (1981) Geothermics: using temperature in hydrocarbon exploration. AAPG Shortcourse Notes 17, pp 1–156
Hacquebard DA (1977) Rank of coal as an index of organic metamorphism for oil and gas in Alberta. In: Deroo G, Powell TG, Tissot B (eds) The origin and migration of petroleum in the Western Canadian sedimentary basin, Alberta. Geol Surv Can Bull 262: 11–22

Haenel R (1979) A critical review of heat flow measurements in sea and lake bottom sediments. In: Cermak V, Rybach L (eds) Terrestrial heat flow in Europe. Springer, Berlin Heidelberg New York, pp 49–73
Hanbaba P, Jüntgen H (1969) Zur Übertragbarkeit von Laboratoriums-Untersuchungen auf geochemische Prozesse der Gasbildung aus Steinkohle und über den Einfluß von Sauerstoff auf die Gasbildung. In: Schenck PA, Havenaar I (eds) Advances in organic geochemistry 1968. Pergamon Press, Oxford, pp 459–471
Haq BU, Boersma A (1978) Introduction to marine micropaleontology. Elsevier, New York, 376 pp
Haq BU, van Eysinga FWB (1987) Geologic time table, 4th edn. Elsevier, Amsterdam
Haq BU, Hardenbol J, Vail PR (1987) Chronology of fluctuating sea levels since the Triassic. Science 235: 1156–1167
Harland WB, Cox AV, Lewellyn PG, Pickton CAG, Smith AG, Walters R (1982) A geological time scale. Cambridge University Press, Cambridge, pp 1–131
Harland WB, Armstrong RL, Cox AV, Craig LE, Smith AG, Smith DG (1990) A geologic time scale 1989. Cambridge University Press, Cambridge, pp 1–263
Helgeson HC (1968) Evaluation of irreversible reactions in geochemical processes involving minerals and aqueous solutions. I. Thermodynamic relations. Geochim Cosmochim Acta 32: 853–877
Helwig JA (1985) Origin and classification of sedimentary basins. 17th Annu Offshore Technol Conf, Houston, OTC 4843, pp 21–32
Hermanrud C (1993) Basin modelling techniques – an overview. In: Doré AG et al. (eds) Basin modelling: advances and applications. NPF Spec Publ 3. Elsevier, Amsterdam, pp 1–34
Hess JC, Lippolt HJ (1986) 40Ar/39Ar ages of tonstein and tuff sanidines: new calibration points for the improvement of the Upper Carboniferous time scale. Isotope Geosci 59: 143–154
Hooper ECD (1991) Fluid migration along growth faults in compacting sediments. J Petrol Geol 14: 161–180
Houseknecht DW (1987) Assessing the relative importance of compaction processes and cementation to reduction of porosity in sandstones. AAPG Bull 71: 633–642
Hower J, Eslinger EV, Hower ME, Perry EA (1976) Mechanism of burial metamorphism of argillaceous sediments. 1. Mineralogical and chemical evidence. GSA Bull 87: 725–737
Hubbert MK, Rubey WW (1959) Role of fluid pressure in mechanics in overthrust faulting. I. Mechanics of fluid-filled porous solids and its application to overthrust faulting. Bull Geol Soc Am 70: 115–166
Hubbert MK, Rubey WW (1961) Role of fluid pressure in mechanics in overthrust faulting. I. Mechanics of fluid-filled porous solids and its application to overthrust faulting. Reply to discussion by Francis Birch. Bull Geol Soc Am 72: 1445–1452
Huck G, Karweil J (1955) Physikalisch-chemische Probleme der Inkohlung. Brennstoff-Chemie 36: 1–11
Isacks BL, Oliver JE, Sykes LR (1968) Seismology and the new global tectonics. J Geophys Res 73: 5855–5899
Jankowsky W (1962) Diagenese und Ölinhalt als Hilfsmittel für die strukturgeschichtliche Analyse des Norwestdeutschen Beckens. Z Dtsch Geol Ges 114: 452–460
Jendrzejewski JP, Zarillo GA (1972) Late Pleistocene paleotemperature oscillations defined by silicoflagellate changes in a sub-antarctic deep-sea core. Deep Sea Res 19: 327–329
Jones PB, Linsser H (1986) Computer synthesis of balanced structural cross-sections by forward modeling. AAPG Bull 70: 605
Jordan TE (1981) Thrust loads and foreland basin development, Cretaceous, western United States. AAPG Bull 65: 2506–2520
Kalkreuth W, McMechan ME (1989) Coalification patterns in Jurassic-Lower Cretaceous strata (Minnes, Bullhead and Fort St. John Groups), Rocky Mountain foothills and foreland, east-central British Columbia and adjacent Alberta. Contrib to Can Coal Geosci Geol Surv Canada, pp 68–79
Kappelmeyer O, Haenel R (1974) Geothermics with special reference to application. Gebrüder Bornträger, Berlin, pp 1–240

Kharaka Y, Mariner R (1989) Chemical geothermometers and their application to formation waters from sedimentary basins. In: Naeser N, McCulloh T (eds) Thermal history of sedimentary basins. Methods and case histories. Springer, Berlin Heidelberg New York, pp 99–117

Kharaka YK, Gunter WD, Aggarwal PK, Perkins EH, DeBraal JD (1988) SOLMINEQ88: a computer program for geochemical modeling of water-rock interactions. Rep 88–4227, USGS Water-Resour Inv, pp 1–420

Kooi H, Cloetingh S (1989) Some consequences of compressional tectonics for extensional models of basin subsidence. In: Poelchau HS, Mann U (eds) Geologic modeling – aspects of integrated basin analysis and numerical simulation. Geol Rundsch 78: 183–195

Larsen G, Chilingar GV (1979) Diagenesis in sediments and sedimentary rocks. Developments in sedimentology 25A. Elsevier, Amsterdam

Larter S (1988) Some pragmatic perspectives in source rock geochemistry. Mar Petrol Geol 5: 194–204

Larter S (1989) Chemical models of vitrinite reflectance evolution. In: Poelchau HS, Mann U (eds) Geologic modeling – aspects of integrated basin analysis and numerical simulation. Geol Rundsch 78: 349–359

Lawrence DT, Doyle M, Aigner T (1990) Stratigraphic simulation of sedimentary basins: concepts and calibration. AAPG Bull 74: 273–295

Leischner K (1994) Kalibration simulierter Temperaturgeschichten von Sedimentgesteinen. Ber Forschungszentrum Jülich 2909, pp 1–309

Leischner K, Welte DH, Littke R (1993) Fluid inclusions and organic maturity parameters as calibration tools in basin Modelling. In: Doré AG et al. (eds) Basin modelling: advances and applications. NPF Spec Publ 3. Elsevier, Amsterdam, pp 161–172

Lessenger MA, Cross TA (1995) Estimating accuracies and uncertainties of stratigraphic prediction. AAPG Bull 79: 1240

Lewis CR, Rose SC (1970) A theory relating high temperatures and overpressures. J Petrol Technol 22: 11–16

Leythaeuser D, Poelchau HS (1991) Expulsion of petroleum from type III kerogen source rocks in gaseous solution: modeling of solubility fractionation. In: England WA, Fleet AJ (eds) Petroleum migration. Geol Soc Lond, Spec Publ 59, pp 33–46

Leythaeuser D, Welte DH (1969) Relation between distribution of heavy n-paraffins and coalification in carboniferous coals from the Saar district, Germany. In: Schenck PA, Havenaar I (eds) Advances in organic geochemistry, 1968. Pergamon Press, New York, pp 429–442

Leythaeuser D, Schaefer RG, Radke M (1987) On the primary migration of petroleum. Proc 12th World Petrol Congr. Wiley, Chichester, pp 227–236

Lichtner PC (1988) The quasi-stationary state approximation to coupled mass transport and fluid-rock interaction in a porous medium. Geochim Cosmochim Acta 52: 143–165

Lopatin NV (1971) Temperature and geologic time as factors in coalification. Akad Nauk SSSR Izvestiya, Seriya Geologicheskaya 3: 95–196 (in Russian)

Mackenzie AS, McKenzie D (1983) Isomerization and aromatization of hydrocarbons in sedimentary basins formed by extension. Geol Mag 12: 417–470

Magara K (1976) Thickness of removed sediments, paleopore pressure and paleotemperature, southwestern part of western Canada Basin. AAPG Bull 60: 554–565

Magara K (1978) Compaction and fluid migration – practical petroleum geology. Developments in petroleum science 9. Elsevier, Amsterdam

Magara K (1986) Geological models of petroleum entrapment. Elsevier, Amsterdam, pp 1–328

Magoon LB (1988) The petroleum system – a classification scheme for research, exploration, and resource assessment. In: Magoon LB (ed) Petroleum systems of the United States. USGS Bull 1870: 2–15

Magoon LB, Claypool GE (1983) Petroleum geochemistry of the North Slope of Alaska: time and degree of thermal maturity. In: Bjorøy M et al. (eds) Advances in organic geochemistry 1981. Wiley, New York, pp 28–38

Magoon LB, Dow WG (1994) The petroleum system. In: Magoon LB, Dow WG (eds) The petroleum system – from source to trap. AAPG Mem 60. Am Assoc Petrol Geol, Tulsa, pp 3–24

Masters JA (1979) Deep Basin gas trap, western Canada. AAPG Bull 63: 152–181
Masters JA (1984) Lower Cretaceous oil and gas in western Canada. In: Masters JA (ed) Elmworth – case study of a Deep Basin gas field. AAPG Mem 38: 1–33
McBride EF (1984) Rules of sandstone diagenesis related to reservoir quality. Trans Gulf Coast Assoc Geol Soc, 34th Annu Meet, Shreveport, Louisiana, pp 137–139
McDonald MG, Harbaugh AW (1988) A modular three-dimensional finite-difference groundwater flow model. Techniques of water-resources investigations of the United States Geological Survey, Book 6, Ch A1. Scientific Software Group, Washington, DC
McIntyre A, Kipp NG, Bé AWH, Crowley TP, Kellogg T, Gardner JV, Prell W, Ruddiman WF (1976) Glacial North Atlantic 18 000 years ago. A CLIMAP reconstruction. In: Cline RM, Hays JD (eds) Geol Soc Am Mem 145, pp 43–76
McKenzie DP (1978) Some remarks on the development of sedimentary basins. EPSL 40: 25–32
McKenzie DP (1981) The variation of temperature with time and hydrocarbon maturation in sedimentary basins formed by extension. EPSL 55: 87–98
Menning M (1989) A synopsis of numerical time scales 1917–1986. Episodes 12: 3–5
Meyer AJ, Landais P, Brosse E, Pagel M, Carisey JC, Krewdl D (1989) Thermal history of the Permian formations from the Breccia Pipes area (Grand Canyon region, Arizona). In: Poelchau HS, Mann U (eds) Geologic Modeling – aspects of integrated basin analysis and numerical simulation. Geol Rundsch 78: 427–438
Nedkvitne T, Bjørlykke K (1992) Secondary porosity in the Brent Group (Middle Jurassic), Huldra Field, North Sea: implication for predicting lateral continuity of sandstones? J Sed Petrol 62: 23–34
Neglia S (1979) Migration of fluids in sedimentary basins. AAPG Bull 63: 573–597
Neugebauer HJ (1994) An integrated dynamic-thermal approach to basin development. In: Welte DH (ed) Basin analysis and reservoir studies. JÜL Ber 2882. Forschungszentrum Jülich GmbH, Jülich, pp A4.1–8
Neugebauer HJ, Walzebuck JP (1987) A modelling theory for cratonic basins and their hydrocarbon accumulations. 12th Petroleum Congr, vol 2. Wiley, New York, pp 9–17
Neuser RD, Richter DK, Vollbrecht A (1989) Natural quartz with brown/violet cathodoluminescence – genetic aspects evident from spectral analysis. Zentralbl Geol Paläont I: 919–930
Nunn JA (1994) Free thermal convection beneath intracratonic basins: thermal and subsidence effects. Basin Res 6: 115–130
Ondrak R (1992) Mathematical simulation of water-mineral interaction and fluid flow with respect to the diagenetic evolution of sandstone. Proc 7th Int Symp Water-Rock Interactions, Park City, Utah
Ortoleva P, Merino E, Moore C, Chadam J (1987) Geochemical self-organization. I. Reaction-transport feedbacks and modelling approach. Am J Sci 287: 979–1007
Oxburgh ER, Turcotte DL (1974) Thermal gradients and regional metamorphism in overthrust terrains with special reference to the eastern Alps. Schweizer Mineral Petrogr Mitt 54: 641–662
Palmer AR (1983) The decade of North American geology 1983 geologic time scale. Geology 11: 503–504
Perlmutter MA, Matthews MD (1990) Global cyclostratigraphy – a model. In: Cross TA (ed) Quantitative dynamic stratigraphy. Prentice Hall, Englewood Cliffs, pp 233–260
Philippi GT (1965) On the depth, time and mechanism of petroleum generation. Geochim Cosmochim Acta 29: 1021–1049
Poelchau HS (1974) Holocene silicoflagellates of the North Pacific: their distribution and use for paleotemperature determination. Diss, Univ Calif, San Diego, pp 1–165
Poelchau HS, Zwach C (1994) Basin simulation and diagenetic models: Albian Cadotte Sandstone, Alberta Deep Basin, Canada. Ber Forschungszentrums Jülich, Jül-2882, pp B1.1–142
Posamentier HW, Vail PR (1988) Sequences, systems tracts and eustatic cycles. AAPG Bull 72: 237
Potter PE (1967) Sand bodies and sedimentary environments: a review. AAPG Bull 51: 337–365
Pytte AM, Reynolds RC (1989) The thermal transformation of smectite to illite. In: Naeser ND, McCulloh TH (eds) Thermal history of sedimentary basins – methods and case histories. Springer, Berlin Heidelberg New York, pp 133–140

Quigley TM, MacKenzie AS, Gray JR (1988) Kinetic theory of petroleum generation. Proc Conf on Migration of hydrocarbons in sedimentary basins, Bordeaux, June 1987. Editions Technip, Paris, pp 649–665

Quinlan G (1987) Models of subsidence mechanisms in intracratonic basins, and their applicability to North American examples. In: Beaumont C, Tankard AJ (eds) Sedimentary basins and basin-forming mechanisms. Mem 12, Can Soc Petroleum Geol, pp 463–481

Radke M, Welte DH (1983) The Methylphenanthrene Index (MPI): a maturity parameter based on aromatic hydrocarbons. In: Bjorøy M et al. (eds) Advances in organic geochemistry 1981. Wiley, Chichester, pp 504–512

Ratcliffe EH (1960) The thermal conductivity of ocean sediments. J Geophys Res 65: 1535–1541

Rieke HH, III, Chilingarian GV (1974) Compaction of argillaceous sediments. Developments in sedimentology 16. Elsevier, Amsterdam, pp 1–424

Robertson EC (1979) Thermal conductivity of rocks. U.S. Geol. Survey, Open File Rpt 79-356, 31 pp

Robinson AG, Gluyas JG (1992) Duration of quartz cementation in sandstones, North Sea and Haltenbanken basins. Mar Petrol Geol 9: 324–327

Rose W (1983) A note on the role played by sediment bedding in causing permeability anisotropy. J Petrol Technol 2: 330–332

Rubey WW, Hubbert MK (1959) Role of fluid pressure in mechanics in overthrust faulting. 2. Overthrust belt in geosynclinal area of western Wyoming in light of fluid-pressure hypothesis. Bull. Geol. Soc. Am. 70: 167–206

Rullkötter J, Marzi R (1988) Natural and artificial maturation of biological markers in a Toarcian shale from northern Germany. In: Mattavelli L, Novelli L (eds) Advances in organic geochemistry 13. Pergamon, Oxford, pp 639–645

Salvador A (1995) Chronostratigraphic and geochronometric scales in COSUNA stratigraphic correlation charts of the United States. AAPG Bull 69: 181–189

Sass JH, Lachenbruch AH, Munroe RJ (1971) Thermal conductivity of rocks from measurements on fragments and its application to heat flow determinations. J Geophys Res 76: 3391–3400

Schaefer RG, Schenk HJ, Hardelauf H, Harms R (1990) Determination of gross kinetic parameters for petroleum formation from Jurassic source rocks of different maturity levels by means of laboratory experiments. In: Béhar F, Durand B (eds) Advances in organic geochemistry. Organic geochemistry 16. Pergamon Press, Oxford, pp 115–120

Scherer M (1987) Parameters influencing porosity in sandstones: a model for sandstone porosity prediction. AAPG Bull 71: 485–491

Schmoker JW, Halley RB (1982) Carbonate porosity vs. depth: a predictable relation for south Florida. AAPG Bull 66: 2561–2570

Scholle PA (1977) Geological studies on the COST No. B-2 well, U.S. Mid-Atlantic outer continental shelf area. US Geol Surv Circ 71

Schopper JR (1982) Porosity and permeability. In: Hellwege KH (ed) Landolt-Börnstein. Numerical data and functional relationships in science and technology, vol 1. Physical properties of rocks. Springer, Berlin Heidelberg New York, pp 184–303

Sclater JG, Christie PAF (1980) Continental stretching: an explanation of the post-mid-Cretaceous subsidence of the central North Sea basin. J Geophys Res 85: 3711–3739

Sekiguchi K (1984) A method for determining terrestrial heat flow in oil basinal areas. Tectonophysics 103: 67–79

Shackleton NJ, Opdyke ND (1973) Oxygen isotope and paleomagnetic stratigraphy of equatorial Pacific core V28-238: oxygen isotope temperatures and ice volumes on a 10^5 year and 10^6 year scale. Quat Res 3: 39–55

Silver C (1968) Principles of gas occurrence, San Juan basin. Natural Gases in North America AAPG Mem, 9. pp 946–960

Sippel RF (1968) Sandstone petrology, evidence from luminescence petrography. J Sed Petrol 38: 530–554

Smith JE (1971) The dynamics of shale compaction and evolution of pore fluid pressures. Math Geol 3: 239–263

Somerton WH (1992) Thermal properties and temperature-related behavior of rock/fluid systems. Developments in petroleum science 37. Elsevier, Amsterdam, pp 1–260

Steckler MS, Watts AB (1978) Subsidence of the Atlantic-type continental margin off New York. EPSL 41: 1–13

Stephenson LP (1977) Porosity dependence on temperature: limits on maximum possible effect. AAPG Bull 61: 407–415

Stephenson R (1990) Beyond first-order thermal subsidence models for sedimentary basins? In: Cross TA (ed) Quantitative dynamic stratigraphy. Prentice Hall, Englewood Cliffs, pp 113–125

Stoneley R (1981) Petroleum: the sedimentary basin. In: Tarling DH (ed) Economic geology and geotectonics. Wiley, New York, pp 51–71

Sweeney JJ (1990) BASINMAT: Fortran program calculates oil and gas generation using a distribution of discrete activation energies. Geobyte 5: 37–43

Sweeney JJ, Burnham AK (1990) Evaluation of a simple model of vitrinite reflectance based on chemical kinetics. AAPG Bull 74: 1559–1570

Sweeney JJ, Burnham AK, Braun RL (1987) A model of hydrocarbon generation from type I kerogen: application to Uinta Basin, Utah. AAPG Bull 71: 967–985

Thompson KFM (1988) Gas-condensate migration and oil fractionation in deltaic systems. Mar Petrol Geol 5: 237–246

Tilley BJ (1988) Diagenesis and pore water evolution in Cretaceous sedimentary rocks of the Alberta Deep Basin. Diss, Univ Alberta, Edmonton, pp 1–224

Tillman RW, Almon WR (1979) Diagenesis of Frontier formation offshore bar sandstones, Spearhead Ranch Field,Wyoming. SEPM Special Publ 26, pp 337–378

Tissot B (1969) Premières données sur les mécanismes et la cinétique de la formation du pétrole dans les sédiments. Simulation d'un schéma réactionnel sur ordinateur. Rev Inst Fr Pét 24: 470–501

Tissot B, Espitalié J (1975) L'évolution thermique de la matière organique des sédiments: application d'une simulation mathématique. Rev Inst Fr Pét 30: 743–777

Tissot BP, Pelet R (1971) Nouvelles données sur les mèchanismes de genèse et de migration du pétrole, simulation mathématique et application à la prospection. 8th World Petrol Congr, vol 2. Wiley, London, pp 35–46

Tissot BP, Pelet R, Ungerer P (1987a) Thermal history of sedimentary basins, maturity indices and kinetics of oil and gas generation. Bull AAPG 71: 1445–1466

Tissot BP, Welte DH, Durand B (1987b) The role of geochemistry in exploration risk evaluation and decision making. Proc 12th World Petrol Congr, Houston, 1987, vol 2. Wiley, London, pp 99–112

Ungerer P, Pelet R (1987) Extrapolation of the kinetics of oil and gas formation from laboratory experiments to sedimentary basins. Nature 327: 52–54

Ungerer P, Burrus J, Doligez B, Chénet PY, Bessis F (1990) Basin evaluation by integrated two-dimensional modeling of heat transfer, fluid flow, hydrocarbon generation, and migration. AAPG Bull 74: 3

Vail PR, Mitchum RM, Jr, Todd RG, Widmier JM, Thompson S, III, Sangree JB, Bubb JN, Hatlelid WG (1977) Seismic stratigraphy and global changes of sea level. In: Payton CE (ed) Seismic stratigraphy – applications to hydrocarbon exploration. AAPG Mem 26, pp 49–212

Van Wagoner JC, Mitchum RM, Campion KM, Rahmanian VD (1990) Siliciclastic sequence stratigraphy in well logs, cores, and outcrops. AAPG Methods in Exploration 7. Am Assoc Petrol Geol, Tulsa, pp 1–55

Van Eysinga FWB (1975) Geological time table. Elsevier, Amsterdam

Van Hinte JE (1978) Geohistory analysis – application of micropaleontology in exploration geology. AAPG Bull 62: 201–222

Vasseur G, Velde B (1993) A kinetic interpretation of the smectite-to-illite transformation. In: Doré AG et al. (eds) Basin modelling: advances and applications. NPF Spec Publ 3, Elsevier, Amsterdam, pp 173–184

Vassoyevich NB, Korchagina Y, Lopatin NV, Chernyshev VV (1969) Principal phase of oil formation. Moscov Univ Vestnik 6: 3–27 (in Russian). Int Geol Rev 12:1276–1296 (1970) (Engl transl)

Visher GS (1984) Exploration stratigraphy. Pennwell Publ, Tulsa, pp 1–334

Vogt M (1990) Ein vektorrechnerorientiertes Verfahren zur Berechnung großräumiger Multikomponenten-Transport-Reaktionsmechanismen im Grundwasserleiter. Fortschr-Ber. VDI, Reihe 15, vol 80. VDI-Verlag, Düsseldorf, pp 1–176

Walker RG (1984) Facies models. Geoscience Canada, Reprint Series, Geol Assoc Canada, Toronto, Ontario, 317 pp

Waples D (1980) Time and temperature in petroleum formation: application of Lopatin's method to petroleum exploration. AAPG Bull 64: 916–926

Weast RC (ed) (1974) CRC handbook of chemistry and physics. CRC Press, Cleveland

Weber KJ (1982) Influence of common sedimentary structures on fluid flow in reservoir models. J Petrol Technol 34: 665–672

Weiss HM (1985) Geochemische und petrographische Untersuchungen am organischen Material kretazischer Sedimentgesteine aus dem Deep Basin, Westkanada. Doctoral Diss, RWTH Aachen, pp 1-261

Welte DH (1965) Relation between petroleum and source rock. AAPG Bull 49: 2246–2268

Welte DH (1988) Migration of hydrocarbons – facts and theory. Proc Conf on Migration of hydrocarbons in sedimentary basins, Bordeaux, June 1987. Editions Technip, Paris, pp 393–413

Welte DH, Yalçin MN (1987) Formation and occurrence of petroleum in sedimentary basins as deduced from computer-aided basin modelling. In: Kumar SP, Dwivedi P, Banerjie V, Gupta V (eds) Petroleum geochemistry and exploration in the Afro-Asian region. Balkema, Rotterdam, pp 17–23

Welte DH, Yalçin MN (1988) Basin modelling – a new comprehensive methodology in petroleum geology. In: Matavelli L, Novelli L (eds) Advances in organic geochemistry 13. Pergamon, Oxford, pp 141–151

Welte DH, Yükler MA (1981) Petroleum origin and accumulation in basin evolution – a quantitative model. AAPG Bull 65: 1387–96

Wilson M (1993) Magmatism and the geodynamics of basin formation. Sed Geol 86: 5–29

Woodside W, Messmer JH (1961) Thermal conductivity of porous media. II. Unconsolidated sands. J Appl Phys 32: 1688–1699

Wygrala BP (1988) Integrated computer-aided basin modeling applied to analysis of hydrocarbon generation history in northern Italian oil field. In: Mattavelli L, Novelli L (eds) Advances in organic geochemistry 13. Pergamon, Oxford, pp 187–197

Wygrala BP (1989) Integrated study of an oil field in the southern Po basin, northern Italy. Diss, Univ Köln, Berichte Kernforschungsanlage Jülich, no 2313, pp 1–217

Wygrala BP, Yalçin MN, Dohmen L (1990) Thermal histories and overthrusting – application of numerical simulation technique. Org Geochem 16: 267–285

Yükler A, Cornford C, Welte DH (1978) One-dimensional model to simulate geologic, hydrodynamic, and thermodynamic development of a sedimentary basin. Geol Rundsch 67: 960–979

Yükler MA (1979) Sensitivity analysis of groundwater flow systems and an application to a real case. In: Gill D, Merriam DF (eds) Geomathematical and petrophysical studies in sedimentology, computery and geology. Pergamon Press, Oxford, pp 33–49

Yükler MA (1987) How essential is quantitative basin modeling in petroleum exploration? Turkish Assoc Petrol Geol: 392–404

Zoback MD, Byerlee JD (1975) Permeability and effective stress. AAPG Bull 59: 154–158

Zwach C (1995) Diagenesis and temperature history of the Cadotte Sandstone, Alberta Deep Basin, Canada: integration of reservoir quality analysis and basin modeling. Ber Forschungszentr Jülich, Jül-Ber 3082, pp 1–173

Zwach C, Poelchau HS (1993) Porosity reduction in the Albian Cadotte Sandstone, Alberta Deep Basin, Canada: small scale vs. large scale transport processes in diagenesis. Geofluids 93, Torquay, pp 220–222

Zwach C, Poelchau HS, Hantschel T, Welte DH (1994) Simulation with contrasting pore fluids: can we afford to neglect hydrocarbon saturation in basin modeling? Conf on Basin modelling, London Geol Soc, Petroleum Group

Chapter 2
Thermal History of Sedimentary Basins

Chapter 2: Overview and Insights

Temperature is the most important single parameter to be studied. It is the driving force for petroleum generation and many other chemical and transport processes. Basin modeling provides a unique opportunity to investigate temperature regimes in space and time. A prerequisite of meaningful temperature reconstruction are adequate temperature sensitive calibration data, such as those derived from organic geochemical maturation parameters, fluid inclusions, fission tracks, or illite crystallinity. This calibration of thermal histories is discussed in this chapter.

The wholistic approach of basin modeling allows analysis of the timing and sequence of events in great detail. Thereby the reconstruction of temperature history and the application of a modified form of chemical reaction kinetics play the decisive role. The thermal history of a basin can be described as a time-dependent energy balance which is controlled, on the one hand, by heat entering and leaving the basin and, on the other, by the ability of the sedimentary fill in the basin to transmit and store heat.

Accordingly, the physical and mathematical aspects of heat transfer and storage under the changing ambient conditions of a basin are treated. The thermophysical material parameters of the sedimentary rocks filling the basin and related changes induced during the evolution of a basin are discussed along with the input of thermal energy at the basin floor and heat losses at the surface. The above changes of the ambient conditions in the subsurface are largely controlled by four geological processes, i.e., by deposition of sediments, periods of nondeposition, processes related to the deformation of the basin fill, and erosion of sediments. These processes are identical with the main processes which must be accounted for when constructing the conceptual geological model for the basin simulation procedure as described in Chapter 1.

For the reconstruction of thermal histories via numerical simulation the conceptual model can provide the only geological framework; the points of reference (calibration), must be derived from organic or inorganic temperature sensitive parameters.

The thermal history of a given rock unit at any one point in the basin is determined by the combined effects of all the variables influencing heat transfer and storage. However, as with many complex interrelationships of this nature, sensitivity studies and/or case histories can single out dominant geological effects which have a decisive influence on temperature histories. Therefore this chapter includes case histories explaining the effect of very low sea bottom temperatures upon the heat budget of underlying rock strata, showing the influence of magmatism, the role of salt as an excellent thermal conductor, and the interplay of subsidence and uplifting upon an individual temperature history curve.

Thermal History of Sedimentary Basins

M.N. Yalçın[1], R. Littke[2], and R.F. Sachsenhofer[3]

2.1 Introduction

Geoscientists have always been interested in temperature values, as the crucial role of temperature in various geological, geochemical, and geophysical phenomena was recognized relatively early. However, this interest was long limited to aspects such as temperature distribution in the earth's crust and mantle, melting temperatures of different rocks, equilibrium temperatures of various metamorphic mineral assemblages, and temperature dependency of some geophysical parameters. Almost all of these are related to thermal conditions in the deeper parts of the earths crust. Low temperature fields at shallow depth have been of less concern. It is interesting to recognize that all these studies sought to define either the present distribution of temperature or the maximum temperature reached at some time in the geological past. In other words, temperature distribution as a steady state case was the subject of interest.

After basic research in the field of petroleum generation clearly showed that hydrocarbons are formed from kerogen in the source rocks when they are exposed to thermal stress over time (Philippi 1965; Welte 1966; Tissot and Welte 1984), many attempts were made to determine the effect of time and temperature on the source rocks. Further research in organic geochemistry led to the so-called "maturity concept" of immature, mature, and overmature source rocks. Many parameters have been developed and are widely applied, including vitrinite reflectance, sterane and hopane isomerization, T_{max} values, methylphenantrene index, spore coloration index, and thermal alteration index, and these help to assess the maturity of organic matter (Oudin 1984; Tissot and Welte 1984). The different approaches to maturity measurements are used mainly to determine whether a source rock is generating or has already generated oil and gas, or whether it is still immature. In other words, the main interest lies not in determining the past but the present status. With

[1]Department of Geology, University of Istanbul, 34850 Avçilar-Istanbul, Turkey and Department of Earth Sciences, TÜBITAK, Marmara Research Centre, P.O. Box 21, 41470 Gebze-Kocaeli, Turkey
[2]Institut für Erdöl und Organische Geochemie (ICG-4), Forschungszentrum Jülich GmbH, 52425 Jülich, Germany
[3]Institut für Geowissenschaflen, Montan-Universität Leoben, 8700 Leoben, Austria

Welte et al. (eds)
Petroleum and Basin Evolution
© Springer-Verlag Berlin Heidelberg 1997

the help of maturity parameters and other geochemical measurements and mainly in an empirical way, zones of oil and gas generation and of immature, mature, and overmature source rocks are determined. However, application of the maturity concept in different basins has shown that zones of oil and gas generation are shifted on the maturity scale (Yuekler and Kokesh 1984; Yalçın and Welte 1988) depending on heating rates even for the same organic matter. In other words, the correlation between maturation indices and steps of hydrocarbon generation is not uniform and is therefore imprecise for quantitative evaluation of petroleum generation. Additionally, present maturity values of mature and overmature source rocks do not indicate when oil and gas generation began. Therefore timing, a crucial parameter for evaluating hydrocarbon potential, cannot be determined by maturity values. Finally, maturity values do not allow quantitative determination of the amount of petroleum generated. The consequence is that an adequate prediction of petroleum generation in source rocks is not possible with the help of maturity parameters alone, since reactions leading to an increase in maturity and to the generation of oil and gas are controlled by different kinetics.

It is generally accepted that kinetics of the conversion of kerogen to petroleum can be described by a model based on first-order reaction kinetics obeying the Arrhenius equation, as shown by Jüntgen (1964) in connection with pyrolysis reactions, by Jüntgen and van Heek (1968) in connection with gas release from coal, and by Tissot (1969) in connection with petroleum formation from kerogen. Further studies have shown that the Arrhenius law is also applicable on a geological time scale (Jüntgen and Klein 1975; Tissot and Espitalie 1975; Ungerer 1984). The application of the kinetic approach to hydrocarbon generation requires both appropriate kinetic parameters and an accurate temperature history. Therefore a realistic temperature history of the relevant source rock is as important as accurate kinetic parameters of the particular organic matter in order to evaluate petroleum prospects.

A source rock is a part of a particular sedimentary basin, which is normally formed by very complex and interrelated processes as a multiparameter and dynamic system. Consequently, when reconstructing the temperature history of a source rock the evolution of a basin as a whole must be considered (Yalçın and Welte 1988). Thermal history of sedimentary basins can be described as a time-dependent energy balance problem, which is controlled, on the one hand, by heat entering and leaving the system, and on the other, by the ability of the system to transmit and to store heat. Therefore an accurate determination of temperature history is possible only if the heat transfer process is treated as a real physical dynamic case under given boundary conditions of a sedimentary basin and its evolution over time.

Temperature is possibly the most critical parameter since it affects not only hydrocarbon generation but also many physical properties of sediments and fluids. Consequently many processes during basin evolution, such as compaction and fluid flow, are controlled directly by the temperature itself. Therefore a realistic reconstruction of temperature history is crucial for the modeling of basin evolution and for understanding the very complex processes and their interactions during basin development.

This chapter discusses the physics and mathematical treatment of heat transfer processes. Aspects of heat transfer in sedimentary basins, initial and boundary conditions, and relevant parameters controlling heat transfer are then explained. Finally, the reconstruction of thermal history of sedimentary basins with the help of a computer-aided integrated modeling approach is presented, demonstrating the role of various parameters and boundary conditions on the temperature distribution.

2.2 Fundamental Concepts of Heat Transfer

Heat is defined as energy transfered by virtue of a temperature difference or gradient and is vectorial in the sense that it flows from regions of higher temperature to regions of lower temperatures. Heat transfer occurs by conduction, convection, and radiation (Rohsenow and Hartnett 1973).

Temperature is a measure of the disordered kinetic energy of the molecules constituting a body. If the temperature is not uniform in a body, a thermal gradient exists, and heat flows in direction of lower temperatures until a uniform temperature distribution is reached. A measure of the temperature or thermal gradient is dT/dh, where dT is the temperature difference between the points h1 and h2. (h2 − h1) is then dh. A uniform temperature distribution means that in the body a steady-state or an equilibrium condition is established. The situation prior to the equilibrium is defined as a transient or disequilibrium condition and is characterized by a continuous change in temperature as a function of time. This change is caused by the flow of heat. However, an equilibrium can also be reached if rate of energy (heat) input into the body and rate of heat output from the body have the same values.

Amount of heat energy is defined by heat flux density (q), which is the measure of heat flow per unit time and per unit area. Heat flux density is also vectorial and is proportional to the temperature gradient (grad T), where the proportionality factor is the thermal conductivity (K):

$$q = -K \text{ grad } T \qquad (2.1)$$

Thermal conductivity is a property of the body through which heat is transferred by conduction. For a given amount of heat it defines the change in temperature per unit length. The consequence is that high thermal conductivity values result in low thermal gradients and vice versa. The SI unit of thermal conductivity is W/m K. Other commonly used units are mcal cm^{-1} s^{-1} deg^{-1} and kcal m^{-1} h^{-1} deg^{-1}. The SI unit of heat flux density is mW/m^2. Units such as HFU (heat flow unit) and cal cm^{-2} s^{-1} are also frequently used.

Any material has the ability to store heat. The capability of a body to store heat is a function of its specific heat. The amount of heat required to increase the temperature of 1 g of material by 1 °C is defined as specific heat (Kappelmeyer and Haenel 1974). It can be defined as:

$$Q = c\, m(T2 - T1) \qquad (2.2)$$

where Q is the amount of heat, m is the mass, T1 and T2 relevant temperatures, and c is the specific heat. The product of mass and specific heat is the heat capacity, which defines the energy required to heat a body of mass (m) by 1 °C (Kappelmeyer and Haenel 1974). The SI unit of specific heat is J kg^{-1} deg^{-1}.

Another thermal property of materials is thermal diffusivity (κ), which is a measure of penetration of temperature changes into a solid material. This can be expressed for a homogeneous medium in terms of thermal conductivity (K), specific heat (c) and density (ρ) as given in equation (2.3):

$$\kappa = \frac{K}{\rho \cdot c} \tag{2.3}$$

High thermal diffusivity results in a fast and deep penetration of temperature changes. The SI unit of the thermal diffusivity is m^2/s.

Conduction is one of the heat transfer mechanisms in which heat is transferred by a physical contact from points at higher temperatures to those at lower temperatures in a body. This process takes place at the molecular level. In this way an equibilirium is reached when all the atoms or molecules have a certain average energy, which means that in different parts of the body one has the same temperature.

Convection is the motion of deformable media by which heat is transferred in the course of relative displacement of portions of a heated body (Kappelmeyer and Haenel 1974). Convection is connected with mass flow (solid, fluid, or gas). Two types of convection are distinguished, namely natural and forced convection. The former is induced by density differences caused by thermal disequilibrium. In case of forced convection the mass movement is caused by external influences.

Radiation is the heat transfer mechanism in which heat radiates in the form of electromagnetic waves from a hot body. This process is significant only at temperatures above 500 °C (MacDonald 1959).

The process of heat transfer can be adequately described and mathematically treated only if the initial and boundary conditions are known. Initial conditions define the temperature distribution at a chosen time, generally at the beginning of the mathematical treatment (t = 0). Boundary conditions describe the thermal state at the boundaries of the body under consideration during the entire time of heat transfer. Boundary conditions may be constant or may change as a function of time and space. A mathematical treatment of the heat transfer process requires an expression of the initial and boundary conditions in mathematical formulae.

2.3 Heat Transfer Equation

The differential equation given by Carslaw and Jaeger (1959) allows the treatment of heat transfer in a homogeneous and isotropic solid under steady-state conditions and with thermal conductivity independent of temperature:

$$\frac{\partial^2 T}{\partial x^2}+\frac{\partial^2 T}{\partial y^2}+\frac{\partial^2 T}{\partial z^2}=0 \tag{2.4}$$

where T = temperature and x, y, and z are the coordinates. Equation 2.4 indicates that heat is neither acquired nor lost in the any given unit volume. For transient conditions the differential Eq. (2.4) must be extended to:

$$\frac{\partial}{\partial x}\left(K_x\frac{\partial T}{\partial x}\right)+\frac{\partial}{\partial y}\left(K_y\frac{\partial T}{\partial y}\right)+\frac{\partial}{\partial z}\left(K_z\frac{\partial T}{\partial z}\right)=\rho c\frac{\partial T}{\partial t} \tag{2.5}$$

where K_x, K_y, and K_z are thermal conductivities, ρ is density, c is specific heat. It describes that heat gain or loss in the unit volume is balanced by temperature change multiplied by volumetric heat capacity (Gretener 1981). For isotropic conditions this equation can be rewritten as:

$$\frac{\partial^2 T}{\partial x^2}+\frac{\partial^2 T}{\partial y^2}+\frac{\partial^2 T}{\partial z^2}=\frac{\rho c}{K}\frac{\partial T}{\partial t}=\frac{1}{\kappa}\frac{\partial T}{\partial t} \tag{2.6}$$

in order to show that the conductive heat transfer under transient conditions is controlled by the thermophysical parameters such as thermal conductivity, heat capacity, density, and thermal diffusivity of a given material. Details of the conductional heat transfer equation can be found in the relevant literature (Rohsenov and Hartnett 1973; Kappelmeyer and Haenel 1974; Eckert and Drake 1987).

As the temperature distribution in a body can also be influenced by convection, a complete heat transfer equation should have a convective term. Assuming that convection takes place by water flow, the additional effects of water flow can be considered with the help of a source or sink term (S) which is added to the conductive heat transfer equation (2.5) as given below:

$$\frac{\partial(K_x\partial T)}{\partial x^2}+\frac{\partial(K_y\partial T)}{\partial y^2}+\frac{\partial(K_z\partial T)}{\partial z^2}=\rho c\frac{\partial T}{\partial t}+S \tag{2.7}$$

and:

$$S=c_w\rho_w\left(V_x\frac{\partial T}{\partial x}+V_y\frac{\partial T}{\partial y}+V_z\frac{\partial T}{\partial z}\right) \tag{2.8}$$

where, V_x, V_y, V_z are the velocity of water, ρ_w the density of water, and c_w the specific heat of water.

The conductive and convective terms of the heat transfer equation were combined by Stallman (1963) to determine the flow velocities of groundwaters:

$$\begin{aligned}&\frac{\partial}{\partial x}K_s\frac{\partial T}{\partial x}+\frac{\partial}{\partial y}K_s\frac{\partial T}{\partial y}+\frac{\partial}{\partial z}K_s\frac{\partial T}{\partial z}\\&-c_w\rho_w\left(V_x\frac{\partial T}{\partial x}+V_y\frac{\partial T}{\partial y}+V_z\frac{\partial T}{\partial z}\right)=c_s\rho_s\frac{\partial T}{\partial t}\pm R\end{aligned} \tag{2.9}$$

where K_s is the thermal conductivity (assumed to be "isotropic"); c_s is the specific heat, ρ_s is the density of sediments; c_w and ρ_w are the specific heat and the density of water, respectively; T is the temperature, V_x, V_y, V_z are the flow velocities in directions x, y, z; and R is a source/sink term, which can be used, for example, for the heat input by radiogenic heat generation.

As shown by Sharp and Domenico (1976); Yuekler et al. (1978); Welte and Yuekler (1981), Stallman's equation can handle the heat transfer adequately. All of the physical parameters of the heat flow equation are temperature and pressure dependent. Hence these dependencies must be taken into account when temperature changes are to be calculated precisely.

2.4 Heat Transfer in Sedimentary Basins

Heat transfer in sedimentary basins is an energy balance problem. Heat entering and leaving the basin and heat storage capacity are the controlling parameters. Thermal history of the basin can therefore be reconstructed if the heat transfer Eq. (2.9) can be applied under conditions of a sedimentary basin development.

Quantitative treatment of heat transfer requires the precise definition of the complex system of "sedimentary basin" in the light of the parameters of the heat transfer equation. These are thermal conductivity, specific heat, heat capacity, density of the system, flow velocities of different mass flows (normally fluids) within the system, and initial and boundary conditions of the heat transfer process. However, these parameters of the conductive and convective heat transfer change as a function of both space and time. For example, thermal conductivities vary spatially due to three-dimensional variations in lithologies and porosities. On the other hand, there is a continuous change in physical entities over time due to changes in sedimentary fill resulting from deposition, compaction, erosion, etc. All of these time- and space-dependent variations must be considered in solving the heat transfer equation. Additionally, boundary conditions may also change in space and time. Therefore not only relevant parameters but also many physical processes which influence heat transfer must be considered. Thus one must treat heat transfer in sedimentary basins in a wholistic manner.

2.4.1 Heat Transfer in Sedimentary Basins by Conduction

Heat conduction takes place as a diffuse process by which the kinetic energy is transferred by interatomic or intermolecular collision in a medium where a temperature disequilibrium exists. The amount of heat transmitted by conduction is proportional to the temperature gradient and to the proportionality factor, thermal conductivity, as given in Eq. (2.1). Consequently an accurate treatment of heat conduction in sedimentary basins requires first an appropriate definition of the thermophysical parameters of the sedimentary fill. This definition must consider both spatial and temporal variations. Spatial variations are due to lithology, porosity, and temperature; temporal variations are due to the changes in porosity and temperature caused by burial and compaction. Since conduction is the dominant heat transfer mechanism in sedimentary basins, a realistic definition of thermal conductivity and heat capacity is of great importance.

Thermal Conductivity of Sediments

Sedimentary rocks are porous, and pores are usually filled with formation fluids. Consequently, for the thermal conductivity determination of sediments a mixed material of at least two components, namely matrix and pore fill must be considered. Matrix is generally polymineralic and in certain cases the pore fill is also a mixture of oil and water, water and gas or all of them. Pore fluids play a crucial role for the so-called "bulk thermal conductivity" or "effective thermal conductivity", because pore fluids have a remarkably low thermal conductivity compared to matrix minerals. The thermal conductivity of sediments can be predicted if the lithology and pore filling fluid or gas are known. The empirical relation given in (2.10) can be used for the determination of bulk conductivities:

$$K_s = K_r(K_w/K_r)^{\phi} \tag{2.10}$$

where, K_s, K_r, and K_w are the thermal conductivities of sediment, grain (matrix) and water, respectively, and ϕ is the porosity of the sediment. However, Eq. (2.10) is valid only if K_w/K_r is neither very small nor very large (Palciauskas 1986; Ungerer et al. 1990).

The effective medium theory, which enables the prediction of properties of a composite material if the volume fractions of the individual components and their properties are known, can also be used for the definition of sediment thermal conductivities (Palciauskas 1986). The basic result of the theory is:

$$K_s^{-1} = \sum_{i=1}^{n} 3V_i(2K_s + K_i)^{-1} \tag{2.11}$$

where, K_s is the sediment thermal conductivity, K_i is the conductivities of individual components, and V_i is the volume fractions. This theory is based on the assumption that the composite material is homogeneous and isotropic, which can be justified if volume is sufficiently larger than the grain size. Since the volume of sediments and the size of grains in a basin fullfil this condition, this rule is valid for sediments (Kappelmeyer and Haenel 1974).

Allen and Allen (1990, p. 287) compared the approaches for conductivity prediction for a mixture of quartz and water in which quartz occupies 70% of sediment volume and water 30%. Thermal conductivities of quartz and water were taken as 13 and 1.7 mcal cm^{-2} s^{-1} $°C^{-1}$, at T = 100 °C, respectively. The bulk conductivities were predicted as 7 and 8 mcal cm^{-2} s^{-1} $°C^{-1}$, with Eqs. (2.10) and (2.11), respectively.

Most sedimentary rocks are anisotropic in terms of thermal conductivity (Gretener 1981). However, the effect of this difference is negligible, as only a minor amount of heat is conducted horizontally due to the absence of significant lateral thermal gradients.

Morin and Silva (1984) investigated effects of high pressure and high temperature on thermal conductivity of sediments and found that the dependency on pressure is negligible. However, there is an obvious dependency on temperature. In general thermal conductivity decreases with increasing temperature; the effect is less for lithologies with low to moderate thermal conductivities and greater for well conducting lithotypes, such as quartzite, and salt (Kappelmeyer and Haenel 1974). Hence, a correction of conductivity values is necessary for sediments at greater depths.

The dependency of thermal conductivities upon porosity is much more important for the calculation of heat transfer in sedimentary basins because the changes can be by a factor of 2–4, depending on the lithotype. Figure 2.1 presents the dependency of some major lithotypes, where the values are computed with Eq. (2.10) without considering changes in conductivity due to temperature. If both porosity and temperature dependency are taken into account, the increase in conductivity values due to pore volume reduction is compensated after a depth of approximately 3000 m by the decrease in thermal conductivities due to higher temperatures for most major lithotypes such as shale, limestone, and sandstone. However, in case of salt and quartzite the dependency on temperature is dominant so that the effect of pore volume reduction is overwhelmed, as shown in Fig. 2.2.

Burial of sediments is of course generally associated with compaction (porosity reduction) and with temperature increase which, in addition to different lithotypes, causes variations of thermal conductivity with depth and time. These changes must be taken into account when heat conduction is calculated.

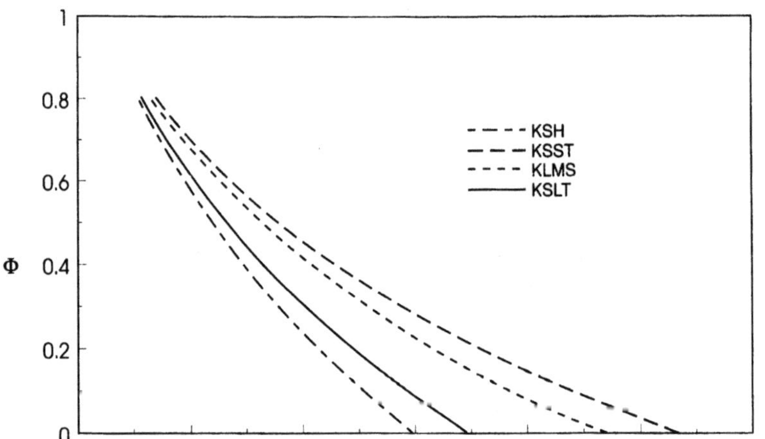

Fig. 2.1. Changes in thermal conductivity of the sediments (grains+pores) shale (*Ksh*), sandstone (*Ksst*), limestone (*Klms*), and siltstone (*Kslt*) due to the varying porosities. Thermal conductivities of the pure lithotypes and water (*Kw*) used for the calculation are: *Ksh* = 1.98 W/m°C; *Ksst* = 3.16 W/m °C; *Klms* = 2.83 W/m °C; *Kslt* = 2.22 W/m °C; *Kw* = 0.60 W/m °C

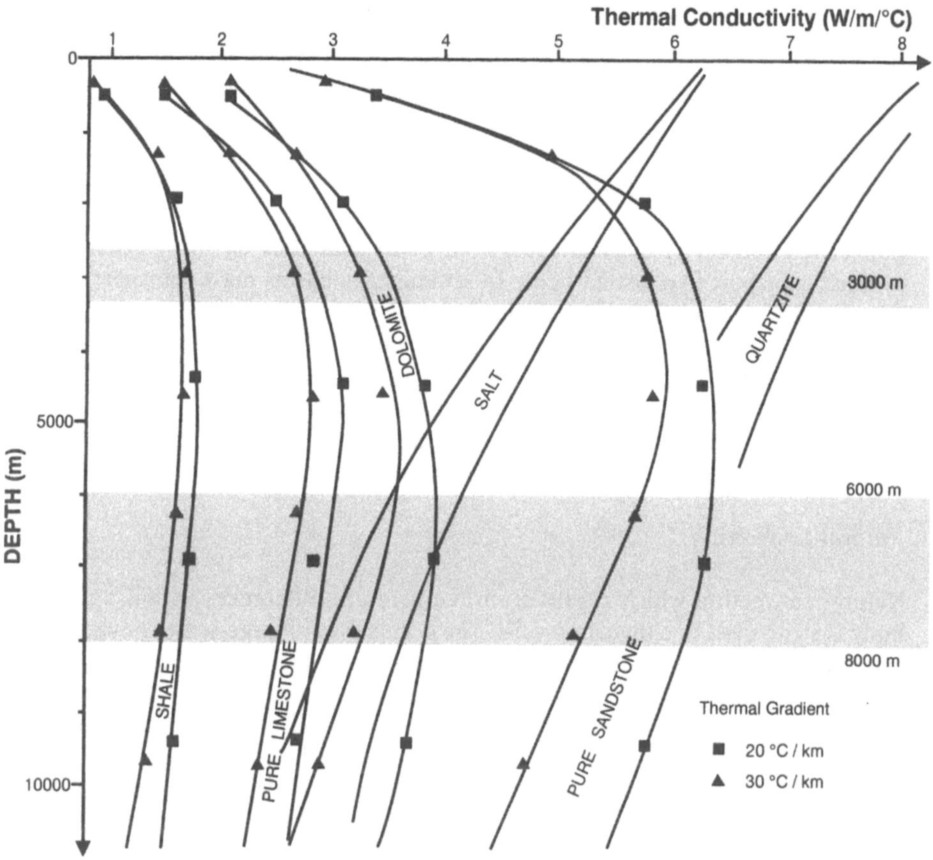

Fig. 2.2. Vertical thermal conductivity of porous rocks during burial, with two different thermal gradients. (Ungerer et al. 1990)

Specific Heat of Sediments

The capability of a body to store heat is expressed in terms of its specific heat. It gives the energy used or liberated in a change in temperature in terms of mass. Accordingly, it has an obvious effect on heat conduction under transient temperature conditions. It does not play a role under steady-state conditions where temperatures remain fixed.

As in the case of thermal conductivity for the specific heat prediction of porous sedimentary rocks, one must know the ratio and specific heat of the matrix and pore filling material. A weighted averaging can then be applied to find the specific heat of the sediment. Porous sediments have a high specific heat since water, the most common pore filling material, has an extremely high specific heat value (c = 4.2 Ws/g K at T = 20 °C) compared to the mean value of rocks (c = 0.8 Ws/g K at T = 20 °C). Specific heat of sedimentary rocks increases as a function of increasing temperatures. Mongelli et al. (1982) measured specific heat of eight different rocks at temperatures ranging from 20 ° to 250 °C and found an average increase of 19%. An increase of 10% was observed in plagioclasic micaschist and 35% in phyllite.

2.4.2 Heat Transfer in Sedimentary Basins by Convection

Convective transfer of thermal energy occurs by mass flow or more generally by mass transport in a heated body. In sedimentary basins mass transport can be achieved by fluid flow, by magmatic intrusions or volcanic eruptions, by tectonically induced displacement of rock volumes (faulting, overthrusting, etc.), or by synsedimentary mass-transport phenomena (debris flow, gravity sliding, etc.). This kind of solid-state convection requires a special consideration and is discussed later. Convection associated with fluid flow in sedimentary basins includes natural and forced convection.

Natural Convection

Natural convection which results from temperature differences within a fluid body is structured in convection cells. The occurrence of large-scale convective cell water systems in sedimentary basins is still controversial. According to Bjoerlykke et al. (1988, 1993), these are very rare because the fluid currents are interrupted by impermeable units. On the other hand, large-scale convectional cells are interpreted as the main mechanism by which water observed in ultradeep wells reaches the upper crust. Therefore more detailed work is necessary before natural convection can be totally excluded as an additional mechanism of heat transfer in sedimentary basins.

Forced Convection

Another mode of convectional heat transfer is due to the forced convection. Possible mechanisms of forced convection in sedimentary basins are: (1) expulsion of formation water during compaction, (2) release of confined water from sediments, which are hydrodynamically overpressured, and (3) flow of groundwater induced by pressure differences due to an artesian system. Compaction-driven water flow theoretically can occur in sedimentary basins continuously. The amount of water released from a 500-m-thick sandy shale sequence with a porosity of 15% (initial porosity 50%) is 350 million m^3/km^2 surface area. Therefore it could be expected that the compaction-driven water transfers a large amount of heat. However, studies of Bethke (1985) and Hermanrud (1986) have clearly shown that compaction-driven fluids move too slowly to be effective in altering temperature distribution in a basin. This is the case for both vertical and lateral flow. In addition to the very slow movement of water, only 0.01–0.05 mm/year, it is significant that most of the formation water is expelled while sediments are buried only to a few hundred meters, where temperatures are only 10 °–15 °C warmer than at the surface. Consequently the amount of heat stored in water is generally very limited. Thermal energy transferred by compaction-driven water flow is therefore negligible, unless it is concentrated in a very narrow zone. Release of confined water in sediments of overpressured zones may occur at a geologically fast expulsion rate. However, under realistic geological boundary conditions, duration and areal extent are insufficient to cause thermal anomalies affecting the heat distribution in sedimentary basins (Hermanrud 1986; Wygrala 1989).

Temperatures in sedimentary basins may also be affected by the convective heat transfer induced by groundwater flow through regional aquifers. This kind of water flow takes place in so-called hydrodynamic or artesian systems. Pressure gradients caused by pieziometric/topographic conditions lead to flow of water from recharge areas, which are topographically high, to discharge areas located in down-dip areas. Hydrodynamic water flow, which is controlled by the amount of water supply, the permeabilities within the aquifer, and topographic variations, plays a major role in heat transfer in sedimentary basins as discussed in detail by Smith and Chapman (1983) and Woodbury and Smith (1985). Effects of regional groundwater flow in artesian systems and its role in heat flow distribution have been investigated in many areas of the world. Examples are the Williston Basin in the United States (Gosnold and Fisher 1986), the Alberta Basin in Canada (Majorowicz et al. 1984), the Eromanga Basin, a part of the Great Artesian Basin in Australia (Chapman 1981), and the North Sea (Andrews-Speed et al. 1984). Each of these studies demonstrates that temperatures in the basins are affected considerably by groundwater flow. Temperatures are low in recharge areas due to the cold meteoric water entering and penetrating the aquifer. Water which is heated in deeper parts of the basin during transport results in anomalously higher temperatures at discharge areas. Whereas the cited examples are of regional nature, the same mechanism may also cause local disturbance of the temperature field in areas where an upward movement of thermal water is observed. Almost all thermal springs are related to fault and fracture zones enabling the upward movement of waters and causing local disturbances of the regional thermal field. Although the effects are local, thermal water springs are a good indication of deep-seated thermal anomalies. Some fault or fracture zones also act as sites of recharge. Hence an opposite effect, namely cooling of the system, would be the result. The general conclusion is that natural (free) convection and water flow forced by compaction and release of overpressuring do not play a crucial role in heat transfer. Only forced convection related to pressure gradients in hydrodynamic groundwater flow systems has a significant effect and must be taken into account for the reconstruction of temperature histories in sedimentary basins.

2.4.3 Boundary Conditions of Heat Transfer in Sedimentary Basins

In order to compute heat transfer in basin fill, physical and thermal conditions at the base and the top of the sedimentary basin must be considered. The lower boundary condition is determined by the heat flux density, which controls heat input into a basin. The upper boundary is determined by temperatures prevailing at the surface of the sedimentary package. Depending on the depositional environment, either the sediment/water interface (sea bottom) temperatures or subaerial surface temperatures define the upper boundary condition.

Heat Flux Density (Heat Flow)

The heat flux density is a measure of heat flow per unit time and unit area. It is often referred to as "heat flow". Sources of heat from the earth's interior are in general the upper mantle and the lower crust. Heat flow from the lower crust stems from the decay of radioactive isotopes, such as uranium, thorium, and potassium and is called the radiogenic heat flow. The remaining portion, called subcrustal heat flow, is related to the cooling of the earth. The relative proportions of these two components depend on many factors. However, the main control is thickness of the crust, as shown by Bodri and Bodri (1985; Fig. 2.3). However, Royden (1986) cautioned as to this oversimplified relationship between crustal thinning and heat flow increase and gave examples in which the increase is much higher than expected. Another correlation of heat flow is with tectonic age, where heat flow of tectonically young areas are much higher than the old cratons (Vitorello and Pollack 1980).

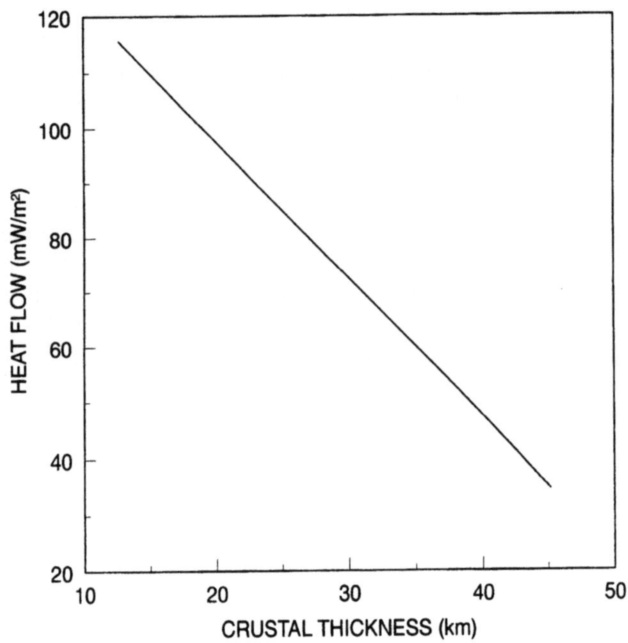

Fig. 2.3. General correlation of heat flow and crustal thickness for tectonic structures such as young Cenozoic tectonic formations, tectonic structures, with thin crust at the subsided peripheral parts of platforms and ancient platform structures of Precambrian age. (Bodri and Bodri 1985)

Heat transfer mechanisms between upper mantle, lower crust, and into the basin (upper crust) are yet unclear and sometimes controversial. Transfer by radiation can be excluded. However, whether heat transfer by conduction is really the dominant mechanism, or whether convection can play a decisive role under certain circumstances (Welte 1995) has yet to be investigated in detail. Theoretical concepts of thermal processes and events in the upper mantle and lower crust must be linked convincingly with an increasing data base on the temperature history of sedimentary basins. Heat input into a sedimentary basin from its base must be entirely controlled by radiogenic and subcrustal heat flow. However, in areas where a basin is formed on a sedimentary basement, which is frequently the case, effects of this basement can drastically modify the lithospheric heat flow. As has been established, in the sedimentary basin fill a major heat transfer mechanism is thermal conduction, and differences in thermal conductivity in space and time within the basin fill are therefore the most important factors controlling thermal irregularities (Welte 1995). Differences in thermal conductivities can be caused by primary effects of lithologic properties (shale, sand, salt, etc.) or induced by geologic processes, such as subsidence and sedimentation, generation and migration of hydrocarbons (especially gas) or overpressuring. Considering the entire heat budget of a sedimentary basin, fluid flow is only of minor importance and responsible for rather local temperature anomalies. For regional temperature anomalies within a basin, deep-reaching groundwater flow or water flow caused by hydrodynamics of overpressured strata are the main factors (Majorowicz et al. 1984, 1986; Andrews-Speed et al. 1984; Luheshi and Jackson 1986).

Measured regional heat flow values reveal important variations. Low values of 30–40 mW/m^2 are typical for old cratons such as the Baltic and Canadian shields and for oceanic trenchs. High heat flow values of 85–120 mW/m^2 are measured in oceanic ridge or active volcanic areas. Regions of extensional tectonics such as the Red Sea area and the basin and range areas of the western United States also have high heat flow values. The overall average value for heat flow is given by Lee (1963) as 63 mW/m^2. According to Kappelmeyer and Haenel (1974), the average for oceans is 61 mW/m^2 and for all continents 60 mW/m^2. Regional variations in surface heat flow, as compiled by Allen and Allen (1990), is given in Table 2.1. Their "worldwide" average value is higher, 69.6 mW/m^2, than those cited above. Although it is risky to generalize, an assignment of heat flow values (value ranges) to basin types is possible. A basin classification based on regional plate tectonic concepts, as proposed, for example, by Kingston et al. (1983a,b), is suitable for such an assignment. Some general results, as outlined in Fig. 2.4, have emerged: (1) ocean ridges, intra-arc basins, active rift areas have higher than normal heat flow; (2) stable cratonic areas, fore-arc basins have lower than average heat flow values; (3) young orogenic areas (suture zones) have high heat flow values.

If the actual aim is the reconstruction of temperature history of sedimentary basins, then not only the present values of heat flow but also the ancient values (paleoheat flow) must be known. Several attempts have been made to estimate paleoheat flow. The approach of McKenzie (1978, 1981) is a typical example, where the temporal changes in heat flow is calculated for a basin formed by

Table 2.1. Regional variations in surface heat flow. (Allen and Allen 1990)

	mW/m^2	HFU
Continents	56.6	1.35
Africa	49.8	1.19
North America	54.4	1.30
Australia	63.6	1.52
Oceans	78.2	1.87
North Pacific	95.4	2.28
Indian Ocean	83.3	1.99
South Atlantic	59.0	1.41
Worldwide	69.6	1.67

extension. The period of stretching and thinning of the lithosphere and uprising of the asthenosphere is associated with an increase in heat flow, which is quantitatively determined as a function of stretching factor (β) and time (Fig. 2.5). The so-called thermal subsidence period following the termination of stretching is represented by a decrease in heat flow over time due to the relaxation of lithospheric isotherms to their original position before stretching. Several modifications to the McKenzie model and also new geodynamic models have been proposed for the definition of basal heat flow as a function of time (Sclater and Christie 1980; Wernicke 1985; Royden 1986; Burrus and Bessis 1986). Although these models contributed much to the understanding of crustal and mantle thermal phenomena, they are still not sufficient to provide reliable data on heat flow variations with time, as our present knowledge of geodynamic processes is not precise enough (Tissot et al. 1987, Welte 1995). Consequently the history of the basal heat flow must be considered a "best estimate" parameter, which can be adjusted with the help of other, in-

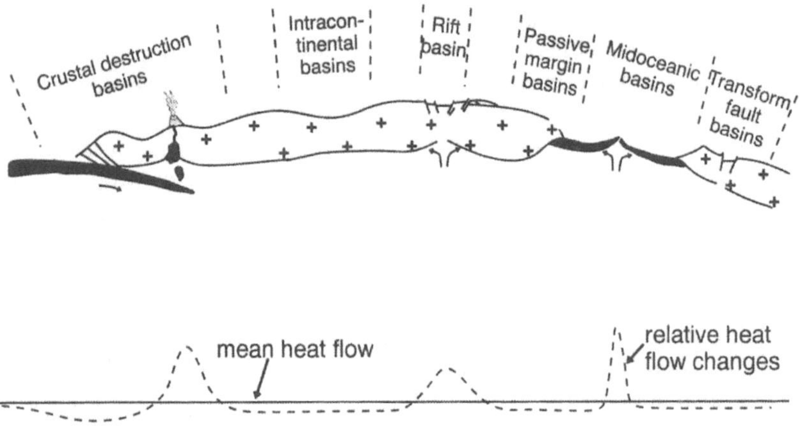

Fig. 2.4. Relative changes in heat flow in different types of sedimentary basins in respect to plate tectonics. (Yalçın and Welte 1988)

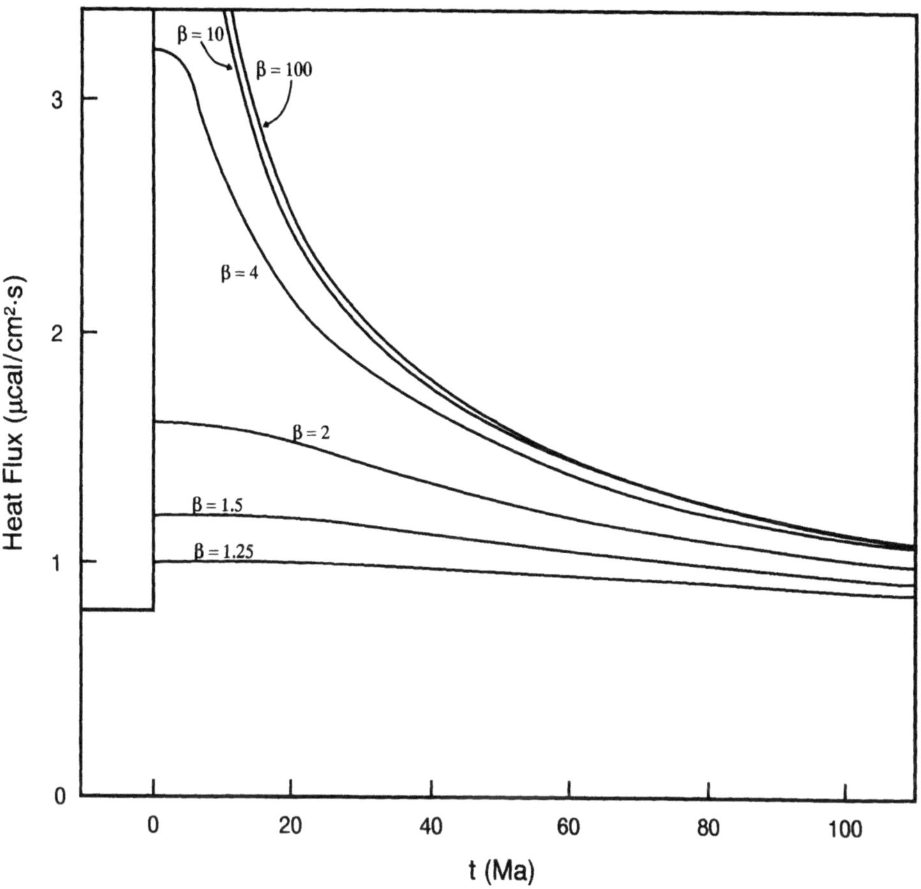

Fig. 2.5. Changes in heat flux as a function of time for various values of β. (McKenzie 1978)

dependent, temperature-sensitive measurements. The estimation must include a consideration of regional tectonic evolution, resulting basin type, heat flow of recent basins with similar characteristics, type and thickness of the actual crust, and the present heat flow to obtain a realistic paleoheat flow value. Calibration of the paleoheat flow estimation, that is, the adjustment of it with the help of independent measurements, is demonstrated in Section 2.5.3.

Surface Temperatures

The upper boundary condition of heat transfer in sedimentary basins is given by the temperatures either at the subaerial surface or at the sediment/water interface (sea floor). Whereas the annual mean ground surface temperatures can be obtained from mean air temperatures, temperatures at sediment/water interface are additionally affected by respective water depth, paleogeographic position of the basin, and global oceanic currents. Effects of local phenomena such as meteoric cold water discharge into the sea or small-scale currents are ignored since their effects are not distinguishable. Definition of the surface and/or sea bottom temperatures requires consideration of long-term variations in surface temperatures during geological time. Although an accurate determination of those variations is still not possible, facies studies of sedimentary units, fossil fauna and flora, ratio of oxygen isotopes, and plate tectonic reconstruction of continents and oceans allow the derivation of temperature variations during the geological past (Meyerhoff 1970; Schwarzbach 1974; Frakes 1979; Frakes et al. 1994; Habicht 1979; Smith et al. 1981; Parrish et al. 1982; Ziegler et al. 1984).

Approximation of the surface and/or sediment/water interface temperatures for a given sedimentary basin requires the following:

- The paleolatitude position of the basin through geological time using paleogeographic maps such as those by Smith et al. (1981) or by the Paleogeographic Atlas Project (Ziegler 1987).
- The mean annual paleoair temperatures as derived from maps of global paleotemperature distribution for a given geological time slice. Wygrala (1989) prepared a chart of mean surface paleotemperatures vs. latitude vs. geological time from the Carboniferous to the recent past (Fig. 2.6) from which a time-temperature curve can easly be derived.
- Contemporary surface temperatures. Corrected mean air temperatures can be used directly as the surface temperatures for periods of erosion or continental deposition. Due to the complex interactions between air and the surface caused by vegetation and soil/rock type, but also by local climatic conditions, no clear rules exist for this matter; addition or subtraction of 1–2 °C from the mean air temperatures is recomended by Haenel (1971) and Clauser (1984), respectively.
- Contemporary sediment/water interface temperatures. Here, effects of water depth, paleogeographic position of the basin and global oceanic currents must be taken into account; except for some shallow marine depositional enviroments (approximately water depths up to 60–200 m), the relevant sea bottom temperatures are determined by the coldest water temperature encountered within the specific marine system. In open marine systems, for example, the relevant value is that of the polar waters, at present 2–4 °C, but remarkably higher during preglaciation periods (Frakes 1979). The present Mediterranean Sea is a typical example of a closed marine system where sea bottom temperatures are not affected by global oceanic currents, and where sea bottom temperatures are therefore not less than 13 °C even at depths exceeding 2000 m (Wygrala 1989).

Despite all of these considerations it cannot be asserted that the determination of surface and sea floor temperatures is entirely correct. However, sensitivity of temperature history to the upper boundary condition can be tested, and a calibration can be carried out (see Sect. 2.5).

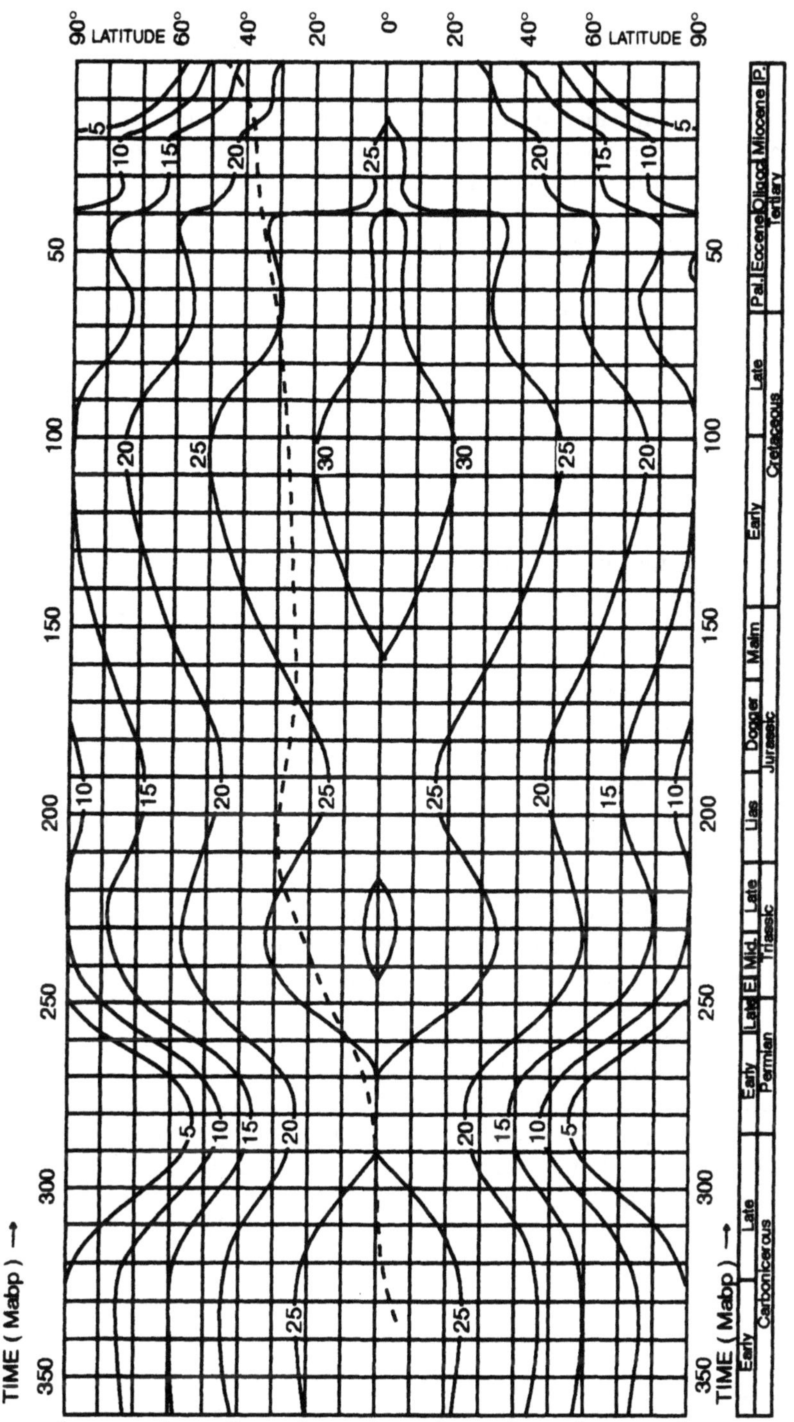

Fig. 2.6. Surface temperatures vs. latitude vs. geological time (Wygrala 1989). Global mean temperatures are based on Frakes (1979). Geological age scale is taken from Harland et al. (1989)

2.4.4 Other Factors Affecting Thermal History of Sedimentary Basins

In addition to the distribution of heat by conduction and convection, other parameters and processes also have an effect on the thermal history of sedimentary basins. Parameters related to depositional and tectonic processes, such as thickness, lithology, erosion, and thrusting, are discussed in Section 2.5. Here only radiogenic heat production and thermal refraction are treated briefly. Heat released from chemical reactions, such as from oxidation of sulfides, and heat related to the frictional energy are of minor importance and are omitted.

Radiogenic Heat Generation in Sedimentary Basins

As mentioned above, radioactive decay contributes significantlly to heat generation in the crust. Most sedimentary rocks contain radioactive minerals (elements). Despite relatively low concentrations heat generation by radioactive decay in sediments may affect the temperatures in sedimentary basins. Therefore the amount of heat generated by radioactive decay in sediments is included as an additional term to the heat transfer Eq. (2.9). All naturally occurring radioactive isotopes generate heat. However, only the series of uranium, thorium and potassium isotope 40 K contribute significantly to heat generation (Kappelmeyer and Haenel 1974; Rybach 1986). Radioactive heat generation (A) of a sediment can be calculated after Eq. (2.12) with the help of heat generation constants if the concentrations of uranium (C_U), thorium (C_{Th}) and potassium (C_K) are known (Schmucker 1969):

$$A = \rho(3.35_{c_K} + 9.79_{c_U} + 2.64_{c_{Th}})10^{-5} \tag{2.12}$$

where A is expressed in $\mu W/m^3$, ρ in kg/m^3, C_K in percentage, and C_U and C_{Th} in ppm. Depending on the concentration, heat production varies with lithology. Results of studies by Rybach (1976), Haack (1982), and Rybach and Cermák (1982), are summarized by Rybach (1986; see Table 2.2). Heat generation is generally low in evaporites and carbonates, medium in sandstones, and high in silt- and claystones. Pereira et al. (1986) have calculated an average heat generation of 0.63 $\mu W/m^3$ in the fine sandy and silty continental shelf sediments of Brazil, which correlates with the values of Table 2.2. They also found a positive correlation between fineness of grain size and heat generation. Rybach (1986) demonstrated the affect of heat production on the geothermal field by a simple, one-dimensional, purely conductive model equibilirium. He calculated temperature for any depth (h_z) using the equation:

$$T(h) = T_o + \frac{q + AH}{K} h_z - \frac{A}{2K} h_z^2 \tag{2.13}$$

where T_o is surface temperature, q heat flow at the base, A is the average radioactive heat production, H is the thickness, and K is the average thermal conductivity of sediments. Results of different A and K values are shown in Fig. 2.7. Effects of heat generation are directly proportional to the amount of heat generated and to depth. As heat generated by internal sources is dissipated

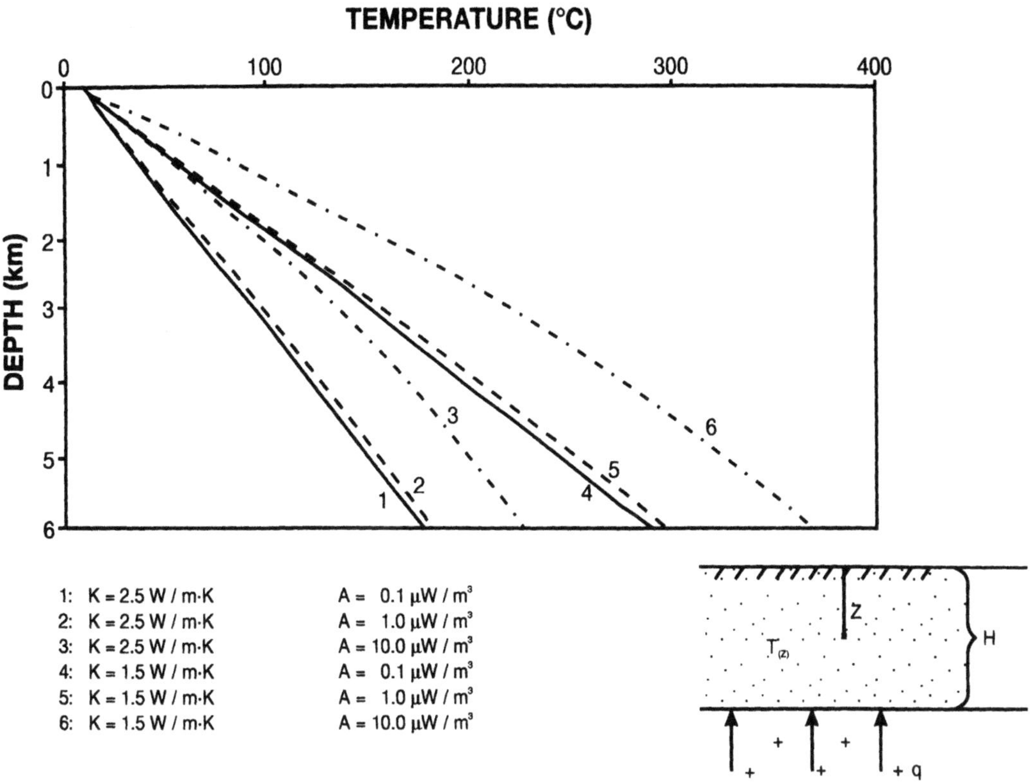

Fig. 2.7. Effect of heat generation in sediment (A) and thermal conductivity (K) on the temperature-depth profile. Temperature curves are calculated with the given K and A values and from Eq. (2.13) for H = 6 km and with q = 70 mW/m^2 and T_o = 10 °C (Rybach 1986). (Reprinted from Rybach L Amount and significance of radioactive heat source in sediments, pp 311–322, 1986, with kind permission from Editions Technip, Paris)

Table 2.2. Average contents of heat producing radioelements in sedimentary rocks. (Rybach 1986)

Rock type	U (ppm)	Th (ppm)	K (%)	Th/U	Density[a] (10^3 kg/m^3)	Heat generation (μW/m^3)
Carbonates					2.6	
Limestone	2.0	1.5	0.3	0.75		0.62
Dolomite	1.0	0.8	0.7	0.80		0.36
Evaporites						
Salt	0.02	0.01	0.1	0.50	2.2	0.012
Anhydrite	0.1	0.3	0.4	3.0	2.9	0.090
Shales, siltstones	3.7	12.0	2.7	3.2	2.4	1.8
Black shales	20.2	10.9	2.6	0.54	2.4	5.5
Sandstones					2.4	
Quartzite	0.6	1.8	0.9	3.0		0.32
Arkose	1.5	5.0	2.3	3.3		0.84
Graywacke	2.0	7.0	1.3	3.5		0.99
Deep sea sediments	2.1	11.0	2.5	5.2	1.3	0.74

[a]Broad average since density strongly depends on porosity. (Reprinted from Rybach L Amount and significance of radioactive heat source in sediments, pp 311–322, 1986, with kind permission from Editions Technip, Paris)

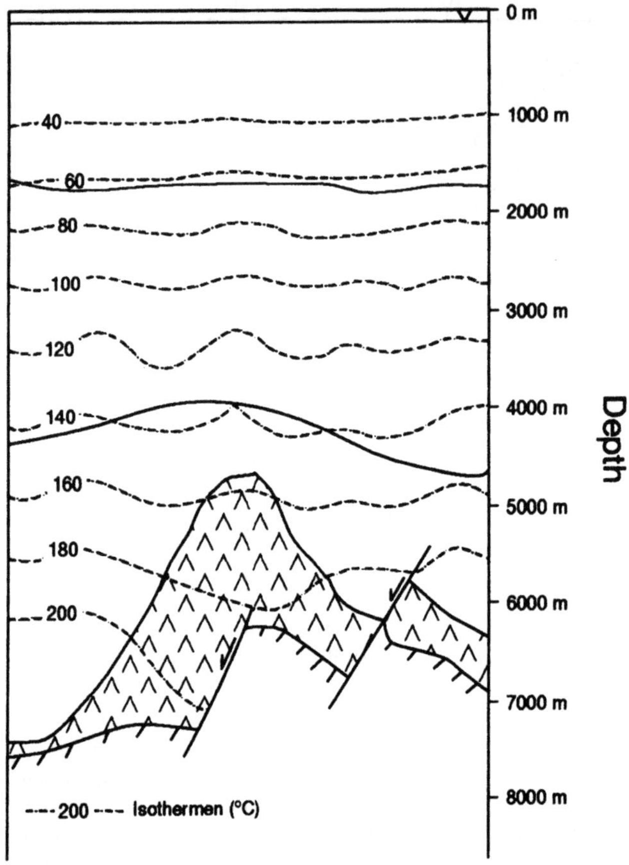

Fig. 2.8. Effect of a salt diapir on the temperature distribution. Note higher temperature gradient on the top of the salt diapir. (Modified from a simulated cross section in the Northern Central Graben Area; Heynisch et al. 1987)

by conduction, the temperature increase with time (i.e., depth) is also controlled by thermal conductivity.

A general statement on the affects of radioactive heat generation in sedimentary basins is difficult. However, if a basin is deep enough (>5 km), if the major lithologies are fine clastics, and if the basin fill is not very young (>10 Ma), one must consider the internal radioactive heat generation in modeling the temperature distribution.

Thermal Refraction

Conductive heat transfer principally occurs vertically as the isotherms are in general horizontal. However, if within the basin lateral conductivity contrasts exist, heat can also be conducted laterally. Lateral contrast of conductivity in sedimentary basins are either related to diapiric structures or basement highs.

Faulting can also result in formation of lateral conductivity changes if two lithotypes of varying conductivities are brought together. Lateral heat conduction leads to a deflection of isotherms depending on the thermal conductivity of the disturbing body. Salt diapirs act as chimneys and carry positive temperature anomalies in their roof (Fig. 2.8) due to very high conductivities of salt. In shale diapirs the opposite is observed. Basement highs, generally formed by crystalline rocks with higher conductivities than the sediments, act as salt diapirs.

The conclusion is that structures such as diapirs, faults, and basement highs may give rise to remarkable temperature irregularities in sedimentary basins as a result of thermal refraction. Lateral changes in lithofacies or any other lateral/horizontal inhomogeneity in the sedimentary rock sequence causing lateral differences in thermal conductivity, such as thinning or thickening of the sedimentary strata, changes in porosity, or saturation of pore space with oil or gas, result in temperature disturbances, i.e., isotherms are not horizontal. Consequently to consider lateral conductivity differences a realistic calculation of heat transfer must be carried out in two or three dimensions.

2.5 Reconstruction of Thermal History in Sedimentary Basins

The reconstruction of thermal history in sedimentary basins is still an area with problems which require further research and development (Hermanrud 1993; Lerche 1993). The basic problem is that sedimentary basins are not static but experience a dynamic evolution through geological time, and thermal energy is in principle transported steadily within the basin. Furthermore, no measurable parameter can be directly converted to paleotemperatures (Tissot et al. 1987). However, several methods and/or concepts contribute to the assessment of temperature conditions in the geological past:

- Methods based on maturation indices of organic matter, for example, degree of coalification, vitrinite reflectance (Karweil 1956, 1975; Lopatin 1971; Waples 1980; Kettel 1981).
- Concepts based on mineralogical changes due to the diagenesis, for example, clay mineral transformations, and illite crystallinity (Burst 1969; Kübler 1967; Pytte and Reynolds 1989).
- Methods based on fluid-inclusion analysis, for example, homogenization temperatures (Burrus 1987; Leischner 1992; Roedder and Bodnar 1980).
- Geothermometers based on specific chemical reactions, such as stable isotope equilibration (Hoefs 1987) and the SiO_2-Na-K-Ca thermometer of Ellis and Mahon (1977).
- Fission track analysis as outlined by Gleadow et al. (1983) is another method of temperature determination. This is similar to fluid inclusion thermometry in that a specific temperature value can be determined. This method gives a record of the temperature, and concurrently the time that the system has experienced (Green et al. 1989). This approach was used by Huntsberger and Lerche (1987) to determine paleoheat flux.

- For radiometric systems, such as K-Ar, Rb-Sr, and U-methods, which are closed at different temperatures, a combination of radiometric age determinations can be used for thermal history reconstructions as indicated by Buntebarth and Stegena (1986).

Considering the limitations of the various approaches noted above it seems that methods based on physical concepts and numerical simulation are the most suitable for reconstructing the thermal history of sedimentary basins. In other words, the temperatures within a basin must be calculated by solving the heat transfer equation given in Eq. (2.9). As stated above, thermal history is linked to the basin evolution, which is a dynamic, complex, and a multi-parameter system with boundary conditions changing through time. Additionally, processes and parameters which control the temperature distribution are closely interrelated and influence each other. The consequence is that calculation of temperatures during basin evolution requires an integrated and quantitative approach and a suitable modeling method. Either of two main modeling techniques, namely forward and inverse modeling, can be applied to predict temperature history. Inverse modeling is used mainly to determine paleoheat flux, either with the help of measured vitrinite reflectance values (Lerche et al. 1984) or of fission track analysis results (Huntsberger and Lerche 1987). Forward modeling is applied to problems of hydrocarbon generation prediction (Tissot and Espitalie 1975) and to model geological processes during basin evolution (Yükler et al. 1978; Welte and Yükler 1981; Welte and Yalçın 1985; Yalçın 1991). The latter (computer-aided basin modeling) seems the most suitable method to handle dynamic, complex and integrated systems in a deterministic manner.

2.5.1 Reconstruction of Thermal History by Computer-Aided Basin Modeling

Reconstruction of the thermal history of sedimentary basins by computer-aided basin modeling is demonstrated with the help of hypothetical and real case histories. A brief description of the modeling approach (PDI basin modeling system) is given below to furnish the reader with the necessary background information.

The PDI system uses a *forward modeling* approach, for example, processes during basin evolution are modeled from past to present employing assumed starting conditions according to the geological principle that "the present is the key to the past." This is a *deterministic model*. Therefore results of the model runs are unique, and no probabilities are involved. A *numerical modeling* technique is applied, and hence numerical values of all input parameters are required. This can cause some problems in parametric studies of multivariable systems. Finally, routines of the system are constructed in such a way that *dynamic modeling*, for example, modeling of non-steady-state systems is possible, whereas static modeling can be applied only to equilibrium systems. The close interrelationship between different processes which must be simulated require an integrated approach, as "determinism" could otherwise not be ensured. Equations used in the system are formulated in such a way that almost all are coupled with each other via common parameters. Hence the simulation represents a *fully integrated* basin modeling approach. The following processes can be simulated: (a) deposition, (b) nondeposition, (c) erosion, (d) faulting, (e) thrusting/overthrusting, (f) halokinetics, (g) compaction, (h) pressure build-up, (i) heat transfer, (j) maturation of organic matter, (k) oil and gas generation, and (l) expulsion of oil.

Most of these processes affect the thermal history and vice versa. Processes such as maturation of organic matter and petroleum generation are, on the other hand, controlled by the temperature history. During the simulation of basin evolution heat flux density at the base of the system and surface (and/or sea floor temperatures) are used as the thermal boundary conditions. Thermal conductivity, specific heat, density, compressibility, permeability, initial porosity of sediments of different lithotypes, and viscosity and density of pore fluids are important parameters employed during the calculation procedures of the modeling routine.

One starts with modeling at a given time in the geological past and simulates the temporal development until the present. As a consequence the input data must be prepared in an appropriate way so that the entire temporal development is represented by the data. This requires a systematic consideration of the geological history using all available data. However, knowledge on basin history is often limited, and some of the paleoparameters can be estimated only using indirect approaches. Estimation of paleoheat flow and paleoclimatic conditions or erosional amounts are typical examples. Therefore prior to the mathematical model a so-called conceptual model must be designed. The conceptual model forms the basis for the model input (Welte and Yalçın 1985) and must be calibrated by comparing the first simulation results with observed data. In most cases an iterative approach is needed to reach an acceptable fit between simulated and observed results. The main point during the construction of the conceptual model is definition of the geochronological units. Structuring of the input data is based on this temporal subdivision in term of events. Time periods which are not represented by physically existing units, such as erosional periods, must also be identified and included. The event-based structured input data consist of absolute geological age and duration, present and eroded original thickness of deposited units, lithology, present porosity, water depths, heat flow at the base of the system, and surface or sediment/water interface temperatures at the top. Additional input data are the physical properties of different lithotypes and formation water, and for purposes of hydrocarbon generation and expulsion the type and amount of organic matter.

The typical outputs of the model for each event consists of thickness, porosity, pore pressure, temperature, maturity, level of hydrocarbon generation, amount of generated hydrocarbons, and the expulsion efficiency as a function of time. Almost all of these parameters can be used directly or indirectly for calibration purposes by comparing the values calculated by the simulation for the present time with measured data. After the calibration of the conceptual model a sensitivity analysis can be performed by changing selected input parameters within geologically justifiable ranges and by simulating again the basin evolution with these values. The sensitivity of results (outputs) to systematically changed input data can be tested by this procedure.

Other details of the system and its modeling aspects are provided by Welte and Yalçın (1985, 1988), Yalçın and Welte (1988), Wygrala (1989) and Yalçın (1991). Some case histories simulated with the system are demonstrated by Yalçın et al. (1988, 1994), Novelli et al. (1988) and Wygrala (1988). Also, see Poelchau et al. (Chap. 1, this Vol.) for further discussion of numerical modeling procedures.

2.5.2 Controls of Thermal History

The PDI basin modeling system has been used to demonstrate the effects of various processes and parameters on thermal history. For this purpose some hypothetical cases of one-dimensional simulations are first carried out. To visualize the effects of different parameters on thermal history a case history is simulated as a reference well. Other wells are then simulated in which only one parameter is changed. Comparing the computed temperature values in each case, reference well and the well with the changed parameter, the effect of the relevant parameter is discussed.

The reference case represents an uninterrupted deposition of a 2400-m-thick sandy shale sequence over 120 million years (Ma) with a sedimentation rate of 20 m/Ma. Forty events are distinguished, each of 3 Ma duration. Boundary conditions remained constant over the entire time. Heat flow at the base was chosen to be 1.2 HFU and sediment/water interface temperatures 14 °C. The water depth was set at 100 m. Simulation was carried out in time steps of 1 Ma. The complete form of the input data is given in Table 2.3. Results are demonstrated using various diagrams (Figs. 2.9–2.11) displaying temperature vs. depth at present, temperature vs. time of the lowermost unit, and a corrected burial history diagram with temperatures. Figures 2.9–2.11 refer to the reference well. Computed values show that temperature at the base of the sequence increases to 82.5 °C at a depth of approximately 2500 m, where the actual thickness of sediments is 2400 m (Fig. 2.9). The difference of 100 m is the water depth at present. Due to the uniform boundary conditions an almost constant geothermal gradient of 28 °C/1000 m is established. For the same reason temperatures increase very gradually with a constant rate of 0.6 °C/Ma (Fig. 2.10), and as shown by the burial history diagram where isotherms remain horizontal during the entire period of deposition (Fig. 2.11).

Rate of Burial and Sedimentation

The rate of deposition may effect the temperature distribution even if all other parameters remain constant. Extremely fast sedimentation rates can lead to decreased temperatures if the sedimentation is faster than the rate of thermal

Table 2.3. Input data of the reference well

Event no., name	Lithology	End-begin (Mabp)	Thickness (m)	Porosity (%)	Water depth (m)	SWI temp. (°C)	Heat flow (HFU)
40 Layer 40	Sandy shale	0–3	60	44	100	14	1.20
39 Layer 39	Sandy shale	3–6	60	38	100	14	1.20
38 Layer 38	Sandy shale	6–9	60	34	100	14	1.20
37 Layer 37	Sandy shale	9–12	60	32	100	14	1.20
36 Layer 36	Sandy shale	12–15	60	30	100	14	1.20
35 Layer 35	Sandy shale	15–18	60	29	100	14	1.20
34 Layer 34	Sandy shale	18–21	60	27	100	14	1.20
33 Layer 33	Sandy shale	21–24	60	26	100	14	1.20
32 Layer 32	Sandy shale	24–27	60	25	100	14	1.20
31 Layer 31	Sandy shale	27–30	60	24	100	14	1.20
30 Layer 30	Sandy shale	30–33	60	24	100	14	1.20
29 Layer 29	Sandy shale	33–36	60	23	100	14	1.20
28 Layer 28	Sandy shale	36–39	60	22	100	14	1.20
27 Layer 27	Sandy shale	39–42	60	22	100	14	1.20
26 Layer 26	Sandy shale	42–45	60	21	100	14	1.20
25 Layer 25	Sandy shale	45–48	60	20	100	14	1.20
24 Layer 24	Sandy shale	48–51	60	20	100	14	1.20
23 Layer 23	Sandy shale	51–54	60	19	100	14	1.20
22 Layer 22	Sandy shale	54–57	60	19	100	14	1.20
21 Layer 21	Sandy shale	57–60	60	19	100	14	1.20
20 Layer 20	Sandy shale	60–63	60	18	100	14	1.20
19 Layer 19	Sandy shale	63–66	60	18	100	14	1.20
18 Layer 18	Sandy shale	66–69	60	17	100	14	1.20
17 Layer 17	Sandy shale	69–72	60	17	100	14	1.20
16 Layer 16	Sandy shale	72–75	60	17	100	14	1.20
15 Layer 15	Sandy shale	75–78	60	16	100	14	1.20
14 Layer 14	Sandy shale	78–81	60	16	100	14	1.20
13 Layer 13	Sandy shale	81–84	60	16	100	14	1.20
12 Layer 12	Sandy shale	84–87	60	15	100	14	1.20
11 Layer 11	Sandy shale	87–90	60	15	100	14	1.20
10 Layer 10	Sandy shale	90–93	60	15	100	14	1.20
9 Layer 09	Sandy shale	93–96	60	14	100	14	1.20
8 Layer 08	Sandy shale	96–99	60	14	100	14	1.20
7 Layer 07	Sandy shale	99–102	60	14	100	14	1.20
6 Layer 06	Sandy shale	102–105	60	14	100	14	1.20
5 Layer 05	Sandy shale	105–108	60	14	100	14	1.20
4 Layer 04	Sandy shale	108–111	60	13	100	14	1.20
3 Layer 03	Sandy shale	111–114	60	13	100	14	1.20
2 Layer 02	Sandy shale	114–117	60	13	100	14	1.20
1 Layer 01	Sandy shale	117–120	60	13	100	14	1.20

equilibration. As a consequence, sediments remain cooler although they are buried. However, the sedimentation rate must exceed values of 1000 m/Ma, as shown by Doligez et al. (1986) for this effect to be significant. The cooling effect of extremely high sedimentation rates can be compensated by a decreased thermal conductivity induced either by overpressuring or by the deposition of low thermal conductivity lithologies. Overpressuring is always associated with high sedimentation rates, which results in turn in high porosity values, i.e., a

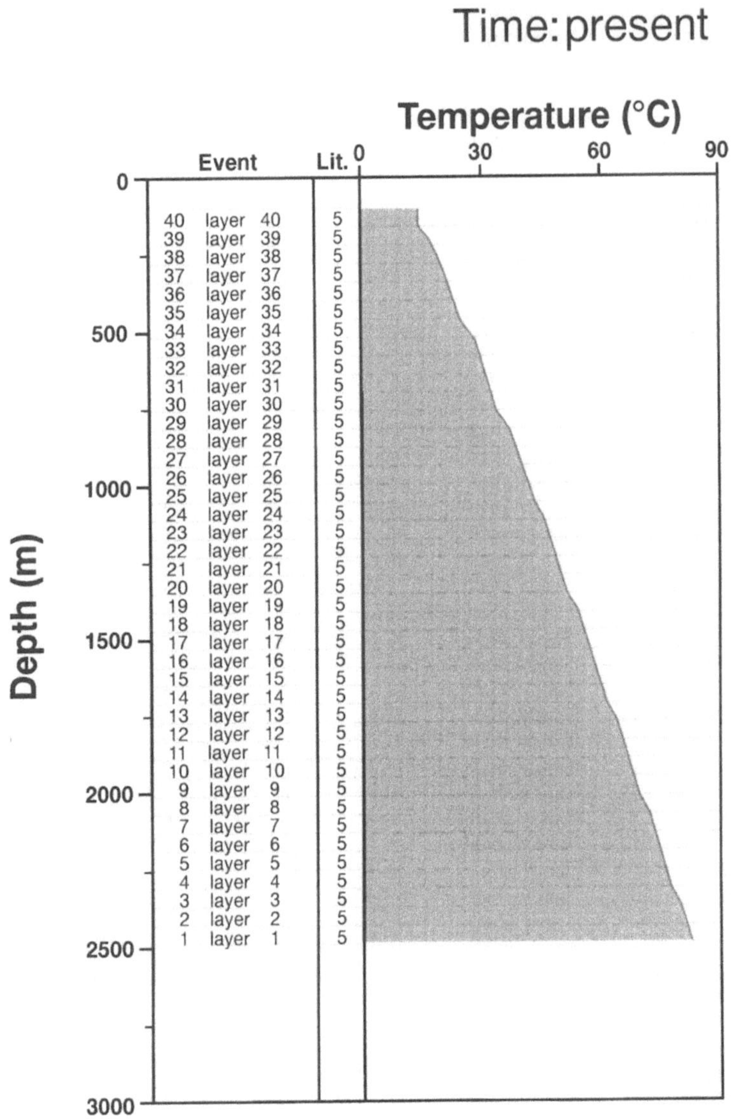

Fig. 2.9. Simulated temperature vs. depth trend of the reference well

high water content. The thermal effect of overpressuring was observed when the sedimentation rate was increased to 300 m/Ma. In this case temperature at the base of the sequence increased to 93 °C due to a decrease in thermal conductivity resulting from high porosities of the overpressured system (Fig. 2.12, line A).

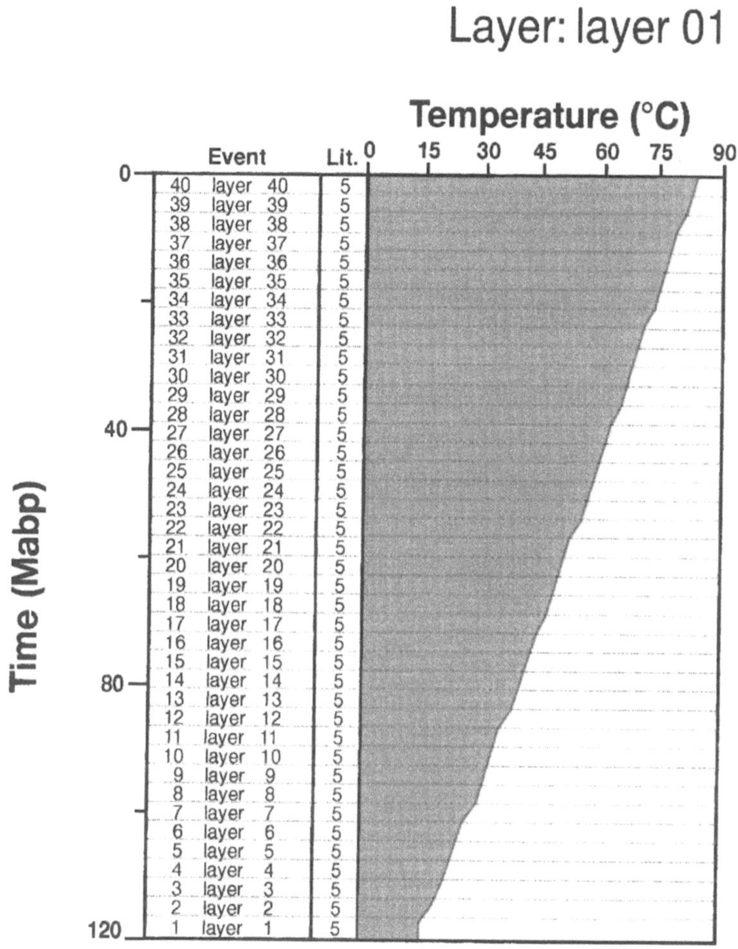

Fig. 2.10. Temporal development of temperature in the reference well

Lithology and Thermophysical Properties

Effects of various lithotypes and associated differences in thermophysical parameters are demonstrated where the lithotype of the reference well is changed from sandy shale to a pure shale and alternatively to salt. Respective values of thermal conductivity and specific heat used during the simulations are listed in Table 2.4. Whereas deposition of shale resulted in a temperature of 88 °C at the base of the sequence at 2500 m depth, in the case of salt the temperature could rise to only 38 °C (Fig. 2.12, lines B and C, respectively). This is due to the very high thermal conductivity of salt. The lower thermal conductivity and the higher porosity values of the pure shale with a resulting decrease in thermal conductivity are the cause of a higher temperature profile than in the reference well (see Yalçın 1991).

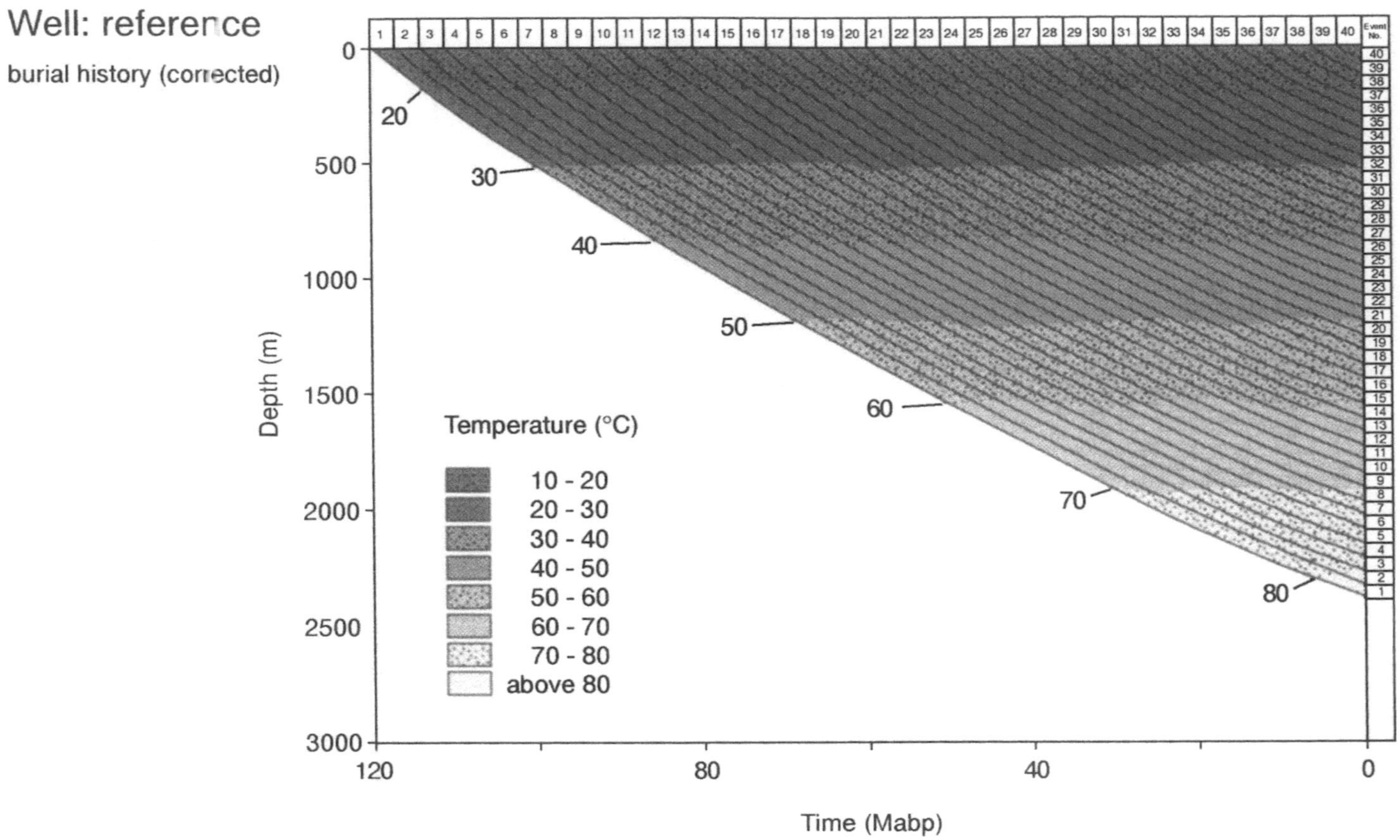

Fig. 2.11. Burial history diagram of the reference well with relevant isotherms. Note the almost uniform development of the geothermal gradient

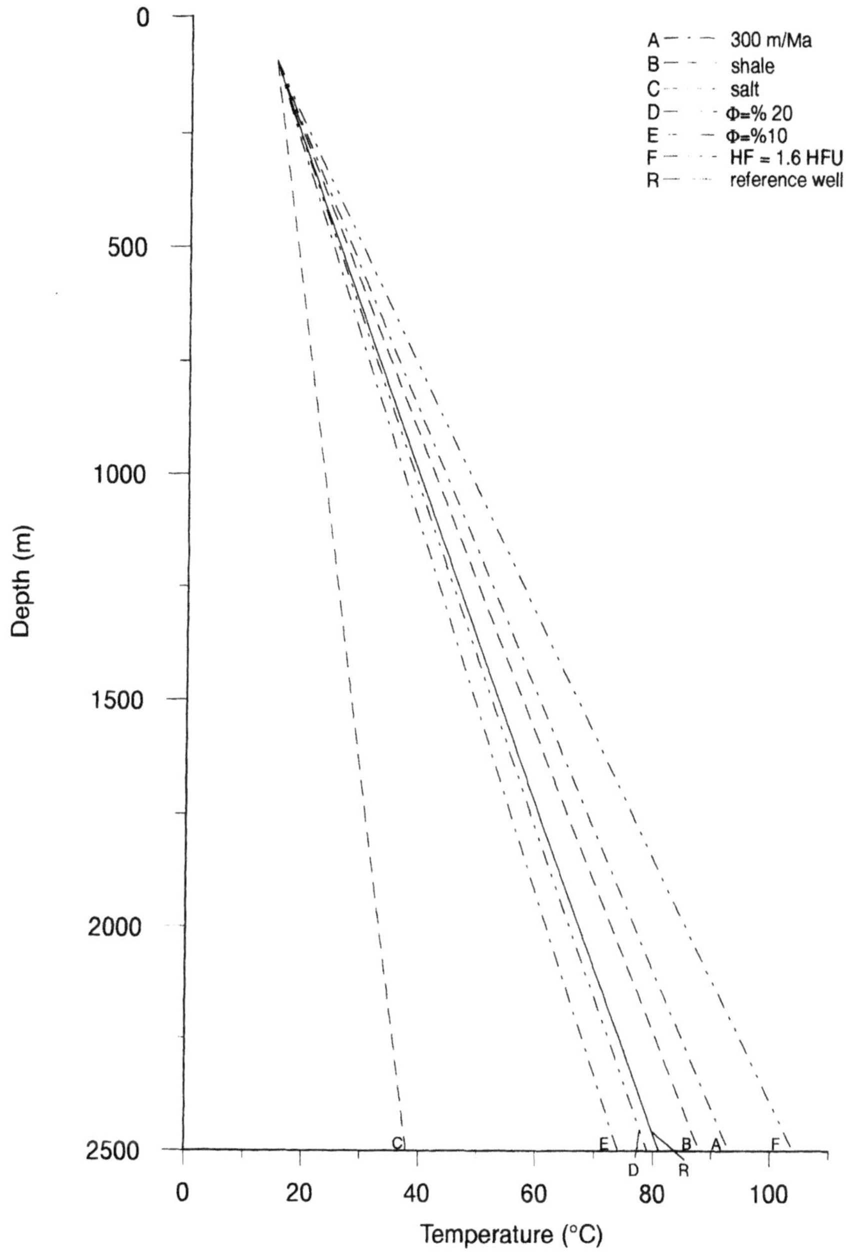

Fig. 2.12. Various temperature vs. depth trends which are calculated with only one varying parameter with otherwise constant parameters. Effects of different lithologies, porosities, heat flux densities, and sedimentation rates are demonstrated

Table 2.4. Thermal properties of the lithotypes sandy shale, shale, and salt which have been used for the simulation of case histories

	Specific heat (cal/g °C)		Thermal conductivity (W/m °C)	
Lithotype	At 20 °C	At 100 °C	At 20 °C	At 100 °C
Sandy shale	0.205	0.248	2.32	2.12
Shale	0.213	0.258	1.98	1.91
Salt	0.206	0.212	5.69	4.76

Porosity and Type of Formation Fluid

To show the effect of porosity the reference well is simulated in such a way that porosities remained constant, both as a function of time and depth during deposition. Two cases are simulated, one with 10% porosity and another with 20%. A higher porosity, for example, higher water content, diminishes the thermal conductivity of the sedimentary sequence. This results in a decrease in conducted heat and hence a lower heat loss at the surface. Consequently, the temperature could rise to 79 °C in the well exhibiting higher porosities, whereas it remained at 74 °C in the well with a porosity of 10% (Fig. 2.12 lines D and E, respectively).

The type of the formation fluid also influences the temperature history. Changes caused by salinity differences are negligible. However, when formation water is replaced by oil and/or gas, the effect on the bulk thermal conductivity may be remarkable, as thermal conductivities of oil and gas differ considerably from those of water. Because thermal conductivities of oil and gas are lower, the bulk conductivity of a hydrocarbon saturated sediment drops. Normally only reservoirs, and to a certain degree carrier beds, are saturated with oil or gas. In such a case the thermal properties of the respective sequence must be specifically defined. Zwach (1995) showed such an effect in the Alberta Deep Basin.

Heat Flux Density (Heat Flow)

The effect of heat flow on thermal history is obvious since it defines the energy input into the system. Therefore temperature and heat flow are directly proportional, as also indicated in Fig. 2.13, where the present temperature distribution of the reference well is compared with another well. In this well heat flow is increased from 1.2 to 1.6 HFU and temperature at the base increased accordingly from 81 °C in the reference well to 104 °C in the other well (Fig. 2.13, line F). The relationship between heat flow and temperatures is also demonstrated where a simulation with varying heat flow histories is carried out. For this simulation everything except the heat flow is taken as it was in the reference well. In the other well heat flow was constant as 1.2 HFU during the first 21 Ma of the basin evolution. For the next 21 Ma it increased to 1.3 HFU and then gradually up to 1.8 HFU during the next 21 Ma. For a period of 9 Ma it remained at this level and then decreased gradually to 1.3 HFU. Input data

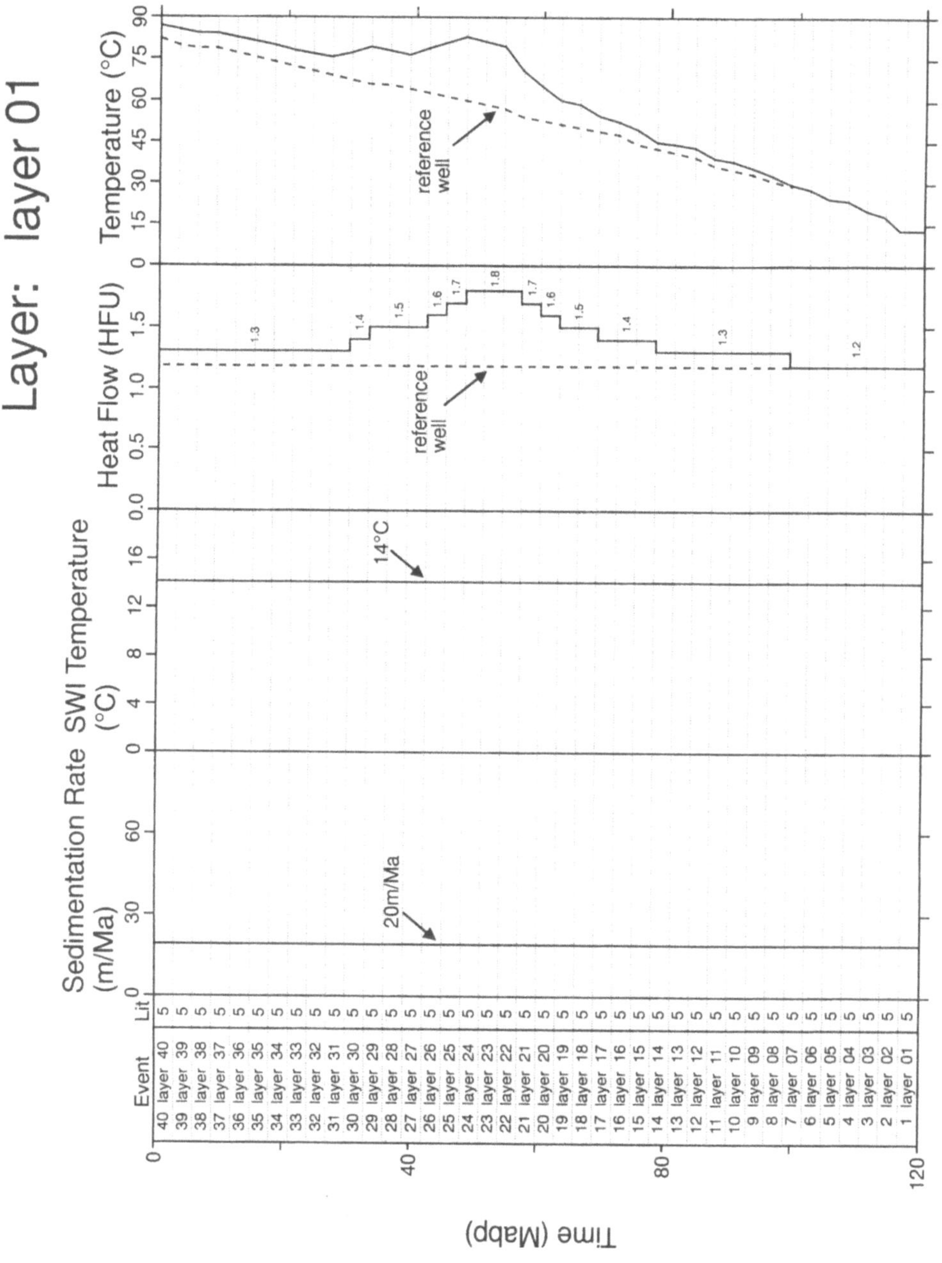

Fig. 2.13. Effects of a variable heat flux density on the temperature history. Other simulation parameters are kept as those of the reference well

used for this simulation and resulting temperature history of the lowermost (layer 1) are shown in Fig. 2.13. The direct relationship between changes in heat flow and temperature of the sediment is clearly demonstrated.

Surface and/or Sea Floor Temperatures

The effect of surface or sea floor (sediment/water interface) temperatures on thermal history is similar to those of heat flow. It can be generally assumed that temperatures increase with rising surface temperatures and vice versa. For example, an increase in surface temperature from 14° to 19 °C will result in a proportional shift of the temperature/depth curve toward higher temperature values. Another simulation with varying sediment/water interface temperatures is also performed. Some of the input parameters of this simulation are shown in Fig. 2.14. While heat flow and sedimentation rate remained constant over the entire simulation period of 120 Ma, sediment/water interface temperatures were varied. Results of the simulation showed that changes in sediment/water interface (SWI) temperature of only 1 °C affects the temperature history of the lowermost unit as indicated by changing gradients of temperature increases (Fig. 2.14). At 36, 15, and 9 Ma before the present (Mabp) the SWI temperatures were dropped from 22 to 20, from 19 to 15 and from 15 to 13 °C, respectively. The result was that the temperatures of the lowermost unit decreased although it was buried continously deeper. This is a typical example that shows the effect of a given parameter being overwhelmed by another in a manner which produces unexpected but valid results. As discussed below, such effects are not an exception in computing thermal histories rigorously based on relevant physical and physicochemical laws. Geological experience and frequently used "scientific intuition," neither of which can quantify complex, interrelated geological processes, may often suggest different answers.

Type of Geological Process

One can distinguish between the following processes: deposition, nondeposition, erosion, and processes related to deformation of basin fill.

Deposition is always associated with basin subsidence. Naturally temperatures are expected to rise with increasing burial, which is generally the case. However, other parameters or contemporaneous processes may indeed have an effect in the opposite way. That is, temperatures in a given layer may be lowered although it is buried deeper. As demonstrated above, effects of the upper and lower boundary conditions, sea floor temperatures and the heat flux densities are of prime importance. In addition, the effects of lithology, sedimentation rate, and porosity reduction (i.e., compaction) also affect the temperature history, sometimes in an unexpected manner. Thus, as shown above, in some cases despite burial the entire system or parts of may actually become cooler.

During *nondepositional* periods and without further subsidence a sedimentary basin loses thermal energy at the surface. Unless additional heat is generated in the basin fill, the system cools toward a state of equilibrium. This effect is demonstrated with the help of the reference well where three non-

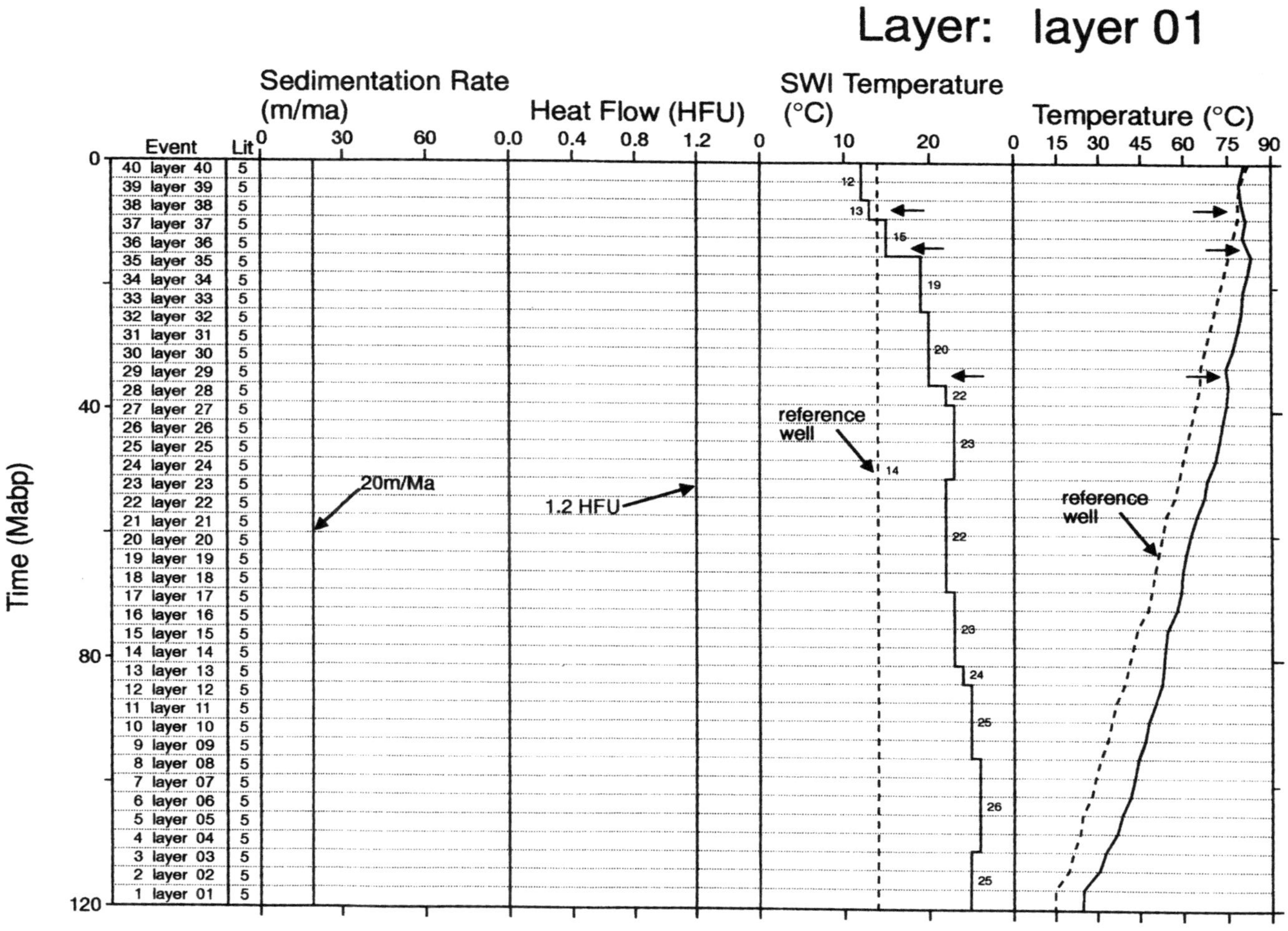

Fig. 2.14. Effects of variable SWI (sea bottom) temperatures on the temperature history. Other simulation parameters are kept as those of the reference well

depositional periods are added (Fig. 2.15). The first two periods have the same duration of 18 Ma. However, the first nondeposional period began after a deposition with a sedimentation rate of 40 m/Ma, whereas the second ensued after a rate of 20 m/Ma. The last nondepositional period lasted only 3 Ma. In all three cases the temperatures decreased within the first 3 Ma following the initiation of the nondeposition periods. They then remained constant until the next depositional period began. It is interesting to note that cooling was faster during the nondepositional period after the rapid deposition (phase I), and much slower following the period with lower sedimentation rates (phase II). During phase III cooling started even before the onset of nondeposition due to a decrease in the sedimentation rate from 40 to 20 m/Ma during the event just prior to nondeposition (Fig. 2.15). Both the sedimentation rate and the type of lithology plays a role. The sandy shale used for the simulation has moderate thermal conductivity but a relatively high heat capacity. These resulted in a long reequilibration time of the temperature distribution.

Erosion is also associated with cooling and occurs during the entire erosional period, as shown in Fig. 2.16. There is a direct correlation between the amounts of cooling and erosion. Temperature dropped from 78 to 42 °C during the erosion phase I, where during 9 Ma, 810 m were eroded. During the erosion phase II, also lasting 9 Ma, 360 m is eroded, with a corresponding temperature decrease of only 13 °C. Erosion leads not only to a decrease in temperatures but also to changes in the geothermal gradient. Figure 2.17 presents the burial history diagram of the hypothetical well and the corresponding change in the isotherms during geological history. Prior to the erosion phase I the average geothermal gradient was 34 °C/1000 m. At the end of the erosional period the value dropped to 26 °C/1000 m. A similar trend can be observed during the erosion phase II as indicated by the deepening of isotherms (Fig. 2.17).

Whereas erosion leads only to vertical changes in basin stratigraphy, most *structural deformation* is associated with lateral displacement. Particularly faulting, thrusting, and diapirs cause drastic changes in the continuity of layers, both in vertical and lateral direction. The displacement of sediments affects both the conductive heat transfer and the water flow in the system. A certain amount of heat can be transferred by such displacements if the movement is rapid enough. However, such movements are generally too slow, and the amount of heat which is transported by mass convection is negligible except in cases of rapid overthrusting (Wygrala et al. 1990; Yalçın 1991).

If along a cross section only a single point is considered, normal and *growth faulting* often results in a thickness reduction of the relevant unit. The thickness reduction is normally compensated by the increased sedimentation rate on the downthrown block of synsedimentary (growth) faults and by enhanced erosion on the upthrown block of a normal fault. In both cases the configuration of the layers prior to faulting changes drastically, and the conductive heat transfer in vertical and lateral directions is affected. Figure 2.18 presents a cross section from the Central Graben area of the North Sea illustrating the formation of a synsedimentary fault. The fault was active during the deposition of units 5 to 17 as indicated by thickness variations. Figure 2.19A shows the

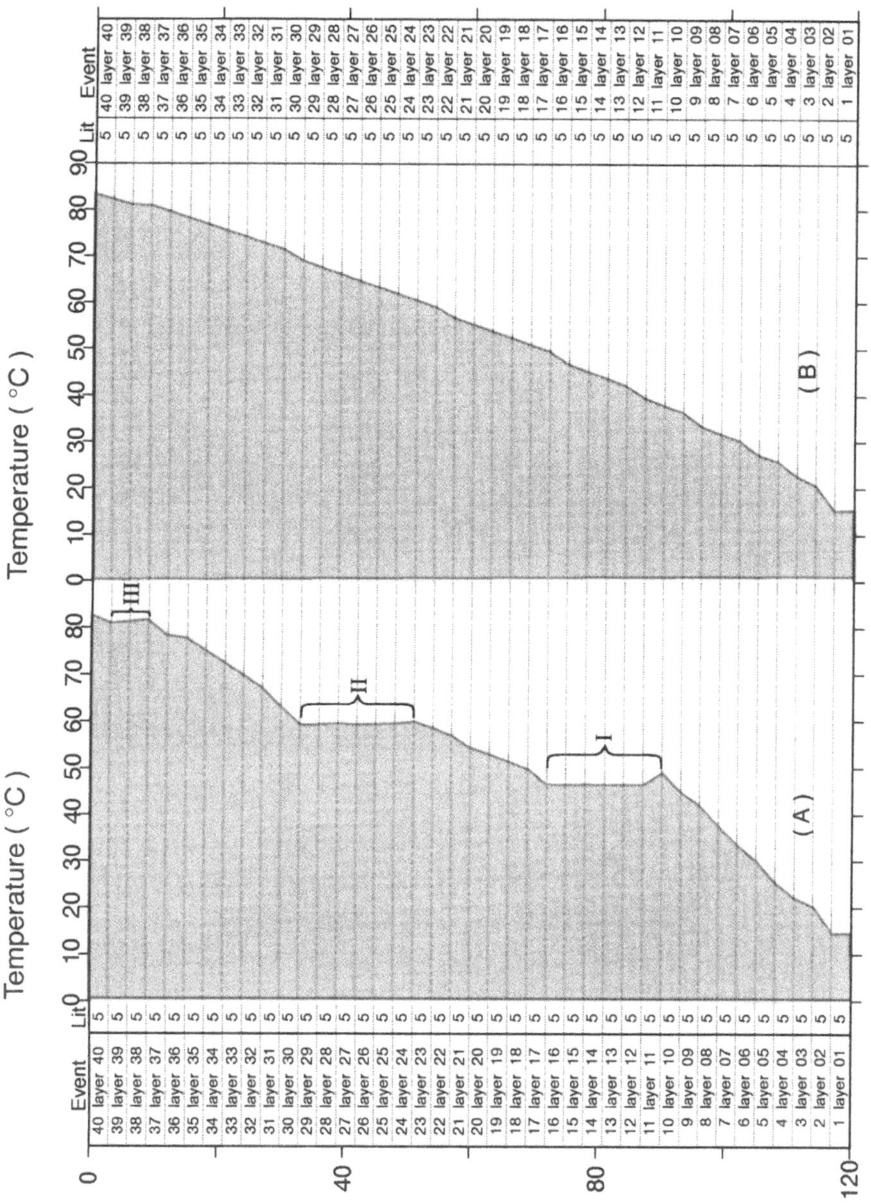

Fig. 2.15. A Temperature history of the lowermost unit in a well similar to the reference well where, however, three nondepositional periods (I, II, III) are integrated. **B** For comparison, the temporal development in the reference well

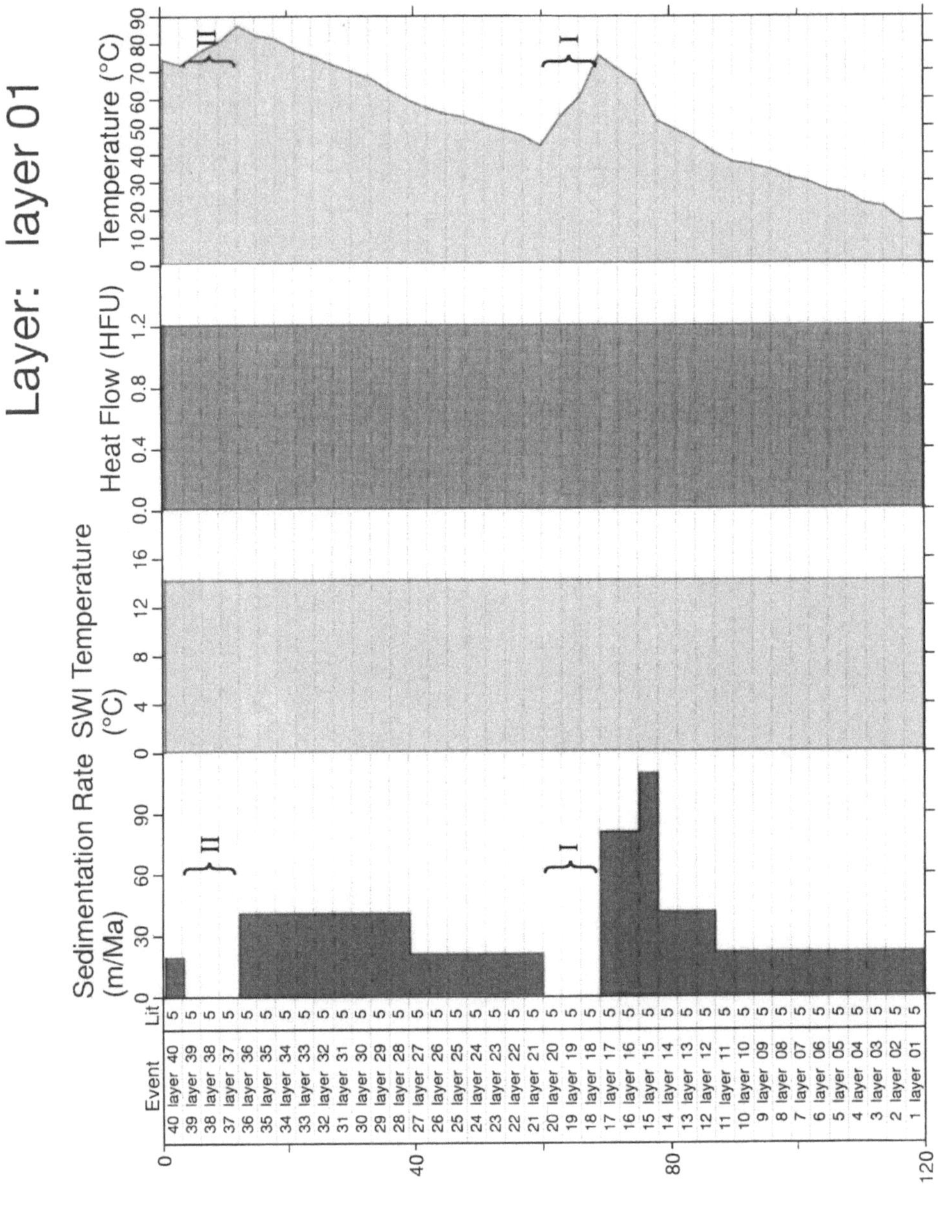

Fig. 2.16. Cooling caused by erosional periods (I, II) demonstrated in a well. Some of the input data are also shown

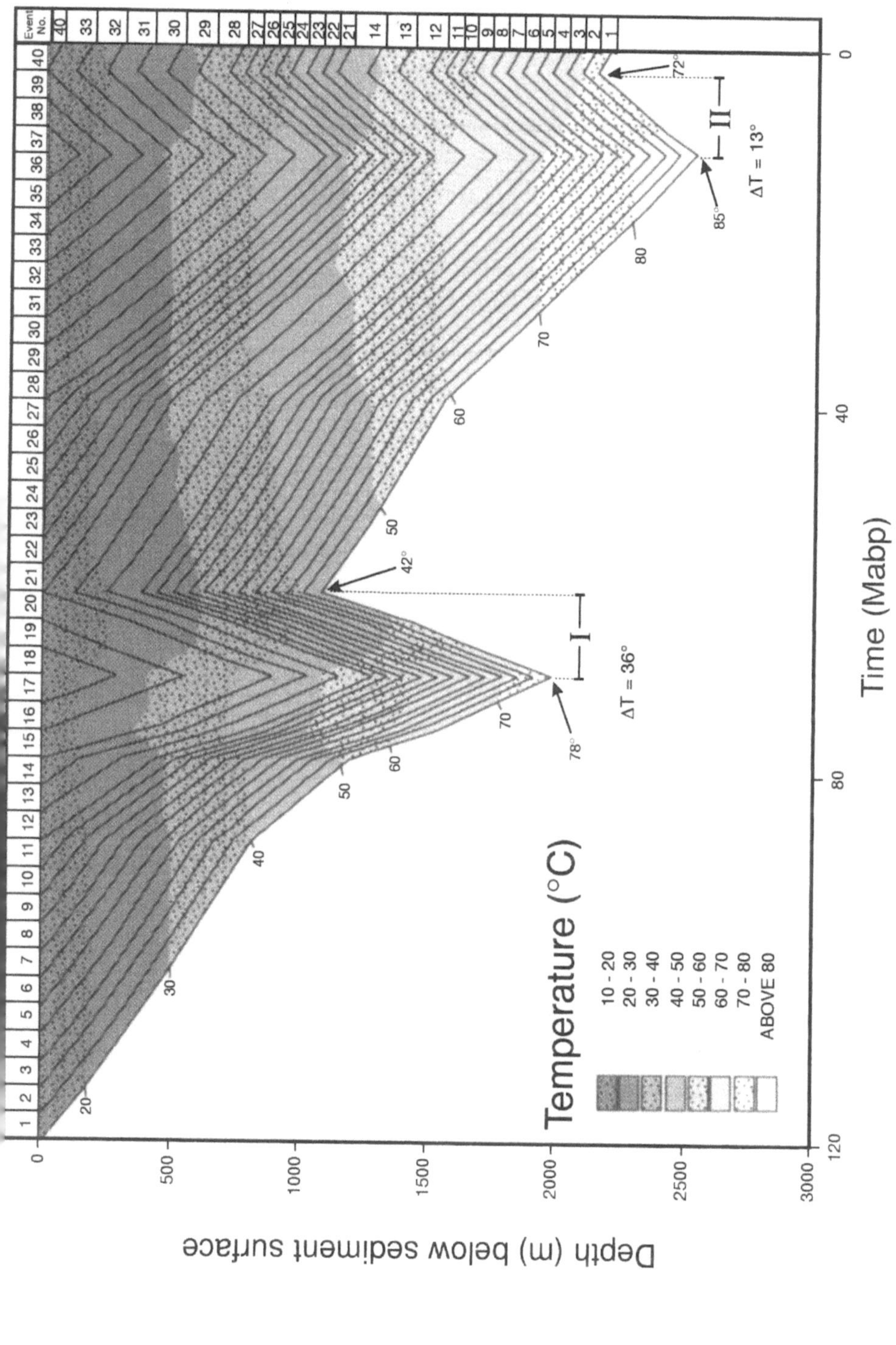

Fig. 2.17. Burial history diagram of the well with two erosional periods (I, II) where isotherms are also shown. Note that the geothermal gradient becomes lower during the erosional periods

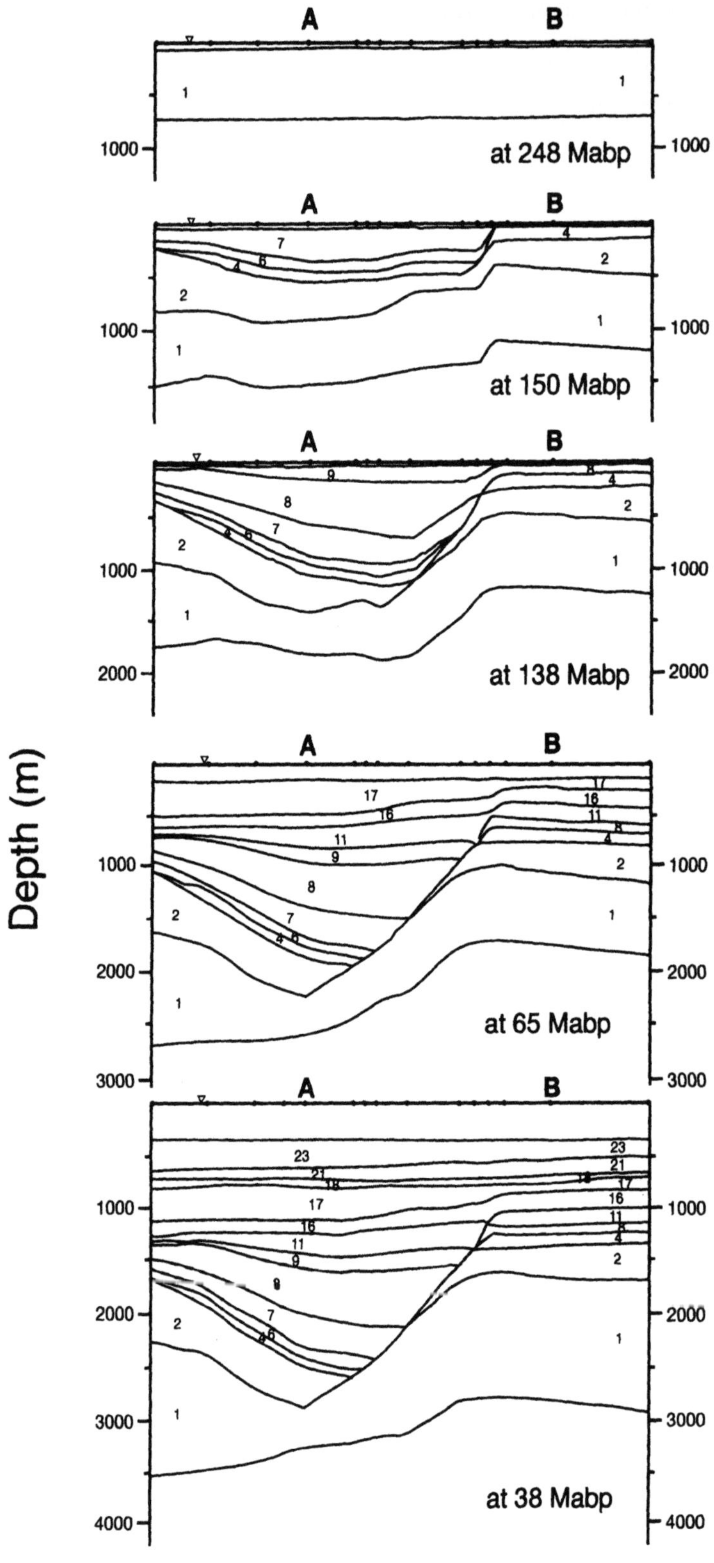

Depth (m)
A
B
at 248 Mabp
at 150 Mabp
at 138 Mabp
at 65 Mabp
at 38 Mabp
1000
2000
3000
4000

temperature history of the lowermost unit at the point A, as affected by faulting. For comparison, the temperature history of point B, which is not affected by faulting, is given in Fig. 2.19B. It is obvious that the same unit may have at two relatively close points remarkably different temperature histories.

The effect of *salt diapirs* on temperature distribution in under- and overlying rock units is much greater than through faulting. This is caused by the very high thermal conductivity of salt. The result is depressed temperatures in the underlying and increased temperatures in the overlying rock units. The formation of a salt diapir and associated compaction and heat transfer processes are shown with the help of a series of paleocross-sections from the North Sea (Fig. 2.20). The isotherms displayed in the three paleosections and the present-day cross section are the result of a numerical simulation of heat transfer. It can be seen how the movement of the salt disturbed the temperature distribution during the entire geological history.

Overthrusting has also a great effect on the temperature distribution and history within a basin since generally a warmer (preheated) rock sequence is emplaced on top of a colder sequence. However, the assumption that a sawtooth-shaped temperature profile is established, as suggested by Oxburgh and Turcotte (1974), is unlikely. A sawtooth-shaped temperature profile could be created only if the thrust sheet emplacement occurred instantaneously, which of course is geologically unrealistic. To arrive at a realistic appraisal of the effects of overthrusting on the history of temperature distributions, it is necessary to look into the details of this geological process. Aspects of the overthrust duration, the thermal properties of the rock sequence, and heat flow must be analyzed. This has been achieved by Wygrala et al. (1990) with the help of a special software routine applied to a hypothetical fault model (Figs. 2.21 and 2.22). First, a sequence consisting of ten layers, each 500 m thick, is deposited, followed by a hiatus to ensure temperature equilibration prior to overthrusting. Then a 5000-m-thick thrust sheet with a basal temperature of 120 °C is moved over a rock sequence with a thrust plane angle of 45 °. The modeling is carried out excluding compaction and assuming that frictional heating is negligible. By assuming variable rates of overthrusting, lithotypes, and heat flow values it was possible to analyze the effect of the duration of the overthrusting, thermal properties, and heat flow. Results show that thrusting rate is a major factor for the resulting temperature distribution. A rate of 0.5 cm/Ma caused only a general depression in thermal gradients and not an inversion in the temperature trend (Fig. 2.21). An increase in the thrusting rate to 5 cm/Ma caused an S-shaped temperature profile at the point of maximum thickness (Fig. 2.22). Development and curvature of the S-shaped temperature curve is also controlled by the basal temperatures of the overthrust sheet prior to overthrusting. When overthrusting stops, the so-called "reequilibration"

Fig. 2.18. Cross sections showing the formation of a synsedimentary fault in the Central Graben Area, North Sea. Note that the lowermost layer is composed of salt, and that salt movement also participated to the fault formation. Numbers indicate layers. (Modified from Heynisch et al. 1987)

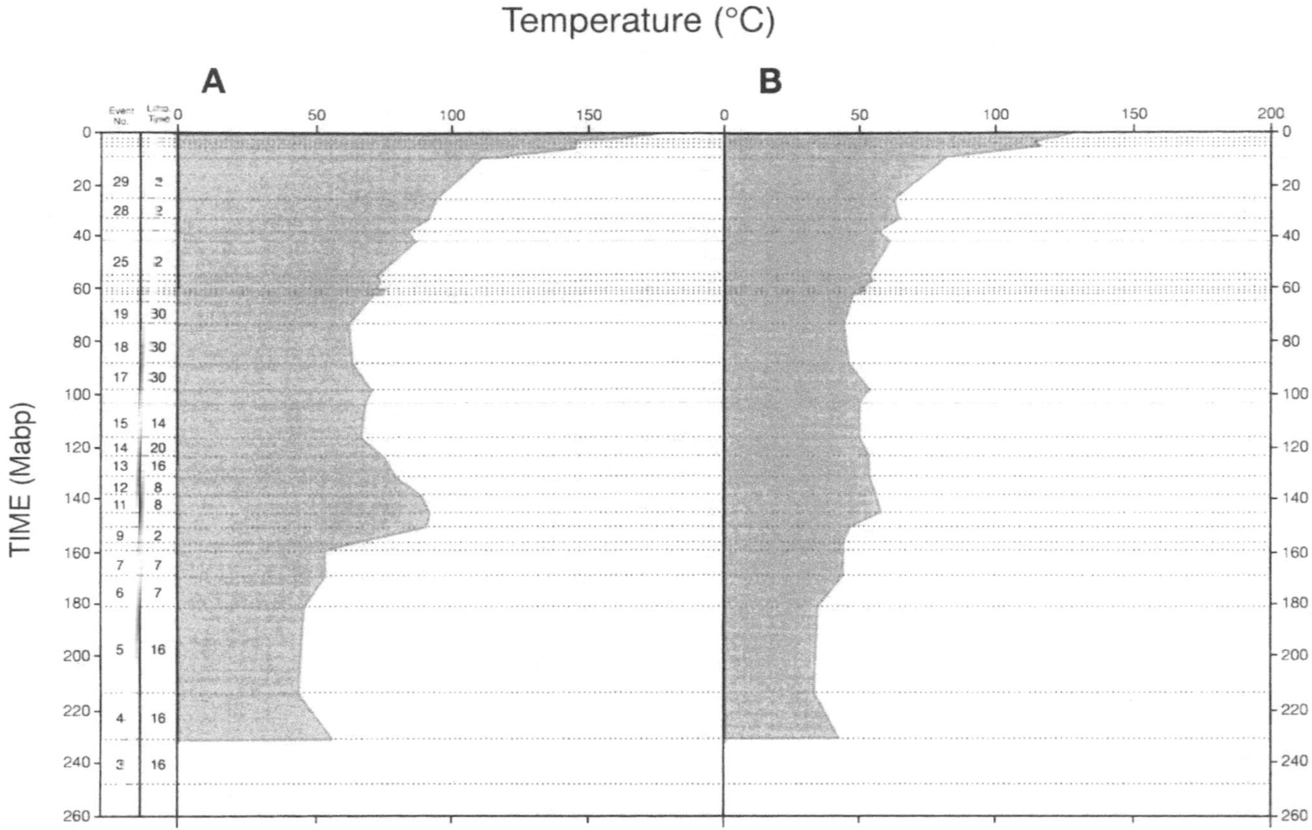

Fig. 2.19. Temperature history of layer no. 3 at points A and B on the cross sections shown in Fig. 2.18. **A** Affected by faulting. **B** Not affected by faulting

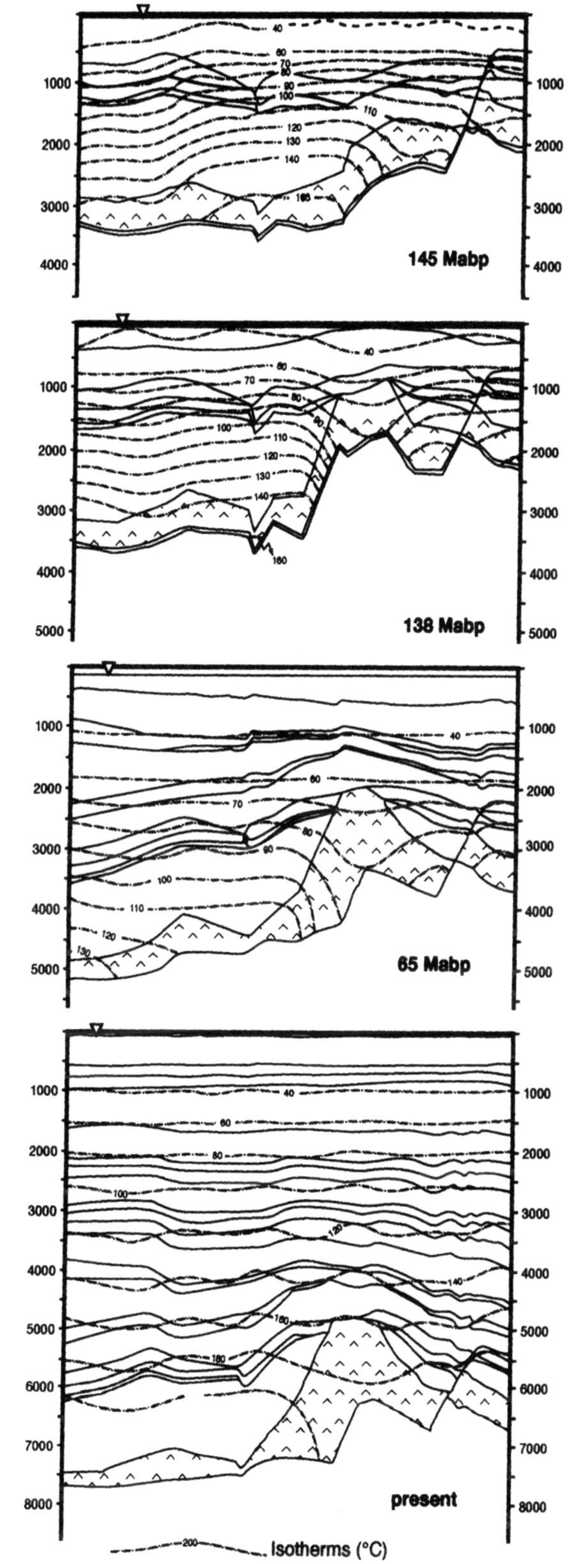

Fig. 2.20. Cross sections showing the development of a salt diapir and its effects on the temperature distribution. See text for details. (Modified from a modeling study in the Central Graben area; Heynisch et al. 1987)

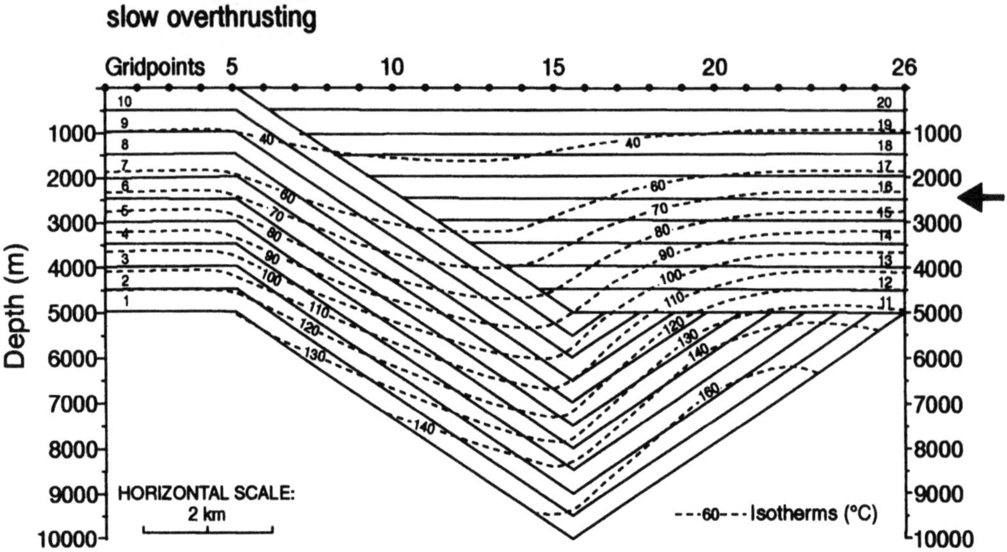

Fig. 2.21. Temperature distribution within the overthrust sheet and overthrusted units after a slow (0.5 cm/year) overthrusting. (Wygrala et al. 1990)

phase begins, during which the disturbed temperatures tend to reach balanced normal values. Sensitivity runs indicate that within the first 1 Ma about half of the difference between initial and final temperatures at the base of the entire sequence has been reached. Complete equilibrium requires approximately 7–

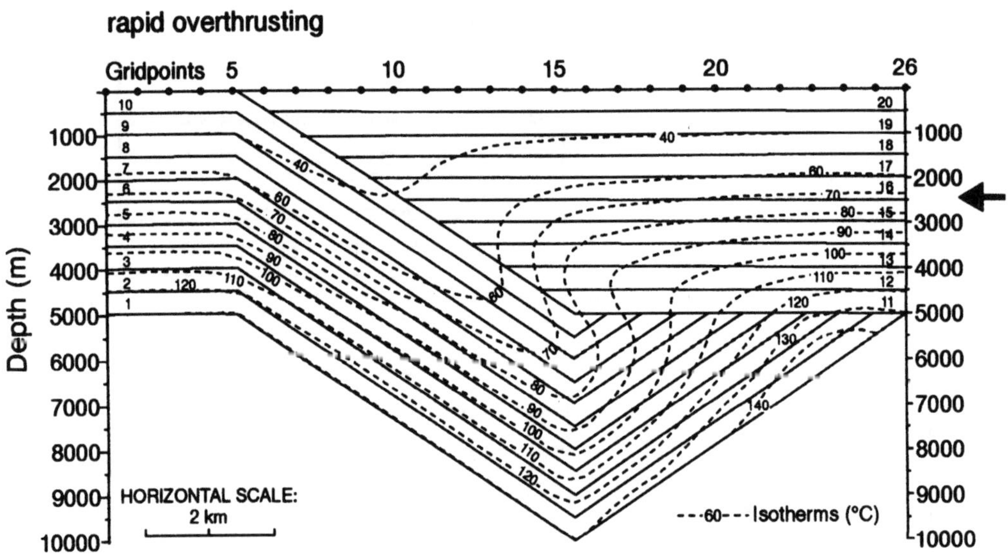

Fig. 2.22. Temperature distribution within the overthrust sheet and overthrusted units after a rapid (5 cm/year) overthrusting. (Wygrala et al. 1990)

10 Ma (Fig. 2.23). Here the equilibrium temperatures are controlled by the thermal conductivity (lithology) of the sequence and by the heat flow at the base. Rates of equilibration are a function of heat capacity of the system.

As briefly demonstrated, changes in basin stratigraphy induced by tectonic movements such as faulting, thrusting, and halokinesis have a remarkable effect on the heat transfer and temperature distribution. Consequently these kind of processes should definitely be considered during the reconstruction of temperature history of sedimentary basins.

2.5.3
Calibration of Thermal History

The conceptual model of basin evolution which forms the basis for the input must be calibrated by comparing simulation results with measured/observed data. Since almost all of the measurable parameters are either a direct result of the temperature history or were at least influenced by it, there are usually enough data from temperature-sensitive parameters to ensure a reasonable calibration. Information useful for calibration is derived mostly from wells. The relevant data are:

- Thickness of different layers
- Porosity distribution in mechanically compacted units
- Pressures
- Temperatures
- Maturity distribution derived from different maturation indicators such as vitrinite reflectance, sterane and hopane isomerization, aromatization, etc.
- Temperature values as determined by fluid inclusion and other geothermometric methods
- Geochemical data indicating status of hydrocarbon generation, such as amount of extract, pyrolysis data, etc.

Parameters such as thickness, porosity, and pressure are of secondary importance since they are coupled with temperature indirectly. Calibration of temperature history requires parameters which are strongly dependent to temperature history. These are primarly the present-day temperatures, temperatures determined with the help of geothermometers, and maturities. However, present-day temperature distribution is of only limited value since over geological time heat transfer is dynamic, and the present-day situation is determined only by thermal phenomena of the last few million years. Nevertheless, it allows a definition of the recent heat flux density, and this is of great help for "backstripping" of the heat flow history. This, however, is valid only if the basin type has not changed during the last few million years. The value of geothermometric methods is also limited unless timing (geological dating) is available for the temperatures determined (Leischner et al. 1993). As a consequence a suite of maturity indicators provides the most practical and available set of parameters for calibration of temperature history. Similarly, data showing the status of hydrocarbon generation can also be helpful.

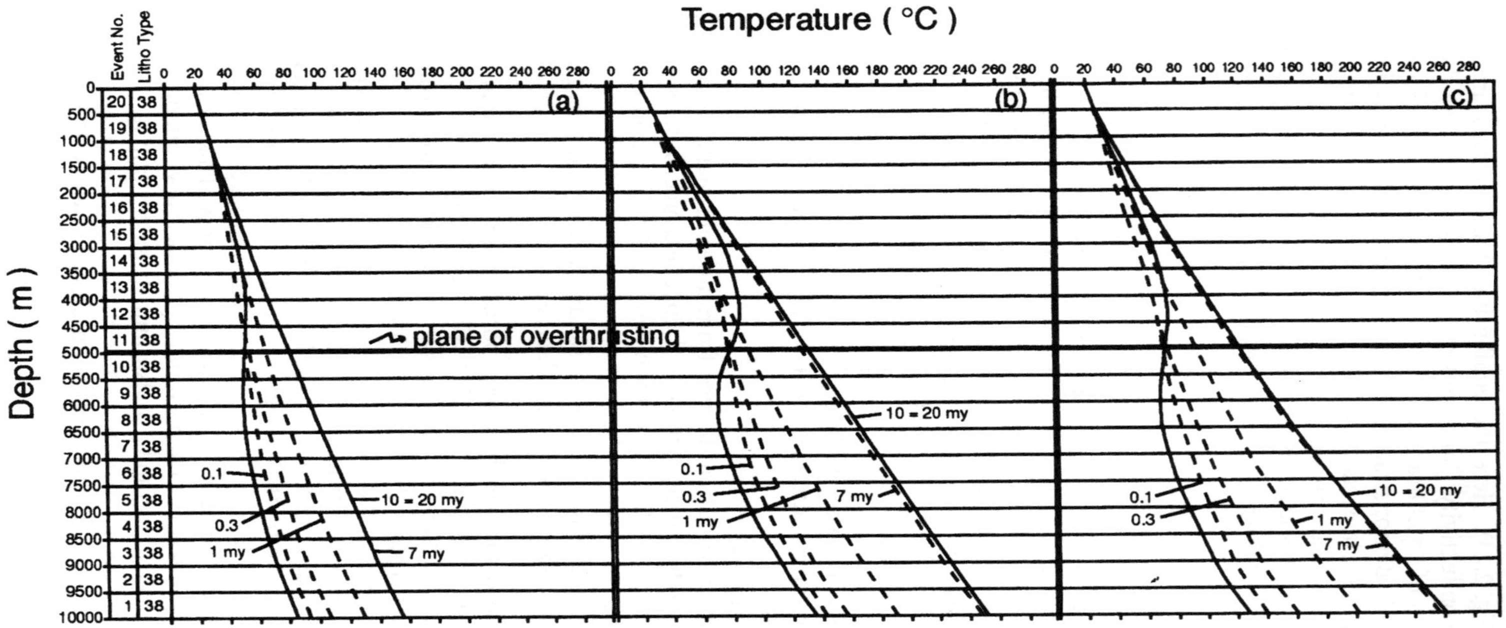

Fig. 2.23a–c. Temperature vs. depth at overthrust step after overthrusting (during reequilibrium) sensitivity to lithology and different heat flow values. **a** Dolomite over dolomite with a basal heat flow of 1.0 HFU. **b** Shale over shale with a basal heat flow of 1.0 HFU. **c** Dolomite over dolomite with a basal heat flow of 1.6 HFU. (Wygrala et al. 1990)

The use of several calibration parameters would be the ideal and most comprehensive manner of thermal history calibration. As emphasized by Leischner et al. (1993), the combination of two independent sets of calibration parameters, for example, one derived from maturation indices and another one from geothermometry, is required for a reliable temperature history reconstruction.

The calibration process is based on modification of the conceptual model, as permitted by relevant field and laboratory data, and on the adjustment of the respective parameters on the input data. However, original input such as thickness of the layers, type, age, and duration of various events, timing of tectonic movements, depositional environment and related water depths, basin type, and thermophysical properties of different lithologies are based on real observations and measurements and therefore cannot be significantly modified for calibration purposes. Consequently only sea bottom temperatures, heat flux, timing and duration of erosional events, and original thickness of eroded units are input data which can be adjusted within geologically acceptable ranges to establish a fit between calculated and measured parameters. Of course here are adjustment limitations even for these parameters. For example, a heat flux density of 2.5 HFU could not be selected for forearc basins, which usually have much lower heat flow values; or a sea bottom temperature of 18 °C in a 1200 m deep open ocean would be unrealistic since sea water temperatures drop very rapidly with increasing water depth (Fig. 2.24). Thermal calibration has only a small degree of freedom, since physical, chemical, and geological data and principles place severe constraints on the procedure.

Although organic maturity indicators have a high sensitivity to thermal history, an acceptable match for all the available calibration parameters should be obtained for rigorous modeling. In certain cases a measured maturity trend can be simulated with more than one version of a conceptual model. Consideration of additional calibration parameters narrows the possible paths and enhances the validity of the conceptual model. A typical example of the value of independent calibration parameters was presented by Leischner et al. (1993). A well in the Lower Saxony Basin in Germany, with a well-known burial history (Fig. 2.25), was simulated to achieve an improved calibration of the temperature history. A suite of organic maturity parameters including Rock-Eval pyrolysis, vitrinite reflectance, and methylphenantrene index calculations was used. These were combined with inorganic temperature indicators such as fluid inclusion measurements, fission-track studies, and illite crystallinity. The combined data were used finally to calibrate the respective conceptual model of basin evolution.

In this well vitrinite reflectance values increase from 0.55% R_o at approximately 500 m depth to 2.0% R_o at 3300 m (Fig. 2.26). This trend of vitrinite reflectance is in good agreement with values of MPI and T_{max} values. Thus the reliability of the maturity trend is ensured. Fluid inclusions were found in anhydrite veins within the Zechstein and Upper Carboniferous and in authigenic and detritial quartz from sediments of Late Carboniferous Age. Large fluid inclusions in anhydrite veins located presently at depths of 1860, 2244,

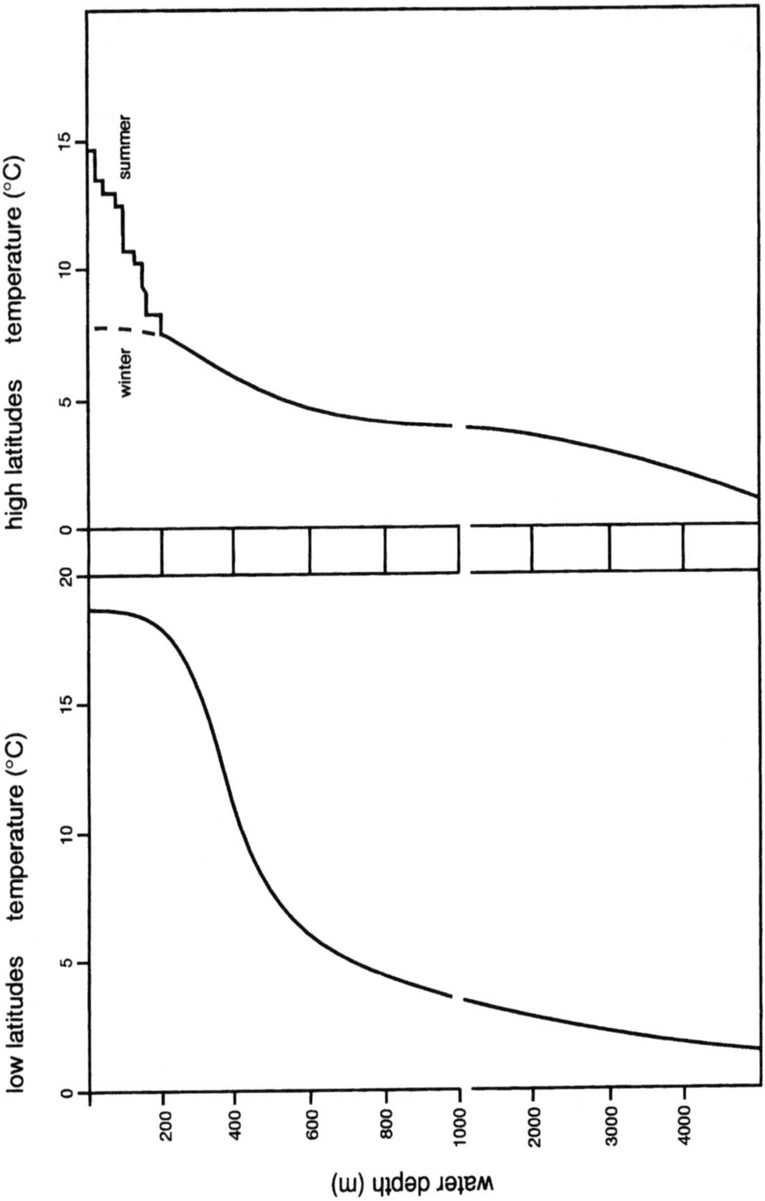

Fig. 2.24. Changes in water temperatures in open oceans as a function of depth. (Smith 1982)

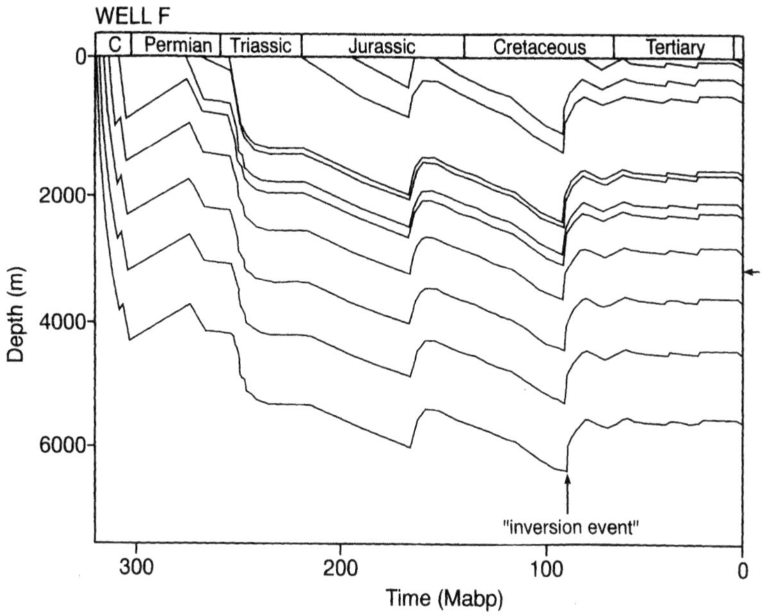

Fig. 2.25. Simplified burial history diagram of well F. *C*, Upper Carboniferous. (Leischner et al. 1993). (Reprinted from Leischner et al., Fluid inclusions and organic maturity parameters as callibration tools in basin modelling, NPF Spec Publ 3, 1993, pp 161–172, with kind permission from Elsevier Science - NL, Sara Burgerhartstraat 25, 1055 KV Amsterdam, The Netherlands)

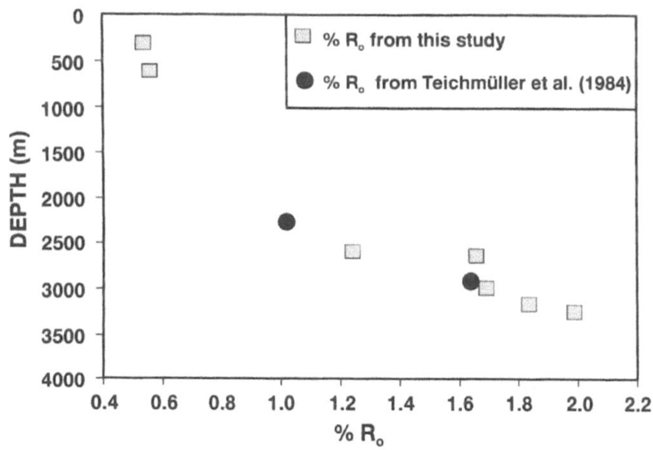

Fig. 2.26. Vitrinite reflectance values measured in well F by Leischner (1994; *gray squares*) and by Teichmüller et al. (1984; *black circles*). (Reprinted from Fortschr Geol Rheinl Westfalen, 32, Teichmüller et al., Inkohlung und Erdgas - eine neue Inkohlungskarte der Karbonoberflaeche in Nordwestdeutschland, pp 11–24, 1984 with kind permission from Elsevier Science - NL, Sara Burgerhartstraat 25, 1055 KV Amsterdam, The Netherlands)

and 2500 m were stretched and/or decrepitated so that homogenization temperatures provide information on maximum paleotemperatures as discussed by Burruss (1987) and Prezbindowski and Tapp (1991). Relevant maximum paleotemperatures were determined as 140, 161, and 175 °C at 1860, 2244, and 2500 m, respectively. Another group of fluid inclusions in authigenic quartz from an Upper Carboniferous sandstone at a depth of 3215 m indicated trapping temperatures in the range of 190–200 °C. These are presumed to be the maximum paleotemperatures since the fluid inclusions were formed during Turonian time when maximum burial was attained. Thus paleotemperatures between 140 and 200 °C were obtained as maximum temperatures for the depth interval from 1860–3215 m (Fig. 2.27).

As shown by the burial history diagram of the simulated well (Fig. 2.25) the basin evolution is characterized by several phases of subsidence interrupted by periods of uplift and erosion. A major uplift and erosion event took place during Late Cretaceous and is termed the "inversion event" of the Lower Saxony Basin. Maximum burial was attained just before this inversion about 88 Ma before the present. At approximately the same time the magmatic bodies of Bramsche and Vlotho intruded. These intrusions and the corresponding hydrothermal events caused anomalously high vitrinite reflectance values mainly in the southern parts of the basin as indicated by the isoreflectance lines (Fig. 2.28) on maps depicting the top Carboniferous (Teichmüller et al. 1984). The simulated well is located in the westernmost part of the basin so that the effect of the magmatic intrusions of Bramsche and Vlotho would not be expected. However, some indications for hydrothermal events have been observed (Leischner 1994) and some effects therefore cannot be excluded.

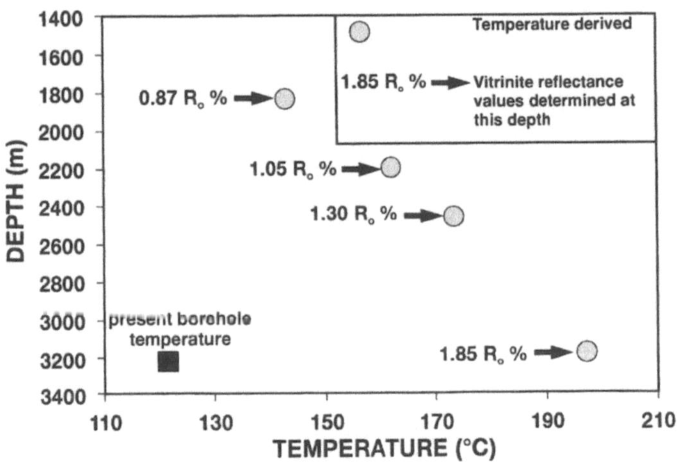

Fig. 2.27. Maximum temperatures derived from fluid inclusions versus depth for well F. The corresponding vitrinite reflectance values for the examined samples are included. (Leischner et al. 1993)

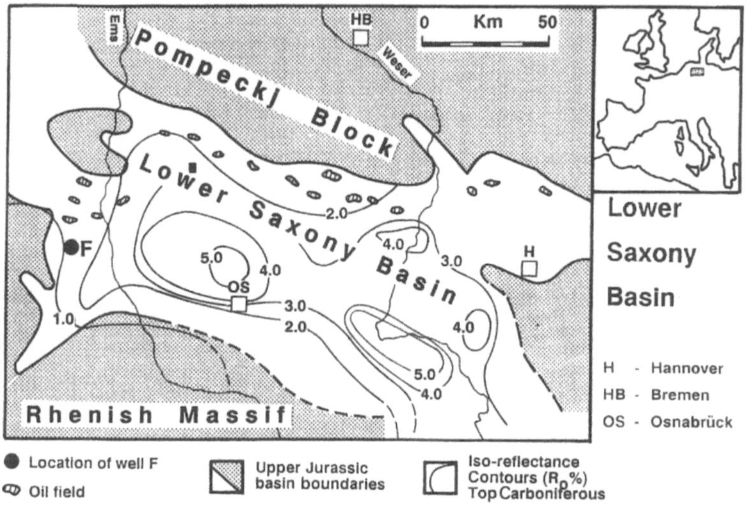

Fig. 2.28. The Lower Saxony Basin and the location of well F. The intrusive complex of Bramsche and accompanying intrusive bodies are outlined by the isoreflectance lines. (Reprinted from Leischner et al., Fluid inclusions and organic maturity parameters as callibration tools in basin modelling, NPF Spec Publ 3, 1993, pp 161–172, with kind permission from Elsevier Science - NL, Sara Burgerhartstraat 25, 1055 KV Amsterdam, The Netherlands)

Vitrinite reflectance values and paleotemperature data (maximum temperatures from fluid inclusions) were used for the thermal calibration of the conceptual model. A constant heat flow of 1.5 HFU resulted in an acceptable fit between measured and calculated vitrinite reflectance data (Fig. 2.29) using a modified time-temperature index (TTI) correlation after Waples (1980). In order to match the present borehole temperatures a very recent increase in heat flow to 1.7 HFU was necessary, which was justified by upward migration of hot fluids causing heat anomalies (Schulz 1989). However, a remarkable discrepancy was observed, as indicated by the isotherms, between paleotemperatures derived by modeling and paleotemperatures determined by fluid inclusions. This difference was about 40–50 °C for maximum temperatures reached during Turonian (Fig. 2.29). Further attempts of recalibration showed that a fit for both calibration parameters, vitrinite reflectance and fluid inclusion homogenization temperatures, was not possible because a heat pulse with a maximum heat flow value of 2.25 HFU during the Turonian, which was necessary to achieve the maximum temperatures, always led to an overestimation of vitrinite reflectance data as calculated by the TTI method. Therefore a recalculation of the vitrinite reflectance data was performed using the Easy-R_o model after Sweeney and Burnham (1990). As shown in Fig. 2.30, this led to an acceptable match between the calculated and measured vitrinite reflectance data, and a fit between determined and calculated maximum temperatures was also achieved. The heat pulse during Turonian can be justfied since it was probably related to the intrusion of the magmatic body of Lingen-

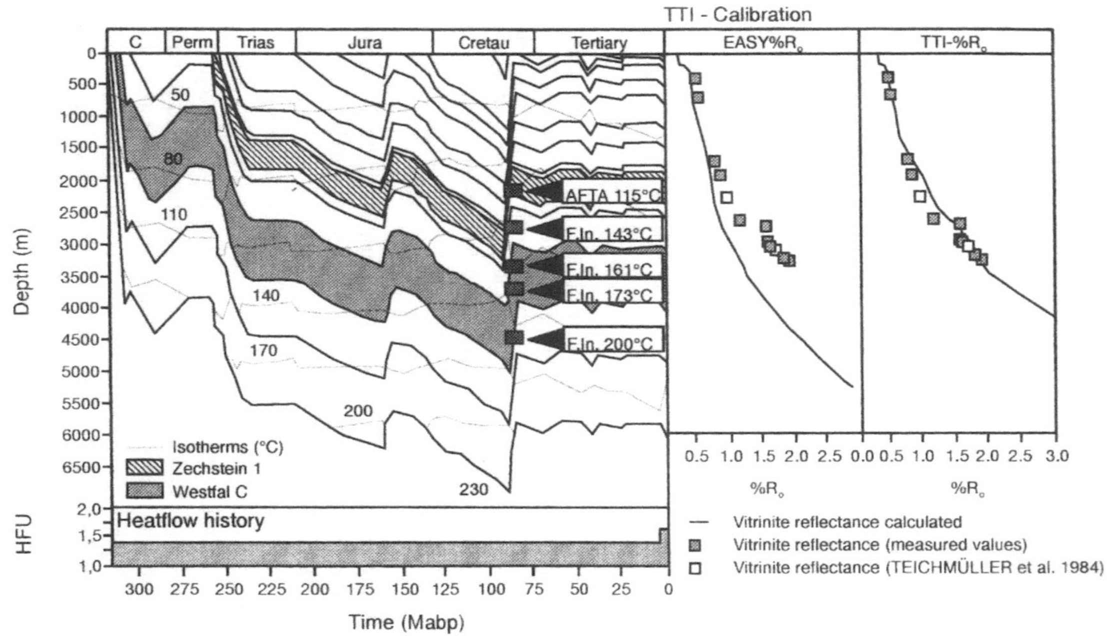

Fig. 2.29. Simulated burial and temperature history using the TTI correlation for vitrinite reflectance calculation. Note that a constant heat flow of 1.5 HFU is required for the entire burial history to achieve a fit between measured and calculated vitrinite reflectance values. The temperatures indicated on the burial history diagram are the temperatures derived from fluid inclusions. They do not agree with the isotherms constructed after the calibration with TTI method. (Leischner 1994)

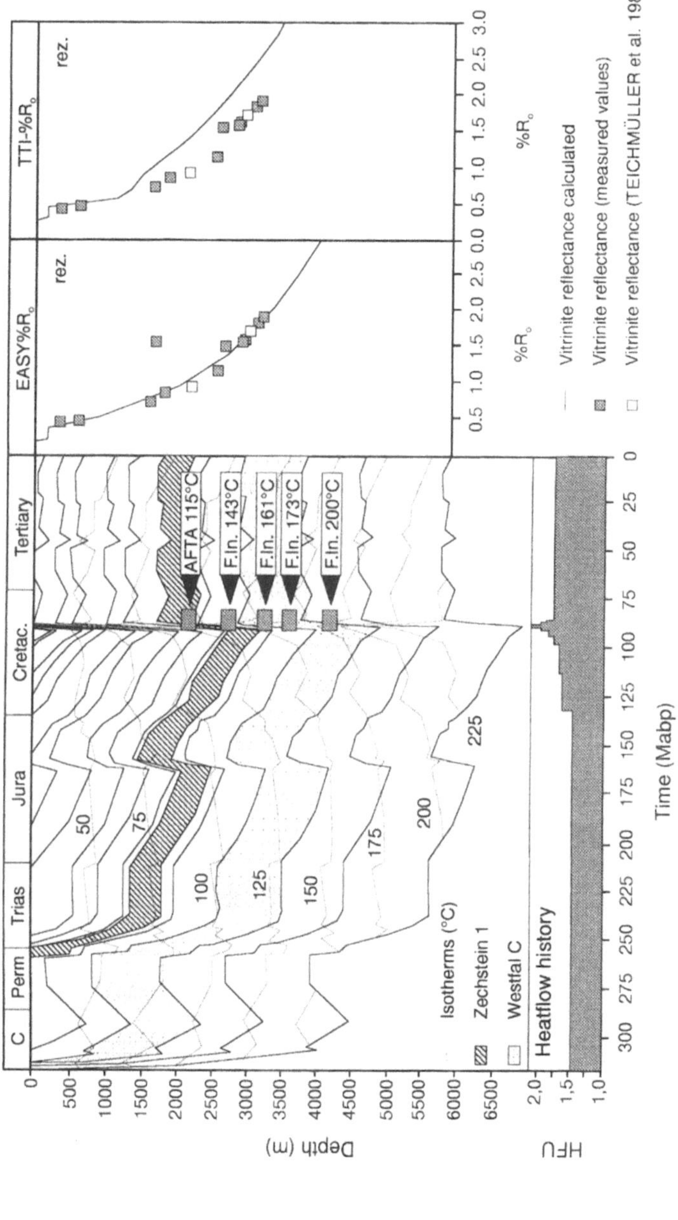

Fig. 2.30. Recalibrated temperature history for well F using the paleotemperatures determined by fluid inclusion and the Easy%-R_o approach. (Leischner 1994)

Bad Bentheim, a western relative of the well-known contemporenous Bramsche and Vlotho plutons.

As demonstrated above, the use of at least two independent calibration tools is recommended for a realistic and reliable temperature history reconstruction. It should also be noted that the effect of time is overemphasized and leads to an overestimation of vitrinite reflectance values if they are calculated using the TTI approach. The Easy-R_o model, on the other hand, is more temperature sensitive and generally provides a more accurate vitrinite reflectance calculation (Leischner et al. 1993).

Nevertheless, an additional calibration procedure is demonstrated below with the help of another case history, where vitrinite reflectance values are calculated with the TTI method to show details of the general approach of reconstructing and calibrating thermal history. Using the data from a particular well and the information on the geological history of the Haltenbanken area of the North Sea, including thickness, lithology, porosity, water depth, sea floor temperatures, heat flux, duration and type of 27 events are listed in Table 2.5. Except for sea floor temperatures, heat flux, erosional amounts, and porosities of some layers all parameters are based on observations and mea-

Table 2.5. Input data of the well in the Haltenbanken area

Event no., name	Lithology	Thickness (m)	Porosity (%)	Water depth (m)	SWI Temp. (°C)	Heat flow (HFU)
27 Quaternary	Shale	286.0	33.0	273.0	4.0	1.30
26 U. Pliocene	Shale	831.0	25.0	50.0	4.0	1.30
25 U. Miocene	Shale silt	311.0	23.0	200.0	4.0	1.30
24 L. Miocene	Shale silt	70.0	21.0	350.0	4.0	1.30
23 Oligocene	Shale sand	67.0	20.0	350.0	4.0	1.30
22 U. Eocene	Shale silt	133.0	18.0	500.0	9.0	1.30
21 M.-L. Eocene	Shale silt	180.0	17.0	500.0	12.0	1.30
20 U. Paleocene	Silt tuff.	45.0	17.0	200.0	13.0	1.30
19 U. Paleocene	Shale	28.0	15.0	500.0	13.0	1.30
18 L. Paleocene	Shale	67.0	15.0	500.0	13.0	1.30
17 U. Maastricht	Shale	0.0	–	150.0	18.0	1.30
16 Campanian	Shale	600.0	12.0	150.0	19.0	1.30
15 Sant-Turonian	Shale	163.0	12.0	500.0	17.0	1.30
14 Cenomanian	Shale	233.0	12.0	500.0	18.0	1.30
13 Albian	Shale	194.0	12.0	500.0	20.0	1.30
12 Aptian	Shale	0.0	–	200.0	24.0	1.35
11 Hauterivian	Marl	16.0	15.0	150.0	24.0	1.40
10 Valanginian	Shale carb.	20.0	12.0	0.0	25.0	1.45
09 Berr.-Tithonian	Shale carb.	69.0	12.0	75.0	24.0	1.45
08 Kimmeridgian	Shale silt	0.0	–	50.0	25.0	1.35
07 Oxfo.-Bathonian	Shale silt	84.0	12.0	50.0	24.0	1.20
06 Bathonian	Sand silty	–20.0	–	0.0	23.0	1.10
05 Bajo.-Aalenian	Sand silty	259.0	12.0	25.0	21.0	1.00
04 Toarcian	Siltstone	152.0	11.0	25.0	20.0	0.95
03 Pliensbachian	Sand silty	211.0	10.0	25.0	19.0	0.95
02 Sine-Hettangian	Shale carb.	299.0	9.0	1.0	20.0	0.90
01 Rhaetgian	Salt	100.0	5.0	0.0	22.0	0.90

surements. Values and temporal distribution of sea bottom temperatures are derived from the constraints of paleoclimates, paleolatitudes, and water depths as discussed previously. Erosional amounts are estimated and are so small that they hardly affect simulation results. The trend of the temporal distribution of heat flux is determined in the light of the tectonic evolution of the northern North Sea which was affected by rifting and subsequent thermal subsidence during Jurassic and Cretaceous (Mo et al. 1989). Absolute values of heat flux are estimated considering items discussed earlier. Porosities of reservoir horizons were available as measured values. For other layers they were determined by log interpretation and/or by the consideration of a relevant porosity/depth trend. Thicknesses of individual layers were also used for calibration purposes. Measured temperature and vitrinite reflectance values are shown in Fig. 2.31.

Results of the simulation are illustrated in Fig. 2.31; calculated thicknesses and porosities of the individual layers and the observed values are listed in Table 2.6. Comparison clearly indicates that the simulation is unsuccessful. The calculated thicknesses are in general greater than the measured values, e.g. the simulated sequence is undercompacted. Calibration can be performed ei-

Table 2.6. Comparison of measured and calculated thickness and porosity values of individual layers of the well in the Haltenbanken area

Event no., name	Thickness (m)		Porosity (%)	
	Measured	Calculated	Measured	Calculated
27 Quaternary	286	304	33	38
26 U. Pliocene	831	871	25	28
25 U. Miocene	311	325	23	27
24 L. Miocene	70	73	21	24
23 Oligocene	67	71	20	25
22 U. Eocene	133	139	18	22
21 M.-L. Eocene	180	188	17	21
20 U. Paleocene	45	47	17	21
19 U. Paleocene	28	29	15	19
18 L. Paleocene	67	70	15	18
17 U. Maastricht	–	–	–	–
16 Campanian	600	623	12	16
15 Sant-Turonian	163	177	12	18
14 Cenomanian	233	252	12	19
13 Albian	194	206	12	17
12 Aptian	–	–	–	–
11 Hauterivian	16	17	15	19
10 Valanginian	-20	–	–	–
09 Berr.-Tithonian	49	55	12	21
08 Kimmeridgian	–	–	–	–
07 Oxfo.-Bathonian	84	89	12	17
06 Bathonian	-20	–	–	–
05 Bajo.-Aalenian	239	259	12	19
04 Toarcian	152	157	11	14
03 Pliensbachian	211	224	10	16
02 Sine-Hettangian	299	300	9	9
01 Rhaetgian	100	100	5	5

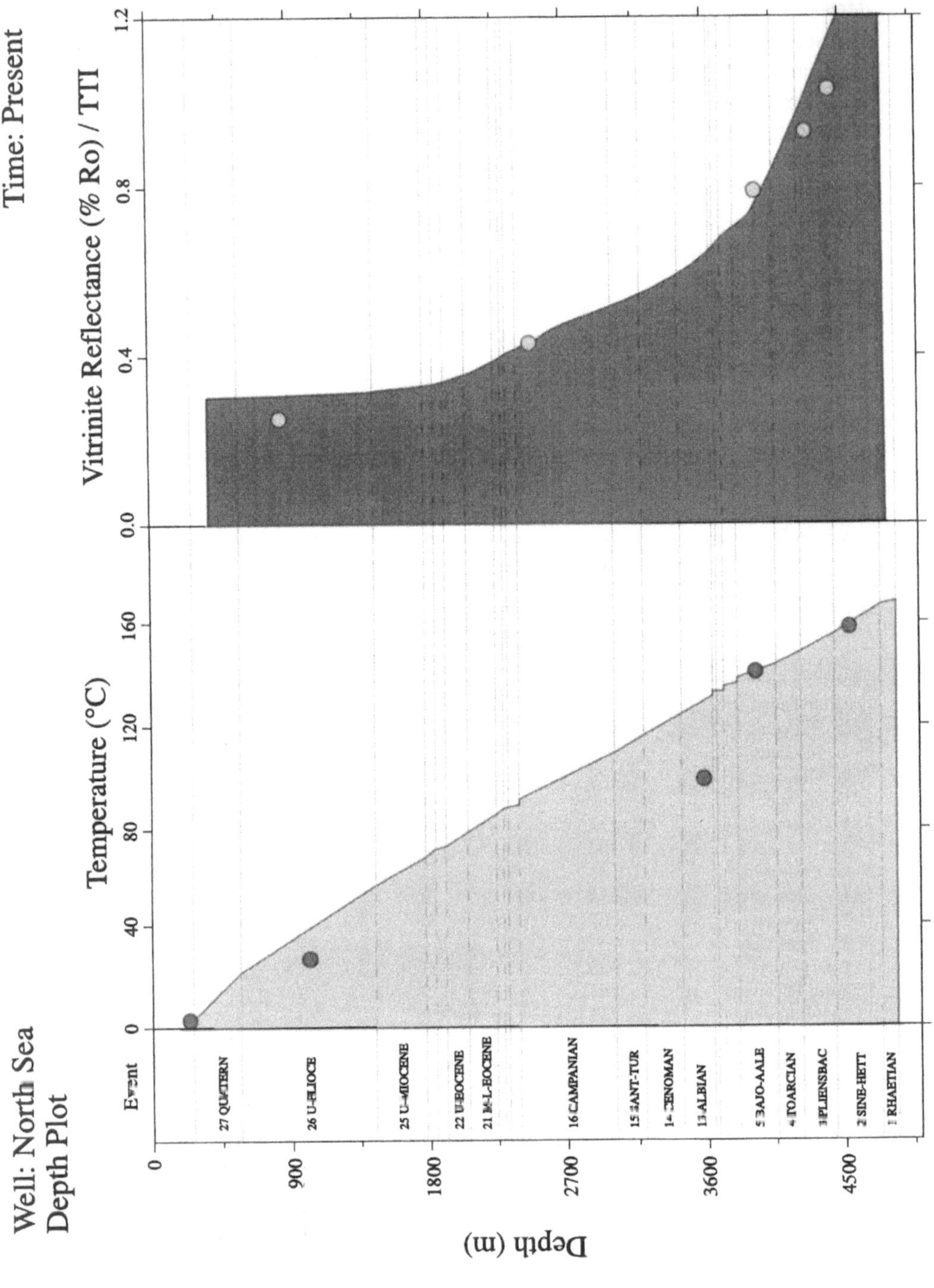

Well: North Sea
Depth Plot
Time: Present
Temperature (°C)
Vitrinite Reflectance (% Ro) / TTI
Depth (m)
Event
0
900
1800
2700
3600
4500
40
80
120
160
0.0
0.4
0.8
1.2
16 CAMPANIAN
13 ALBIAN
4 TOARCIAN
2 SINE-HETT
1 RHAETIAN

Table 2.7. Comparison of measured and calculated thickness and porosity values of individual layers of the well in the Haltenbanken area after the adjustment of porosity values after the first simulation

Event no., name	Thickness (m)		Porosity (%)	
	Measured	Calculated	Measured	Calculated
27 Quaternary	286	282	38	38
26 U. Pliocene	831	848	29	30
25 U. Miocene	311	308	29	28
24 L. Miocene	70	70	25	25
23 Oligocene	67	67	26	25
22 U. Eocene	133	132	22	22
21 M.-L. Eocene	180	178	21	20
20 U. Paleocene	45	45	21	21
19 U. Paleocene	28	28	19	19
18 L-Paleocene	67	67	19	18
17 U. Maastricht	-	-	-	-
16 Campanian	600	597	16	16
15 Sant-Turonian	163	164	18	18
14 Cenomanian	233	234	18	18
13 Albian	194	192	18	18
12 Aptian	-	-	-	-
11 Hauterivian	16	16	21	20
10 Valanginian	-20	-	-	-
09 Berr.-Tithonian	49	52	16	20
08 Kimmeridgian	-	-	-	-
07 Oxfo.-Bathonian	84	84	18	18
06 Bathonian	-20	-	-	-
05 Bajo.-Aalenian	239	238	20	20
04 Toarcian	152	151	15	15
03 Pliensbachian	211	211	16	16
02 Sine-Hettangian	299	299	10	9
01 Rhaetian	100	100	5	5

ther by increasing the compressibilities or by decreasing the decompacted thicknesses. The latter is preferred since in the numerical system porosities are defined as the bulk porosities created by mechanical compaction, whereas, measured porosity values normally reflect the effective porosities often reduced by cementation. Therefore, for a better match between measured and calculated thicknesses and porosities, the present-day porosities were increased, that is the decompacted thicknesses are reduced. Results of the simulation with these new porosities are demonstrated in Fig. 2.32 and Table 2.7. For thickness and porosity the match can be considered as acceptable, but the calculated maturities are still too high.

Fig. 2.31. Calculated (*lines*) vs. measured (*dots*) temperature and maturity values of the well in the Haltenbanken area. Notice high maturity values at greater depths and a greater total thickness which indicate an unsuccessful simulation

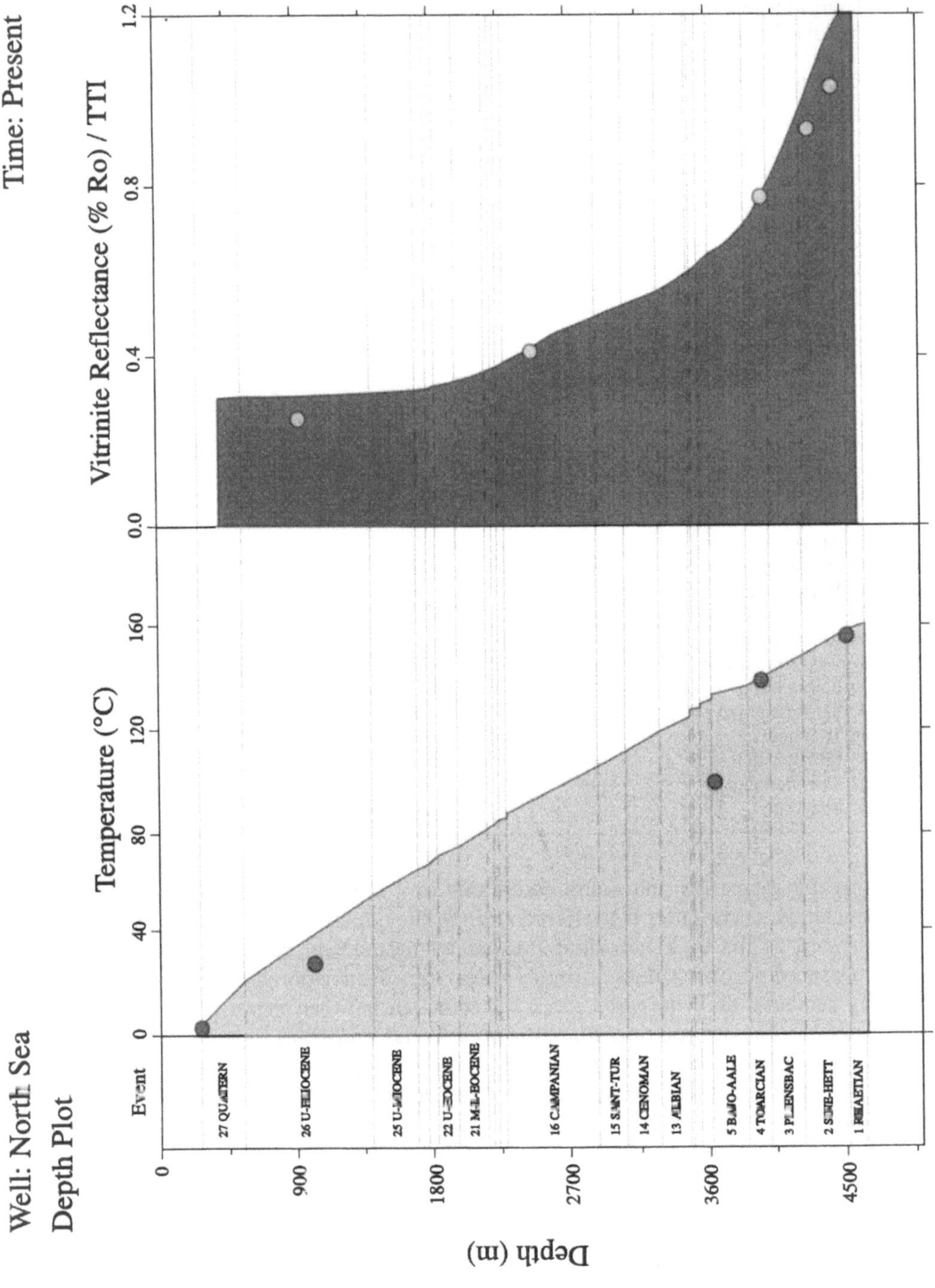
Well: North Sea
Depth Plot
Time: Present
Temperature (°C)
Vitrinite Reflectance (% Ro) / TTI
Depth (m)
Event
27 QUATERN
26 U-PLIOCENE
16 CAMPANIAN
15 SANT-TUR
14 CENOMAN
13 ALBIAN
5 BAJO-AALE
4 TOARCIAN
2 SINE-HETT
1 RHAETIAN
0
900
1800
2700
3600
4500
40
80
120
160
0.0
0.4
0.8
1.2

To obtain a better fit to maturity parameters the heat flux values must be decreased. However, this would result in a mismatch between calculated and measured temperature values which are in a good agreement. As thermal equilibrium requires a relatively short time, the present temperature trend observed in the well can be simulated with an appropriate heat flux value persisting only during the last few million years. Consequently the respective value of the last event is retained and the heat flux values of all events except the last one are decreased by 0.1 HFU. This abrupt increase in heat flow during the last 2 Ma has been required by almost all maturation models in the North Sea (see Jensen and Doré 1993). Since other indications of tectonism are unrecognized, transient conditions caused by glaciation, overpressuring and fluid flow, which cannot be handled completely and accurately by the existing modeling algorithms, are considered as the most probable for conditions requiring an arbitrary increase in heat flux (Yalçın 1991; Jensen and Doré 1993). The adjusted input data given in Table 2.8 led to a successful match, as shown in Fig. 2.33.

Table 2.8. Calibrated Input data of the well in the Haltenbanken area

Event no., name	Lithology	End-begin (Mabp)	Thickness (m)	Porosity (%)	Water depth (m)	SWI temp. (°C)	Heat flow (HFU)
27 Quaternary	Shale	0–2	286	38	273	4	1.30
26 U. Pliocene	Shale	2–4	831	29	50	4	1.20
25 U. Miocene	Shale silt	4–11	311	29	200	4	1.20
24 L. Miocene	Shale silt	11–25	70	25	350	4	1.20
23 Oligocene	Shale sand	25–38	67	26	350	4	1.20
22 U. Eocene	Shale silt	38–42	133	22	500	9	1.20
21 M.-L. Eocene	Shale silt	42–55	180	21	500	12	1.20
20 U. Paleocene	Shale tuff.	55–58	45	21	200	13	1.20
19 U. Paleocene	Shale	58–60	28	19	500	13	1.20
18 L. Paleocene	Shale	60–65	67	19	500	13	1.20
17 U. Maastricht	Shale	65–67	0	–	150	18	1.20
16 Campanian	Shale	67–83	600	16	150	19	1.20
15 Sant Turonian	Shale	83–91	163	18	500	17	1.20
14 Cenomanian	Shale	91–103	233	18	500	18	1.20
13 Albian	Shale	103–113	194	18	500	20	1.25
12 Aptian	Shale	113–123	0	–	200	24	1.30
11 Hauterivian	Marl	123–131	16	21	150	24	1.35
10 Valanginian	Shale carb.	131–138	-20	–	0	25	1.40
09 Berr.-Tithonian	Shale carb.	138–147	69	16	75	24	1.40
08 Kimmeridgian	Shale silt	147–156	0	18	50	25	1.25
07 Oxfo.-Bathonian	Shale silt	156–172	84	18	50	24	1.15
06 Bathonian	Sand silt	172–175	-20	–	0	23	1.05
05 Bajo.-Aalenian	Sand silt	175–187	259	20	25	21	0.95
04 Toarcian	Siltstone	187–194	152	15	25	20	0.90
03 Pliensbachian	Sand silty	194–202	211	16	25	19	0.90
02 Sine-Hettangian	Shale carb.	202–215	299	10	1	20	0.85
01 Rhaetgian	Salt	215–216	100	5	0	22	0.85

Fig. 2.32. Results of simulation with corrrected porosities. The total thickness is now acceptable. However, maturities are still too high

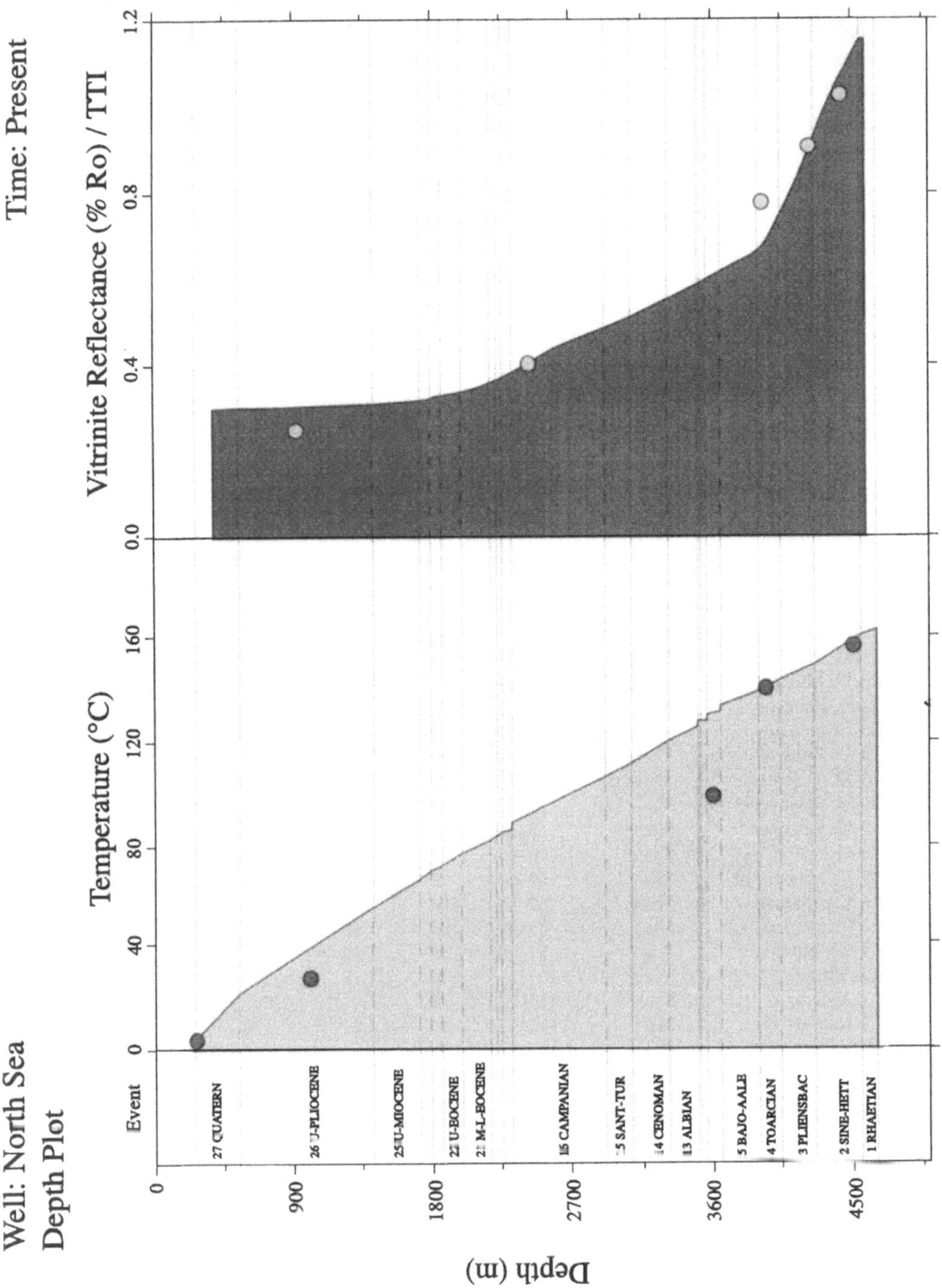

Well: North Sea
Depth Plot
Time: Present
Temperature (°C)
Vitrinite Reflectance (% Ro) / TTI
Depth (m)
Event
27 QUATERN
26 U-PLIOCENE
25 U-MIOCENE
22 U-EOCENE
21 M-L-EOCENE
15 CAMPANIAN
15 SANT-TUR
14 CENOMAN
13 ALBIAN
5 BAJO-AALE
4 TOARCIAN
3 PLIENSBAC
2 SINE-HETT
1 RHAETIAN

It is important to recognize that calibration of the thermal history by modifying the conceptual model and adjusting the relevant input parameters should not be generalized because regional and even local aspects of the geological evolution play a more important role than general assumptions. Therefore during the calibration procedure each basin part or even each well should be considered separately.

2.6 Thermal History of Sedimentary Basins: Case Histories

Thermal histories of six basins are presented below to demonstrate and emphasize the close relationship between different parameters and effects of the dynamics of basin evolution to thermal history. For this purpose usually one or two wells or a cross section from each basin are simulated. Only the final results, that is, results achieved with a previously calibrated conceptual model, are presented. The selected basins are:

- Cambay Basin, India
- San Joaquin Basin, California, USA
- Adana Basin, Turkey
- Styrian Basin, Austria
- Zonguldak Basin, Turkey
- Northwest German Basin

For each basin the geological history and differentiation of events, followed by the input data used for the simulation, are briefly presented. Critical and interesting aspects of the simulated temperature history are then discussed in the light of the relevant parameters of the conceptual model. Most of the examples are taken from published studies.

2.6.1 Cambay Basin, India

The Cambay Basin is an intracratonic fault-bounded graben in the northwestern part of India (Fig. 2.34). A thick sequence of siliciclastic Cenozoic sedimentary rocks, with a thickness of over 7 km in the deepest part of the basin, was deposited on the Deccan Trap floor basalts. A thick sedimentary sequence, Olpad Formation, comprising conglomerates, siltstones, and claystones was deposited during late Paleocene. The latest Paleocene was a non-depositional period. During early Eocene to early Oligocene the deposition of the Cambay Shale, Kalol Formation, Tarapur Shale and Anklesvar Formation took place. These units are generally composed of argillaceous and clastic rocks. Thereafter the basin experienced a long break in sedimentation extending to the early Miocene. This was followed by the deposition of Kathana,

◄

Fig. 2.33. Results of simulation with an adjusted heat flow trend. The match between measured and calculated values are now acceptable, for example, the simulation is successful

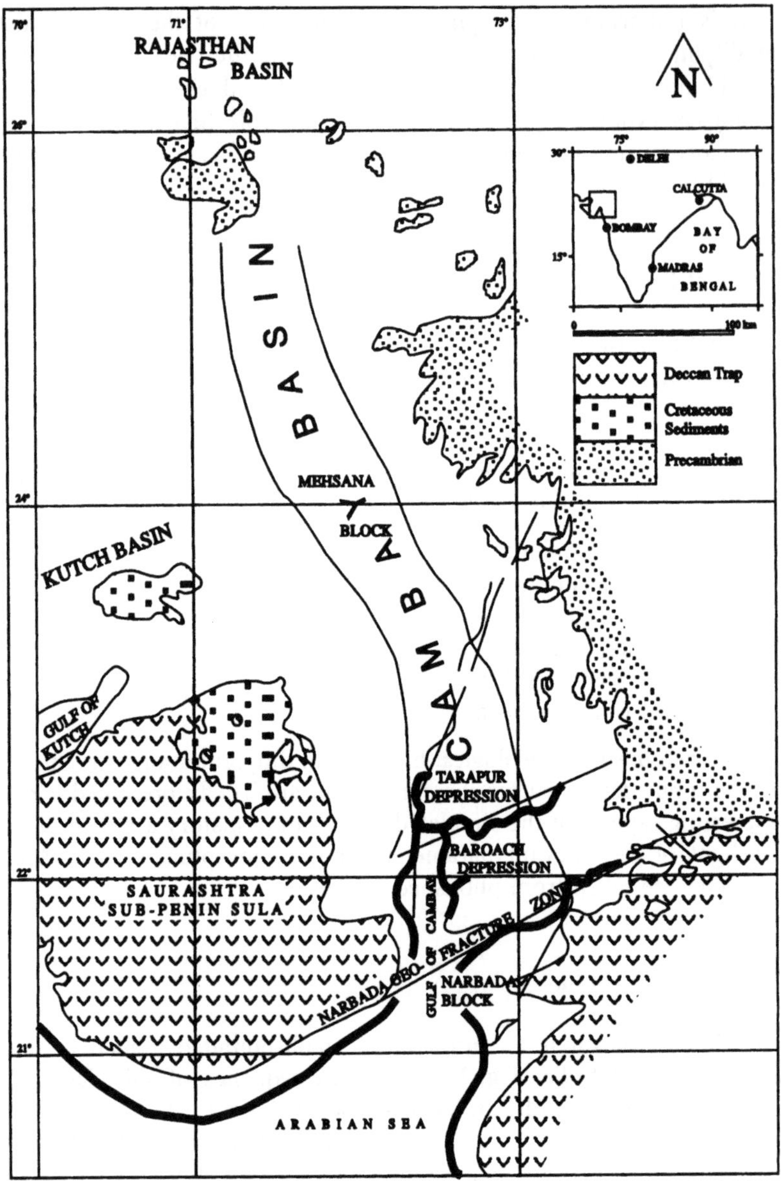

Fig. 2.34. Location map of the Cambay Basin. (Yalçın et al. 1988). (Reprinted from Kumar, Ruby K, Dwivedi P, Banerjee V, Gupta V (eds), Petroleum geochemistry and exploration in Afro-Asian region – proceedings of the first international conference, Dehra Dun, India, 25–27 November 1985, 1987, 558 pp., A.A. Balkema, P.O. Box 1675, Rotterdam, The Netherlands)

Babaguru, Kand, and Jhagadia Formations above the Tarapur Shale during early and middle Miocene. Babaguru and Jhagadia Formations are arenaceous and Kand and Kathana Formations argillaceous. This deposition was followed by another break in sedimentation during late Miocene. During Pliocene and thereafter the entire basin received sediments under marine and brackish water conditions (Broach Formation). Quaternary sediments of Jambusar Formation and the Gujarat Alluvium are the youngest units in the basin.

The entire sedimentary sequence was subdivided into 18 units: 15 chronostratigraphic (depositional) and 3 geochronological (erosional and/or nondepositional) units (Fig. 2.35). Lateral lithofacies changes are common in almost all depositional units. Fifteen different types of lithologies have been distinguished in terms of sand, shale, and coal ratios. The lithological composition of each unit is also shown in Fig. 2.35. Maximum thickness of the units and estimated water depths and sea bottom temperatures are listed in Table 2.9. Heat flux density was determined as constant and uniform at 1.1–1.3 HFU except for small areas where an increase of up to 1.7 HFU was indicated.

Some interesting aspects of the thermal history of the Cambay Basin are be explained below using time-temperature diagrams of selected units at one site located in the Broach depression (Fig. 2.34). The temperature course of Olpad, Cambay Shale I, and Kalol Formations at this point are analyzed. Immediately after its deposition the Olpad Formation reached a temperature of about 175 °C at the base because of a very high sedimentation rate and rapid burial. During the next 4 Ma cooling took place due to uplift and nondeposition. Deposition of Cambay Shale units I–V (events 3–7) caused a rapid temperature increase

Table 2.9. Paleobathymetric estimates, paleotemperatures at sediment water interface and maximum thickness of the events used for the simulation of basin evolution of Cambay Basin

Event no., name	Water depth (m)	Temperature (°C)	Maximum thickness (m)
18 Gujarat Alluvium	5	30	100, 150
17 Jambusar Formation	5, 10	30, 29	500
16 Broach Formation	5, 10	30, 29	200
15 Unconformity III	0	30	-
14 Jhagadis Formation	10	29	400
13 Kand Formation	20	28	600
12 Babaguru Formation	10	28	250
11 Kathana Formation	20	28	300
10 Unconformity II	0	28	-
09 Tarapur Shale	10, 40, 100, 200	28, 27, 20, 10	300
08 Kalol Formation	10, 40	28, 27	400
07 Cambay Shale V	10, 80	27, 22	50
06 Cambay Shale IV	10, 50	27, 26	200
05 Cambay Shale III	10, 40, 100	27, 26, 22	50
04 Cambay Shale II	10, 20, 60	27, 27, 24	350
03 Cambay Shale I	10, 40, 120	27, 26, 20	1000
02 Unconformity I	0	26	-
01 Olpad Formation	5	26	1000

(Reprinted from Kumar, Ruby K, Dwivedi P, Banerjee V, Gupta V (eds), Petroleum geochemistry and exploration in Afro-Asian region – proceedings of the first international conference, Dehra Dun, India, 25–27 November 1985, 1987, 558 pp., –. A.A. Balkema, P.O. Box 1675, Rotterdam, The Netherlands)

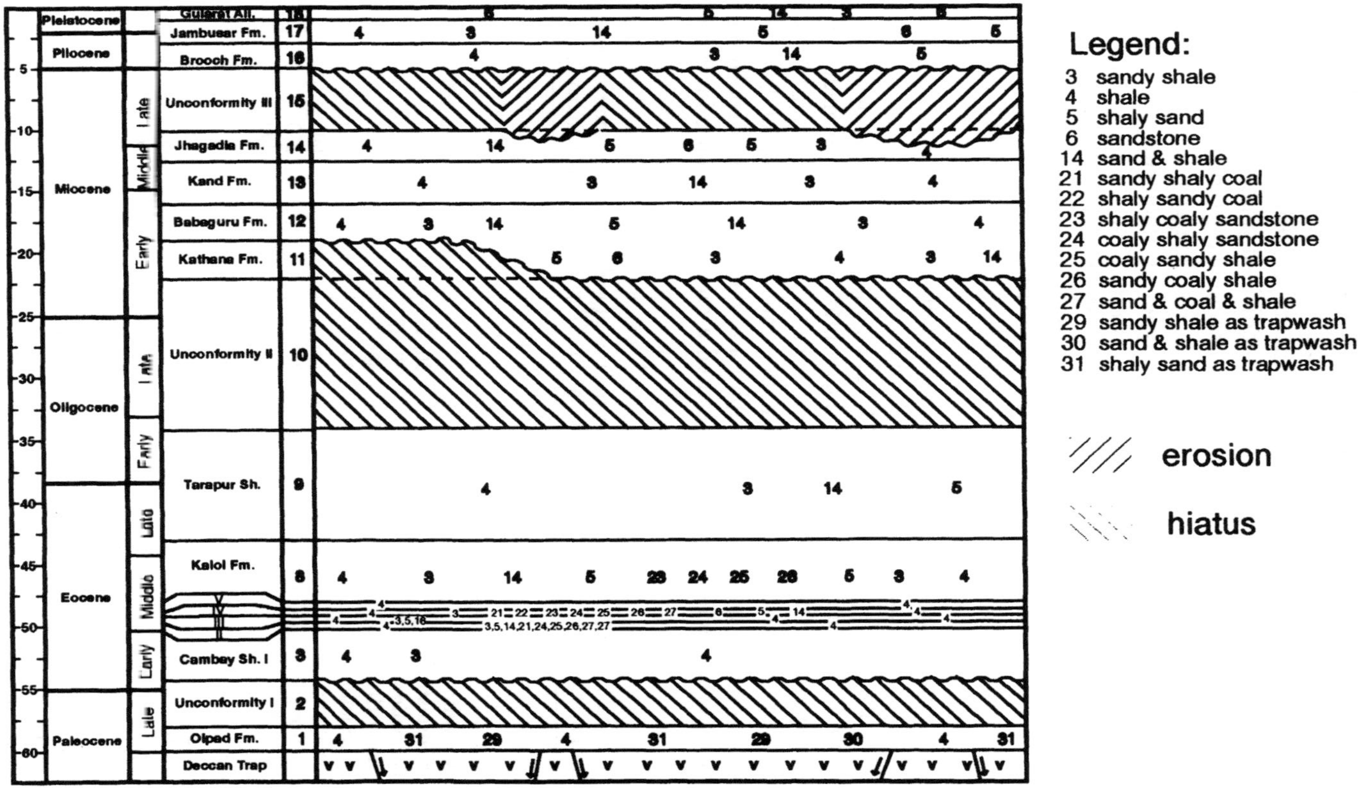

Fig. 2.35. Time-rock synopsis of Cambay Basin. (Yalçın et al. 1988). (Reprinted from Kumar, Ruby K, Dwivedi P, Banerjee V, Gupta V (eds), Petroleum geochemistry and exploration in Afro-Asian region – proceedings of the first international conference, Dehra Dun, India, 25–27 November 1985, 1987, 558 pp., A.A. Balkema, P.O. Box 1675, Rotterdam, The Netherlands)

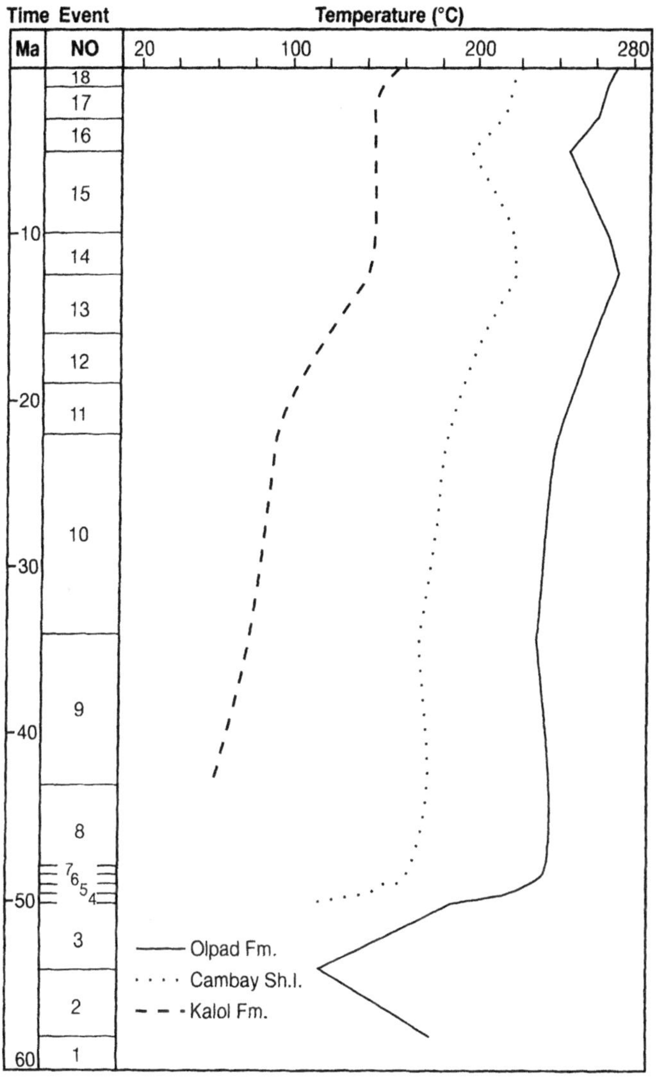

Fig. 2.36. Temporal development of temperature in the Olpad formation, Cambay shale I, and Kalol formation at point A in Broach depression. (Reprinted from Kumar, Ruby K, Dwivedi P, Banerjee V, Gupta V (eds), Petroleum geochemistry and exploration in Afro-Asian region - proceedings of the first international conference, Dehra Dun, India, 25-27 November 1985, 1987, 558 pp., -. A.A. Balkema, P.O. Box 1675, Rotterdam, The Netherlands)

both in Olpad Formation and in the Cambay Shale I (Fig. 2.36). The rapid increases are related to relatively high sedimentation rates and rapid burial. The deposition of the Kalol Formation caused only a slight temperature increase, whereas the deposition of the Tarapur shale (event 9) resulted in a slight decrease in temperature inspite of a deposition of nearly 1000 m. The reason for this unusual cooling effect is the extremely low sea bottom temperatures of 10 °C during the deposition of Tarapur Shale. During the next 12 Ma, which is a nondepositional period (event 10), temperatures increased slightly because the surface temperatures had risen to about 27 °C and therefore less heat was lost at the surface. Deposition of the Jhagadia Formation (event 14) led to a decrease in temperatures of Olpad Formation and Cambay Shale I, but a slight temperature increase took place in the Kalol Formation. This is a typical indication of the system dynamics, demonstrating that stratigraphic units can be affected differently by the same boundary conditions. In this case the critical parameter was the difference in heat capacity of the respective units, which is the main control of the thermal equilibriation time. The following unconformity period (events 15) was characterized by further cooling until interrupted by the depositional events in Pliocene-Quaternary. The present-day temperatures are 270, 220, and 155 °C at the base of Olpad, Cambay Shale I, and Kalol Formations, respectively. Details of the thermal history of Cambay Basin are presented by Yalçın et al. (1988).

2.6.2
San Joaquin Basin, California, USA

The San Joaquin Basin is located in the southern extension of the Great Valley in California (Fig. 2.37). The Great Valley sediments were deposited in a forearc basin between a trench in the west and a magmatic arc in the east. Deposition in the San Joaquin Basin started with Upper Cretaceous (Santonian) sediments. The growth of the forearc basin continued into the Campanian and led to the accumulation of several thousand feet of submarine fan deposits (Forbes, Sacremento, and Lathrop Formations). Very fast deposition and rapid progradation to the west caused a filling up of the basin and a gradual replacement of the deep marine sedimentation by slope and shelf deposition during late Campanian and Maastrichtian (Sawtooth, Tracy-Starkey, Ragged Valley, Blewet, and Moreno Formations). By the end of Cretaceous the forearc basin was almost completely filled. Consequently Paleocene Garzas Sandstone and Lodo Formations are shallow marine deposits. The truncation of the Paleocene units at the base of Middle Eocene Domengine sands indicates uplift and erosion during early Eocene. Regional subsidence at the beginning of late Eocene caused an extensive marine transgression resulting in the deposition of marine shaly lithologies of the Kreyenhagen Formation until the end of early Oligocene. Late Oligocene was an erosional period in the northern part of the basin. The deposition during the last 20 Ma was strongly controlled by the movement of the San Andreas transform fault system and related wrench tectonism. During this period mainly shallow marine to nonmarine units (Vaqueros, Temblor, St. Margarita, McLure and Tulare Formations) were deposited. Only in the southernmost part of the basin is Lower Miocene represented by deep to moderately deep marine units. The time-rock synopsis

→

Fig. 2.37. Location map of San Joaquin Basin. (Welte et al. 1985)

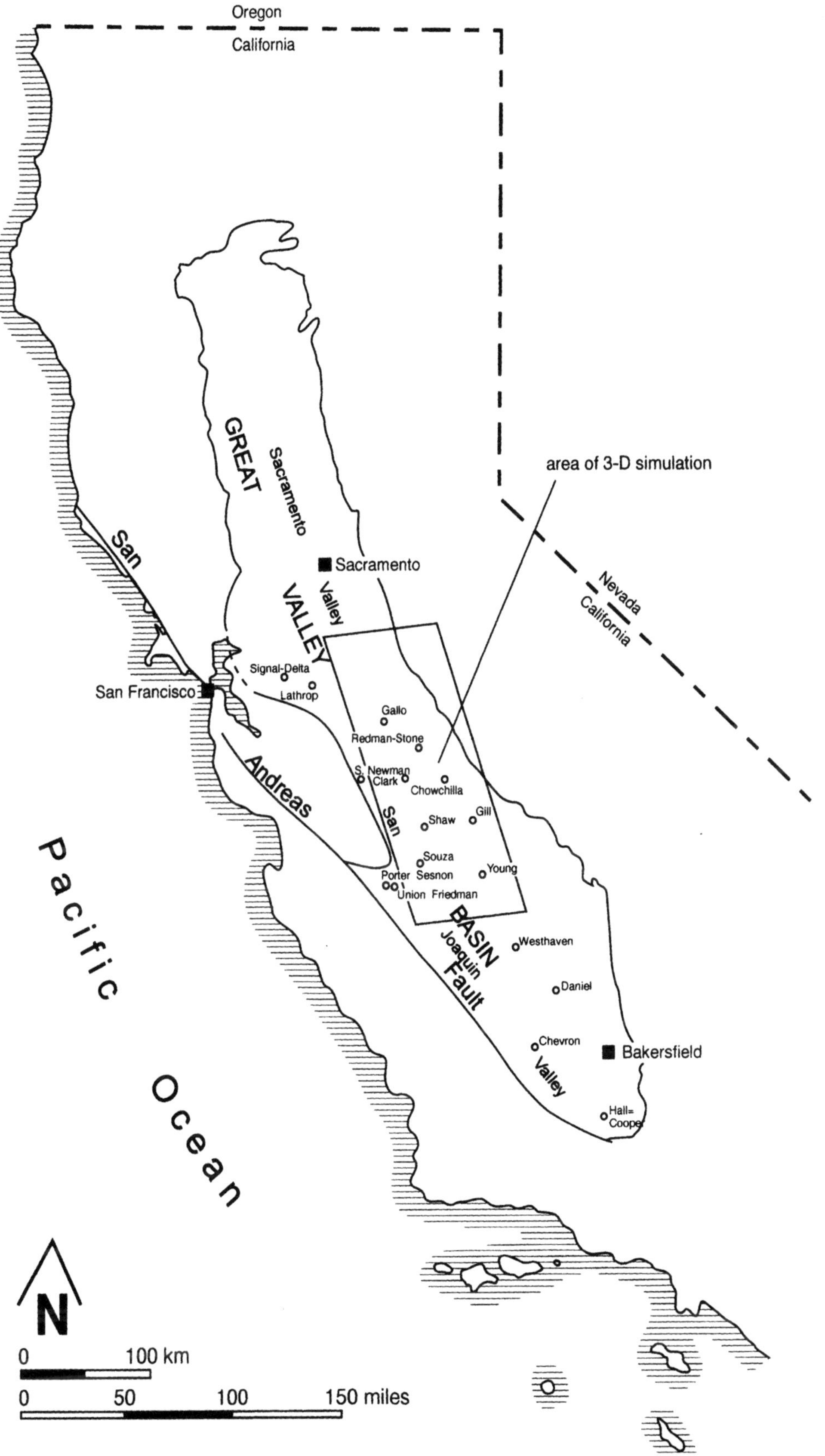

Oregon
California
area of 3-D simulation
GREAT
Sacramento
VALLEY
Valley
Sacramento
San
Andreas
Fault
Nevada
California
Signal-Delta
Lathrop
San Francisco
Gallo
Redman-Stone
S. Newman
Clark
Chowchilla
Shaw
Gill
San
Souza
Young
Porter
Sesnon
Union
Friedman
BASIN
Joaquin
Westhaven
Daniel
Chevron
Valley
Bakersfield
Hall=
Cooper
Pacific
Ocean
N
0
100 km
0
50
100
150 miles

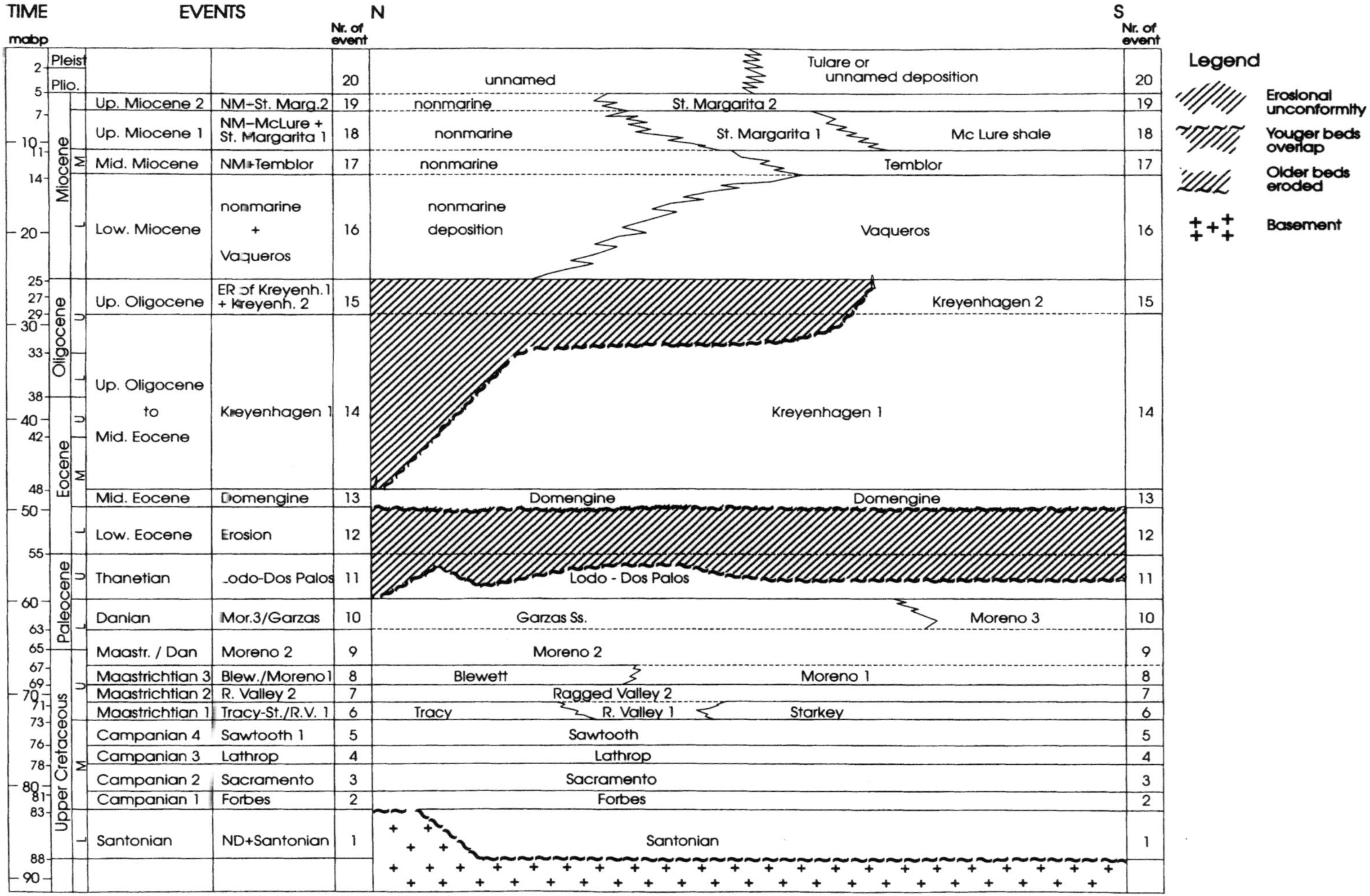
TIME
mabp
EVENTS
Nr. of event
N
S
Legend
Erosional unconformity
Youger beds overlap
Older beds eroded
Basement
Pleist
Plio.
Miocene
Oligocene
Eocene
Paleocene
Upper Cretaceous
Up. Miocene 2
Up. Miocene 1
Mid. Miocene
Low. Miocene
Up. Oligocene
Up. Oligocene to Mid. Eocene
Mid. Eocene
Low. Eocene
Thanetian
Danian
Maastr. / Dan
Maastrichtian 3
Maastrichtian 2
Maastrichtian 1
Campanian 4
Campanian 3
Campanian 2
Campanian 1
Santonian
NM–St. Marg.2
NM–McLure + St. Margarita 1
NM+Temblor
nonmarine + Vaqueros
ER of Kreyenh.1 + Kreyenh. 2
Kreyenhagen 1
Domengine
Erosion
Lodo-Dos Palos
Mor.3/Garzas
Moreno 2
Blew./Moreno1
R. Valley 2
Tracy-St./R.V. 1
Sawtooth 1
Lathrop
Sacramento
Forbes
ND+Santonian
unnamed
Tulare or unnamed deposition
nonmarine
St. Margarita 2
St. Margarita 1
Mc Lure shale
Temblor
nonmarine deposition
Vaqueros
Kreyenhagen 2
Kreyenhagen 1
Domengine
Lodo - Dos Palos
Garzas Ss.
Moreno 3
Moreno 2
Blewett
Moreno 1
Ragged Valley 2
Tracy
R. Valley 1
Starkey
Sawtooth
Lathrop
Sacramento
Forbes
Santonian

Table 2.10. Paleobathymetric estimates and paleotemperatures at sediment/water interface during different events in San Joaquin Basin

Event no., name	Water depth (feet)	Temperature (°C)
20 Tulare/nonmarine	10, 15	18, 18
19 St.Margarita 2 nonmarine	10, 60	18, 17
18 St.Margarita 1 nonmarine	10, 60	17, 17
17 Temblor nonmarine	10, 30	17, 17
16 Vequeros nonmarine	10, 60	17, 17
15 Kreyenhagen 2	0, 600, 1500	17, 15, 11
14 Kreyenhagen 1	600, 1500, 3000	15, 11, 7
13 Domengine	0, 200	17, 16
12 Erosion of Lodo	0	16
11 Lodo	300	14
10 Garzas/Moreno 3	450, 1800, 2400	14, 8, 8
09 Moreno 4	600, 1800, 2400	12, 8
08 Blewett/Moreno 1	600, 1800	12, 8
07 Regged-Valley 2	450, 1500	14, 8
06 Tracy-Starkey/Ragged-Valley 1	450, 1500	14, 8
05 Sawtooth	600, 1500, 1800	10, 7, 7
04 Lathrop	450, 1800	11, 6
03 Sacramento	3000	5
02 Forbes	450, 1800, 2400	11, 6, 6
01 Santonian	0, 60, 450, 1800	15, 14, 13, 7

summarizes the stratigraphic evolution and shows the absolute geological time of different units (Fig. 2.38).

Twenty events were distinguished. Eighteen of these are depositional, and lateral lithofacies changes are common within them. A wide spectrum of clastic rocks ranging from shale and siltstone to coarse sandstone define the lithotypes. Estimates of paleowater depths and sea floor/surface temperatures are shown in Table 2.10. Considering the geological history and the basin type a relatively low and constant heat flow value of 1.10 HFU (Fig. 2.39), typical for forearc basins, was used which gave good agreement between calculated and observed calibration parameters in all the wells at the western flank of the basin. However, simulation of the key wells at the eastern flank showed that the constant and low heat flow values did not lead to an acceptable match. Further calibration and sensitivity tests indicated that the heat flow values must have been much higher than 1.10 HFU in the geological past and lower during recent times. Values up to 1.6 HFU in the past and 0.90 HFU in more recent periods were determined (Fig. 2.39). Such high heat flow values are likely caused by magmatic or volcanic events. However, there is no evidence of such events at the surface or within the sequence. Therefore deep seated magmatic bodies which probably formed in connection with the Sierra Neveda magmatic arc are postulated. Evidence for such intrusions is detected on the gravity map

◄

Fig. 2.38. Time-rock synopsis of the northern part of the San Joaquin Basin

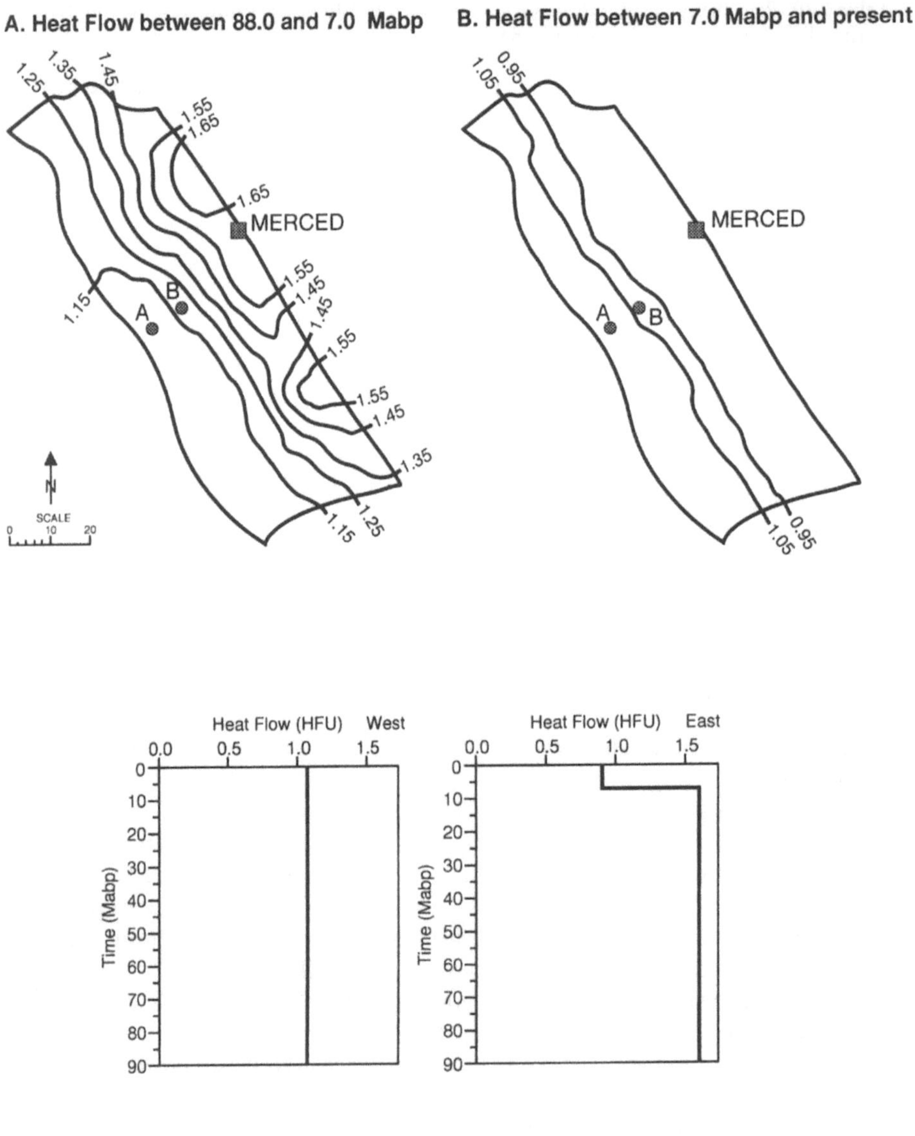

Fig. 2.39. Changes in heat flow with time in San Joaquin basin. The drastic changes at 7 Mabp are due to the effect of the San Andreas fault. (Welte 1989)

of the basin as positive gravity anamolies which coincide with areas of past high heat flow. Furthermore, the time of the rapid heat flow decrease coincides with the abrupt change in the crustal configuration at about 7–8 Mabp caused by the northward migration of the Mendocino triple junction which converted the subduction margin into a transform margin. Consequently subduction, associated magmatic activity, and the high heat flux in the sedimentary sequence ceased.

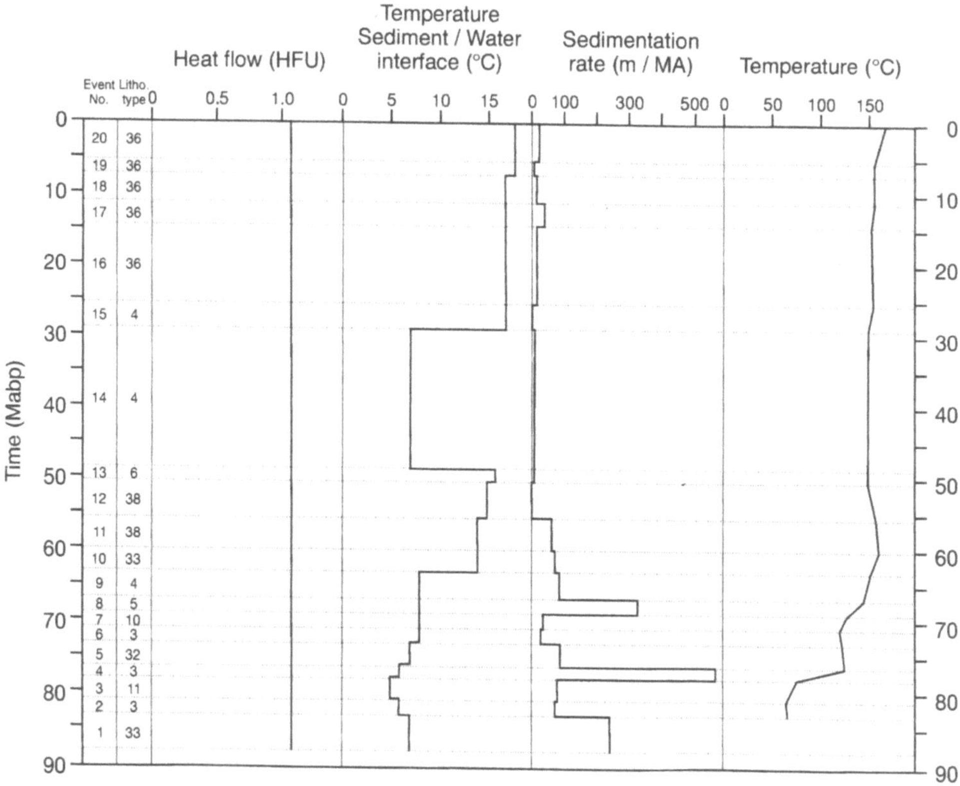

Fig. 2.40. Temporal distribution of heat flow, sediment/water interface temperature, sedimentation rate, and temperature of the layer Santonian at point A. (See Fig. 2.39 for the location of point A)

Using this heat flow configuration and other data the thermal history of the San Joaquin Basin was reconstructed as a three-dimensional study (Welte et al. 1985). General trends of the thermal history are demonstrated here with the help of the relevant input data and the temperature-time curves of the Santonian at two selected grid points A and B (Fig. 2.39). Point A is located in the western part of the basin, where the heat flow remained constant at its initial value of 1.10 HFU (Fig. 2.40). Point B is located in a basin area with a higher heat flow of 1.20 HFU until 7 Mabp, after which it was reduced to 1.00 HFU due to the above mentioned changes from a subduction into a transform margin (Fig. 2.41). The temperature-time curves of Santonian at these points indicate clearly that major increases in temperature are controlled by the rapid deposition of Lathrop and Blewett/Moreno formations (events 4 and 8). At point A even with the slightly lower heat flow than point B, higher temperatures are reached because Santonian was buried deeper at this point by Lathrop and Blewett/Moreno formations. At point A as a consequence of this rapid increase the temperature at the base of Santonian started to drop at 60 Mabp

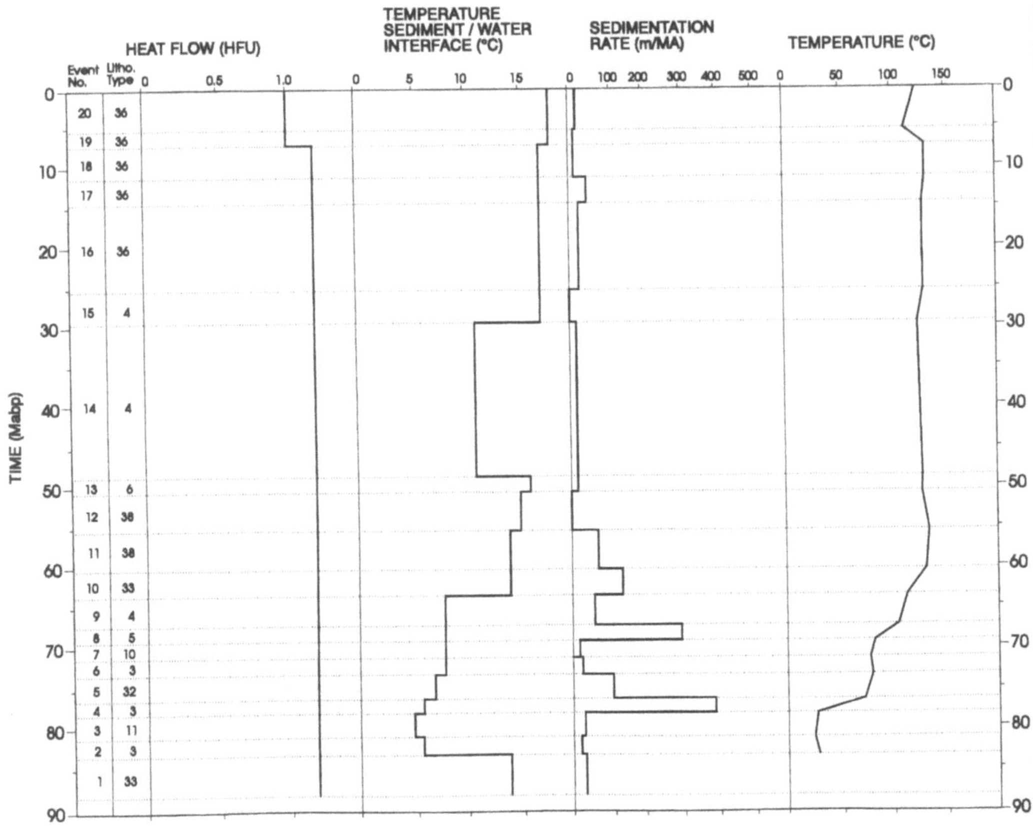

Fig. 2.41. Temporal distribution of heat flow, sediment/water interface temperature, sedimentation rate, and temperature of the layer Santonian at point B. (See Fig. 2.39 for the location of point B)

although burial was continuing and SWI temperatures and heat flow were remaining unchanged (Fig. 2.40). At point B cooling started as expected 5 Ma later, that is at 55 Mabp, when an uplift and erosion took place (Fig. 2.41). Thereafter temperatures of the Santonian remained more or less constant until the period of regional heat flow decrease at 7 Mabp, which caused a slight drop in temperature at point B (Fig. 2.41). At point A, where the heat flow remained unchanged, temperature increased slightly, as expected (Fig. 2.40). Other details of the thermal history of the San Joaquin Basin are given by Welte et al. (1985), Yalçın and Welte (1988), Welte and Yalçın (1988) and Welte (1989).

2.6.3
Adana Basin, Turkey

The Adana Basin is located in southern Turkey between the Taurus Mountains in the north and Cyprus in the south (Fig. 2.42). The basin formation was

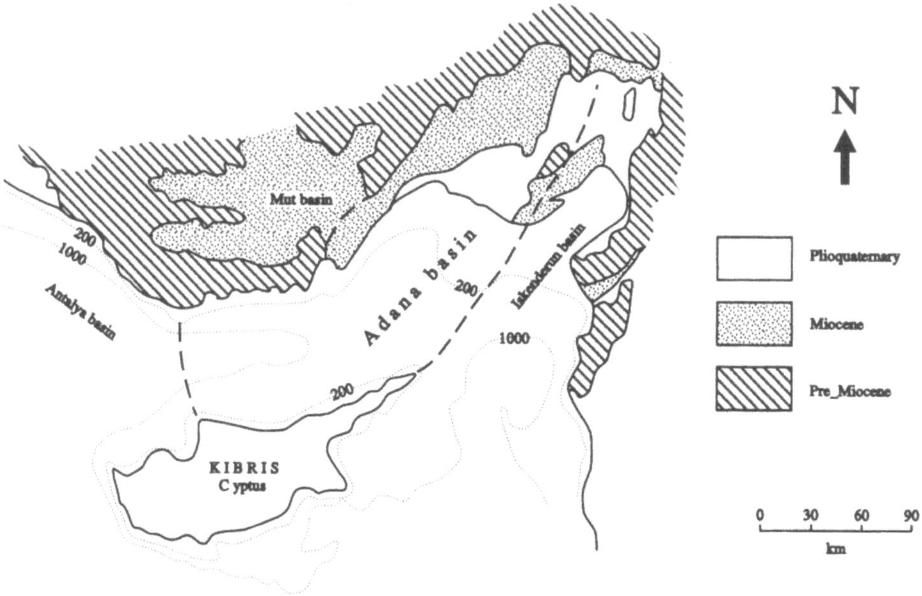

Fig. 2.42. Location map of the Adana Basin

induced by a complicated combination of wrench faulting and crustal flexuring in Early Miocene (Şengör and Yilmaz 1981). The fault activity continued until recent times and controlled, together with other large-scale geological phenomena such as the Messinian Salinity Crisis, the deposition within the basin. The sedimentary sequence consists of three sedimentary cycles of Miocene, Pliocene, and Pliocene-Quaternary ages (Yalçın and Görür 1984). Twenty-three chronological units (events) of different duration are distinguished. Except for one nondepositional event all others represent depositional units. The sequence consists mainly of detritial sediments such as shale, siltstone, sandy shale, and a thick evaporitic unit of Messinian age. A simulation beginning at 17 Mabp with input data from the Kuransa well are shown in Fig. 2.43. The heat flow increase during the last 3 Ma which was determined after several calibration runs fits to the tectonic evolution of the area because it was effected by an extensional stress regime during this time. Additionally, a volcanic activity is also observed which normally leads to an increase in the heat flux density.

The analysis of the simulated temperature history of the lowermost unit in the Kuransa well indicates some interesting aspects. Namely, during the deposition of the evaporitic unit (event 13) the temperature increased only insignificantly, although the burial was more than 400 m and sea bottom temperature was relatively high. This unusual behavior is the result of the very high thermal conductivity of evaporitic rocks leading to an enhanced heat loss

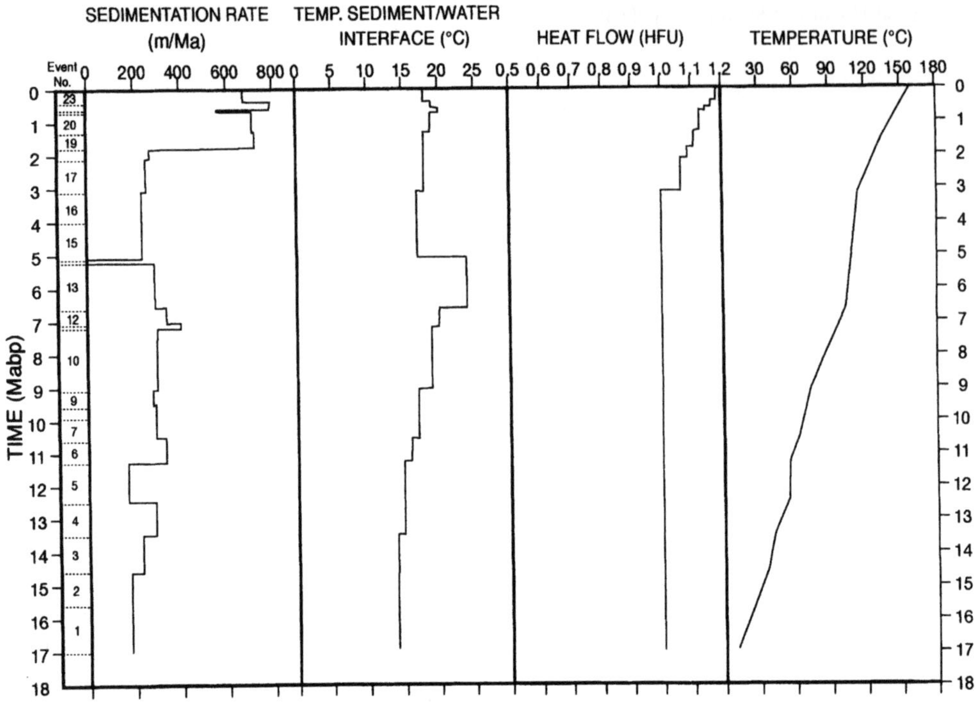

Fig. 2.43. Sedimentation rate, sediment/water interface temperatures, heat flow, and temperature history of the lowermost unit in the Kuransa well

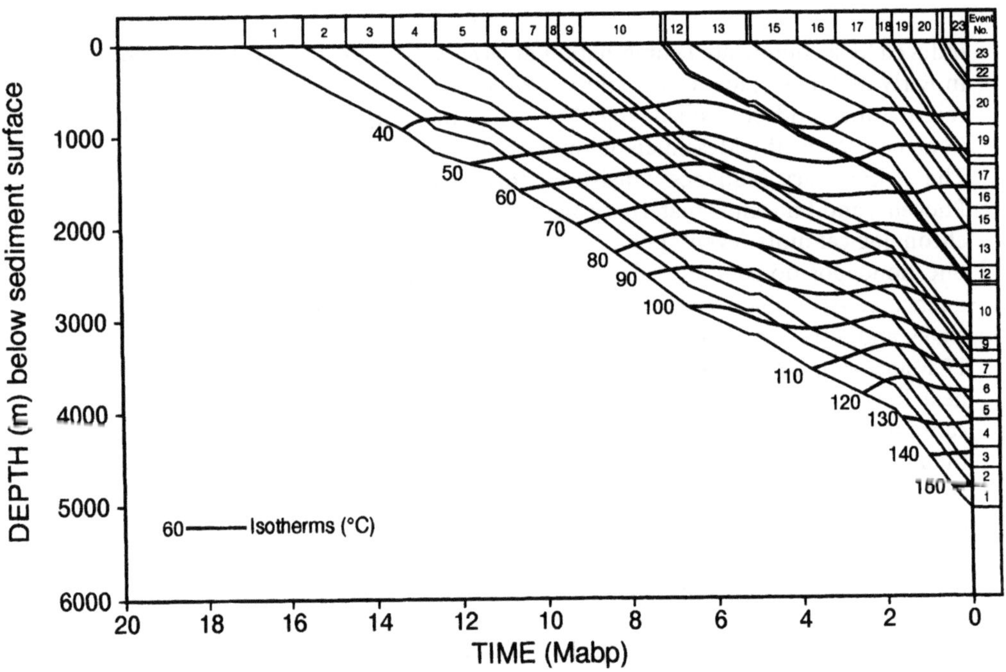

Fig. 2.44. Burial history diagram of the Kuransa well in Adana Basin with isotherms showing the temperature development as function of time and space

from the sediment surface. Figure 2.44 shows that not only the lowermost unit in the well but the entire sequence was effected by the deposition of evaporites. Although temperatures increased slightly during this period, the geothermal gradient decreased, as indicated by dipping of isotherms. The change in the geothermal gradient both as a function of time and depth is a good indication of basin dynamics. Details of the temperature history of the Adana Basin are presented by Yalçın (1988, 1990, 1991).

2.6.4 Styrian Basin, Austria

The more than 4-km-deep Neogene Styrian Basin is located at the eastern margin of the Alps and forms part of the Pannonian Basin System (Fig. 2.45). It

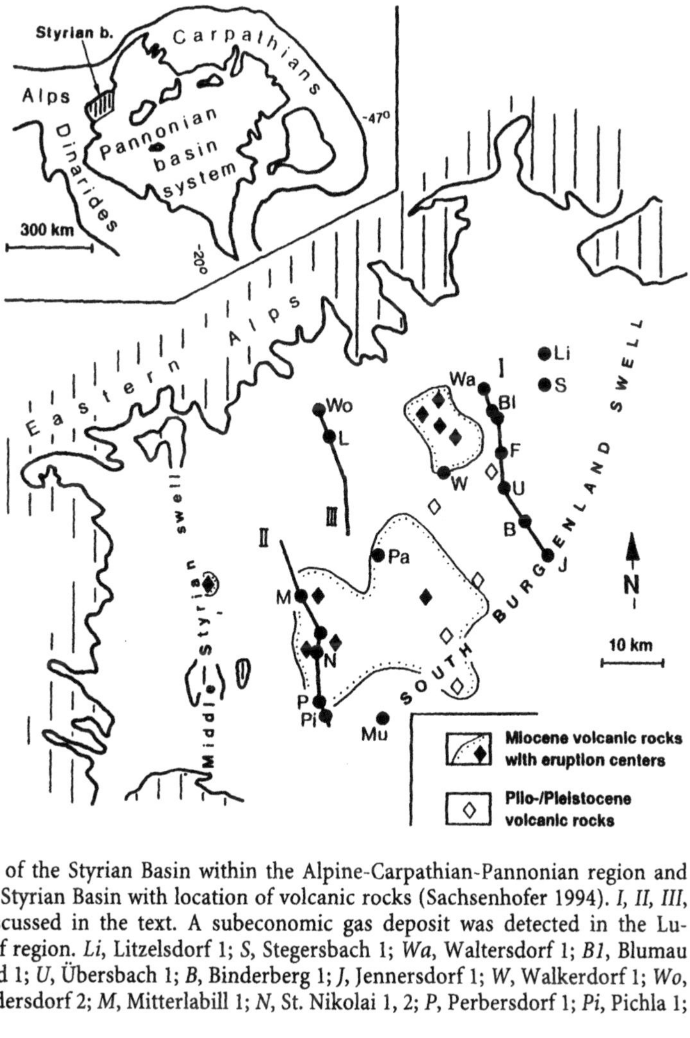

Fig. 2.45. Position of the Styrian Basin within the Alpine-Carpathian-Pannonian region and sketch map of the Styrian Basin with location of volcanic rocks (Sachsenhofer 1994). *I, II, III,* Cross sections discussed in the text. A subeconomic gas deposit was detected in the Ludersdorf-Wollsdorf region. *Li,* Litzelsdorf 1; *S,* Stegersbach 1; *Wa,* Waltersdorf 1; *B1,* Blumau 1, 1a; *F,* Fürstenfeld 1; *U,* Übersbach 1; *B,* Binderberg 1; *J,* Jennersdorf 1; *W,* Walkerdorf 1; *Wo,* Wollsdorf 1; *L,* Ludersdorf 2; *M,* Mitterlabill 1; *N,* St. Nikolai 1, 2; *P,* Perbersdorf 1; *Pi,* Pichla 1; *Mu,* Mureck 1

represents an extensional structure on top of a crustal wedge, which was moving eastward during the final stages of the Alpine orogeny. Eastward extrusion was a consequence of continental escape and extensional collapse within the Eastern Alps. The evolution of the Styrian Basin can be subdivided into an early Miocene (Ottnangian to Karpatian) synrift and a middle to late Miocene postrift phase of subsidence (Ebner and Sachsenhofer 1995; Fig. 2.46). During the synrift phase thick clastic limnic/fluviatile and marine sediments were deposited. The climax of extension during the synrift phase favored the ascent of andesitic magmas which are most probably related to subduction along the Carpathian front. Magmatic activity continued after the end of the synrift phase into the early Badenian. Today the voluminous shield volcanoes are almost totally buried by younger sediments. An unconformity separates lower and middle Miocene sediments and is interpreted as the transition from the synrift to the postrift stage. Sedimentation during the postrift stage is controlled by middle Miocene (early Badenian and Sarmatian) transgressions and a subsequent regression with a lowering of salinity. During this time period intercalations of sandy and shaly sediments were deposited. Algal reefs and rhodolith platforms developed along the basin margins and along the

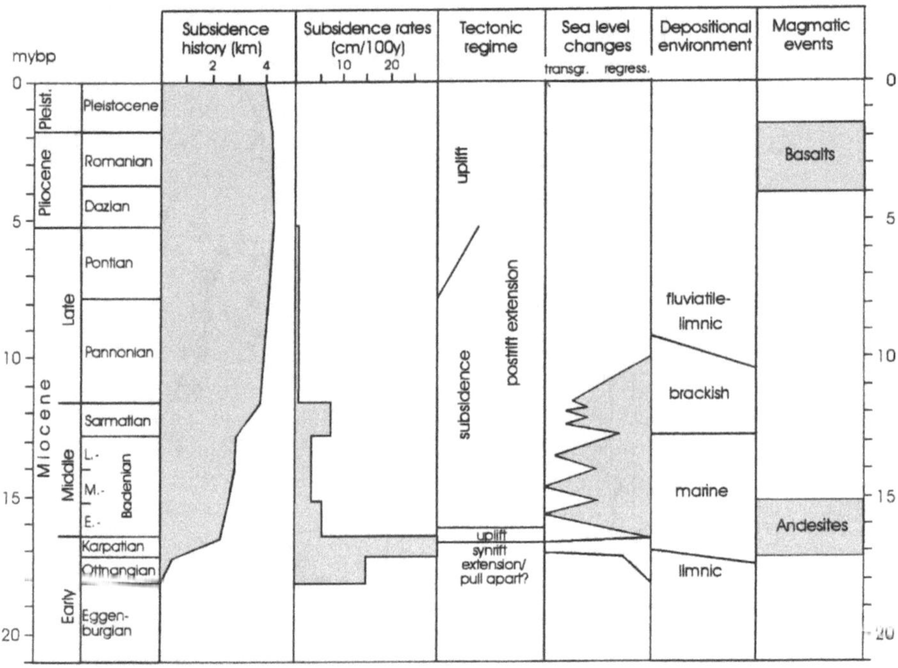

Fig. 2.46. Summary of the Neogene evolution of the Styrian Basin (Ebner and Sachsenhofer 1995). The geological time scale follows Steininger et al. (1990). (Reprinted from Tectophysics, 242 (1–2), 133–150, Ebner F, Sachsenhofer RF, Paleogeography, subsidence and thermal history of the Neogene Styrian Basin (Pannonian basin system, Austria) 1995, with kind permission from Elsevier Science – NL, Sara Burgerhartstraat 25, 1055 KV Amsterdam, The Netherlands)

slopes of volcanic islands during early and middle Badenian times. In Pliocene times a basin inversion resulted in the erosion of a few hundred meters of sediment. Uplift was accompanied by a second volcanic phase producing basalts in Plio-/Pleistocene times.

The evolution of the Styrian Basin and the reconstruction of the thermal history were studied in five individual wells and along two cross sections (I, II; Fig. 2.45) using a two-dimensional finite-element software system. Hydrocarbon migration was studied along a third cross section (III; see Chap. 7, this Vol.). A relatively complex geological evolution and periods of volcanic activity, influenced the temporal and spatial distribution of heat flow substantially. Therefore a careful calibration of the heat flow history was necessary. The heat flow evolution along cross section II which crosses a Karpatian (Nikolai) and an early Badenian (Mitterlabill) volcano are discussed here as an example (Fig. 2.47a). A total of 30 events and 50 gridpoints including 5 calibration wells were used to define the geological model. Several calibration runs were necessary to reach a fit between measured and calculated data, which are presented in Fig. 2.49. The temporal and spatial distribution of the heat flow is shown in Fig. 2.47b. The simulation of the basin evolution with this heat flow

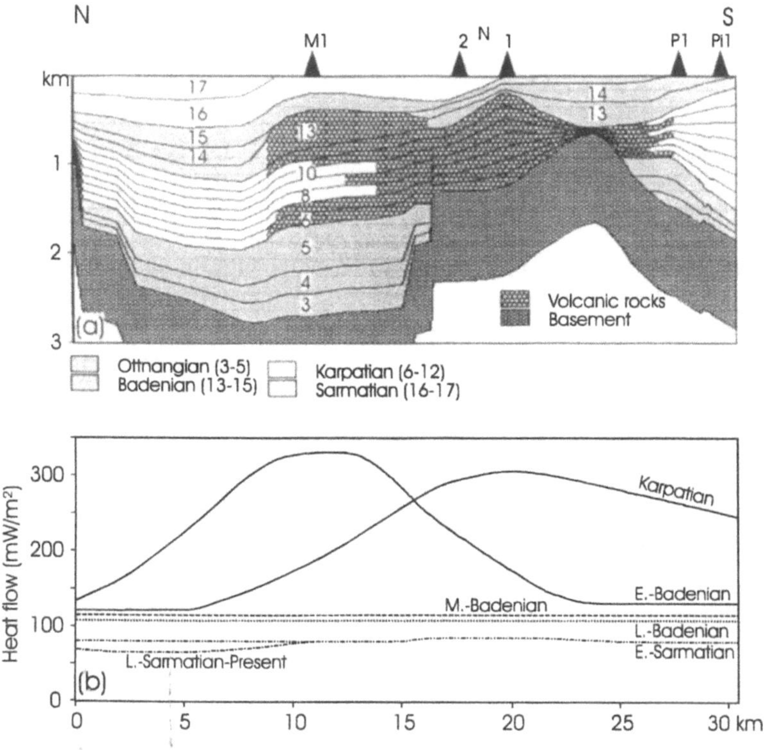

Fig. 2.47.a Stratigraphy along cross section II with position of calibration wells. *M*, Mitterlabill; *N*, St. Nikolai; *P*, Perbersdorf; *Pi*, Pichla. **b** Heat flow model applied to cross section II (Sachsenhofer 1994)

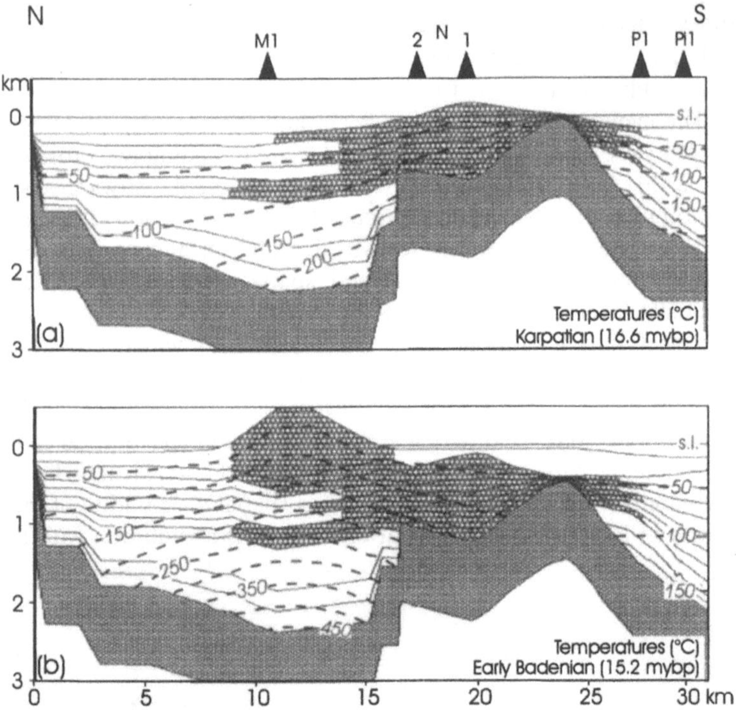

Fig. 2.48. Calculated temperature distribution along cross section II for Karpatian (**a**) and early Badenian times (**b**). *s.l.*, Sea level. (Sachsenhofer 1994)

data enabled the reconstruction of thermal evolution along the cross section. The calculated temperature distribution during Karpatian and early Badenian is shown in Fig. 2.48. The great effect of the volcanic activity on the heat flow pattern and the resulting temperature distributions are shown in Figs. 2.47 and 2.48. The Karpatian heat flow was extremely elevated in vicinity of the Karpatian (Nikolai) volcano resulting in an updoming of the isotherms. In conformity with moving magmatic activity, the heat flow maximum was shifted northward to the Mitterlabill area, which became a center of volcanic activity in early Badenian times. A good fit between measured and calculated vitrinite reflectance data in this area can be established only assuming extremely high heat flow (330 mW/m^2) and erosion of 100–450 m thick early Badenian andesitic rocks. Determination of the exact values of heat flow and erosion, which are necessary to establish a good fit, also depends on the physical properties (particularly thermal conductivities) of the eroded rocks. These parameters cannot be measured. Furthermore, the eroded thickness would be overestimated if heat is transfered from the nearby volcanic vent or from the overlying andesites laterally and with resulting enhancement of vitrinite reflectance in shallow Karpatian sediments below the andesites. In any case the erosion of thick andesitic rocks (and the shallow water depth of the Badenian

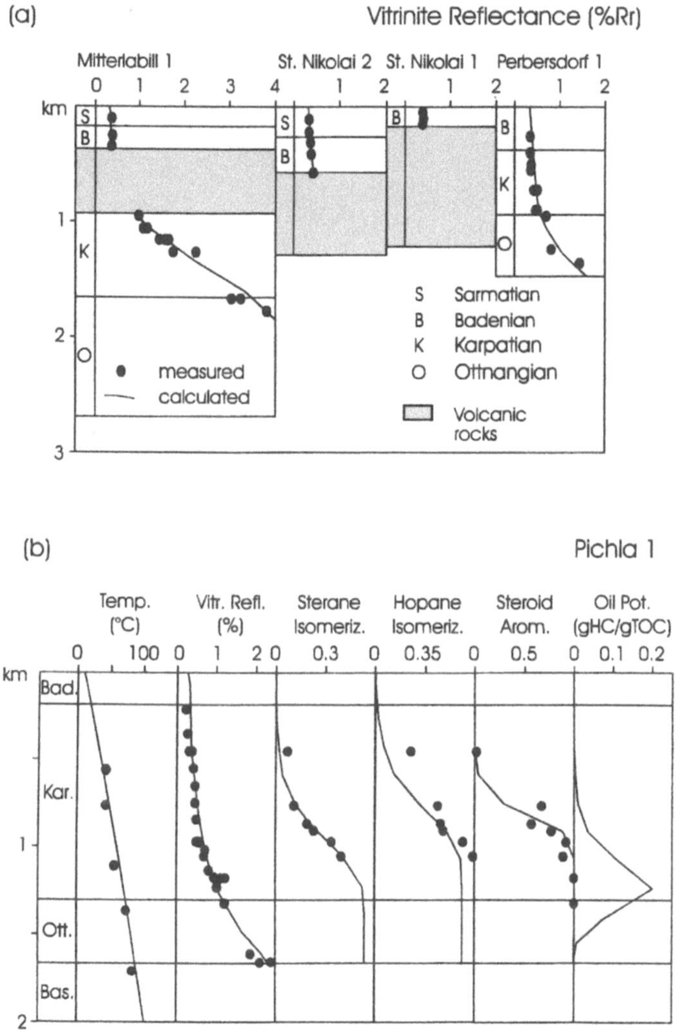

Fig. 2.49. a Comparison of measured and calculated vitrinite reflectance data of four wells along cross section II. **b** Comparison of measured and calculated temperature data and maturity parameters (vitrinite reflectance, sterane isomerization, hopane isomerization, steroid aromatization) of the Pichla 1 well. Oil potential has been calculated using kinetic parameters for Karpatian sediments (Sachsenhofer 1994)

Sea) indicate that the Mitterlabill volcano must have been raised significantly (500 m) above the early Badenian sea level (Fig. 2.48b). Erosion occurred during the middle Badenian, which is represented by a stratigraphic gap in the Mitterlabill 1 well. After the end of the magmatic activity heat flow decreased along the whole cross section and has been 55–85 mW/m^2 since late Sarmatian times.

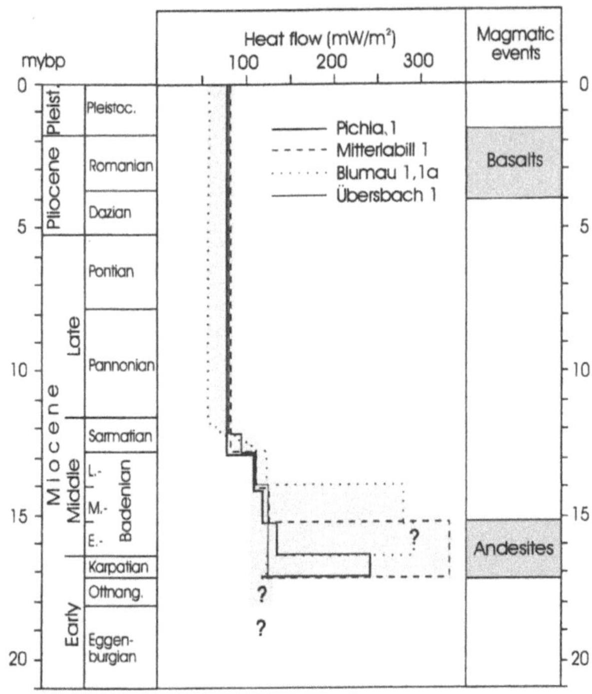

Fig. 2.50. Heat flow history of the Styrian Basin. Different lines indicate the heat flow evolution of selected wells. Pichla 1, Mitterlabill 1, and Blumau 1, 1a wells are situated in vicinity of Miocene volcanoes. (Modified after Sachsenhofer 1994)

The heat flow history of the Styrian Basin is summarized in Fig. 2.50. Extremely elevated heat flows during Karpatian to Badenian times are most probably a consequence of shallow magma chambers. In the southern part of the basin (e.g., Pichla area), heat flow decreased immediately after the Karpatian, whereas it remained high until early or even middle Badenian times near the Mitterlabill well and close to the northeastern volcano (e.g., Blumau area). This is a consequence of northward shifting of magmatic activity. In the Pichla area heat flow decreased immediately after the end of magmatic activity whereas in the Blumau area this decrease took place 1 Ma after the end of magmatic activity. This difference may indicate the presence of a larger magma chamber below the northeastern volcano, which required a longer time to cool down. Modeling showed that the effect of raised heat flow in vicinity of the volcanoes disappeared at a distance of about 10 km and heat flow approached background values of about 120 mW/m^2 (Fig. 2.47b). Since late Sarmatian, heat flow along the entire cross section has been generally in the range of 55–85 mW/m^2. This relatively high heat flow is probably due to a thinned crust beneath the Styrian Basin. Namely, present thickness of the crust is about 25–30 km (Aric et al. 1987). Present formation temperatures in some wells in the

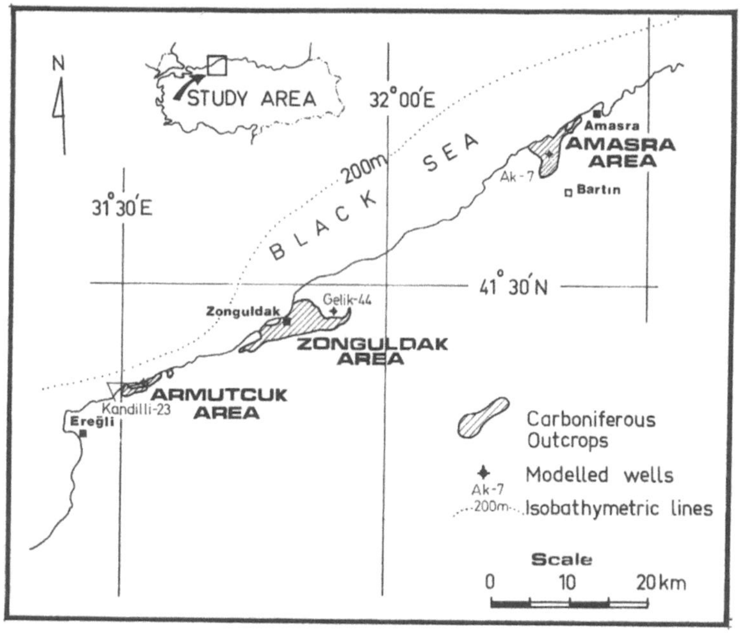

Fig. 2.51. Location map of the Zonguldak Basin also showing sites of the modeled wells

easternmost part of the basin indicate a slight heat flow increase (5–10 mW/m^2) during Pliocene or Pleistocene. Whether this is a consequence of hydrodynamic effects or due to Plio-/Pleistocene basaltic volcanism is yet not known. In any case this latter volcanic phase had only little effect on regional heat flow patterns. Perhaps this is a result of the great depth of the relevant magma chambers (50–80 km; Kurat et al. 1980). More details of the thermal history of the Styrian Basin are given by Sachsenhofer (1994).

2.6.5 Zonguldak Basin, Turkey

The Zonguldak Basin which is located in northwestern Turkey on the Black Sea coast (Fig. 2.51) is formed on a Hercynian continental sliver (Okay et al. 1994). Basin development started at the end of Visean when a carbonate platform was converted in to a deltaic basin due to a tectonic uplift to the north. Deposition started with prodelta Namurian sediments which were followed by a progradational Westphalian sequence of delta-plain sediments which include coal seams. At the end of Westphalian the entire basin was uplifted and eroded. Continental clastics in the eastern part of the basin represent a new depositional period which commenced in late Permian and continued until middle Liassic. After a second period of uplift and erosion the basin area was flooded by a shallow sea during early Malm and platform-type carbonates were deposited. The formation of the Black Sea Basin as an extensional back-arc basin by rifting led in the Zonguldak area to a second major depositional period. This period commenced in Hautrivian and continued up to middle Eocene. Different types of sediments representing synrift, postrift, volcanic arc and passive continental margin enviroments were deposited. Collision of the Hercynian continental sliver with the Sakarya continent in the south also affected the Zonguldak area and resulted in a second major uplift and erosion period which began during middle Eocene. As a result of the ongoing uplift in many parts of the basin the entire sequence of the major second depositional period is eroded. Hence even Carboniferous coal-bearing units can be observed in outcrops at certain locations. The geological evolution of the study area is outlined in a time-rock synopsis of the basin (Fig. 2.52).

Three wells (Kandilli-23, Gelik-44, and Ak-7) from the western, middle, and eastern parts of the basin (Fig. 2.51), respectively, were simulated to determine the temperature history of coal-bearing units and to calculate gas generation in given coal seams (Yalçın et al 1994). In each well the base of Namurian Alacaagzi formation and the present topographic surface are taken as the lower and upper spatial boundaries. Consequently basin evolution is simulated from 333 Mabp to present time. Depending on the well, up to 29 events are designated for the simulation. Many of these events are erosional representing the two major uplift period at the end of Carboniferous and in the Eocene. The original thickness at the eroded units was first estimated using data from regional geology and from wells in surrounding areas. These thicknesses were then verified during the calibration procedure. One of the deep wells in the middle part of the basin, where both for the Carboniferous and Cretaceous a good vitrinite reflectance trend was determined, contributed much to the calibration. With the help of this well not only the original thickness of the eroded units but also the heat flow history could be constructed as demonstrated by Yalçın (1995). Heat flow values were constant until late Jurassic to early Cretaceous except for a slight increase during the late Westphalian due to small-scale volcanic activity. As a consequence of the tectonic stretching during early Cretaceous related to the opening of the Black Sea heat flow values began to increase. Arc volcanism represented by the pyroclastics and volcanics of the Yemisli Cay formation (Fig. 2.52) caused continued increase in heat flow which reached a peak during Coniacian. Thereafter heat flow values decreased gradually to their present value. This calibrated temporal distribution of heat flow and other input data such as sedimentation rate, water depths, and SWI temperatures are shown in Fig. 2.53. The thermal evolution of the basin is demonstrated with the help of temperature vs. time diagrams of the Namurian Alacaagzi formation at the 3 wells (Fig. 2.54). The first temperature increase was related to rapid burial during Westphalian. Uplift and erosion after Westphalian caused a cooling which lasted until the next deposition (burial) which commenced in the western part of the basin either in late Jurassic (Gelik-44) or early Cretaceous (Kandilli-23). In the east (Ak-7) the deposition

→

Fig. 2.52. The time-rock synopsis of the Zonguldak basin

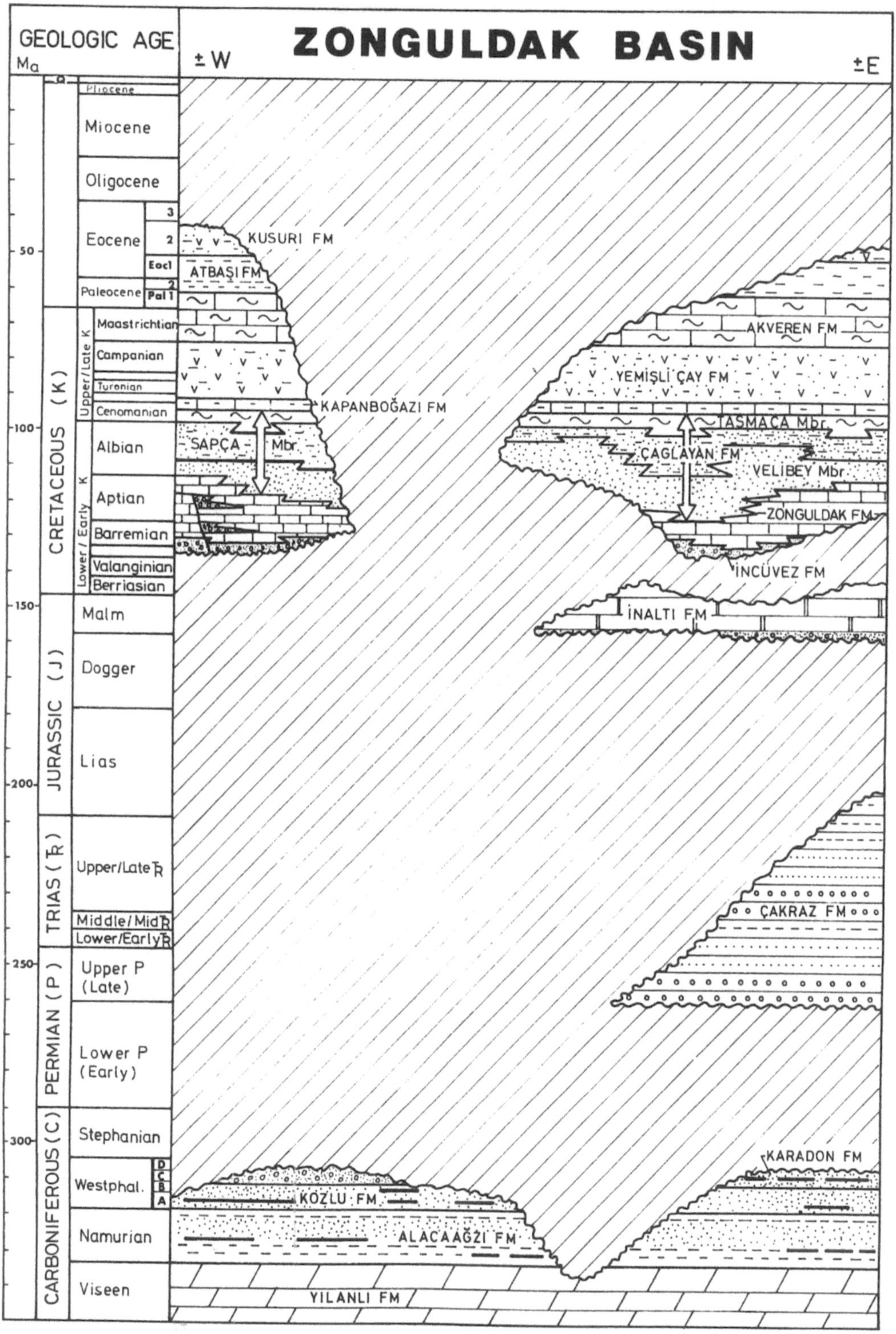

GEOLOGIC AGE
Ma
ZONGULDAK BASIN
± W
± E
Pliocene
Miocene
Oligocene
Eocene
Paleocene
Maastrichtian
Campanian
Turonian
Cenomanian
Albian
Aptian
Barremian
Valanginian
Berriasian
Malm
Dogger
Lias
Upper/Late Ŗ
Middle/Mid Ŗ
Lower/Early Ŗ
Upper P (Late)
Lower P (Early)
Stephanian
Westphal.
Namurian
Viseen
CRETACEOUS (K)
JURASSIC (J)
TRIAS (Ŗ)
PERMIAN (P)
CARBONIFEROUS (C)
Upper / Late K
Lower / Early K
50
100
150
200
250
300
KUSURI FM
ATBAŞI FM
KAPANBOĞAZI FM
SAPÇA Mbr
AKVEREN FM
YEMİŞLİ ÇAY FM
TAŞMACA Mbr
ÇAĞLAYAN FM
VELİBEY Mbr
ZONGULDAK FM
İNCÜVEZ FM
İNALTI FM
ÇAKRAZ FM
KARADON FM
KOZLU FM
ALACAAĞZI FM
YILANLI FM

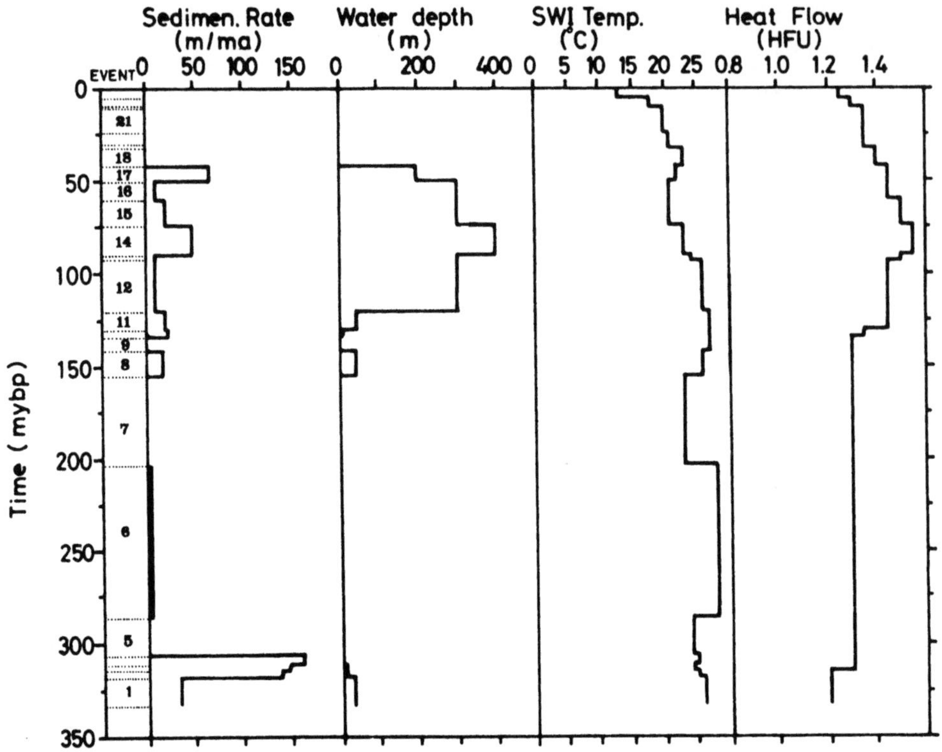
Sedimen. Rate (m/ma)
Water depth (m)
SWI Temp. (°C)
Heat Flow (HFU)
EVENT
Time (mybp)
21
18
17
16
15
14
12
11
9
8
7
6
5
1

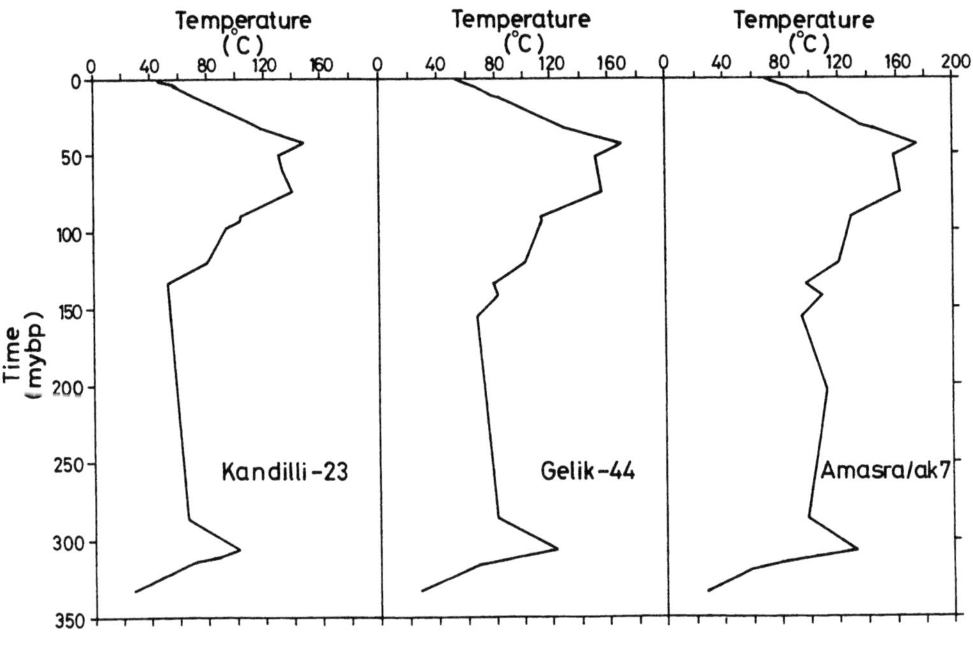
Temperature (°C)
Temperature (°C)
Temperature (°C)
Time (mybp)
Kandilli-23
Gelik-44
Amasra/ak7

of the Upper Permian- Liassic Çakraz formation led to an interruption of this cooling (Fig. 2.54). The highest temperatures were reached at the end of the second major deposition period at 42 Mabp as a consequence of maximum burial coincident with high heat flow. Relevant values at the base of the Namurian Alacaagzi formation in the wells Kandilli-23, Gelik-44, and Ak-7 are 150, 171, and 175 °C, respectively (Fig. 2.54).

2.6.6 Northwest German Basin

The Northwest German Basin is composed of Paleozoic, Mesozoic, and Cenozoic sediments, which occur in variable thickness in different parts of the basin. The maximum thickness probably exceeds 10 km at Hamburg. The general stratigraphy and lithology is summarized in Fig. 2.55 (modified from Barnard and Cooper 1983). With respect to petroleum generation the Upper Carboniferous is considered the major source rock unit for gas. This formation is several kilometers thick and composed of sandstones, siltstones, and shales with interlayered coal seams. The latter are restricted to the late Namurian (Namurian C) and early Westphalian (Westphalian A, B, and C). For the simulation of petroleum generation in this region the temperature history of the latter units is most important, although a contribution of petroleum from deeper, more mature formations cannot be ruled out. However, information on the older Paleozoic is sparse due to its great depth; therefore the older Paleozoic units are not considered as source rocks at this stage.

The Palaeozoic strata in this area reached their maximum temperatures only recently (during Late Tertiary or Quaternary); therefore vitrinite reflectance and other maturation profiles generally reflect mainly the Cenozoic burial and temperature history. Nevertheless, the earlier temperature evolution is of great importance for the gas generation in this region because it determines how much gas generation potential is preserved. Information on this early temperature history, however, cannot be obtained from maturation data but must be constructed from geological information such as the geotectonic position of the basin, volcanic activity, etc.

One transect through the study area for which a thermal reconstruction was performed is presented in Fig. 2.56. Temperature reconstructions are based on the burial and heat flow histories of the Namurian and Westphalian, which are different for the northern and southern parts of the transect. One major difference between the two areas is the greater thickness of Autunian volcanoclastic rocks in the north (up to 600 m) which contrasts with no Autunian volcanoclastics in the south. This difference affects the heat flow model. It is thought that volcanic centers existed close to the northern part of the transect,

Fig. 2.53. The temperal distribution of sedimentation rate, water depth, sediment/water interface temperature, and heat flow which are used for the modeling of the well Ak-7 as input

Fig. 2.54. Temperature vs. time diagrams of the Namurian Alacaagzi formation in the modeled wells

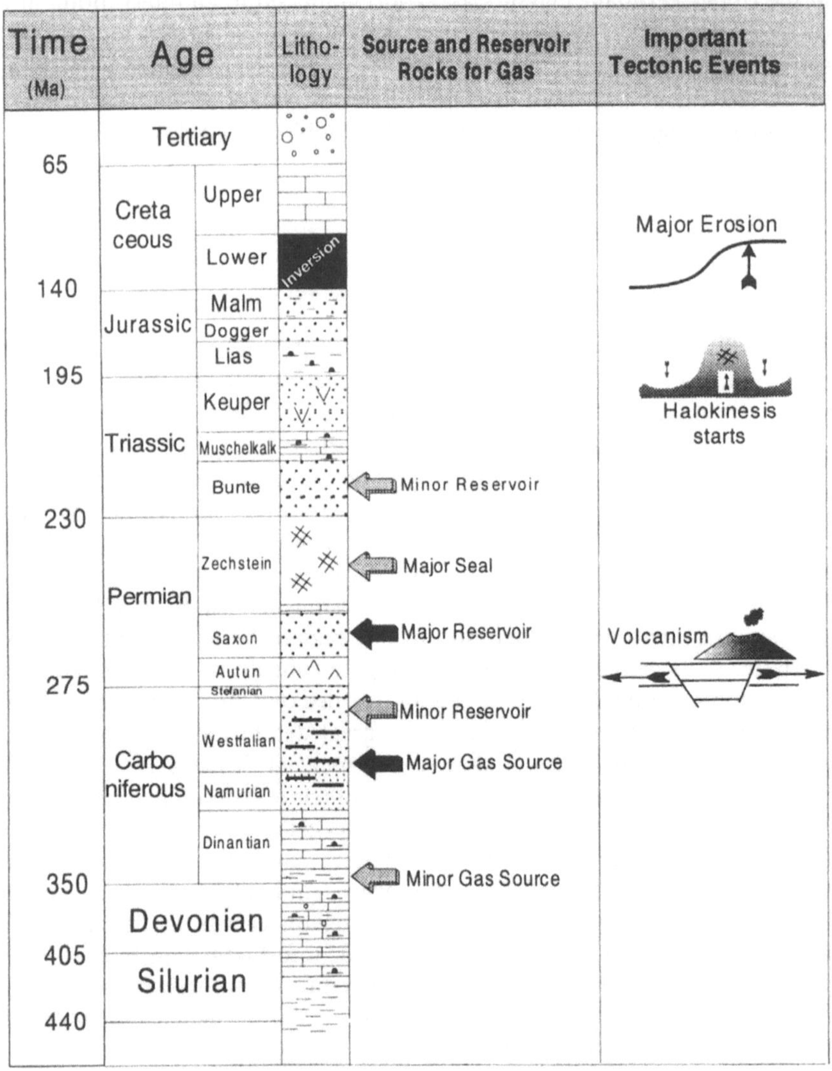

Fig. 2.55. Schematic lithological and stratigraphic column for the study area with major gas source and reservoir rocks and some important tectonic events. (Modified after Barnard and Cooper 1983)

which led to very high heat flows (up to 150–200 mW/m^2) there. For the south, lower heat flows (but still above the crustal average of 60 mW/m^2) are thought to have persisted during the Autunian. Therefore the early Permian is regarded as the time of first gas generation from the Carboniferous source rocks along the entire profile, but it is thought that the rate of gas generation was much higher in the north than in the south. This expectation is due not only to the greater heat flows in the north but also to the greater burial depth (Fig. 2.57).

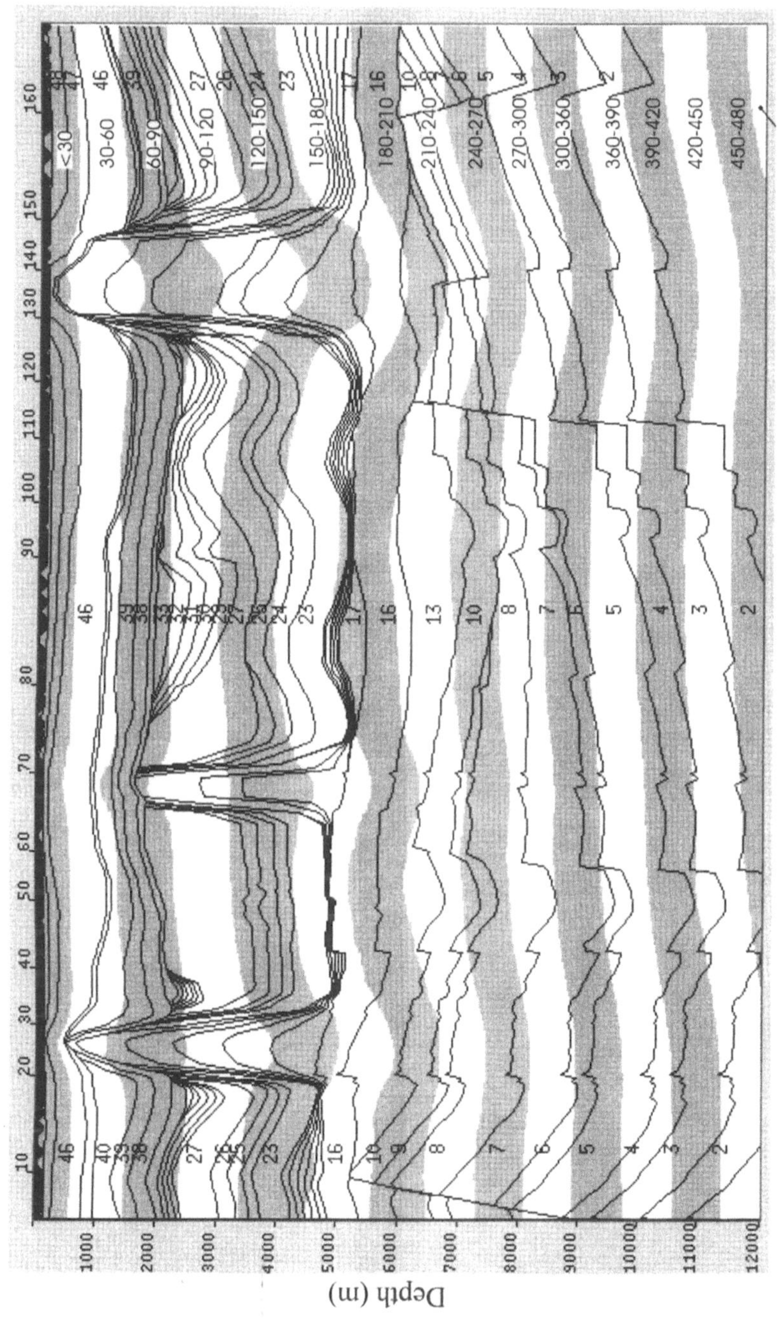

Fig. 2.56. General stratigraphy and temperature along a north/south profile in northwest Germany. The transect is about 100 km long and situated south of Hamburg

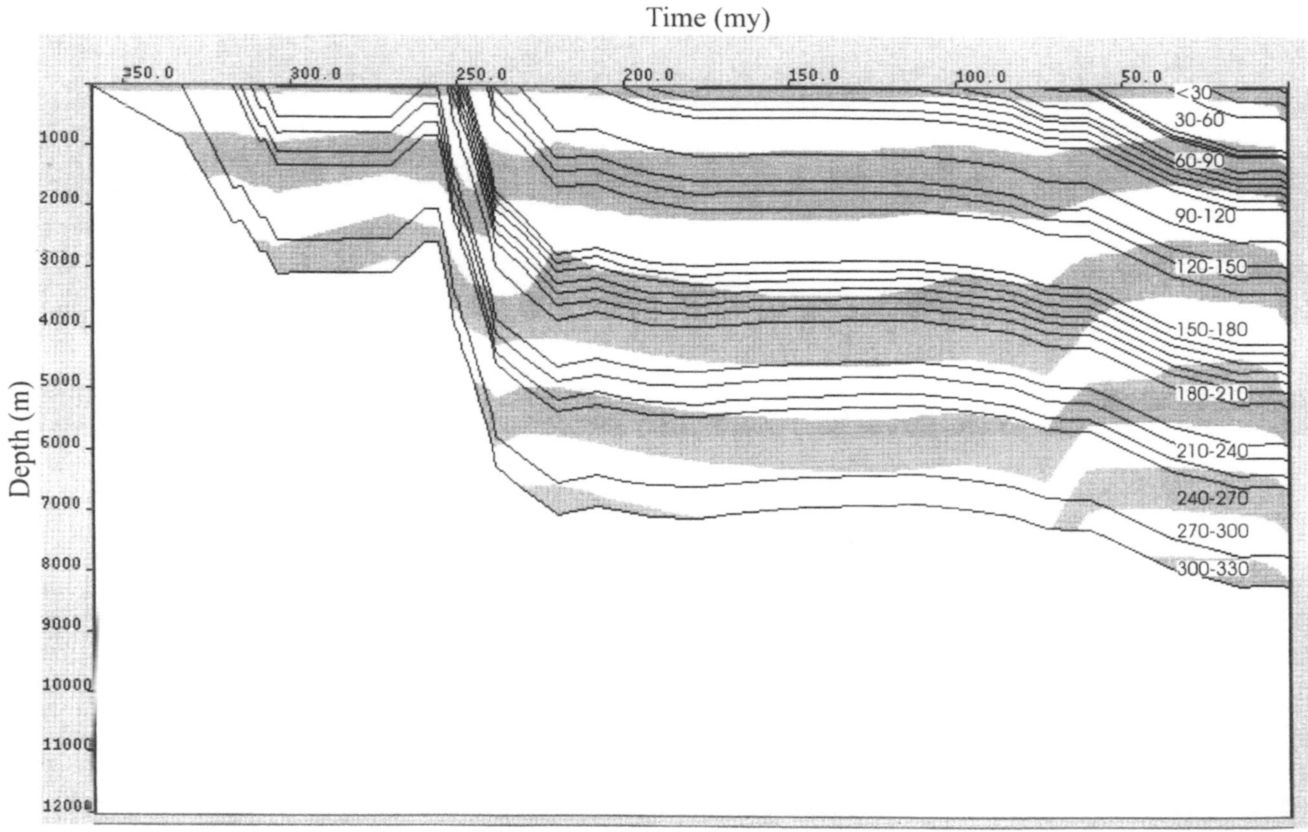

Fig. 2.57. Burial histories for two sites in the northern (**a**) and southern (**b**) part of the transect shown in Fig. 2.56. The relatively high heat flows during the Autunian are due to assumed high heat flows in relation to the volcanic activity at that time

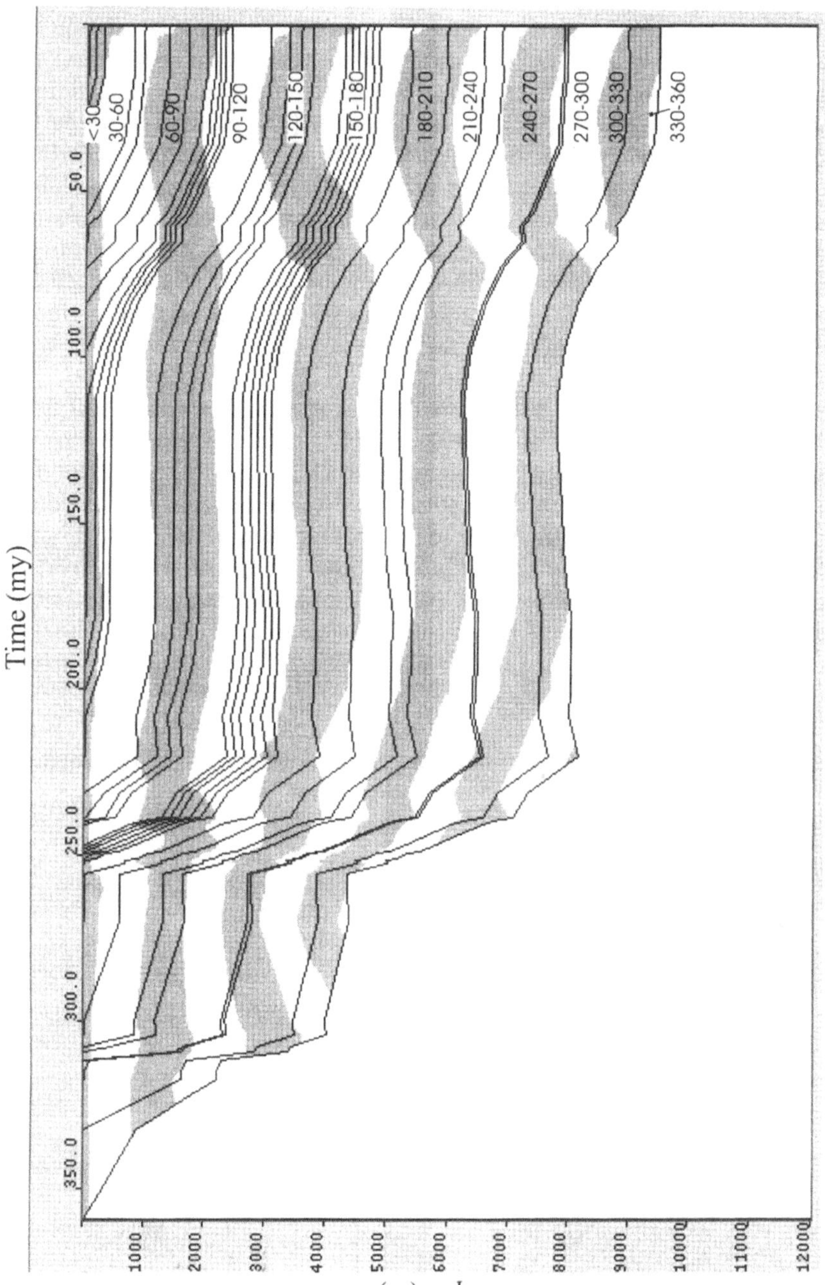
Time (my)
50.0
100.0
150.0
200.0
250.0
300.0
350.0
Depth (m)
1000
2000
3000
4000
5000
6000
7000
8000
9000
10000
11000
12000
<30
30-60
60-90
90-120
120-150
150-180
180-210
210-240
240-270
270-300
300-330
330-360

Fig. 2.57b

For the Mesozoic and Cenozoic times constant heat flows of about 60 mW/m^2 are thought to have persisted, although it can be argued that slightly elevated heat flows may have occurred during some periods. There are, however, no indications for very high heat flows; for example, there are no indications of strong volcanic activity or rifting. Whether or not the North Sea rifting has influenced the study area remains speculative at present. In any case organic maturity parameters bear no imprint of any high temperature events during the Mesozoic or Cenozoic but are a function of the Neogene heat flows and burial depth. The present-day burial is the greatest which the sediments experienced, but the burial depth reached during the Jurassic was not much less (Fig. 2.57). Therefore the Triassic/Jurassic is regarded as the second period of gas generation and migration and the Cenozoic as the third period. More details of the gas accumulation history are discussed in Chapter 7.

Present-day temperature profiles for several wells indicate that heat flows have recently been slightly lower than 60 mW/m^2, and that average geothermal gradients for the upper crust are in the range of 27 °C/km (Fig. 2.58). Not only present-day temperatures but also vitrinite reflectance values are a function of the present depth of burial, as shown in Fig. 2.58. In view of the fact that burial depth during the Jurassic differed little from those at present, this observation is regarded as an additional proof of low or average (but not high!) heat flows during the Mesozoic. With the burial and heat flow history described above, it was possible to obtain good fits between measured and calculated vitrinite reflectance values.

One special aspect of the temperature simulation along the transect is the temperature disturbance below and above salt domes (Fig. 2.56). Generally

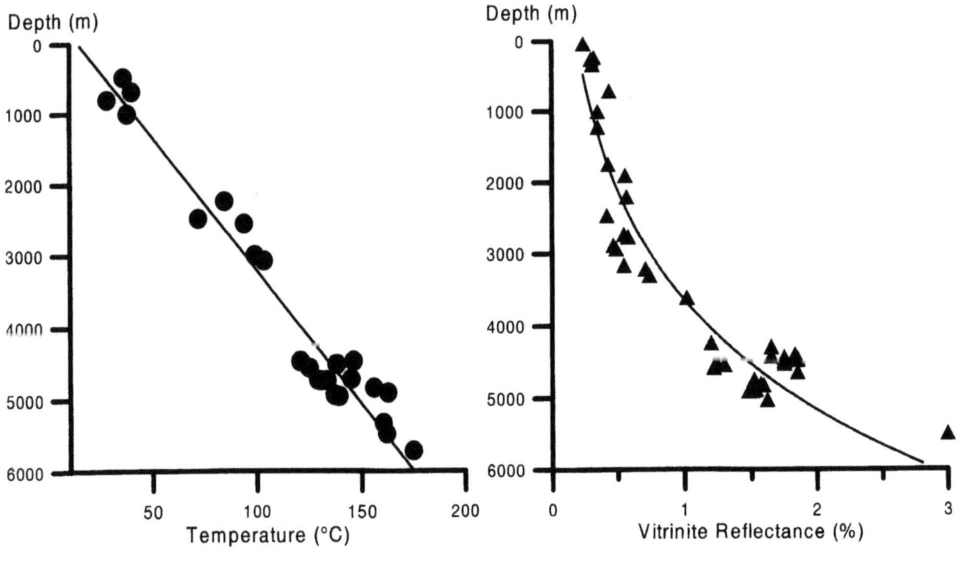

Fig. 2.58. Plots of corrected present-day borehole temperatures (**A**) and vitrinite reflectance values (**B**) versus depth for several wells in the study area. *Symbols*, various wells

temperatures below the salt domes are lower than typical for the respective depth level, whereas those above the salt domes are higher. This phenomenon is due to the high heat conductivity of salt compared to other lithologies. The difference in temperature between areas directly below salt domes and those adjacent to them may exceed 10 °C for the uppermost Carboniferous units according to the simulations (Neunzert et al. 1996). This difference would be large enough to significantly reduce gas generation below the salt domes or, in other words, to preserve a higher gas generation potential in the Carboniferous units below the salt domes. Whether this temperature effect is as large as calculated or smaller due to higher rates of lateral, possibly convective, heat transfer remains to be tested in the future.

2.7
Concluding Remarks

The thermal history of sedimentary basins is a time-dependent energy balance process. In accordance with the physics of the heat transfer phenomena this process is affected by almost every parameter involved in basin evolution. These include: type of geological process such as deposition, nondeposition, erosion, faulting, salt movement, overthrusting; sedimentation rates, original thicknesses, duration, and timing of events; lithotype and properties of deposited units; type and properties of the formation fluids; bathymetry during basin development; paleoclimatic conditions, and heat flow regime. The thermal history of a basin is determined by the combined effects of all these parameters. No single parameter can be identified as the most important one for the entire basin and for the entire time of basin evolution. A particular parameter may be dominant only in a part of the basin area and during a limited time period. Consequently methods which emphasize only a single parameter may lead to erroneous results. Misleading conclusions may result, such as burial always causing a temperature increase and temperature always rising when the heat flow increases. It is even possible that different levels within a sequence can be effected in very different ways by the same boundary conditions. Therefore when reconstructing the temperature history of sedimentary basins, the evolution of the basin as a whole must be considered. For these reasons basin modeling, the numerical simulation of basin evolution which allows an integrated and synergistic, i.e., interactive approach, is probably the most accurate method for a realistic temperature history reconstruction.

References

Allen PA, Allen JR (1990) Basin analysis, principles and applications. Blackwell, Oxford, 451 pp

Andrews-Speed CP, Oxburgh ER, Cooper BA (1984) Temperatures and depth-dependent heat flow in western North Sea. AAPG Bull 68:1764–1781

Aric K, Gutdeutsch R, Klinger G, Lenhardt W (1987) Seismological studies in the eastern Alps. In: Flügel HW, Faupl P (eds) Geodynamics in the Eastern Alps. Deuticke, Wien, pp 325–333

Barnard PC, Cooper BS (1983) A review of geochemical data related to the northwest European gas province. In: Brooks J (ed) Petroleum geochemistry and exploration of Europe. Pergamon Press, Oxford, pp 19–33

Bethke CM (1985) A numerical model of compaction driven groundwater flow and heat transfer and its application to the paleohydrology of intracratonic sediment basins. J Geophys Res 90: 6817–6828

Bjoerlykke K (1993) Fluid flow in sedimentary basins. Sediment Geol 86: 137–158

Bjoerlykke K, Mo A, Palm E (1988) Modelling of thermal convection in sedimentary basins and its relevance to diagenetic reactions. Mar Petrol Geol 5: 338–351

Bodri L, Bodri B (1985) On the correlation between heat flow and crustal thickness, vol 120. Elsevier, Amsterdam, pp 69–81

Buntebarth G, Stegena L (1986) Paleogeothermics: evaluation of geothermal conditions in the geological past. Lecture Notes in Earth Sciences 5. Springer, Berlin Heidelberg New York, 234 pp

Burrus J, Bessis F (1986) Thermal modeling in the Provencal Basin (NW-Mediterranean). In: Burrus J (ed) Thermal modeling in sedimentary basins. Editions Technip, Paris

Burruss RC (1987) Diagenetic paleotemperature from aqueous fluid inclusions. Miner Mag 51: 477–481

Burst JP (1969) Diagenesis of Gulf Coast clayey sediments and its possible relation to petroleum migration. AAPG Bull 53: 73–77

Carslaw HS, Jaeger JC (1959) Conduction of heat in solids, 2nd edn. Oxford University Press, Oxford, 510 pp

Chapman RE (1981) Geology and water. Developments in applied earth sciences, 1. Nijhoff-Junk, The Hague, 228 pp

Clauser C (1984) A climatic correction on temperature gradients using surface-temperature series of various periods. Tectonophysics 103: 33–46

Doligez B, Bessis F, Burrus J, Ungerer P, Chenet PY (1986) Integrated numerical simulation of the sedimentation, heat transfer, hydrocarbon formation and fluid migration in a sedimentary basin: the Themis model. In: Burrus J (ed) Thermal modeling in sedimentary basins. Editions Technip, Paris

Ebner F, Sachsenhofer RF (1995) Paleogeography, subsidence and thermal history of the Neogene Styrian basin (Pannonian basin system, Austria). Tectonophysics 242(1–2): 133–150

Eckert ERG, Drake RM (1987) Analysis of heat and mass transfer. Springer, Berlin Heidelberg New York, 806 pp

Ellis AJ, Mahon WAJ (1977) Chemistry and geothermal systems. Academic Press, New York, 392 pp

Frakes LA (1979) Climates throughout geologic time. Elsevier, Amsterdam, 310 pp

Frakes LA, Probst J-L, Ludwig W (1994) Latitudinal distribution of paleotemperature on land and sea from Early Cretaceous to Middle Miocene. CR Acad Sci Paris 318,II: 1209–1218

Gleadow AJW, Duddy IR, Lovering JF (1983) Fission track analysis: a new tool for the evolution of thermal histories and hydrocarbon potential. APEA J 23: 93–102

Gosnold WD, Fisher DW (1986) Heat flow studies in sedimentary basins. In: Burrus J (ed) Thermal modeling in sedimentary basins. Editions Technip, Paris

Green PF, Duddy IR, Gleadow AJW, Lovering JF (1989) Apatite fission track analysis as a peleotemperature indicator for hydrocarbon exploration. In: Naeser ND, McCulloh TH (eds) Thermal history of sedimentary basins – methods and case histories. Springer, New York, pp 181–195

Gretener PE (1981) Geothermics: using temperature in hydrocarbon exploration. AAPG Education Course Note Series 17

Haack U (1982) Radioactivity of rocks. In: Angenheister G (ed) Physical properties of rocks, vol 1b. Springer, Berlin Heidelberg New York, pp 433–481

Habicht JKA (1979) Paleoclimate, paleomagnetism, and continental drift. AAPG Stud Geol 9: 31

Haenel R (1971) Bestimmungen der terrestrischen Waermestromdichte in Deutschland. Z Geophys 37: 119–134

Harland WB, Armstrong RL, Cox AV, Craig LE, Smith AG, Smith DG (1989) A geological time scale. Cambridge University Press, Cambridge, 263 pp
Hermanrud C (1986) On the importance to the petroleum generation of heating effects from compaction-derived water: an example from the northern North Sea. In: Burrus J (ed) Thermal modeling in sedimentary basins. Editions Technip, Paris
Hermanrud C (1993) Basin modelling techniques - an overview. In: Doré AG, Augustson JH, Hermanrud C, Stewart DJ, Sylta Q (eds) Basin modelling: advances and applications. NPF Spec Publ 3. Elsevier, Amsterdam, pp 1–34
Heynisch S, Yalçın MN, Wygrala BP, Dohmen L, Messner J, Welte DH (1987) Two-dimensional modeling of the profile 56T 8606-205, 1205 in the Northern Central Graben Area, IES GmbH (unpublished)
Hoefs J (1987) Stable isotope geochemistry, 3rd edn. Springer, Berlin Heidelberg New York, 241 pp
Huntsberger TL, Lerche I (1987) Determination of paleo heat-flux from fission scar tracks in apatite. J Petrol Geol 10(4): 365–394
Jensen RP, Doré AG (1993) A recent Norwegian shelf heating event - fact or fantasy. In: Doré AG, Augustson JH, Hermanrud C, Stewart DJ, Sylta Q (eds) Basin modelling: advances and applications. NPF Spec Publ 3. Elsevier, Amsterdam, pp 85–106
Jüntgen H (1964) Reaktionskinetischen Überlegungen zur Deutung von Pyrolyse-Reaktionen. Erdöl Kohle Erdgas Petrochem17: 180–186
Jüntgen H, Klein J (1975) Entstehung von Erdgas aus kohligen Sedimenten. Erdöl Kohle-Erdgas Petrochem 28: 6573
Jüntgen H, van Heek KH (1968) Gas release from coal as a function of rate of heating. Fuel 47: 103–117
Kappelmeyer O, Haenel R (1974) Geothermics with special reference to application. Gebrueder Borntraeger, Berlin, 240 pp
Karweil J (1956) Die Metamorphose der Kohlen vom Standpunkt der physikalischen Chemie. Z Dtsch Geol Ges 107: 132–139
Karweil J (1975) The determination of paleotemperatures from the optical reflectance of coaly particles. In: Alpern B (ed) Petrographie de la matiere organique des sediments. CNRS, Paris, pp 195–203
Kettel D (1981) Maturitätsberechnung für das nordwestdeutsche Oberkarbon - ein Test verschiedener Methoden. Erdöl-Erdgas 97, 11: 395–404
Kingston DR, Dishroon CP, Williams PA (1983a) Global basin classification system. AAPG Bull 67: 2175–2193
Kingston DR, Dishroon CP, Williams PA (1983b) Hydrocarbon plays and global basin classification. Bull AAPG 67: 2194–2198
Kübler B (1967) La cristallinite de l'illite et les zones tout a fait superieures du metamorphisme: etages tectoniques. Colloq Neuchatel, pp 105–122
Kurat G, Palme H, Spettel B, Baddenhausen H, Hofmeister H, Palme C, Wanke H (1980) Geochemistry of ultramafic xenoliths from Kapfenstein, Austria: evidence for a variety of upper mantle processes. Geochim Cosmochim Acta 44: 45–60
Lee WHK (1963) Heat flow data analysis. Rev Geophys 1: 449–479
Leischner K (1994) Kalibration simulierter Temperaturgeschichten von Gesteinen mit organischen Reifeparametern und anorganischen Temperaturindikatoren. Dissertation, University of Bochum, Ber. Forschungszentrum Jülich, 2909: 309pp
Leischner K, Welte DH, Littke R (1993) Fluid inclusions and organic maturity parameters as calibration tools in basin modelling. In: Doré AG, Augustson JH, Hermanrud C, Stewart DJ, Sylta Q (eds) Basin modelling: advances and applications. NPF Spec Publ 3. Elsevier, Amsterdam, pp 161–172
Lerche I (1993) Theoretical aspects of problems in basin modelling. In: Doré AG, Augustson JH, Hermanrud C, Stewart DJ, Sylta Q (eds) Basin modelling: advances and applications. NPF Spec Publ 3. Elsevier, Amsterdam, pp 35–65
Lerche I, Yarzab RF, Kendall CGStC (1984) Determination of paleoheat flux from vitrinite reflectance data. AAPG Bull 68: 1704–1717
Lopatin NV (1971) Temperature and geologic time as factors in coalification. Akad Nauk SSSR Izvestiya, Seriya Geologicheskaya 3: 95–196 (in Russian)

Luheshi MN, Jackson D (1986) Conductive and convective heat transfer in sedimentary basins. In: Burrus J (ed) Thermal modeling in sedimentary basins. Editions Technip, Paris
MacDonald GFJ (1959) Calculations on the thermal history of the earth. J Geophys Res 64(11): 1967–2000
Majorowicz JA, Jones FW, Lam HL, Jessop AM (1984) The variability of heat flow both regional and with depth in southern Alberta, Canada: Effect of groundwater flow?: Tectonophysics 106: 1–29
Majorowicz JA, Jones FW, Jessop AM (1986) Geothermics of the Williston Basin in Canada in relation to hydrodynamics and hydrocarbon occurrences. Geophysics 51(3): 767–779
Mathieu Y (1984) Estimation des conductivites thermiques de quelques roches en fonction de leur enfouissement: Inst Fr Pet Rep 32199: 39
McKenzie DP (1978) Some remarks on the development of sedimentary basins. Earth Planet Sci Lett 40: 25–32
McKenzie DP (1981) The variation of temperature with time and hydrocarbon maturation in sedimentary basins formed by extension. Earth Planet Sci Lett 55: 87–98
Meyerhoff AA (1970) Continental drift: implications of paleomagnetic studies, meteorology, physical oceanography, and climatology. AAPG Bull 78: 1–51
Mo ES, Havik T, Throndsen T, Kjeller P, Andresen P, Backström SA (1989) A dynamic deterministic model of hydrocarbon generation in the Midgard field drainage area offshore Mid-Norway. Geol. Rundsch 78/1: 305–317
Mongelli F, Loddo M, Tramacere A (1982) Thermal conductivity, diffusivity and specific heat variation of some Travale field (Tuscany) rocks versus temperature. Tectonophysics 83: 33–43
Morin R, Silva AJ (1984) The effects of high pressure and high temperature on physical properties of ocean sediments. J Geophys Res 89: 511–526
Neunzert GH, Gaupp R, Littke R (1996) Absenkungs- und Temperaturgeschichte paläozoischer und mesozoischer Formationen im Nordwestdeutschen Becken. Z.Dt. Geol. Ges. 147/2 (in press)
Novelli L, Welte DH, Mattavelli L, Yalçın MN, Cinelli D, Schmitt KJ (1988) Hydrocarbon generation in southern Sicily – a three dimensional computer aided basin modeling study. In: Mattavelli L, Novelli L (eds) Advances in organic geochemistry 1987. Org Geochem 13: 141–151
Okay AI, Şengör AMC, Görür N (1994) Kinematic history of the opening of the Black Sea and its effect on the surrounding regions. Geology 22: 267–270
Oudin JL (1984) Thermal maturation indices in geochemistry. In: Durand B (Ed) Thermal phenomena in sedimentary basins. Technip, Paris, pp 117–125
Oxburgh ER, Turcotte DL (1974) Thermal gradients and regional metamorphism in overthrust terrains with special reference to the eastern Alps. Schweiz Mineral Petrogr Mitt 54: 641–662
Palciauskas VV (1986) Models for thermal conductivity and permeability in normally compacting basins. In: Burrus J (ed) Thermal modeling in sedimentary basins. Editions Technip, Paris, pp 323–336
Parrish JT, Ziegler AM, Scotese CR (1982) Rainfall patterns and the distributions of coals and evaporites in the Mesozoic and Cenozoic. Palaeogeogr Palaeoclimatol Palaeoecol 40: 67–101
Pereira EB, Hamza VM, Furtado VV, Adams JAS (1986) U, Th and K content, heat production and thermal conductivity of Sao Paulo, Brazil, continental shelf sediments: a reconnaissance work. Chem Geol 58: 217–226
Philippi GT (1965) On the depth, time and mechanism of petroleum generation. Geochim Cosmochim Acta 29: 1021–1049
Prezbindowski DR, Tapp JP (1991) Dynamics of fluid inclusion alteration in sedimentary rocks: a review and discussion. Org Geochem 17: 131–142
Pytte AM, Reynolds RC (1989) The thermal transformation of smectite to illite. In: Naeser ND, McCulloh TH (eds) Thermal history of sedimentary basins – methods and case histories. Springer, New York, pp 133–140
Roedder E, Bodnar RJ (1980) Geologic pressure determinations from fluid inclusion studies. Annu Rev Earth Planet Sci 8: 263–301

Rohsenow WM, Hartnett JP (1973) Handbook of heat transfer. McGraw-Hill, New York
Royden L (1986) A simple method for analyzing subsidence and heat flow in extensional basins. In: Burrus J (ed) Thermal modeling in sedimentary basins. Editions Technip, Paris, pp 49–73
Rybach L (1976) Radioactive heat production in rocks and its relation to other petrophysical parameters. Pure Appl Geophys 114: 309–317
Rybach L (1986) Amount and significance of radioactive heat sources in sediments. In: Burrus J (ed) Thermal modeling in sedimentary basins. Editions Technip, Paris, pp 311–322
Rybach L, Cermak V (1982) Radioactive heat generation in rocks. In: Angenheister G (ed) Physical properties of rocks, vol 1b. Springer, Berlin Heidelberg New York, pp 353–371
Sachsenhofer RF (1994) Petroleum generation and migration in the Styrian Basin (Pannonian Basin system, Austria): an integrated organic geochemical and numeric modelling study. Mar Petrol Geol 11: 684–701
Schmucker U (1969) Conductivity anomalies, with special reference to the Andes. In: Runcorn SK (ed) The application of modern physics to the earth and planetary interiors. Wiley-Interscience, London, pp 125–138
Schulz R (1989) Temperaturverteilung in Nordwestdeutschland (Abstract). Nachrichten Dtsch Geolog Ges 41: 72
Schwarzbach M (1974) Das Klima der Vorzeit - eine Einführung in die Paläoklimatologie, 3. Aufl. Enke, Stuttgart
Sclater JG, Christie PAF (1980) Continental stretching; an explanation of the Post-Mid-Cretaceous subsidence of the central North Sea basin. J Geophys Res 85: 3711–3739
Sharp JM Jr, Domenico PA (1976) Energy transport in thick sequences of compacting sediments. Geol Soc Am Bull 87: 390–400
Smith AG, Hurley AM, Briden JC (1981) Phanerozic paleocontinental world maps. Cambridge Univ Press, Cambridge, 162 pp
Smith DG (1982) The Cambridge encylopedia of earth sciences. Cambridge University Press, Cambridge, 496 pp
Smith L, Chapman DS (1983) On the thermal effects of groundwater flow, 1. Regional scale systems. J Geophys Res 88: 593–608
Stallman RW (1963) Computation of ground water velocities from temperature data. In: Methods of collecting and interpreting ground water data. US Geol Surv, Water Supply Pap 1544-H: 36–46
Steininger FF, Bernor RL, Fahlbusch V (1990) European marine/continental chronological correlations.-In: Lindsay EH, Fahlbusch V, Mein P (eds) European Neogene mammal chronology. Plenum, New York, pp 15–46
Sweeney JJ, Burnham AK (1990) Evaluation of a simple model of vitrinite reflectance based on chemical kinetics. AAPG Bull 74: 1559–1570
Teichmüller M, Teichmüller R, Bartenstein H (1984) Inkohlung und Erdgas – eine neue Inkohlungskarte der Karbonoberflaeche in Nordwestdeutschland. Fortschr Geol Rheinl Westfalen 32: 11–34
Tissot B (1969) Premières donnés sur les mécanismes et la cinétique de la formation du pétrole dans les sédiments. Simulation d'un schème reactionnel sur ordinateur. Rev Inst Fr Pétrol 24: 470–501
Tissot B, Espitalié J (1975) L'évolution thermique de la matière organique des sédiments: Applications d'une simulation mathématique. Rev Inst Fr Pétrol 30: 743–777
Tissot B, Welte DH (1984) Petroleum formation and occurrence. Springer, Berlin Heidelberg New York, 699 pp
Tissot BP, Pelet R, Ungerer P (1987) Thermal history of sedimentary basins, maturation indices, and kinetics of oil and gas generation. AAPG Bull 71: 1445–1466
Ungerer P (1984) Models of petroleum formation. How to take into account geology and chemical kinetics. In: Durand B(ed) Thermal phenomena in sedimentary basins. Editions Technip, Paris, pp 235–246
Ungerer P, Burrus J, Doligez B, Chenet PY, Bessis F (1990) Basin evaluation by integrated two-dimensional modeling of heat transfer, fluid flow, hydrocarbon generation, and migration. AAPG Bull 74, 3: 309–335

Vitorello I, Pollack HN (1980) On the variation of continental heat flow with age and the thermal evolution of continents. J Geophys Res 85, B2: 983–995
Waples DW (1980) Time and temperature in petroleum formation: application of Lopatin's method to petroleum exploration. AAPG Bull 64: 916–926
Welte DH (1966) Kohlenwasserstoffgenese in Sedimentgesteinen. Untersuchungen über den thermischen Abbau von Kerogen unter besonderer Berücksichtigung der n-Paraffinbildung. Geol Rundsch 55: 131–144
Welte DH (1989) The changing face of geology and future needs. Geologische Rundschau 78/1: 7–20
Welte DH (1995) The German-Norwegian Geoscientific Cooperation: a first summary report on an integrated study of several northern European basins. (unpublished)
Welte DH, Yalçın MN (1985) Formation and occurrence of petroleum in sedimentary basins as deduced from computer-aided basin modeling. Int Conf on Petroleum geochemistry and exploration in the Afro-Asian region, Dehra Dun, Nov1985. Key Note Pap, pp 1–21
Welte DH, Yalçın MN (1988) Basin modeling – a new method in petroleum geology. In: Advances organic geochemistry 1987. Mattavelli L, Novelli L (eds) Org Geochem 13: 141–152
Welte DH, Yuekler MA (1981) Petroleum origin and accumulation in basin evolution – a quantitative model. AAPG Bull 65: 1387–1396
Welte DH, Yalçın MN, Heynisch S, Schmitt KJ, Wygrala B (1985) Computer-aided basin study in the San Joaquin Basin, California, USA. Final Report, Integrated Exploration Systems (IES), Jülich, FRG (unpublished)
Wernicke B (1985) Uniform-sense normal simple shear of the continental lithosphere. Can J Earth Sci 22: 108–125
Woodbury AD, Smith L (1985) On the thermal effects of three-dimensional groundwater flow. J Geophys Res 90: 759–767
Wygrala BP (1988) Integrated computer-aided basin modeling applied to analysis of hydrocarbon generation history in a northern Italian oil field. In: Mattavelli L, Novelli L (eds) Advances in organic geochemistry 1987. Org Geochem 13: 187–197
Wygrala BP (1989) Integrated study of an oil field in the southern Po Basin, northern Italy. Berichte der Kernforschungsanlage Jülich - No 2313, ISSN 0366–0885, 217 pp
Wygrala BP, Yalçın MN, Dohmen L (1990) Thermal histories and overthrusting – application of numerical simulation technique. Advances in organic geochemistry 1989. Org Geochem 16: 267–285
Yalçın MN (1988) Numerical simulation of the geologic evolution, the thermal history and the hydrocarbon generation potential of the Adana Basin (South Turkey). AAPG Bull 72: 1031–1032 (Abstr)
Yalçın MN (1990) Computer-aided basin modelling in hydrocarbon exploration. 8th Petroleum Congr of Turkey, Proc Geology, Turkish Assoc Petrol Geol, pp 228–239 (in Turkish with English Abstr)
Yalçın MN (1991) Basin modelling and hydrocarbon exploration. J Petrol Sci Eng 5: 379–398
Yalçın MN (1995) Contribution of the Kozlu-K20/G well to the computer-aided modelling studies in the Zonguldak basin. In: Yalçın MN, Gürdal G (eds) Zonguldak hardcoal basin research wells-I: Kozlu-K20/G. Spec Publ of TÜBITAK, MAM, pp 173–196 (in Turkish with English Abstr)
Yalçın MN, Görür N (1984) Sedimentological evolution of the Adana Basin. In: Tekeli O, Göncüoglu MC(eds) Proc Int Symp on the Geology of the Taurus Belt, pp 65–172
Yalçın MN, Welte DH (1988) The thermal evolution of sedimentary basins and significance for hydrocarbon generation. Bull Turkish Petrol Geol, Ankara 1: 12–26
Yalçın MN, Welte DH, Kumar SR, Misra KN, Mandal SK, Balan KC, Mehrotra KL, Lohar BL (1988) Three-dimensional computer-aided basin modeling of Cambay Basin, India. A case history of hydrocarbon generation. In: Kumar RK, Dwivedi P, Banerjie V, Gupta V (eds) Petroleum geochemistry and exploration in the Afro-Asian region. AA Balkema, Rotterdam, pp 417–450
Yalçın MN, Schenk HJ, Schaefer RG (1994) Modelling of gas generation in coals of the Zonguldak basin (northwestern Turkey). Int J Coal Geol 25: 195–212

Yuekler MA, Kokesh F (1984) A review of models in petroleum resource estimation and organic geochemistry. In: Brooks J, Welte DH (eds) Adv Org Geochem 1: 69–113

Yuekler MA, Cornford C, Welte DH (1978) One-dimensional model to simulate geologic, hydrodynamic and thermodynamic development of a sedimentry basin. Geol Rundsch 67: 960–979

Ziegler AM (1987) Paleogeographic atlas project – current activities. Univ of Chicago (unpublished)

Ziegler AM, Hulver ML, Lottes AL Schmachtengerg WF (1984) Uniformitarianism and palaeoclimates: inferences from the distribution of carbonate rocks. In: Berenchley PJ (ed) Fossils and climate. John Wiley & Sons, Chichester, pp 3–25

Zwach C (1995) Diagenesis and temperature history of the Cadotte Sandstone, Alberta Deep Basin, Canada: integration of reservoir quality analysis and basin modeling. Thesis, University of Kiel, Germany. Berichte des Forschungszentrum Jülich, Germany, No 3082, 173 pp

Chapter 3
Maturation and Petroleum Generation

Chapter 3: Overview and Insights

The maturation concept was originally based on a suite of empirical observations documenting thermally induced changes in naturally occurring organic matter. These changes range from an increase in vitrinite reflectance, or progressive colour changes of pollen-grains, when assessing maturation by means of a microscope, to chemical structural changes on a molecular level, when for instance analysing aromatic-type molecules such as phenanthrenes or benzothiophenes. In all these cases it was evident that the observed advances in maturity of organic matter could be linked directly to an increasing thermal stress the sample material had experienced. Parallel and subsequently to this source rock maturation concept the concept of an "oil window" was established.

More detailed and more specific geochemical analyses investigating different molecular compound groups and chemical structural features of rock samples of different maturity and numerous oils supported the oil window concept and the existence of maturation sequences among source rocks and oils alike. Molecular parameters, such as the ratio of *n*-hexane to methylcyclopentane or the methylphenanthrene index, were elaborated to show maturation progress in bitumen or liquid oil. The establishment of these hydrocarbon internal maturity parameters was an important step to directly compare and relate oils of a given maturity level to a source rock of corresponding maturity as indicated by vitrinite reflectance. Such geochemical studies also revealed, that averaging the yields of components, like certain alkylphenanthrenes or alkyldibenzothiophenes, at a given maturity, permits the definition of a C_{15+} hydrocarbon generation profile for a source rock.

The different means to assess maturity of kerogen down to the structural, molecular level are discussed and compared in this chapter. It is important to know that the refined analyses of structural chemical changes on a molecular level finally opened up the means for predictive kinetic considerations with respect to source rock maturation and petroleum generation.

Maturation and Petroleum Generation

M. Radke[1], B. Horsfield[1], R. Littke[1], and J. Rullkötter[2]

3.1 Introduction

It is now firmly established that crude oil and most natural gas, collectively termed petroleum, are generated from kerogen in sedimentary source rocks. The organic origin of crude oil is beyond doubt based on optical activity (Oakwood et al. 1952; Hills and Whitehead 1966) and isotopic composition (Silverman 1964). The chemical structure of biological markers in ancient sediments and crude oils compared to that of living cell constituents (Calvin 1969; Albrecht and Ourisson 1971; Tegelaar et al. 1989a), and regularities in crude oil composition according to sedimentary environments (Tissot and Welte 1984) further confirm an organic origin. However, it was uncertain for a long time at what depth petroleum forms in the earth. The discovery of hydrocarbons in Recent sediments by Smith (1952) gave support to a shallow origin for oil. Baker (1960) and Meinschein (1961) noted that the amount of hydrocarbons in Recent sediments could account for known oil reserves. However, Stevens (1956) found only a few simple aromatic hydrocarbons in Recent sediments as compared to the numerous complex aromatic hydrocarbons in ancient sediments and crude oils. Other authors (Emery and Hoggan 1958; Dunton and Hunt 1962; Hunt 1975) noted the abundance of light hydrocarbons (C_4–C_{13}) in petroleums and their absence in young sediments. It was thus argued that petroleum must form at greater burial depths. Importantly, Bray and Evans (1961) and later workers (Brooks and Smith 1967; Leythaeuser and Welte 1969; Allan and Douglas 1977) observed that hydrocarbon distributions gradually evolved in going from Recent sediments to ancient sediments to crude oils, suggesting that the origin of crude oil is irrefutably tied to the maturation of sedimentary organic matter. Owing to this relationship, maturity assessments of petroleum source rocks and crude oils are to be considered crucial in petroleum exploration studies (for a review, see Brooks 1981).

[1]Institut für Erdöl und Organische Geochemie (ICG-4), Forschungszentrum Jülich GmbH, 52425 Jülich, Germany
[2]Institut für Chemie und Biologie des Meeres (ICBM), Universität Oldenburg, Carl-von-Ossietzky-Str. 9-11, 26111 Oldenburg, Germany

Welte et al. (eds)
Petroleum and Basin Evolution
© Springer-Verlag Berlin Heidelberg 1997

3.2
Maturation: Definition and Driving Force

Maturation is a technical term commonly used in petroleum geochemistry to address thermally induced changes in the nature of organic matter during catagenesis. It may refer to the entire source rock, which is said to gain maturity when heated sufficiently. Maturation summarizes kerogen conversion processes including petroleum generation the "gross" kinetical aspects of which are treated by Schenk et al. (Chap. 4). The driving force of all processes involved is their negative free energy or Gibbs function ($-\Delta G$), which is the difference in free energy between the reactants at the initial (immature) state and the products at the final (mature) state (see textbooks of physical chemistry, e.g., Atkins 1990). The states are dependent on temperature (T), pressure (p), and volume (V). The Gibbs function is defined by Eq. (3.1), where ΔH and ΔS are, respectively, the differences in enthalpy and entropy of the system between these states. This equation refers to an isothermal change at constant pressure:

$$\Delta G = \Delta H - T\Delta S \tag{3.1}$$

A similar equation applies to an isothermal change at constant volume. Since ΔS is always a positive quantity, ΔG must decrease when the temperature is raised at constant pressure. The dependency of ΔG on temperature is given by the Gibbs-Helmholtz equation, which takes the form of Eq. (3.2) when applied to a chemical reaction. As far as the reaction is confined to condensed (liquid and/or solid) phases, the pressure effect on ΔG can generally be neglected:

$$\delta(\Delta G/T)/\delta T = -\Delta H/T^2 \tag{3.2}$$

A heat flow (W) within the system results in a local entropy production (ϑ) per second (see Sommerfeld 1965 and references therein), as defined by the expression (3.3):

$$\vartheta = -(W/T^2)\ \mathrm{grad}\ T \tag{3.3}$$

where grad T is the temperature gradient.

Chemical reactions such as kerogen cracking certainly contribute to the local entropy production in a source rock, but not to the heat flow. On the contrary, the local heat flow is reduced because most reactions taking place in the source rock presumably are endothermic, that is, the difference in enthalpies between reactants and products is positive. Thus, it appears that the term "thermal stress," which is sometimes used in petroleum geochemistry, means the local entropy production rather than the local heat flow.

Temperature was considered by Philippi (1965) and others (Louis and Tissot 1967; Vassoyevich et al. 1969; Price 1983) to be of overriding importance in generating petroleum from organic matter enclosed in source rocks. The effect of temperature is emphasized when speaking of thermal maturation or thermal evolution of kerogens. However, maturation is ultimately controlled by entropy rather than temperature. It is the increase in entropy of the whole system that hinders primary cracking products from recombining with residual kerogen, hence allowing petroleum generation to proceed. The dominant role of entropy is not always recognized although it had been accentuated quite a long time ago, for example, by Emden (1938). In his popular article on heating, this prominent physicist observed: "In the huge manufactory of natural processes, the principle of entropy occupies the position of manager, for it dictates the manner and method of the whole business, whilst the principle of energy merely does the book-keeping, balancing credits and debits."

In statistical thermodynamics, the entropy is defined by the expression (3.4), where k is the Boltzmann constant, Eq. (3.5) and Ω is the thermodynamic probability of the state (see Sommerfeld 1965 and references therein).

$$S = k \ln \Omega \tag{3.4}$$

$$k = R/N_A = 1.38066 \times 10^{-23} JK^{-1} \tag{3.5}$$

where R is the gas constant, and N_A the Avogadro constant.

In contrast to Eq. (3.1), which characterizes the system at a macroscopic scale, Eq. (3.4) refers to a microscopic scale and can consequently be applied to kerogen maturation at a molecular level. During thermal evolution the entropy of the kerogen decreases owing to increasing regularity (decreasing thermodynamic probability) of its structure as described in detail below. The decrease in entropy of the kerogen is overcompensated by an increase in entropy of the products released. Because the molecules attain a more disordered (thermodynamically more probable) state in the petroleum generated, their entropy is higher than that of the precursor entities in the kerogen.

3.3
The Phenomenon of Petroleum Generation

As regards the actual precursors of petroleum, only kerogen decomposition can be considered as being quantitatively significant in forming the bulk of reservoired petroleum (Abelson 1963); of the 6×10^{14} tons of organic matter in the earth's crust 95% is in the form of kerogen (Welte 1970). The decarboxylation of fatty acids and the dehydroxylation and reduction of fatty alcohols also contribute (Cooper and Bray 1963; Kvenvolden 1970; Welte and Waples 1973) but only to a minor degree. Any quantitative or qualitative assessment of source rock potential must therefore consider the amount and type of kerogen that is or was present in the source rock before it generated petroleum.

Generated petroleumlike compounds are thought to result from a multitude of quasi-irreversible, first-order (assumed) thermal cracking reactions (Huck and Karweil 1955; Hanbaba and Jüntgen 1969; Tissot 1969). The progressive loss of smaller molecules from the macromolecular kerogen structure proceeds according to bond strengths with weaker bonds breaking before stronger ones. Production of alkanes requires hydrogen to be transferred from kerogen to intermediates of the cracking process, leaving behind a hydrogen-depleted residue. Kerogen at an optimum stage of liquid hydrocarbon generation is said to be mature.

The early part of catagenesis is dominated by oil generation from kerogen and accumulation within the pore and fracture system of the source rock. Initial extracts are rich in polar components of high molecular weight, whereas those generated later contain higher proportions of both aromatic and particularly saturated hydrocarbons of lower molecular weight (Louis and Tissot 1967; Connan 1974; Allan and Douglas 1974; Powell et al. 1978; Powell 1978). For this reason polar fractions are viewed by some authors as an intermediate in the conversion of kerogen to petroleum (Louis and Tissot 1967; Tissot 1969), although most models of petroleum generation nowadays assume a fixed number of parallel, rather than sequential reactions (see Schenk et al. 1990, Chap. 4). Normal alkanes generated from the kerogen overwhelmingly swamp other saturated hydrocarbons in the same boiling range preserved from early diagenesis.

Generally, maturity of bitumen (C_{15+}-soluble organic matter) is discussed with the understanding that the maturity of the organic matter recovered by solvent extraction of a given rock sample corresponds to that of the associated kerogen. However, "initial oil" not originating from kerogen (Cooles et al. 1986) may obscure the early generation products (Radke and Willsch 1994), hence leading to erroneous conclusions. Furthermore, the thermal evolution of the (primary) C_{15+}-soluble organic matter remaining in place may differ from that of the respective precursor entities in the kerogen (e.g., Requejo et al. 1992; Requejo 1994). Possible redistribution of bitumen among source and reservoir rocks at different stages of their geological history (see Chap. 7), i.e., depletion by primary migration or enrichment by oil impregnation, also complicates the interpretation of bitumen maturation data. Unless otherwise stated, the following discussions refer to samples in which virtually all the C_{15+}-soluble organic matter is autochthonous. That the extent of hydrocarbon depletion is not generally known seems uncritical concerning the molecular maturity parameters discussed below. At least the commonly used aromatic hydrocarbon parameters are unaffected by depletion when petroleum is expelled as a bulk phase (Leythaeuser et al. 1988; Radke and Willsch 1994).

It is difficult to define clearly what oil maturity means. Some petroleum geochemists believe that the term "immature oil" is a contradiction. They replace it by "early mature oil" because, as they say, if oil was released from the source rock, it must be mature. When an oil originates from different source rocks, not all having the same maturity, bulk parameters indicate an average maturity, whereas molecular parameters may lead to conflicting conclusions. For example, based on biomarkers that are present only in the immature portion of the oil, the whole oil is classified as immature, although the major part may be mature or postmature. The same applies to oils derived from the same source rock, but released at different maturation levels.

Studies in western Canada, the Paris Basin, and the Hils syncline area of northwestern Germany show that substantial amounts of light hydrocarbons (C_1–C_7) are produced throughout oil generation (Tissot et al. 1972; Monnier et al. 1983; Schaefer and Littke 1988). This associated gas is considered to be a primary breakdown product of kerogen. At higher levels of catagenesis (generalized as R_r >1.2%) gas concentrations continue to increase, with an increasing preference for methane. Conversely, the yield of C_{15+} components decreases, resulting in a bell-shaped curve with values falling to or below the initial amounts at the base of the mature zone (Albrecht and Ourisson 1969; Vassoyevich et al. 1969; Le Tran et al. 1974; Powell et al. 1978). This curve is commonly used in definition of the "oil window" (Pusey 1973) although the terms "oil generation window" and "liquid window" are also in common usage. The onset of intense oil generation (top of the oil window) is clearly revealed by a drastic increase in C_{15+} hydrocarbon yields, as seen in the Paris Basin (Tissot et al. 1974) and several other sedimentary basins (see Connan 1974). Complete bell-shaped depth profiles for C_{15+} hydrocarbons have been elucidated for all three kerogen types I, II, and III by Tissot et al. (1978), Le Tran et al. (1974) and Albrecht et al. (1976) who studied the Uinta Basin in the United States, the Aquitaine Basin in France, and the Douala Basin in Cameroon, respectively. The inflexion in the C_{15+} hydrocarbon curve was originally explained (but has never been demonstrated explicitly) in terms of oil breaking down to give gas. However, oil cracking accounts only to a limited extent for the decrease in yields of C_{15+} hydrocarbons which are intrinsically rather stable (Mango 1991) and have been detected in sediments of very high maturity (Price et al. 1981). With oil-prone source rocks, the decrease in yields of C_{15+}-soluble organic matter beyond 0.8% R_r is in fact mostly due to petroleum expulsion (Rullkötter et al. 1988). Other phenomena such as facies changes and, in the case of permeable zones, fractionation and displacement phenomena may blur the generation profile and hence complicate determination of the oil window (Albrecht and Ourisson 1969; Larter 1988; Claypool and Mancini 1989).

3.4 Kerogen Maturation

3.4.1 Petrography: Vitrinite, Other Macerals, and Microscopic Approaches

A term comparable to "maturation" used in coal petrography is "coalification," which follows peat diagenesis and leads to an increase in coal rank from subbituminous through bituminous and anthracite to meta-anthracite stages. These later stages are not normally addressed when speaking of maturation as they are beyond the maximum maturity at which oil persists. It is interesting to note that White (1915) drew attention to the fact that the limits of oil occurrence can be inferred from the rank of associated coal beds. Subsequently, the principal phase of oil formation corresponding to the oil window was defined by Vasso(y)evich et al. (1967, 1969) in terms of (Russian) coal ranks, i.e., between D (German *Flammkohle*) and Zh (German *Fettkohle*). Thus coal rank appears to be the first maturity indicator applied in petroleum geochemistry.

Vitrinite reflectance was originally used to measure accurately and rapidly the rank of coals. The observation that vitrinite particles are ubiquitous in sedimentary rocks led to an intensified use of vitrinite reflectance for other purposes. It was increasingly applied (1) as a maturity parameter predicting the stage of oil generation mainly from macerals other than vitrinite and (2) as a calibration tool for numerical simulations of temperature histories in sedimentary basins (Lopatin 1971; Waples 1980).

Organic matter in rich petroleum source rocks is mainly composed of liptinite derived from phytoplankton (alginite; e.g., Hutton et al. 1980; Jankowski and Littke 1986; Littke et al. 1988). Since alginite is preserved only under favorable conditions, its occurrence is restricted mostly to relatively thin source rock intervals. This is one major reason why changes in the optical properties of liptinite, especially of alginite, cannot generally be used quantitatively as maturity indicators in thick stratigraphic columns. Optical maturity parameters other than vitrinite reflectance may nevertheless provide important and reliable qualitative information on maturity levels.

Only organic petrographic methods allow the comparison of optical properties of the same types of identifiable organic particles at different maturation stages, whereas geochemical methods commonly measure properties of the mixture of all types of organic constituents present in a rock.

Organic petrography using incident light microscopy on polished whole rock blocks is derived from coal petrography (Teichmüller 1986). Maceral groups can easily be identified in coals because organic particles are generally large. Clay and other minerals, which have a low reflectivity similar to liptinite, do not interfere with the proper identification of macerals in this instance. In source rocks, however, composed mainly of a mineral matrix, it is difficult to distinguish the three maceral groups by their reflectivity. Thus, a major weakness of organic petrographic maturity evaluation is the subjectivity of identification of the correct maceral group or maceral used for optical measurements.

3.4.2 Maturity-Related Changes of Optical Properties of Macerals

Upon maturation, vitrinites lose volatile products such as water, carbon dioxide, organic acids, and hydrocarbons (van Krevelen 1993; Littke et al. 1989). These chemical changes are accompanied by changes of physical properties. Most importantly, there is an increase in vitrinite reflectance which according to the Fresnel-Beer equation (3.6) is a function of the absorption coefficient (k), the refractive index of the vitrinite particle (n), and the refractive index of the overlying medium (n_o; usually oil with $n_o = 1.518$; see Ting 1981 for more details).

$$R = [(n - n_o)^2 + n^2k^2]/[(n + n_o)^2 + n^2k^2] \quad (3.6)$$

The applicability of vitrinite reflectance as a maturity parameter predicting the stage of oil and gas generation depends on the extent to which the decrease in entropy of vitrinite due to loss of volatile products is correlated with the decrease in entropy of liptinite, which is the major source of liquid hydrocarbons. In nonisothermal heating experiments under ambient pressure this correlation was bad because most petroleum generation from alginite took place at temperatures clearly below those at which vitrinite reflectance started

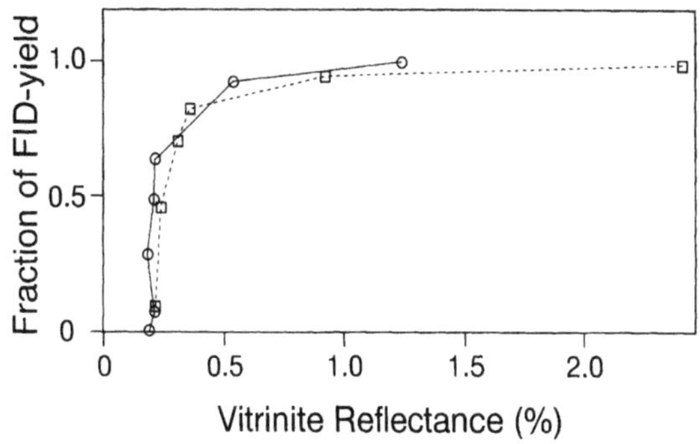

Fig. 3.1. Fraction of total FID yields (hydrocarbons and related substances) during pyrolysis of vitrinite as a function of vitrinite reflectance at heating rates of 0.1 (*solid lines*) and 20 °C/min (*dashed lines*). (Reprinted from Schenk et al., Structural modifications of vitrinite and alginite concentrates during pyrolytic maturation at different heating rates. A combined infrared, ^{13}C NMR and microscopial study. In: Durand B, Béhar F (eds) Advances in organic geochemistry 1989. Pergamon Press, Oxford, 1990, pp 943–950, with kind permission from Elsevier Science Ltd, The Boulevard, Langford Lane, Kidlington OX5 1GB, UK)

to increase (Fig. 3.1; Schenk et al. 1990). In contrast, case studies on alginite-rich source rocks of the Kimmeridge Clay and Posidonia Shale formations suggest that an excellent correlation does exist under geological conditions. Certainly the onset of intense C_{15+} hydrocarbon generation was consistently detected at 0.50%R_r despite variations in geothermal history among the study areas (Radke and Willsch 1994). Thus heating experiments as those mentioned above cannot serve as a model for the maturation of vitrinite in nature. The observation that oil generation from type II kerogen generally starts at 0.5%R_r and ends at 1.3%R_r does not mean that this must be true for all sedimentary basins. High heating rates supposedly result in different reflectance-generation relationships (Yalçın and Welte 1988; Mukhopadhyay 1992). Therefore geological information on burial and temperature history may help interpret vitrinite reflectance data to predict the state of oil generation correctly.

An additional problem is the fact that it is not yet known to what extent factors other than time and temperature, such as host rock facies or pore fluids, affect vitrinite reflectance during catagenesis. In comparison with depth-reflectance trends defined by measurements on coal samples, those determined on vitrinite particles from clastic and carbonate rocks usually show the same general trend but more scatter than in coals (Fig. 3.2). In most cases mean vitrinite reflectance values measured on rocks of different petrographic composition but from the same narrow depth interval do not differ by more than 15% of the value and may be attributed to different botanical vitrinite precursors or early diagenetic processes (Buiskol Taxopeus 1983).

"Suppressions" of vitrinite reflectance in oil shales reported by Hutton and Cook (1980) and other authors were attributed to an incorporation of bitumen from external sources, i.e., from liptinites ubiquitously present in the oil shales. However, as no significant difference in reflectance exists between data measured on extracted and nonextracted (bitumen-impregnated) specimens of the same sample (Wenger and Baker 1987), this explanation seems to be pre-

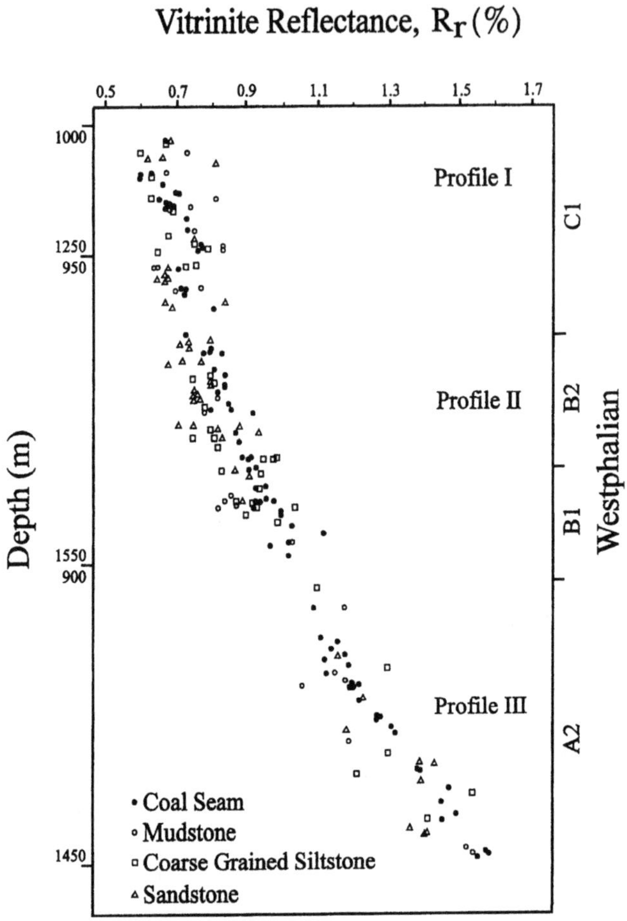

Fig. 3.2. Increase in mean random vitrinite reflectance (%R_r) with depth in three cores of different maturity levels, all drilled in the northern part of the Ruhr area, Germany. *Vertical scale*, depth of the sample in the individual core. The profiles are put together according to their stratigraphic age. (Scheidt and Littke 1989)

mature. What is more important, Schenk et al. (1990) demonstrated that not even the release of the bulk of volatile products from vitrinite upon pyrolysis at 300–350 °C significantly changes reflectivity. The reflectance increase above 350 °C is due mainly to the rearrangement of aromatic units (Fig. 3.1). Accordingly, differences between reflectance values of dispersed vitrinites in clastic rocks are unlikely to depend on bitumen impregnation.

A different usage of reflectance data was initiated by basin history simulations during the last two decades. Lopatin (1971) developed a mathematical formalism describing maturity as an exponential function of temperature and a linear function of time. Considering Lopatin's method, Waples (1980) introduced the time-temperature index as a new maturity parameter. TTI values were originally calibrated against measured vitrinite reflectance and thermal alteration index values of 402 samples from 31 wells. According to this calibration, a doubling of the rate of the chemical reactions involved in thermal maturation with every 10 °C rise in temperature was assumed. Slightly modified equations are still widely used in quantitative maturation simulations. Their potential to adequately model the evolution of vitrinite reflectance has in fact been demonstrated in many cases.

Theoretical considerations confirm, however, that TTI calculations use a wrong temperature dependence in connection with the Arrhenius law. Quigley et al. (1987) observed that TTI values do not provide an accurate measure of maturity. For example, the extent of oil generation is significantly underestimated with geological heating rates exceeding 1 °C/Ma. Considering the release of phenols from vitrinite under laboratory and natural conditions, Larter (1989) was the first to arrive at a sound chemical model of vitrinite reflectance evolution. This and other models (e.g., Sweeney and Burnham 1989) for vitrinite reflectance prediction need to be further tested and improved, however, before the TTI method can definitely be deemed outdated (Leischner et al. 1993).

Reflectance of zooclasts and organic particles other than vitrinite were used in the past to evaluate maturity, especially in pre-Devonian rocks and in other sedimentary rocks in which no vitrinite is present. Reflectance of liptinite macerals, for example, sporinite, does not change significantly at early maturation levels corresponding to vitrinite reflectance values below 0.8% (e.g., Dormans et al. 1957; Littke 1987). Above this level liptinite reflectance increases more rapidly than vitrinite reflectance. The major disadvantage of liptinite reflectivity as a maturity parameter is the difference in optical properties between the major liptinite macerals, i.e., reflectance of resinite, sporinite, alginite, cutinite etc. differ greatly from each other (Teichmüller 1982). In summary, reflectance of liptinite can serve as reliable maturity indicator only where the same type of liptinite occurs in a thick stratigraphic section or where the same type of liptinite occurs in one particular source rock in different areas.

Bertrand (1990) summarized data on the reflectance of zooclasts and found that "chitinozoans and graptolites show similar reflectivity and values slightly lower than vitrinite, while scolecodents show significantly lower values than that of vitrinite." Zooclast reflectance is regarded as a valuable maturity indicator in Palaeozoic rocks.

Solid bitumen (compare pyrobitumen, migrabitumen, natural tar; see Curiale 1986 for chemical information) is a ubiquitous constituent of sedimentary rocks. The potential of its reflectance as a maturity indicator has been discussed by Jacob (1989). Solid bitumen reflectance was claimed to be an alternative to vitrinite reflectance, although the data scatter is generally greater at low levels of maturation (< 1% vitrinite reflectance). Reflectance measurements on solid bitumen are more meaningful with mature to overmature sedimentary rocks where the data display less scatter. These should be interpreted with caution, however, since Bertrand (1990) observed discrepancies in mean reflectance of solid bitumen among various lithologies (sandstone, shale, and limestone). Also, regression lines between solid bitumen and vitrinite reflectivity were variable among these lithologies. Further studies and more data are hence needed before solid bitumen reflectance can be regarded as a reliable maturity parameter.

The observation of progressive color changes of liptinite macerals in transmitted light with increasing maturation led to the early development of carbonization measurements on pollen-grains and spores applied in petroleum exploration (Gutjahr 1966; see Staplin [1977] for historical review). Based on these observations, the thermal alteration index and level of organic metamorphism (Hood et al. 1975) were established as maturity parameters. One major problem of these parameters is that different liptinites and even different types of spores differ in their optical properties; accordingly, "each type of organic matter should have its own scale calibrated to hydrocarbon analysis" (Staplin 1977). Furthermore, grain thickness greatly affects transmission, and description of color was until recently a subjective evaluation rather than a physical measurement. Nevertheless, coloration of palynomorphs and other particles, such as conodonts (e.g., Nöth 1991) and ostracods (e.g., Ainsworth et al. 1990), was successfully used as a rapid, though rough, estimate of thermal maturation in many sedimentary rocks. Translucency measurements on one specific pollen genus (*Carya*; Eocene-Recent) have been related by Lerche and McKenna (1991) to thermal history through first-order time-temperature integrals.

The development of a practicable method providing chemical information by means of spectroscopy on individual macerals, for example, a microprobe for organic particles, is still missing (see Blob et al. 1988 for discussion). The most widely used spectroscopic and microscopic method for characterizing organic particles in rocks is fluorescence spectroscopy (see Teichmüller 1986 for historical review; Lin and Davis 1988). Unfortunately, it reveals almost no chemical information (Pradier et al. 1990). It was noted, however, that most brightly fluorescing organic particles are hydrogen-rich (oil-prone) immature liptinites, whereas most weakly fluorescing particles are hydrogen-lean macerals. At increasing maturation stages, fluorescence of liptinites becomes weaker and is shifted toward longer wavelengths.

From fluorescence spectra a number of parameters were deduced as a measure of maturity. In this context the most widely used parameters are the wavelength of maximum fluorescence intensity (λ_{max}), the spectral red/green quotient (Q), the fluorescence intensity (I), and the alteration of green fluorescence (546 nm) during irradiation. More complex and sophisticated measurements (parameters) were also presented by Hagemann and Hollerbach (1981), and Michelsen and Khavari Khorasani (1990). Whereas liptinite fluorescence undisputedly provides a rapid, though rough, estimate of maturity and can be used in almost all (except overmature) sedimentary sequences, its applicability as a quantitative maturity parameter is severely restricted by the great difference in fluorescence properties of different types of liptinite.

3.4.3
Model for Kerogen Maturation: Evolution of Physical Structure

Detailed X-ray diffraction studies on various macerals have revealed cognate structures. However, differences do exist in regularity, size, and the nature of the packing of the layers. At medium rank hydroaromatic rings are relatively more abundant in exinites than in vitrinites. With increasing rank the extent of fully aromatized areas ("clusters") mainly consisting of flat six-membered rings was shown to increase. The flat arrangement of structural units remained imperfect where incorporation of twisted five-membered rings resulted in buckled layers (Cartz and Hirsch 1960).

Increasing regularity in the orientation of aromatic structural units with increasing thermal evolution has been observed by high-resolution transmission electron microscopy (Oberlin et al. 1974). In an investigation of Precambrian kerogens this technique proved superior to X-ray measurements since structural ordering was discovered for samples where no or little indications of crystallinity were obtained by powder X-ray diffraction methods (Buseck et al. 1988). Graphitization of anthracite occurs in nature at temperatures of 300–500 °C, whereas in the laboratory at ambient pressure onset of graphitization was observed only above 1000 °C. Temperatures as high as 2000–3000 °C were necessary to convert significant amounts of anthracite into graphite. Creep experiments under constant high confining pressure (500 MPa), deviatoric stress, variable temperature (300–600 °C), and strain (to 33%) have demonstrated a decrease with increasing strain energy in temperatures required for conversion (Ross and Bustin 1990; Ross et al. 1991). Experimental conditions in this case corresponded more closely to those in nature where graphitized anthracites commonly occur in areas of comparable temperatures and strain energies. These experiments underline the importance that stress may have in the development of structural order in kerogen. It certainly loses increasing quantities of entropy with mounting tectonic stress.

3.4.4
Changes in Chemical and Carbon Isotope Composition

The compositional significance of changes in the chemical structure of coal with increasing rank due to dehydration, decarboxylation, and demethanation reactions has been demonstrated by a C–H–O ternary diagram rather than by the binary H/C vs. O/C van Krevelen diagram by Stephens (1979). The ternary diagram was also used when relating compositional changes of kerogens to the release of carbon dioxide, water, methane, and oil (Ujiié 1978).

The natural and artificial thermal evolution of kerogens when studied by transmission infrared spectrometry is similar to that seen on the van Krevelen diagram. One observes a decrease in relative abundance of carbonyl groups and carbon-hydrogen single bonds corresponding to a decrease in the O/C and H/C atomic ratios, respectively (Robin 1975; Rouxhet and Robin 1978; Rouxhet et al. 1980; Ganz and Kalkreuth 1987). Because of its selectivity regarding bond types, e.g., aliphatic C–H, aromatic C–H, C=O, transmission infrared spectrometry more clearly reveals changes in the chemical structure of kerogen, such as depletion in aliphatic chains, increase in aromatic content, and elimination of carbonyl groups.

Solid-state ^{13}C nuclear magnetic resonance spectrometry has been used to specify structural changes of terrestrial organic matter during diagenesis and coalification. Samples taken sequentially through the cross section of a fossilized gymnosperm log with a diameter of 3 m and through a modern species of the same family have provided information about the combined processes of peatification and coalification (Bates and Hatcher 1989). The degradation of cellulosic materials and demethylation of lignin appeared to be early diagenetic processes, as indicated by decreasing carbohydrate and methoxyl contents when going from the center to the periphery of the log. Whereas this investigation demonstrated processes likely to occur during diagenesis and independently of coalification, another study revealed coalification processes directly. Specifically, conventional and dipolar dephasing CP/MAS ^{13}C nuclear magnetic resonance experiments on numerous coals and coal macerals ranging in maturity from lignite to anthracite rank demonstrated an overall increase in hydrogen aromaticity as the carbon aromaticity increased (Fig. 3.3; Wilson et al. 1984). With increasing degree of coalification the aromatic rings obviously are defunctionalized and hydrogen replaces the functional groups faster than cross-linking reactions occur. However, not all the functional groups removed

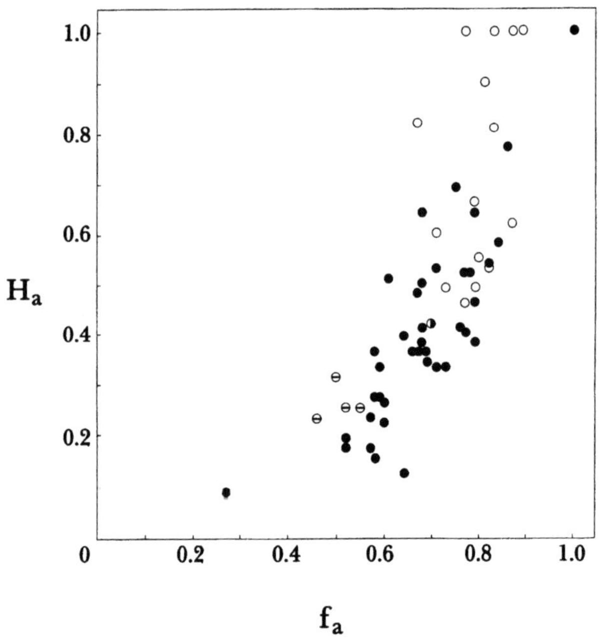

Fig. 3.3. Relationship between the fraction of hydrogen that is aromatic (H_a) and carbon aromaticity (f_a) in coals and coal macerals. *Filled circles*, whole coals, vitrinites or vitrains; *half-filled circle*, vitrinite + inertinite; *open circles*, inertinites or durains; *circles with horizontal line*, exinites. (Wilson et al. 1984)

are replaced by hydrogen. A growing number of radical sites at the kerogen matrix, resulting from homolytic cleavage of chemical bonds during petroleum generation persist, as evidenced by electron spin resonance (ESR). ESR spectroscopy detects free radical concentrations (spin densities) in kerogen isolates. Spin density was claimed to be a useful palaeothermometer (Pusey 1973), but the significance is questionable since a discrepancy in maturation trends between kerogen types II and III and a reversal around 2%R_r was observed (Durand et al. 1977a). Due to the effect of kerogen structure on formation rates and stabilities of free radicals successive attempts to correlate ESR parameters with kerogen maturity were also not fully successful (Shibaoka and Steven 1977; Morishima and Matsubayashi 1978; Marchand and Conrad 1980; Bakr et al. 1988). More recently Bakr et al. (1990) have drawn attention to the influence on ESR measurements of organic molecules that remain trapped within the kerogen matrix upon treatment with moderately polar organic solvents commonly used in source rock extraction. These problems are overcome when trapped molecules are removed by treatment of the kerogen with pyridine. As the increase in spin density with thermal evolution of kerogen is related to breakage of chemical bonds during the release of smaller molecules from the kerogen matrix, ESR measurements have proved useful when simulating petroleum generation in pyrolysis experiments. For example, deviations in oil generation thresholds among different kerogens were inferred from the temperatures at which spin densities increased rather abruptly (Aizenshtat et al. 1986; Bakr et al. 1991).

Due to preferential breakage of ^{12}C–^{12}C bonds, oil and gas should be isotopically lighter than the source kerogen, which hence is expected to become isotopically heavier during catagenesis. However, only in a few cases were variations in the carbon isotope composition of kerogen clearly related to maturity. For example, kerogens of the Milligan shale became isotopically heavier with approach to the eastern margin of the Idaho batholith, where the $\delta^{13}C$ value of the sample nearest to the contact was 20‰ higher than that of samples not affected by the batholith (Baker and Claypool 1970). Within certain uniform kerogen series, $\delta^{13}C$ values seemed to increase by up to 1‰ during catagenesis, but changes in carbon isotope composition were insignificant with most kerogens investigated (Redding et al. 1980; Clayton 1991). As long as oil is the major product being released from kerogen, the isotopic effect is small and obscured by variations in the isotopic composition of the starting materials. Only in the overmature stage when substantial quantities of methane are formed in a single-step reaction from kerogen fractionation becomes significant and residual kerogen becomes isotopically heavier (Redding et al. 1980).

3.4.5 Pyrolysis Characterization

Pyrolysis has been defined as "a chemical degradation reaction that is induced by thermal energy alone" (Ericsson and Lattimer 1988). As reviewed by Horsfield (1984), the technique is well-suited to analyzing macromolecular sedimentary organic matter fractions, with both simple "bulk-flow" (e.g.,

Rock-Eval) and sophisticated analytical systems being employed. At their simplest level, pyrolysates can be divided into bulk components such as hydrocarbons and related compounds (S_2) which are measured by flame ionization detection, and carbon dioxide (S_3) which is measured by thermal conductivity detection (Espitalié et al. 1977; also see Barker 1974; Gransch and Eisma 1970; Le Tran and van der Weide 1969). S_2 yield, and its carbon-normalized parameter, the hydrogen index, is considered a measure of petroleum generation potential. Changes in these quantities as a result of maturation have been well documented.

Type I kerogens give the highest hydrogen indices at any given level of maturity. Alginite occurring in coals, generally thought synonymous with type I kerogen, yields higher proportions of high molecular weight "oil-like" compounds than do other kerogen types (Larter et al. 1977). Type II kerogens are compositionally equivalent to liptinite coal macerals, and have a lower hydrogen index than does type I kerogen. So-called "marine" organic matter, sporinite, cutinite and resinite fall in this category. Type III kerogens, equivalent to vitrinite in coals and dispersed ubiquitously in clastic sediments, have the lowest inherent petroleum-generation yields and the highest proportions of short-chain, mainly gas-forming, precursors. The gas from type II kerogen is "wetter" than that from type III kerogens, that is it contains higher proportions of C_2–C_4 components relative to methane (Horsfield and Douglas 1980). This finding is corroborated by field data (e.g., Powell 1978).

The absolute amount of petroleum that can be generated by a given source rock is a function of its hydrogen index value when immature versus that when mature, its original organic carbon content, thickness and lateral extent (Cooles et al. 1986; Larter 1985; Goff 1984), all of which are determined by the depositional setting of the source rock. Yields can be calculated in one of two ways. One is to use an algebraic mass balance scheme, exploiting routine geochemical measurements for mature source rocks and their immature equivalents (Cooles et al. 1986). The other is to apply a kinetic model that has been calibrated from laboratory pyrolysis measurements.

The fundamental assumption in the algebraic model is that kerogen consists of a reactive and inert part, that the reactive part is thought to be a true measure of the potential yield of petroleum per unit carbon for a given source rock, and that the inert carbon fraction measured in the laboratory (C_R of Gransch and Eisma 1970) is proportionately representative of inert carbon in nature. Petroleum masses are calculated by measuring inert carbon in the mature sample, assuming the ratio of reactive versus inert carbon then figuring the yield of petroleum by normalization to inert carbon. As pointed out by Dembicki and Pirkel (1985), it may be valid to assume that a rock unit, 10-ft-thick and containing 10% organic carbon, *generates* the same as one that is 1000-ft-thick and containing 0.1% organic carbon (same facies, lateral extent, maturation history) but this does not mean that they *expel* the same amount. The dispersed nature of the hydrocarbons in the thicker interval would make migration difficult. Clearly, generation models alone cannot infer the amount of petroleum leaving the source rock and entering the carrier system; both generation and migration must be considered together in this regard (see Chap. 7).

Kinetic models are the second approach for determining degrees of petroleum generation and are dealt with in detail in Chapter 6. Briefly, commonly used kinetic models determine the extent of petroleum generation using pyrolysis data from immature samples (Ungerer and Pelet 1987; Burnham et al. 1988). The progressive loss of kerogen substituents both here and in nature is assumed to proceed according to bond strengths. A good example of this is afforded by the case of sulfur-rich organic matter (types I-S and II-S) whose weak carbon-sulfur bonds break before stronger carbon-carbon bonds. Therefore, these kerogens are more thermally labile than their low-sulfur type II counterparts (Tissot et al. 1987; Baskin and Peters 1992; di Primio and Horsfield 1995). This means that under uniform programmed pyrolysis conditions and for most given kerogen types (type I kerogen is an exception) there is a progressive increase in T_{max} (the temperature at which the maximum of the S_2 pyrolysis peak occurs) as a function of increasing maturity because the more labile components on the front of the S_2 peak are stripped away first (Girling 1963; Barker 1974; Espitalié et al. 1977). This is manifested by higher mean values of activation energy (Schaefer et al. 1990). At very high levels of maturity (R_r >ca. 2.0%), a rapid increase in T_{max} may occur (Peters et al. 1980) that coincides with a reversal in ESR spin density (Retkofsky et al. 1968), and a decrease in the intensity of aromatic C-H infrared spectral vibrations (Rouxhet and Robin 1978) associated with aggregation of aromatic stacks.

When examined in more detail, it is clear that kerogens yield exceedingly complex mixtures when pyrolysed, and that abundances can be related to maturation. Numerous compound classes have been found including hydrocarbons, ketones, alcohols, nitriles and thiols, as represented by cyclic and acyclic, saturated and unsaturated carbon skeletons (Rovere et al. 1983; Wilson et al. 1983). Pyrolysis-gas chromatography using flame ionization detection, affords a rapid and convenient means of studying some of the most common and major pyrolysate components, such as the one- and two-ring aromatic hydrocarbons, certain phenols, straight-chained paraffins, straight-chained olefins, and isoprenoid hydrocarbons. Pyrolysis-gas chromatography-mass spectrometry has been used to gain insight into the minor components, such as biomarkers. The relative and absolute abundances of pyrolysate components have been used to infer biochemical contributions, degree and type of diagenetic modification, and level of catagenesis in numerous study areas (e.g., Romováček and Kubát 1968; Giraud 1970; Chaffee et al. 1983; Nip et al. 1986; Senftle et al. 1986; Curry and Simpler 1988). Importantly, major resolved species in high-temperature pyrolysates appear to be proportionally representative of structural moieties in the kerogen as a whole rather than of atypical part-structures, at least as far as the maturity range R_r = 0.27–1.04% is concerned. This is the case despite their absolute yields being low. On the premise that kerogen composition directly controls the types and yields of volatile products generated during natural maturation, it can be concluded that the abundances and distributions of resolved pyrolysis products give clues as to the bulk compositions of natural petroleums, such as paraffinicity and aromaticity. Chapter 6 (this Vol.) provides a detailed overview of pyrolysate components and their origins. Some points relating to maturation are repeated here in an abbreviated form.

Aromatic hydrocarbons and phenolic compounds predominate the pyrolysis products of humic vitrinitic coals and type III kerogens deposited in coal-forming environments. Benzene, toluene, C_2- and C_3-alkylbenzenes, C_1- and C_2-alkylnaphthalenes, phenol, cresols, and xylenols are major components (Holden and Robb 1958; Girling 1963; van de Meent et al. 1980; Allan and Larter 1983). Their presence may in some cases indicate the presence of preserved biopolymeric benzenoid structures (lignins, tannins). Coals of increasing rank yield pyrolysates that are progressively enriched in total and low molecular weight aromatic compounds (Romováček and Kubát 1968; McHugh et al. 1976; Larter and Douglas 1980). Contemporaneously, the simple phenols decrease in abundance with increasing rank (van Graas et al. 1980; Senftle et al. 1986), a feature that has been used to build kinetic models for vitrinite reflectance prediction (Larter 1989). Moieties yielding aromatic hydrocarbons and phenols on pyrolysis are less abundant in hydrogen-rich kerogen types, though yields can be variable. For instance, type I kerogen from the Green River Shale of the Uinta Basin and alginites in boghead coals generate very little resolved aromatic pyrolysate whereas this is more pronounced for type I kerogen from the Green River Shale (Laney Member) of the Washakie Basin (Horsfield et al. 1994; Horsfield 1989 and references therein). Type II kerogen from the Toarcian Shales (Germany and France), Kimmeridge Clay (UK) and the Bakken Shale (USA) generate relatively high abundances of resolved aromatic compounds at low maturity. Yields actually decrease with progressive maturation, resulting in a relative increase in *n*-alkenes and *n*-alkanes in the C_{10+} fraction (Horsfield and Düppenbecker 1991; Muscio and Horsfield 1996; van Graas et al. 1981; Solli et al. 1985). Similar findings apply to the closed system pyrolysates of Kap Stewart Formation and Posidonia Shale kerogens (Horsfield et al. 1989; Düppenbecker and Horsfield 1990). These findings appear to be at odds with the statements made above regarding the representativity of pyrolysates for determining kerogen structure because increasing maturity is usually associated with increasing aromaticity. Further investigation is warranted.

High sulfur kerogens from clay-poor depositional settings yield abundant sulfur compounds on pyrolysis whereas kerogens deposited in freshwater or in the presence of excess reactive iron yield only small quantities (Eglinton et al. 1990). The major resolved sulfur compounds dominating the C_{6+} fraction in kerogen pyrolysates are the alkylthiophenes and alkylated benzothiophenes (Sinninghe Damsté et al. 1989). Sulfur-containing products from type I kerogens are dominated by the 2-alkylthiophenes and those from type II-S kerogens by the 2,5-dialkylthiophenes whereas those from coals and type III kerogens consist predominantly of branched isomers such as 2,4- and 3,4-dialkylthiophenes (Eglinton et al. 1992). With increasing maturity the ratio of 2-methylthiophene/2,5-dimethylthiophene decreases, as revealed by the microscale sealed vessel pyrolysis of Posidonia Shale kerogen (transformation ratios > 40%). Changes in the relative abundances of 2,3-dimethylthiophene, 2-ethyl-

5-methylthiophene and 2,3,5-trimethylthiophene have also been documented for the natural maturation (R_r = 0.48–1.45%) of these sediments (Muscio et al. 1991). Maturation sequences from the Monterey Formation, Kimmeridge Clay and Mahakam Delta, supplemented by simulated maturation results on corresponding immature samples, showed decreasing thiophene ratios (2,3-dimethylthiophene/[*o*-xylene + *n*-non-1-ene]) and increasing proportions of branched versus linear isomers (Eglinton et al. 1990). An increase in the alkylbenzothiophene/alkylthiophene ratio was also reported, although quantitatively the formation of benzothiophenes is minor compared to the overall fall in sulfur content.

The principal acyclic isoprenoid hydrocarbon identified in high-temperature kerogen pyrolysates is prist-1-ene (Larter et al. 1979). This may be transformed to prist-2-ene when clays are present (Regtop et al. 1986). Isoprenoid structures yielding prist-1-ene on pyrolysis appear to be more labile than are *n*-alkenes and *n*-alkanes, but the disappearance of isoprenoid structures varies from case to case (R_r = 0.8–1.1%) as reported by Øygard et al. (1988), Curry and Simpler (1988) and Larter et al. (1983). The decrease is correlated with the appearance of pristane in associated rock extracts with increasing maturation (van Graas et al. 1981). The Pristane Formation Index (pristane/[pristane+pristenes]) has been formulated as a maturity parameter (Goossens et al. 1988). Although prist-1-ene is thought likely to originate from kerogen-bound tocopherols (Goossens et al. 1984), its kinetics of degradation requires that it is bound in more than one form in the kerogen (Burnham 1989).

The average chain length of *n*-alkanes in coals and siliciclastic sediments of increasing maturity decreases, ostensibly due to the instability of the higher homologues (Bray and Evans 1961; Philippi 1965; Leythaeuser and Welte 1969). The precursors of these normal alkanes can be cracked from the kerogen structure mainly as *n*-alk-1-enes and *n*-alkanes during laboratory pyrolysis. With the exception of humic coals, resinites (Snowdon 1980; Horsfield et al. 1983; van Aarssen et al. 1991), and some unusual alginites (Horsfield et al. 1992), *n*-alkenes and *n*-alkanes represent the visually dominant species in almost all pyrograms. Odd carbon number-dominated *n*-alkane and even carbon number dominated *n*-alkene distributions in the C_{22+} range of type II kerogens with land plant admixture (van de Meent et al. 1980) and in the C_9–C_{18} region of *Gloeocapsomorpha* alginite (Reed et al. 1986; Douglas et al. 1991) suggest that *n*-alkyl pyrolysate distributions are related to parent structures and not random chain scission. Aliphatic biopolymers in plant cuticular membranes, now thought to be a major precursor of aliphatic structures in kerogens (Tegelaar et al. 1989b), yield *n*-alkanes with a pronounced odd carbon number predominance even when artificially matured to high levels of thermal stress. Thus, waxy crude oils containing *n*-alkanes with an odd carbon number predominance are not necessarily generated at early levels of maturation (Tegelaar et al. 1989a) as classical geochemistry once inferred (see Bray and Evans 1961). Interestingly, long-chain hydrocarbons, some exhibiting strong periodicity, are also rather predominant components of the analytical pyrolysates of some overmature kerogens (Dungworth and Schwartz 1972; McKirdy et al. 1980; Jackson et al. 1984; Horsfield 1989). These observations suggest that certain biologically derived compositional characteristics such as chain length distribution might sometimes be discernible from the analysis of overmature sediments, allowing original oil potential to be inferred. Equally remarkable is the essentially uniform C_{2+} *n*-alkyl chain length distribution displayed by a boghead coal, sporinite, Talang Akar coal, and the Ordovician Glenwood Shale on closed system pyrolysis despite there being a 15-fold difference in *n*-alkyl yield (Horsfield et al. 1989; Horsfield 1990). This points to the presence of precursors with a statistically fixed distribution of chain lengths. By way of corroboration, the *n*-alkyl chain length distribution of analytical pyrolysates from High Volatile Bituminous rank vitrinites, sporinites and whole coals (R_r = 0.5–1.2%) do not show a progressive depletion in longer-chain homologues with increasing maturity (McHugh et al. 1976; Larter and Douglas 1980).

Extrapolating these findings to natural geological settings, it can be concluded that many kerogens, especially algal kerogens, generate petroleums with a fixed C_{2+} *n*-alkane bulk composition throughout catagenesis, and that progressive shortening of average *n*-alkyl chain length, concomitant with a change from inherently oil-prone to inherently gas-prone character, does not always occur during the maturation of kerogen. Examples of chain-shortening are afforded by the early maturation history of the Green River Shale and Posidonia Shale kerogens, and for the Kap Stewart Formation (Greenland) as revealed by simulation experiments (Horsfield et al. 1989; Horsfield and Düppenbecker 1991).

3.5 Bitumen and Petroleum: Geochemical Maturation

3.5.1 Maturation Changes in Bulk Properties and Gross Composition

Changes with increasing maturity in bulk properties of petroleum and in the gross composition of C_{15+}-soluble organic matter that have proved useful tools for thermal maturation studies include a decrease in wetness of natural gas, an increase in API gravity (3.7) of crude oil, and an increase in the relative abundance of C_{15+} hydrocarbons in total extract:

$$\mathrm{API} = [141.5/(\text{specific gravity } 60/60\ ^\circ\mathrm{F})] - 131.5 \tag{3.7}$$

The percentage of wet gas $[(C_2 - C_4)/(C_1 - C_4)\times 100]$ in a large number of canned cuttings samples from single lithostratigraphic units of Canadian Arctic Basins generally was extremely variable with depth (Monnier et al. 1983). Although the data were at best semiquantitative, increasing generation of liquid hydrocarbons from kerogen with transition from immature to marginally mature zones and from marginally mature to mature zones was indicated by the proportion of wet gas surpassing 30% and 60%, respectively. Overmature zones were identified by a decline in the proportion of wet gas from above 30% to below 30% as liquid hydrocarbons were cracked to gas (Snowdon and Roy 1975; Powell 1978).

The API gravity (3.7) is a bulk parameter that allows the maturity of crude oils to be assessed in the most convenient way. For example, variations in API gravity of 13°–45° (specific gravity = 0.98–0.80 g/ml) in an extensive series of Gulf of Suez oils were attributed to maturities ranging from immature to mature (Rohrback 1983). The API gravity showed positive correlation with the percentage of hydrocarbons < C_{15} and the saturate-to-aromatic ratio, whereas correlation with sulfur content was negative. A positive correlation of API gravity with the saturate-to-aromatic ratio has been demonstrated also for a suite of North Sea oils with API gravities mostly in the range of 25°–45° (specific gravity = 0.90–0.80 g/ml). In addition, API gravity was shown to be controlled by the asphaltene content. As a rule of thumb, 15% asphaltene content in a normal crude oil reduces API gravity by 10° (Cornford et al. 1983).

With siliciclastic rocks, the relative abundance of C_{15+} hydrocarbons in total extract generally increases rather regularly from 30% at early thermal evolution to 60% at the peak of the oil generation curve. Thus, it may be used as a maturity indicator within this maturity interval, although the large data scatter allows for only a rough assessment (Foscolos et al. 1976; Powell et al. 1978; Héroux et al. 1979). Despite the limited precision, a tentative calibration of relative C_{15+} hydrocarbon abundance in total dichloromethane extracts against $\%R_r$ proved useful for differentiating oil-impregnated from nonimpregnated shales (Radke and Welte 1983). This was an important point in the development of hydrocarbon-internal maturity parameters because correlation with $\%R_r$ was meaningful only with nonimpregnated samples.

3.5.2
Maturation Changes in Molecular Distributions of Hydrocarbons

Changes with maturity in the composition of petroleum or C_{15+}-soluble organic matter can be defined more specifically on the molecular level. A number of concentration ratios of saturated and aromatic hydrocarbons, dibenzothiophenes, and porphyrins commonly used as maturity indicators have been introduced mainly on an empirical basis. As far as genetic relationships among the key compounds were postulated, evidence for a direct conversion of a given precursor into the related product has rarely been presented. Generally, isomer distributions seem to be controlled by preferential thermal degradation of the less stable isomers rather than by isomerization reactions, such as epimerization of isoprenoid chiral centers or shift of methyl groups on aromatic rings.

Maturity indices based on the molecular composition of light hydrocarbons rely on an assumed greater stability of *n*-alkanes as compared to the isomeric alkylcycloalkanes. For example, the ratio of *n*-hexane to methylcyclopentane was shown to be sensitive to maturity changes, as evidenced by correlation with conventional parameters, such as $\%R_r$ (Jonathan et al. 1975). Using a large data base Thompson (1979) was able to develop paraffinicity indices (e.g., "heptane value") which have found wide application in organic maturation studies. Schaefer and Littke (1988) determined light hydrocarbon distributions in a great number of Lower-Toarcian shales from the Hils syncline area, northwestern Germany. They found a relationship between maturity indicators based on C_7 paraffin/naphthene ratios and $\%R_r$ in the range of 0.5–1.5%. Also, ratios of certain positional or stereoisomers, such as the ratio of 2-methyl- to 3-methylhexane or of *trans*- to *cis*-1,2-dimethylcyclopentane, were shown to depend on maturity of the organic matter. This latter ratio that was mentioned first by Hunt (1975) has proved a useful tool for maturity assessment of strata bearing type III kerogen and coal in the Elmworth gas field, Western Canada Basin (Schaefer et al. 1984). The gradual decrease in this ratio from 7.5 at $0.7\%R_r$ to 1.0 at $1.3\%R_r$ appeared to be, at least partly, thermodynamically controlled, as indicated by a similar decrease with temperature of the equilibrium constant for the 1,2-dimethylcyclopentane isomerization (Fig. 3.4). Calculation of the equilibrium constants suffered from uncertainties of the available thermochemical data. Standard enthalpies and entropies of formation referred to the ideal gas state and were obtained by an incremental method rather than determined experimentally.

In the earlier days of organic geochemistry attempts to assess the maturity of source rocks and crude oils concentrated on C_{15+} saturated hydrocarbons. Those fractions are generally representative of a larger proportion of the C_{15+}-soluble organic matter and are more amenable to analysis by gas chromatography than aromatics or NSO compounds. The carbon preference index which is a measure of the predominance of odd-numbered over even-num-

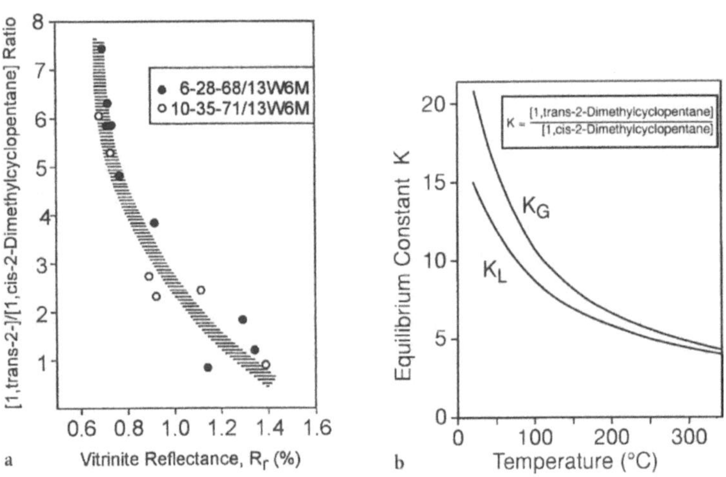

Fig. 3.4. **a** Variation of the 1,*trans*-2-dimethylcyclopentane/1,*cis*-2-dimethylcyclopentane ratio with mean random vitrinite reflectance for rock samples from two boreholes in the Elmworth area, western Canada. **b** Equilibrium constant for 1,2-dimethylcyclopentane isomerization in the gas phase (K_G) and in the liquid phase (K_L), as a function of temperature. (Reprinted from Schaefer et al., Geochemistry of low molecular weight hydrocarbons in two exploration wells of the Elmworth gas field (Western Canada Basin). In: Schenck PA, de Leeuw JW, Lijmbach GWM (eds) Advances in organic geochemistry 1983. Pergamon Press, Oxford, 1984, pp 695–701, with kind permission from Elsevier Science Ltd, The Boulevard, Langford Lane, Kidlington OX5 1GB, UK)

Fig. 3.5. Relationship between carbon preference index and mean random vitrinite reflectance for coals of different origins. (Durand et al. 1977b)

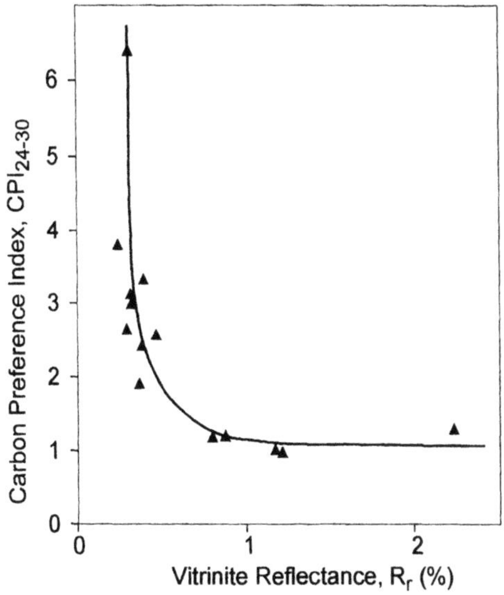

bered long-chain *n*-alkanes commonly was applied in organic maturation studies (Bray and Evans 1961, 1965; Philippi 1975). For coal samples with 0.2–1.7%R_r, the decrease in the carbon preference index of *n*-alkanes (carbon numbers > 23) from 6 to 1 was most pronounced below 0.9%R_r (Fig. 3.5; Durand et al. 1977b; Allan and Douglas 1977; Radke et al. 1980).

Pristane in immature Messel Shale comprises solely the 6(S),10(R) isomer that has the same configuration as phytol in zooplankton (Patience et al. 1978). Stereospecifity is lost with increasing maturity, as indicated by a decrease in relative abundance of the 6(S),10(R) isomer to 80% in Green River Shale (Maxwell et al. 1972). Mackenzie et al. (1980) showed that pristane isomer ratios changed gradually with increasing depth in a suite of Toarcian shales from Paris Basin until an equimolar mixture of the biogenic 6(S),10(R) configuration and the newly formed 6(S),10(S) and 6(R),10(R) isomers was reached at about 1000 m depth; a similar isomerization trend was apparent for phytane in the same sample series. A gradual decrease in the relative abundance of 6(S),10(R) pristane from 65% to 50% (measured on extractable alkane/alkene fractions) was observed in rocks of the Permian Kupferschiefer formation with approach to the Krefeld High in northwestern Germany. This decrease corresponds to enhanced maturity due to an intrusive body (Püttmann and Eckardt 1989).

The observed increase in the ratio of 20S/(20S + 20R) 5α,14α,17α–24-ethylsteranes in sediments with increasing thermal stress has been proposed to involve the reversible isomerization of the biologically inherited 20R configuration to the 20S configuration occurring only in geological systems, until an equilibrium is reached (Fig. 3.6b; Mackenzie et al. 1980). The same applies to the change in the ratio of 5α,14α,17α (20R + 20S) to 5α,14β,17β (20R + 20S) steranes (Fig. 3.6c; Mackenzie et al. 1980). The precursor/product relationship of these apparent geochemical reactions for a long time was considered to be

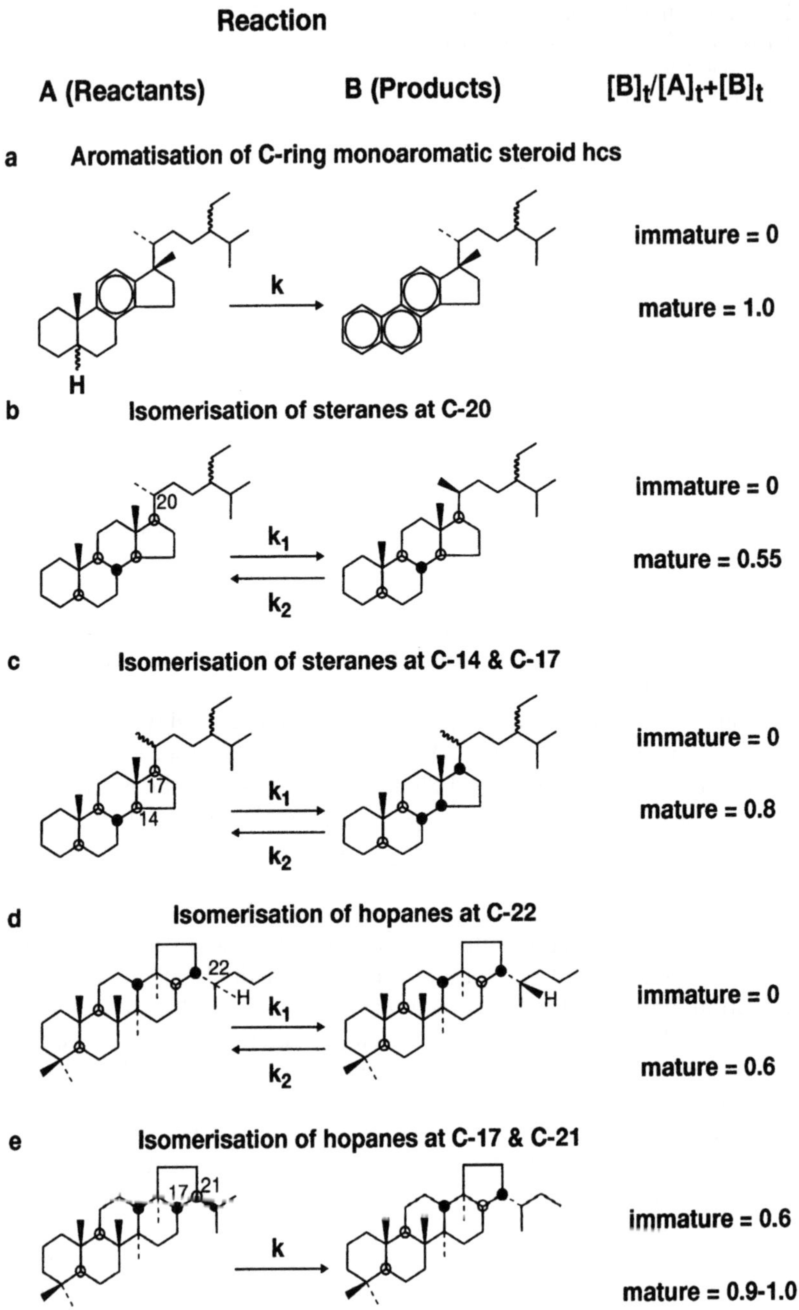

Fig. 3.6a–e. Biological marker reactions in sediments with apparent precursor/product relationships together with starting and end ("equilibrium") values for the corresponding biological marker compound ratios. (Mackenzie et al. 1985b)

sufficiently well established for the isomer ratios to be used as thermal maturation indicators (see Mackenzie and Maxwell 1981, Mackenzie et al. 1982a, Mackenzie 1984 for overviews) and empirical application in petroleum geochemistry was successful in many cases (e.g., Seifert and Moldowan 1981; Mackenzie et al. 1983; Hong et al. 1986; Li et al. 1987). However, recent hydrous pyrolysis experiments using oil shale samples showed that there may not be such a simple precursor/product relationship for sterane isomers, as indicated in Fig. 3.6b,c (Lewan et al. 1986; Rullkötter and Marzi 1989; Peters et al. 1990; Marzi and Rullkötter 1992).

As demonstrated schematically by the progress of sterane isomer ratios with increasing thermal stress in Fig. 3.7a, a reversal to lower values of the 20S/(20S + 20R) sterane isomer ratio of around 0.4 occurs at higher maturity levels (Rullkötter and Marzi 1989; Marzi and Rullkötter 1992). In contrast to the expected and theoretically predicted equilibrium value of 0.54 (van Graas et al. 1982), the recursion agrees with low values detected in natural sediment sequences (e.g., Mackenzie et al. 1985b; Snowdon et al. 1987). In the past these

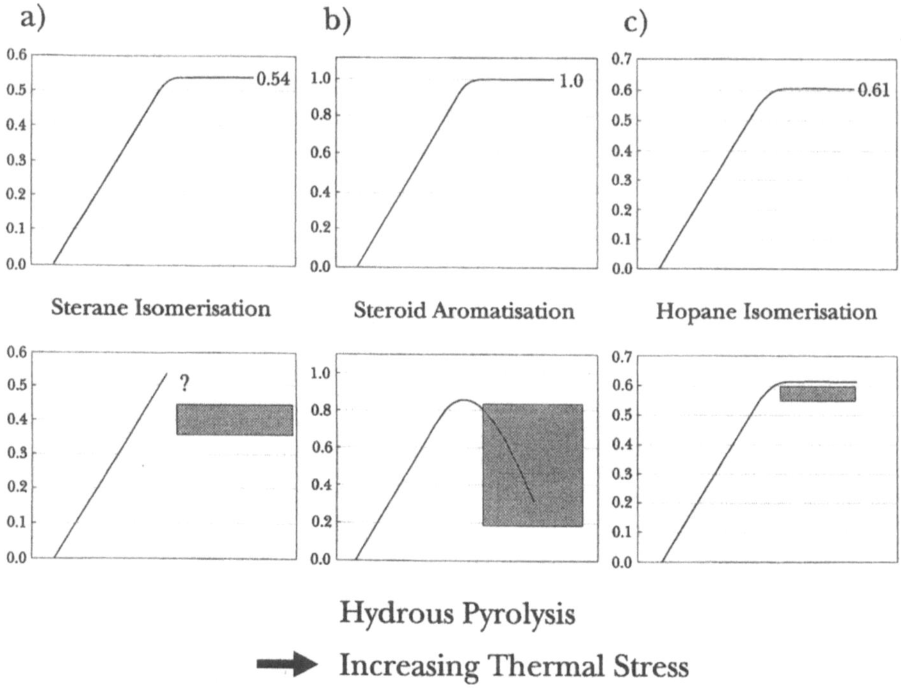

Fig. 3.7a–c. Predicted (*above*) and observed (hydrous pyrolysis; *below*) extent of biological marker reactions with the x-axis showing the thermal stress in arbitrary units. **a** Isomerization of C_{29} steranes at C-20. **b** Transformation of C-ring monoaromatic into ABC ring triaromatic steroid hydrocarbons. **c** Isomerization of 17α-homohopanes at C-22. (Marzi and Rullkötter 1992)

were often interpreted in terms of contamination with less mature material. Laboratory experiments with pure sterane standards failed to bring about sterane isomerization and this added to the doubts about simple product/precursor relationships raised by hydrous pyrolysis results and natural observations (Peakman and Maxwell 1988; Abbott et al. 1990). The experimental trend shown in Fig. 3.7a should therefore be interpreted in a way that the initial increase in the 20S/(20S + 20R) sterane isomer ratio is due to the fact that newly formed steranes (from functionalized precursors in the bitumen fraction and from kerogen) are a mixture of 20S and 20R isomers which add to the 20R steranes formed earlier during diagenesis without affecting the chiral center at C-20. At higher levels of thermal stress, sterane formation competes with sterane destruction. In this later process, the 20S isomer is apparently destroyed more rapidly than the 20R isomer. This happens at a stage during catagenesis when the absolute concentrations of biological markers in sediments decrease dramatically (Rullkötter et al. 1984). As a consequence, the sterane isomer ratio operates as a reliable maturation parameter only in the ascending portion of the trend shown in Fig. 3.7a.

Considering this modified view of sterane isomerization, Marzi (1989) attempted to reassess the kinetics of this apparent geochemical reaction. He used his hydrous pyrolysis results in combination with those natural data of Mackenzie and McKenzie (1983) from the North Sea and the Pannonian Basin which he could relate to the ascending part of the isomer ratio trend and determined new values for activation energy and frequency factor (Fig. 3.8). Although these new values are numerically very much different from those of

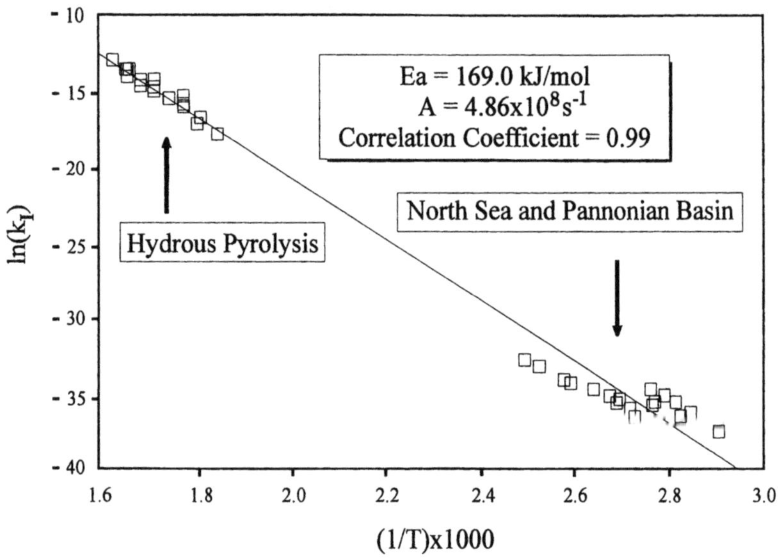

Fig. 3.8. Arrhenius diagram for determination of kinetic data of sterane isomerization based on hydrous pyrolysis experiments and selected natural samples from the North Sea and the Pannonian Basin (Mackenzie and McKenzie 1983). (Marzi and Rullkötter 1992)

Mackenzie and McKenzie (1983), a sensitivity analysis showed that both sets of data react in a similar way when applied to the geothermal histories of sedimentary basins (Marzi et al. 1990). This is in contrast to kinetic data for apparent sterane isomerization determined by Suzuki (1984) and Sajgó and Lefler (1986) which do not seem to be geologically meaningful (Marzi et al. 1990). Also, the kinetic data of Abbott et al. (1990) based on the more sophisticated view of new formation and destruction of steranes cannot be applied to geological systems but seem to be valid only for the laboratory experiments performed (Marzi 1992; Abbott et al. 1992). Despite the fact that sterane isomerization at C-20 is apparently not a geochemical reaction with a straightforward precursor/product relationship, empirical application of the (pseudo) kinetic data for sterane isomerization has led to novel conclusions with respect to the geothermal histories of sedimentary basins (e.g., Mackenzie and McKenzie 1983; Marzi and Rullkötter 1992; Rullkötter et al. 1992). In the case of the Michigan Basin (Fig. 3.9) both measured and calculated sterane isomerization values showed that there has been higher geothermal heat flow in the past than assumed from geophysical modeling (Marzi and Rullkötter 1992).

Sterane isomer ratios obviously cannot be applied as maturation indicators in carbonates, siliceous or phosphatic sediments and evaporites, or crude oils

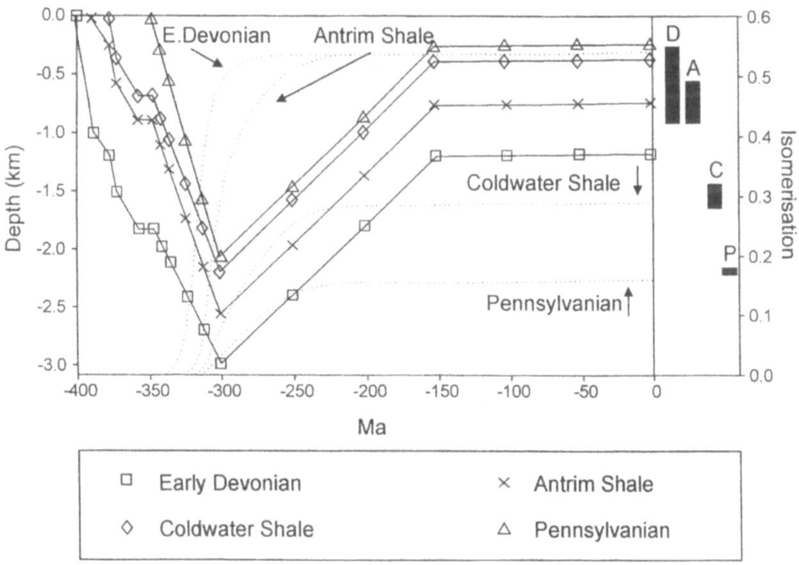

Fig. 3.9. Late Palaeozoic subsidence of the center of the Michigan Basin (*solid lines*) and calculated evolution of sterane isomerization (*dotted lines*) based on a modified geothermal scenario (higher heat flow and deeper subsidence during the Carboniferous) relative to that based on geothermal modeling. Ranges of measurements (*bars*) on natural samples are shown for comparison. *D*, Early Devonian; *A*, Antrim Shale; *C*, Coldwater Shale; *P*, Pennsylvanian coal. In contrast to that put forward by Nunn et al. (1984), oil generation in the Michigan Basin occurred early in Devonian and older strata and was even possible in late Devonian sediments. (Marzi and Rullkötter 1992)

derived from these sources. As an example, oils and asphalts from the Dead Sea area were found to have uniform sterane isomer ratios despite large differences in thermal maturation as evident from other maturity parameters such as steroid aromatization (Rullkötter et al. 1985). Ten Haven et al. (1986) found that specific steroid precursor molecules with a double bond at C-7, biosynthesized by organisms thriving in (hypersaline) environments where deposition of carbonates and evaporites occurs, are responsible for early isomerization at C-14, C-17, and C-20 in the steroid molecules. A facies dependence of sterane isomerization at C-14 and C-17 in the early catagenesis stage was also found in Lower Toarcian carbonate-rich shales and marls and adjacent mudstones from the Hils syncline in northern Germany (Fig. 3.10; Rullkötter and Marzi 1988). In the same sample series, sterane isomerization at C-20 showed no facies dependence (Fig. 3.10).

Ubiquitous hopanes of bacterial origin in sediments and crude oils are another class of biological marker hydrocarbons that undergo systematic changes as a function of thermal stress in sediments. The hopane isomer ratios in general are more dependent on the source of organic matter than sterane isomer ratios, and several of the apparent hopane isomerization reactions (Fig. 3.6d,e) are also complete before oil generation really starts (Seifert and Moldowan 1980). Moretanes (17β,21α-hopanes) are common constituents of several land plant species and thus may be directly incorporated into sediment. This complicates their use as maturity indicators in the case where the organofacies of the sample series investigated varies. For a series of coals, ten Haven et al. (1989) have studied moretane/17α-hopane transformations in some detail. A slight facies dependence of the 17α-hopane/moretane ratio is obvious for the Liassic sample series from northern Germany.

Isomerization in the side-chain of extended 17α-hopanes (C_{31}–C_{35}; Fig. 3.6d) is complete at the onset of catagenesis (Mackenzie et al. 1980; Mackenzie 1984). In hydrous pyrolysis experiments on Lower Toarcian shales from northern Germany the 22S/(22S + 22R) ratios of extended hopanes unlike the steroid hydrocarbons do not show a reversal in their trend at high thermal stress (Fig. 3.7c; Marzi and Rullkötter 1992). In a similar experiment Peters et al. (1990) found a reversal in the trend of 22S/(22S + 22R) ratios of 17α-homohopanes at high temperatures (> 340 °C) for the phosphatic member of the Monterey Formation but not for the siliceous member. They could not exclude problems in determining this ratio precisely with the low amounts of hopanes remaining at these high temperatures. There is also no facies dependence of this ratio in the core sequence of Lower Toarcian and adjacent rocks from northern Germany (Fig. 3.10).

In contrast to this, the ratio of 22,29,30-*trinor*-neohopane (Ts) to 22,29,30-*trinor*–17α-hopane (Tm) shows a strong facies dependence (Fig. 3.10). This hopane parameter is known to be both source and maturity dependent (Seifert and Moldowan 1978), with the neohopane component being the more stable isomer. Again, the use of this parameter is limited to sequences with uniform organofacies characteristics.

Steroid hydrocarbons with ring A or B aromatized are formed during early diagenesis (Hussler et al. 1981; Hussler and Albrecht 1983) but are relatively

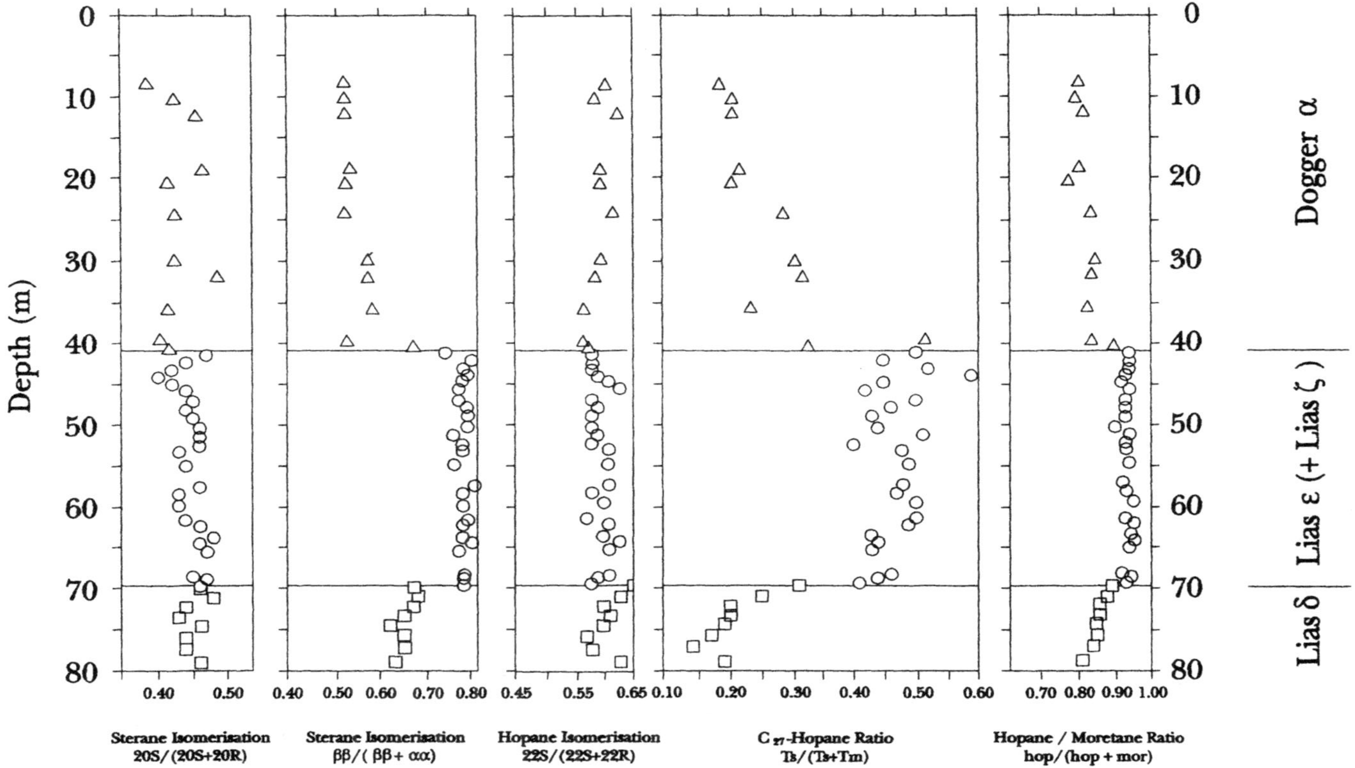

Fig. 3.10. Biological marker compound ratios as a function of depth in borehole samples of Lower Toarcian carbonate-rich shales and marls and adjacent mudstones from the Hils syncline, northern Germany. Sterane isomerization at C-14 and C-17 [ββ/(ββ+αα)], C_{27} hopane ratios and 17α-hopane/moretane ratios show distinct facies dependence, whereas sterane isomerization at C-20 and isomerization of 17α-homohopanes at C-22 do not (Reprinted from Rullkötter J, Marzi R., Natural and artificial maturation of biological markers in a Toarcian shale from northern Germany. In: Mattavelli L, Novelli L (eds) Advances in organic geochemistry 1987. Pergamon Press, Oxford, 1988, pp 639–645, with kind permission from Elsevier Science Ltd, The Boulevard, Langford Lane, Kidlington OX5 1GB, UK)

unstable and disappear again when diagenesis proceeds further (Rullkötter and Welte 1983). They are followed by C-ring monoaromatic steroid hydrocarbons (Riolo and Albrecht 1985; Riolo et al. 1985; Moldowan and Fago 1986) at a later diagenetic stage. They persist well into the catagenesis zone where they are gradually transformed into ABC ring triaromatic steroid hydrocarbons (Mackenzie et al. 1981; Ludwig et al. 1981). This apparent transformation (Fig. 3.6a) was introduced as a maturation indicator based on a tri-/(tri- + mono-) aromatic steroid hydrocarbon ratio by Mackenzie et al. (1981). It was later modified by Moldowan and Fago (1986) due to the identification of new series of (rearranged) monoaromatic steroid hydrocarbons and enhanced gas chromatographic resolution.

The tri-/(tri- + mono-)aromatic steroid hydrocarbon ratio, as the sterane isomer ratio, shows a reversal in its trend with increasing thermal stress both in hydrous pyrolysis experiments (Fig. 3.7b; Rullkötter and Marzi 1989; Peters et al. 1990) and in natural sample series (Mackenzie et al. 1985b). This indicates that at elevated temperatures the preferential thermal destruction of the more labile triaromatic species becomes more important than the conversion of mono- into triaromatic steroid hydrocarbons. It should be noted, however, that at this level the absolute concentration of aromatic steroid hydrocarbons has decreased by about three orders of magnitude compared to the early catagenesis stage (Rullkötter et al. 1984; Marzi and Rullkötter 1992).

The aromatic steroid hydrocarbon maturation parameter seems to be more widely applicable than the sterane isomer ratios, i.e., steroid aromatization is less dependent on organic facies but reacts virtually only to thermal stress. This has been demonstrated in the maturity assessment of carbonate-derived asphalts and crude oils from the Dead Sea area, Israel (Rullkötter et al. 1985), from a series of 50 crude oils from various sources (Riolo et al. 1986), and a series of sulfur-lean and sulfur-rich crude oils from the Monterey Formation in California (Rullkötter and Orr 1989). In the latter case maturity assessment using sterane isomer ratios failed for sulfur-rich oils but was applicable to sulfur-lean Californian oils, whereas the steroid aromatization parameter was able to discriminate within the entire set of oils according to maturity differences (Fig. 3.11).

Besides mono- and triaromatic steroid hydrocarbons with intact carbon skeletons (mainly C_8–C_{10} side-chains) there are the corresponding lower pseudohomologs with short side-chains (C_2–C_3) in sediments and crude oils. Although this has never been proven, the idea is that the short side-chain species derive from their higher pseudohomologs by thermal cleavage of the side-chain (Mackenzie et al. 1981). Empirically the ratio of short to long side-chain aromatic steroids was found to be a maturation parameter particularly in the higher maturity range (Seifert and Moldowan 1978; Mackenzie et al. 1981, 1985b; Shi et al. 1982; Riolo et al. 1986). Recently, Lichtfouse et al. (1990) found a new series of 3- and 4-methyl triaromatic steroids (C_{21}–C_{23}) in marine shales from the Paris Basin. The concentration of 3-methyl triaromatic steroids systematically increased with maturity relative to the concentration of the corresponding 4-methyl isomers. Based on the assumption that this is due to a thermally induced methyl group shift from C-4 to C-3 in ring A of the

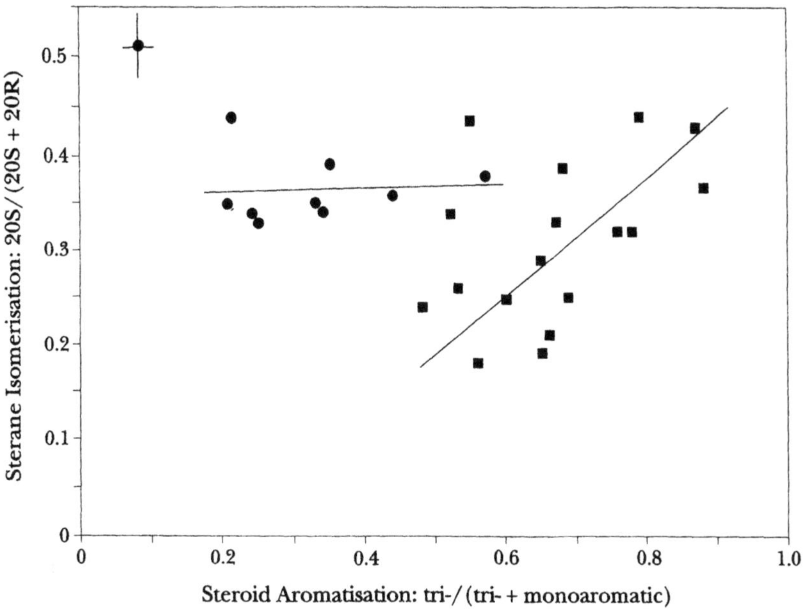

Fig. 3.11. Cross plot of sterane isomerization and steroid aromatization for crude oils from the Monterey Formation. Sulfur-rich oils (*circles*) exhibit uniform sterane isomerization values whereas sulfur-lean oils (*squares*) can be differentiated by both parameters. (Rullkötter and Orr 1989)

triaromatic steroids, Lichtfouse et al. (1990) proposed the isomer ratio as a maturation parameter in the higher maturity range. A detailed discussion of various maturity parameters based on aromatic steroid hydrocarbons, also including stereoisomer ratios, has been published by Riolo et al. (1986).

Systematic changes in isomer distributions of methylated aromatic hydrocarbons and sulfur heterocycles in the soluble organic matter of sedimentary rocks and in crude oils are increasingly applied for assessment of thermal maturation levels (Radke 1987). Generally the abundance of relatively unstable isomers tends to decrease at the higher levels. In some instances (phenanthrenes, chrysenes, perylenes, dibenzothiophenes) the degree of alkylation of the nonalkylated (parent) compound seems to increase up to the level of maximum hydrocarbon generation at 0.9%R_r, but the evidence for such trends is less compelling.

A decrease in abundance of the less stable α-type isomers, 1-methylnaphthalene (1-MN) and 1-ethylnaphthalene (1-EN), relative to the more stable β-type isomers (2-MN and 2-EN) with increasing rank was observed for extractable aromatics of bituminous coals between 0.8 and 1.0%R_r (MNR, ENR: Radke et al. 1982b). Likewise, the abundance of the less stable α,α-type dimethylnaphthalenes (1,5-DMN or 1,8-DMN) relative to the more stable β,β- and α,β-type isomers was used as an indicator of increasing maturity of crude oils and the organic extracts of sedimentary rocks including coals with 0.7–1.5%R_r (DNR: Radke et al. 1982b 1984; Alexander et al. 1984, 1985). This approach has been extended to ratios of α,β,β-type and β,β,β-type trimethylnaphthalenes, but the dynamic range of the derived parameters (TNR 1: Alexander et al. 1985; TNR 2: Radke et al. 1986) apparently is narrower than that of DNR and ends around 1.0%R_r (Weiß 1985; Radke et al. 1994).

On the other hand, owing to boiling points higher than those of MNs and DMNs, TMNs generally are affected to a minor extent by evaporation losses during migration and accumulation of an oil (volatility segregation) or during subsequent sampling, storage, and analysis. When volatiles are lost, β-type isomers, due to the greater volatility, are removed preferentially, thus decreasing MNR and DNR. Where erratic evaporation losses account for a large scatter within a narrow depth interval of MNR and DNR, maturity assessment should rely on maximum rather than mean values.

When compared to TMNs, monomethyl- and dimethylphenanthrenes are even less volatile and hence not normally affected by evaporation losses. Case histories have demonstrated that aromatic hydrocarbon parameters based on phenanthrenes, such as the methylphenanthrene index (MPI 1), were often equivalent and sometimes superior to %R_r (Radke et al. 1982a; Radke 1988; Farrington et al. 1988). For example, in Cretaceous rocks of the Deep Basin, western Canada, distributions of extractable phenanthrenes were related to %R_r between 0.65 and 1.35% (Fig. 3.12; Radke et al. 1982a; Welte et al. 1984; Weiß 1985). Since the empirical relationship was supported by data from other basins, a vitrinite reflectance equivalent (R_c) based on the MPI 1 has been introduced (Radke and Welte 1983). With type III organic matter, R_c seemed to be valid only beyond the onset of intense C_{15+} hydrocarbon generation commonly assumed at 0.65%R_r. A greater dynamic range including immature to

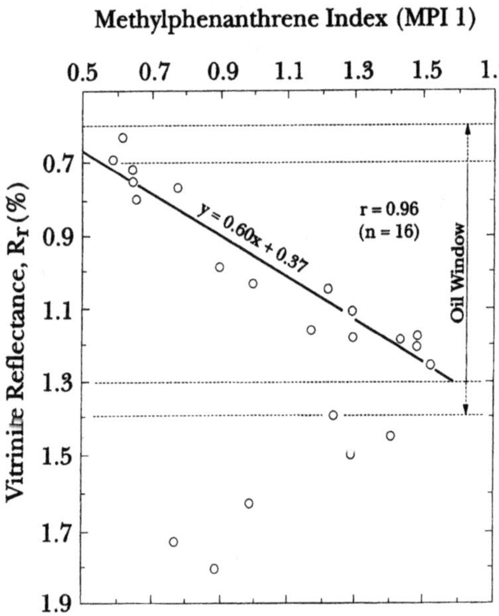

Fig. 3.12. Relationship between MPI 1 and mean random vitrinite reflectance for rock samples from a borehole in the Elmworth area, western Canada. (Radke et al. 1982a)

marginally mature stages was indicated for Miocene coals with 0.40–0.64%R_r of the Mahakam Delta, Indonesia by a linear increase in the MPI 1 with depth (Garrigues et al. 1988a). Regression analysis of MPI 1 data for Australian rock samples of Carboniferous, Permian, Jurassic, and Tertiary ages containing terrestrial organic matter suggested that calibration against %R_r generally may be extended to 0.5%R_r (Boreham et al. 1988). Discrepancies among MPI 1 and %R_r or T_{max} of certain oil-prone source rocks, however, left the significance of maturity parameters below 0.7%R_r based on phenanthrenes open to doubt (Radke et al. 1986; Radke 1987; Cassani et al. 1988; Lu and Kaplan 1989).

In Tertiary and Jurassic rocks of the Upper Rhine Graben erratic MPI 1 variations at 0.30–0.67%R_r were suspected to arise from migrated oil (Radke and Welte 1983). In order to improve understanding of these unexpected aromatic hydrocarbon distributions in sediments of apparently low maturity, a wider range of extractable aromatic hydrocarbons were investigated in four boreholes of the Upper Rhine Graben, three of which had been studied for maturity indicators based on methyl- and dimethylphenanthrenes (Radke and Welte 1983; Radke et al. 1991). Alkylated phenanthrenes showed a regular thermal evolution with depth only for the borehole Kork 1 (Fig. 3.13), as indicated by an increase in MPR and MPI 1. In organic-matter-lean immature sediments of the other boreholes alkylated compounds generally were outranked by nonalkylated (parent) compounds. Phenanthrene, fluoranthene, and pyrene along with minor amounts of monomethyl- and dimethylphenanthrenes presumably were associated with organic matter that had been severely altered in high-temperature and/or oxidative conditions before it was finally incorporated into the sediments. During catagenesis these PAHs retained a pyrolyticlike distribution until, depending on type and amount of unaltered autochthonous organic matter in the same samples, alkylated aromatic hydrocarbons released from kerogen eventually became more abundant. The MPI 1 was a useful maturity indicator only where those latter hydrocarbons prevailed. In some of the organic-matter-lean samples where an enhanced abundance of methylated relative to parent compounds was attributed to migrated oil, the MPI 1 was higher than expected and believed to reflect the maturity of the oil.

The potential of the MPI 1 for assessing oil maturities has been tested with more than 200 crude oils of various origins (Radke 1987, 1988). A preliminary classification into immature, mature, postmature, and overmature categories was refined when influences of the organic facies were taken into account. Four modes of the R_c frequency distribution were tentatively attributed to types I, I/II, II/III, and III/II of the source kerogens (Fig. 3.14), because the modes at 0.75%, 0.82%, and 0.92%R_c corresponded to maxima of the hydrocarbon generation curves generally assumed at the respective %R_r values. The first mode at 0.65%R_c, however, was lower than expected for oils derived from type I kerogen-bearing strata, which likely were the source beds of oils from the Songliao Basin. This discrepancy may be due to the fact that calibration of MPI 1 against %R_r was based on humic coals and type III kerogen-bearing rocks. As the impact of organic facies on the MPI 1 is reduced with increasing maturity, R_c values are particularly useful for assessing the maturity of postmature oils and condensates, which generally is outside the dynamic range of biomarker

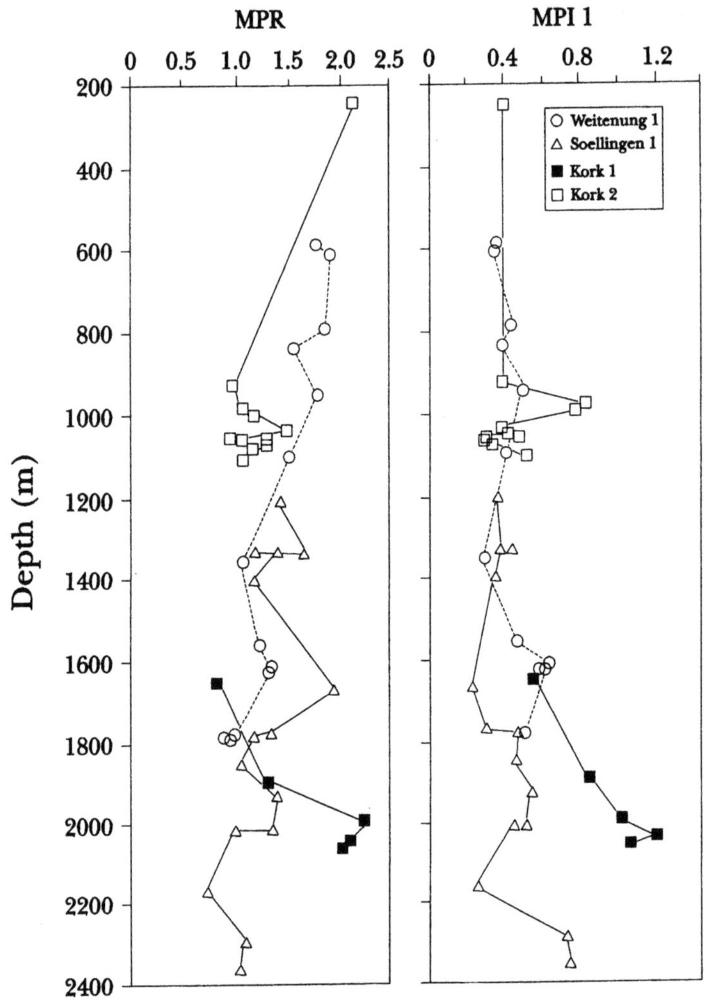

Fig. 3.13. Variation in MPR and MPI 1 with depth for rock samples from four boreholes in the Upper Rhine Graben. (Radke et al. 1991)

parameters. The MPI 1 parameter has been successfully used in numerous oil studies that are mostly unavailable from the open literature. Published applications include maturity assessment of 130 crude oils representing 107 oil pools in the Cooper and Eromanga basins of Australia (Tupper and Burckhardt 1990), 60 crude oils from different regions of the former Soviet Union (Ivanov and Golovko 1992), and 60 crude oils from three Java Sea (sub-)basins (Radke et al. 1994).

Maturity parameters based on alkylated biphenyls rely on an abundance decrease with maturity of 2-methylbiphenyl or 2,3-dimethylbiphenyl (2,3-DMBP). Lower stability of isomers with a methyl group in *ortho* (2) position as compared to those with methyl groups in *para* (4) or *meta* (3,5) positions has

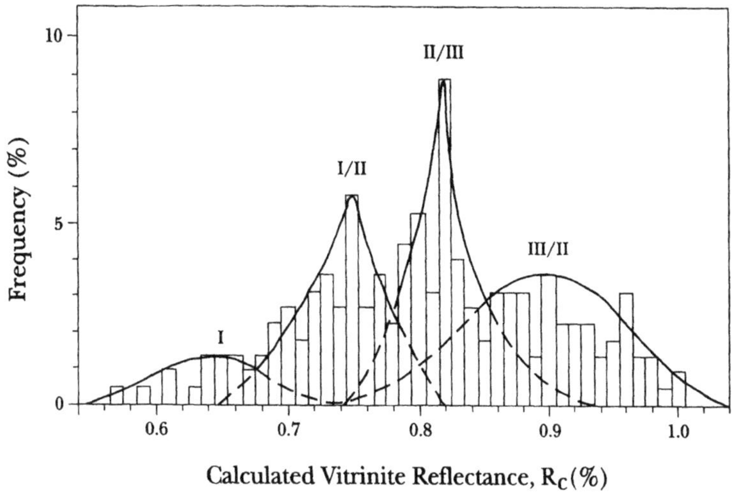

Fig. 3.14. Frequency distribution of R_c for crude oils of different origins. (Radke 1987)

been confirmed by pyrolysis experiments. Case histories of two wells drilled in the Carnavon Basin, Australia revealed increasing trends with temperature of the 3,5-DMBP/2,3-DMBP ratio. They fit to the trend curves predicted using kinetic parameters determined in the laboratory (see Sect. 3.5.5; Fig. 3.15; Alexander et al. 1986, 1988; Cumbers et al. 1987; Kagi et al. 1990).

Alkylated aromatic hydrocarbons the distribution of which is controlled by maturity, but which have not found wide application include chrysenes, perylenes, and benzenes. For a series of Miocene coals with 0.40–0.64%R_r from the Mahakam Delta, Indonesia, β-type 2-methyl- and 3-methylchrysenes, believed to be the most stable isomers, showed a regular increase in relative concentrations with depth. Also, the sum of methylchrysenes increased relative to the parent chrysene, but the correlation with depth was less certain (Garrigues et al. 1988a). Consecutive alkylation of perylene during early thermal evolution of organic matter at temperatures > 30 °C was indicated in deep-ocean sediments of Quaternary and Miocene ages by an increase in C_1–C_3 perylenes relative to the nonalkylated compound (Louda and Baker 1984). A series of Spanish crude oils representing a wide range of maturities, as indicated by bulk parameters and biomarker ratios, revealed a decrease in abundance of long-chain *ortho* *n*-alkyltoluenes relative to the more stable *meta* isomers with increasing maturity (Albaigés et al. 1986).

3.5.3
Maturation Changes in Molecular Distributions of Heterocompounds

The effect of maturation on sulfur aromatics has been demonstrated for Middle East crude oils where the abundance relative to alkylbenzenes of alkylbenzo[*b*]thiophenes decreased as maturity increased (Joly et al. 1974). The benzothiophene-to-dibenzothiophene ratio introduced by Ho et al. (1974) relies on the lower stability of alkylbenzo[*b*]thiophenes as compared to alkyldi-

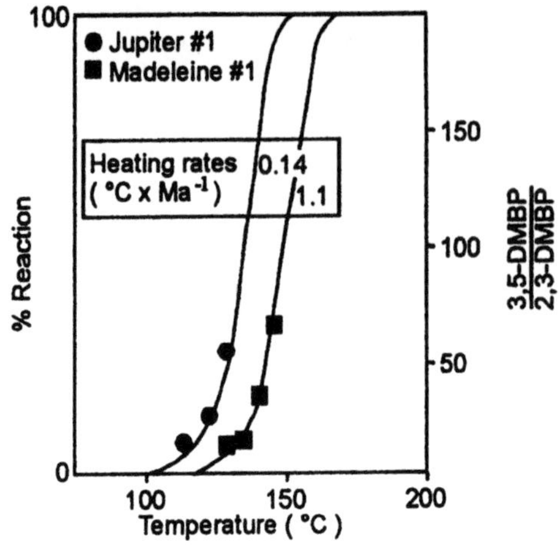

Fig. 3.15. Curves showing theoretical progress of reaction with increase in temperature based on laboratory-determined kinetic parameters for the cyclization of 2,3-dimethylbiphenyl and the heating rates of the sediments in boreholes Jupiter #1 and Madeleine #1. *Solid circles and squares*, 3,5-DMBP/2,3-DMBP values measured for sediment samples. (Reprinted from Alexander et al., 2,3-Dimethylbiphenyl: Kinetics of its cyclisation reaction and effects of maturation upon its relative concentration in sediments. In: Mattavelli L, Novelli L (eds) Advances in organic geochemistry 1987. Pergamon Press, Oxford, 1988, pp 833–837, with kind permission from Elsevier Science Ltd, The Boulevard, Langford Lane, Kidlington OX5 1GB, UK)

benzothiophenes. It was applied to assess the maturity of a great number of crude oils which were roughly grouped into immature, altered, and mature categories. Among alkyldibenzothiophenes, 1-methyldibenzothiophene is characterized by relatively low stability, and this led to the development of the methyldibenzothiophene ratio (3.8):

$$\text{MDR} = [\text{4-MDBT}]/[\text{1-MDBT}] \tag{3.8}$$

The MDR has proved a useful maturity indicator, especially for oil-prone source rocks (Radke et al. 1986). Correlation with $\%R_r$ and T_{max} was fair for various type I/II, type II, and type III kerogen-bearing rock samples (Fig. 3.16; Radke 1988). A similar indicator, the dibenzothiophene index (3.9):

$$\text{DBTI} = ([\text{2-MDBT}] + [\text{3-MDBT}])/2\times[\text{1-MDBT}] \tag{3.9}$$

has been applied to assess relative maturities of a limited number of crude oils from the Tarragona Basin, Spanish Mediterranean, and western Canada, for example, Cold Lake and Pembina oil fields (Albaigés et al. 1986; Payzant et al. 1989). Calibration against $\%R_r$ and T_{max} has not been carried out with rock samples in this instance.

Certain petroporphyrins were employed in very sensitive maturity indicators which were mainly based on (1) the decrease in deoxophylloerythroetioporphyrin (DPEP) abundance relative to ETIO-type porphyrins and (2) variations in the average molecular weight and carbon number distributions of vanadyl porphyrins (Didyk et al. 1975; Baker and Louda 1986). The effect (1) was first suggested by Corwin (1959) and substantiated later by Morandi and

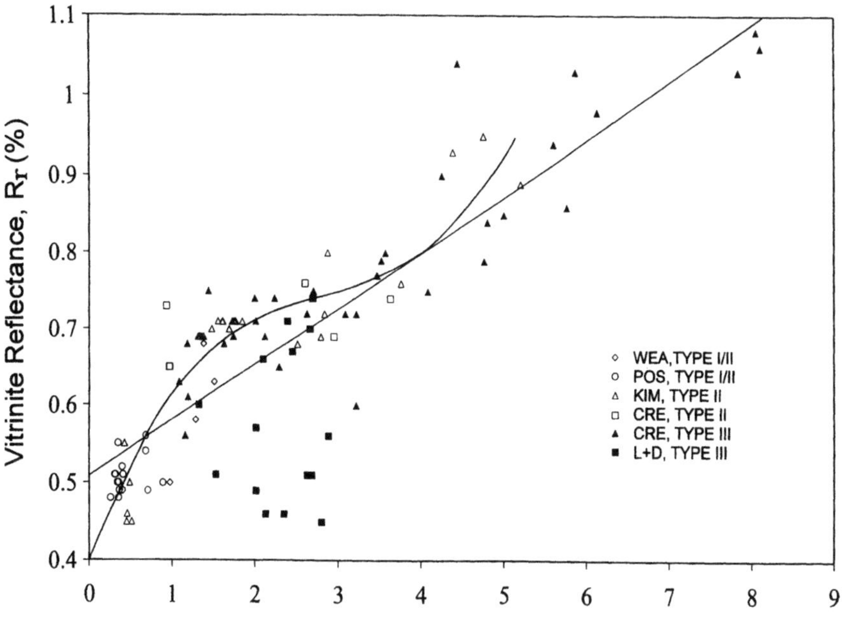

a

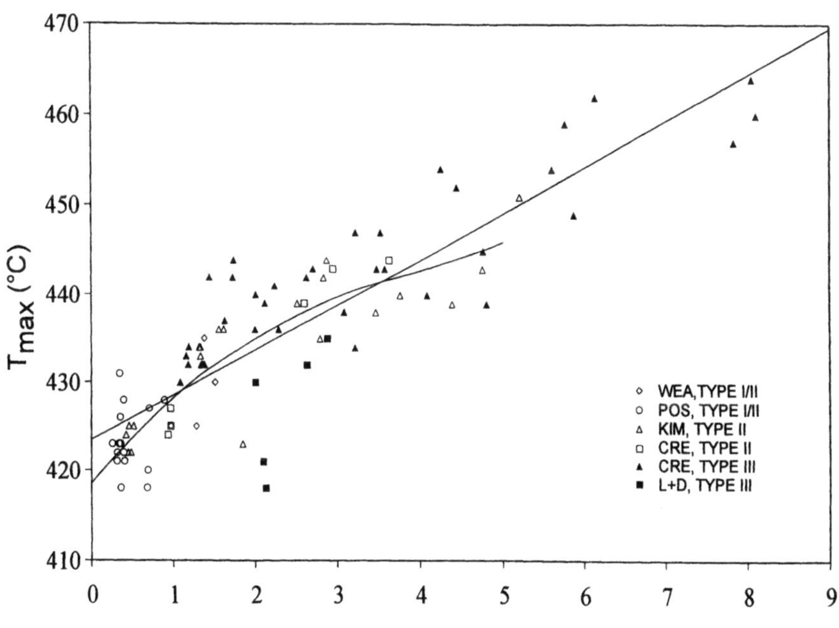

b

Fig. 3.16. Relationship between MDR and mean random vitrinite reflectance (**a**) and T_{max} (**b**) for rock samples of different origins. (Radke 1988)

Jensen (1966) who observed that the ratio of DPEP to ETIO porphyrins was lower in the shale oil derived from retorting of Green River Shale than in the starting material. This change likely was due to thermal destruction of DPEP, rather than to a conversion of DPEP to ETIO porphyrins by opening of the isocyclic five-membered ring, as indicated by heating experiments on artificial vanadyl DPEP and vanadyl ETIO porphyrin mixtures (Burkova et al. 1980). The ETIO porphyrin survived a 10-h heat treatment at 220 °C, whereas most of the DPEP was degraded.

From the variation with depth in carbon-normalized yields of both porphyrin types in El Lajjun Shale (Jordan) it was deduced that changes in the DPEP to ETIO porphyrin ratio not only were due to the higher thermal stability of the ETIO porphyrins but also to generation of those compounds from kerogen (Barwise and Roberts 1984). The percentage of deoxophylloerythroetioporphyrins (DPEP) in total porphyrins from Permian Marl Slate (Durham, England) sediments exhibited high sensitivity to maturity at 0.3–0.7%R_r. Thus, it is an attractive alternative to %R_r for maturity assessment of samples with high porphyrin contents, such as algal-rich sediments, where %R_r generally is difficult to measure (Barwise and Park 1983). A related vanadyl porphyrin maturity parameter (3.10) has been introduced by Sundararaman et al. (1988):

$$PMP = C_{28}E/(C_{28}E + C_{32}D) \tag{3.10}$$

where $C_{28}E$ is ETIO porphyrin with 28 carbon atoms, and $C_{32}D$ is DPEP porphyrin with 32 carbon atoms

Using pyrolysis experiments those authors were able to scrutinize earlier models for porphyrin conversions during catagenesis. According to their results, the thermal evolution of vanadylporphyrin distributions is controlled by the generation of ETIO porphyrins that would increasingly dilute preexisting DPEP porphyrins. The potential of the PMP parameter for assessing the maturity of marine source rocks and crude oils has been demonstrated with hydrous pyrolysis experiments and several field examples (Sundararaman and Raedeke 1993).

The effect (2) arises from alkylation-dealkylation processes and/or formation of polyalkyl porphyrins from bacteriochlorophylls (Chicarelli et al. 1987; Ocampo et al. 1987 and references therein). Variations in the extent of alkylation beyond C_{32} are measured by an alkylation index (Baker et al. 1967; Baker and Palmer 1978). When plotted against the percentage of DPEP, the alkylation index reaches a maximum greater than 1.0 around 60% DPEP (Baker and Louda 1986). The ETIO porphyrin ratios, $C_{28}E'/C_{27}E'$ and $C_{29}E'/C_{28}E'$, proved useful in maturity assessment of postmature marine crude oils and rock extracts (Gallango and Cassani 1992).

3.5.4 Maturation Changes in Carbon Isotope Composition

Because during petroleum generation isotopically heavy oil is mixed with an isotopically light fraction of initial oil possibly derived from bacteria, $\delta^{13}C$ of the generated oil increases by 0.5‰ on an average (Clayton 1991). Since this change occurs essentially during the first 20% of petroleum generation it likely reflects total swamping by generated oil of initial oil. This interpretation agrees with that of Schoell et al. (1983) and Schoell (1984) who attributed variations in the isotopic composition of organic extracts from sedimentary rocks of the Mahakam Delta, Indonesia, to mixing of isotopically light initial soluble organic matter with isotopically heavier generated oil.

3.5.5 Thermochemistry, Kinetics, and Mechanisms of Molecular Transformations

The isomerization of one pristane stereoisomer [(6R, 10S) = *meso*-pristane] was brought about in the laboratory under free radical conditions by heating an intimate mixture of the substrate, a solvent-extracted shale, and sulfur at temperatures between 500 and 530 K (227–257 °C). The activation parameters of the reaction were derived from the temperature dependence of the rate constant by use of the Arrhenius expression. The preexponential factor and activation energy were found to be $2.1\times10^7 s^{-1}$ and 120 kJ mol^{-1} (29 kcal mol^{-1}). Similar experiments were carried out with the 5α(H) and 5β (H) isomers of (20R)- and (20S)-17β-methyl-18-norcholesta-8,11,13-triene to yield (20R)- and (20S)-15β-methyl-18,19-dinorcholesta-1,3,5(10),6,8,11,13-heptaene.

This aromatization of monoaromatic steroids was more temperature-dependent than the configurational isomerization of pristane, as indicated by the higher activation energy of 145 kJ mol^{-1} (35 kcal mol^{-1}) of the former reaction. The frequency factor of 6.7×10^{12} s^{-1} also was higher (Abbott et al. 1985a,b). Kinetic studies on the aromatization of rearranged ring-C monoaromatic steroid hydrocarbons have shown that under the same experimental conditions the rate constant is lower for the rearranged than for the nonrearranged isomers (Abbott and Maxwell 1988).

Methylated naphthalenes and phenanthrenes were shown to undergo rearrangement reactions when heated in the presence of acidic catalysts or clays (Oku and Yuzen 1975; Béarez 1985). For example, when heated at 40–70 °C in the presence of trifluoroacetic acid, 1,8-DMN rearranged very slowly into 1,7-DMN without forming by-products. The reaction followed a first-order rate equation (activation energy: 22.1 kcal mol^{-1}, frequency factor: $7.2\times10^9 s^{-1}$; Oku and Yuzen 1975). However, substantial proportions of by-products were produced when naphthalene, methylnaphthalenes, dimethylnaphthalenes, and dibenzothiophene were subjected to a 48-h hydrous pyrolysis at 235–335 °C in the presence of montmorillonite, illite or chlorite. The material that was recovered by solvent extraction corresponded to less than 40% of the starting material, while the residual material appeared to be strongly bound to the clay matrix. It was inferred that intramolecular methyl-shift was only one out of a large number of possible reactions to explain the increase in β-type relative to α-type isomers (Burkow et al. 1990). Inorganic catalysis is unlikely to affect the thermal evolution of alkylphenanthrene relative distributions in shales, as indicated by artificial maturation of Kimmeridge Clay samples (Garrigues et al. 1990). When heated at 270–500 °C for 45 h, changes in the distribution of extractable C_0–C_2 phenanthrenes were similar among rock and nonextracted kerogen samples isolated therefrom, both matching that of naturally matured samples.

The methyl group at the *ortho* (2) position of biphenyl, although thermally relatively unstable, is not shifted because it enters into a cyclization reaction to give fluorene at elevated temperatures. In a kinetic study on the conversion of 2,3-dimethylbiphenyl, 1-methylfluorene was in fact the only significant reaction product (Alexander et al. 1988). For determination of the kinetic parameters, solutions of the substrate in Dexsil were heated without a catalyst in sealed ampoules at five temperatures between 460 and 500 °C for appropriate times in the range 40 min–21 h. As expected, the reaction was first order [activation energy: 181 kJ mol^{-1}(43 kcal mol^{-1}), frequency factor: $8.2\times10^{7} s^{-1}$]. With naturally matured sediments, however, the ratio of 2,3-dimethylbiphenyl to 1-methylfluorene did not change with maturity in a systematic way. Because 1-methylfluorene obviously originated also from other sources, it was unsuitable as a reference when monitoring 2,3-dimethylbiphenyl conversion in sediments. Instead, the 3,5-dimethylbiphenyl finally chosen for this purpose proved inert under conditions that resulted in complete conversion of 2,3-dimethylbiphenyl to 1-methylfluorene. A more detailed investigation of *ortho*-methylbiphenyls included mono-, di-, and trimethyl derivatives that were partially deuterated in the benzylic position. It showed that the cyclization reaction involved a thermally induced homolytic cleavage of the benzylic C-H bond as the rate determining step (Kagi et al. 1990). Regarding application to maturity assessment of sedimentary rocks, it is important to note that the reaction is uncatalyzed. Lithological variations do not affect the reaction rates in this instance.

Of the semiempirical quantum mechanical methods for calculating thermochemical data of molecules as a function of molecular geometry, molecular mechanics (MM; see Kao and Allinger 1977) and a method based on the zero-differential overlap approximation (complete neglect of differential overlap or CNDO; see Pople and Beveridge 1970; Murrel and Harget 1972) have been applied in organic geochemistry. Energies of molecules at their ground states were obtained by the MM method, while the total energy as a function of H–C···H distances was assessed by CNDO calculations. An improved version (MM2, CNDO-2) that differed from the original method by modified parameterization was employed in each case. Accordingly, van Graas et al. (1982) used the MM2 method for calculating heats of formation (gaseous standard state) of 13 cholestane isomers and predicted the extent to which isomerization reactions can proceed during maturation. Thermodynamic stabilities of various alkylated, dealkylated, and rearranged 17α- and 17β-hopane isomers were likewise assessed by Kolaczkowska et al. (1990). They observed that calculated equilibrium ratios of certain isomers generally agreed with those found in thermally mature crude oil samples. To calculate relative differences in free energy of 32 cholestene isomers increments of the empirical MM2 force field had to be adjusted for sp^2 hybridized carbon atoms. From these results it follows that double bond isomerization via secondary carbocations is unlikely (de Leeuw et al. 1989).

The MM2 method has also been used to verify the greater stability of β-type when compared to α-type methylphenanthrenes, thus to substantiate the chemical basis of the MPI 1 (Garrigues et al. 1990). Because for 2-MP and 3-MP calculated enthalpies of formation of 41.9 kcal mol^{-1} are less positive than for

1-MP and 9-MP (43.8 kcal mol^{-1}) a negative heat of isomerization of –1.9 kcal mol^{-1} (gaseous standard state) is inferred. Correspondingly, based on published heats of combustion, a negative heat of isomerization of –5.8 kcal mol^{-1} was assessed for 9,10-DMP to yield 2,7-DMP (Radke et al. 1982b). As long as entropies of formation of the compounds of interest are unknown, free energies of formation ($\Delta G^{\circ}_{f,T}$), cannot be determined. Predicting thermodynamic equilibrium distributions is therefore critical with methylated phenanthrenes. This is in contrast to dimethyl-, trimethyl- and tetramethylnaphthalenes where thermodynamic equilibrium distributions were based on $\Delta G^{\circ}_{f,T}$ (I) of each isomer (3.11) using an increment method (Garrigues et al. 1988b) or the MM3 force field (van Duin et al. 1993). However, the results should be considered with caution also in this instance. It must be kept in mind that the thermochemical data refer to the gaseous standard state, which will never occur in a geological system:

$$\Delta G^{\circ}_{f,T}(I) = RT\,\ln[\sum_{i=1}^{N_I} \exp(\Delta G^{\circ}_{f,T,i})/RT] \tag{3.11}$$

where N_I is the number of isomers.

Considering $\Delta G^{\circ}_{f,T}$ data, Garrigues et al. (1988b) observed that the theoretical DMN distribution agreed with that of extractable DMNs in an ancient sediment reported earlier by Alexander et al. (1985). At the given temperature of 344 K (71 °C), DMNs apparently were formed under conditions close to thermodynamic equilibrium. Deviations seen for TMNs were attributed to enhanced production of specific isomers from biological precursors (Garrigues et al. 1988b). Likewise, disagreement between the theoretical thermodynamic equilibrium distribution of DMNs at 527 °C and that of low-temperature coal tar was attributed to an influence of the organic source material. That is, the distribution of DMN isomers in the tar was believed to reflect the structure of the coal from which it was formed (Karr et al. 1967).

A novel approach to a better understanding of chemical reactions taking place in geological conditions concentrated on the epimerization of hydrocarbon chiral centers. The epimerization model assumed that three carbon atoms next to the chiral center were attached to the rigid kerogen network, and thus were immobile, while the central carbon atom could move (Fig. 3.17). The chiral center was viewed as part of a "microuniverse" of atoms belonging to different molecular entities of the organic matrix, which surrounded the chiral center ("epimerization siton"). By providing or binding a proton, entities on opposite sides of the chiral center presumably assist in the vibrational displacement of the central carbon atom ("assisted vibrational displacement mechanism"; Costa Neto 1983, 1988; Costa Neto and Nakayama 1987). A double-well potential energy function for the successive displacement of the central carbon atom was obtained by the second version of the CNDO method (CNDO-2). Displacement occurs when a proton bound to the central carbon atom is released while a new chemical bond is being established to another proton, provided by the kerogen matrix, which approaches the chiral center from the opposite side. The total energy at the bipyramidal symmetrical con-

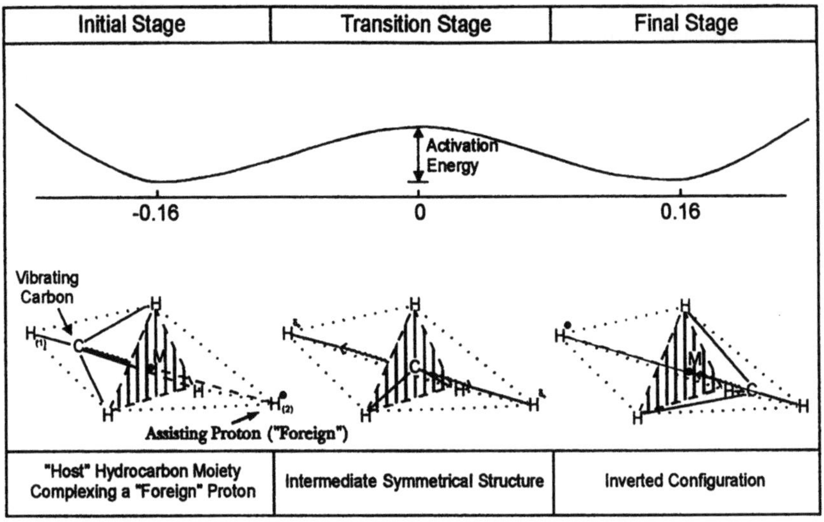

Fig. 3.17. Structural changes of a simplified "epimerization siton" by the "assisted vibrational displacement mechanism" and total energy function for this reaction. (Costa Neto 1988)

figuration of the transition state of a simplified "epimerization siton" ($CH_4 \cdots H^+$ complex) as low as 2.36 kcal mol^{-1} suggested epimerization is induced by a proton (H^+) rather than a hydride ion (H^-) which would result in a much higher energy barrier of 7.35 kcal mol^{-1}. Confining pressure should reduce the distances between atoms within the "epimerization siton" and hence the height of the energy barrier. As an important implication of the "siton" concept it appears that kinetic approaches to organic maturation neglecting pressure are meaningless (Costa Neto 1991). However, the relevance of pressure as a kinetic parameter is more obvious when activation volumes ($V^\ddagger$) rather than atomic distances are considered (Dominé 1989). The extent to which the rate constant (k) is modified by pressure can be inferred from the following relationship:

$$(\delta \ln k/\delta\, p)T = -\Delta V^\ddagger/RT \tag{3.12}$$

Dominé (1991) has shown that the rate of alkane pyrolysis may decrease by as much as 1000-fold in the range 21–1500 MPa, which would correspond to 2–150 km burial depth at hydrostatic pressure.

3.5.6
Relationships Among Various Maturity Indicators

A compilation and correlation of the major maturation indicators commonly applied in hydrocarbon exploration in the late seventies was published by Héroux et al. (1979). In addition to mineralogical and optical parameters, the various organic geochemistry parameters of kerogen, bitumen, and gas dis-

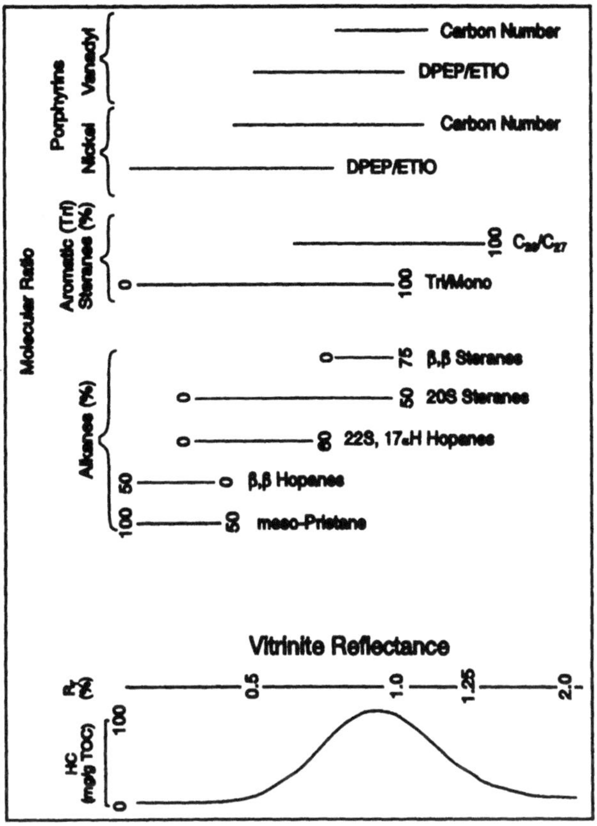

Fig. 3.18. Ranges of individual molecular measurements of thermal maturation relative to a downhole hydrocarbon generation curve and a vitrinite reflectance scale. (Mackenzie and Maxwell 1981)

cussed earlier in this chapter were related to maturity in terms of coalification and hydrocarbon generation stages. Mackenzie and Maxwell (1981) provided a semiquantitative assessment of ranges of individual molecular parameters of thermal maturation relative to a downhole hydrocarbon generation curve and a vitrinite reflectance scale (Fig. 3.18). Due to the different kinetic constants of the chemical reactions involved, such a calibration of molecular and bulk geochemical parameters would be valid only for a specific geothermal regime. Although this approach was *not* advised in general by the authors, some general information can be derived from Fig. 3.18. It is obvious that most biological marker reactions are complete before the maximum of hydrocarbon generation. On the other hand, biological markers allow maturity assessment in a range where vitrinite reflectance measurements are least reliable ($< 0.7\%R_r$). Aromatic steroid hydrocarbon parameters, including side-chain cleavage, and porphyrin ratios extend somewhat further into the higher maturity range.

If the kinetic data of geochemical reactions, i.e., their temperature dependence, are sufficiently different, the progress of these reactions in comparison can be used to reconstruct the geothermal development in a sedimentary basin as a function of geological time. Initial assessment of the kinetic behaviour of steroid aromatization and sterane or hopane isomerization by Mackenzie et al. (1982b) led to the observation that the rate of steroid aromatization is more temperature dependent and thus occurs at a faster relative rate at higher temperatures than sterane or hopane isomerization. Kinetic data derived from natural observations and laboratory experiments (steroid aromatization only) were then applied to sedimentary basins with an assumed more quiescent history (e.g., North Sea, Paris Basin) and areas with high geothermal heat flow (Pannonian Basin, Lower Saxony Basin) where distinctly different trends of biological marker ratios (Fig. 3.19) clearly distinguished between extremes (McKenzie et al. 1983; Mackenzie and McKenzie 1983). Further areas of successful application of this approach were the Alberta Foreland Basin (Beaumont et al. 1985) and the Nova Scotia margin (Mackenzie et al. 1985a).

Reassessment of kinetic data of steroid aromatization and sterane/hopane isomerization by Marzi (1989) revealed that the differences are much smaller than previously determined (Mackenzie and McKenzie 1983). As a consequence, sensitivity analysis showed that a difference in geothermal heating rate by a factor of 10 leads only to a minor distinction of isomerization/aromatization trends, (Fig. 3.20; Marzi and Rullkötter 1992) which in view of the experimental errors involved in the experimental determination of biological marker ratios is too small to be useful for application to the reconstruction of geothermal histories of sedimentary basins based on biological markers alone. Under these circumstances it remains unclear at present why previous applications of isomerization/aromatization plots and of the (pseudo) kinetic data of apparent geochemical reactions were successful.

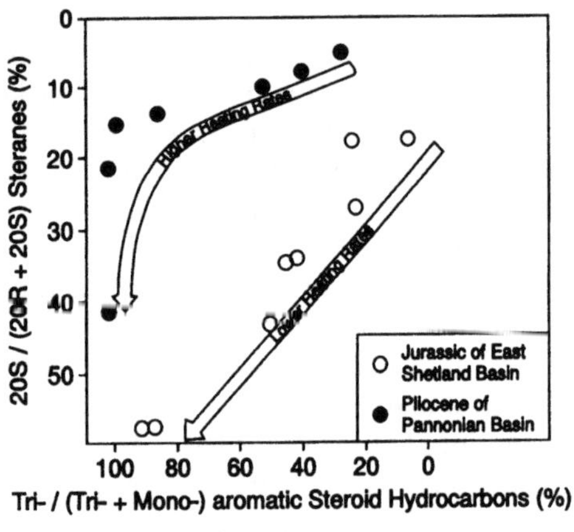

Fig. 3.19. Comparison of the variation of the extent of steroid aromatization relative to the extent of sterane isomerization at C-20 for an assumed quiescent geothermal regime (North Sea) and an area of high heat flow (Pannonian Basin). (Mackenzie 1984)

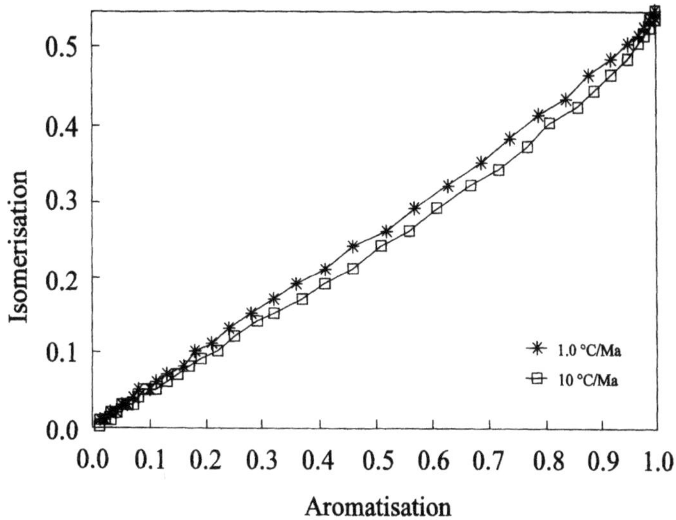

Fig. 3.20. Simulated progress of sterane isomerization and steroid aromatization using kinetic data determined by a combination of hydrous pyrolysis and natural samples and assuming heating rates of 1 and 10 °C Ma^{-1}. In contrast to earlier kinetic assessment of these two geochemical reactions the difference of temperature effect is only minor. (Marzi and Rullkötter 1992)

In a study on the geochemistry of greater Ekofisk crude oils Hughes et al. (1985) observed that the sensitivity to thermal stress increases in the order C_{20-21}/C_{26-28} triaromatic sterane ratio < Ts/Tm ≈ MPI 1 < sterane isomerization. Maturity of the Palaeozoic Alum Shale was assessed in different regions of Sweden by combined use of biomarker ratios [20S/(20S + 20R) steranes, 22S/(22S + 22R) homohopanes, moretane/hopane] and methylphenanthrene parameters (phenanthrene distribution fraction, F_1; % R_r predicted according to Kvalheim et al. 1987). Those parameters consistently indicated that the bitumen from central Sweden is more mature than that from eastern Sweden. Since biomarker abundance in the extracts, as the H/C atomic ratios of the kerogen, were inversely proportional to whole-rock uranium concentrations, maturity indicators have been determined only for samples with relatively low uranium contents (Dahl et al. 1989).

Relationships have been observed between the disappearance of smectite and (1) oil accumulations in the Gulf Coast region of Texas, (2) vitrinite reflectance for samples from Cameroon, and (3) organic extracts, adsorbed gases, and %R_r for samples from Labrador (Kübler 1980). Correlation between illite crystallinity and %R_r was statistically significant in Carboniferous shales of the Ouachita Mountains. Thus illite crystallinity was considered potentially useful for the maturity assessment of samples with vitrinite contents insufficient for %R_r determination (Guthrie et al. 1986). Although the smectite to illite reaction seems to be influenced by both temperature and time (Velde and Espitalié 1989), the kinetic factors are basically unknown, and it is not clear how to apply these results to polytype transformations. However, for the

Smackover Formation of the Manila Embayment, southwestern Alabama, the illite polytype 1M-2M transition has proved a good local, qualitative maturity indicator (Wilcoxon et al. 1990).

Very low-grade metamorphism in external parts of the Central Alps has been investigated by combined use of illite crystallinity, $\%R_r$, and fluid inclusion data (Frey et al. 1980). Correlation among the three parameters was good in some areas, but the relationships were not generally valid due to variable effects of the tectonic history.

References

Abbott GD, Maxwell JR (1988) Kinetics of the aromatization of rearranged ring-C monoaromatic steroid hydrocarbons. In: Mattavelli L, Novelli L (eds) Advances in organic geochemistry 1987. Pergamon Press, Oxford. Org Geochem 13: 881–885

Abbott GD, Lewis CA, Maxwell JR (1985a) Laboratory models for aromatization and isomerization of hydrocarbons in sedimentary basins. Nature 318: 651–653

Abbott GD, Lewis CA, Maxwell JR (1985b) The kinetics of specific organic reactions in the zone of catagenesis. Philos Trans R Soc Lond A 315: 107–122

Abbott GD, Wang GY, Eglinton TI, Home AK, Petch GS (1990) The kinetics of biological marker release and degradation processes during hydrous pyrolysis of vitrinite kerogen. Geochim Cosmochim Acta 54: 2451–2461

Abbott GD, Petch GS, Wang GY (1992) Reply to comment by R. Marzi on "The kinetics of sterane biological marker release and degradation process during the hydrous pyrolysis of vitrinite kerogen". Geochim Cosmochim Acta 56: 535–536

Abelson PH (1963) Organic geochemistry and the formation of petroleum. In: Proc 6th World Petr Congr, Frankfurt, vol 1. Frankfurt, pp 397–407

Ainsworth NR, Burnett RD, Kontrovitz M (1990) Ostracod colour change by thermal alteration, offshore Ireland and western UK. Mar Petrol Geol 7: 288–297

Aizenshtat Z, Pinsky I, Spiro B (1986) Electron spin resonance of stabilized free radicals in sedimentary organic matter. Org Geochem 9: 321–329

Albaigés J, Algaba J, Clavell E, Grimalt J (1986) Petroleum geochemistry of the Tarragona Basin (Spanish Mediterranean off-shore). In: Leythaeuser D, Rullkötter J (eds) Advances in organic geochemistry 1985. Pergamon Press, Oxford. Org Geochem 10: 441–450

Albrecht P, Ourisson G (1969) Diagenèse des hydrocarbures saturés dans une série sédimentaire épaisse (Douala, Cameroun). Geochim Cosmochim Acta 33: 138–142

Albrecht P, Ourisson G (1971) Biogenic substances in sediments and fossils. Angew Chem Int Ed Engl 10: 209–286

Albrecht P, Vandenbroucke M, Mandengué M (1976) Geochemical studies on the organic matter from the Douala Basin (Cameroon). I. Evolution of the extractable organic matter and the formation of petroleum. Geochim Cosmochim Acta 40: 791–799

Alexander R, Kagi R, Sheppard P (1984) 1,8-Dimethylnaphthalene as an indicator of petroleum maturity. Nature 308: 442–443

Alexander R, Kagi RI, Rowland SJ, Sheppard PN, Chirila TV (1985) The effects of thermal maturity on distributions of dimethylnaphthalenes and trimethylnaphthalenes in some ancient sediments and petroleums. Geochim Cosmochim Acta 49: 385–395

Alexander R, Cumbers KM, Kagi RI (1986) Alkylbiphenyls in ancient sediments and petroleums. In: Leythaeuser D, Rullkötter J (eds) Advances in organic geochemistry 1985. Pergamon Press, Oxford. Org Geochem 10: 841–845

Alexander R, Fisher SJ, Kagi RI (1988) 2,3-Dimethylbiphenyl: Kinetics of its cyclisation reaction and effects of maturation upon its relative concentration in sediments. In: Mattavelli L, Novelli L (eds) Advances in organic geochemistry 1987, Pergamon Press, Oxford. Org Geochem 13: 833–837

Allan J, Douglas AG (1974) Alkanes from the pyrolytic degradation of bituminous vitrinites and sporinites. In: Tissot B, Bienner F (eds) Advances in organic geochemistry 1973. Éditions Technip, Paris, pp 203–206

Allan J, Douglas AG (1977) Variations in the content and distribution of *n*-alkanes in a series of Carboniferous vitrinites and sporinites of bituminous rank. Geochim Cosmochim Acta 41: 1223–1230

Allan J, Larter SR (1983) Aromatic structures in coal maceral extracts and kerogens. In: Bjorøy M et al. (eds) Advances in organic geochemistry 1981, Wiley, Chichester, pp 534–546

Atkins PW (1990) Physical Chemistry, 4th edn. Oxford University Press, Oxford, 995 pp

Baker DR, Claypool GE (1970) Effects of incipient metamorphism on organic matter in mudrock. AAPG Bull 54: 456–468

Baker EG (1960) A hypothesis concerning the accumulation of sediment hydrocarbons to form crude oil. Geochim Cosmochim Acta 19: 309–317

Baker EW, Louda W (1986) Porphyrins in the geological record. In: Johns RB (ed) Biological markers in the sedimentary record. Elsevier, Amsterdam, pp 125–225

Baker EW, Palmer SE (1978) Geochemistry of porphyrins. In: Dolphin D (ed) The porphyrins, vol 1. Academic Press, London, pp 486–552

Baker EW, Yen TF, Dickie JP, Rhodes RE, Clark LF (1967) Mass spectrometry of porphyrins II. Characterization of petroporphyrins. J Am Chem Soc 89: 3631–3639

Bakr M, Akiyama M, Sanada Y (1991) In situ high temperature ESR measurements for kerogen maturation. Org Geochem 17: 321–328

Bakr M, Akiyama M, Yokomo T, Sanada Y (1988) Radical concentration of kerogen as a maturation parameter. Org Geochem 12: 29–32

Bakr MY, Akiyama M, Sanada Y (1990) ESR assessment of kerogen maturation and its relation with petroleum genesis. Org Geochem 15: 595–599

Barker C (1974) Pyrolysis techniques for source rock evaluation. AAPG Bull 58: 2349–2361

Barwise AJG, Park PJD (1983) Petroporphyrin fingerprinting as a geochemical marker. In: Bjorøy M et al. (eds) Advances in organic geochemistry 1981. Wiley, Chichester, pp 668–674

Barwise AJG, Roberts I (1984) Diagenetic and catagenetic pathways for porphyrins in sediments. In: Schenk PA, de Leeuw JW, Lijmbach GWM (eds) Advances in organic geochemistry 1983. Pergamon Press, Oxford. Org Geochem 6: 167–176

Baskin DK, Peters KE (1992) Early generation characteristics of a sulfur-rich Monterey kerogen. AAPG Bull 76: 1–13

Bates AL, Hatcher PG (1989) Solid-state ^{13}C NMR studies of a large fossil gymnosperm from the Yallourn Open Cut, Latrobe Valley, Australia. Org Geochem 14: 609–617

Béarez C (1985) Transformation catalytique de composés représentatifs de la matière organique sédimentaire sur mineraux naturels et synthétiques. PhD Thesis, Université de Poitiers, 192 pp

Beaumont C, Boutilier R, Mackenzie AS, Rullkötter J (1985) Isomerization and aromatization of hydrocarbons and the paleothermometry and burial history of Alberta Foreland Basin. AAPG Bull 69: 546–566

Bertrand R (1990) Correlations among the reflectances of vitrinite, chitinozoans, graptolites and scolecodents. Org Geochem 15: 565–574

Blob AK, Rullkötter J, Welte DH (1988) Direct determination of the aliphatic carbon content of individual macerals in petroleum source rocks by near-infrared microspectroscopy. In: Mattavelli L, Novelli L (eds) Advances in organic geochemistry 1987. Pergamon Press, Oxford. Org Geochem 13: 1073–1078

Boreham CJ, Crick IH, Powell TG (1988) Alternative calibration of the Methylphenanthrene Index against vitrinite reflectance: Application to maturity measurements on oils and sediments. Org Geochem 12: 289–294

Bray EE, Evans ED (1961) Distribution of *n*-paraffins as a clue to recognition of source beds. Geochim Cosmochim Acta 22: 2–15

Bray EE, Evans ED (1965) Hydrocarbons in non-reservoir-rock source beds. AAPG Bull 49: 248–257

Brooks J (1981) Organic maturation of sedimentary organic matter and petroleum exploration: a review. In: Brooks J (ed) Organic maturation studies and fossil fuel exploration. Academic Press, London, pp 1–37

Brooks JD, Smith JW (1967) The diagenesis of plant lipids during the formation of coal, petroleum and natural gas. I. Changes in the *n*-paraffin hydrocarbons. Geochim Cosmochim Acta 31: 2389–2397

Buiskol Taxopeus JMA (1983) Selection criteria for the use of vitrinite reflectance as a maturity tool. In: Brooks J (ed) Petroleum geochemistry and exploration of Europe. Blackwell, Oxford, pp 295–308

Burkova UN, Ryadova OV, Serebrennikova OW, Titov VI (1980) Geoporphyrin composition as an indication of organic matter transformation. Geokhimiya 9: 1417–1421

Burkow IC, Jørgensen E, Meyer T, Rekdal Å, Sydnes L (1990) Experimental simulation of chemical transformations of aromatic compounds in sediments. Org Geochem 15: 101–108

Burnham AK (1989) On the validity of the Pristane Formation Index. Geochim Cosmochim Acta 53: 1693–1697

Burnham AK, Braun RL, Samoun AM (1988) Further comparison of methods for measuring kerogen pyrolysis rates and fitting kinetic parameters. In: Mattavelli L, Novelli L (eds) Advances in organic geochemistry 1987. Pergamon Press, Oxford. Org Geochem 13: 839–845

Buseck PR, Bo-Jun H, Miner B (1988) Structural order and disorder in Precambrian kerogens. Org Geochem 12: 221–234

Calvin M (1969) Chemical evolution. Clarendon Press, Oxford

Cartz L, Hirsch PB (1960) A contribution to the structure of coals from X-ray diffraction studies. Philos Trans R Soc Lond A 252: 557–602

Cassani F, Gallango O, Talukdar S, Vallejos C, Ehrmann U (1988) Methylphenanthrene maturity index of marine source rock extracts and crude oils from the Maracaibo Basin. In: Mattavelli L, Novelli L (eds) Advances in organic geochemistry 1987. Pergamon Press, Oxford. Org Geochem 13: 73–80

Chaffee AL, Perry GJ, Johns RB (1983) Pyrolysis-gas chromatography of Australian coals. I. Victorian brown coal lithotypes. Fuel 62: 303–310

Chicarelli MI, Kaur S, Maxwell JR (1987) Sedimentary porphyrins: unexpected structures, occurrence and possible origins. In: Filby RH, Branthaver JF (eds) Metal complexes in fossil fuels - geochemistry, characterization, and processing. ACS Symp Ser 344. Am Chem Soc, Washington, pp 40–67

Claypool GE, Mancini EA (1989) Geochemical relationships of petroleum in Mesozoic reservoirs to carbonate source rocks of Jurassic Smackover Formation, southwestern Alabama. AAPG Bull 73: 904–924

Clayton CJ (1991) Effect of maturity on carbon isotope ratios of oils and condensates. Org Geochem 17: 887–899

Connan J (1974) Diagenèse naturelle et diagenèse artificielle de la matière organique à éléments végétaux prédominants. In: Tissot B, Bienner F (eds) Advances in organic geochemistry 1973. Éditions Technip, Paris, pp 73–95

Cooles GP, Mackenzie AS, Quigley TM (1986) Calculation of petroleum masses generated and expelled from source rocks. In: Leythaeuser D, Rullkötter J (eds) Advances in organic geochemistry 1985. Pergamon Press, Oxford. Org Geochem 10: 235–245

Cooper JE, Bray EE (1963) A postulated role of fatty acids in petroleum formation. Geochim Cosmochim Acta 27: 1113–1127

Cornford C, Morrow JA, Turrington A, Miles JA, Brooks J (1983) Some geological controls on oil composition in the U.K. North Sea. In: Brooks J (ed) Petroleum geochemistry and exploration of Europe. Blackwell, Oxford, pp 175–194

Corwin AH (1959) Petroporphyrins. In: Proc 5th World Petrol Congr, New York, vol. 5. New York, pp 119–129

Costa Neto C (1983) Theoretical organic geochemistry. I. An alternative model for the epimerization of hydrocarbon chiral centers in sediments. In: Bjorøy M et al. (eds) Advances in organic geochemistry 1981. Wiley, Chichester, pp 834–838

Costa Neto C (1988) Theoretical organic geochemistry. II. The siton concept and the assisted vibrational displacement mechanism of geochemical reactions in oil shales. An Acad brasil Ci 60(2): 137–148

Costa Neto C (1991) The effect of pressure on geochemical maturation: theoretical considerations. Org Geochem 17: 579–584

Costa Neto C, Nakayama HT (1987) The stratigraphic function for phenol content in the CERI-1 column of the Irati Formation. An Acad brasil Ci 59(4): 319–328

Cumbers KM, Alexander R, Kagi RI (1987) Methylbiphenyl, ethylbiphenyl and dimethylbiphenyl isomer distribution in some sediments and crude oils. Geochim Cosmochim Acta 51: 3105–3112

Curiale JA (1986) Origin of solid bitumens, with emphasis on biological marker results. In: Leythaeuser D, Rullkötter J (eds) Advances in organic geochemistry 1985. Pergamon Press, Oxford. Org Geochem 10: 559–580
Curry DJ, Simpler TK (1988) Isoprenoid constituents in kerogens as a function of depositional environment and catagenesis. In: Mattavelli L, Novelli L (eds) Advances in organic geochemistry 1987. Pergamon Press, Oxford. Org Geochem 13: 995–1001
Dahl J, Chen RT, Kaplan IR (1989) Alum shale bitumen maturation and migration: implications for Gotland's oil. J Pet Geol 12: 465–476
de Leeuw JW, Cox HC, van Graas G, van de Meer FW, Peakman TM, Baas JMA, van de Graaf B (1989) Limited double bond isomerization and selective hydrogenation of sterenes during early diagenesis. Geochim Cosmochim Acta 53: 903–909
Dembicki H, Pirkel FL (1985) Regional source rock mapping using a source potential rating index. AAPG Bull 69: 567–581
Didyk BM, Alturky YIA, Pillinger CT, Eglinton G (1975) Petroporphyrins as indicators of geothermal maturation. Nature 256: 563–565
di Primio R, Horsfield B (1995) Predicting the generation of heavy oils in carbonate/evaporitic environments using pyrolysis methods. In: Grimalt JO, Dorronsoro C (eds) Organic geochemistry: developments and applications to energy, climate, environment and human history. Selected Papers from the 17th Int Meet on Organic geochemistry, Donostia-San Sebastián, The Basque Country, Spain, AIGOA, Donostia-San Sebastián, pp 410–412
Dominé F (1989) Kinetics of hexane pyrolysis at very high pressures. 1. Experimental study. Energy Fuels 3: 89–96
Dominé F (1991) High pressure pyrolysis of *n*-hexane, 2,4-dimethylpentane and 1-phenylbutane. Is pressure an important geochemical parameter? Org Geochem 17: 619–634
Dormans HNM, Huntjens FJ, van Krevelen DW (1957) Chemical structure and properties of coal. XX. Composition of the individual macerals (vitrinites, fusinites, micrinites and exinites). Fuel 36: 321–333
Douglas AG, Sinninghe Damsté JS, Fowler MG, Eglinton TI, de Leeuw JW (1991) Unique distributions of hydrocarbons and sulphur compounds released by flash pyrolysis from the fossilised alga *Gloeocapsomorpha prisca*, a major constituent in one of four Ordovician kerogens. Geochim Cosmochim Acta 55: 275–291
Dungworth G, Schwartz AW (1972) Kerogen isolates from the Precambrian of South Africa and Australia. In: Von Gaertner HR, Wehner H (eds) Advances in organic geochemistry 1971. Pergamon Press, Oxford, pp 699–706
Dunton ML, Hunt JM (1962) Distribution of low-molecular-weight hydrocarbons in Recent and ancient sediments. AAPG Bull 46: 2246–2248
Düppenbecker S, Horsfield B (1990) Compositional information for kinetic modelling and petroleum type prediction. In: Durand B, Béhar F (eds) Advances in organic geochemistry 1989. Pergamon Press, Oxford. Org Geochem 16: 259–266
Durand B, Marchand A, Combaz A (1977a) Étude de kérogènes en résonance paramagnétique électronique. In: Campos R, Goñi J (eds) Advances in organic geochemistry 1975. Enadimsa, Madrid, pp 763–779
Durand B, Niçaise G, Roucaché J, Vandenbroucke M, Hagemann HW (1977b) Étude géochimique d'une série de charbons. In: Campos R, Goñi J (eds) Advances in organic geochemistry 1975. Enadimsa, Madrid, pp 601–631
Eglinton TI, Sinninghe Damsté JS, Kohnen MEL, de Leeuw JW, Larter SR, Patience RL (1990) Analysis of maturity-related changes in the organic sulfur composition of kerogens by flash pyrolysis-gas chromatography. In: Orr WL, White CM (eds) ACS symposium series, 429: Geochemistry of sulfur in fossil fuels. Am Chem Soc, Washington, DC, pp 529–565
Eglinton TI, Sinninghe Damsté JS, Pool W, de Leeuw JW, Eijkel G, Boon JJ (1992) Organic sulphur in macromolecular sedimentary organic matter. II. Analysis of distributions of sulphur-containing pyrolysis products using multivariate techniques. Geochim Cosmochim Acta 56: 1545–1560
Emden R (1938) Why do we have winter heating? Nature 141: 908–909
Emery KO, Hoggan D (1958) Gases in marine sediments. AAPG Bull 42: 2174–2188
Ericsson I, Lattimer RP (1988) Pyrolysis nomenclature. J Anal Appl Pyrolysis 14: 219–221

Espitalié J, LaPorte JL, Madec M, Marquis F, Leplat P, Paulet J, Boutefeu A (1977) Méthode rapide de caractérisation des roches mères, de leur potentiel pétrolier et de leur degré d'évolution. Rev Inst Fr Pét 32: 23–42

Farrington JW, Davis AC, Tarafa ME, McCaffrey MA, Whelan J, Hunt JM (1988) Bitumen molecular maturity parameters in the Ikpikpuk well Alaska North Slope. In: Mattavelli L, Novelli L (eds) Advances in organic geochemistry 1987. Pergamon Press, Oxford. Org Geochem 13: 303–310

Foscolos AE, Powell TG, Gunther PR (1976) The use of clay minerals and inorganic and organic geochemical indicators for evaluating the degree of diagenesis and oil generating potential of shales. Geochim Cosmochim Acta 40: 953–966

Frey M, Teichmüller M, Teichmüller R, Mullis J, Künzi B, Breitschmid A, Gruner U, Schwizer B (1980) Very low-grade metamorphism in external parts of the Central Alps: illite crystallinity, coal rank and fluid inclusion data. Eclogae Geol Helv 73: 173–203

Gallango O, Cassani F (1992) Biological marker maturity parameters of marine crude oils and rock extracts from the Maracaibo Basin, Venezuela. Org Geochem 18: 215–224

Ganz H, Kalkreuth W (1987) Application of infrared spectroscopy to the classification of kerogen-types and the evaluation of source rock and oil shale potential. Fuel 66: 708–711

Garrigues P, de Sury R, Angelin ML, Bellocq J, Oudin JL, Ewald M (1988a) Relation of the methylated aromatic hydrocarbon distribution pattern to the maturity of organic matter in ancient sediments from the Mahakam delta. Geochim Cosmochim Acta 52: 375–384

Garrigues P, Druez O, Rayez JC (1988b) Équilibre thermodynamique et géochimie organique des alkylnaphtalènes: vers un accord de principe? C R Acad Sci Paris. Sér II, 307: 921–926

Garrigues P, Oudin JL, Parlanti E, Monin JC, Robcis S, Bellocq J (1990) Alkylated phenanthrene distribution in artificially matured kerogens from Kimmeridge clay and the Brent Formation (North Sea). In: Durand B, Béhar F (eds) Advances in organic geochemistry 1989. Pergamon Press, Oxford. Org Geochem 16: 167–173

Giraud A (1970) Application of pyrolysis and gas chromatography to geochemical characterisation of kerogen in sedimentary rocks. AAPG Bull 54: 439–451

Girling GW (1963) Evolution of volatile hydrocarbons from coal. J Appl Chem 13: 77–91

Goff JC (1984) Hydrocarbon generation and migration from Jurassic source rocks in the East Shetland Basin and Viking Graben of the northern North Sea. In: Demaison G, Murris RJ (eds) Petroleum geochemistry and basin evaluation. AAPG Mem 35: 273–302

Goossens H, de Lange F, de Leeuw JW, Schenck PA (1988) The pristane formation index, a molecular maturity parameter. Confirmation in samples of the Paris Basin. Geochim Cosmochim Acta 52: 2439–2444

Goossens H, de Leeuw JW, Schenck PA, Brassell SC (1984) Tocopherols as likely precursors of pristane in ancient sediments and crude oils. Nature 312: 440–442

Goossens H, Due A, de Leeuw JW, van de Graaf B, Schenck PA (1988) The Pristane Formation Index, a new molecular maturity parameter. A simple method to assess maturity by pyrolysis/evaporation-gas chromatography of unextracted samples. Geochim Cosmochim Acta 52: 1189–1193

Gransch JA, Eisma E (1970) Characterisation of the insoluble organic matter of sediments by pyrolysis. In: Hobson GG, Speers GC (eds) Advances in organic geochemistry 1966. Pergamon Press, New York, pp 407–426

Guthrie JM, Houseknecht DW, Johns WD (1986) Relationships among vitrinite reflectance, illite crystallinity, and organic geochemistry in Carboniferous strata, Ouchita Mountains, Oklahoma and Arkansas. AAPG Bull 70: 26–33

Gutjahr CCM (1966) Carbonization measurements of pollen-grains and spores and their application. JJ Groen and Zoon, Leiden. 29 pp

Hagemann H, Hollerbach A (1981) Spectral fluorometric analysis of extracts, a new method for the determination of the degree of maturity of organic matter in sedimentary rocks. Bull Centres Rech, Expl-Prod Elf-Aquitaine 5: 635–650

Hanbaba P, Jüntgen H (1969) Zur Übertragbarkeit von Laboratoriums-Untersuchungen auf geochemische Prozesse der Gasbildung aus Steinkohle und über den Einfluß von Sauerstoff auf die Gasbildung. In: Schenck PA, Havenaar I (eds) Advances in organic geochemistry 1968. Pergamon Press, Oxford, pp 459–471

Héroux Y, Chagnon A, Bertrand R (1979) Compilation and correlation of major thermal maturation indicators. AAPG Bull 63: 2128-2144
Hills JR, Whitehead EV (1966) Triterpanes in optically active petroleum distillates. Nature 209: 977-979
Ho TY, Rogers MA, Drushel HV, Koons CB (1974) Evolution of sulfur compounds in crude oils. AAPG Bull 58: 2338-2348
Holden HW, Robb JC (1958) Mass spectrometry of substances of low volatility. Nature 182: 340
Hong Zhi-Hua, Li Hui-Xiang, Rullkötter J, Mackenzie AS (1986) Geochemical application of sterane and triterpane biological marker compounds in the Linyi Basin. In: Leythaeuser D, Rullkötter J (eds) Advances in organic geochemistry 1985. Pergamon Press, Oxford. Org Geochem 10: 433-439
Hood A, Gutjahr CCM, Heacock RL (1975) Organic metamorphism and the generation of petroleum. AAPG Bull 59: 986-996
Horsfield B (1984) Pyrolysis studies and petroleum exploration. In: Brooks J, Welte DH (eds) Advances in petroleum geochemistry, vol 1. Academic Press, London, pp 247-298
Horsfield B (1989) Practical criteria for classifying kerogens: some observations from pyrolysis-gas chromatography. Geochim Cosmochim Acta 53: 891-901
Horsfield B (1990) The rapid characterisation of kerogens according to source quality, compositional heterogeneity and thermal lability. Rev Palaeobot Palynol 65: 357-365
Horsfield B, Douglas AG (1980) The influence of minerals on the pyrolysis of kerogens. Geochim. Cosmochim. Acta 44: 1119-1131
Horsfield B, Düppenbecker SJ (1991) The decomposition of Posidonia Shale and Green River Shale kerogens using Microscale Sealed Vessel (MSSV) pyrolysis. J Anal Appl Pyrol 20: 107-123
Horsfield B, Dembicki H, Ho TTY (1983) Some potential applications of pyrolysis to basin studies. J Geol Soc Lond 140: 431-443
Horsfield B, Disko U, Leistner F (1989) The microscale simulation of maturation: outline of a new technique and its potential applications. Geol Rundsch 78: 361-374
Horsfield B, Bharati S, Larter SR, Leistner F, Littke R, Schenk HJ, Dypvik H (1992) On the atypical petroleum-generating characteristics of alginite in the Cambrian Alum Shale. In: Schidlowski M, Kimberley MM, McKirdy DM, Trudinger PA, Golubic S (eds) Early organic evolution. Implications for mineral and energy resources. Springer, Berlin Heidelberg New York, pp 257-266
Horsfield B, Curry DJ, Bohacs K, Littke R, Rullkötter J, Schenk HJ, Radke M, Schaefer RG, Carroll AR, Isaksen G, Witte EG (1994) Organic geochemistry of freshwater and alkaline lacustrine environments, Green River Shale, Wyoming. In: Øygard K et al. (eds) Advances in organic geochemistry 1993. Org Geochem 22: 415-440
Huck G, Karweil J (1955) Physikalisch-chemische Probleme der Inkohlung. Brennstoff-Chemie 36: 1-11
Hughes WB, Holba AG, Miller DE, Richardson JS (1985) Geochemistry of greater Ekofisk crude oils. In: Thomas BM et al. (eds) Petroleum geochemistry in exploration of the Norwegian Shelf. Norwegian Petroleum Society, Graham & Trotman, London, pp 75-92
Hunt JM (1975) Origin of gasoline range hydrocarbons in the deep sea. Nature 288: 688-690
Hussler G, Albrecht P (1983) C_{27}-C_{29} Monoaromatic anthrasteroid hydrocarbons in Cretaceous black shales. Nature 304: 262-263
Hussler G, Chappe B, Wehrung P, Albrecht P (1981) C_{27}-C_{29} ring A monoaromatic steroids in Cretaceous black shales. Nature 294: 556-558
Hutton AC, Cook AC (1980) Influence of alginite on the reflectance of vitrinite from Joadja, New South Wales, and some other coals and oil shales containing alginite. Fuel 59: 711-714
Hutton AC, Kantsler AJ, Cook AC, McKirdy DM (1980) Organic matter in oil shales. Aust Petr Expl Assoc 20: 44-67
Ivanov VL, Golovko AK (1992) Phenanthrene hydrocarbons in USSR oils. Sib Khim Zh 1: 94-102 (in Russian)
Jackson KS, McKirdy DM, Deckelmann JA (1984) Hydrocarbon generation in the Amadeus Basin, central Australia. APEA J 24: 43-65

Jacob H (1989) Classification, structure, genesis and practical importance of natural solid oil bitumen ("migrabitumen"). Int J Coal Geol 11: 65–79

Jankowski B, Littke R (1986) Das organische Material der Ölschiefer von Messel. Geowiss unserer Zeit 4: 73–80

Joly D, Vasse L, Bordenave ML (1974) Application de méthodes d'analyse physique à la recherche de parenté entre différents pétroles du Moyen-Orient. In: Tissot B, Bienner F (eds) Advances in organic geochemistry 1973. Éditions Technip, Paris, pp 531–547

Jonathan D, L'Hote G, du Rochet J (1975) Analyse géochimique des hydrocarbures légers par thermovaporisation. Rev Inst Fr Pét 30: 65–88

Jones RW (1978) Kerogen maturation and petroleum generation. Nature 275: 567

Kagi RI, Alexander R, Toh E (1990) Kinetics and mechanism of the cyclisation reaction of *ortho*-methylbiphenyls. In: Durand B, Béhar F (eds) Advances in organic geochemistry 1989. Pergamon Press, Oxford. Org Geochem 16: 161–166

Kao J, Allinger NL (1977) Conformational analysis -122. Heats of formation of conjugated hydrocarbons by the Force Field Method. J Am Chem Soc 99: 975–986

Karr C, Estep PA, Chang TCL, Comberiati JR (1967) Identification of distillable paraffins, olefins, aromatic hydrocarbons, and natural heterocyclics from a low-temperature bituminous coal tar. Bur Mines Bull 637: 1–198

Kolaczkowska E, Slougui NE, Watt DS, Maruca RE, Moldowan JM (1990) Thermodynamic stability of various alkylated, dealkylated and rearranged 17α- and 17β-hopane isomers using molecular mechanics calculations. In: Durand B, Béhar F (eds) Advances in organic geochemistry 1989. Pergamon Press, Oxford. Org Geochem 16: 1033–1038

Kübler B (1980) Les premiers stades de la diagenèse organique et de la diagenèse minérale (une tentative d'équivalence). II. Zonéographie par les transformations minéralogiques, comparaison avec la réflectance de la vitrinite, les extraits organiques et les gaz adsorbés. Bull Ver Schweiz Petrol-Geol-Ing 46: 1–22

Kvalheim OM, Christy AA, Telnæs N, Bjørseth A (1987) Maturity determination of organic matter in coals using the methylphenanthrene distribution. Geochim Cosmochim Acta 51: 1883–1888

Kvenvolden KA (1970) Evidence for transformation of normal fatty acids in sediments. In: Hobson GD, Speers GC (eds) Advances in organic geochemistry 1966. Pergamon Press, Oxford, pp 335–366

Landes KK (1967) Eometamorphism, and oil and gas in time and space. AAPG Bull 51: 828–841

Larter SR (1984) Application of analytical pyrolysis techniques to kerogen characterization and fossil fuel exploration/exploitation. In: Voorhees KJ (ed) Analytical pyrolysis – techniques and applications. Butterworth, Guildford, pp 212–275

Larter SR (1985). Integrated kerogen typing and the quantitative evaluation of petroleum source rocks. In: Thomas BM et al. (eds) Petroleum geochemistry in exploration of the Norwegian Shelf. Graham and Trotman, London, pp 269–286

Larter S (1988) Some pragmatic perspectives in source rock geochemistry. Mar Petrol Geol 5: 194–204

Larter SR (1989) Chemical models of vitrinite reflectance evolution. Geol Rundsch 78: 349–359

Larter SR, Douglas AG (1980) A pyrolysis-gas chromatographic method for kerogen typing. In: Douglas AG, Maxwell JR (eds) Advances in organic geochemistry 1979. Pergamon Press, Oxford, pp 579–584

Larter SR, Horsfield B, Douglas AG (1977) Pyrolysis as a possible means of determining the petroleum-generating potential of sedimentary organic matter. In: Jones CER, Cramers CA (eds) Analytical pyrolysis. Elsevier, Amsterdam, pp 189–202

Larter SR, Solli H, Douglas AG (1983) Phytol-containing melanoidins and their bearing on the fate of isoprenoid structures in sediments. In: Bjorøy M et al. (eds) Advances in organic geochemistry 1981. Wiley, Chichester, pp 513–523

Larter SR, Solli H, Douglas AG, de Lange F, de Leeuw JW (1979) Occurrence and significance of prist-1-ene in kerogen pyrolysates. Nature 279: 405–408

Le Tran K, van der Weide BM (1969) Détermination automatique du degré de carbonisation de la matière organique des roches. Bull Centre Rech, Pau SNPA 3: 449–457

Le Tran K, Connan J, van der Weide B (1974) Probèmes relatifs à la formation d'hydrocarbures de d'hydrogène sulfuré dans le bassin sud-ouest Aquitain. In: Tissot B, Bienner F (eds) Advances in organic geochemistry 1973. Éditions Technip, Paris, pp 761–789

Leischner K, Welte DH, Littke R (1993) Fluid inclusions and organic maturity parameters as calibration tools in basin modelling. In: Doré AG, Augustson JH, Hermann C, Stewart DJ, Sylta Ø (eds) Basin modelling: advances and applications. NPF Spec Publ 3. Elsevier, Amsterdam, pp 161–172

Lerche I, McKenna T (1991) Pollen translucency as a thermal maturation indicator. J Petrol Geol 14: 19–36

Lewan MD, Dolcater DL, Bjorøy M (1986) Effects of thermal maturation on steroid hydrocarbons as determined by hydrous pyrolysis of Phosphoria Retort Shale. Geochim Cosmochim Acta 50: 1977–1987

Leythaeuser D, Welte DH (1969) Relation between distribution of heavy *n*-paraffins and coalification in Carboniferous coals from the Saar district, Germany. In: Schenck PA, Havenaar I (eds) Advances in organic geochemistry 1968. Pergamon Press, New York, pp 429–442

Leythaeuser D, Radke M, Willsch H (1988) Geochemical effects of primary migration of petroleum in Kimmeridge source rocks from Brae field area, North Sea. II. Molecular composition of alkylated naphthalenes, phenanthrenes, benzo- and dibenzothiophenes. Geochim Cosmochim Acta 552: 2879–2891

Li Taiming, Rullkötter J, Radke M, Schaefer RG, Welte DH (1987) Crude oil geochemistry of the southern Songliao Basin, People's Republic of China. Erdöl Kohle, Erdgas, Petrochem 40: 337–346

Lichtfouse E, Riolo J, Albrecht P (1990) Occurrence of 2-methyl-, 3-methyl- and 6-methyltriaromatic steroid hydrocarbons in geological samples. Tetrahedron Lett 31: 3937–3940

Lin R, Davis A (1988) A fluorogeochemical model for coal macerals. Org Geochem 12: 363–374

Littke R (1987) Petrology and genesis of Upper Carboniferous seams from the Ruhr region, western Germany. Int J Coal Geol 7: 147–184

Littke R, Baker DR, Leythaeuser D (1988) Microscopic and sedimentologic evidence for the generation and migration of hydrocarbons in Toarcian source rocks of different maturities. In: Mattavelli L, Novelli L (eds) Advances in organic geochemistry 1987. Pergamon, Oxford. Org Geochem 13: 549–559

Littke R, Horsfield B, Leythaeuser D (1989) Hydrocarbon distribution in coals and dispersed organic matter of different maceral compositions and maturities. Geol Rundsch 78(1): 391–410

Lopatin NV (1971) Temperature and geologic time as factors in coalification. Izv Akad Nauk Uzb SSR, Ser Geol 3: 95–106 (in Russian)

Louda JW, Baker EW (1984) Perylene occurrence, alkylation and possible sources in deep-ocean sediments. Geochim Cosmochim Acta 48: 1043–1058

Louis MC, Tissot BP (1967) Influence de la température et de la pression sur la formation des hydrocarbures dans les argiles à kérogène. In: Proc 7th World Petrol Congr, Mexico City, vol. 2. Elsevier, London, pp 47–60

Lu ST, Kaplan IR (1989) Pyrolysis of kerogens in the absence and presence of montmorillonite. II. Aromatic hydrocarbons generated at 200 and 300 °C. Org Geochem 14: 501–510

Ludwig B, Hussler G, Wehrung P, Albrecht P (1981) C_{26}–C_{29} Triaromatic steroid derivatives in sediments and petroleums. Tetrahedron Lett 22: 3313–3316

Mackenzie AS (1984) Application of biological markers in petroleum geochemistry. In: Brooks J, Welte DH (eds) Advances in petroleum geochemistry, vol 1. Academic Press, London, pp 115–214

Mackenzie AS, McKenzie D (1983) Isomerization and aromatization of hydrocarbons in sedimentary basins formed by extension. Geol Mag 120: 417–470

Mackenzie AS, Maxwell JR (1981) Assessment of thermal maturation in sedimentary rocks by molecular measurements. In: Brooks J (ed) Organic maturation studies and fossil fuel exploration. Academic Press, London, pp 239–254

Mackenzie AS, Patience RL, Maxwell JR, Vandenbroucke M, Durand B (1980) Molecular parameters of maturation in the Toarcian shales, Paris Basin, France. I. Changes in the

configurations of acyclic isoprenoid alkanes, steranes and triterpanes. Geochim Cosmochim Acta 44: 1709–1721
Mackenzie AS, Hoffmann CF, Maxwell JR (1981) Molecular parameters of maturation in the Toarcian shales, Paris Basin, France. III. Changes in aromatic steroid hydrocarbons. Geochim Cosmochim Acta 45: 2369–2376
Mackenzie AS, Brassell SC, Eglinton G, Maxwell JR (1982a) Chemical fossils: the geological fate of steroids. Science 217: 491–504
Mackenzie AS, Lamb NA, Maxwell JR (1982b) Steroid hydrocarbons and the thermal history of sediments. Nature 295: 223–226
Mackenzie AS, Maxwell JR, Coleman ML, Deegan CE (1983) Biological marker and isotope studies of North Sea crude oils and sediments. In: Proc 11th World Petrol Congr, London, vol. 2. Wiley, Chichester, pp 45–56
Mackenzie AS, Beaumont C, Boutilier R, Rullkötter J (1985a) The aromatization and isomerization of hydrocarbons and the thermal and subsidence history of the Nova Scotia margin. Philos Trans R Soc Lond A 315: 203–232
Mackenzie AS, Rullkötter J, Welte DH, Mankiewicz P (1985b) Reconstruction of oil formation and accumulation in North Slope, Alaska, using quantitative gas chromatography-mass spectrometry. In: Magoon LB, Claypool GE (eds) Alaska North Slope oil/rock correlation study. Am Assoc Petr Geol, Tulsa. AAPG Stud Geol 20: 319–377
Mango FD (1991) The stability of hydrocarbons under the time-temperature conditions of petroleum genesis. Nature 352: 146–148
Marchand A, Conrad J (1980) Electron paramagnetic resonance in kerogen studies. In: Durand B (ed) Kerogen - insoluble organic matter from sedimentary rocks. Éditions Technip, Paris, pp 243–270
Marzi R (1989) Kinetik und quantitative Analyse der Isomerisierung und Aromatisierung von fossilen Steroidkohlenwasserstoffen im Experiment und in natürlichen Probensequenzen. Berichte der Kernforschungsanlage Jülich, Nr 2264. KFA Jülich, ISSN 0336-0885. 169 pp
Marzi R (1992) Comment on "The kinetics of sterane biological marker release and degradation processes during hydrous pyrolysis of vitrinite kerogen" by Abbott GD, Wang GY, Eglinton TI, Home AK, and Petch GS. Geochim Cosmochim Acta 56: 533–534
Marzi R, Rullkötter J (1992) Qualitative and quantitative evolution and kinetics of biological marker transformations. Laboratory experiments and application to the Michigan Basin. In: Moldowan JM, Albrecht P, Philip RP (eds) Biological markers in sediments and petroleum. Prentice Hall, Englewood Cliffs, pp 18–41
Marzi R, Rullkötter J, Perriman WS (1990) Application of the change of sterane isomer ratios to the reconstruction of geothermal histories: implications of the results of hydrous pyrolysis experiments. In: Durand B, Béhar F (eds) Advances in organic geochemistry 1989. Pergamon Press, Oxford. Org Geochem 16: 91–102
Maxwell JR, Cox RE, Ackman RG, Hooper SN (1972) The diagenesis and maturation of phytol. The stereochemistry of 2,6,10,14-tetramethylpentadecane from an ancient sediment. In: Von Gaertner HR, Wehner H (eds) Advances in organic geochemistry 1971. Pergamon Press, Oxford, pp 277–291
McHugh DJ, Saxby JD, Tardif JW (1976) Pyrolysis-hydrogenation gas-chromatography of carbonaceous material from Australian sediments. I. Some Australian coals. Chem Geol 17: 243–259
McKenzie D, Mackenzie AS, Maxwell JR, Sajgó CS (1983) Isomerisation and aromatisation of hydrocarbons in streched sedimentary basins. Nature 301: 504–506
McKirdy DM, McHugh DJ, Tardif JW (1980) Comparative analysis of stromatolitic and other microbial kerogens by pyrolysis-hydrogenation-gas chromatography (PHGC). In: Trudinger PA, Walter MR, Ralph BJ (eds) Biogeochemistry of ancient and modern environments. Aust Acad Sci and Springer, Berlin Heidelberg New York, pp 187–200
Meinschein WG (1961) Significance of hydrocarbons in sediments and petroleum. Geochim Cosmochim Acta 22: 58–64
Michelsen JR, Khavari Khorasani G (1990) Monitoring chemical alterations of individual oil-prone macerals by means of microscopical fluorescence spectrometry combined with multivariate data analysis. Org Geochem 15: 179–192

Moldowan JM, Fago FJ (1986) Structure and significance of a novel rearranged monoaromatic steroid hydrocarbon in petroleum. Geochim Cosmochim Acta 50: 343–351
Monnier F, Powell TG, Snowdon LR (1983) Qualitative and quantitative aspects of gas generation during maturation of sedimentary organic matter. Examples from Canadian Frontier Basins. In: Bjorøy M et al. (eds) Advances in organic geochemistry 1981. Wiley, Chichester. pp 487–495
Morandi JR, Jensen H (1966) Composition of porphyrins from shale oil, oil-shale, and petroleum by absorption and mass spectroscopy. J Chem Eng Data Ser 11: 81–88
Morishima M, Matsubayashi H (1978) ESR diagrams: a method to distinguish vitrinite macerals. Geochim Cosmochim Acta 42: 537–540
Mukhopadhyay PK (1992) Maturation of organic matter as revealed by microscopic methods: applications and limitations of vitrinite reflectance, and continuous spectral and pulsed laser fluorescence spectroscopy. In: Wolf KH, Chilingarian GV (eds) Diagenesis. III. Developments in sedimentology, 47. Elsevier, Amsterdam, pp 435–510
Murrel JN, Harget AJ (1972) Semi-empirical self-consistent-field molecular orbital theory of molecules. Wiley, New York, 180 pp
Muscio GPA, Horsfield B (1996) Neoformation of inert carbon during the natural maturation of a marine source rock; Bakken Shale, Williston Basin. Energy Fuels 10: 10–18
Muscio GPA, Horsfield B, Welte DH (1991) Compositional changes in the macromolecular organic matter (kerogens, asphaltenes and resins) of a naturally matured source rock sequence from northern Germany as revealed by pyrolysis methods. In: Manning DAC (ed) Organic geochemistry – advances and applications in the natural environment. Manchester University Press, Manchester, pp 447–449
Nip M, Tegelaar EW, Brinkhuis H, de Leeuw JW, Schenck PA, Holloway P.J. (1986) Analysis of modern and fossil plant cuticles by Curie point Py-GC and Curie point Py-GC-MS: recognition of a new, highly aliphatic and resistant biopolymer. In: Leythaeuser D, Rullkötter J (eds) Advances in organic geochemistry 1985. Pergamon Press, Oxford. Org Geochem 10: 769–778
Nöth S (1991) Die Conodontendiagenese als Inkohlungsparameter und ein Vergleich unterschiedlich sensitiver Diageneseindikatoren am Beispiel von Triassedimenten Nord- und Mitteldeutschlands. Boch Geol Geotech Arb 37: 1–169
Nunn JA, Sleep NH, Moore WE (1984) Thermal subsidence and generation of hydrocarbons in Michigan Basin. AAPG Bull 68: 296–315
Oakwood TS, Shriver DS, Fall HH, McAleer WJ, Wunz PR (1952) Optical activity of petroleum. I E Chem 44: 2568–2570
Oberlin A, Boulmier JL, Durand B (1974) Electron microscope investigation of the structure of naturally and artificially metamorphosed kerogen. Geochim Cosmochim Acta 38: 647–650
Ocampo R, Callot HJ, Albrecht P (1987) Evidence for porphyrins of bacterial and algal origin in oil shale. In: Filby RH, Branthaver JF (eds) Metal complexes in fossil fuels – geochemistry, characterization, and processing. ACS Symp Ser 344. Am Chem Soc, Washington, pp 68–73
Oku A, Yuzen Y (1975) Acid-catalyzed rearrangements of polymethylnaphthalenes. J Org Chem 40: 3850–3857
Øygard K, Larter SR, Senftle J (1988) The control of maturity and kerogen type on analytical pyrolysis data. In: Mattavelli L, Novelli L (eds) Advances in organic geochemistry 1987. Pergamon Press, Oxford. Org Geochem 13: 1153–1162
Patience RL, Rowland SJ, Maxwell JR (1978) The effect of maturation on the configuration of pristane in sediments and petroleum. Geochim Cosmochim Acta 42: 1871–1875
Payzant JD, Mojelsky TW, Strausz OP (1989) Improved methods for the selective isolation of the sulfide and thiophenic classes of compounds from petroleum. Energy Fuels 3: 449–454
Peakman TM, Maxwell JR (1988) Early diagenetic pathways of steroid alkenes. In: Mattavelli L, Novelli L (eds) Advances in organic geochemistry 1987. Pergamon Press, Oxford. Org Geochem 13: 583–592
Peters KE, Moldowan JM, Sundararaman P (1990) Effects of hydrous pyrolysis on biomarker thermal maturity parameters: Monterey phosphatic and siliceous members. Org Geochem 15: 249–265

Peters KE, Rohrback BG, Kaplan IR (1980) Laboratory-simulated thermal maturation of recent sediments. In: Douglas AG, Maxwell JR (eds) Advances in organic geochemistry 1979. Pergamon Press Oxford, pp 547–558

Philippi GT (1965) On the depth, time and mechanism of petroleum generation. Geochim Cosmochim Acta 29: 1021–1049

Philippi GT (1975) The deep subsurface temperature controlled origin of the gaseous and gasoline-range hydrocarbons of petroleum. Geochim Cosmochim Acta 39: 1353–1373

Pople JA, Beveridge DL (1970) Approximate molecular orbital theory. McGraw-Hill, New York, 214 pp

Powell TG (1978) An assessment of the hydrocarbon source rock potential of the Canadian Arctic Islands. Geol Survey Can Pap, Calgary, 78–12, 82 pp

Powell TG, Foscolos AE, Gunther PR, Snowdon LR (1978) Diagenesis of organic matter and fine clay minerals: a comparative study. Geochim Cosmochim Acta 42: 1181–1197

Pradier B, Largeau C, Derenne S, Martinez L, Bertrand P, Pouet Y (1990) Chemical basis of fluorescence alteration of crude oils and kerogens. I. Microfluorimetry of an oil and its isolated fractions; relationships with chemical structure. In: Durand B, Béhar F (eds) Advances in organic geochemistry 1989. Pergamon Press, Oxford. Org Geochem 16: 451–460

Price LC (1983) Geologic time as a parameter in organic metamorphism and vitrinite reflectance as an absolute paleogeothermometer. J Petr Geol 6: 5–38

Price LC, Clayton JL, Rumen L (1981) Organic geochemistry of the 9.6 km Bertha Rogers No. 1. well, Oklahoma. Org Geochem 3: 59–77

Pusey WC (1973) The ESR-kerogen method – how to evaluate potential gas and oil source rocks. World Oil 176: 71–75

Püttmann W, Eckardt CB (1989) Influence of an intrusion on the extent of isomerism in acyclic isoprenoids in the Permian Kupferschiefer of the Lower Rhine Basin, N.W. Germany. Org Geochem 14: 651–658

Quigley TM, Mackenzie AS, Gray JR (1987) Kinetic theory of petroleum generation. In: Doligez B (ed) Migration of hydrocarbons in sedimentary basins. Éditions Technip, Paris, pp 131–171

Radke M (1987) Organic geochemistry of aromatic hydrocarbons. In: Brooks J, Welte D (eds) Advances in petroleum geochemistry, vol 2. Academic Press, London, pp 141–207

Radke M (1988) Application of aromatic compounds as maturity indicators in source rocks and crude oils. Mar Petrol Geol 5: 224–236

Radke M, Welte DH (1983) The Methylphenanthrene Index (MPI): A maturity parameter based on aromatic hydrocarbons. In: Bjorøy M et al. (eds) Advances in organic geochemistry 1981. Wiley, Chichester, pp 504–512

Radke M, Willsch H (1994) Extractable alkyldibenzothiophenes in Posidonia Shale (Toarcian) source rocks: relationship of yields to petroleum formation and expulsion. Geochim Cosmochim Acta 58: 5223–5244

Radke M, Schaefer RG, Leythaeuser D, Teichmüller M (1980) Composition of organic matter in coals: Relation to rank and liptinite fluorescence. Geochim Cosmochim Acta 44: 1787–1800

Radke M, Welte DH, Willsch H (1982a) Geochemical study of a well in the Western Canada Basin: relation of aromatic distribution pattern to maturity of organic matter. Geochim Cosmochim Acta 46: 1–10

Radke M, Willsch H, Leythaeuser D, Teichmüller M (1982b) Aromatic components of coal: relation of distribution pattern to rank. Geochim Cosmochim Acta 46: 1831–1848

Radke M, Leythaeuser D, Teichmüller M (1984) Relationship between rank and composition of aromatic hydrocarbons for coals of different origins. In: Schenck PA, de Leeuw JW, Lijmbach GWM (eds) Advances in organic geochemistry 1983. Pergamon Press, Oxford. Org Geochem 6: 423–430

Radke M, Welte DH, Willsch H (1986) Maturity parameters based on aromatic hydrocarbons: Influence of the organic matter type. In: Leythaeuser D, Rullkötter J (eds) Advances in Organic Geochemistry 1985. Pergamon Press, Oxford. Org Geochem 10: 51–63

Radke M, Welte DH, Willsch H (1991) Distribution of alkylated aromatic hydrocarbons and dibenzothiophenes in sediments of the Upper Rhine Graben. Chem Geol 93: 325–341

Radke M, Rullkötter J, Vriend SP (1994) Distribution of naphthalenes in crude oils from the Java Sea: source and maturation effects. Geochim Cosmochim Acta 58: 3675–3689

Redding CE, Schoell M, Monin JC, Durand B (1980) Hydrogen and carbon isotopic composition of coals and kerogens. In: Douglas AG, Maxwell JR (eds) Advances in organic geochemistry 1979. Pergamon Press, Oxford, pp 711–723

Reed J, Illich HA, Horsfield B (1986) Biochemical evolutionary significance of Ordovician oils and their source. In: Leythaeuser D, Rullkötter J (eds) Advances in organic geochemistry 1985. Pergamon Press, Oxford. Org Geochem 10: 347–358

Regtop RA, Crisp PT, Ellis J, Fookes CJR (1986) 1-Pristene as a precursor for 2-pristene in pyrolysates of oil shale from Condor, Australia. Org Geochem 9: 233–236

Requejo AG (1994) Maturation of petroleum source rocks. II. Quantitative changes in extractable hydrocarbon content and composition associated with hydrocarbon generation. Org Geochem 21: 91–105

Requejo AG, Gray NR, Freund H, Thomann H, Melchior MT, Gebhard LA, Bernardo M, Pictroski CF, Hsu CS (1992) Maturation of petroleum source rocks. 1. Changes in kerogen structure and composition associated with hydrocarbon generation. Energy Fuels 6: 203–214

Retkofsky AL, Stark JM, Friedel RA (1968) Electron spin resonance in American coals. Anal Chem 40: 1699–1704

Riolo J, Albrecht P (1985) Novel rearranged ring C monoaromatic steroid hydrocarbons in sediments and petroleum. Tetrahedron Lett 26: 2701–2704

Riolo J, Ludwig B, Albrecht P (1985) Synthesis of ring C monoaromatic steroid hydrocarbons occurring in geological samples. Tetrahedron Lett 26: 2697–2700

Riolo J, Hussler G, Albrecht P, Connan J (1986) Distribution of aromatic steroids in geological samples: their evaluation as geochemical parameters. In: Leythaeuser D, Rullkötter J (eds) Advances in organic geochemistry 1985. Pergamon Press, Oxford. Org Geochem 10: 981–990

Robin P (1975) Caractérisation des kérogènes et de leur évolution par spectroscopie infrarouge. PhD Thesis, Université Catholique de Louvain, 162 pp

Rohrback RG (1983) Crude oil geochemistry of the Gulf of Suez. In: Bjorøy M et al. (eds) Advances in organic geochemistry 1981. Wiley, Chichester, pp 39–48

Romováček J, Kubát J (1968) Characterization of coal substance by pyrolysis-gas chromatography. Anal Chem 40: 1119–1126

Ross JV, Bustin RM (1990) The role of strain energy in creep graphitization of anthracite. Nature 343: 58–60

Ross JV, Bustin RM, Rouzaud JN (1991) Graphitization of high rank coals – the role of shear strain: experimental considerations. Org Geochem 585–596

Rouxhet PG, Robin PL (1978) Infrared study of the evolution of kerogens of different origins during catagenesis and pyrolysis. Fuel 57: 533–540

Rouxhet PG, Robin PL, Niçaise G (1980) Characterization of kerogens and of their evolution by infrared spectroscopy. In: Durand B (ed) Kerogen – insoluble organic matter from sedimentary rocks. Éditions Technip, Paris, pp 163–190

Rovere CE, Crisp PT, Ellis J, Bolton PD (1983) Chemical characterization of shale oil from Condor, Australian. Fuel 62: 1274–1282

Rullkötter J, Leythaeuser D, Horsfield B, Littke R, Mann U, Müller PJ, Radke M, Schaefer RG, Schenk HJ, Schwochau K, Witte EG, Welte DH (1988) Organic matter maturation under the influence of a deep intrusive heat source: a natural experiment for quantification of hydrocarbon generation and expulsion from a petroleum source rock (Toarcian shale, northern Germany). In: Mattavelli L, Novelli L (eds) Advances in organic geochemistry 1987. Pergamon Press, Oxford. Org Geochem 13: 847–856

Rullkötter J, Mackenzie AS, Welte DH, Leythaeuser D, Radke M (1984) Quantitative gas chromatorgraphy-mass spectrometry analysis of geological samples. In: Schenck PA, de Leeuw JW, Lijmbach GWM (eds) Advances in organic geochemistry 1983. Pergamon Press, Oxford. Org Geochem 6: 817–827

Rullkötter J, Marzi R (1988) Natural and artificial maturation of biological markers in a Toarcian shale from northern Germany. In: Mattavelli L, Novelli L (eds) Advances in organic geochemistry 1987. Pergamon Press, Oxford. Org Geochem 13: 639–645

Rullkötter J, Marzi R (1989) New aspects of the application of sterane isomerization and steroid aromatization to petroleum exploration and the reconstruction of geothermal histories of sedimentary basins. Prepr, Div Pet Chem, Am Chem Soc 34: 126–131

Rullkötter J, Orr WL (1989) A comparative study of thermal maturation effects in sulfur-rich and less sulfur-rich crude oils of Tertiary age from California Basins. 14th Int Meet on Organic geochemistry, Paris. Book of Abstracts, no 134

Rullkötter J, Welte DH (1983) Maturation of organic matter in areas of high heat flow: A study of sediments from DSDP Leg 63, offshore California, and Leg 64, Gulf of California. In: Bjorøy M et al. (eds) Advances in organic geochemistry 1981. Wiley, Chichester, pp 438–448

Rullkötter J, Spiro B, Nissenbaum A (1985) Biological marker characteristics of oils and asphalts from carbonate source rocks in a rapidly subsiding graben, Dead Sea, Israel. Geochim Cosmochim Acta 49: 1350–1357

Rullkötter J, Marzi R, Meyers PA (1992) Biological markers in Paleozoic sedimentary rocks and crude oils from the Michigan Basin: reassessment of sources and thermal history of organic matter. In: Schidlowski M, Kimberley MM, McKirdy DM, Trudinger PA, Golubic S (eds) Early organic evolution: implications for mineral and energy resources. Springer, Berlin Heidelberg New York, pp 324–335

Sajgó Cs, Lefler J (1986) A reaction kinetic approach to the temperature-time history of sedimentary basins. In: Buntebarth F, Stegena L (eds) Paleogeothermics. Lecture Notes in Earth Sciences, vol 5. Springer, Berlin Heidelberg New York, pp 119–151

Schaefer RG, Littke R (1988) Maturity-related compositional changes in the low-molecular-weight hydrocarbon fraction of Toarcian shales. In: Mattavelli L, Novelli L (eds) Advances in organic geochemistry 1987. Pergamon Press, Oxford. Org Geochem 13: 887–892

Schaefer RG, Welte DH, Pooch H (1984) Geochemistry of low molecular weight hydrocarbons in two exploration wells of the Elmworth gas field (Western Canada Basin). In: Schenck PA, de Leeuw JW, Lijmbach GWM (eds) Advances in organic geochemistry 1983. Pergamon Press, Oxford. Org Geochem 6: 695–701

Schaefer RG, Schenk HJ, Hardelauf H, Harms R (1990) Determination of gross kinetic parameters for petroleum formation from Jurassic source rocks of different maturity levels by means of laboratory experiments. In: Durand B, Béhar F (eds) Advances in organic geochemistry 1989. Pergamon Press, Oxford. Org Geochem 16: 115–120

Scheidt G, Littke R (1989) Comparative organic petrology of interlayered sandstones, siltstones, mudstones and coals in the Upper Carboniferous Ruhr basin, northwest Germany, and their thermal history and methane generation. Geol Rundsch 78: 375–390

Schenk HJ, Witte EG, Littke R, Schwochau K (1990) Structural modifications of vitrinite and alginite concentrates during pyrolytic maturation at different heating rates. A combined infrared, ^{13}C NMR and microscopical study. In: Durand B, Béhar F (eds) Advances in organic geochemistry 1989. Pergamon Press, Oxford. Org Geochem 16: 943–950

Schoell M (1984) Wasserstoff- und Kohlenstoffisotope in organischen Substanzen, Erdölen und Erdgasen. Geol Jahrb D67: 1–161

Schoell M, Teschner M, Wehner H, Durand B, Oudin JL (1983) Maturity related biomarker and stable isotope variations and their application to oil/source rock correlation in the Mahakam Delta, Kalimantan. In: Bjorøy M et al. (eds) Advances in organic geochemistry 1981. Wiley, Chichester, pp 156–163

Seifert WK, Moldowan JM (1978) Applications of steranes, terpanes and monoaromatics to the maturation, migration and source of crude oils. Geochim Cosmochim Acta 42: 77–92

Seifert WK, Moldowan JM (1980) The effect of thermal stress on source rock quality as measured by hopane stereochemistry. In: Douglas AG, Maxwell JR (eds) Advances in organic geochemistry 1979. Pergamon Press, Oxford, pp 229–237

Seifert WK, Moldowan JM (1981) Paleoreconstruction by biological markers. Geochim Cosmochim Acta 45: 783–794

Senftle JT, Larter SR, Bromley BW, Brown JH (1986) Quantitative chemical characterization of vitrinite concentrates using pyrolysis-gas chromatography. Rank variation of pyrolysis products. Org Geochem 9: 345–350

Shi Ji-Yang, Mackenzie AS, Alexander R, Eglinton G, Gowar AP, Wolff GA, Maxwell JR (1982) A biological marker investigation of petroleums and shales from the Shengli oilfield, the People's Republic of China. Chem Geol 35: 1–31

Shibaoka M, Steven JR (1977) Characterization of kerogen by electron spin resonance. Fuel 56: 458–459

Silverman SR (1964) Investigations of petroleum origin and mechanisms by carbon isotope studies. In: Miller SL, Wasserburg GJ (eds) Isotopic and cosmic chemistry. North-Holland, Amsterdam, pp 92–102

Sinninghe Damsté JS, Eglinton TI, de Leeuw JW, Schenck PA (1989) Organic sulphur in macromolecular sedimentary organic matter. 1. Structure and origin of sulphur containing moieties in kerogen, asphaltenes and coal as revealed by flash pyrolysis. Geochim Cosmochim Acta 53: 873–889

Smith PV (1952) The occurrence of hydrocarbons in recent sediments from the Gulf of Mexico. Science 116: 437–439

Snowdon LR (1980) Resinite – a potential petroleum source in the Upper Cretaceous/Tertiary of the Beaufort-Mackenzie Basin. Can Soc Petrol Geol Mem 6: 509–521

Snowdon LR, Roy KJ (1975) Regional organic metamorphism in the Mesozoic strata of the Sverdrup Basin. Bull Can Pet Geol 23: 131–148

Snowdon LR, Brooks PW, Williams GK, Goodarzi F (1987) Correlation of Canol Formation source rock with oil from Norman Wells. Org Geochem 11: 529–548

Solli H, Schou L, Krane J, Skjetne T, Leplat P (1985) Characterization of sedimentary organic matter using nuclear magnetic resonance and pyrolysis techniques. In: Thomas BM et al. (eds) Petroleum geochemistry in exploration of the Norwegian Shelf, Norwegian Petroleum Society, Graham & Trotman, London, pp 309–317

Sommerfeld A (1965) Thermodynamik und Statistik, 3te Aufl., Vorlesungen über theoretische Physik, Bd. 5. Akademische Verlagsgesellschaft, Leipzig, 338 pp

Stach E, Teichmüller M, Mackowsky MTh, Taylor GH, Chandra D, Teichmüller R (1982) Stach's textbook of coal petrology, 3rd edn. Gebrüder Bornträger, Berlin

Staplin FL (1977) Interpretation of thermal history from color of particulate organic matter – a review. Palynology 1: 9–18

Stephens JF (1979) Coal as a C-H-O ternary system. 1. Geochemistry. Fuel 58: 489–494

Stevens NP (1956) Origin of petroleum – a review. AAPG Bull 40: 51–61

Sundararaman P, Raedeke LD (1993) Vanadyl porphyrins in exploration: maturity indicators for source rocks and oils. Appl Geochem 8: 245–254

Sundararaman P, Biggs WR, Reynolds JG, Fetzer JC (1988) Vanadylporphyrins, indicators of kerogen breakdown and generation of petroleum. Geochim Cosmochim Acta 52: 2337–2341

Suzuki N (1984) Estimation of maximum temperature of mudstone by two kinetic parameters; epimerization of sterane and hopane. Geochim Cosmochim Acta 48: 2273–2282

Sweeney JJ, Burnham AK (1989) Evaluation of a simple model of vitrinite reflectance based on chemical kinetics. AAPG Bull 74: 1559–1570

Tegelaar EW, Matthezing RM, Jansen JBH, Horsfield B, de Leeuw JW (1989a) Possible origin of *n*-alkanes in high-wax crude oils. Nature 342: 529–531

Tegelaar EW, de Leeuw JW, Derenne S, Largeau C (1989b) A reappraisal of kerogen formation. Geochim Cosmochim Acta 53: 3103–3106

Teichmüller M (1982) Origin of petrographic constituents of coal. In: Stach E, Mackowsky MT, Teichmüller M, Taylor GH, Chandra D, Teichmüller R (eds) Stach's textbook of coal petrology. Bornträger, Stuttgart, pp 219–294

Teichmüller M (1986) Organic petrology of source rocks, history and state of the art. In: Leythaeuser D, Rullkötter J (eds) Advances in organic geochemistry 1985. Pergamon Press, Oxford. Org Geochem 10: 581–599

ten Haven HL, de Leeuw JW, Peakman TM, Maxwell JR (1986) Anomalies in steroid and hopanoid maturity indices. Geochim Cosmochim Acta 50: 853–855

ten Haven HL, Littke R, Rullkötter J (1989) Hydrocarbon biological markers in Carboniferous coals of different maturities from the Ruhr area (northwest Germany). Prepr, Div Ret Chem, Am Chem Soc 34: 149–153

Thompson KFM (1979) Light hydrocarbons in subsurface sediments. Geochim Cosmochim Acta 43: 657–672

Ting FTC (1981) Uniaxial and biaxial vitrinite reflectance models and their relationship to palaeotectonics. In: Brooks J (ed) Organic maturation studies and fossil fuel exploration. Academic Press, London, pp 379–392

Tissot BP (1969) Premières données sur les mécanismes et la cinétique de la formation du pétrole dans les sédiments. Simulation d'un schéma réactionnel sur ordinateur. Rev Inst Fr Pét 24: 470–501

Tissot B, Welte DH (1984) Petroleum formation and occurrence, 2nd edn., Springer, Berlin Heidelberg New York, 699 pp

Tissot B, Oudin JL, Pelet R (1972) Critères d'origine et d'évolution des pétroles. Application à l'étude géochimique des bassins sédimentaires. In: Von Gaertner HR, Wehner H (eds) Advances in organic geochemistry 1971. Pergamon Press, Oxford, pp 113–134

Tissot B, Durand B, Espitalié J, Combaz A (1974) Influence of nature and diagenesis of organic matter in formation of petroleum. AAPG Bull 58: 499–506

Tissot B, Deroo G, Hood A (1978) Geochemical study of the Uinta Basin: formation of petroleum from the Green River formation. Geochim Cosmochim Acta 42: 1469–1485

Tissot BP, Pelet R, Ungerer P (1987) Thermal history of sedimentary basins, maturity indices and kinetics of oil and gas generation. AAPG Bull 71: 1445–1466

Tupper NP, Burckhardt DM (1990) Use of the Methylphenanthrene Index to characterise expulsion of Cooper and Eromanga Basin oils. APEA J 30: 373–385

Ujiié Y (1978) Kerogen maturation and petroleum genesis. Nature 272: 438–439, 275: 568

Ungerer P, Pelet R (1987) Extrapolation of the kinetics of oil and gas formation from laboratory experiments to sedimentary basins. Nature 327: 52–54

van Aarssen BGK, de Leeuw JW, Horsfield B (1991) A comparative study of three different pyrolysis methods used to characterise a biopolymer isolated from fossil and extant Dammar resins. J Anal Appl Pyrol 20: 125–139

van de Meent D, Brown SC, Philp RP, Simoneit BRT (1980) Pyrolysis-high resolution gas chromatography and pyrolysis gas chromatography – mass spectrometry of kerogen precursors. Geochim Cosmochim Acta 44: 999–1014

van Duin ACT, Baas JMA, van de Graaf B, de Leeuw JW, Bastow TP, Alexander R (1993) Comparison of experimental and calculated thermodynamic values of alkylnaphthalenes; an approach to recognize maturity changes in source rocks and crude oils. In: Øygard K (ed) Organic geochemistry – Poster Sessions from the 16th Int Meet on Organic geochemistry, Stavanger 1993. Falch Hurtigtrykk, Oslo, pp 194–197

van Graas G, de Leeuw JW, Schenck PA (1980) Analysis of coals of different rank by Curie-point pyrolysis-mass spectrometry and Curie-point pyrolysis-gas chromatography-mass spectrometry. In: Douglas AG, Maxwell JR (eds) Advances in organic geochemistry 1979. Pergamon Press, Oxford, pp 485–494

van Graas G, de Leeuw JW, Schenck PA, Haverkamp J (1981) Kerogen of Toarcian shales of the Paris Basin. A study of its maturation by flash pyrolysis techniques. Geochim Cosmochim Acta 45: 2465–2474

van Graas G, Baas JMA, van de Graaf B, de Leeuw JW (1982) Theoretical organic geochemistry I. The thermodynamic stability of several cholestane isomers calculated by molecular mechanics. Geochim Cosmochim Acta 46: 2399–2402

van Krevelen DW (1993) Coal – typology, physics, chemistry, constitution, 3rd edn., Elsevier, Amsterdam, 979 pp

Vassoevich NB, Visotski IV, Guseva AN, Olenin VB (1967) Hydrocarbons in the sedimentary mantle of the earth. In: Proc 7th World Petr Congr, Mexico City, vol. 2. Elsevier, London, pp 37–45

Vassoyevich NB, Korchagina YuI, Lopatin NV, Chernyshev VV (1969) Principal phase of oil formation. Moscov Univ Vestnik 6: 3–27 (in Russian). Engl trans Int Geol Rev 12: 1276–1296 (1970)

Velde B, Espitalié J (1989) Comparison of kerogen maturation and illite/smectite composition in diagenesis. J Petrol Geol 12: 103–110

Waples DW (1980) Time and temperature in petroleum formation: application of Lopatin's method to petroleum exploration. AAPG Bull 64: 916–926

Weiß HM (1985) Geochemische und petrographische Untersuchungen am organischen Material kretazischer Sedimentgesteine aus dem Deep Basin, Westkanada. PhD Thesis, Rheinisch-Westfälische Technische Hochschule, Aachen, 261 pp
Welte DH (1970) Organischer Kohlenstoff und die Entwicklung der Photosynthese auf der Erde. Naturwissenschaften 57: 17–23
Welte DH, Waples D (1973) Die Bevorzugung geradzahliger *n*-Alkane in Sedimentgesteinen. Naturwissenschaften 60: 516–517
Welte DH, Schaefer RG, Stoessinger W, Radke M (1984) Gas generation and migration in the Deep Basin of Western Canada. Mitt Geol-Paläontol Inst Univ Hamburg 56: 263–285. AAPG Mem 38: 35–47
Wenger LM, Baker DR (1987) Variations in vitrinite reflectance with organic facies - examples from Pennsylvanian cyclothems of the Midcontinent, U.S.A. Org Geochem 11: 411–416
White D (1915) Geology: Some relations in origin between coal and petroleum. J Wash Acad Sci 5: 189–212
Wilcoxon BR, Ferrell RE, Sassen R, Wade WJ (1990) Illite polytype distribution as an inorganic indicator of thermal maturity in the Smackover formation of the Manila embayment, southwest Alabama. Org Geochem 15: 1–8
Wilson MA, Philp RP, Gillam AH, Gilbert TD, Tatle KR (1983) Comparison of the structures of humic substances from aquatic and terrestrial sources by pyrolysis gas chromatography-mass spectrometry. Geochim Cosmochim Acta 47: 497–502
Wilson MA, Pugmire RJ, Karas J, Alemany LB, Woolfenden WR, Grant DM, Given PH (1984) Carbon distribution in coals and coal macerals by cross polarization magic angle spinning carbon-13 nuclear magnetic resonance spectrometry. Anal Chem 56: 933–943
Yalçın MN, Welte DH (1988) The thermal evolution of sedimentary basins and significance for hydrocarbon generation. TAPG Bull 1/1: 12–26

Chapter 4
Kinetics of Petroleum Formation and Cracking

Chapter 4: Overview and Insights

After parameters for maturation assessment of source rocks and petroleum were established, it became clear that two kinds of information were still missing for predictive purposes in petroleum exploration prior to drilling: information about quantities of petroleum generated at a given maturity and the timing of generation. Petroleum explorationists agreed that for exploration strategies in a typical predrilling situation, while exploration still relies on interpreted seismics, a means had to be developed quantitatively to assess possible petroleum generation in source rocks.

Heating experiments with immature kerogen have revealed that petroleum-generating reactions in nature and in the laboratory are quasi-irreversible with (first-order) rate constants obeying the Arrhenius equation. The precursors of petroleum, or the educts for the petroleum generation reactions, cannot be rigorously defined. Similarly the reaction product, petro-leum, cannot be identified in terms of all its thousands of components. Therefore the strict physicochemical approach of classical reaction kinetics of a well-defined precursor-product relationship under specific reaction conditions cannot be applied. Instead geochemists have developed a simplified "gross" kinetic, or "pseudo" kinetic concept, whereby molecular precursors are replaced by "bulk petroleum" potentials which represent fractions of total prod-uct yields. Upon heating in nature or in the laboratory these potentials are said to be transformed into the product "petroleum" by a certain number of parallel pseudoreactions, each of which is kinetically characterized by a reaction order and a rate constant. The rate constants are assumed to obey the Arrhenius equation with activation energies ranging from 40 to 80 kcal/mol and a preexponential or frequency factor in the order of magnitude of the frequencies of molecular vibrations. A typical frequency factor is around 10^{14} s^{-1}.

This gross or pseudokinetic approach very elegantly solves the problem of the multitude of chemical reactions involved during oil and gas generation. This simplification of kinetics also provides the link between the short-time high-temperature reactions in the laboratory, on the one hand, and the long-time low-temperature reactions in sedimentary basins, on the other. In addition, this concept has recently been applied to the generation of methane and gaseous nonhydrocarbon compounds such as molecular nitrogen.

The stability of oil in deeply buried reservoirs or in general the survival of liquid oil under high temperature regimes is another important aspect of reaction kinetics under natural geological conditions. There are indications that the depth range for oil-to-gas conversion is likely to occur at greater depth than depth ranges over which petroleum is generated from kerogen. Naturally this aspect is of interest for geologically early generated and pooled oil which has been buried deeper during basin evolution.

Kinetics of Petroleum Formation and Cracking

H.J. Schenk, B. Horsfield, B. Krooss, R.G. Schaefer, and K. Schwochau[1]

4.1 Introduction

One of the most fundamental problems in basin modeling as related to petroleum exploration is assessing the temporal and spatial limits of petroleum generation in sedimentary basins. It is well known that petroleum is generated from macromolecular sedimentary organic matter as it thermally degrades upon burial. The multitude of chemical reactions involved are unknown in detail (Philippi 1965; Welte 1965) but are recognized to be quasi-irreversible (Huck and Karweil 1955; Hanbaba and Jüntgen 1969; Tissot 1969). The organic components of subsiding sedimentary rocks are generally far away from thermodynamic equilibrium (Dayhoff et al. 1967; Tackach et al. 1987). Consequently, the formation of oil and gas in nature is controlled by chemical reaction kinetics, in particular by non-isothermal kinetics because temperature changes as a function of time under geological conditions (Tissot and Espitalié 1975).

Historically, the idea of describing organic matter maturation in terms of chemical kinetics was developed to account for the close similarity between coal pyrolysis and natural coalification (van Krevelen et al. 1951; Pitt 1962; Jüntgen 1964; Hanbaba and Jüntgen 1969). Comparable concepts for the formation of petroleum from kerogen were elaborated somewhat later and were mainly derived from a kinetic interpretation of geochemical data from source rocks of increasing maturity (Tissot 1969). Nowadays all attempts to establish kinetic models for petroleum generation are essentially based on the following assumptions:

- Petroleum generation is the result of a large number of chemical reactions leading from kerogen to liquid and gaseous products of lower molecular weight and to residues of increasing degree of condensation.
- These reactions are governed by the basic laws of chemical kinetics although these were originally developed for well-defined and relatively simple reactions in the gas phase and in dilute solutions.

[1]Institut für Erdöl und Organische Geochemie (ICG-4), Forschungszentrum Jülich GmbH, 52425 Jülich, Germany

Welte et al. (eds)
Petroleum and Basin Evolution
© Springer-Verlag Berlin Heidelberg 1997

- Laboratory pyrolysis performed at temperatures substantially higher than those occurring under usual geological conditions does not duplicate, but to a certain extent simulates the processes of petroleum formation and yields kinetic parameters that can be incorporated into geothermal histories (Chap. 2) for calculating the timing and the intensity of petroleum generation in sedimentary basins.

This chapter treats the various aspects of petroleum generation kinetics in a pragmatic way with special emphasis on the role of kinetic models as a valuable link between short time–high temperature (laboratory) and long time–low temperature (nature) configurations. For a comprehensive description of reaction mechanisms on a molecular level the reader is referred to the excellent review by Ungerer (1990); possible catalytic effects have recently been discussed by Mango et al. (1994). After a brief introduction of the basic laws of chemical kinetics (Sect. 4.2) there is a more detailed discussion of gross kinetic concepts which are currently used to model bulk petroleum generation (Sect. 4.3). This section furthermore justifies the second assumption listed above, i.e., the application of simple kinetic laws to the decomposition of complex organic solids. Kinetic studies also play an important role in investigating the sources for and generation and accumulation mechanisms of natural gas. There is an increased tendency to extend this research also to nonhydrocarbon gases. Thus the occurrence of nitrogen-rich natural gas accumulations in northern Germany prompted research efforts to establish the generation kinetics of methane and molecular nitrogen (N_2) from coals and related organic matter. First results of this investigation are summarized in Section 4.4.

Section 4.5 is devoted exclusively to the third assumption and summarizes the problems and possible shortcomings arising from the extrapolation of laboratory derived kinetic models to conditions within a naturally subsiding basin. The conversion of oil to gas in petroleum reservoirs (Sect. 4.6) covers the final stage of the natural evolution leading from kerogen to pooled petroleum; intermediate steps such as secondary cracking of primary products in overmature source rocks and during migration are omitted because these phenomena are up to now very poorly understood.

4.2 Concepts of Chemical Kinetics

Chemical reactions are processes which involve the transformation of molecules by cleavage and reformation of chemical bonds. The evolution of the theories of chemical bonding resulted in the establishment of chemical kinetics as a branch of physical chemistry concerned with the investigation of the rates and mechanisms of individual chemical reactions.

Chemical reactions are commonly represented in terms of stoichiometric equations, for instance:

$$n_A A + n_B B \longrightarrow n_C C + n_D D \tag{4.1}$$

where the stoichiometric coefficients n_i denote the number of molecules of constituent i participating in this particular reaction. In the context of petroleum generation generally decomposition reactions of the type

$$n_A A \longrightarrow n_B B + n_C C + \ldots \longrightarrow n_D D + \ldots \tag{4.2}$$

(respectively, the precursor, intermediate steps, and product) are considered. An example of a well-defined decomposition reaction related to organic geochemistry is the thermal decarboxylation of acetic acid:

$$CH_3COOH \longrightarrow CH_4 + CO_2 \tag{4.3}$$

Kharaka et al. (1983) who investigated the kinetics of this reaction under the aspect of the role of carboxylic acids in natural gas formation found that it follows a first-order rate law (see Sect. 4.2.1) with respect to carboxylic acid.

The above stoichiometric formulations establish a precursor (educt)-product relationship for the *overall* reaction. They yield, however, no information on the mechanisms and the rate laws of these reactions. Chemical reactions represented in this way usually involve a number of intermediate steps, so-called *elementary* reactions. These elementary reactions may lead to the formation of short-lived, unstable intermediate compounds which do not appear in the formulation of the overall equation. Each of these elementary reactions may affect the kinetics of the overall reaction. Thus in a series of consecutive elementary reactions the overall reaction rate is determined by the slowest reaction step, whereas in a set of parallel reactions the fastest elementary reaction is rate determining. The assessment of the mechanism of chemical reactions and establishment of the corresponding rate laws is the main objective of chemical kinetics.

4.2.1 Rate Laws and Order of Reactions

The increase or decrease per unit time of the concentration of the molecular species participating in a given reaction is termed the *rate of the reaction* or *conversion rate*. In a well-defined chemical reaction performed under controlled conditions the conversion rate can be expressed by the rate of change of concentration of each compound participating in the reaction. The rate of decrease in the educt concentrations is related, through the stoichiometric coefficients, to the rate of increase in product concentrations.

For the generalized stoichiometric Eq. (4.1) the conversion rate is given by:

$$-\frac{1}{n_A}\cdot\frac{d(A)}{dt} = -\frac{1}{n_B}\cdot\frac{d(B)}{dt} = \frac{1}{n_c}\cdot\frac{d(C)}{dt} = \frac{1}{n_D}\cdot\frac{d(D)}{dt} \tag{4.4}$$

Here the parentheses denote the concentrations of the individual components.

Most theories of chemical kinetics have evolved from the idea of collisions of molecules either in the gas phase or in solutions resulting in a chemical reaction. Evidently, the rate of a reaction:

$$A + B \longrightarrow C \tag{4.5}$$

depends on the probability of a collision between two molecules A and B which, in turn is a function of the concentrations of these molecules. Consequently a rate law for this reaction should have the form:

$$-\frac{d(A)}{dt} = -\frac{d(B)}{dt} = k \cdot (A) \cdot (B) \tag{4.6}$$

where k is the *rate constant* of this reaction and the quantities (A), (B), etc. denote the concentrations of the reacting components.

Similarly, a reaction:

$$A + A \longrightarrow C \tag{4.7}$$

would be expected to proceed according to a rate law of the type:

$$-\frac{d(A)}{dt} = k \cdot (A) \cdot (A) = k \cdot (A)^2 \tag{4.8}$$

Generally, the conversion rate of many chemical reactions can be described by a *rate law* of the following exponential type:

$$-\frac{d(A)}{dt} = k \cdot (A)^x \cdot (B)^y \tag{4.9}$$

By definition, the sum of exponents (x+y+...) in the rate-law equation is termed the *order* of the reaction. Thus, both rate laws in Eqs. (4.6) and (4.8) are of the order 2 (second-order reactions). On the other hand, a decomposition reaction of the kind:

$$A \longrightarrow B + C + \dots \tag{4.10}$$

would be expected to follow a *first-order* rate law involving only the concentration of the educt (A) at a given time:

$$-\frac{d(A)}{dt} = k \cdot (A) \tag{4.11}$$

The order of a chemical reaction cannot be derived from its stoichiometric formula but must be determined experimentally. The common approach to determine the rate law of a chemical reaction is to observe the formation or decay of certain components under well-defined isothermal conditions and to find a rate law that describes the experimental data. More complicated rate laws involving, for instance, noninteger overall reaction orders may result from the combination of several forward and backward elementary reactions. The different techniques to determine rate laws of chemical reactions are referenced in most textbooks of physical chemistry. For a comprehensive treatment of the theories of reaction kinetics the reader is referred to Bamford and Tipper (1969). An introduction to the methods which have been applied in geochemical kinetics has been given by Lasaga (1981a).

4.2.2 Temperature Dependence of Reaction Rates

It is a common experience that chemical reactions are temperature dependent, and that relatively small increases in temperature can result in a considerable acceleration of chemical reactions. Several approaches have been made to quantify this temperature dependence in the kinetic modeling of petroleum generation. A very simplistic approach is based on the "rule of thumb" that a reaction rate approximately doubles with a temperature increase of 10 °C. This rule has no theoretical background but is purely empirical. Put into a mathematical form this relationship reads:

$$k(T_0 + n \cdot 10) = k_0 \cdot 2^n \tag{4.12a}$$

If the temperature, T, is expressed by:

$$T = T_0 + n \cdot 10 \tag{4.13}$$

this relationship can be written as:

$$\ln \frac{k(T)}{k_0} = \frac{(T - T_0)}{10} \cdot \ln 2 \tag{4.12b}$$

where k_0 denotes the reaction rate constant at a reference temperature T_0 and k(T) is the rate constant at the temperature (T). This "rule of thumb" is the basis of the time-temperature index concept established by Lopatin (1971; see Waples 1980) for the reconstruction of thermal histories by means of vitrinite reflectance data. Although this approach is outdated and at best acceptable for low maturity levels of source rocks, it is still applied.

A substantially better approach to quantify the temperature dependence of chemical reactions is the semiempirical Arrhenius equation. This equation, which is essentially based on van't Hoff's theory of thermodynamic equilibrium, involves the concept of an activation energy, i.e., a potential barrier between educt and product compounds that must be overcome during a chemical reaction. According to the Arrhenius theory the temperature dependence of the rate "constant" k is given by the relationship:

$$k(T) = A \cdot \exp\left(\frac{-E}{RT}\right) \tag{4.14a}$$

or, in a logarithmic form:

$$\ln k(T) = \ln A - \frac{E}{R} \cdot \frac{1}{T} \tag{4.14b}$$

Here T is the absolute temperature in Kelvin, E is the activation energy of the reaction (in J/mol or cal/mol), and R is the gas constant (J mol^{-1} K^{-1} or cal mol^{-1} K^{-1}). In a reaction following a first-order kinetic rate law, as considered here, the preexponential factor A has the dimension of a frequency (t^{-1}). It is therefore also termed "frequency factor" and has been related to the vibration frequency (Polanyi and Wigner 1928) of the "activated complex," an intermediate molecular species formed in the transition state of the reaction. The factor RT is a measure of the thermal energy of the system at a given temperature T. If this factor is small compared to the activation energy E, the reaction rate constant k is also small.

Comparison of Eqs. (4.12) and (4.14) immediately reveals the difference between the two formulae. While in the Arrhenius equation (4.14) the rate coefficient k approaches an upper limit (equal to A) with increasing temperature, the rate coefficient of the "rule of thumb" Eq. (4.12) increases infinitely with temperature. The logarithm of k(T) in Eq. (4.12b) is a linear function of temperature (T) whereas in Eq. (14b) a hyperbolic relationship is obtained.

A more sophisticated theory of chemical kinetics, the transition state theory (TST) was developed by Eyring (1935a,b) on the basis of statistical thermodynamics. The TST predicts a similar type of functional relationship for the temperature dependence of reaction rate as the Arrhenius function but, in addition, a temperature dependence of the preexponential factor is obtained:

$$k = A(T) \cdot \exp\left(-\frac{E}{RT}\right) \tag{4.15}$$

The TST yields a good representation of gas phase reactions. However, additional assumptions must be introduced for the treatment of reactions in liquids and solids. A detailed review of the TST and its applications in geochemistry has been given by Lasaga (1981b). The temperature dependence of the preexponential factor as derived by the TST is relatively small and cannot be readily determined with sufficient precision in pyrolysis experiments aimed at the establishment of kinetic parameters for petroleum generation. Therefore the TST approach is not used in geochemical modeling of petroleum formation. It provides, however, the theoretical basis that supports the application of the Arrhenius theory as an approximation for the temperature dependence of petroleum generation reactions.

In summary, two principal functional relationships have been used in the description of the temperature dependence of petroleum generation one of which is empirical and the other one based on well-established reaction kinetic theories. As shown above and elaborated for instance by Wood (1988) and Robert (1985) these two functions are incompatible and yield different results when applied to the formation of petroleum in natural systems under different thermal regimes. From a physicochemical point of view and in terms of applicability the Arrhenius theory is considered as the most adequate approach to account for the temperature dependence of petroleum generation reactions (see Sect. 4.3).

4.2.3 Fundamentals of Non-isothermal Kinetics

To investigate the temperature dependence of reaction rates, kinetic experiments are traditionally performed at a number of different temperatures. The most common procedure in this "isothermal approach" is to prepare an Arrhenius diagram, i.e., a plot of the logarithm of the reaction rate constant (ln k) over the reciprocal of the absolute reaction temperature (1/T). The slope of this Arrhenius plot yields the activation energy (E) and the intercept with the (ln k) axis the preexponential factor (A).

Conducting "non-isothermal" kinetic experiments at a *constant heating rate* (r) is an alternative approach. The rate laws used in this method are the same as for "isothermal kinetics," but due to an additional variable (heating rate) the mathematical evaluation of the experiments is different. In contrast to the isothermal approach, a single experiment with a linear heating rate yields the activation energy and frequency factor of a given reaction. In practice, however, three heating rates are used for geochemical work. The reasons for this are explained in Section 4.3.1. A comprehensive treatment of non-isothermal reaction kinetics has been given by Jüntgen and van Heek (1970) and Koch (1977). These articles cover the experimental methods, mathematical treatment and evaluation of experimental data and give numerous examples of reactions investigated by this method, especially reactions related to coal pyrolysis.

For a first-order reaction conducted under non-isothermal conditions with a constant heating rate the rate equation can be derived as follows:

According to the first-order rate law the rate of generation of product (conversion rate) is proportional to the remaining amount of educt (precursor substance) at a given time:

$$\frac{dm}{dt} = k \cdot (M - m) \tag{4.16}$$

where m is the mass of product generated (mass of educt consumed) at time t, M is the initial mass of educt at time t_o, M–m is the residual mass of educt at time t, and k is the reaction rate constant. Assuming furthermore a linear heating rate:

$$r = \frac{dT}{dt} \tag{4.17}$$

and a temperature dependence of the reaction rate according to the Arrhenius relationship (4.14a) the generation rate as a function of temperature (the "generation curve") is:

$$\frac{dm}{dT} = \frac{dm}{dt} \cdot \frac{dt}{dT} = \frac{dm}{dt} \cdot \frac{1}{r} \tag{4.18}$$

Combination of Eqs. (4.14a), (4.16), and (4.18) yields:

$$\frac{dm}{dT} = \frac{A}{r} \cdot \exp\left(-\frac{E}{RT}\right) \cdot (M - m) \tag{4.19}$$

And upon integration ($T = T_o$ at $t = t_o$) one obtains:

$$M - m = M \cdot \exp\left(-\frac{A}{r} \cdot \int_{T_0}^{T} \exp\left(-\frac{E}{RT}\right) dT\right) \tag{4.20}$$

Combination of Eqs. (4.19) and (4.20) yields the function for a non-isothermal generation curve for a first-order reaction:

$$\frac{dm}{dT} = M\frac{A}{r}\exp\left(-\frac{E}{RT} - \frac{A}{r}J\right)$$
$$J = \int_{T_0}^{T} \exp\left(-\frac{E}{RT}\right) dT \tag{4.21}$$

In a similar way expressions for second- and higher order reactions can be derived.

4.3
Bulk Petroleum Generation

4.3.1
Kinetic Models

The natural or pyrolytic transformation of kerogen into petroleum cannot be exactly described in terms of individual precursor-product relationships. Neither the educt (kerogen) nor the product (petroleum) of this process is a well-defined chemical compound. In particular, the tremendously complex structure of kerogens as exemplified by the models of Burlingame et al. (1969), Yen (1974), Oberlin et al. (1980), and Béhar and Vandenbroucke (1986) requires the application of simplified "gross" kinetic concepts, whereby molecular precursors of oil and gas are replaced by so-called bulk petroleum or gross hydrocarbon potentials M_i which are fractions of total product yields (Tissot and Espitalié 1975). During pyrolysis or natural evolution of the organic matter these potentials are assumed to be more or less simultaneously transformed into the product "petroleum" by a certain number of n parallel pseudoreactions each of which is kinetically characterized by a reaction order and a rate constant k_i. The pyrolytic release of volatile products from coal has been shown to be consistent with the assumption of first-order kinetics (van Krevelen et al. 1951; Pitt 1962; Jüntgen 1964; Jüntgen and Klein 1975; Jüntgen 1984). The extension of this conclusion to the thermal decomposition of sedimentary organic matter in general is supported by theoretical considerations concerning the unimolecular decay of complex molecules (Hinshelwood 1927) and by a statistical treatment of the formation of petroleum from kerogen (Tissot 1969). Additional experimental reasons corroborating the assumption of first-order kinetics are discussed below (The effect of kerogen type and maturity). Thus the rate dM/dt of bulk petroleum generation:

$$\frac{dM}{dt} = \sum_{i=1}^{n} k_i.(M_i - m_i) \tag{4.22}$$

is formally very similar to Eq. (4.16) of Section 4.2, with m_i being the fraction of partial bulk petroleum potential M_i which is already transformed into petroleum at time t.

The rate constants k_i are generally believed to obey the Arrhenius equation (Eq. 14a, Sect. 4.2) with activation energies E_i expected somewhere between 40 and 80 kcal/mol or 167 and 334 kJ/mol (Ungerer et al. 1986) and pre-exponential factors A_i that are predicted to be of the order of magnitude of the frequencies of molecular vibrations (Polanyi and Wigner 1928; Hanbaba and Jüntgen 1969). Immediate evidence for an Arrhenius-like temperature dependence has been provided partly by heating experiments involving the generation of CO_2 and H_2O from solid organic acids (van Heek et al. 1967) and of gaseous hydrocarbons from coal (Hanbaba et al. 1968; Hanbaba and Jüntgen 1969), and partly by the kinetic interpretation of geological observations such as the nonlinear increase of extractable bitumen with increasing burial temperature of source rocks (Tissot 1969) or the inverse relationship between age and temperature of producing formations (Connan 1974). For practical computing reasons the preexponential factors are mostly assumed to be the same for all parallel reactions, with the consequence that these reactions proceed in the order of increasing activation energy with increasing temperature, whatever the temperature range (Tissot and Espitalié 1975; Ungerer 1990). Assuming a linear heating rate of $dT/dt = r$, the integration of Eq. (4.22) yields the temperature-related rate dM/dT of bulk petroleum generation as a function of temperature T:

$$\frac{dM}{dT} = \sum_{i=1}^{n} M_i \frac{A}{r} \exp\left(-\frac{E_i}{RT} - \frac{A}{r} J_i\right)$$
$$J_i = \int_0^T \exp\left(\frac{-E_i}{RT}\right) dT \qquad (4.23)$$

which closely resembles Eq. (4.21) of Section 4.2 with the current approximation of $T_o=0$.

Equations (4.22) and (4.23) presume a limited number of n parallel reactions or a discrete bulk petroleum potential versus activation energy distribution. Such discrete models have been reported in many kinetic studies (Tissot and Espitalié 1975; Ungerer et al. 1986; Ungerer and Pelet 1987; Burnham et al. 1987, 1988; Espitalié et al. 1988; Schaefer et al. 1990; Issler and Snowdon 1990; Forbes et al. 1991; Jarvie 1991; Schenk and Horsfield 1993). These models generally involve the optimization of n+1 parameters, namely n gross hydrocarbon or bulk petroleum potentials M_i associated with the n activation energies E_i and one preexponential factor A. The number of parameters to be optimized can be reduced by assuming an infinite number of parallel reactions with a Gaussian distribution of gross hydrocarbon potential versus activation energy (Pitt 1962; Hanbaba et al. 1968; Campbell et al. 1980; Braun and Burnham 1987, 1990; Burnham et al. 1988; Mackenzie and Quigley 1988; Burnham and Braun 1990). Discrete models, however, fit better to experimental hydrocarbon formation rate versus temperature curves, which, depending on kerogen type and maturity, are more or less asymmetric (Burnham et al. 1987; Burnham et al. 1988; Ungerer 1990).

4.3.2 Model Calibration Against Programmed-Temperature Open-System Pyrolysis

Although some of the above kinetic models were derived from geochemical observations in nature (Tissot and Espitalié 1975; Ungerer et al. 1986), most of them were calibrated against open-system non-isothermal pyrolysis (e.g., Rock-Eval) with various temperature programs. Generally, the preexponential factor A and the gross hydrocarbon potentials M_i associated with the assumed activation energies E_i are optimized by a least-squares iteration method that compares measured and calculated bulk petroleum formation rates at a certain number of temperatures until the corresponding error function (sum of squared differences) presents a well-defined absolute minimum.

The optimization procedure is known to produce artifacts if the error function presents several minima corresponding to very different kinetic parameters (Ungerer 1990). Extremely low or high values of both preexponential factors and activation energies may still reproduce the main features of pyrolysis curves (e.g., temperatures of maximum generation rate) but lead to erroneous results when extrapolated to geological heating rates (Jüntgen and Klein 1975; Welte et al. 1988). The reliability of output parameters can be considerably enhanced if bulk petroleum evolution profiles are measured using at least two heating rates which differ by one to two orders of magnitude. This is because a reasonable starting value of A can be derived from the shift of peak generation temperatures (van Heek and Jüntgen 1968; Ungerer and Pelet 1987; Burnham et al. 1988; Schaefer et al. 1990). Furthermore, the heating rates are recommended to be as low as possible in order to ensure correct temperature measurements (Gregg et al. 1981; Burnham et al. 1988; Schaefer et al. 1990).

The experimental devices which have been used to perform open-system programmed-temperature pyrolysis range from standard Rock-Eval (e.g., Ungerer and Pelet 1987; Burnham et al. 1988) to self-built instruments (e.g., Schaefer et al. 1990). Details are found in the quoted literature. Two important prerequisites, however, should be retained. Care must be taken to place the finely ground kerogen or rock sample in the furnace in such a way as to facilitate heat and mass transfer and to ensure that the decomposition of the organic matter and not the diffusion of products is overall rate controlling (Jüntgen and van Heek 1969; Rajeshwar and Dubow 1982). Furthermore, in most experimental configurations a flow of inert gas (argon or helium) is maintained during pyrolysis in order to transport released products rapidly from the sample to the flame ionization detector. This is done to minimize secondary cracking reactions, the extent of which is dependent upon the residence time of primary products in the hot zone of the oven (van Heek et al. 1967).

The Influence of Kerogen Type and Maturity

Figures 4.1–4.3 compare gross hydrocarbon (bulk petroleum) formation rate versus temperature profiles for selected lacustrine (Fig. 4.1), marine (Fig. 4.2) and terrestrial (Fig. 4.3) source rocks at heating rates of 0.1, 0.7, and 5.0 K/min. Curves defined by dotted lines represent the experimental input data smoothed by appropriate spline functions. Curves defined by solid lines were calculated as best fit from the kinetic parameters of Figs. 4.1–4.6 according to the mathematical model of Schaefer et al. (1990) which assumes 25 parallel reactions with activation energies regularly spaced between 46 and 70 kcal/mol (192 and 293 kJ/mol) and a single preexponential factor. The Green River Shale sample (C_{org}=9.61%) is AP 22 of Burnham et al. (1987). The Toarcian (Posidonia) Shale sample (C_{org}=12.70%) was obtained from the Lower Saxony Basin of northern Germany (Mann 1987). The coal sample was obtained from the South China Sea area of Indonesia. All samples can be classified as immature (R_r~0.5%) with hydrogen indices of 720, 660, and 170 mg/g C_{org}, respectively.

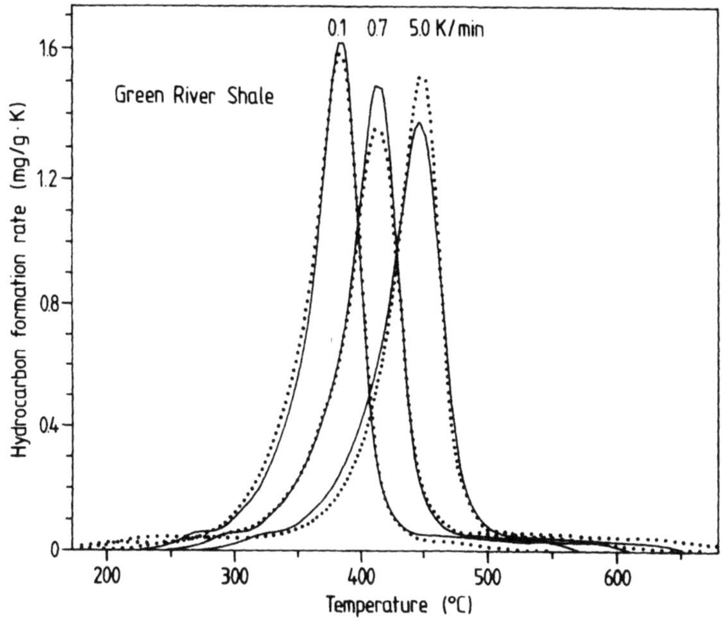

Fig. 4.1. Gross hydrocarbon (bulk petroleum) formation rates (milligram per gram of rock and per degree) from open-system pyrolysis of a Green River Shale sample as a function of temperature (°C) at heating rates of 0.1, 0.7, and 5.0 K/min. (Schenk and Horsfield, unpublished)

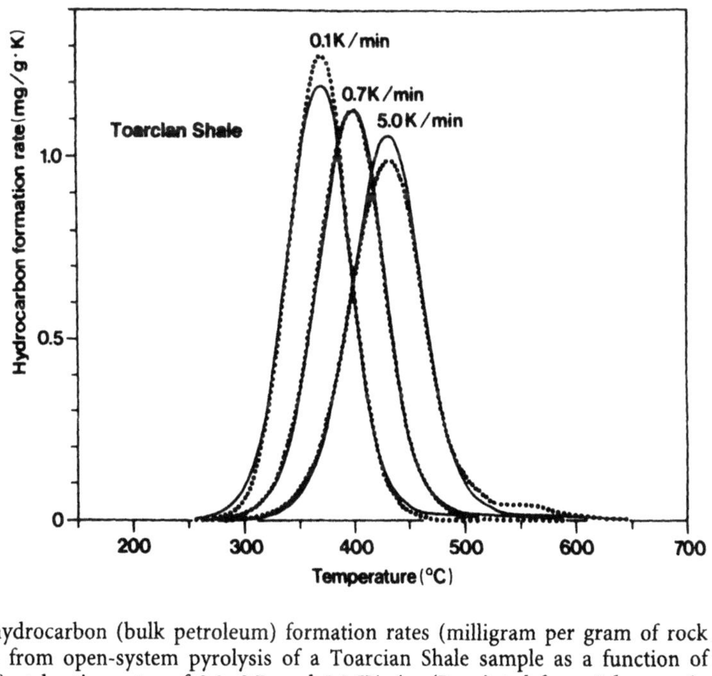

Fig. 4.2. Gross hydrocarbon (bulk petroleum) formation rates (milligram per gram of rock and per degree) from open-system pyrolysis of a Toarcian Shale sample as a function of temperature (°C) at heating rates of 0.1, 0.7, and 5.0 K/min. (Reprinted from Advances in Organic Geochemistry 1989, Schaefer et al., Determination of gross kinetic parameters for petroleum formation from Jurassic source rocks of different maturity levels by means of laboratory experiments, 1990, pp 115–120, with kind permission from Elsevier Science Ltd, The Boulevard, Langford Lane, Kidlington OX5 1GB, UK)

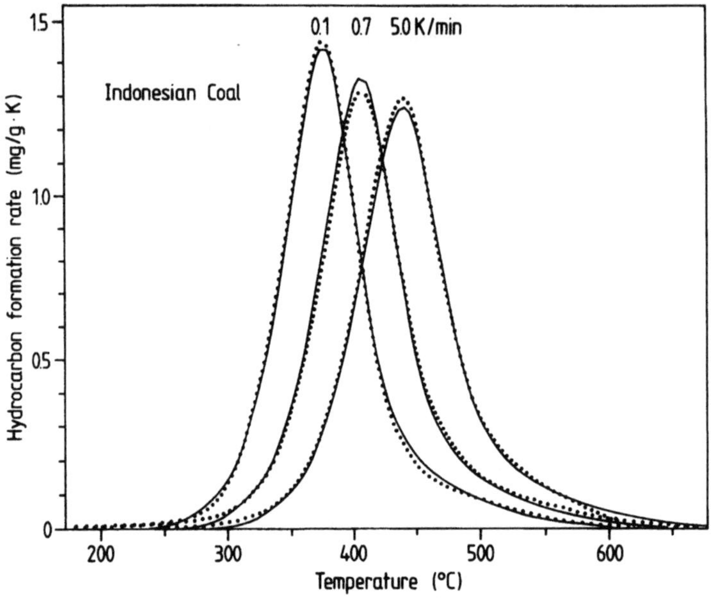

Fig. 4.3. Gross hydrocarbon (bulk petroleum) formation rates (milligram per gram of coal and per degree) from open-system pyrolysis of an Indonesian coal sample as a function of temperature (°C) at heating rates of 0.1, 0.7, and 5 K/min. (Horsfield and Schenk, unpublished)

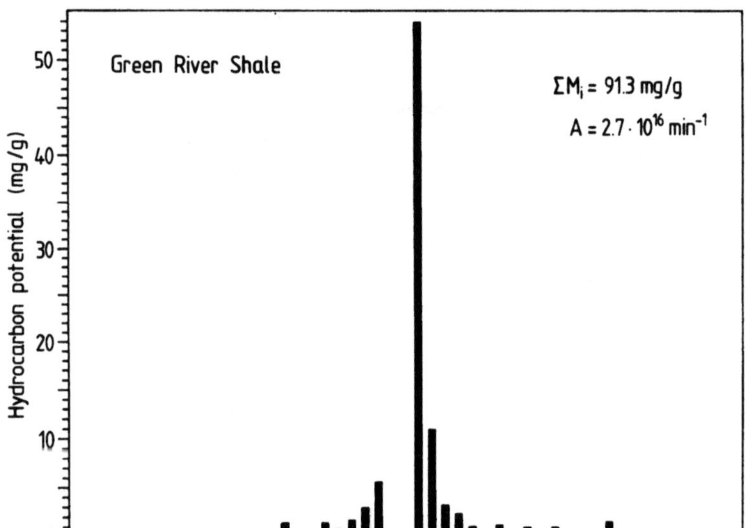

Fig. 4.4. Gross hydrocarbon potential (milligram per gram of rock) versus activation energy (kcal/mol) as derived from open-system pyrolysis of a Green River Shale sample ΣM_i is the total potential (milligram per gram of rock) and A is the preexponential factor (per minute). (Schenk and Horsfield, unpublished)

The observed large temperature shift of pyrolytic product generation with increasing rate of heating is expected from non-isothermal kinetics (Jüntgen and van Heek 1970). The evolution profiles obtained from the lacustrine kerogen (Fig. 4.1) are characterized by a gradual rise of reaction rate up to T_{max} and by a rather steep decline at higher temperatures. This is in excellent agreement with the curve shape predicted by Eq. 21 of Section 4.2 for a single first-order reaction. Consequently, the potential vs activation energy distribution (Fig. 4.4) turns out to be very narrow with one dominant potential at 56 kcal/mol (234 kJ/mol). By contrast, a very broad distribution tailing up to 70 kcal/mol (293 kJ/mol) is obtained for the Indonesian coal (Fig. 4.6) and other coals (Yalçin et al. 1994). A similar type dependence has been reported by Ungerer and Pelet (1987).

While the three distributions described above typify many lacustrine, anoxic marine and fluviodeltaic settings (e.g., Burnham 1990; Sachsenhofer 1994), it should not be construed that this is always the case. For instance, it is now well-known that sulfur-rich oil-prone kerogens such as from the Monterey or Italian source rocks have a broad distribution of energies whose mean lies at appreciably lower values than the range shown in Fig. 4.5 (Tissot et al. 1987; di Primio 1995). Additionally, some horizons of the Toarcian Shales are characterized by a very narrow activation energy distribution that more closely resembles that of the Green River Shale than the Toarcian Shale distribution shown in Fig. 4.5 (Düppenbecker and Horsfield 1990). Another example is

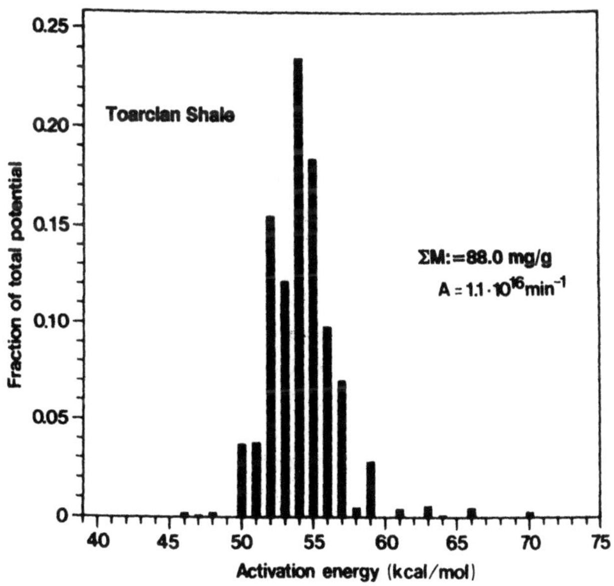

Fig. 4.5. Fraction of total gross hydrocarbon potential versus activation energy (kcal/mol) as derived from open-system pyrolysis of a Toarcian Shale sample. ΣM_i is the total potential (milligram per gram of rock) and A is the preexponential factor (per minute). (Reprinted from Advances in Organic Geochemistry 1989, Schaefer et al., Determination of gross kinetic parameters for petroleum formation from Jurassic source rocks of different maturity levels by means of laboratory experiments, 1990, pp 115–120, with kind permission from Elsevier Science Ltd, The Boulevard, Langford Lane, Kidlington OX5 1GB, UK)

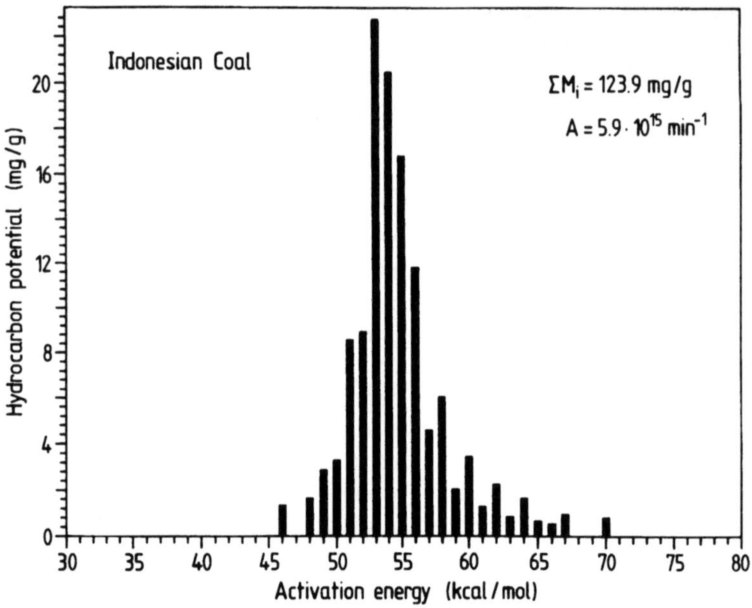

Fig. 4.6. Gross hydrocarbon potential (milligram per gram of coal) versus activation energy (kcal/mol) as derived from open-system pyrolysis of an Indonesian Coal sample. ΣM_i is the total potential (milligram per gram of coal) and A is the preexponential factor (per minute). (Horsfield and Schenk, unpublished)

provided by the Green River Formation in the Washakie Basin of Wyoming. Here, the activation energy distribution for the freshwater Luman Tongue Member resembles the classic Uinta Basin Green River Shale distribution shown in Fig. 4.4 whereas that for the alkaline Laney Shale Member is exceedingly broad, pointing to the occurrence of a heterogeneous rather than homogeneous kerogen structure (Horsfield et al. 1994).

Specific relationships have not yet been clearly established between kerogen structure and kinetic parameters because of the physicochemical complexity of most kerogens. Some important conclusions can nevertheless be drawn. First, it is well known that Green River Shale kerogen from the Uinta Basin has a rather homogeneous composition whereas humic coals are physically and/or chemically more heterogeneous (Rullkötter and Michaelis 1990). This close accordance of compositional and kinetic findings probably provides the most convincing support for the validity of first-order kinetic models; second- or higher order kinetics would result in a broader distribution for type I and a narrower distribution for type III kerogens. Second, heteroelements, especially sulfur, exert a crucial control on the thermal lability of kerogen. This has been most comprehensively documented by Tegelaar and Noble (1994) using a large database from a variety of organic facies. Finally, Schenk et al. (1993) found that the thermal stability of a suite of Green River Shale kerogens (Luman Tongue and Laney Shale Members, Washakie Basin) was directly linked to the proportion of aliphatic carbon that aromatizes on pyrolysis, with highest stabilities being related to high concentrations of alicyclic and bridged moieties. The latter was also associated with high relative abundances of C_2-alkylphenols in the total phenol fraction (Horsfield et al. 1994). The origin of the alicyclic and oxygen functions was not known but formation via diagenetic reactions rather than selective preservation was suggested based on literature review.

Figure 4.7 compares the gross hydrocarbon (bulk petroleum) potential vs activation energy distributions derived from open-system pyrolysis of Toarcian Shale samples of increasing maturity (Schaefer et al. 1990). The corresponding vitrinite reflectance values range from R_r=0.48% (sample WEN) through R_r=0.88% (sample HAR) to R_r=1.45% (sample HAD); sample WEN is identical to the sample of Figs. 4.2 and 4.5. Since the differences in depositional environment between the three samples and their drilling locations are believed to be negligible (Rullkötter et al. 1988) the variations of kinetic parameters are mainly related to the different maturity levels. The most striking changes are observed in going from the immature WEN to the mature HAR stage. The partial transformation of kerogen into oil and gas during natural maturation is evidenced by the net disappearance of pyrolytic potentials available at low activation energies (50–57 kcal/mol = 209–238 kJ/mol). The progressive release of low molecular weight compounds during catagenesis, however, is accompanied by a shift of the remaining pyrolytic potentials to higher activation energies (58–62 kcal/ mol = 242–259 kJ/mol). This shift is essentially linked to the kinetic model which employs a single preexponential

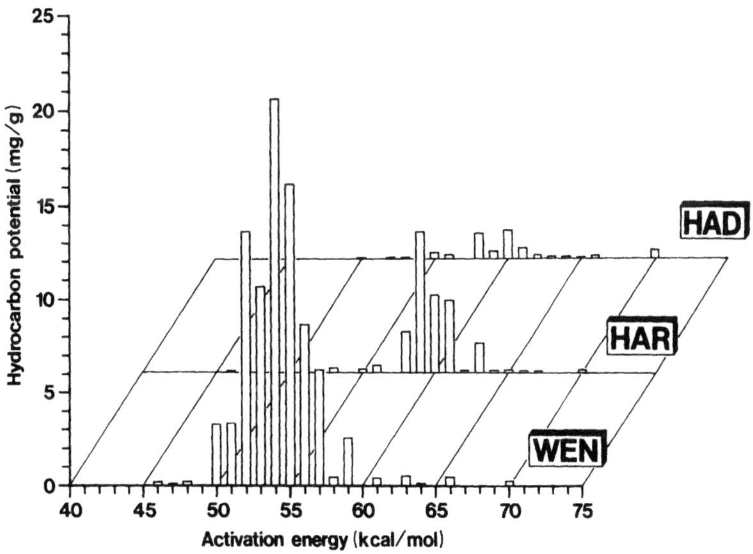

Fig. 4.7. Gross hydrocarbon potential (milligram per gram of rock) versus activation energy (kcal/mol) as derived from open-system pyrolysis of Toarcian Shale samples of increasing maturity. WEN: $R_r = 0.48\%$; HAR: $R_r = 0.88\%$; HAD: $R_r = 1.45\%$. The preexponential factors are $1.1 \cdot 10^{16}$ (WEN), $1.3 \cdot 10^{17}$ (HAR), and $4.4 \cdot 10^{16}$ min^{-1} (HAD). (Reprinted from Advances in Organic Geochemistry 1989, Schaefer et al., Determination of gross kinetic parameters for petroleum formation from Jurassic source rocks of different maturity levels by means of laboratory experiments, 1990, pp 115–120, with kind permission from Elsevier Science Ltd, The Boulevard, Langford Lane, Kidlington OX5 1GB, UK)

factor for all parallel reactions and reflects the increase in this factor by one order of magnitude in going from the WEN to the HAR stage. Probably these high energy potentials can still be released during *pyrolysis* but cannot be mobilized upon further *natural evolution* of the kerogen from the mature HAR to the overmature HAD stage (Witte et al. 1988).

The conclusion that there has been no significant natural generation of oil and gaseous hydrocarbons during this final maturation step is inferred from comparative analytical and spectroscopic studies (Witte et al. 1988) which showed that the total organic carbon content of the shales remained unchanged in going from the HAR to the HAD stage whereas the aliphatic carbon content continued to decrease as a consequence of aromatization reactions within the kerogen. This example illustrates the importance of calibrating laboratory-based kinetic models against pyrolysis of really immature samples (Tissot et al. 1987). Similar effects have recently been observed for a series of coals of increasing maturity (Schenk and Horsfield 1996). In these cases, however, the increase of potentials at high activation energies was much more pronounced and pointed to concomitant structural reorganizations and aromatizations within the coal macromolecule which could be confirmed by infrared spectroscopy. Such structural changes within the residual organic matter leading to an absolute increase in bulk petroleum potentials at higher activation energies are only detectable if both activation energy distributions and preexponential factors are optimized individually for each maturation stage; there is no reason to assume that the model preexponential factors remain unchanged over a broad maturation sequence.

Extrapolation to Geological Heating Rates

Laboratory-derived non-isothermal kinetic models can readily be adapted to the very different time-temperature regimes of geological maturation by a simple readjustment of the parameter "heating rate" in Eq. (4.23). This extrapolation is illustrated in Fig. 4.8 (Schaefer et al. 1990) which compares calculated gross hydrocarbon (bulk petroleum) formation rates from open-system pyrolysis of the immature Toarcian Shale sample WEN (R_r=0.48%) for a laboratory heating rate of 0.1 K/min and two geological heating rates of 10^{-12} K/min (0.53 K/10^6 a) and 10^{-11} K/min (5.3 K/10^6 a). Geological T_{max} values are found between 135 and 155 °C, the onset of bulk petroleum generation between 90 and 105 °C. It must be emphasized that the reliability of these predictions is strongly affected by the accuracy of laboratory experiments. T_{max} temperatures measured during pyrolysis at heating rates r (range 0.1–5.0 K/min) are subjected to an error of at least ± 1 °C. According to the correlation (van Heek and Jüntgen 1968):

$$\ln \frac{r}{T_{max}^2} = \ln \frac{AR}{E} - \frac{E}{R} \cdot \frac{1}{T_{max}}$$

this uncertainty of laboratory T_{max} implies that the dominant activation energy E and the logarithm of the preexponential factor A may both be erroneous by about 2% which, in the worst case, can add up to an error of $\sqrt{(4+4)} \approx 3\%$ or 12 °C for an extrapolated geological T_{max} temperature of 400 K (127 °C). Similar error considerations have been reported by Braun et al. (1991) and Espitalié et al. (1993).

The formalism of non-isothermal kinetics implies that maximum formation rates increase whereas T_{max} values and curve widths decrease with decreasing rate of heating (Jüntgen and van Heek 1968). The immediate practical consequences of these well-known relationships are (1) that high geological heating rates require relatively high paleotemperatures for significant petroleum generation to take place, and (2) that these temperatures must be reconstructed very exactly since main product formation is predicted to occur within rather narrow temperature limits. The latter fact is a driving force for even better basin modeling techniques and refinement in the reconstruction of geological temperature histories (see Chap. 2).

The extrapolation of Fig. 4.8 presumes that laboratory derived kinetic parameters (preexponential factors and potential vs activation energy distributions) remain valid under natural heating conditions; the relevance of this fundamental assumption is discussed in Section 4.5.

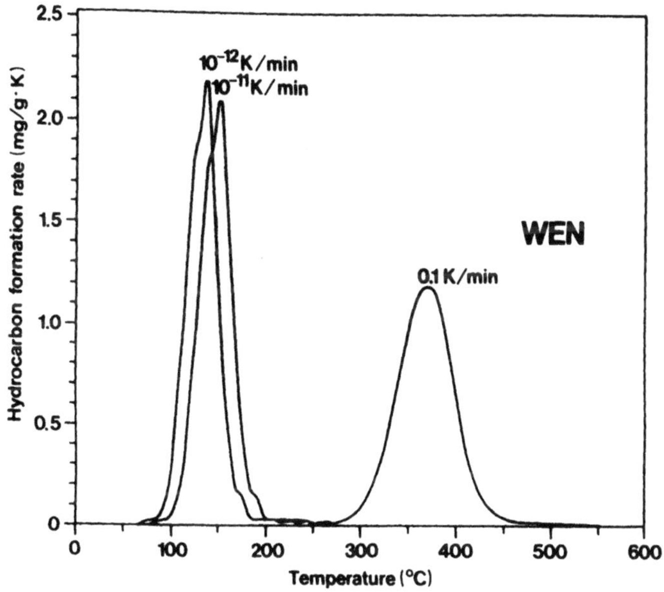

Fig. 4.8. Calculated gross hydrocarbon (bulk petroleum) formation rates (milligram per gram of rock and per degree) as a function of temperature (°C) from open-system pyrolysis of an immature Toarcian Shale sample (sample WEN, $R_r = 0.48\%$) at a laboratory heating rate of 0.1 K/min and two geological heating rates of 10^{-12} K/min (0.53 K/10^6a) and 10^{-11} K/min (5.3 K/10^6 a). (Reprinted from Advances in Organic Geochemistry 1989, Schaefer et al., Determination of gross kinetic parameters for petroleum formation from Jurassic source rocks of different maturity levels by means of laboratory experiments, 1990, pp 115–120, with kind permission from Elsevier Science Ltd, The Boulevard, Langford Lane, Kidlington OX5 1GB, UK)

4.3.3 Closed Versus Open-System Configurations

The most common way of gathering kinetic data for gross petroleum generation is by programmed-temperature open-system pyrolysis. Closed-system pyrolysis is less convenient to perform because cumulative yields cannot be measured in a continuous way. Furthermore, most kinetic studies under confined conditions have been based on isothermal hydrous pyrolysis (Lewan 1985; Barth et al. 1989; Castelli et al. 1990; Hunt et al. 1991). Previous comparisons of kinetic parameters derived from Rock-Eval-like and hydrous pyrolysis revealed significant differences that affect the predicted temperature ranges of main product formation (Burnham et al. 1987; Burnham et al. 1988). However, besides the fact that under closed-system conditions a certain overlap of primary petroleum generation with secondary petroleum cracking reactions can never be avoided, the observed differences are perhaps linked to factors other than the open or closed nature of the experiments per se. It is well known that the Rock-Eval signal is equated to total volatile oil and gas yield, whereas hydrous pyrolysis kinetics are currently related to the oil fraction that is expelled from the residual rock and bitumen mixture by rather ill-defined transport processes which probably interfere with the observed "expelled oil" formation kinetics. These arguments are further supported by recent oil evolution experiments from a self-purging reactor (Burnham 1991). Whereas the latter experimental technique can be considered as being intermediate between rigorously open- and closed-system simulations, the measured temperature ranges for oil evolution agree essentially with those calculated from open-system kinetic parameters. Substantial discrepancies are observed, however, with predictions using the hydrous pyrolysis kinetics reported by Lewan (1985), partly because single activation energies rather than activation energy distributions were calculated from the hydrous pyrolysis data (Lewan 1985).

The effectiveness of laboratory-derived kinetic models must finally be judged according to their ability to accurately predict the timing of bulk petroleum generation within a given geothermal history. It is quite obvious that such predictions should depend exclusively on the type of the organic material on one hand and on the geological conditions on the other hand. Any strong dependence on calibration methods would seriously decrease the predictive value of kinetic models because neither open nor closed experimental configurations can exactly reproduce the degree of confinement of naturally subsiding source rocks.

In a recent study Schenk and Horsfield (1993) have compared bulk petroleum generation from an immature Toarcian Shale sample (sample WEN of Sect. 4.3.2) under open- and closed-system conditions using directly comparable non-isothermal input data in both cases. Open-system pyrolysis was performed according to the procedure of Schaefer et al. (1990) whereas closed-system experiments were carried out by use of the microscale sealed vessel (MSSV) technique of Horsfield et al. (1989). Closed-system pyrolysis of a kerogen does not allow the continuous registration of hydrocarbon formation rates as a function of temperature that constitute the most common data base for kinetic modeling of non-isothermal simulation experiments. Only cumulative yields can be determined after discrete temperature intervals. But the quality of a kinetic fit can better be judged with respect to differential formation rate curves. In order to compare closed- and open-system configurations under non-isothermal conditions a differentiation (versus temperature) of such discrete evolution profiles had to be performed. This procedure, which provides a suitable link between closed-system experiments and non-isothermal kinetic modeling, is illustrated in Fig. 4.9. The differentiated curves were calculated from closed-system kinetic parameters which were very similar

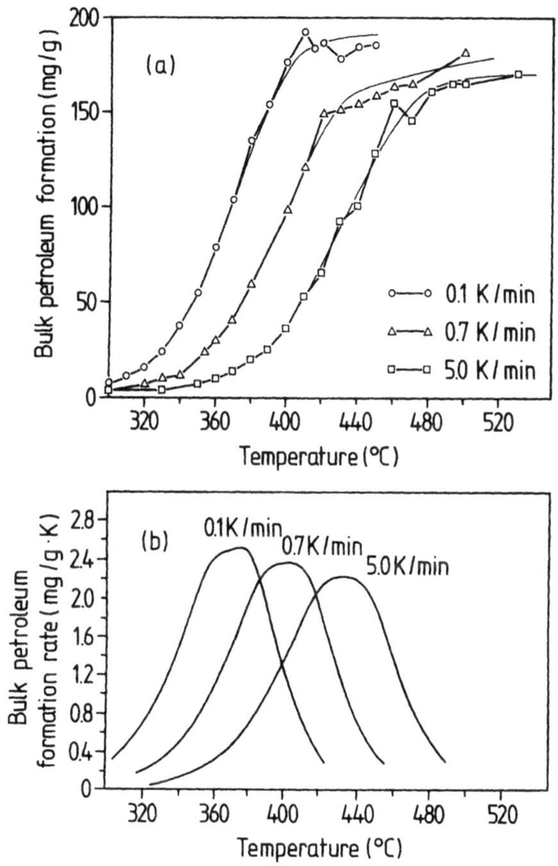

Fig. 4.9. **a** Cumulative formation of bulk petroleum (milligram per gram of kerogen concentrate) during closed-system pyrolysis (microscale sealed vessels) of Toarcian Shale concentrates with increasing temperature (°C) at heating rates of 0.1, 0.7, and 5.0 K/min. *Thick lines*, connection between experimental data points; *thin lines*, spline functions smoothing the discrete data sets. The concentrates were prepared from sample WEN of Section 4.3.2. **b** Bulk petroleum formation rates (milligram per gram of kerogen concentrate and per degree) from closed-system pyrolysis (microscale sealed vessels) of Toarcian Shale concentrates as a function of temperature (°C) at heating rates of 0.1, 0.7, and 5.0 K/min. The curves were calculated as best fit to the first derivatives of the spline functions of a using kinetic parameters closely similar to those of Fig. 4.5. (Schenk and Horsfield 1993)

to the corresponding open-system parameters (see Fig. 4.5); the preexponential factors coincidentally turned out to be identical.

Figure 4.10 compares the gross hydrocarbon (bulk petroleum) formation rates calculated from the open and closed-system kinetic parameters for laboratory and geological heating rates of 0.1 K/min and 10^{-11} K/min (5.3 K/10^6 a), respectively. For the two configurations the temperature ranges of bulk petroleum generation turn out to be virtually identical. This result is an immediate consequence of the identical preexponential factors and similar activation energy distributions, but the intrinsic reason is probably the irreversible nature of petroleum generating reactions, the rates of which remain unaffected by the build-up of considerable product concentrations in the closed MSSV configurations. Pyrolysate yields from the open- and closed-systems differ by a factor of about two. Reduced MSSV yields may partly be due to the formation of coke by secondary cracking reactions which cannot be excluded under confined conditions and the extent of which is difficult to assess in naturally subsiding source rocks (Durand 1985). However, different product yields do not affect the major and essential outcome of this closed-open comparison, namely that bulk petroleum generation is simulated equally well by either closed- or open-system pyrolysis, and that the kinetics derived from the latter, which is more convenient to perform, may be used to calculate the timing of petroleum formation within a given thermal history (Horsfield et al. 1993).

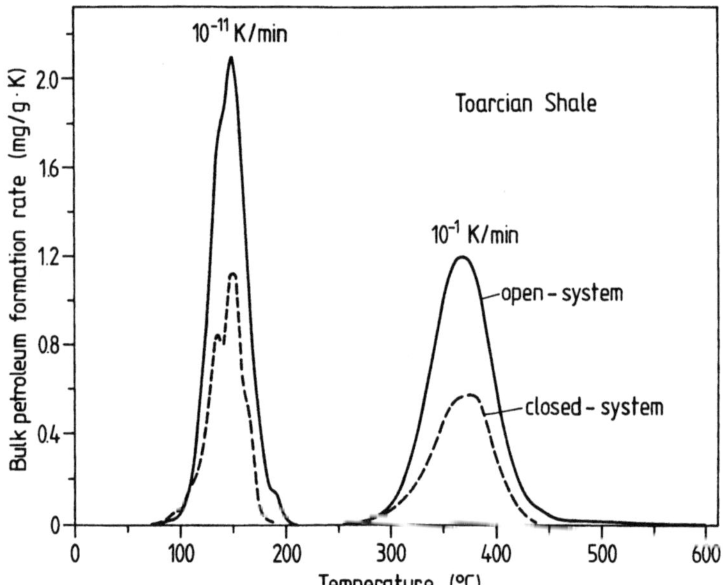

Fig. 4.10. Calculated gross hydrocarbon (bulk petroleum) formation rates (mg per g of rock and per degree) as a function of temperature (°C) from open- and closed-system pyrolysis of an immature Toarcian Shale sample (sample WEN of Sect. 4.3.2) at heating rates of 0.1 K/min (laboratory) and 10^{-11} K/min (nature). For comparison the closed system rates have been reduced by a factor of 12.7/56.4 according to the increase in organic carbon content from 12.7% (rock) to 56.4% (kerogen concentrate). (Schenk and Horsfield 1993)

4.4 Generation of Methane and Molecular Nitrogen from Coals

Natural gas accumulations in the Permian Rotliegend and the Triassic Buntsandstein formations of the North German basin exhibit high concentrations of molecular nitrogen (N_2) ranging from 50% to 90%. Various theories have been proposed to explain these natural gas "anomalies" (see Krooss et al. 1993, 1995), but the formation mechanism still remains disputed. Natural gas in northern Germany is sourced by thick strata of Carboniferous coals and dispersed organic matter. Coals represent an important reservoir of fixed nitrogen in the lithosphere, with concentrations ranging between 0.4 and 3 wt%. It is therefore conceivable that methane and molecular nitrogen in the gas reservoirs originated from a common source. Generation potentials of methane (c. 150 m^3/t organic matter) and nitrogen (7 m^3/t organic matter) show clearly that, assuming simultaneous formation and accumulation, N_2 contents in natural gases derived from coals as a common source cannot exceed values of about 5 vol%. Therefore, if nitrogen-rich natural gas accumulations are derived from organic matter, their formation requires a mechanism which results in substantial enrichment of nitrogen over methane. Two hypotheses have been proposed to explain this enrichment: (1) migration-related fraction-ation processes resulting in a "chromatographic" separation of methane and nitrogen and (2) generation of nitrogen and methane at different stages of sedimentary diagenesis, catagenesis, or metamorphism.

The latter mechanism involves "fractional generation" of methane and nitrogen. To test this hypothesis it is necessary to establish the generation kinetics of these two gases from coals. The only previous experimental work of this kind was performed by Klein and Jüntgen (1972). A reevaluation of this work (Krooss et al. 1993) indicated that phases of preferential generation of nitrogen may occur during the thermal maturation of coals. On the other hand, it showed significant shortcomings in the assessment of the kinetic parameters.

Figure 4.11 shows the generation curves for methane and nitrogen from a low-rank bituminous coal (R_m=0.70%) of the Ruhr area of Germany (Krooss et al. 1993). These generation curves were obtained with a constant heating rate of 2 °C/min from room temperature to 1100 °C (see Lillack 1992, for experimental details). It can be noted that methane and nitrogen generation does not occur simultaneously during the experiment. Methane formation starts between 300° and 400 °C and in this experiment reaches its maximum rate at about 500 °C. The generation rate then decreases, and no further methane is generated above 850 °C. The onset of measurable nitrogen evolution occurs only above 700 °C. The nitrogen generation curve exhibits its first maximum at 780 °C. The most conspicuous feature, however, is the strong increase in the nitrogen generation rate at extremely high temperatures. This observation agrees with the results of Klein (1971) who pyrolysed coals of different rank from the same area. Boudou and Espitalié (1995) performed pyrolysis experiments up to even higher temperatures and found that using a heating rate of 30 K/min the second nitrogen peak maximum for semianthracites occurred around 1200 °C. These experimental results indicate that a significant portion

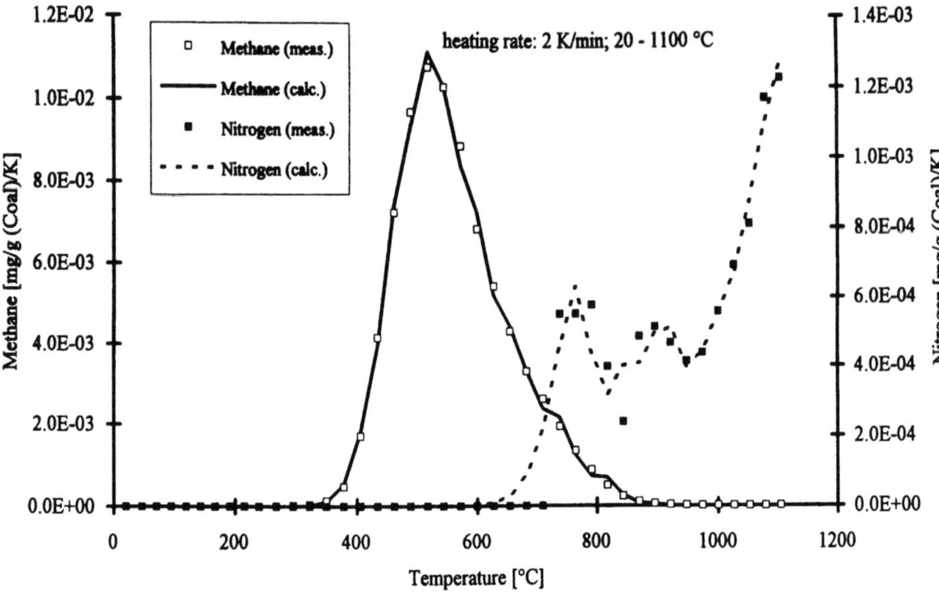

Fig. 4.11. Formation rates (mg · g^{-1} coal · K^{-1}) of methane and molecular nitrogen as a function of temperature (°C) during open-system pyrolysis of a humic coal from Northern Germany at a heating rate of 2 K/min. (Krooss et al. 1993)

of the nitrogen in the coals is very refractory and also under natural conditions requires excessive temperatures to be liberated.

Based on the experimental generation curves activation energy distributions for the generation of methane and nitrogen from coal were determined using a nonlinear least-squares fit procedure described by Schaefer et al. (1990). The underlying assumptions for this evaluation were that the generation reactions of both gases can be described by a set of parallel first-order reactions with Arrhenius type temperature dependence and a single frequency factor.

The calculated generation curves resulting from this evaluation are also shown in Fig. 4.11. The corresponding kinetic parameters are given in Krooss et al. (1993). A better constraint on the preexponential factor can be obtained by performing pyrolysis experiments at different heating rates.

Figure 4.12 shows an application of the experimental kinetic parameters to a simple scenario of a constant geological heating rate of $4K/10^6$ a. According to this scenario, formation of natural gas with significant concentrations of molecular nitrogen occurs only at temperatures in excess of 280 °C. Temperatures in excess of 300 °C are required to generate gas with N_2 as a major component. Geological studies indicate that these temperature levels may well be reached by deeply buried Carboniferous strata in the northern part of the North German basin. As stated above, cumulative trapping of the generated gas reduces or even wipes out the compositional effects of fractional gas generation. One supposition of this model is therefore that the present reservoir gases correspond only to the gas supplied "most recently" on a geological time scale and

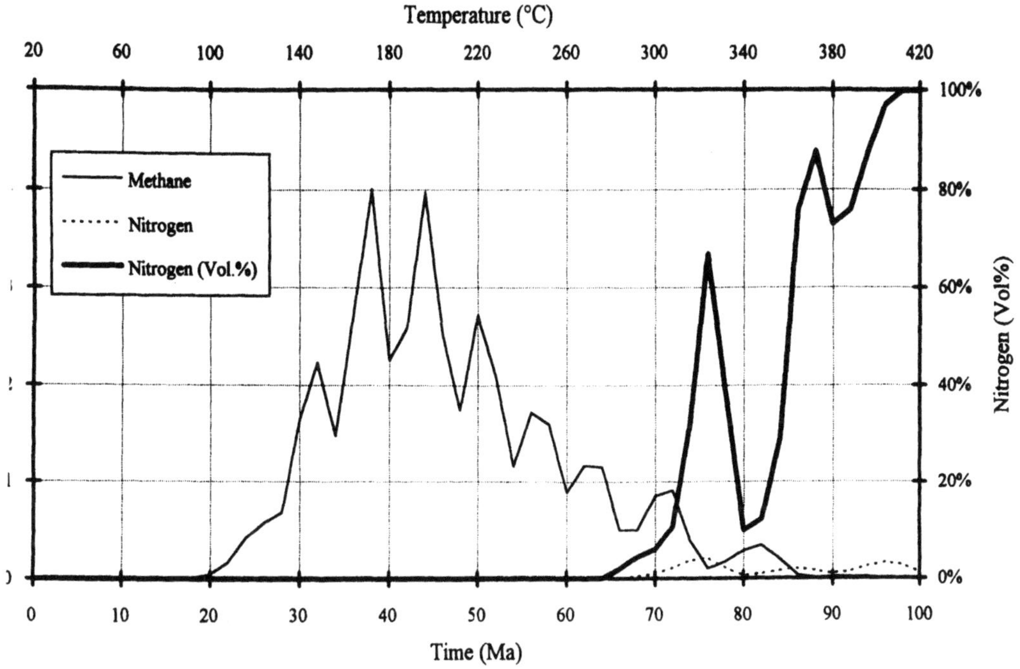

Fig. 4.12. Generation rates ($m^3/t \cdot 10^6$ a) of methane and molecular nitrogen from a humic coal of northern Germany and nitrogen content (vol%) of methane-nitrogen mixtures vs. time (10^6 a) and temperature (°C) at a geological heating rate of 4 K/10^6 a. (Krooss et al. 1993)

therefore represent only a small fraction of the total generated gas. This concept which takes into account the dynamic nature of natural gas accumulations is supported by gas balance calculations performed by Jurgan et al. (1983). According to this study the gas in place in Rotliegend reservoirs of eastern Lower Saxony corresponds to only 0.1% of the gas generated within the corresponding drainage area. It is therefore reasonable to consider the present day reservoirs, on a geological time scale, as containers which have been flushed by large amounts of gas and which represent probes of the gas that has been supplied most recently.

At present the details of the formation mechanism and evolution of gas accumulations are far from being understood. One important point to remember from the experiments described above is that the products are from open-system pyrolysis and therefore refer only to primary gases. No account has been made of the higher molecular weight pyrolysis products, which can also be considered a potential secondary source of hydrocarbon and non-hydrocarbon gases. This aspect is discussed in the following section. Newly established kinetic parameters for methane and nitrogen generation from coals in combination with basin modeling are expected to provide improved insight into the sources, composition, and migration pathways of natural gas and the dynamics and longevity of gas reservoirs on a geological time scale.

4.5
The Problems of Predicting Petroleum Generation Rates and Compositions in Nature

Since the generation of oil and gas from kerogen is controlled by chemical kinetics, the incorporation of appropriate kinetic models into geothermal histories, both individually optimized to the source rock of interest, should be a promising tool for calculating the timing of petroleum formation and reducing the risk of petroleum exploration. Kerogen, however, is a tremendously complex material, the thermal alteration of which must be anticipated to proceed through a sequence of competing parallel and consecutive reactions involving the cleavage of carbon-carbon and carbon-hydrogen bonds, cyclization and ring-opening reactions, aromatizations, decarboxylations and other eliminations of heteroelements. Gross kinetic modeling of kerogen thermal decomposition (Sect. 4.3.1) replaces the multitude of actually occurring but unknown reactions by a finite number of parallel first-order pseudoreactions (reaction channels), characterized by different initial potentials (partial yields) and activation energies. After appropriate calibration (Sects. 4.3.2, 4.3.3) this system reproduces the overall generation of oil and gas from a source rock during laboratory pyrolysis and allows the calculation of gross petroleum yields for any temperature and heating rate. If this artificial scheme remains valid for geological heating rates, the global amount of petroleum generated from a source rock up to a defined maturation stage can be rather unambiguously derived from the subsidence conditions and paleotemperatures according to Fig. 4.13.

The pyrolytic degradation of an immature kerogen under programmed-temperature conditions results in the continuous release of hydrocarbons and related FID detectable species, whose formation rate reaches a maximum somewhere between 400 and 500 °C, and the concomitant formation of inert carbonaceous materials. Gross kinetic analysis of product evolution curves, measured at various feasible heating rates, yields partial hydrocarbon generating potentials M_i as a function of activation energies E_i generally in the form of a discrete distribution somewhere between 40 and 80 kcal/mol (167–335 kJ/mol) as well as preexponential factors. The sum of partial potentials is equal to the area of each evolution curve. Assuming these parameters to be independent of heating rate, the hydrocarbon formation curves can easily be calculated for the geological maturation of the same source rock assuming, for simplicity, that an average geological heating rate (GEOL.dT/dt) can be reasonably derived from the reconstruction of the time-temperature history of the basin. The drastic decrease in heating rate from about 0.1 K/min (laboratory) to about 10^{-11} K/min (geology) results in a shift of peak generation temperature down to 100–200 °C and in a strong reduction of curve width. The calculated "geological" dM/dT versus T curve defines the temperature range of primary bitumen generation under natural heating conditions, and each paleotemperature (GEOL.T_E) is related to a partial curve area (black in Fig. 4.13) representing the kinetically predicted amount of oil and gas produced from the source rock up to this maturation stage.

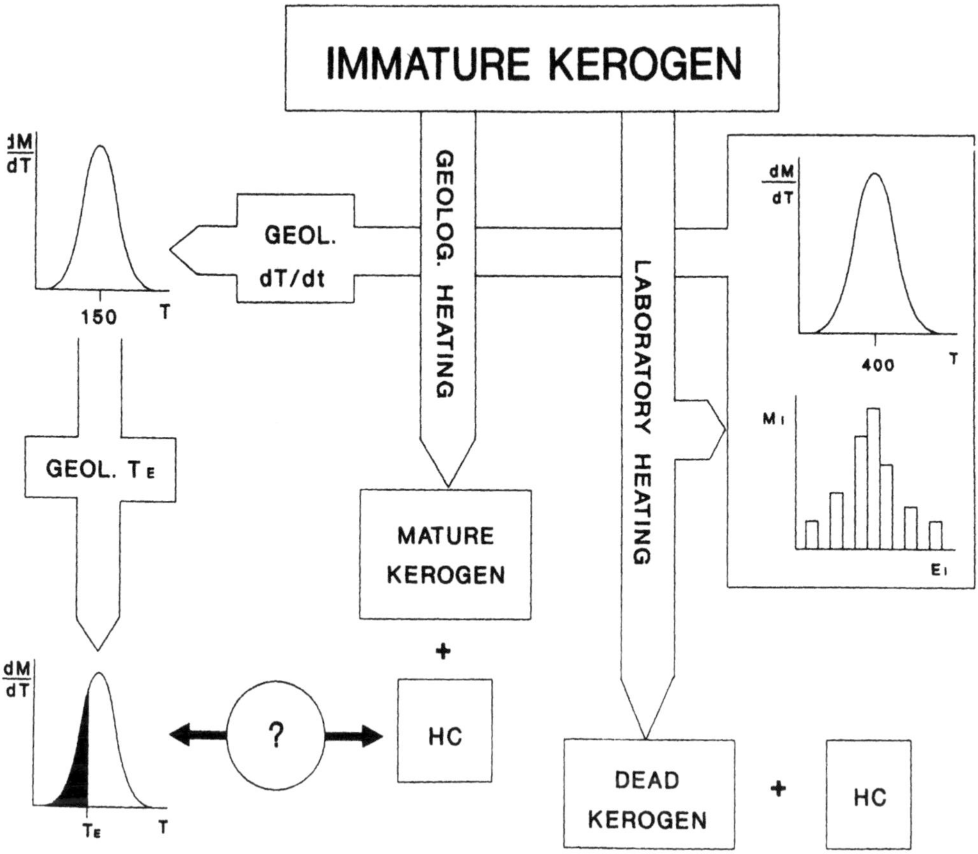

Fig. 4.13. Possible procedure for testing the relevance of extrapolating from laboratory kinetics to geological systems.

The relevance of extrapolating laboratory-based kinetic models to geological heating rates that are lower by 10 orders of magnitude is very controversial (Snowdon 1979). Several obstacles preclude kinetic models from being checked by comparing predicted (Fig. 4.13) and actually observed petroleum generation. Firstly, samples are not usually available from deep generative depressions; most wells are drilled on structural highs. Secondly, and even under special circumstances where petroleum generation has occurred at shallow depth (Rullkötter et al. 1988), petroleum expulsion efficiencies for source rocks are high (Cooles et al. 1986), and secondary migration pathways are tortuous, making unequivocal assessments of yields almost impossible. Nevertheless, by extrapolation to geological heating rates of 10^{-11}–10^{-12} K/min, some kinetic models have succeeded in predicting natural maturation trends for coals (Ungerer and Pelet 1987; Forbes et al. 1991) and kerogens (Sweeney et al. 1987; Issler and Snowdon 1990) or explaining the absence of gaseous hydrocarbons in certain coal seams (van Heek et al. 1971) and in a deep petroleum reservoir (Horsfield et al. 1992). The isothermal generation of oil from Green River kerogen exhibits no change of kinetic parameters within the temperature range of 500–209 °C (Freund and Kelenen 1989). The elegance of these predictions is surprising because the bulk composition of unaltered petroleum, a product of kerogen degradation under geological conditions, is fundamentally different to that of kerogen pyrolysate, produced under laboratory conditions, irrespective of crude oil class, kerogen type or analytical configuration (Larter and Horsfield 1993). Crude oils are undoubtedly hydrocarbon-rich systems, whereas pyrol-ysates from either hydrous or anhydrous systems contain a much higher proportion of polar compounds (Urov 1980; Castelli et al. 1990). There is therefore a fundamental flaw in assuming that "petroleum" generated in nature and by laboratory pyrolysis are compositionally alike. This difference has probably nothing to do with fractionation because expulsion efficiencies associated with mature oil-prone source rocks are exceedingly high (Cooles et al. 1986; Larter 1988). Also, most mature source rock extracts contain a high proportion of hydrocarbons (Philippi 1965; Powell 1978). This fundamental difference in composition limits the detailed chemical significance of "kinetic" models of "petroleum" or "hydrocarbon" generation when they are calibrated using nonhydrocarbon rich pyrolysate evolution curves. Examining these findings in terms of a sequential K→B and B→O model (K, kerogen; B, polar bitumen; O, crude oil), Larter and Horsfield (1993) concluded that the kinetic parameters for the two reactions were closely similar and that a crossover occurred on the Arrhenius diagram so that the rate-controlling step under laboratory pyrolysis conditions is the bitumen to oil conversion whereas, in the subsurface, the kerogen to bitumen reaction is the rate controlling step.

An analogous finding is provided by the observation that the natural evolution of sedimentary organic matter is marked by a rather fast and selective elimination of oxygen functionalities prior to the release of hydrocarbons whereas during pyrolysis both atomic H/C and O/C ratios decrease simultaneously (Boudou et al 1984; Schenk et al. 1990). The apparent difference between natural and artificial maturation in this case is a consequence of lower activation energies and preexponential factors for the formation of H_2O and CO_2 as compared to the generation of hydrocarbons (Jüntgen and Klein 1975).

There is an additional ramification from the above considerations that concerns the prediction of petroleum composition, including gas-oil ratio. While attempts have been made to determine the kinetic parameters for the generation of specific compounds from coal and kerogen (Hoering and Abelson 1963; Wall and Smith 1987; Jüntgen 1984; Esser and Schwochau 1991), the distributed activation energy models of petroleum generation outlined above do not provide any information on the composition of the petroleum formed. As a first step at integrating the generation of compound classes into kinetic models, Espitalié et al. (1988) defined "primary cracking" kinetic parameters for four "petroleum" fractions, i.e., C_1, C_2–C_5, C_6–C_{15} and C_{15+} pyrolysate, using a modified Rock-Eval. An alternative approach, outlined by Düppenbecker et al. (1991), involved predicting the transformation ratio using kinetic modeling and thereafter assessing gas versus oil yield by a laboratory-determined transformation ratio vs. gas/oil ratio curve (Düppenbecker and Horsfield 1990). Inferring the gas-oil ratio of natural petroleum from the quotient of C_1–C_5/C_6+ products in kerogen pyrolysates (Mackenzie and Quigley 1988; Espitalie et al. 1988), even after data manipulation (Düppenbecker and Horsfield 1990) appears to be conceptually erroneous because a high proportion of the C_{6+} fraction consists of polar components from the K→B reaction. The distribution of gas versus oil in this reaction is not necessarily the same as in the B→O reaction (Horsfield and Düppenbecker 1991), the likely outcome being an overestimation of oil potential. In the case of coals, for instance, this means that secondary gas generating potential may be underestimated (Horsfield and Idiz, unpublished). Because of these considerations, "product prediction" is an area of ongoing research.

The practical use of extremely powerful mathematical models including the generation and degradation of more than 60 product species besides a large number of physical variables (Braun and Burnham 1990; Burnham and Braun 1990) is severely limited by the comparably poor knowledge of relevant geological parameters. The elaboration of more and more sophisticated kinetic models increases the already existing gap between chemical and mathematical possibilities, on the one hand, and the geological state of the art, on the other. As far as predicting the timing and intensity of petroleum generation is concerned, there is more need for reliable data on geological parameters such as the calibration and reconstruction of temperature histories, hydrocarbon migration versus formation rates and average residence times of primary bitumen components in the hot zones of sedimentary rocks. To delve deeper and deeper into the complexities of laboratory pyrolysis does not seem to be a priority task at this stage. However, the opposite is true as far as predicting petroleum compositions is concerned because of the uncertain relationship between products produced under laboratory heating rates and those formed in sedimentary basins.

4.6
The Conversion of Oil to Gas in Petroleum Reservoirs

Crude oils show a progressive increase in gas-oil ratio, API gravity, and hydrocarbon content with increasing maturity (Holmquest 1965), and oil occurrence ultimately gives way to that of gas and condensates (Landes 1967). The present day temperature of deepest oil occurrence can lie anywhere between 90 and 200 °C depending on the area and geological history in question (see Andreev et al. 1968; Evans and Staplin 1972; Pusey 1973; Evamy et al. 1978; Zieglar and Spotts 1978; Price 1980; Momper and Williams 1984) and is directly related to heat flow history, the type and stability of crude oil, displacement phenomena or combinations thereof. Crude oils are commonly considered to break down via hydrogen transfer reactions to yield gas and pyrobitumen (Bailey et al. 1974; Connan et al. 1975), the conversion being assumed to proceed via systems of first-order constituent reactions (Tissot et al. 1974; Quigley et al. 1987; Ungerer et al. 1988; Welte et al. 1988; Braun and Burnham 1990; Béhar et al. 1991). Kinetic models of oil degradation are commonly used, in addition to petroleum generation models (e.g., Ungerer and Pelet 1987), to help predict the occurrence of oil versus gas in time and space.

Because there are no well-constrained geological case histories (Quigley et al. 1987), these oil degradation models are calibrated almost exclusively using hydrous pyrolysis or other closed-system experimental approaches.

This section shows how closed-system pyrolysis of a North Sea crude oil (well 33/9-14), using programmed rather than isothermal heating, was used to obtain kinetic constants for oil to gas conversion and outlines how the results were calibrated using the case history of a well (2/4-14 reservoir) in the 2/4 Block, offshore Norway (Horsfield et al. 1992).

All heating experiments were carried out using MSSV pyrolysis, the basic elements being (1) three different heating rates for producing cumulative gas generation curves, (2) differentiation (see Sect. 4.3.3) to convert these data into a format resembling that used in open-system petroleum generation experiments, and from this (3) calculation of a gas potential versus activation energy distribution and a single preexponential factor for pseudoreactions of gas generation (see Schaefer et al. 1990). The generation of total gas (C_1–C_4) with increasing temperature for the three heating rates is shown in Fig. 4.14. Importantly, the displacement of gas generation curves to progressively lower temperatures with decreasing heating rate is in accordance with the kinetic modeling approach (see Jüntgen and van Heek 1970). Additionally, high temperatures were required to bring about the complete conversion of oil to gas, even using low heating rates, indicating that oil to gas conversion in nature is likely to take place deeper than the depth range over which petroleum is generated from kerogen (see Schaefer et al. 1990).

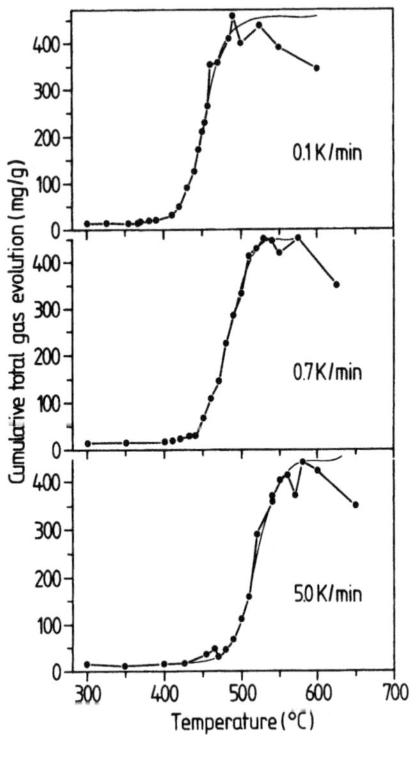

Fig. 4.14. Cumulative evolution of total gas (milligram per gram) in MSSV pyrolysates with increasing temperature (°C) at three heating rates (0.1, 0.7, and 5.0 K/min). *Thin lines*, spline functions smoothing the experimental discrete data points (connected by *solid lines*). (Reprinted from Advances in Organic Geochemistry 1991, Horsfield et al., An investigation of the in-reservoir conversion of oil to gas: Compositional and kinetic findings from closed-system programmed-temperature pyrolysis, 1992, pp 191–204, with kind permission from Elsevier Science Ltd, The Boulevard, Langford Lane, Kidlington OX5 1GB, UK)

The gas generation curves result from a very large number of individual chemical reactions, each possessing its own kinetic parameters. While it is not possible to identify these individual reactions, we can nevertheless identify some of the processes accompanying gas generation, many of which, such as increasing paraffinicity, decreasing average boiling point, the high wetness of generated gases, and the interpreted decrease in asphaltene content, match the compositional features of naturally matured crude oils (Bailey et al. 1974; Neumann et al. 1981; Cornford et al. 1983; Tissot and Welte 1984; Claypool and Mancini 1989). Based on a comparison with these compositional changes, the 2/4-14 petroleum appears to have originated at a stage where a maximum of 15% of the in-place oil had been converted to gas.

The best-fit between measured and calculated gas formation rate curves, shown in Fig. 4.15, was obtained with the rather narrow gas potential versus activation energy distribution shown in Fig. 4.16, the principal distribution being between 66 and 70 kcal/mol. The preexponential factor was found to be $6.69 \cdot 10^{17}$ min^{-1}. The application of these parameters to geological heating rates is illustrated in Fig. 4.17. An onset of oil cracking between 160 and 190 °C is indicated for heating rates in the range 0.53 to 5.3 K/10^6 a. According to these figures, severe oil cracking is not likely to have taken place in the 2/4-14 reservoir as a result of exposure to temperatures around 165 °C. These results are consistent with compositional considerations. Namely, the composition of the 2/4-14 liquid petroleum bears a resemblance to artificially matured 33/9-14 crude oil at low degrees of conversion (<15%).

Numerical modeling using these kinetic parameters and geological input data specific to the 2/4-14 reservoir predicted that the degree of oil conversion

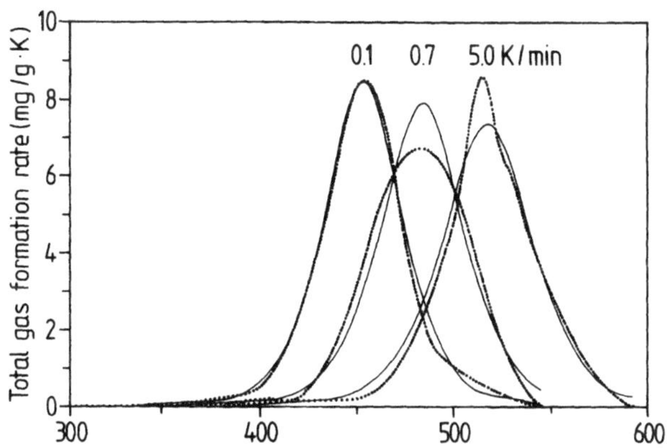

Fig. 4.15. Total gas formation rates ($mg \cdot g^{-1} \cdot K^{-1}$) as a function of temperature (°C) for three heating rates (0.1, 0.7, and 5.0 K/min). *Dotted curves*, those resulting from differentiation of the spline functions of Fig. 4.14; *solid curves*, those calculated from the kinetic parameters of Fig. 4.16 by the iteration method outlined in Section 4.3.2. (Reprinted from Advances in Organic Geochemistry 1991, Horsfield et al., An investigation of the in-reservoir conversion of oil to gas: Compositional and kinetic findings from closed-system programmed-temperature pyrolysis, 1992, pp 191–204, with kind permission from Elsevier Science Ltd, The Boulevard, Langford Lane, Kidlington OX5 1GB, UK)

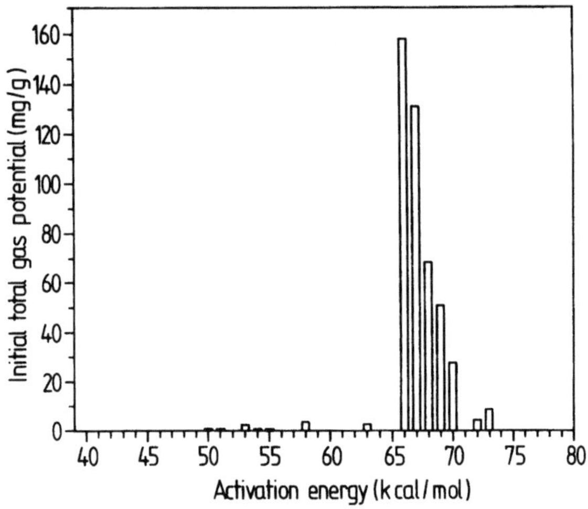

Fig. 4.16. Distribution of initial total gas potentials (milligram per gram) versus activation energies (kcal/mol). The best-fit preexponential factor is A=6.69 · 10^{17} min^{-1}. (Reprinted from Advances in Organic Geochemistry 1991, Horsfield et al., An investigation of the in-reservoir conversion of oil to gas: Compositional and kinetic findings from closed-system programmed-temperature pyrolysis, 1992, pp 191–204, with kind permission from Elsevier Science Ltd, The Boulevard, Langford Lane, Kidlington OX5 1GB, UK)

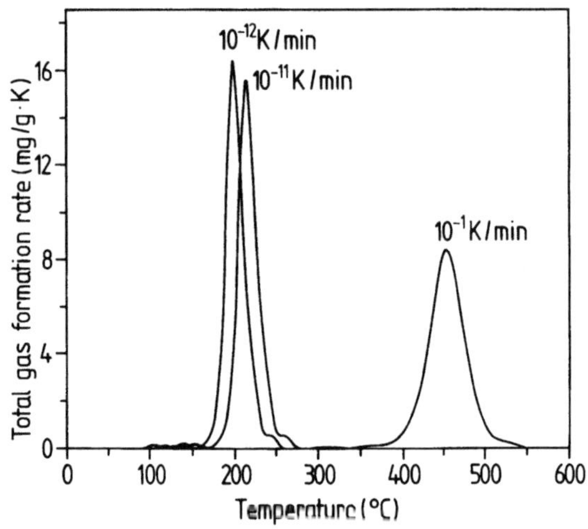

Fig. 4.17. Extrapolation of total gas formation rate ($mg \cdot g^{-1} \cdot K^{-1}$) versus temperature (°C) from a laboratory (10^{-1} K/min) to typical geological heating rates of 10^{-12}–10^{-11} K/min (0.53 and 5.3K/10^6 a). (Reprinted from Advances in Organic Geochemistry 1991, Horsfield et al., An investigation of the in-reservoir conversion of oil to gas: Compositional and kinetic findings from closed-system programmed-temperature pyrolysis, 1992, pp 191–204, with kind permission from Elsevier Science Ltd, The Boulevard, Langford Lane, Kidlington OX5 1GB, UK)

should be very low at the well location (Andresen et al. 1993). For modeling, the well had been divided into 38 events extending from the Permian (248 Ma) to the present day, there being a long period of nondeposition in the Late Triassic/Early Jurassic but no periods of major erosion. The heat flow history showed a peak of 1.45 heat flow units (HFU) in the Jurassic, decreasing to a constant value of 1.15 HFU during the Cretaceous and Tertiary and a rise to 1.30 HFU in the Quaternary. Judging by the results of Schaefer et al. (1990), who reported that petroleum generation from kerogen is approx. 90% complete for temperatures in the range 170–190 °C and heating rates between 0.53 and 5.3 °C Ma^{-1} (see Fig. 4.8), it appears that oil generation and destruction are separated from one another in time and space, rather than gradually merging from one to the next as the classical oil window concept might infer (see Tissot and Welte 1984). This point has previously been illustrated by Sweeney et al. (1987).

The stability of oil might have been overestimated by the kinetic model in quantitative terms because compositional considerations indicate that as much as 15% of the original petroleum may have been converted into gas. The high stability predicted by the kinetic model may in part be related to the fitting of the spline function to the gas generation curves. To date we have optimized the whole of each gas generation profile. While this approach has been successfully applied to oil generation studies using open-system pyrolysates (e.g., Schaefer et al. 1990), it may not be ideally suited for assessing the onset of oil to gas cracking. Determining the onset, rather than peak, of oil to gas cracking is very important *in the general case* because, depending on P-V-T conditions, only a small conversion may be required to result in the formation of a separate gas phase and the displacement of the liquid phase to shallower structures. Optimizing the fit for the first third of gas generation profile may therefore prove critical in evaluating oil versus gas prospectivity. Sensitivity analysis revealed that the predicted high stability of the 33/9-14 oil is not tracable to analytical artefacts and errors.

If the rate of oil to gas cracking depended strongly on pressure, it would be invalid to extend low pressure experimental studies to geological conditions. This aspect must therefore also be considered. The maximum pressure attained in these experiments was estimated to be 3 MPa whereas the pore pressure in the 2/4-14 reservoir is estimated to be 65 MPa. If increased pressure were to lead to a reduction in the rate of catagenetic change, as reported by Brooks et al. (1971), Dominé (1991) and Price and Wenger (1992), low pressure experiments would overestimate the rate of oil cracking. Conversely, applying the results of Fabuss et al. (1964) on hexane cracking, cracking rates may be faster by 5- to 15-fold at 65 MPa in comparison to 3 MPa. Horsfield et al. (1992) found no evidence to suggest that pressure has reduced the rate of oil conversion. In fact, judging by the high stability of oil predicted by the kinetic model on the one hand and evidence of approx. 15% conversion from compositional data on the other, the converse might be true, with pressure having slightly accelerated the rate of oil cracking. Unfortunately, given the uncertainties in geological input data to the kinetic model (temperature and burial histories) and sample representativity inherent to such studies, presumptions regarding reaction mechanisms and the unknown role of potential mineral catalysts, it was difficult to determine whether pressure has or has not played a role in oil to gas conversion at the 2/4-14 location.

References

Andreev PF, Bogomolov AI, Dobrayanskii AF, Kartsev AA (1968) Transformation of petroleum in nature. Pergamon Press, London

Andresen P, Mills N, Schenk HJ, Horsfield B (1993) The importance of kinetic parameters in modelling oil and gas generation - a case study in 1-D from well 2/4-14. In: Doré AG, Augustson JH, Hermanrud C, Stewart DJ, Sylta Ø (eds) Basin modelling: advances and applications. Norwegian Petroleum Society (NPF) Spec Publ 3. Elsevier, Amsterdam, pp 563–571

Bailey NJL, Evans CR, Milner CWD (1974) Applying petroleum geochemistry to search for oil: examples from Western Canada Basin. Bull Am Assoc Petrol Geol 58: 2284–2294

Bamford CH, Tipper CFH (1969) Comprehensive chemical kinetics. vol 2. The theory of kinetics. Elsevier, Amsterdam

Barth T, Borgund AE, Hopland AL (1989) Generation of organic compounds by hydrous pyrolysis of Kimmeridge oil shale. Bulk results and activation energy calculations. Org Geochem 14: 69–76

Béhar F, Vandenbroucke M (1986) Représentation chimique de la structure des kérogènes et des asphaltènes en fonction de leur origine et de leur degré d'évolution. Rev Inst Fr Pét 41: 173–188

Béhar F, Ungerer P, Kressmann S, Rudkiewicz JL (1991) Thermal evolution of crude oils in sedimentary basins: experimental simulation in a confined system and kinetic modeling. Rev Inst Fr Pét 46: 151–181

Boudou JP, Espitalié J (1995) Molecular nitrogen from coal pyrolysis. Kinetic modelling. In: Rice D, Schoell M (eds) Sources of natural gas formation. Chem Geol 126: 319–333

Boudou JP, Durand B, Oudin JL (1984) Diagenetic trends of a Tertiary low-rank coal series. Geochim Cosmochim Acta 48: 2005–2010

Braun RL, Burnham AK (1987) Analysis of chemical reaction kinetics using a distribution of activation energies and simpler models. Energy Fuels 1: 153–161

Braun RL, Burnham AK (1990) Mathematical model of oil generation, degradation and expulsion. Energy Fuels 4: 132–146

Braun RL, Burnham AK, Reynolds JG, Clarkson JE (1991) Pyrolysis kinetics for lacustrine and marine source rocks by programmed micropyrolysis. Energy Fuels 5: 192–204

Brooks JD, Hesp WR, Rigby D (1971) The natural conversion of oil to gas in sediments in the Cooper Basin. APEA J 11: 121–125

Burlingame AL, Haug PA, Schnoes HK, Simoneit BR (1969) Fatty acids derived from the Green River Formation Oil Shale by extractions and oxidations - a review. In: Schenk PA, Havenaar I (eds) Advances in organic geochemistry 1968. Pergamon Press, Oxford, pp 85–129

Burnham AK (1990) Pyrolysis kinetics and composition for Posidonia Shale. Rep UCRL-1D-105871, Lawrence Livermore National Laboratory, 12 pp

Burnham AK (1991) Oil evolution from a self-purging reactor: kinetics and composition at 2 °C/min and 2 °C/h. Energy Fuels 5: 205–214

Burnham AK, Braun RL (1990) Development of a detailed model of petroleum formation, destruction and expulsion from lacustrine and marine source rocks. In: Durand B, Béhar F (eds) Advances in organic geochemistry 1989. Org Geochem 16: 27–39

Burnham AK, Braun RL, Gregg HR (1987) Comparison of methods for measuring kerogen pyrolysis rates and fitting kinetic parameters. Energy Fuels 1: 452–458

Burnham AK, Braun RL, Samoun AM (1988) Further comparison of methods for measuring kerogen pyrolysis rates and fitting kinetic parameters. In: Mattavelli L, Novelli L (eds) Advances in organic geochemistry 1987. Org Geochem 13: 839–845

Campbell JH, Gallegos G, Gregg M (1980) Gas evolution during oil shale pyrolysis. 2. Kinetic and stoichiometric analysis. Fuel 59: 727–732

Castelli A, Chiaramonte MA, Beltrame PL, Carniti P, Del Bianco A, Stroppa F (1990) Thermal degradation of kerogen by hydrous pyrolysis. A kinetic study. In: Durand B, Béhar F (eds) Advances in organic geochemistry 1989. Org Geochem 16: 75–82

Claypool GE, Mancini EA (1989) Geochemical relationships of petroleum in Mesozoic reservoirs to carbonate source rocks of Jurassic Smackover Formation, southwestern Alabama. Bull Am Assoc Petrol Geol 73: 904–924.

Connan J (1974) Time-temperature relation in oil genesis. AAPG Bull 58: 2516–2521

Connan J, LeTran K, van der Weide BM (1975) Alteration of petroleum in reservoirs. Proc 9th World Petroleum Congr, Tokyo, vol 2, pp 171–178

Cooles GP, Mackenzie AS, Quigley TM (1986) Calculation of petroleum masses generated and expelled from source rocks. In: Leythaeuser D, Rullkötter J (eds) Advances in organic geochemisrtry 1985. Org Geochem 10: 235– 245

Cornford C, Morrow JA, Turrington A, Miles JA, Brooks J (1983) Some geological controls on oil composition in the U.K. North Sea. In: Brooks J (ed) Petroleum geochemistry and exploration of Europe. Blackwell, Oxford, pp 175–194

Dayhoff MO, Lippincott ER, Eck RV, Nagarajan G (1967) Thermodynamic equilibrium in prebiological atmospheres of C,H,O,N,P,S and Cl. Rep NASA SP-3040, National Biomedical Research Foundation, Silver Spring, Maryland, 259 pp

di Primio R (1995) The generation and migration of sulphur rich petroleums in a low maturity carbonate source rock sequence from Italy. PhD Thesis, University of Cologne

Dominé F (1991) High pressure pyrolysis of n-hexane, 2,4-dimethylpentane and 1-phenylbutane. Is pressure an important geochemical parameter. Org Geochem 17: 619–634

Düppenbecker S, Horsfield B (1990) Compositional information for kinetic modelling and petroleum type prediction. In: Durand B, Béhar F (eds) Advances in organic geochemistry 1989. Org Geochem 16: 259–266

Düppenbecker SJ, Dohmen L, Welte DH (1991) Numerical modelling of petroleum expulsion in two areas of the Lower Saxony Basin, (northern Germany). In: England WA, Fleet A (eds) Proc of the Meet on Petroleum migration, Geol Soc Lond, Spec Publ 59, pp 47–64

Durand B (1985) Diagenetic modification of kerogens. Philos Trans R Soc Lond A 315: 77–90

Espitalié J, Ungerer P, Irwin H, Marquis F (1988) Primary cracking of kerogen. Experimenting and modeling C_1, C_2–C_5, C_6–C_{15} and C_{15+} classes of hydrocarbons formed. In: Mattavelli L, Novelli L (eds) Advances in organic geochemistry 1987. Org Geochem 13: 839–845

Espitalie J, Margues F, Drouet S (1993) Critical study of kinetic modelling parameters. In: Doré AG, Augustson JH, Hermanrud C, Stewart DJ, Sylta Ø (eds) Basin modelling, advances and applications. Norwegian Petroleum Society (NPF) Spec Publ 3. Elsevier, Amsterdam, pp 233–242

Esser W, Schwochau K (1991) Evolution of individual hydrocarbons (C_1–C_4) by non-isothermal pyrolysis of petroleum source rocks. J Anal Appl Pyrolysis 22: 61–71

Evamy BD, Harembourne, Kamerling P, Knapp WA, Molloy FA, Rowlands PH (1978) Hydrocarbon habitat of Tertiary Niger Delta. Bull Am Assoc Petrol Geol 62: 1–39

Evans CR, Staplin FL (1972) Regional facies of organic metamorphism. CIM Spec Vol 11: 517–521

Eyring H (1935a) The activated complex in chemical reactions. J Chem Phys 3: 107–120

Eyring H (1935b) The activated complex and the absolute rate of chemical reactions. Chem Rev 17: 65–82

Fabuss BM, Duncan DA, Satterfield CN (1964) Thermal cracking of pure saturated hydrocarbons. Adv Pet Chem Refining 3: 156–201

Forbes PL, Ungerer PM, Kuhfuss AB, Riss F, Eggen S (1991) Compositional modeling of petroleum generation and expulsion. Application to a local mass balance in the Smorbukk Sor field, Haltenbanken area, Norway. AAPG Bull 75: 873–893

Freund H, Kelenen SR (1989) Low-temperature pyrolysis of Green River kerogen. AAPG Bull 73: 1011–1017

Gregg ML, Campbell JH, Taylor JR (1981) Laboratory and modelling investigation of a Colorado oil shale block heated to 900 °C. Fuel 60: 179–188

Hanbaba P, Jüntgen H (1969) Zur Übertragbarkeit von Laboratoriums-Untersuchungen auf geochemische Prozesse der Gasbildung aus Steinkohle und über den Einfluß von

Sauerstoff auf die Gasbildung. In: Schenck PA, Havenaar I (eds) Advances in organic geochemistry 1968. Pergamon Press, Oxford, pp 459–471

Hanbaba P, Jüntgen H, Peters W (1968) Nicht-isotherme Reaktionskinetik der Kohlepyrolyse. Teil II: Erweiterung der Theorie des Gasabspaltung und experimentelle Bestätigung an Steinkohlen. Brennstoff-Chemie 49: 368–376

Hinshelwood CN (1927) On the theory of unimolecular reactions. Proc R Soc (A) 113: 230–233

Hoering TC, Abelson PH (1963) Hydrocarbons from kerogen. Carnegie Inst Wash Year Book 62: 229–234

Holmquest HJ (1965) Deep pays in Delaware and Val Verde Basins. In: Young A, Galley JE (eds) Fluids in sub-surface environments. AAPG Mem 4: 257–279

Horsfield B, Düppenbecker SJ (1991) The decomposition of Posidonia Shale and Green River Shale kerogens using microscale sealed vessel (MSSV) pyrolysis. J Anal Appl Pyrol 20: 107–123

Horsfield B, Disko U, Leistner F (1989) The microscale simulation of maturation: outline of a new technique and its potential applications. Geol Rundsch 78: 361–374

Horsfield B, Schenk HJ, Mills N, Welte DH (1992) An investigation of the in-reservoir conversion of oil to gas: compositional and kinetic findings from closed-system programmed-temperature pyrolysis. In: Eckardt CB, Maxwell JR, Larter SR, Manning DAC (eds) Advances in organic geochemistry 1991. Org Geochem 19: 191–204

Horsfield B, Düppenbecker SJ, Schenk HJ, Schaefer RG (1993) Kerogen typing concepts designed for the quantitative geochemical evaluation of petroleum potential. In: Doré AG, Augustson JH, Hermanrud C, Stewart DJ, Sylta Ø (eds) Basin modelling; advances and applications. Norwegian Petroleum Society (NPF) Spec Publ 3. Elsevier, Amsterdam, pp 243–249

Horsfield B, Curry DJ, Bohacs K, Littke R, Rullkötter J, Schenk HJ, Radke M, Schaefer RG, Carroll AR, Isaksen G, Witte EG (1994) Organic geochemistry of freshwater and alkaline lacustrine sediments in the Green River formation of the Washakie Basin, Wyoming, USA. In: Telnaes N, van Graas G, Øygard K (eds) Advances in organic geochemisrtry 1993. Org Geochem 22: 415–440

Huck G and Karweil J (1955) Physikalisch-chemische Probleme der Inkohlung. Brennstoff-Chemie 36: 1–11

Hunt JM, Lewan MD, Hennet RJ-C (1991) Modeling oil generation with time-temperature index graphs based on the Arrhenius equation. AAPG Bull 75: 795–807

Issler DR, Snowdon LR (1990) Hydrocarbon generation kinetics and thermal modelling, Beaufort-MacKenzie basin. Bull Can Petrol Geol 38: 1–16

Jarvie DM (1991) Factors affecting Rock-Eval derived kinetic parameters. Chem Geol 93: 79–99

Jüntgen H (1964) Reaktionskinetische Überlegungen zur Deutung von Pyrolyse-Reaktionen. Erdöl Kohle-Erdgas-Petrochem 17: 180–186

Jüntgen H (1984) Review of the kinetics of pyrolysis and hydropyrolysis in relation to the chemical constitution of coal. Fuel 63: 731–737

Jüntgen H, Klein J (1975) Entstehung von Erdgas aus kohligen Sedimenten. Erdöl Kohle-Erdgas-Petrochem 28: 65–73

Jüntgen H, van Heek KH (1968) Gas release from coal as a function of the rate of heating. Fuel 47: 103–117

Jüntgen H, van Heek KH (1969) Fortschritte der Forschung auf dem Gebiet der Steinkohlenpyrolyse. Brennstoff Chemie 50: 172–178

Jüntgen H, van Heek KH (1970) Reaktionsabläufe unter nicht-isothermen Bedingungen. In: Davison A, Dewar MJS, Hafner K, Heilbronner E, Hofmann U, Niedenzu K, Schäfer K, Wittig G (eds) Fortschritte der chemischen Forschung. Topics Curr Chem 13: 601–699

Jurgan H, Devay L, Block M, Kettel D, Mattern G (1983) Erdgasmigration und Lagerstättenbildung am Beispiel der Erdgasfelder Ost-Niedersachens. BMFT-FB-T 83–153, 78 pp

Kharaka YK, Carothers WW, Rosenbauer RJ (1983) Thermal decarboxylation of acetic acid: implications for origin of natural gas. Geochim Cosmochim Acta 47: 397–402

Klein J (1971) Untersuchungen zur Abspaltung von Wasserdampf, CO, CO_2, N_2 und H_2 bei der nichtisothermen Steinkohlenpyrolyse unter inerter und oxidierender Atmosphäre. Thesis, RWTH Aachen, 117 pp

Klein J, Jüntgen H (1972) Studies on the emission of elemental nitrogen from coals of different rank and its release under geochemical conditions. In: Gaertner HR v, Wehner H (eds) Advances in organic geochemistry 1971. Pergamon Press, Oxford, pp 647–656

Koch E (1977) Non-isothermal reaction analysis. Academic Press, London, 607 pp

Krooss BM, Leythaeuser D, Lillack H (1993) Nitrogen-rich natural gases. Qualitative and quantitative aspects of natural gas accumulation in reservoirs. Erdöl Kohle-Erdgas-Petrochem 46: 271–276

Krooss BM, Littke R, Müller B, Frielingsdorf J, Schwochau K, Idiz EF (1995) Generation of nitrogen and methane from sedimentary organic matter: implications on the dynamics of natural gas accumulations. In: Rice D, Schoell M (eds) Sources of natural gas. Chem Geol 126: 291–318

Landes KK (1967) Eometamorphism and oil and gas in time and space, part 1. AAPG Bull 51(6): 828–841

Larter SR (1988) Some pragmatic perspectives in source rock geochemistry. Mar Petrol Geol 5: 194–204

Larter SR, Horsfield B (1993) Determination of structural components of kerogens by the use of analytical pyrolysis methods. In: Engel MH, Macko SA (eds) Organic geochemisrtry. Plenum Press, New York, pp 271–287

Lasaga AC (1981a) Rate laws of chemical reactions. In: Lasaga AC, Kirkpatrick RJ (eds) Kinetics of geochemical processes. Mineralogical Society of America, Rev Mineral 8: 1–68

Lasaga AC (1981b) Transition state theory. In: Lasaga AC, Kirkpatrick RJ (eds) Kinetics of geochemical processes. Mineralogical Society of America, Rev Mineral 8: 135–169

Lewan MD (1985) Evaluation of petroleum generation by hydrous pyrolysis experimentation. Philos Trans R Soc Lond A 315: 123–134

Lillack H (1992) Untersuchungen zur Beeinflussung der pyrolytischen Gasbildung aus Kerogen durch Muttergesteinsmineralien. Dissertation, RWTH Aachen, 193 pp

Lopatin NV (1971) Temperature and geologic time as factors in coalification. Akad Nauk SSSR Izv Ser Geol 3: 95–106 (in Russian)

Mackenzie AS, Quigley TM (1988) Principles of geochemical prospect appraisal. AAPG Bull 72: 399–415

Mango FD, Hightower JW, James AT (1994) Role of transition-metal catalysis in the formation of natural gas. Nature 368: 536–538

Mann U (1987) Veränderung von Mineralmatrix und Porosität eines Erdölmuttergesteins durch einen Intrusivkörper (Lias epsilon 2–3: Hils-Mulde, NW-Deutschland). Facies 17: 181–188

Momper JA, Williams JA (1984) Geochemical exploration in the Powder River Basin. In: Demaison G, Murris RJ (eds) Petroleum geochemistry and basin evaluation. AAPG Mem 35: 181–191

Neumann HJ, Paczynska-Lahme B, Severin D (1981) Composition and properties of petroleum. In: Beckmann H (ed) Geology of petroleum 5. Wiley, Chichester, 137 pp

Oberlin A, Boulmier JL, Villey M (1980) Electron microscopic study of kerogen microtexture. Selected criteria for determining the evolution path and evolution stage fo kerogen. In: Durand B (ed) Kerogen. Insoluble organic matter from sedimentary rocks. Editions Technip, Paris, pp 191–241

Philippi GT (1965) On the depth, time and mechanism of petroleum generation. Geochim Cosmochim Acta 29: 1021–1049

Pitt GJ (1962) The kinetics of the evolution of volatile products from coal. Fuel 41: 267–274

Polanyi M, Wigner E (1928) Über die Interferenz von Eigenschwingungen als Ursache von Energieschwankungen und chemischer Umsetzungen. Z Physik Chem A 139: 439–452

Powell TG (1978) An assessment of the hydrocarbon source rock potential of the Canadian arctic islands. Geol Surv Can Pap, pp 78–12

Price LC, Wenger LM (1992) The control of pressure on petroleum generation, maturation, and thermal destruction as delineated by hydrous pyrolysis. In: Eckardt CB, Maxwell JR, Larter SR, Manning DAC (eds) Advances in organic geochemistry 1991. Org Geochem 19: 141–160

Price LC (1980) Shelf and shallow basin oil as related to hot-deep origin of petroleum. J Petrol Geol 3: 91–116

Pusey WC III (1973) The ESR-kerogen method: a new technique of estimating the organic maturity of sedimentary rocks. Petrol Times 77: 21–26

Quigley TM, Mackenzie AS, Gray JR (1987) Kinetic theory of petroleum generation. In: Doligez B (ed) Migration of hydrocarbons in sedimentary basins. Editions Technip, Paris, pp 649–666

Rajeshwar K, Dubow J (1982) On the validity of a first-order kinetics scheme for the thermal decomposition of oil shale kerogen. Thermochim Acta 54: 71–85

Robert P (1985) Histoire géothermique et diagenèse organique. Bulletin des Centres de Recherches Exploration-Production Elf-Aquitaine, Mem 8, Pau

Rullkötter J, Michaelis W (1990) The structure of kerogen and related materials. A review of recent progress and future trends. In: Durand B, Béhar F (eds) Advances in organic geochemistry 1989. Org Geochem 16: 829–852

Rullkötter J, Leythaeuser D, Horsfield B, Littke R, Mann U, Müller PJ, Radke M, Schaefer RG, Schenk HJ, Schwochau K, Witte EG, Welte DH (1988) Organic matter maturation under the influence of a deep intrusive body: a natural experiment for quantitation of hydrocarbon generation and expulsion (Toarcian Shale, northern Germany). In: Mattavelli L, Novelli L (eds) Advances in organic geochemistry 1987. Org Geochem 13: 847–856

Sachsenhofer RF (1994) Petroleum generation and migration in the Styrian Basin (Pannonian Basin system, Austria): an integrated geochemical and numerical modelling study. Mar Petrol Geol 11: 684–701

Schaefer RG, Schenk HJ, Hardelauf H, Harms R (1990) Determination of gross kinetic parameters for petroleum formation from Jurassic source rocks of different maturity levels by means of laboratory experiments. In: Durand B, Béhar F (eds) Advances in organic geochemistry 1989. Org Geochem 16: 115–120

Schenk HJ, Horsfield B (1993) Kinetics of petroleum generation by programmed-temperature closed versus open system pyrolysis. Geochim Cosmochim Acta 57: 623–630

Schenk HJ, Horsfield B (1996) Kinetics of petroleum generation: applications and limitations. Zentralbl Geol Paläont Teil I (in press)

Schenk HJ, Witte EG, Littke R, Schwochau K (1990) Structural modifications of vitrinite and alginite concentrates during pyrolytic maturation at different heating rates. In: Béhar F, Durand B (eds) Advances in organic geochemistry 1989. Org Geochem 16: 943–950

Schenk HJ, Horsfield B, Witte EG (1993) Organic geochemistry of freshwater and alkaline lacustrine environments Green River Formation, Wyoming. Part III: Comparative characterization of selected shale samples by various spectroscopic and pyrolysis techniques including kinetic measurements. In: Øygard K (ed) Poster sessions from the 16th Int Meet on Organic geochemistry, Stavanger 1993. Falch Hurtigtrykk, Oslo, pp 324–327

Snowdon LR (1979) Errors in extrapolation of experimental kinetic parameters to organic geochemical systems. AAPG Bull 63: 1128–1134

Sweeney JJ, Burnham AK, Braun RL (1987) A model of hydrocarbon generation from type I kerogen: application to Uinta Basin, Utah. AAPG Bull 71: 967–985

Tackach NE, Barker C, Kemp MK (1987) Stability of natural gas in the deep surface: thermodynamic calculation of equilibrium composition. AAPG Bull 71: 322–333

Tegelaar EW, Noble RA (1994) Kinetics of hydrocarbon generation as a function of the macromolecular structure of kerogen. In: Telnaes N, van Graas G, Øygard K (eds) Advances in organic geochemisrtry 1993. Org Geochem 22: 543–574

Tissot B (1969) Premières données sur les mécanismes et la cinétique de la formation du pétrole dans les sédiments. Simulation d'un schéma réactionnel sur ordinateur. Rev Inst Fr Pét 24: 470–501

Tissot B, Espitalié J (1975) L'évolution thermique de la matière organique des sédiments: application d'une simulation mathématique. Rev Inst Fr Pét 30: 743–777

Tissot B, Welte DH (1984) Petroleum formation and occurrence. Springer, Berlin Heidelberg New York, 699 pp

Tissot B, Espitalié J, Deroo G, Tempere C, Jonathan D (1974) Origin and migration of hydrocarbons in the eastern Sahara (Algeria). In: Tissot B, Bienner F (eds) Advances in organic geochemistry 1973. Éditions Technip, Paris, pp 315–334

Tissot BP, Pelet R, Ungerer Ph (1987) Thermal history of sedimentary basins, maturation indices and kinetics of oil and gas generation. AAPG Bull 71: 1445–1466

Ungerer P (1990) State of the art of research in kinetic modelling of oil formation and expulsion. In: Durand B, Béhar F (eds) Advances in organic geochemistry 1989. Org Geochem 16: 1–25

Ungerer P, Pelet R (1987) Extrapolation of the kinetics of oil and gas formation from laboratory experiments to sedimentary basins. Nature 327: 52–54

Ungerer P, Espitalié J, Marquis F, Durand B (1986) Use of kinetic models of organic matter evolution for the reconstruction of paleotemperatures. Application to the case of the Gironville well (France). In: Burrus J (ed) Thermal modeling in sedimentary basins. Editions Technip, Paris, pp 531–546

Ungerer P, Béhar F, Villalba M, Heum OR, Audibert A (1988) Kinetic modelling of oil cracking. In: Mattavelli L, Novelli L (eds) Advances in organic geochemistry 1987. Org Geochem 13: 857–868

Urov KE (1980) Thermal decomposition of kerogens, mechanisms and analytical application. J Anal Appl Pyrol 1: 323–338

van Heek KH, Jüntgen H (1968) Bestimmung der reaktionskinetischen Parameter aus nicht-isothermen Messungen. Ber Bunsenges Phys Chem 72: 1223–1231

van Heek KH, Jüntgen H, Peters W (1967) Nicht-isotherme Reaktionskinetik der Kohlepyrolyse. I. Theoretische und experimentelle Grundlagen, Voruntersuchungen an Carbonsäuren. Brennstoff-Chemie 48: 163–194

van Heek KH, Jüntgen H, Luft KF, Teichmüller M (1971) Aussagen zur Gasbildung in frühen Inkohlungsstadien auf Grund von Pyrolyseversuchen. Erdöl Kohle-Erdgas-Petrochem 24: 566–572

van Krevelen DW, van Heerden C, Huntjens FJ (1951) Physicochemical aspects of the pyrolysis of coal and related organic compounds. Fuel 30: 253–259

Wall GC, Smith SJC (1987) Kinetics of the production of individual products from the isothermal pyrolysis of seven Australian oil shales. Fuel 66: 345–349

Waples DW (1980) Time and temperature in petroleum formation: application of Lopatin's method to petroleum exploration. AAPG Bull 64: 916–926

Welte DH (1965) Relation between petroleum and source rock. AAPG Bull 49: 2246–2268

Welte DH, Schaefer RG, Yalçin MN (1988) Gas generation from source rocks: aspects of a quantitative treatment. Chem Geol 71: 105–116

Witte EG, Schenk HJ, Müller PJ, Schwochau K (1988) Structural modifications of kerogen during natural evolution as derived from ^{13}C NMR, IR spectroscopy and Rock-Eval pyrolysis of Toarcian Shales. In: Mattavelli L, Novelli L (eds) Advances in organic geochemistry 1987. Org Geochem 13: 1039–1044

Wood DA (1988) Relationships between thermal maturity indices calculated using the Arrhenius equation and Lopatin method: implications for petroleum exploration. AAPG Bull 72: 115–134

Yalçin MN, Schenk HJ, Schaefer RG (1994) Modelling of gas generation in coals of the Zonguldak basin (northwestern Turkey). Int J Coal Geol 25: 195–212

Yen TF (1974) A new structural model of oil shale kerogen. Prepr.-Div Fuel Chem Am Chem Soc 9: 109–114

Zieglar DL, Spotts JH (1978) Reservoir and source-bed history of Great Valley, California. Bull Am Assoc Petrol Geol 62: 813–826

Chapter 5
Deposition of Petroleum Source Rocks

Chapter 5: Overview and Insights

Judging from their contributions to world petroleum reserves, marine silled basins, anoxic continental shelves, and progradational submarine fans are among the most important settings in which source rocks are deposited.

Typical examples for source rock type sediments from marine silled basins are sediments in the Black Sea and Cenomanian/Turonian black shales of the early Atlantic Ocean. In these sediments the organic matter, as seen in the microscope, is mainly unstructured and without primary biological features. A strong microbial degradation of the original biomass is the likely cause for this phenomenon. Consequently the petroleum generation potential of these kerogens is, relatively speaking, not particularly high. These sediments are frequently laminated and consist of marlstones and shales. Following the shape of the basin there can be concentric patterns of the organic matter concentration.

Black shales deposited in the marine shallow environment of anoxic continental shelves are quite common in the Palaeozoic and Mesozoic. Typical representatives are Carboniferous black shales in Europe, northern Africa, and North America. One of the best studied examples is the Mesozoic, Lower Toarcian Posidonia Shale in northwestern and central Europe. The type II kerogen is derived from a mixture of phytoplankton, zooplankton and bacteria. The degradation of the primary biomass is less severe than in deep water environments. Hence petroleum generation potential is usually higher than in source rocks of deep marine basins. Shallow marine source rocks originating on anoxic shelves may extend over vast areas. In terms of lithology they often consist of shales intermixed with siltstones. Frequently they are marly or even consist of real carbonates. In the iron-deficient marine carbonate sequences sulfur is incorporated into the organic matter during diagenesis. This process results in a sulfur-rich kerogens.

Progradational submarine fans accumulate in deep marine water close to a continent. Recent examples are the Mississippi, the Indus, and the Bengal fans. These submarine fans are characterized by higher organic matter concentrations than other deep sea sediments. Generally their organic matter content is mainly of terrigenous origin and the petroleum generating potential is relatively low. Compared to source rocks from silled basins, especially with respect to source rocks from shallow marine shelves, oil-prone source rocks within submarine fans are of restricted areal extent. They consist of shales, siltstones, marlstones, and turbidites. The overall importance of submarine fans as petroleum source is limited.

Source rock facies, as described above for different geological settings, strongly influences the composition and structural make-up of kerogen. Further, recognition of source rock facies is important in the interpretation and prediction of the areal and geometrical distribution of source rock occurrence. Consequently source rock facies is a key element in the definition of the petroleum kitchen in a basin.

Deposition of Petroleum Source Rocks

R. Littke[1], D.R. Baker[2], and J. Rullkötter[3]

5.1 Introduction

It has been long recognized that fine-grained sediments of both siliciclastic and carbonate compositions are the principal source rocks of petroleum. The common coincidence of significant, and often high, content of organic matter within such sediments, as first comprehensively documented by Trask and Wu (1930), provided the initial clue and support for this now generally accepted opinion.

Subsequent studies by petroleum geochemists and organic petrographers have largely confirmed and greatly expanded and quantified this view. Currently the assessment and evaluation of the petroleum-generating capacity, i.e., "potential" of a source rock, is based on the quantity and quality of its organic matter content. Chemical analysis of the amount (concentration) of total organic carbon generally provides the basis for quantitative assessment, although estimates of volumetric amounts of organic matter by petrographic methods serve as well. Qualitative evaluation is more complex and depends considerably on the interpretive views of petroleum generation. From a chemical viewpoint hydrogen-rich organic matter is considered to have the greatest potential for oil generation, whereas rocks with terrigenous organic matter of low hydrogen content are viewed as gas prone. Methods to study the quality of organic matter with respect to petroleum generation are discussed in the textbooks of Durand (1980), Stach et al. (1982), and Tissot and Welte (1984).

The deposition of sediments rich in organic matter (C_{org} roughly greater than 1%) is usually restricted to marine and lacustrine subaquatic environments in which organic matter is produced faster than it can be destroyed (Tourtelot 1979). Recent marine environments generally have in common that they are situated near land, i.e., less than a few hundred kilometers away from at least one coastline (Kruijs and Barron 1990), and that bottom waters are

[1]Institut für Erdöl und Organische Geochemie (ICG-4), Forschungszentrum Jülich GmbH, 52425 Jülich, Germany
[2]Department of Geology and Geophysics, The Wiess School of Natural Sciences, Rice University, Houston, 77251 TX, USA
[3]Institut für Chemie und Biologie des Meeres (ICBM), Universität Oldenburg, Carl-von-Ossietzky-Str. 9-11, 26111 Oldenburg, Germany

Welte et al. (eds)
Petroleum and Basin Evolution
© Springer-Verlag Berlin Heidelberg 1997

oxygen depleted compared to surface waters. Similarly, contemporary lakes which accumulate significant amounts of organic matter also are characterized by oxygen-depleted bottom waters, for example, Lake Tanganyika (Huc 1988a,b). Subaerial accumulation of organic matter is restricted to special settings such as raised bog peats (mires) which grow in humid climates and are considered to be important coal precursors (Smith 1962).

In contrast to marine and lacustrine sediments, deltaic and fluvial sediments and coals generally do not contain much organic matter derived from aquatic organisms. However, they often contain sizeable quantities of tissues of higher plants which are less easily destroyed by oxidation and bacterial degradation (Tissot and Welte 1984). This type of organic material is less hydrogen rich and less oil prone than the aquatic type.

In addition to organic matter content and quality, stratigraphic considerations are of importance in source rock assessment and evaluation. Specifically, the thickness and areal extent, and hence the volume, of a source rock unit place important constraints on its ultimate petroleum potential. Similarly, vertical and regional variations in the concentration and type of organic matter within a source rock unit clearly affects its capacity for oil generation. Often source rock assessment is only based on a local-scale data base, i.e., a few isolated samples, or perhaps a single well or outcrop profile. Clearly, adequate evaluation should be carried out on a basin scale involving a data base combining regional geochemical and stratigraphic analysis.

A potential source rock may also be judged from the standpoint of properties which govern its hydrocarbon expulsion effectiveness. Petroleum migration and accumulation greatly depend on source rock characteristics. As discussed in an early publication by Ronov (1958) and in Tissot and Welte (1984), there is a lower limit of total organic carbon values of about 0.5%. Rocks with lower percentages of organic carbon may not be able to expel the generated oil, possibly due to the adsorption of the hydrocarbon molecules on minerals. Especially, organic matter-lean source rocks within thick shale sequences cannot expel petroleum efficiently unless oil is cracked to more mobile gas, or unless fracturing provides additional avenues for petroleum migration. Whether migration occurs mainly in vertical or horizontal direction also depends on the source rock properties. For example, fractures seem to develop more often parallel to the bedding plane in shaly source rocks than in carbonate source rocks, in which fractures cut bedding at high angles (Littke et al. 1988). Espitalié et al. (1987) concluded for the Toarcian shales of the Paris basin that the petroleum migration occurred laterally inside the source rock. For rocks of similar organic matter content, the physical distribution of the kerogen, for example, disseminated vs. concentrated in larger entities such as lenses and layers, may result in different expulsion properties (see also Chap. 7, this Vol.). Further, small-scale lithological variations, such as laminations of siltstone and sandstone, may provide permeable pathways which enhance the expulsion process (Littke 1993, pp. 126–129).

Although it is evident that the ultimate behavior of a rock as a petroleum source depends on its long and often complex history of generation and expulsion, these processes and resultant products are substantially affected by primary features and properties which are determined by the depositional history of the sediment. The objective of source rock depositional studies, and the focus of this chapter, is to establish and elucidate the controls and factors which govern the formation of sedimentary rocks which have qualities and properties considered to be essential for petroleum generation, i.e., source rock potential. Clearly, the parameters of organic matter concentration and character coupled with stratigraphic aspects are the most crucial. Hence, depositional controls of organic matter accumulation and factors influencing the qualitative nature of entombed organic matter is emphasized. Concrete understanding of the depositional history of petroleum source rocks, as stated by Demaison et al. (1984), permits us to predict the location and distribution of potential oil source beds in sedimentary basins. Clearly, such "predictive source bed stratigraphy" (Demaison et al. 1984), especially when applied on a basin scale, should be a significant aid to petroleum exploration.

5.2 Production and Preservation of Organic Matter

5.2.1 The Debate

The opinion that bottom water "stagnation," i.e., oxygen deficiency, anoxia, has played an important role in the preservation of organic matter and hence in the formation of organic matter-rich sediments, for example, black shales, has been a part of sedimentological thinking for many years. However, only recently, when large amounts of stratigraphic, sedimentological, and geochemical observational data on modern oceanic sediments [e.g., Deep Sea Drilling Project (DSDP), Ocean Drilling Program (ODP)] and correlative oceanographic data became available and could be evaluated and interpreted, was it possible to place some constraints on the causes for the development and persistence of anoxia in depositional environments. In particular the synthesis by Demaison and Moore (1980) has proven to be a most influential analysis of the problem and has provided a landmark for the evaluation of the role of anoxia in source rock deposition. Currently two contrasting models (Pedersen and Calvert 1990; Demaison 1991; Pedersen and Calvert 1991) are used to explain the deposition of subaquatic, fossil, organic matter-rich sediments. These are called below the "preservation model" and "productivity model" (Fig. 5.1; Thunell et al. 1984).

Emphasis on the accumulation of organic matter in sediments and sedimentary rocks has led to a generalized (and somewhat simplified) view that organic matter *productivity* and its *preservation* are the two dominant controls of the deposition of source rock facies. This viewpoint was emphasized by Curtis (1980) in her statement that "many geographic settings are favorable for high organic productivity, but preservation of organic materials requires a setting where rapid oxidation cannot take place...." The relative importance of these two general controls is currently being actively debated. For example, Stein (1986a) in his discussion of black shale deposition in the Mesozoic

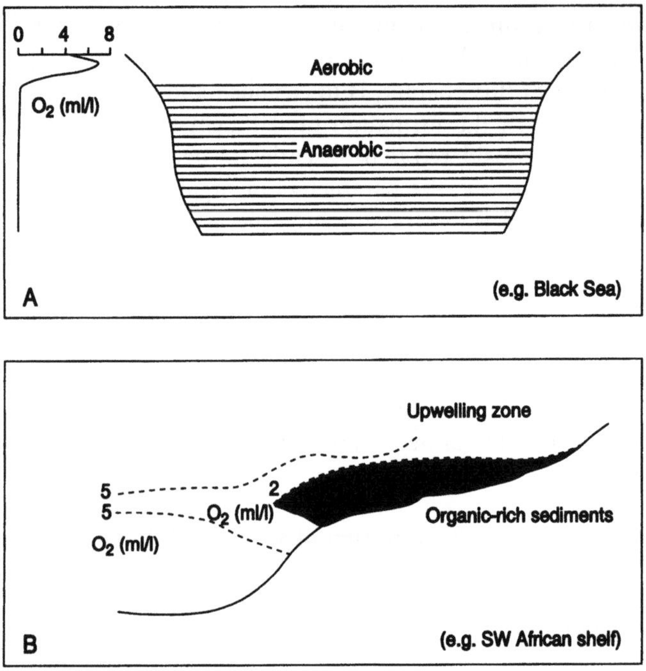

Fig. 5.1A,B. Generalized models for the deposition of organic matter-rich sediments. **A** Stagnation model for anoxic silled basins. **B** Productivity model for upwelling regions. (Modified from Littke and Welte 1992; Thunell et al. 1984)

Atlantic Ocean notes that "the genesis of these sediments can be explained by (1) high flux rates of marine organic matter in areas of high oceanic productivity, (2) high flux rates of terrigenous organic matter, and/or (3) increased preservation of organic matter in areas of anoxic deep-water environments as well as due to rapid burial of organic matter..." and that "which of these different mechanisms is more or less important for the origin of black shales is still open and controversial." Recently, enhanced preservation of organic matter adsorbed to mineral surfaces has been discussed as an additional, and possibly important, control on organic matter accumulation in sediments (Collins et al. 1995).

Advocates of the importance of productivity, i.e., the supply of organic matter, including Calvert (1987), Pedersen and Calvert (1990), and Bailey (1991) argue cogently that "sporadic temporal and spatial increases in primary production, reflecting changes in the behavior and/or state of the ocean-atmosphere system... constitute a more tenable explanation for the occurrence of modern and Quaternary carbon-rich sediments and Cretaceous black shales," and that "consequently, the fundamental control on the accumulation of carbon-rich facies in the oceans and marginal seas is not the presence or absence of anoxia." In this view suitable preservation conditions, for example, low-oxygen bottom waters, "are the result of the high production" (Calvert 1987, p. 146) and due to the high oxygen demand of the settling organic matter.

The "productivity model" (Fig. 5.1B) is based on high biological productivity in surface waters, as recently observed in upwelling areas, for example, offshore Peru (Thiede and Suess 1983; Suess and Thiede 1983; Miller 1989; Pedersen and Calvert 1990; Summerhayes et al. 1992). Essential for this model are high nutrient supply and sufficient sunlight. The great masses of decaying organic matter inherent in high bioproductivity situations also cause oxygen deficiency in the bottom water of these environments. Whether this oxygen deficiency supports organic matter preservation is a matter of current scientific debate (Pedersen and Calvert 1990, 1991; Demaison 1991).

It is, however, not entirely proven whether the same conceptual settings also created organic matter-rich aquatic sediments in the past. For example, it is known that enhanced atmospheric CO_2 concentration leads to enhanced marine bioproductivity (Kruijs and Barron 1990). Such conditions may have favored the deposition of organic matter in the past. Also, transgression of the sea may ultimately enhance accumulation of organic matter in shelf regions (Wenger and Baker 1986), whereas times of regression may promote organic matter accumulation in prograding deltaic fans in deep water. For the understanding of organic matter deposition such time-dependent geological processes are probably of great importance.

The "preservation model" assumes deposition under a stratified water mass of which the bottom part is anoxic or suboxic (Table 5.1; Tyson and Pearson 1991). Contemporaneous examples are the Black Sea and Lake Tanganyika (Huc 1988a). Essential for the validity of this model is a small rate of water exchange, especially vertically; high rates would ultimately result in an oxygenation of bottom waters. It is generally thought that small rates of water exchange are favored in silled basins (Fig. 5.1A; Demaison and Moore 1980). Limited degradation of organic matter in the absence of oxygen rather than high bioproductivity is the principal distinction of the "preservation model." It should be noted, however, that most coastal areas or shelf seas are characterized by higher than average bioproductivity (Huc 1988a; Kruijs and Barron 1990), i.e., the model is generally applied to sedimentary rocks for which medium to high bioproductivity rates can be assumed.

Table 5.1. Terminology for low oxygen regimes and the resulting biofacies. (Tyson and Pearson 1991)

Oxygen (ml/l)	Environments	Biofacies	Physiological regime
8.0–2.0	Oxic	Aerobic	Normoxic
2.0–0.2	Dysoxic	Dysaerobic	Hypoxic
2.0–1.0	Moderate		
1.0–0.5	Severe		
0.5–0.2	Extreme		
0.2–0.0	Suboxic	Quasi-anaerobic	
0.0 (H_2S)	Anoxic	Anaerobic	Anoxic

The authors suggest that the boundary between suboxic and dysoxic conditions marks the limit for bioturbation (see Savrda and Bottjer 1991) and nitrate reduction.

Those who advocate the greater importance of the enhancement of organic matter entombment generally rely on special environmental and depositional conditions which are believed to promote preservation. The most common and accepted conditions are bottom-water *anoxia* and *sedimentation* (burial) *rate*. For the latter the concept is an old and simple one, specifically, that preservation of organic matter is enhanced due to rapid burial and consequently removal (isolation) from the multitude of alteration reactions which occur at and near the depositional interface, i.e., oxidation, bioturbation, and bacterial mineralization, including aerobic and anaerobic processes.

The relative importance of the rate of sediment accumulation has been rather rigorously demonstrated by Müller and Suess (1979), who show that the organic concentration of sediments from a wide variety of *oxic* oceanic settings roughly doubles with each tenfold increase in the rate of sediment accumulation. Similar studies by others (Heath et al. 1977; Ibach 1982; Stein 1986a,b; Bralower and Thierstein 1987; Littke et al. 1991c) also recognize this relationship. However, Tyson (1987, p. 50) notes that where good biostratigraphic data (i.e., time control) are available a correlation of *low* sedimentation rates and the deposition of marine petroleum source rocks (MPSR) is indicated. He cautions that "increased rates of sedimentation should not automatically be considered the key to MPSR formation." In this regard excessive sedimentation rates are thought by many to cause a diminishment of organic matter accumulation, i.e., a dilution effect; however, with some exceptions (Brumsack 1980; Jones 1983; Ibach 1982) the recognition of this effect does not seem unequivocally and generally established.

Most workers (see, for example, the review by Fleet and Brooks 1987, pp. 12–13) take a balanced view of the relative importance of productivity versus preservation controls of source rock deposition. After all, it is trivially obvious that the accumulation of an organic matter-rich sediment requires both a supply (i.e., production) of organic precursors and the survival (i.e., preservation) of some portion of those precursors during the vicissitudes of sedimentation history. Thus, the focus of the controversy should be on an understanding of the precise conditions and processes which govern the productivity and preservation of organic matter.

5.2.2 Some Observations

Primary production of organic matter in essence refers to the rate of fixation of carbon by photosynthetic organisms. Estimates of the total annual phytoplankton production in the modern world ocean vary somewhat according to definition, but center around 20–30 gigatons (10^9 tons) of carbon (Platt and Subba Rao 1975; Koblents-Mishke 1977; Berger et al. 1989b). Maps of surface ocean productivity show that zones of high organic matter production are concentrated on continental margins whereas the central open oceans are mostly characterized by low productivity numbers (see Berger 1989 for an overview). Derived from these maps, about one-quarter of the total production is delivered by slightly less than 10% of the ocean area (Berger et al. 1989b), i.e., mainly coastal waters. Because of changing land mass distributions and climatic conditions both magnitude and global pattern of oceanic organic matter production are likely to have changed considerably during the geological history of the earth (Arrhenius 1952).

Müller and Suess (1979) used a relationship between organic carbon content of sediments and sedimentation rate to obtain Pleistocene productivity data for open ocean conditions. After calibrating with data for the present ocean, they arrived at an empirical equation for paleoproductivity (PaP, in g Cm^{-2} y^{-1}):

$$PaP = C\delta(1 - \phi)/0.003\ S^{0.3}$$

where C is the percentage of organic carbon, δ is the dry sediment density, ϕ is the porosity (as a fraction of the volume), and S the bulk sedimentation rate. They thus suggested that productivity can simply be estimated from the organic matter content of a sediment using a fraction of the sedimentation rate for correction. Paleoproductivity calculation must be regarded, however, as an estimation rather than an exact quantification procedure because several important parameters are neglected. For example, different classes of phytoplankton have vastly different labilities upon oxic and anoxic decay, as demonstrated by the often excellent preservation of green algae such as *Bottryococcus* and *Tasmanales* or of dinoflagellate cysts and by the limited degree of preservation of organisms with siliceous and carbonate shells such as coccoliths (e.g., Littke et al. 1991b). In addition, total organic carbon contents in sediments depend on factors such as oxygen and sulfate availability during early diagenesis, which can hardly be incorporated into this type of formula. Nevertheless, some useful refinements and applications have been published on paleoproductivity estimates.

It has been shown that the organic carbon accumulation rate is a function of carbon flux near the sea floor, which in turn is related to the productivity and the water depth. Using flux information of Betzer et al. (1984), Stein (1986a, 1991) derived the following more complex empirical relationship for estimating paleoproductivity:

$$PaP = 5.31[C(\delta_{WB} - 1.026\phi)]^{0.71} S^{0.07} D^{0.45}$$

with δ_{WB} = wet bulk density and D = water depth. Values calculated this way are linearly related, although not numerically identical to values obtained from another empirical equation by Sarnthein et al. (1987, 1988):

$$PaP = [15.9342C \cdot S \cdot \delta(1 - \phi)]^{0.6590} \cdot S_{B-C}{}^{-0.7143} \cdot D^{0.3174}$$

where $S_{B\text{-}C}$ is the organic-carbon-free bulk sedimentation rate.

Applying their simple relationship to sediments from offshore northwestern Africa, Müller and Suess (1979) found that Pleistocene productivity in the ocean was about the same as today during interglacial stages but higher by a factor of up to 3 during glacial times, due mainly to more intense upwelling or mixing of water masses. In a high-resolution stratigraphic study for the last 500 000 years in the same area, Sarnthein et al. (1987, 1988) found the same range of productivity fluctuations in upwelling sediments with their more sophisticated equation; the productivity amplitudes were much less pro-

Table 5.2. Estimates of paleoproductivity for different time intervals at upwelling site 658 and nonupwelling site 659 offshore West Africa. (Modified after Stein et al. 1989)

Depth (mbsf)	Age (Ma)	Wet-bulk-density ($g\ cm^{-3}$)	Porosity (%)	Mean sed. rate ($cm\ 1000\ a^{-1}$)	Total organic carbon content (%)	Marine organic carbon content (%)	Paleoproductivity I	Paleoproductivity II
Site 658								
0–100	0–0.7	1.65	68	14.7	a: 1.61	1.05	190	150
					b: 1.26	0.82	160	130
100–174	1.6–2.45	1.60	62	7.2	a: 2.13	1.38	240	210
					b: 1.81	1.18	220	190
174–240	2.45–3.1	1.60	62	10.8	a: 1.99	1.29	240	200
					b: 1.67	1.09	210	170
240–295	3.1–3.6	1.73	56	10.8	a: 2.54	1.65	320	260
					b: 1.49	0.97	220	180
Site 659								
0–50	0–1.6	1.51	68	3	0.22	0.15	40	50
50–135	1.6–4.5	1.60	64	3	0.15	0.10	30	40
135–230	4.5–8.5	1.74	56	0.3	0.07	0.05	20	30

For the time interval from 0–2.45 Ma at site 658, a is interpreted as glacial and b as interglacial stages based on different carbonate contents. Paleoproductivity calculations are based on: I, Stein (1986a); II, Sarnthein et al. (1987).

nounced in oceanic "deserts" further offshore, but approximately with the same cycle.

Table 5.2 compares average paleoproductivity data over larger time intervals for sediments from within (site 658) and from outside a coastal upwelling cell (site 659) offshore West Africa (Stein et al. 1989). Average values vary little over the last several million years and are similar to present-day productivity data. Productivity is higher by a factor of 5 to 10 in the upwelling area. Total organic carbon data were corrected for terrigenous contributions based on organic petrographic observations. Thus, productivity data are for marine organic matter only. In the upwelling area values calculated by the method of Stein (1986a) are slightly higher than those based on the relationship of Sarnthein et al. (1987) whereas the opposite was observed for the low-productivity area at site 659. Similar inconsistencies of different techniques applied to Pleistocene sediments were observed by Mix (1989).

As indicated by Stein (1991), the equations of Müller and Suess (1979), Stein (1986a) and Sarnthein et al. (1987, 1988) describe the relationship between surface-water productivity and organic carbon accumulation specifically for oxic environments. For anoxic environments Bralower and Thierstein (1987) from comparison of accumulations rates of (marine) organic carbon and primary production rates in recent anoxic settings suggested a relationship

Fig. 5.2. Paleoproductivity (in $gC\ m^{-2}\ y^{-1}$) calculated for times of mudstone/limestone deposition (**A**) and times of black shale deposition (**B**) for DSDP sites in the eastern Atlantic Ocean. *Underlined values*, calculated using the paleoproductivity equation modified by Bralower and Thierstein (1987) for anoxic deep-water conditions. (Stein 1986a)

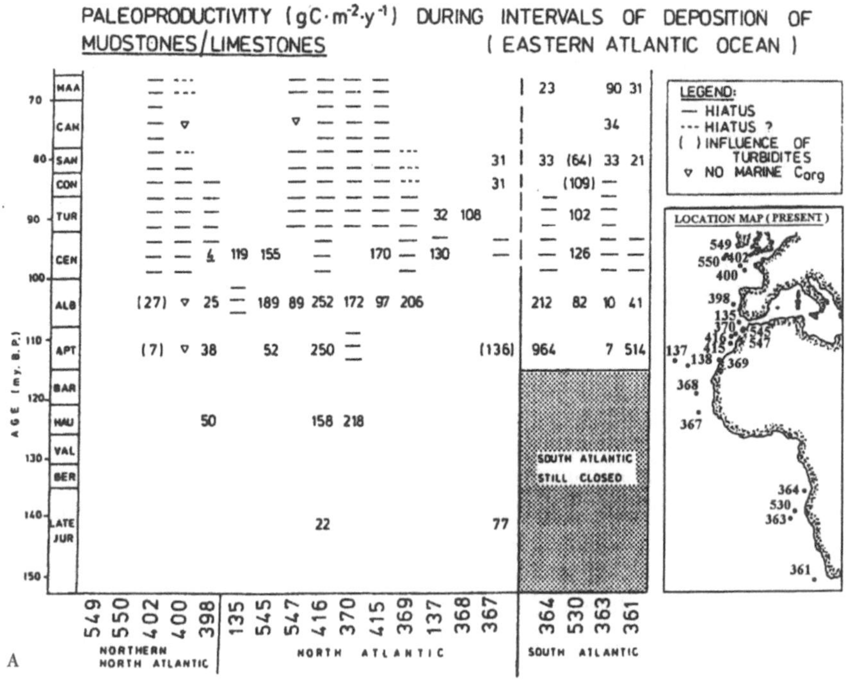
PALEOPRODUCTIVITY (gC·m-2·y-1) DURING INTERVALS OF DEPOSITION OF
MUDSTONES/LIMESTONES (EASTERN ATLANTIC OCEAN)
AGE (m.y. B.P.)
MAA
CAM
SAN
CON
TUR
CEN
ALB
APT
BAR
HAU
VAL
BER
LATE JUR
LEGEND:
— HIATUS
--- HIATUS ?
( ) INFLUENCE OF TURBIDITES
▽ NO MARINE Corg
LOCATION MAP (PRESENT)
SOUTH ATLANTIC STILL CLOSED
549 550 402 400 398
NORTHERN NORTH ATLANTIC
135 545 547 416 370 415 369 137 368 367
NORTH ATLANTIC
364 530 363 361
SOUTH ATLANTIC
A

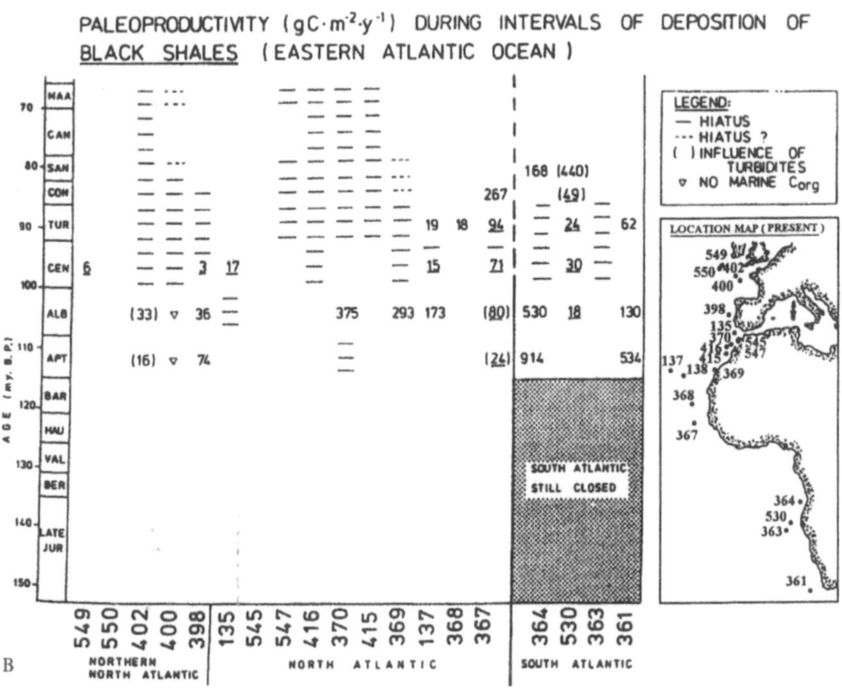
PALEOPRODUCTIVITY (gC·m-2·y-1) DURING INTERVALS OF DEPOSITION OF
BLACK SHALES (EASTERN ATLANTIC OCEAN)
AGE (m.y. B.P.)
MAA
CAM
SAN
CON
TUR
CEN
ALB
APT
BAR
HAU
VAL
BER
LATE JUR
LEGEND:
— HIATUS
--- HIATUS ?
( ) INFLUENCE OF TURBIDITES
▽ NO MARINE Corg
LOCATION MAP (PRESENT)
SOUTH ATLANTIC STILL CLOSED
549 550 402 400 398
NORTHERN NORTH ATLANTIC
135 545 547 416 370 415 369 137 368 367
NORTH ATLANTIC
364 530 363 361
SOUTH ATLANTIC
B

implying that at least 2% of the primary organic carbon is preserved in the sediments:

$$PaP = 5 \cdot C \cdot S \cdot (\delta_{WB} - 1.026\phi)$$

This equation was used by Stein (1986a) to calculate paleoproductivity data in the Mesozoic Atlantic Ocean (Fig. 5.2). Considering particularly the difficulties of obtaining sufficiently accurate sedimentation rate data, a careful preliminary interpretation shows that off northwestern Africa productivity was low during the Late Jurassic but distinctly increased during Hauterivian times. Maximum productivity values, approaching those in the modern ocean, occurred during Aptian/Albian times, probably caused by intensified upwelling off northwestern Africa. Interestingly, paleoproductivity was relatively low at the time of deposition of black shales during the Cenomanian/Turonian boundary event. For mudstone/chalk lithologies similar paleoproductivity values were calculated as for black shales (Fig. 5.2).

The difficulties in assessing the productivity of ancient oceans have been summarized by Berger et al. (1989b) and Thierstein (1989). Particularly the mid-Cretaceous, which is characterized by global enrichment of sediments in organic carbon, represents a major challenge to environmental modeling because of the possible lack of polar glaciation, differences in orbital forcing, and changes in the biosphere which indicate a planetary state much different from that in the Late Neogene. The influence of orbital forcing on carbonate and organic carbon accumulation during the mid-Cretaceous has recently been described and discussed by Jendrzejewski (1995).

The empirical relationships for paleoproductivity assessment illustrate how organic matter accumulation is related to primary productivity through factors such as organic carbon flux through the water column and bulk sedimentation rate. In addition, it is clear that reduced oxygen levels in the water column enhance organic matter preservation. Thus, organic carbon-rich sediments and sedimentary rocks are likely to be formed in most cases by the mutually supporting effects of both anoxia (static or dynamic) and productivity, and the question of assigning a single controlling factor (Pederson and Calvert 1990) appears too restrictive. Extremes can be found in both directions, for example, the Holocene sediments of the Black Sea where anoxic water conditions alone are insufficient to produce black shales, compared to the equatorial upwelling area where enhanced primary productivity in the photic zone is not reflected in elevated organic carbon concentrations in the surface sediments due to the antagonistic effect of the deep oxic water column.

Stein (1986b, 1990) presented a simple diagram of sedimentary organic carbon concentrations vs. sedimentation rates showing the effects of open-marine oxic and anoxic conditions (Fig. 5.3A). Field A, enclosed by the diagonal lines, represents the sedimentation rate controlled preservation of organic matter under oxic marine conditions, whereas the hatched area B refers to anoxic (or strongly oxygen-depleted) depositional environments. At low sedimentation rates the hatched area B signifies stagnant conditions like in the Black Sea, whereas in the stippled area A′ where fields A and B overlap, oxygen depletion is controlled by the oxygen-consuming effect of high primary pro-

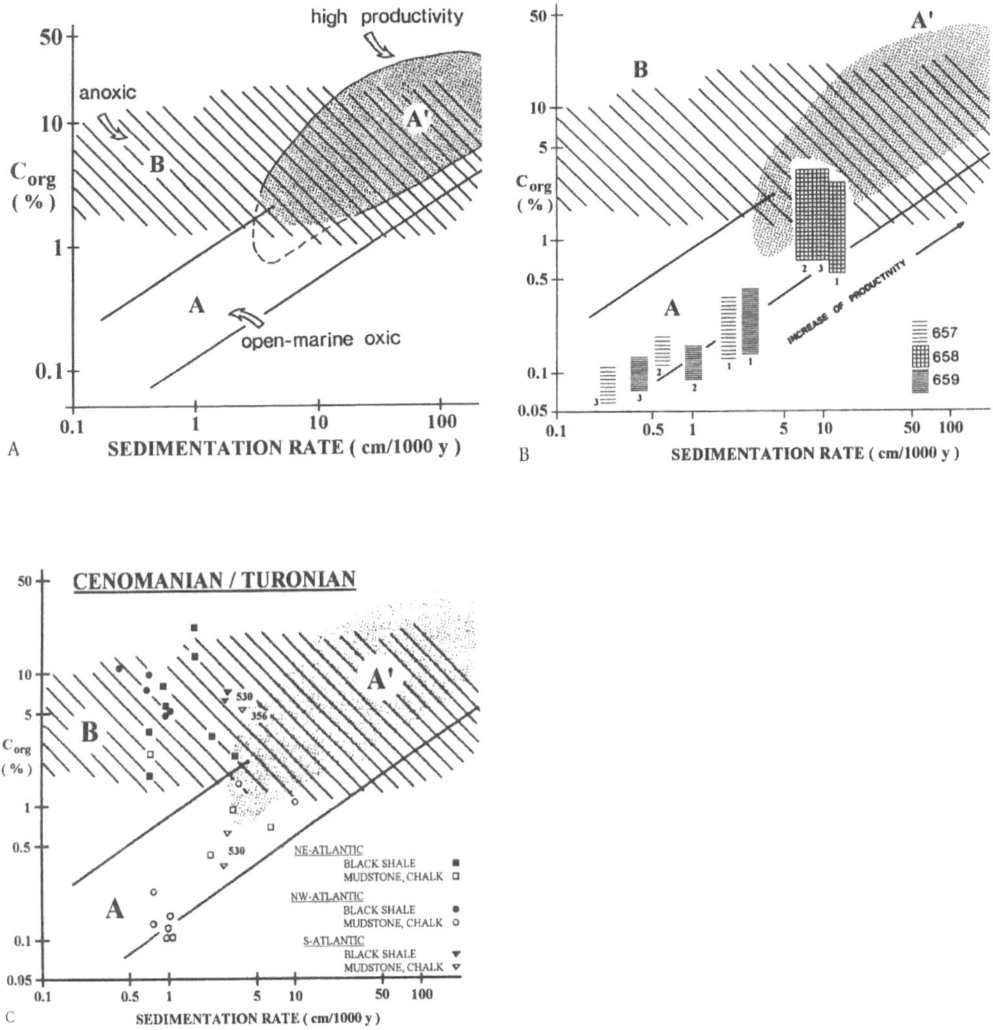

Fig. 5.3. A Correlation between (marine) organic carbon and sedimentation rates. The distinction between fields A, A′, and B is based on data derived from Recent to Miocene sediments deposited in normal open-ocean environments (*field A*), upwelling high-productivity areas (*field A′*; *stippled area*, coastal upwelling and the open area at the lower end of field A′ indicating equatorial upwelling data) and anoxic environments (*field B*). From Stein (1990; modified after Stein 1986b). **B** Organic carbon vs. sedimentation rate diagram applied to sediments from ODP sites 657–659 offshore West Africa. Site 657: *1*, Pliocene-Pleistocene (0–4.5 Ma); *2*, latest Miocene (6–7 Ma); *3*, late Miocene (7–8.5 Ma). Site 658: *1*, Pleistocene (0–0.7 Ma); *2*, latest Pliocene (1.6–2.45 Ma); *3*, late Pliocene (2.45–3.6 Ma). Site 659: *1*, Pliocene-Pleistocene (0–4.5 Ma); *2*, late Miocene (4.5–8.5 Ma); *3*, late Miocene 8.5–10 Ma. From Stein et al. (1989). **C** Organic carbon vs. sedimentation rate diagram applied to Cenomanian/Turonian sediments from the Atlanic Ocean. (Stein 1986b)

ductivity as in upwelling areas. In Fig. 5.3B this diagram is applied to sediments from offshore northwestern Africa (ODP sites 657–659; see Table 5.2) and shows that only the sediments from within the current (and ancient) upwelling zone fall into the anoxic field A′, whereas in all other cases organic matter preservation is controlled by sedimentation under oxic conditions.

Figure 5.3C shows the same relationship for Cenomanian/Turonian black shales (closed symbols) and mudstones/chalks (open symbols) from the Atlantic Ocean (Stein 1986b) based on data compiled by Stein et al. (1986). All black shales at the Cenomanian/Turonian boundary, which represent the Atlantic Ocean expression of a world-wide event (e.g., Herbin et al. 1986), appear to be deposited under stagnant oxygen-depleted bottom-water conditions. This finding puts constraints on the paleoceanographic assessment of the depositional environment at the time of this event, particularly with respect to oceanic circulation and exchange of water masses.

Thus, the organic carbon vs. sedimentation rate diagram appears to be a useful tool in understanding the depositional conditions of organic matter-rich sediments. It should be emphasized again, however, that before applying this diagram a distinction must be made between the content of marine and terrigenous organic matter using a suitable technique such as microscopic maceral analysis, and that only the marine portion must be considered (Stein et al. 1986).

5.3 Transport of Organic Particles

Most of the biologically produced organic matter is transported to its depositional site by wind or water. Exceptions include plants in swamps and raised bogs and benthic algal and bacterial life forms (microbial mats) that can be incorporated into sediments at the site of growth. During transport the degradation of the organic matter begins and the most labile parts of the organic matter are remineralized. This chemical degradation is enhanced by mechanical breakdown and a decrease in particle size causing higher surface/volume ratios. Although the initial transport of organic particles preserved in sediments may occur by wind, the following discussion is restricted to transport in water.

The water transport of organic matter follows the same principles as the transport of mineral grains (e.g., von Engelhardt 1973). A measure of vertical transport is the determination of the sinking velocity (vs) of an ideal spheric particle according to Stokes' law

$$vs = [(\varphi_2 - \varphi_1) \cdot g \cdot D^2]/18\eta$$

where φ_2 and φ_1 are the densities (kg m^{-3}) of the particle and water, g is the force of gravity (m s^{-2}), D is the particle diameter (m) and η is the dynamic viscosity (kg m^{-1} s^{-1}). According to the above equation, the sinking velocity of terrigenous organic particles (vitrinite precursors) of 100 μm diameter is about 1.6 mm/s in freshwater. In a 1000 m deep lake such a particle would reach the sediment surface within 7 days, if stationary conditions are assumed (Table 5.3). Terrigenous organoclasts of equal shape and density but only 10 μm in diameter would need almost 2 years to sink to the bottom, and algal particles that have a lower density than terrigenous particles sink slower. It should be noted that real sinking velocities depend on the grain shape. For example, von Engelhardt (1973) showed that the sinking velocity of quartz spheres of 10–100 μm diameter is two orders of magnitude greater than that of muscovite plates of equal diameter. Many organic particles are not spherical or well rounded. The typical shape of terrigenous organic particles in young, open-marine sediments is irregularly cylindrical, with the longest axis being about twice the length of the shortest axis (Littke et al. 1991c). Therefore lower sinking velocities than calculated in Table 5.3 must be assumed for organic particles. Furthermore, compared to freshwater, sinking velocities in sea water are lower due to its greater dynamic viscosity. A more rapid transportation of organic remains is possible if they are incorporated in large fecal pellets (Tourtelot 1979), which are known to settle faster than small particles. Degens and Ittekott (1987) strongly advocate the transfer of organic particles in fecal pellets "which are jetted to the sea floor at velocities of about 500 cm/day." Even higher velocities can be assumed according to Stokes' law for large fecal pellets or other organic matter-rich aggregates (Table 5.3).

In addition to such theoretical calculations on the transport of particles, sinking rates of aggregated and single organoclasts have also been presented and discussed in the past (e.g., Riley 1970), but there are only few data on grain size distributions of organic particles in sediments. Littke et al. (1991c) determined the grain sizes of vitrinites and inertinites (i.e., allochthonous, terrigenous organic particles) in marine sediments in the central Indian Ocean and found that there are virtually no particles larger than 20 μm (Fig. 5.4). Particles of up to 30 μm in diameter were observed only in some pre-Maestrichtian sediments (site 755) deposited in a paleogeographic position close to the Cretaceous Kerguelen-Heard plateau, i.e., close to a major island. Such

Table 5.3. Sinking velocity, travel time (through 1000 m water), and lateral transport distance (at a current velocity of 1 mm/s and 100 m water depth) for spheric particles in nonturbulent water. (See von Engelhardt 1973)

Particle type	Sinking velocity (m/s)	Travel time (1000 m water)	Lateral transport (m)
Terrigenous organoclast Ø=100 μm, ρ=1300 kg/m^3	1.6×10^{-3}	7.1 days	62.5
Terrigenous organoclast Ø=10 μm, ρ=1300 kg/m^3	1.6×10^{-5}	1.9 years	6.25×10^{3}
Fecal pellet Ø=1 mm, ρ=1300 kg/m^3	1.6×10^{-1}	2 hours	6.25×10^{-1}
Quartz grain Ø=10 μm, ρ=2600 kg/m^3	8.7×10^{-5}	133 days	1.15×10^{3}

Densities of coal macerals usually vary between 1.1 and 1.7 g/cm^3 (van Krevelen 1961).

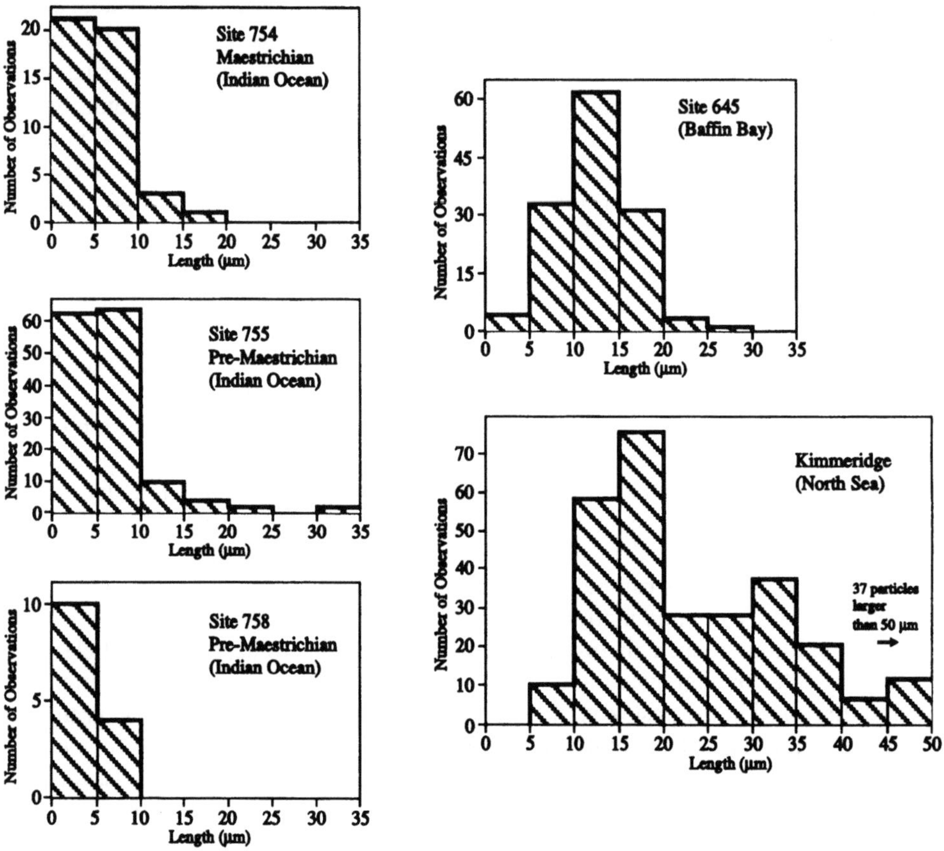

Fig. 5.4. Histograms of length of terrigenous organic particles (vitrinites and inertinites) at sites 754 and 755 (Broken Ridge, Indian Ocean), 758 (Ninetyeast Ridge, Indian Ocean), 645 (Baffin Bay, i.e., near land, see Stein et al. 1989), and in the Kimmeridge Clay in the North Sea (Brae Field; see Leythaeuser et al. 1988). The particle size is related to the distance of transport. (Littke et al. 1991c)

small grain sizes seem to be typical for open marine sediments, providing no turbidite transport (see Degens et al. 1986) affected deposition. In contrast, shallow marine or deep water, continent-near deposits such as the Kimmeridge Clay from the Brae Field, North Sea, contain particles of 50 µm diameter and greater (Fig. 5.4). Even larger grain sizes are found in fluvial-deltaic sediments deposited under humid climatic conditions. For example, Upper Carboniferous siltstones and sandstones of the Ruhr area contain abundant plant fragments of several millimeters length (Scheidt 1988).

The observations on grain sizes of allochthonous organic particles in different environments indicate that transport of organoclasts leads to mechanical grain size reduction and sorting of grains according to size with increasing distance of transport or transport energy in very much the same way as well established for mineral grains. In general, large transport distances of orga-

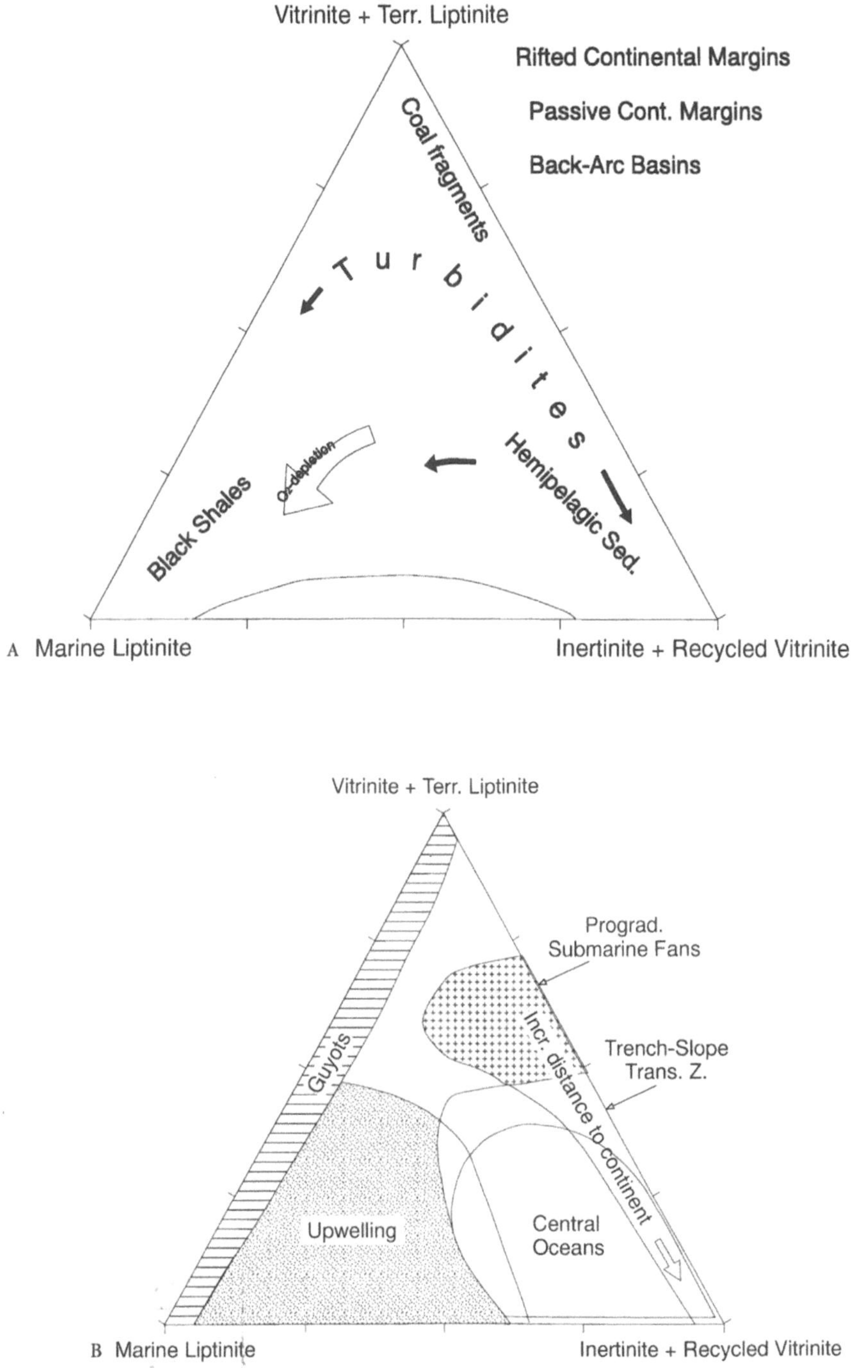

Fig. 5.5. A,B Percentages (of total macerals) of vitrinite, inertinite, and liptinite from different deep sea environments. (Littke and Sachsenhofer 1994)

noclasts are favored by their low density compared to mineral grains. This also leads to large residence times of organoclasts in the water.

While large transport distances of terrigenous organoclasts compared to terrigenous mineral matter lead to a relative enrichment of the former, it is unlikely that transport sorting causes deposition of layers rich in terrigenous organic matter at distal sites in the oceans because more and more autochthonous siliceous and calcareous marine particles are admixed. It is also not expected that liptinite-rich sediments evolve from transport sorting, although liptinites (hydrogen-rich organic particles) have a lower density than vitrinites and inertinites. Liptinites are, however, more susceptible to degradation than, for example, inertinites and therefore if transportation is of long duration less well preserved. This is the reason why in most marine sediments deposited under oxic bottom waters terrigenous organic particles (resedimented vitrinites and inertinites) predominate over marine particles (alginite; Littke and Sachsenhofer 1994).

Generally, with increasing distance of transport, the inertinite/vitrinite ratio increases, indicating a greater degree of chemical degradation. In many coals as well as in many lacustrine, fluvial, deltaic, and shallow marine deposits, vitrinite is the most abundant organoclast (e.g., Horsfield et al. 1988), but in most open marine deposits inertinite predominates (Fig. 5.5; Littke and Sachsenhofer 1994). The latter is known to be the product of a more severe degradation of terrigenous organic matter, whereas vitrinite derives from a less severe degradation of the same principle precursors (Stach et al. 1982; Styan and Bustin 1983). It is an open question why at least some small vitrinite is preserved in open ocean sediments derived from slow deposition under well-oxygenated bottom water, i.e., under conditions favorable to a strong degradation of organic matter.

5.4 Deep Marine Silled Basins

Demaison and Moore (1980) identified anoxic silled basins as a major setting for marine oil source rocks and discussed the Black Sea as a typical contemporary example. In anoxic silled basins the major control on deposition of organic carbon-rich sediments is anoxicity of bottom water and pore water leading to high ratios of preserved over produced organic carbon (Bralower and Thierstein 1987). Although they acknowledge the importance of several other source rock depositional controls, for example, primary biological productivity, biochemical alteration of dead organic matter, transport time and mechanism, and the effect of sediment particle size and sedimentation rate, clearly their view places considerable importance on the existence of anoxic conditions and its resultant enhancement of organic matter preservation. Simply stated, they note that "anoxic conditions occur where the natural demand for oxygen in water exceeds the supply," or more explicitly, where degradation of organic matter depletes the oxygen supply, which results in anoxia unless associated circulation of aerated water masses renews the oxygen supply. Because of common restrictions to water circulation, for example, stratification (including thermoclines, pycnoclines, and haloclines) and other impediments to the dynamics of water circulation, often the oxygen supply cannot keep up with oxygen demand, and oxygen depletion and anoxia result. Their analysis of both modern and ancient depositional settings in which organic matter-rich sediments, i.e., potential source rocks, have accumulated led to a fourfold classification of anoxic depositional models: (1) anoxic lakes, (2) anoxic silled basins, (3) anoxic layers with upwelling, and (4) anoxic open oceans. In each model the interplay of oxygen demand via organic matter degradation and oxygen supply via water circulation, both governed by the specific conditions of the depositional site, is recognized as the pertinent control of the development and persistence of associated anoxia. Similarly, the importance of organic matter preservation due to anoxic environmental conditions is acknowledged by many others as the major control of the accumulation of organic matter-rich sediments (see for example, Tyson 1987; Summerhayes 1981; Bralower and Thierstein 1987; Stein 1986a,b; Schlanger and Jenkyns 1976; Zimmerman et al. 1987).

One typical feature of silled basins is water stagnation. Usually the upper part of the water is oxygen-rich whereas the bottom water is anoxic or dysoxic (Table 5.1) and characterized by low bioproductivity. Generally surface and bottom waters mix only slowly, i.e., there are no strong currents or waves disturbing the water. Deposition usually takes place below storm wave base which can reach as deep as 200 m (Galloway and Hobday 1983) but is usually much shallower (Tyson and Pearson 1991). Stagnation is often enhanced by contrasting physical or chemical properties of bottom and surface water such as temperature and salinity. Accordingly, the boundary between surface and bottom water is called thermocline or halocline, respectively.

In the case of the Black Sea, marine water derived from the Mediterranean is overlain by less saline water of fluvial origin (Danube, Don, Dnieper). The resultant halocline marks also the boundary between oxic and anoxic water. When the anoxic bottom waters were established about 7000 years ago (see Degens et al. 1978 for revision of age data), maximum organic carbon contents of the sediments increased from 0.7% to 20% (Demaison and Moore 1980). Most sediments in the 40-cm-thick, laminated black shale layer deposited between 7000 and 3000 years before the present contain between 2% and 7% organic carbon (Glenn and Arthur 1984). Sediments deposited after 3000 years (before the present) are laminated coccolith oozes and contain only 1–5% organic carbon (Shimkus and Trimonis 1974), although the extension of the bottom water anoxic zone is even greater than during the deposition of the laminated black shales (Deuser 1974; Pedersen and Calvert 1990). The smaller organic carbon percentage can, by analogy to observations on ancient black shales (Littke et al. 1991b), be tentatively explained as a dilution effect by coccoliths (see Müller and Blaschke 1969), which are most enriched in the uppermost Black Sea sediments.

In contrast to numerous observations on oxic deep sea deposits in which sedimentation rate and organic carbon percentage are positively correlated (Müller and Suess 1979), a negative correlation exists for Black Sea sediments (Demaison and Moore 1980) to an extent that "fields of high organic matter content correspond to areas of low primary production and vice versa" (Fig. 5.6; Huc 1988a,b; Shimkus and Trimonis 1974). This type of observation led Stein (1986a) to suggest that organic matter-rich sedimentary rocks may be assigned to either anoxic or oxic bottom water conditions by plotting sedimentation rate versus organic carbon percentage (see Fig. 5.3A). According to this reasoning, anoxic deposits are characterized by high organic carbon percentages and low to high sedimentation rates whereas other deposits only contain high concentrations of organic matter, if sedimentation rates are also high (Fig. 5.6). It should be noted, however, that in the case of the Black Sea as the "classic" silled anoxic basin, sedimentation rates during black shale deposition were also high (about 20–30 cm/1000 a) due to the deposition of great masses of terrigenous material (Degens et al. 1978). In such a case oxic and anoxic palaeo-bottom water conditions can be differentiated only if marine and terrigenous sedimentation rates are calculated, and if the marine sedimentation rate is plotted versus organic carbon percentages. Also, high-resolution stratigraphy is a prerequisite for the adaption of the classification scheme proposed in Fig. 5.3A because only exact stratigraphic and absolute age information allows calculation of sedimentation rates accurately.

Furthermore, studies by Huc (1988a,b) suggest that in anoxic silled basins, the highest concentration of organic carbon occurs in the deepest parts of the basins leading to a concentric pattern of organic matter concentration. This is thought to be an additional criterion for identifying anoxic basins. With regard to ancient deep-sea black shales, those of Cenomanian/Turonian age are probably the best studied (Dean and Arthur 1989). Most of these black shales were sampled in the realm of the Atlantic Ocean, but also at other sites. Therefore they are assumed to be the product of an oceanic anoxic event, witnessing the presence of anoxic bottom waters in large parts of the ocean 91 million years ago (Schlanger et al. 1987). Most favorable conditions for black

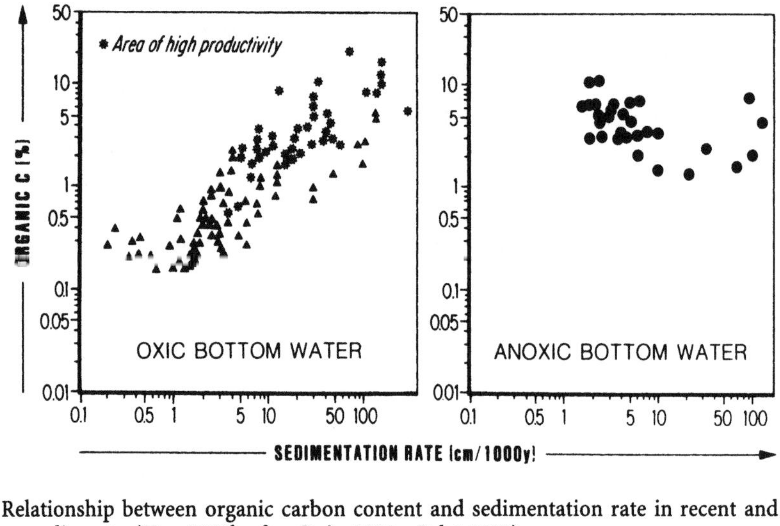

Fig. 5.6. Relationship between organic carbon content and sedimentation rate in recent and Quaternary sediments. (Huc 1988b after Stein 1986a; Pelet 1983)

shale sedimentation existed at that time in parts of the oceanic basin which were closed (silled) with small rates of water circulation such as in the young, opening Atlantic Ocean. Organic carbon percentages are highly variable in these sediments and reach maximum values greater than 20%. However, the petroleum generation potential of the kerogen is not particularly high, i.e., hydrogen index (HI) values do not generally exceed 600 mg hc/g C_{org} and are often much lower. Petrographically the organic matter is mainly unstructured without any primary biological aspects even when studied under high magnifications by transmission electron microscopy or scanning electron microscopy. A strong microbial degradation of the original marine biomass is the likely cause of these maceral characteristics and possibly also of the relatively low HI values compared to shallow marine and lacustrine petroleum source rocks (Littke and Sachsenhofer 1994).

In summary, deep, marine, silled basins are characterized by a small water exchange between surface and bottom waters. Often there is a narrow zone within the water column in which the physical and chemical characteristics such as temperature, density, salinity as well as oxygen and hydrogen sulfide concentrations change abruptly. Below this narrow zone anoxic bottom waters and pore waters persist in which organic matter degradation is restricted to anaerobic processes (probably mainly to microbial sulfate reduction; see Chap. 5.7). Sediments in these environments are often rich in organic carbon, similar to those from lakes with small water circulation.

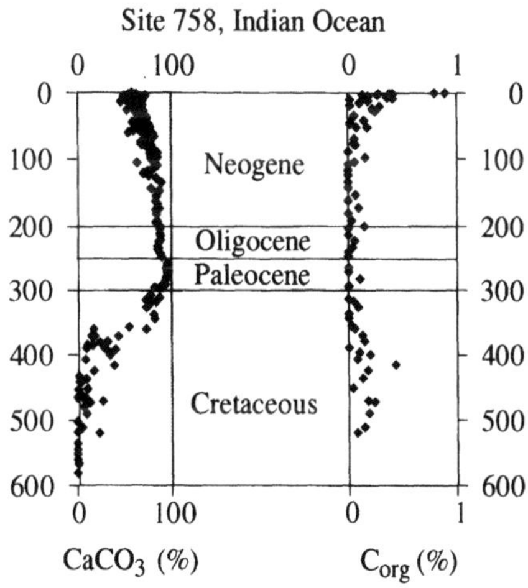

Fig. 5.7. Carbonate and organic carbon percentages in Cretaceous to Holocene sediments of ODP site 758, northern Ninetyeast Ridge, Indian Ocean. The decrease in carbonate and increase in C_{org} in the upper 100 m is an effect of the enhanced detrital supply to the Bengal fan (Ganges and Brahmaputra) which itself is a response to the ongoing rise of the Himalayan mountains. In other words, the source of organic particles became more proximal with increasing progradation of the Bengal fan. Higher C_{org} values in the Cretaceous are affected by supply of terrigenous organic particles in the then smaller Cretaceous Indian Ocean. (Littke et al. 1991c)

5.5
Progradational Submarine Fans

Progradational submarine fans are sedimentary sequences which accumulate in deep marine water close to a continent and are usually fed by river discharge. Recent examples include the Mississippi fan, Indus fan, and Bengal fan.

C_{org} and carbonate concentration profiles through the distal Bengal fan on the northern Ninetyeast Ridge in the Indian Ocean (Littke et al. 1991c; Weissel et al. 1991; site 758 of the ODP) demonstrate the evolution from a purely pelagic towards a mixed marine/continental sedimentation (Fig. 5.7). The Palaeocene and Oligocene pelagic deposits are almost free of organic matter (C_{org} less that 0.2%) as typical for open ocean deposits (Degens and Mopper 1976) and consist of about 80% carbonate derived from planktonic and benthic organisms. With decreasing depth and increasing progradation of the Bengal fan there is a decrease in carbonate content and an increase in C_{org} values. This evolution is enhanced by the ongoing rise of the Himalayan mountain range which caused an increasing sediment discharge by the Ganges and Brahmaputra rivers into the Indian Ocean. The deposited organic matter is mainly of terrigenous origin and hydrogen-poor (Table 5.4). In the central Indian Ocean, C_{org} values and organic carbon accumulation rates are positively correlated with sediment accumulation rates (Littke et al. 1991c), as previously established for other oxygenated open ocean sediments by Müller and Suess (1979).

Table 5.4. Organic matter characteristics of progradational submarine fan deposits.

	Distance (km)	Water depth (m)	C_{org} (%)	HI (mg hc/g org_C)	Origin
Site 615 Lower Mississippi Fan	450	3284	0.7	76 (89)	Terrestrial (vitrinite)
Site 620 Middle Mississippi Fan	250	2612	1.0	98 (107)	Terrestrial (vitrinite)
Site 720 Middle Indus Fan	850	4045	0.7	52 (108)	Terrestrial
Site 721 Distant Indus Fan+ Upwelling Deposits	850	1945	1.0	189 (400)	Marine (+ terrestrial)
Proximal Indus Fan	100	550	2.8	286 (415)	Marine (+ terrestrial)
Site 758 Distant Bengal Fan	1800	2924	0.4	45 (78)	Terrestrial (inertinite)

Distance, distance between river (delta) mouth and depositional site; water depth, present water depth; C_{org}, mean organic carbon percentage; HI, mean and maximum (in parentheses) hydrogen index value; Origin, origin of organic particles based on microscopy and/or geochemical results. Sites 615 and 620 according to Marzi and Rullkötter (1986); site 720 (upper 150 m, Pleistocene), and 721 (upper 100 m Pleistocene and Late Pliocene), according to Prell et al. (1989), site 758 (upper 30 m, Pleistocene and Late Pliocene), according to Peirce, Weissel et al. (1989).

This positive correlation is generally interpreted to be an effect of a higher organic matter preservation rate due to a more rapid burial. Müller and Suess (1979) concluded that "as a first approximation, one can expect a doubling of sedimentary organic carbon contents with each 10-fold increase in sedimentation rate."

Compared to the distal Bengal fan, C_{org} values are slightly higher (0.4–1.1%) in samples from the lower and middle Mississippi fan recovered during DSDP leg 96 in the Gulf of Mexico (Marzi and Rullkötter 1986). As in the Bengal fan, organic matter is mainly of terrigenous origin and HI values are low (less than 110 mg hc/g C_{org}). The greater HI values compared to the distal Bengal fan (Table 5.4) are probably the result of better preservation of terrestrial organic clasts due to shorter transport distances (about 250–450 km versus 1800 km between delta and depositional site). Accordingly, the ratio vitrinite/inertinite is high in the Mississippi fan (Marzi and Rullkötter 1986) and small in the distal Bengal fan (Littke et al. 1991c).

Results on organic matter characteristics in the middle Indus fan (ODP site 720; Prell et al. 1989) show strong similarities to the results obtained on the Mississippi and Bengal fans (Table 5.4). C_{org} values generally vary between 0.5% and 1% and the organic matter is hydrogen poor (HI values less than 110 mg hc/g C_{org}), as typical for a strong terrigenous input. Rapid sedimentation by turbidity currents (Prell et al. 1989) probably enhanced preservation of terrigenous organic matter. However, at the adjacent ODP site 721 on Owen Ridge, which is overlain by an upwelling area, turbidites of the Indus fan did not reach the site of deposition, and organic matter is mainly of marine origin and more hydrogen rich (Prell et al. 1989). Similarly, sites on the Indus fan which are situated more proximal to the Pakistanian coastline are affected by the extended oxygen minimum zone of the Arabian Sea and characterized by high contents of marine-derived, hydrogen-rich kerogen (Table 5.4).

In summary, submarine fans are characterized by higher C_{org} concentrations than most other deep sea sediments. The organic matter is, however, generally hydrogen poor and mainly of terrigenous origin. Its character seems to be strongly affected by transport distance and intensity of alteration. Preservation of organic matter is enhanced by rapid turbidity sedimentation. It should be noted that locally favorable conditions for the preservation of marine organic matter may occur within submarine fans, for example, due to the presence of an ideal topography or hypersaline brines (see Kennicutt et al. 1986).

5.6 Upwelling Areas

Upwelling areas are regions in which great masses of oceanic surface waters move offshore and are replaced by oceanic deep water. The upwelling process is driven by either atmospheric or ocean forcing mechanisms such as wind, coastal currents, and the Coriolis force (Suess and Thiede 1983). Upwelling rates are calculated from surface wind stress and the Coriolis force (Kruijs and Barron 1990) and range from close to 0 to more than 20 cm per day. The upwelling deep waters are often enriched in nutrients such as nitrate, phosphate, and silicate (Baturin 1983; Bishop 1989) thus favoring high primary productivity rates (greater than 180 g C m^{-2} a^{-1}; see Berger et al. 1989a,b) in surface waters and oxygen shortage in waters below the photic zone (Fig. 5.1). Areas of high primary productivity occupy only a very small proportion of the modern oceans (Berger 1989) and are often related to upwelling. Baturin (1983) estimated that the total area of recent strong upwelling is only 500 000 km^2 or about 1‰ of the world ocean. Prominent upwelling cells are situated offshore the coasts of northwestern Africa, southwestern Africa, Peru, northwestern North America, Oman, and in the equatorial Pacific. The processes influencing sedimentation in upwelling areas are described in great detail by Thiede and Suess (1983), Suess and Thiede (1983), and Summerhayes et al. (1992). Typical features of sediments underlying upwelling areas include (1) high organic carbon (2–20%), (2) high biogenic silica (5–70%), (3) elevated phosphorus (0.2–1%) contents, (4) high rates of biogenic sedimentation (up to 0.5 mm/a), and (5) the occurrence of coprogenic material (Baturin 1983). Petrographically the upwelling sediments often contain predominantly unstructured organic matter with only a small contribution of algal remains resembling the primary biomass and only little terrigenous organic matter (Littke and Sachsenhofer 1994; Lückge et al., 1996). This observation, together with the rather low hydrogen indices, is taken as an argument for a restructuring of the primary biomass due to degradation processes before and shortly after deposition.

Organic carbon accumulation in upwelling areas does not depend solely on primary productivity related to upwelling but also on the interaction of the upwelling system with the topography of the shelf and continental slope (Reimers and Suess 1983; Kruijs and Barron 1990). This is supported by Baturin (1983) who suggested that unfavorable conditions for accumulation of organic matter-rich sediments persist where there occur steep shelf platform morphologies, high energy wave climates, strong bottom currents, high rates of terrigenous sedimentation, bioturbation by bottom fauna, and repeated transgressive/regressive cycles (erosional events). Three examples from the upwelling areas, offshore northwestern Africa, Oman, and Peru, illustrate the influence of various factors on organic matter accumulation.

An upwelling area below which sediments are not particularly enriched in organic carbon is the northwestern African shelf at ODP site 658 (Table 5.5; Stein and Littke 1990), where organic carbon contents generally vary between 1% and 3% and do not exceed 4%. Furthermore, HI values from Rock-Eval pyrolysis are relatively low (Stein et al. 1989). This is explained by the occurrence of greater amounts of terrigenous organic matter than at other upwelling sites (Littke and Sachsenhofer 1994). In other upwelling areas the bulk of the organic matter stems from bacterially degraded phytoplankton, and terrigenous organic matter contributes only a small proportion to the total organic matter (Summerhayes 1983; ten Haven et al. 1990).

Higher organic carbon contents than offshore northwestern Africa are reported from the upwelling area of offshore Oman (Tables 5.5, 5.6; Prell et al. 1989). Figure 5.8 shows an idealized cross section through the continental margin off Oman with the position of six different wells drilled by the ODP.

Table 5.5. Average organic carbon percentages, average HI, bulk sediment (ARBS), and organic carbon accumulation rates (AROC)

Area	C_{org} (%)	HI (mg hc/g C_{org})	ARBS ($g\ m^{-2}\ yr^{-1}$)	AROC ($g\ m^{-2}\ yr^{-1}$)
Site 658 Northwestern Africa	2.0	261	120	2.0
Site 723 Oman margin	3.1	340	183	5.7
Site 679 Peru margin	4.0	455	65	2.6

From sediments on the continental margins at site 658 (Pliocene to recent; Stein and Littke 1990), at site 723 (Pleistocene to recent; Prell et al. 1989), and site 679 (Quaternary; ten Haven et al. 1990).

One of the sites (725) is situated at the very top of the oxygen-minimum zone which underlies the surface water. Two sites are just within the oxygen minimum zone (723 and 728), two are on the Owen Ridge and just below the oxygen minimum zone (731 and 722), and one is outside the upwelling area in the deep sea and affected by the Indus fan (720). Organic carbon contents and HI values of the uppermost sediments from these sites are summarized in Table 5.6 (after Prell et al. 1989; ten Haven and Rullkötter 1991). Organic carbon percentages and HI values are greatest in sediments deposited within the oxygen minimum zone, i.e., at water depths which exceed 350 m but are less than 2000 m. Even there only about 1–3% of the primary organic carbon production in surface waters is preserved in the sediments; more than 97% was lost due to remineralization (Lückge et al., 1996). Shelf sediments deposited in relatively shallow water like those drilled at site 725 (311 m) contain little organic matter which is characterized by low HI values. Most favorable conditions for organic matter deposition occur where sedimentation takes place in the upper part of the oxygen minimum layer, for example, at site 723 at a water depth of 807 m.

Table 5.6. Depth intervals (meters below sea floor, mbsf), stratigraphic age, average organic carbon contents, average HI, accumulation rates for bulk sediment (ARBS), and accumulation rates for organic carbon (AROC) for the uppermost intervals drilled at ODP sites offshore Oman (see Fig. 5.8). (Prell et al. 1989; ten Haven and Rullkötter 1991)

Site	Depth (mbsf)	Stratigraphic age	C_{org} (%)	HI (mg hc/g C_{org})	ARBS ($g\ m^{-2}\ yr^{-1}$)	AROC ($g\ m^{-2}\ yr^{-1}$)
725	0–38	Holocene-Pleistocene	0.56±0.53 (35)	110 (2)	144	2.5
723	0–40	Holocene-Pleistocene	2.72±1.39 (41)	298±94 (26)	183	5.7
728	0–65	Holocene-Pleistocene	1.60±0.59 (11)	219±100 (11)	40	0.6
731	0–50	Holocene-Pleistocene	0.65±0.37 (8)	141±53 (5)	38	0.2
722	0–53	Holocene-Pleistocene	0.97±0.32 (8)	243±41 (6)	32	0.3
720	0–48	Holocene-Pleistocene	0.56±0.54 (13)	56±34 (10)	231	1.4

Note that average C_{org} and HI values are tabulated for the depth interval only, while accumulation rates are calculated for the entire Holocene-Pleistocene section (see also Littke 1993).

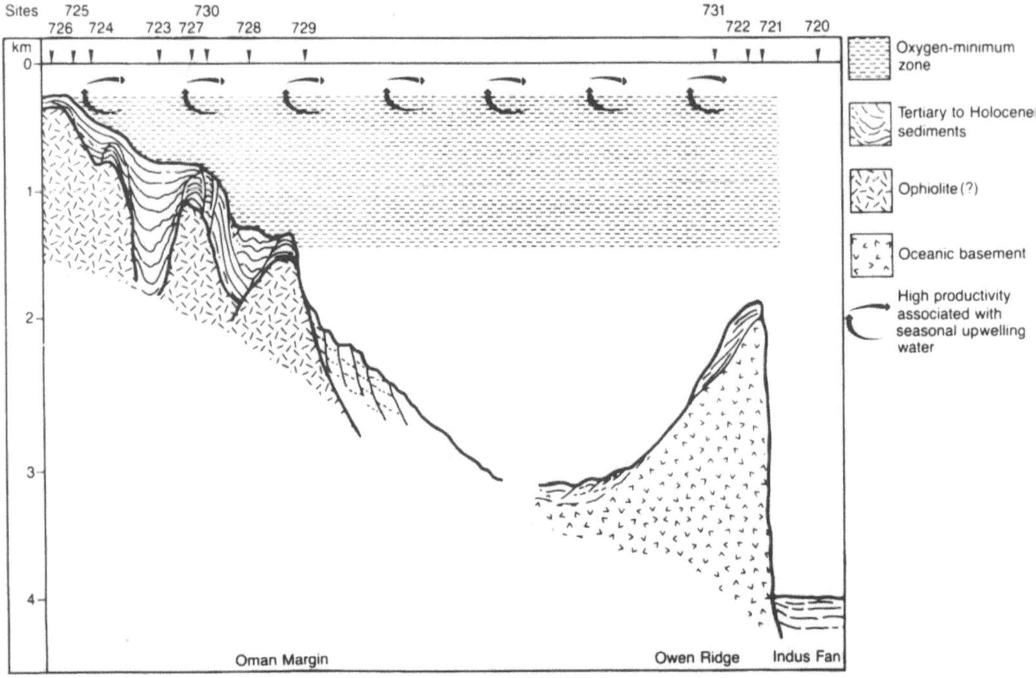

Fig. 5.8. Idealized transect of the continental margin off Oman showing major geological and oceanographic features and locations of sites discussed in text (vertical scale exaggerated). (Spaulding 1991)

For these sediments the average HI value is about 300 mg hc/g C_{org}. This value is greater than those established for the other sites on the Oman continental margin but still lower than for many ancient petroleum source rocks. Favorable conditions for the deposition of organic matter-rich sediments also occur where the sediment/water interface is within the lower part of the oxygen minimum zone or slightly deeper as at site 722 (about 2000 m). Sediments deposited in the deep sea as at site 720 (about 4000 m water depth) contain lower average organic carbon contents and have low HI values. The occurrence of mass flow or turbidity current sediments at these deep locations causes few exceptionally high organic carbon values and is responsible for the high standard deviation recorded in Table 5.6.

The upwelling system offshore Peru serves as another example of the complexity of sedimentation patterns in these depositional environments. In upwelling areas the distribution of organic carbon-rich deposits is not only controlled by primary productivity but also by current activity and ocean floor morphology (Reimers and Suess 1983). This is evident from Figure 5.9, which compares primary productivity rates in surface water and organic carbon percentages of the uppermost sediments for the Peruvian coast. In general, bioproductivity is most enhanced and rather homogeneously distributed along the coastline and becomes successively smaller towards the open ocean. Organic carbon percentages of surface sediments are exceptionally high in two small areas between 11° and 14° S, but lower towards the north, although upwelling intensity and bioproductivity are in the same range. Reimers and Suess (1983) explain this difference by shallower water depths between 7° and 10° S, which promote continuous reworking by bottom currents and inhibit sedimentation.

Furthermore, the data from the Peruvian upwelling system nicely illustrate that high C_{org} contents do not necessarily coincide with high organic matter accumulation rates. Table 5.7 compares C_{org} and HI values and accumulation rates of a site at about 11° S (area of high C_{org} contents) and of a site at about 14° S (area of high accumulation rates). Offshore Peru, high C_{org} percentages coincide with high HI values, whereas high accumulation rates do not coincide with high HI values. These data further support that in oxygen-deficient bottom water environments the general rule of a positive correlation between organic carbon percentages and sediment accumulation rates finds its exception (Fig. 5.3).

Comparisons between the occurrences of upwelling, on the one hand, and high primary productivity, on the other, reveal that: (1) upwelling is the most common but not the only cause for high bioproductivity rates; another source of nutrients is river discharge (van der Zwaan and Jorissen 1991); (2) many upwelling areas (82% according to Kruijs and Barron 1990) are not associated with high productivity rates or high C_{org} percentages of underlying sediments, especially in the open ocean; it is only along coastlines that a strong positive correlation between upwelling and high productivity and C_{org} percentages (Kruijs and Barron 1990) is observed; this may also be due to the additional nutrient supply from land and to favorable topography of the sea floor.

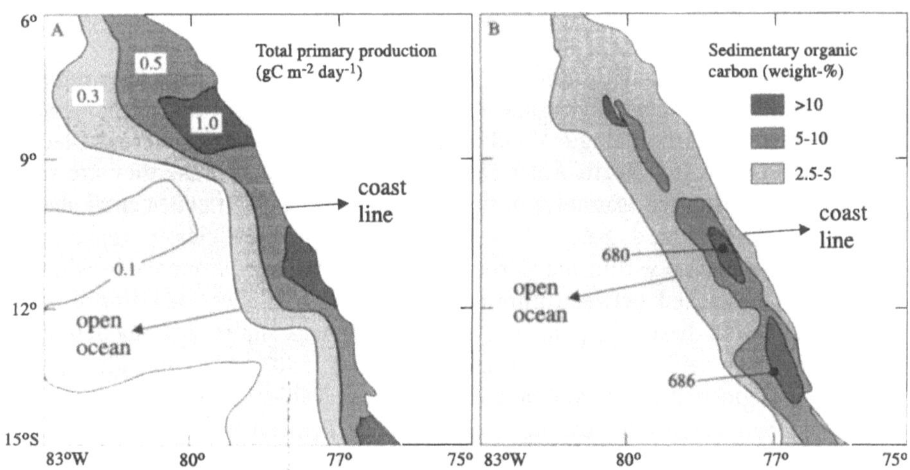

Fig. 5.9. Total primary production (**A**) and sedimentary organic carbon in surface sediments (**B**) in the upwelling area off Peru. (Redrawn after Reimers and Suess 1983)

Table 5.7. Depth intervals (meters below sea floor), age intervals (million years, Myr), average organic carbon contents, average HI, accumulation rates for bulk sediment (ARBS), and accumulation rates for organic carbon (AROC) for the youngest sediments in Holes 680B and 686B, offshore Peru

Site	Water depth (m)	Depth interval (mbsf)	Age (Ma)	C_{org} (%)	HI (mg hc/g org_C)	ARBS (g cm^{-2} 1000 yr^{-1})	AROC (g cm^{-2} 1000 yr^{-1})
680 (11 °S)	260	0–41	0–0.62	5.34±2.57(47)	506±177(46)	3.7	0.20
686 (14 °S)	450	0–160	0–1.41	2.32±1.36(8)	334±78(8)	19.6	0.45

Accumulation rates are corrected for the hiatus between 0.62 and 1.37 Myr at site 686 (Wefer et al. 1990). C_{org} and HI data are from Suess et al. (1988; site 686) and Emeis and Morse (1990; site 680), age information is from Wefer et al. (1990) and sediment densities are from Suess et al. (1988). C_{org} was measured by the Rock-Eval method.

According to Summerhayes (1983), "massive, short-period production rather than steady state high productivity controls the accumulation of organic matter in upwelling regimes, by overwhelming the system's ability to recycle organic matter as food in the water column." Other factors in addition to upwelling and primary productivity also greatly affect organic carbon accumulation and organic carbon percentages in sediments.

The above features of deep sea sediments as well as those of other deep sea sample series (Littke and Sachsenhofer 1994) reveal that most of the organic matter-rich sediments from the deep sea have a predominance of unstructured organic matter and rather low HI values in common. These observations indicate – due to the scarcity of terrigenous organic matter – that degradation of the primary marine biomass is strong in the deep sea settings, even if environmental conditions allow the accumulation of much organic matter.

5.7 Anoxic Continental Shelves

Black shales which are deposited in shallow marine environments are common in the Palaeozoic and Mesozoic ancient geological record, but there is no well-documented modern analogue (Hallam 1981, p. 90). Late Carboniferous black shales of Europe and North America are discussed here because they are regarded as well-studied examples of these organic matter-rich continental shelf deposits (Wenger and Baker 1986; Heckel 1991). They clearly represent transgressive phases within the Carboniferous sedimentary sequence in which they are sandwiched between nonmarine and marine facies including limestones above. As these thin (usually 1 m and less) black shales lack any benthic fossils and any evidence of bioturbation (O'Brian 1990), it is concluded that they were deposited under anoxic conditions (see Table 5.1).

A geochemical profile through one of these Pennsylvanian black shales (maturity corresponds to 0.54% vitrinite reflectance) is shown in Figure 5.10. Organic carbon contents are high (17%) at the base of the black shale and decrease toward the top. HI values are also greatest at the base (320 mg hc/g C_{org}) but not particularly high if compared to other immature black shales.

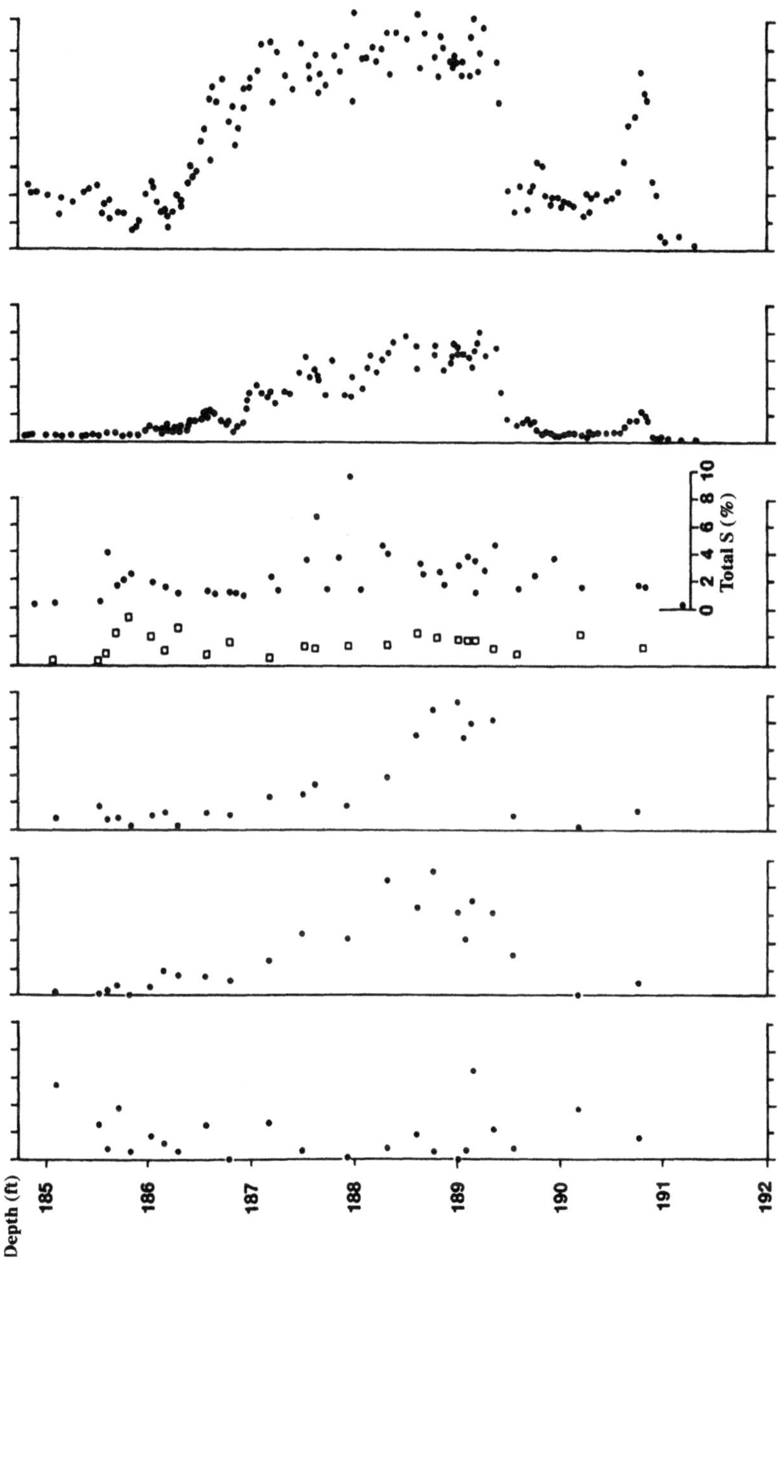

Fig. 5.10. Geochemical and petrological profile through a core of the Little Osage Shale (Upper Carboniferous) from a well in northwestern Oklahoma, United States (R_r = 0.54%). Carb., carbonate (Part of the data are from Wenger and Baker 1986)

This pattern of an abrupt increase in C_{org} and HI values at the base of black shales and a more gradual decrease towards the top was also observed in other black shales (e.g., Piasecki and Stemmerik 1991). For the Pennsylvanian black shales, maceral analysis revealed that the organic particles are a mixture of fluorescing alginite probably derived from marine phytoplankton and of brightly reflecting, weakly or nonfluorescing particles. The presence of the latter material is expected to cause the relatively low HI values of these black shales. Low HI values are generally typical for terrigenous organic matter, and the "bright particles" were initially proposed to be derived from peat clasts (Wenger and Baker 1986). However, molecular geochemical data and organic petrographic observations (such as the lack of phenols in pyrolysis products and the lack of trimacerites) do not support a terrigenous *organic-clast origin*; an explanation as hydrogen-poor *humic precipitate* is more realistic. The precipitation of humic acids is favored in environments where water mixing, for example, mixing of saline water with freshwater, occurs (Swanson and Palacas 1965; Lyons et al. 1984). Such a mixing is certainly not improbable during the time of Pennsylvanian black shale deposition.

The organic richness of many thin shallow marine sedimentary rocks which occur in coal-bearing fluviodeltaic strata suggests a common controlling mechanism and depositional model. Wenger and Baker (1986) and Wenger (1987) suggested that transgressive-regressive cycles are the major control of the accumulation of the Pennsylvanian black shales of the North American mid-continent. Rapid transgression of epicontinental seas over laterally adjacent widespread peat swamps and emergent delta-plain surfaces, formed during the previous regressive cycle, resulted in leaching of flooded sediments and an enhanced influx of nutrients and humic materials to the marine environment. The nutrients stimulated high algal productivity and the terrestrially derived organic material provided an additional sink for oxygen which enhanced anoxia. As transgression proceeded to its maximum, the swamps and delta plain were progressively flooded and diminished in areal extent, the influx of nutrients and humic material decreased, productivity dwindled, intensity of anoxia and preservation of organic matter decreased, and black shale deposition terminated. Subsequent regression led to the reestablishment of delta-plain surfaces and coexisting swamp environments creating the conditions appropriate for the deposition of a succeeding black shale during the next cycle of transgression (see Wenger and Baker 1986). This evolution is summarized in Fig. 5.11 and is in accordance with the geochemical data (Fig. 5.10) which suggest that black shale deposition starts more abruptly than it ends. Similarly, evidence for the importance of eutrophication of shallow seas by riverine inflow of nutrients has recently been published by van der Zwaan and Jorissen (1991) who concluded that "the chance of anoxia is highest during periods of high sea level, leading to large shelf areas."

One of the best-studied of all shallow marine Mesozoic petroleum source rocks is the Lower Toarcian Posidonia Shale, which extends over an area of more than 300 000 km^2 in northwestern and central Europe (Table 5.8). The Posidonia Shale is the source for economic oil accumulations in northern Germany (Wehner et al. 1989), in the Paris basin (Espitalié et al. 1987), and in

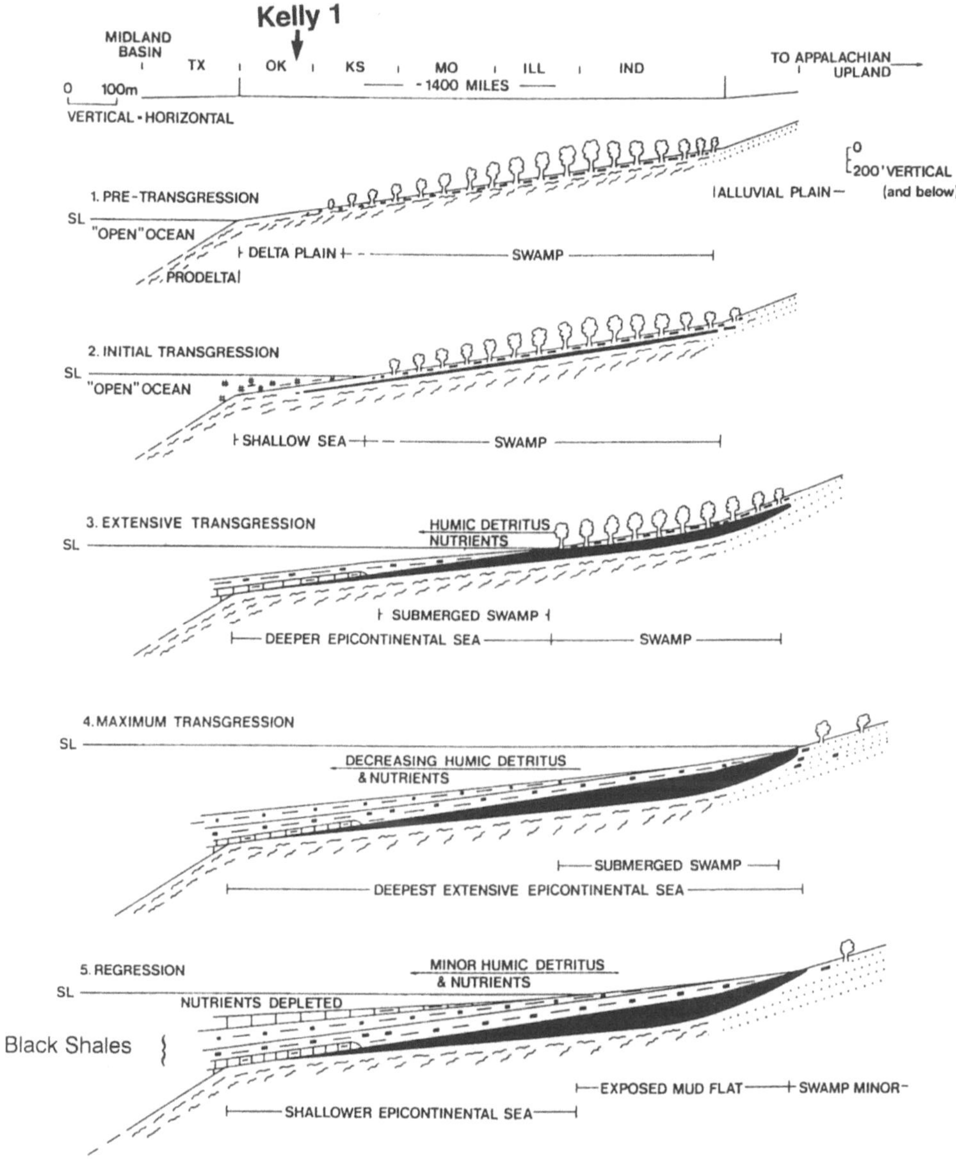

Fig. 5.11. Depositional model for transgressive Upper Carboniferous black shales. Inflow from distal, flooded land areas and the leaching of the former land surface probably provided the nutrients for enhanced algal productivity and humic matter which substantially intensified anoxia and led to high organic matter contents in the sediments (see Fig. 5.10). (Wenger and Baker 1986; Wenger 1987)

Table 5.8. Regional comparisons of the Toarcian black shales of central and northern Europe

Region	Thickness (m)	Lithology	C_{org} (wt/%)	Carbonate (wt%)	HI (mg hc/g C_{org})	Remarks	Reference
Eastern England and North Sea	10–33	Shale	5–15 av.<11.4	2–5	300–600	Limestone and shell beds	Jenkyns (1985), Morris (1980)
Northwestern Germany	16–40	Laminated shale and marlstone	av. 11	35–61	600–700	Upper shale, lower marlstone units	Littke and Rullkötter (1987), Littke et al. (1988)
Southwestern Germany	5-20	Shale and marlstone; partly laminated	9–14	av. 40	700	$\delta^{13}C$-excursion	Küspert (1982)
			av. 9			Widespread limestones	Urlichs (1977)
						"Anoxic biomarkers"	Moldowan et al. (1986)
Paris Basin	10–30	–	5–14	10–30	-	–	Durand et al. (1972)
	–	–	5–11	20–40	-	Homogeneous org. matter	Huc (1977)
	65	Shale-marlstone	4–7	-	500–700	–	Espitalié et al. (1987)
Tethys	8–80	Marlstone	1–2	<50	200–500	Restricted basins in Greece	Baudin et al. (1990)
	–	Laminated shale	av. 1.3	-	200–300	Italian Alps	Jenkyns (1985)

parts of the Upper Rhine Valley (Welte 1979). The organic matter from the Posidonia Shale was defined as the typical type II kerogen (Tissot and Welte 1984). Petrogaphically it consists of a mixture of mostly alginite with an admixture of bituminite and terrigeneous macerals (Littke et al. 1991a,b), although bituminite may be predominant at some locations (Teichmüller and Ottenjann 1977). Geochemically the organic matter is characterized by a high petroleum generation potential, as documented by HI values in excess of 600 mg hc/g C_{org}. This type II kerogen, derived from a mixture of phytoplankton, zooplankton, and bacteria, is typical of many high-quality petroleum source rocks which were deposited in shallow marine, reducing environments (Tissot and Welte 1984). Apparently the degradation of the primary, lipid-rich, autochthonous biomass is less severe than in deep-water environments.

Nevertheless, part of the metabolizable organic matter is consumed during early diagenesis in this type of environment. This early diagenetic degradation is indicated by common high percentages of sulfur in the form of iron sulfides (pyrite) in shallow marine black shales. Pyrite framboids derived from bacterial sulfate reduction and organic matter oxidation are recognized as ubiquitous constituents of petroleum source rocks. Sulfur concentrations and sulfur/organic carbon ratios in shallow marine source rocks are usually higher than in lacustrine black shales due to the limited sulfate availability in freshwater compared to seawater (Berner 1984), and also higher than in organic carbon-rich deep water sediments, which are often rich in siliceous and carbonate material, but lean in clay minerals and iron. Therefore this important early diagenetic process of carbon consumption and pyrite formation is most important in shallow marine sediments.

The general equations describing the formation of iron sulfides and the related consumption of organic matter (after Leventhal 1983; Berner 1984) are:

$$SO_4^{-2} + 2CH_2O \rightarrow 2HCO_3^- + H_2S$$

$$3H_2S + 2FeOOH \rightarrow 2FeS + S^\circ + 4H_2O$$

$$FeS + S^\circ \rightarrow FeS_2$$

According to the first equation, 2 mol organic carbon is consumed to produce 1 mol reduced sulfur. The compositional pathway for this process is shown in Fig. 5.12. The projection of the alteration pathway from a specific sample point gives an approximation of the organic carbon content of the original sediment. This estimate is a minimum, because it is assumed that 100% of the H_2S-sulfur generated in the sediments was fixed as pyrite (see Vetö et al. 1994), and that there was no loss of organic carbon by other processes such as aerobic oxidation and methanogenesis. For the Posidonia Shale a loss of (at least) 17–32% of the original organic carbon occurred due to bacterial sulfate reduction (Fig. 5.12; see Littke et al. 1991b for further explanation). In the case of the Posidonia Shale the calculated carbon loss is negatively correlated to HI values, indicating that a loss of kerogen quality can occur due to bacterial sulfate

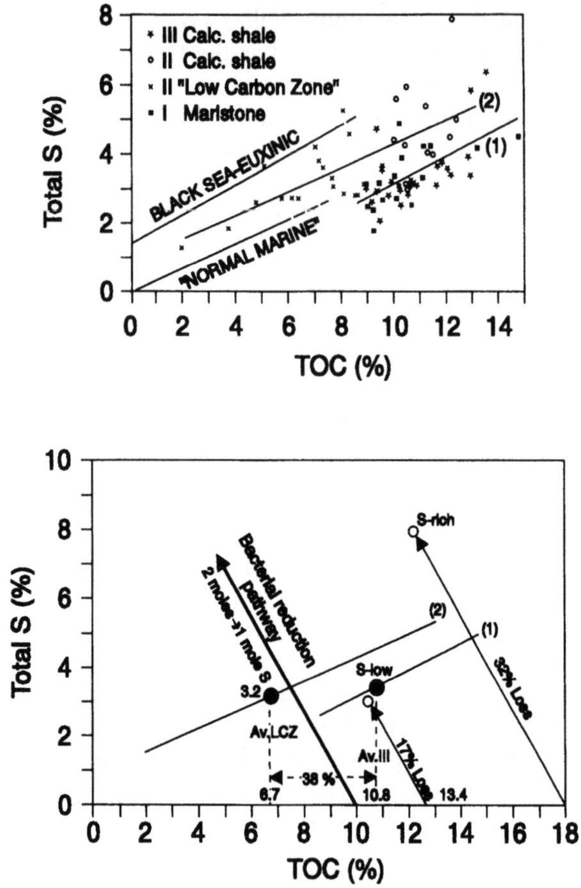

Fig. 5.12. Total sulfur and organic carbon weight percentages of samples from different lithological units within the Posidonia Shale in northern Germany (*above*) and reconstructions of "original" organic carbon values following the bacterial reduction pathway. *Regression lines 1, 2,* calculated for the upper plus lower part and the central part of the Posidonia Shale profile investigated. Carbon loss (17%, 32%) was calculated for samples characterized by low and high sulfur/organic carbon ratios, respectively. Average organic carbon values of the "low carbon zone" within the Posidonia Shale and of the upper unit I are 6.7% and 10.8%, respectively. (Littke et al. 1991b)

reduction. Based on the reasoning above, Lallier-Vergès et al. (1993) calculated a *sulfate reduction index*, which is the quotient of total organic carbon plus carbon loss over total organic carbon. The authors found that in their sample series (upwelling sediments from offshore Oman) HI and sulfate reduction index are positively correlated, because sulfur is incorporated into the organic matter which therefore becomes nonmetabolizable. In contrast, the Posidonia Shale which is characterized by a negative correlation does not contain much organic sulfur. In this case the degradation of the substrate (organic matter) causes a loss of hydrocarbon generation potential which is not compensated by a preservation effect due to transformation of metabolizable into non-metabolizable, sulfur-rich organic matter.

The calculation of organic carbon lost during sulfate reduction allows the treatment of siliciclastic and carbonate-bearing petroleum source rocks as a three-component system, composed of carbonates, silicates, and organic matter only. This three-component composition can be plotted in ternary diagrams (Fig. 5.13). In the case of the Posidonia Shale all samples from one location plot along one trend line, but for different locations the position of the trend line, i.e., the inclination, may be different. For all locations, however, a (theoretical) silicate-free Posidonia Shale would contain almost no organic matter according to this diagram. The latter observation indicates that the carbonate microfossils (mainly coccoliths) in this source rock did not significantly contribute to the preserved kerogen.

Other shallow marine, clastic petroleum source rocks which are rich in hydrogen-rich alginite or bituminite include the Cambrian Alum Shale of Scandinavia (Horsfield et al. 1992), Cambrian and Ordovician shales of Saskatchewan, Canada and the Upper Devonian Bakken Shale of the Williston Basin, Canada (Stasiuk 1994), the Upper Jurassic Kimmeridge Clay (Littke 1993), and Bazhenov formations (Peters et al. 1994) as the major source rocks for oil in the North Sea and in the West Siberian Basin, the Devonian Sunbury Shale of Kentucky (Crisp et al. 1987), and the Permian Irati Formation of southern Brazil (Mello et al. 1988).

Organic petrological studies on shallow marine carbonate source rocks are rare compared to those on siliciclastic black shales. One special feature of many carbonate source rocks is the concentration of organic particles in sty-

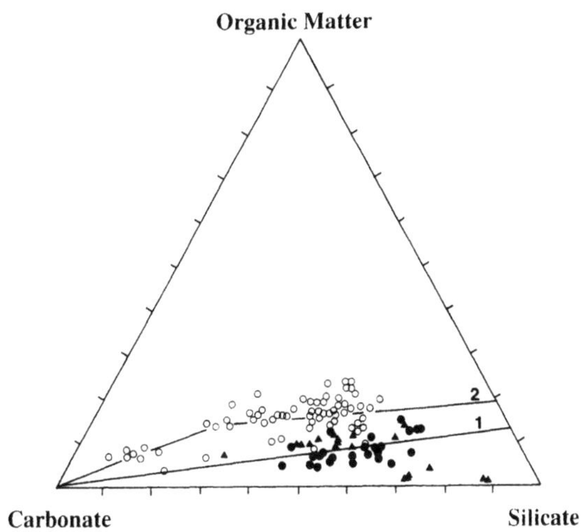

Fig. 5.13. Triangular plot of original sediment composition of Posidonia Shale from northern and southern Germany. Organic carbon percentages are the sum of measured organic carbon contents and the organic carbon loss due to sulfate reduction (see Fig. 5.12). *Open circles*, samples from northern Germany; *dark symbols*, samples from southern Germany. The maturity corresponds to about 0.5% vitrinite reflectance. (Littke et al. 1991a)

lolites and solution seams. Examples for this diagenetic enrichment are described for three important petroleum source rocks: the Permian Bone Springs Formation in Texas (see photos in Littke 1993), Triassic carbonates from Italy (Leythaeuser et al. 1995), and the Jurassic Lower Smackover Formation (Sassen et al. 1987) which extends from Texas to Florida. Except for the latter formation, alginites are the predominant macerals in these carbonates. Usually the same maceral compositions are observed in the carbonate and the intercalated solution seams or stylolites. However, these macerals occur finely disseminated and in low concentrations in the carbonate and in very high concentrations in the solution seams or stylolites. Based on transmitted light studies on kerogen concentrates Sassen et al. (1987) showed that the organic matter of the Lower Smackover carbonates consists mainly of algal-derived amorphous kerogen. While the average organic carbon content of the Smackover Formation is only 0.5%, the mean value for the stylolites is 10%, and the maximum value is 63%.

Due to the limited iron availability in shallow marine carbonate sequences, sulfur incorporation into the organic matter during early diagenesis is a common phenomenon. One example is Upper Cretaceous carbonates west of the Dead Sea. The incorporated organic matter is mainly of algal origin with some terrigenous contribution and was deposited under reducing conditions (Spiro et al. 1983). Activity of sulfate-reducing bacteria induced sulfur incorporation in the organic matter to a vast extent (up to 12%!; Dinur et al. 1980). Such sulfur-rich kerogen produces a petroleum rich in asphaltenes, i.e., an "immature oil," at low levels of thermal maturity (Rullkötter et al. 1984; Tannenbaum and Aizenshtat 1985).

In summary, shallow marine source rocks may be rich in oil-prone organic matter and extend over vast areas. The degree of organic matter degradation is often less severe than in the case of deep marine source rocks. However, a significant organic carbon loss due to degradation reactions (sulfate reduction and organic matter oxidation) affects even those source rocks which were deposited under anoxic waters. Furthermore, in shallow-marine carbonate source rocks the distribution of organic matter is often diagenetically altered by processes such as pressure solution and stylolitization.

5.8 Evaporitic Environments

Until recently, relatively few detailed geochemical and petrographic studies dealt with the organic matter in evaporites and carbonates. This is surprising in view of the fact that biomass production can be extremely high in evaporitic environments, for example, 1810 mg C_{org} m^{-2} per day in Solar Lake of the Sinai (Cohen et al. 1977), and degradation of organic matter by methanogenic bacteria is inhibited (Hite and Anders 1991). The following discussion is restricted to evaporitic sequences in which chemical precipitation of carbonate and other minerals have occurred, and which are characterized by elevated organic carbon contents.

Kenig et al. (1990) and Kenig and Huc (1990) described organic matter from modern carbonates deposited in a hypersaline lagoon in Abu Dhabi. The lagoon provides three principal sources of sedimentary organic matter: microbial mats, mangroves, and seagrass (Fig. 5.14 and Table 5.9). The latter grows in the central part of the lagoonal area, where C_{org} values reach 1.3% in surface sediments. Decreasing C_{org} values with increasing depth of burial could indicate that preservation of seagrass in the upper centimeters of the sediments is poor, but other explanations are also possible (Kenig et al. 1990). Mangroves grow at shallower water depth in the intertidal zone (Fig. 5.14). One mangrove soil contains 8.2% C_{org} and consists of root and leaf (cuticle) tissues (Kenig and Huc 1990) at greater depth of burial, C_{org} values are lower (0.5–4.6%; Table 5.9). Organic matter in the supratidal sabkha environment is derived from microbial mats. Corresponding sediments consist of interlayered detrital (storm) deposits and organic mats with C_{org} values between 1.6% (seaward) and 4.5% (landward; Kenig and Huc 1990). The organic matter of all three principal sedimentary environments is hydrogen-rich (HI = 400–750 mg hc/g C_{org}).

While C_{org} rich sediments accumulate in the sabkha and intertidal zone of the Abu Dhabi lagoon, the adjacent shallow marine carbonates (ooliths) are poor in organic carbon (0.1%, Kenig et al. 1990). Higher C_{org} percentages (0.2–0.4%) and considerable hydrocarbon quantities were reported for ooliths from the British Jurassic (Ferguson 1987), and much higher C_{org} values were found by Ferguson and Ibe (1982; 2.3% C_{org}) for recent ooids. At least with respect to high C_{org} values in ancient oolithic limestones, it is noted that (part of) the

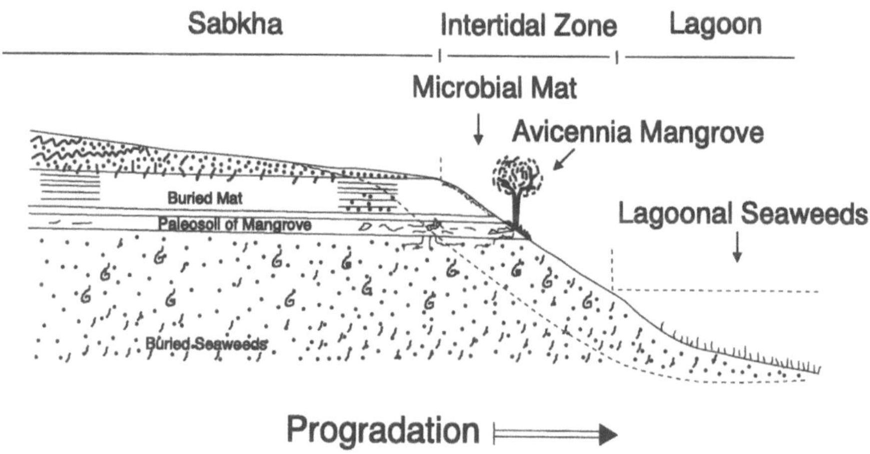

Fig. 5.14. Schematic section through the Abu Dhabi lagoon coastline (see text for geochemical characteristics). (Kenig and Huc 1990). (Reprinted from Kenig F, Huc AY, Incorporation of sulfur into recent organic matter in a carbonate environment, 1990, pp 170–185, with kind permission from Elsevier Science Ltd, The Boulevard, Langford Lane, Kidlington 0X5 1GB, UK)

Table 5.9. Organic matter characteristics of three evaporitic deposits

Locality	Lithology	C_{org}	Geochemical data	Reference
Hypersaline lagoon (Abu Dhabi)	Lagoonal mats (seagrass)	0.3–1.0	HI=550–710	Kenig et al. (1990)
	Coastal soils (mangroves)	0.5–4.6	HI=400–550	
	Sabkha (microbial mats)	0.5–2.7	HI=510–675	
	Shallow marine oolithes	0.1	-	
Modern saline (Spain)	Carbonate	3.1–3.4	Highly branched acyclic isoprenoids	Barbé et al. (1990)
	Gypsum	0.5	C_{15}-C_{18} and C_{25}-C_{33} *n*-alkanes	
	Halite	0.25	Dihydrophytol, phytanic acid	
Oligocene evaporites (France)	Marlstone	0.5–5.0	HI=230–580, H/C=0.9-1.5	Blanc-Valleron et al. (1991)
	Anhydrite	<0.5	–	
	Halite	<0.5	–	

organic matter may be an oil residue, i.e., oolithes can serve as petroleum reservoirs (Horsfield et al. 1991).

Barbé et al. (1990) reported on a modern Spanish salina (Fig. 5.15 and Table 5.9). The carbonates deposited in the pond at relatively low levels of evaporation contain 3.1–3.4% C_{org}. Sulfates and halite, precipitated at higher levels of evaporation, contain much less organic carbon (0.5% and 0.25%, respectively). This is mainly a dilution effect. Accumulation rates of organic carbon are high for sulfates and halites because these sediments are deposited at extremely high sedimentation rates and form the bulk of the total evaporitic sequence. Barbé et al. (1990) demonstrated that the molecular composition of

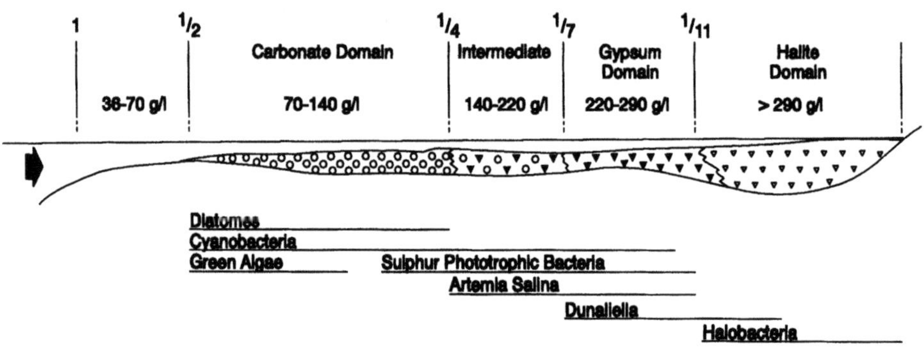

Fig. 5.15. Schematic section through a modern saline in Spain with evaporation ratios, salt concentrations and occurrence of different groups of biota. (Barbé et al. 1990). (Reprinted from Barbé et al., Characterization of model evaporitic environments through the study of lipid components, 1990, pp 815–828, with kind permission from Elsevier Science Ltd, The Boulevard, Langford Lane, Kidlington 0X5 1GB, UK)

compound classes (sterols, lipids, hydrocarbons) differs considerably between the different facies as a response to the variability of produced biomass. For example, only the hydrocarbon fractions of carbonates are characterized by a dominance of highly branched acyclic isoprenoids, which may be derived from diatoms or green algae (see Table 5.9, Fig. 5.15). These compounds are absent or very minor in gypsum, which in turn contains abundant sterenes.

Evaporite systems formed in subaqueous (barred) basins may be a more favorable habitat. High bioproductivity occurring in the photic zone and normal marine waters above the deeps may furnish a substantial organic matter-flux to the stagnant and briny bottom waters of these deep basins. Anoxic and euxinic conditions can extend well above the sediment-water interface and persist over long durations of time, and widespread sapropelic sediments such as the well-known Neogene Mediterranean sapropels can be deposited (ten Haven et al. 1987).

Ancient evaporitic sediments of Oligocene age from the Mulhouse Basin (France) were studied in detail by a joint European organic geochemical research group. First results (Blanc-Valleron et al. 1991) reveal that halites and anhydrites are poor in organic matter (less than 0.5%; Table 5.9), but that organic matter accumulation rates were high due to high sedimentation rates. Interlayered thin marlstones do, however, contain up to 5% C_{org}, and the organic matter is of variable composition (H/C = 0.9–1.5, HI = 230–580). Saturated hydrocarbons are characterized by high concentrations of phytane, pristane and other isoalkanes. Interestingly, the yield of C_{org}-normalized solvent extract from these evaporitic rocks is extremely high, if compared to other immature sediments (Fig. 5.16, Hofmann 1992). Organic geochemical aspects

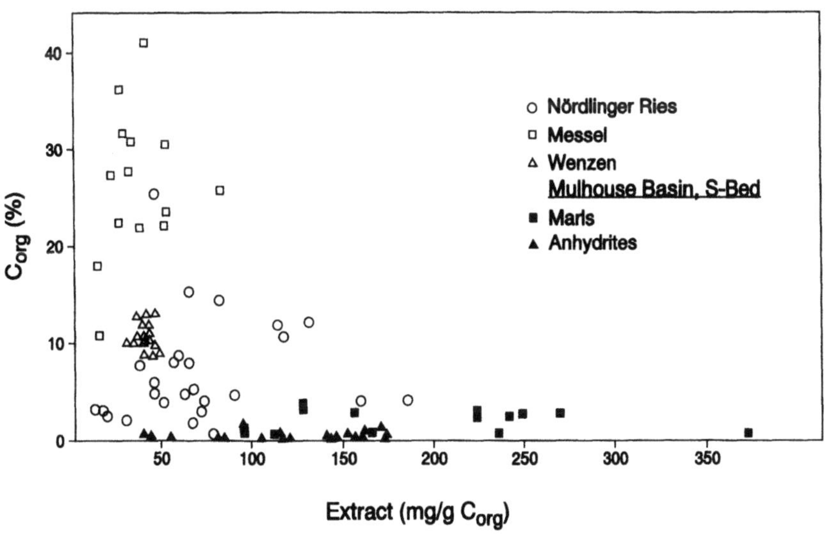

Fig. 5.16. Organic carbon percentages (C_{org}) versus solvent extract yields of samples from an evaporitic environment (Mulhouse basin, S-bed; Hofmann 1992) in comparison to other immature oil shales. (Littke et al. 1988)

of the evaporites in the Mulhouse Basin were published in 1993 and mainly molecular geochemical aspects of the organic matter in several evaporitic sequences were presented in 1995, both in a special volume of *Organic Geochemistry*.

In summary, in evaporitic sequences high organic matter concentrations can be expected only in carbonates and anhydrites, as well as in intercalated shales. In halites (and potassium salts), C_{org} values are low due to the dilution of organic matter by rapid precipitation of minerals. Organic matter in evaporitic sequences is often hydrogen rich; this may be due to an inhibited degradation before and after deposition. Geochemical differences between various evaporitic settings are more obvious than similarities. Even relatively well-established organic indicators of highly saline depositional environments such as low pristane/phytane ratios (ten Haven et al. 1988), predominance of even numbered over odd numbered *n*-alkanes (Welte and Waples 1973), and the occurrence of gammacerane (Mello et al. 1988) do not seem to be of general validity (de Leeuw and Sinninghe Damsté 1990).

5.9 Lakes

Lakes are known to be suitable settings for deposition of rich petroleum source rocks. According to the kerogen classification scheme of Espitalié et al. (1977), the most hydrogen-rich type I kerogens (H/C approx. 1.5 or more, O/C lower than 0.1) are derived mainly from lacustrine depositional environments (Tissot and Welte 1984).

The organic richness of lake deposits is extremely variable and depends upon the nutrient supply by inflowing rivers, water circulation in the lake, and the stability of water stratification. A mapping of organic richness based on organic carbon measurements on recent sediments from Lake Tanganyika was presented by Huc (1988a,b). His results demonstrated that – similar to the situation in silled marine basins such as the Black Sea – the most organic matter-rich sediments were deposited in the deepest part of the lake. The distribution pattern is, however, complex and testifies to the fact that not only water depth affects organic richness.

With respect to the depositional controls on the characteristics of organic matter in lacustrine oil shales, a detailed comparative study was undertaken by Horsfield et al. (1994) on two members of the Eocene Green River Shale from the southwestern United States. The Laney Shale and the Luman Tongue member were deposited under vastly different environmental conditions (open versus closed lake, humid versus arid climate, freshwater versus alkaline lake) and these environmental conditions greatly affected the chemical characteristics of the organic matter. For example, hydrogen richness and hydrocarbon generation potential as well as the amount of terrigenous organic matter differ greatly between the two members. Only the sediments from the alkaline lake stage contain predominantly algal-derived kerogen (alginite) and yield much hydrocarbons upon pyrolysis. The freshwater lake stage, however, produced sediments rich in terrigenous organic matter (vitrinite) which is characterized by a low to medium hydrocarbon generation potential only.

Two other Tertiary oil shales, the Eocene Messel Shale and the Miocene Nördlinger Ries Shale, both from southern Germany, serve further to document the impact of variable environmental conditions on the organic matter in oil shales. In the Eocene Messel Shale organic carbon values vary between 7% and 40% (Rullkötter et al. 1988; Table 5.10). Palaeoenvironment reconstructions indicate that this spectacular organic richness in the 200-m-thick sequence is the consequence of the combination of ideal environmental factors. A warm and humid climate and an inflowing river with nutrients fostered plant growth at the margins and in the photic zone of the lake (Franzen and Michaelis 1988). On the other hand, water circulation caused by river in- and outflow did not inhibit the development of stratification. This is obvious from the rather regular interlayering of algal-rich laminae (often about 0.01 mm thick) and clay-rich laminae (often about 0.1 mm thick) which indicates a seasonal deposition typical of stratified lakes. Lack of oxygen at the sediment/water interface is also indicated by absence of bioturbation. The tectonic setting of the lake in the developing graben system of the Upper Rhine valley permitted deposition of a thick lacustrine sequence.

A different pattern of C_{org} contents characterizes the Nördlinger Ries Shale which was deposited as a lacustrine filling of a meteoric impact crater of Miocene age (Jankowski 1981). In contrast to the Messel Shale, its filling history was not influenced by tectonic processes, i.e., the basin became shallower with ongoing deposition. C_{org} values vary from less than 2% to 25% in this sequence (Table 5.10), which reaches a maximum thickness of more than 200 m in the central part of the crater. Organic matter-rich intervals (C_{org} greater than 8%) are relatively rare and restricted to several dark green layers, which are usually less than 10 cm thick. These layers are concentrated in the lower half of the sequence, testifying to the fact that a critical water depth of about 100 m was necessary to allow deposition of such organic carbon-rich

Table 5.10. Organic matter characteristics of deposits from an Eocene freshwater lake (Messel) and a Miocene hypersalinar lake (Nördlinger Ries)

cality, age	Lithology	C_{org}	S	HI (mg hc/g org_C)
꞉ssel, W Germany Ꜫocene)	Finely laminated, fine-grained shales	28.3±6.1 (7–40)[a]	1.0±0.6 min (0.2–2.4)	553±43 (470–640)[a]
rdlinger Ries, ɔuthern Germany ʎiocene)	Extremely variable finely laminated shales, marlstones, dolomites and others[b]	6.9±5.3 (1.6–25)[c]	2.4±1.2 min (0.5–4.8)	513±176 (177–920)[c]

[a]Rullkötter et al. (1988).
[b]Jankowski (1981).
[c]Rullkötter et al. (1990).

sediments. Also, great differences between C_{org} values in marginal and central parts of the basin were recognized (Rullkötter et al. 1990).

The type of organic matter deposited in the two lakes is also different and clearly depends on the environmental conditions in the adjacent land areas. In the case of the Messel Shale remains of higher land plants were fluvially transported into the lake. Accordingly, the Messel Shale kerogen is composed of about 20% terrigenous higher plant material and 80% autochthonous material derived from subaquatic organisms (Jankowski and Littke 1986; Rullkötter et al. 1988). In the Nördlinger Ries Shale, terrigenous organic particles such as vitrinite and inertinite are generally less abundant, especially in the central part of the deposit (Rullkötter et al. 1990). This is probably due to the lack of an effective transport mechanism, i.e., the Ries was *a closed lake* in contrast to *the open lake* Messel. Therefore terrigenous particles were mainly deposited at the rims of the basin.

Another important environmental factor influencing kerogen composition is water chemistry. The Ries lake had variable, often anomalously high salinities (Jankowski 1981). Organic matter input into the sediments changed with changes in salinity, leading to substantial variability in organic petrographic and bulk geochemical characteristics, for example, in HI values (Table 5.10). The fact that the three most hydrogen-rich samples of the Nördlinger Ries Shale (HI greater than 800 mg hc/g C_{org}) are characterized by a fluorescing, structureless organic groundmass (Rullkötter et al. 1990) may indicate an origin from bluegreen algae, which are relatively resistant to hypersaline conditions (Jankowski 1981: 135; Barbé et al. 1990). No salinity changes affected the Messel Shale, which contains a remarkably homogeneous kerogen on a petrological and bulk geochemical basis. Its rather uniform, intermediate HI values (Table 5.10), normally indicating type II kerogen, are the effect of a mixture of aqueous (type I) and terrigenous (type III) organic matter.

Also, ionic composition of the water in the two lakes was very different. Lake Messel was rich in iron, whereas the Ries lake, completely surrounded by Malmian carbonates, was dominated by alkaline and alkaline earth ions (Na, K, Ca, Mg) and poor in iron (Jankowski 1981). This difference in iron availability greatly affected the sulfur content of the organic matter. In the Messel lake little sulfate was available, and sulfur was mainly fixed in pyrite. In the Ries lake more sulfate was available, and fixation of sulfur in pyrite was inhibited due to the low iron concentration. A great part of the available sulfur was therefore bound in organic matter, leading to anomalously high atomic S_{org}/C_{org} ratios (up to 0.08, Rullkötter et al. 1990). The Nördlinger Ries data show that hypersaline lakes may produce sediments as rich in sulfur as marine sediments (Fig. 5.17). In this case the sulfur versus C_{org} plot cannot be used to differentiate between marine and lacustrine sediments (Berner 1984).

In the Nördlinger Ries deposits extract yields are much higher than would be expected for immature organic matter at very shallow depth (Fig. 5.16). In most samples, phytane is the most abundant single aliphatic compound and phytane/pristane ratios are extremely high (greater than 10) as typical of sediments deposited under hypersaline conditions (ten Haven et al. 1988). The "aromatic hydrocarbon" fraction is dominated by thiophenes, thiolanes, and

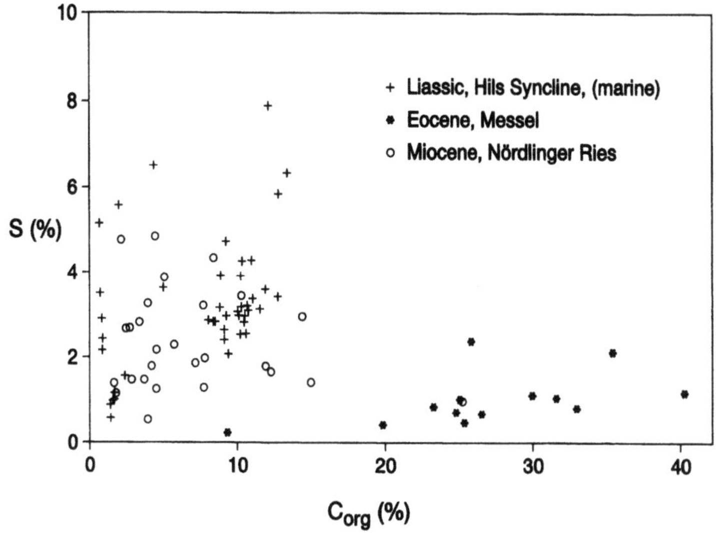

Fig. 5.17. Organic carbon percentages (C_{org}) versus sulfur percentages for three different, immature oil shales. *Crosses*, Lower Toarcian Posidonia Shale from the Wenzen borehole; *circles*, hypersaline, lacustrine Nördlinger Ries deposit of Miocene age; *stars*, Eocene, lacustrine Messel Shale

other unidentified sulfur compounds (Rullkötter et al. 1990), thus confirming the early diagenetic sulfur incorporation into the kerogen. The high sulfur content of the organic matter in the Nördlinger Ries sediments (Fig. 5.17) is believed to cause early petroleum generation, i.e., petroleum generation at very low temperatures, because sulfur-carbon bonds are more easily cracked than carbon-carbon bonds (Baskin and Peters 1992). Similar conclusions were drawn for the diatomaceous Monterey formation in California which is also characterized by sulfur-rich organic matter (Baskin and Peters 1992).

In summary, lakes are potential settings for organic matter-rich sediments with hydrogen-rich kerogen. Due to their smaller size compared to most marine basins, their deposits are strongly influenced by environmental factors such as river inflow and outflow and water chemistry. Accordingly, variability of organic richness, kerogen type, and bitumen yield and composition are great both in vertical sections and horizontal profiles.

5.10 Fluviodeltaic Coal-Bearing Sequences

Fluvial and deltaic sediments are among the most important source rocks for natural gas (e.g., Lutz et al. 1975; Masters 1984; Rice et al. 1989; Littke et al. 1995; Lopatin et al. 1995) but are also reported as source rocks for oil, especially in the Australo-Indonesian realm (Thomas 1982; Thompson et al. 1985; Durand and Paratte 1983; Risk and Rhodes 1985; Shanmugam 1985; Khavari-Khorasani 1987; Horsfield et al. 1988; Hvoslef et al. 1988). Fluvial and deltaic sequences are highly variable with respect to concentration of organic matter.

This heterogeneity is partly due to the fact that preservation of organic matter is strongly affected by differences in climate and exposure to either subaerial or subaquatic conditions after deposition. In fluvial and deltaic sequences the bulk of the organic matter consists of parts of higher land plants, most of which are deposited close to but not exactly at the place of plant growth (Scheihing and Pfefferkorn 1984). As growth of higher plants is inhibited in arid regions, deposition of organic matter-rich fluvial and deltaic sediments is usually restricted to humid climates. Furthermore, source rocks are not expected to occur in pre-Devonian fluviodeltaic rocks because the evolution of higher land plants only started during the late Silurian.

Coal-bearing fluvial and deltaic sequences deposited in humid climates contain the highest concentration of organic matter in rocks both on a meter scale (50–99% C_{org} in coals) and on a basin-wide (kilometer) scale. Figure 5.18 compares organic matter percentages for five coal-bearing basins according to a compilation by Scheidt and Littke (1989). In most of these basins more organic matter is fixed in coals than finely dispersed in clastic sediments. The average amount of total organic matter ranges from 4 to 12 vol%, and the ratio of coal to dispersed organic matter from 0.7 to 2.5% over intervals of several hundred to more than a thousand meters thickness.

Total amounts of dispersed organic matter are different in different lithologies, as exemplified for the Ruhr basin in Figure 5.19 (Scheidt and Littke 1989). For this area a decrease in organic matter content with increasing grain size was found, for example, mudstones on an average contain more organic matter than siltstones and sandstones. More than 2 vol% dispersed organic matter on an average is fixed in gray mudstones, one of the major lithologies. In contrast, the various siltstone and sandstone facies contain less than 2 vol% and often only 1 vol% dispersed organic matter. The only exception are rare conglomeratic sandstones with more than 2 vol% dispersed organic matter in the form of coal pebbles and barks of twigs and stems. Baker (1962) and Huc et al. (1986) reported similar values for shale samples from other coal-bearing sequences, i.e., the Cretaceous Douala basin (1–3% C_{org}), the Miocene Mahakam delta (1–4% C_{org}), the Westphalian in the Paris basin (1–5% C_{org}), and the Upper Carboniferous of the Cherokee group of Kansas and Oklahoma, United States (1–5% C_{org} in gray shales). Few clastic rocks in coal-bearing basins contain more than 10% C_{org}. Exceptions seem to be black shales (see Fig. 5.19) derived from marine incursions (Wenger and Baker 1986: up to 20% C_{org}) and mudstone partings within coal seams (Littke and ten Haven; 1989: up to 32% C_{org}). As discussed above, the organic richness of the former group is explained by high algal production and the occurrence of humic precipitates fostered by the flooding of swamps, whereas the richness of the latter is partly due to the in situ growth and preservation of roots after clastic deposition (Scheidt 1988).

In fluvial and deltaic sequences, terrigenous organic matter usually predominates over aquatic organic matter (e.g., Scheihing and Pfefferkorn 1984; Scheidt and Littke 1989; Smyth 1989), although the latter occurs in interbedded (1) transgressive marine deposits (e.g., Wenger and Baker 1986) and (2) lacustrine deposits such as rare sapropelic coals (e.g., Stach et al. 1982). Terri-

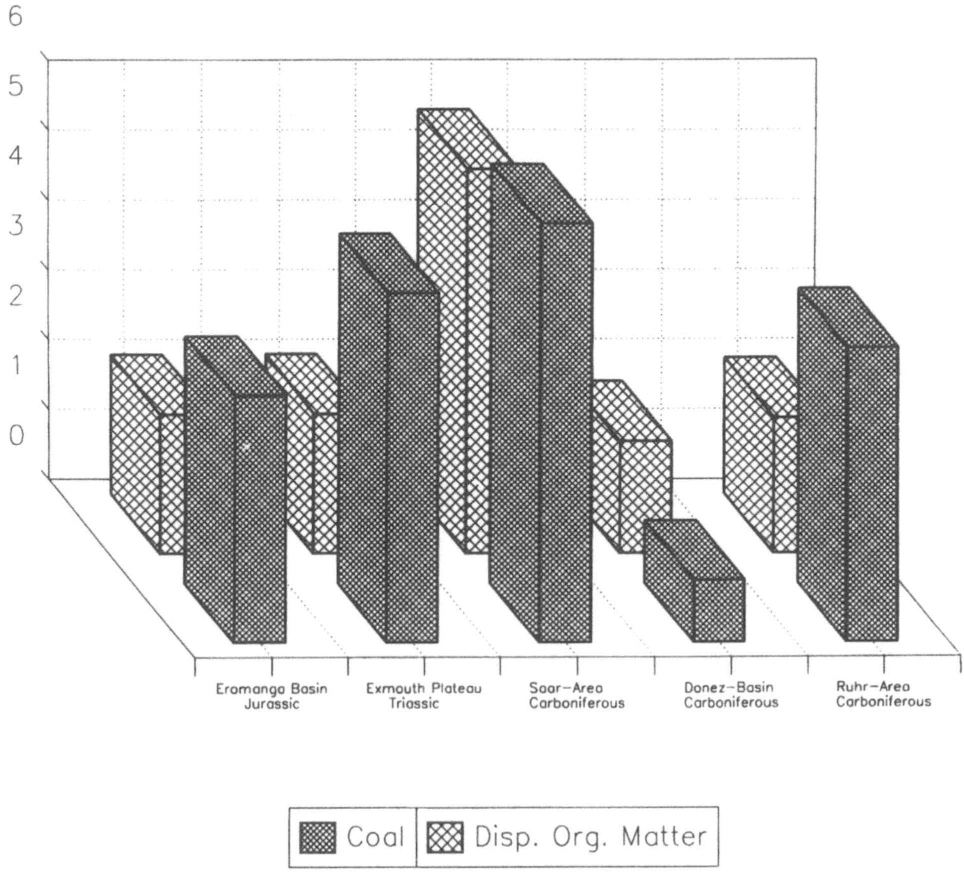

Fig. 5.18. Volume percentages of organic matter bound as coals seams and as dispersed organic matter in different coal-bearing basins. (Compilation by Scheidt and Littke 1989, based on literature data).

genous organic matter is known to contain less hydrogen and more oxygen than aquatic organic matter (Pelet 1983; Tissot and Welte 1984). Accordingly, in immature fluvial and deltaic rocks, H/C ratios of kerogen are generally lower than 1 and O/C ratios greater than 0.2 (Boudou et al. 1984; Huc et al. 1986).

Hydrogen Index values of coals are generally in the range of 150–300 mg hc/g C_{org} at immature or marginally mature stages (Table 5.11); similar values are typical for handpicked vitrinites from coals (Littke et al. 1989) and suggest an intermediate hydrocarbon generation potential. Interestingly, HI values of kerogen in mudstones and siltstones interbedded with the coals are generally lower (Table 5.11), although the petrographic composition of the organic matter is roughly similar (Scheidt and Littke 1989). This difference is tentatively explained by an inhibited expulsion of high molecular weight hydrocarbons generated from the coals. The resultant low expulsion efficiency causes

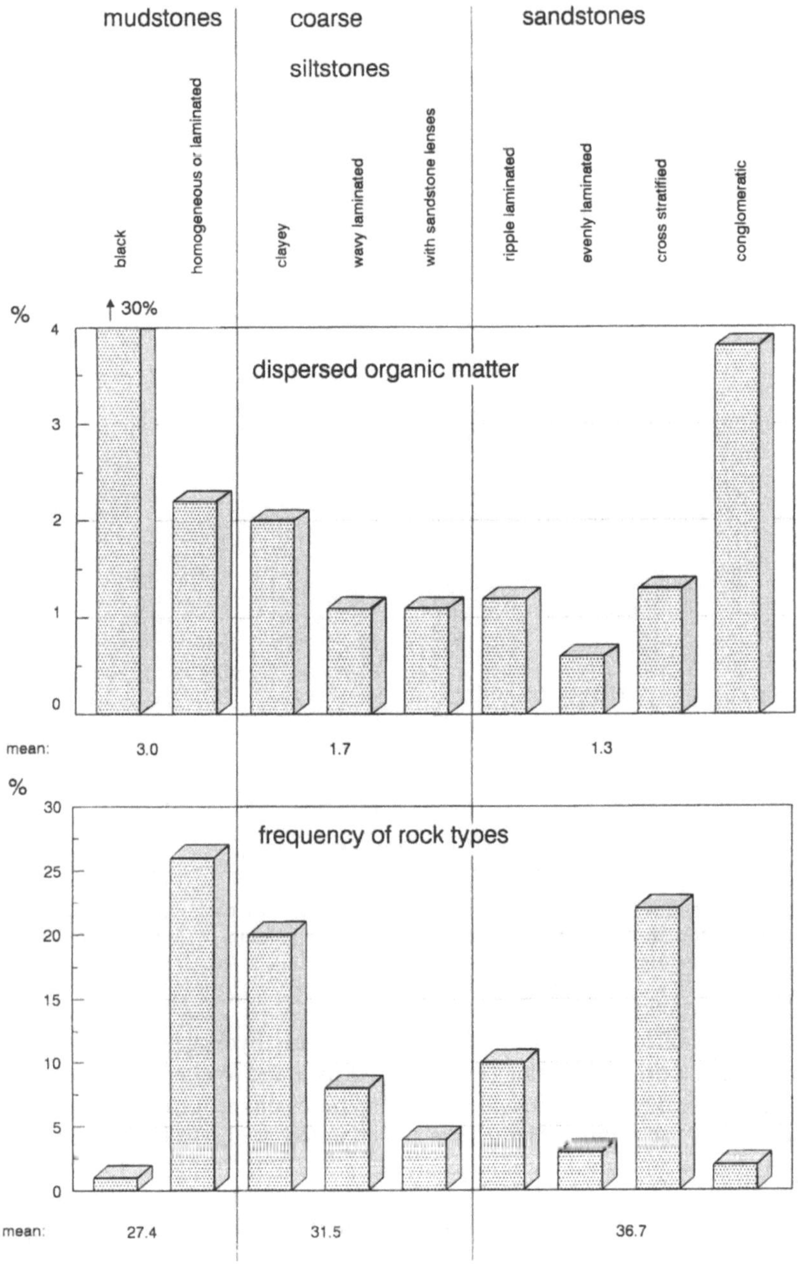

Fig. 5.19. Volume percentages of organic matter in different clastic rocks of the Carboniferous Ruhr basin (*above*) and percentage of these rock types in this sedimentary sequence (*below*). The grain size of the sediments as well as the average grain size of the organic clasts increases from left to right. The volume percentage of organic matter in black shales (about 30%) is offscale. (Scheidt and Littke 1989)

Table 5.11. Organic matter characteristics of coal-bearing strata (coals, mudstones, siltstones). (Horsfield et al. 1988; Ramanampisua et al. 1990; Littke et al. 1989)

	Area	Age	R_r (%)	C_{org} (%)	HI (mg hc/g C_{org})	Major macerals
Talang akar coals	Java	Tertiary	0.3–0.8	59.9	299 (374)	V>L>I
Malagasy coals	Madagascar	Carboniferous-Permian	0.7–0.8	59.9	182 (284)	V~I>L
Malagasy mudstones	Madagascar	Carboniferous-Permian	0.7–0.8	14.4	143 (250)	I~L>V
Ruhr coals	Germany	Carboniferous	0.7	75.2	234 (315)	V>I>L
Ruhr mudstones and siltstones	Germany	Carboniferous	0.7	6.6	93 (191)	V>I>L

R_r, Mean vitrinite reflectance; HI, hydrogen index, mean (maximum in parentheses); V, vitrinite; I, inertinite; L, liptinite.

a preservation of hydrogen until gas is generated by cracking of the trapped bitumen at more elevated maturity stages (Littke and Leythaeuser 1993).

Petrographically, organic matter in fluvial and deltaic deposits is usually composed of a mixture of different macerals. Vitrinite is the most frequent maceral group in most coals (Stach 1982), although in several coal-bearing strata such as the Permian basins of the southern hemisphere inertinite prevails (Chandra and Taylor 1982; Smyth 1989; Ramanampisoa et al. 1990). Liptinite rarely forms more than 50% of coals, although there are exceptions (e.g., Horsfield et al. 1988). Figure 5.20 provides a compilation of average vitrinite, inertinite, and liptinite percentages of coals from different basins. Average liptinite contents vary between 5 and 35%, average vitrinite contents between 34% and 88%, and average inertinite contents between 1 and 50%. As "the chemical and botanical precursors of inertinite are mainly the same as those for vitrinites, namely cellulose and lignin from the cell walls of plants" (Teichmüller 1982), and vitrinite is thought to reflect better preserved organic matter than inertinite, the ratio vitrinite/inertinite is used as an indicator for the degree of pre- and syndepositional degradation of plant particles (Shibaoka and Smyth 1975; Diessel 1987; Littke 1985, 1987).

The petrographic composition of organic particles in clastic rocks is either similar to that of associated coal seams (Scheidt and Littke 1989), or vitrinite is less abundant (Smyth 1989; Ramanampisoa et al. 1990). In most basins inertinite seems to be enriched in clastic rocks (Fig. 5.20) relative to coals. In the Ruhr basin this enrichment is less obvious than in other basins and has been established only for siltstones, whereas mudstones are enriched in liptinite on an average. This is explained by transport sorting leading to a relative enrichment of the lightest maceral group liptinite in mudstones deposited at lowest current velocities. The predominance of vitrinite is explained by the great percentage of vitrinite roots in the clastic rocks of the Ruhr basin. It should be noted, however, that the variability of petrographic composition in each rock type is enormous, for example, liptinite proportions (vol% of organic

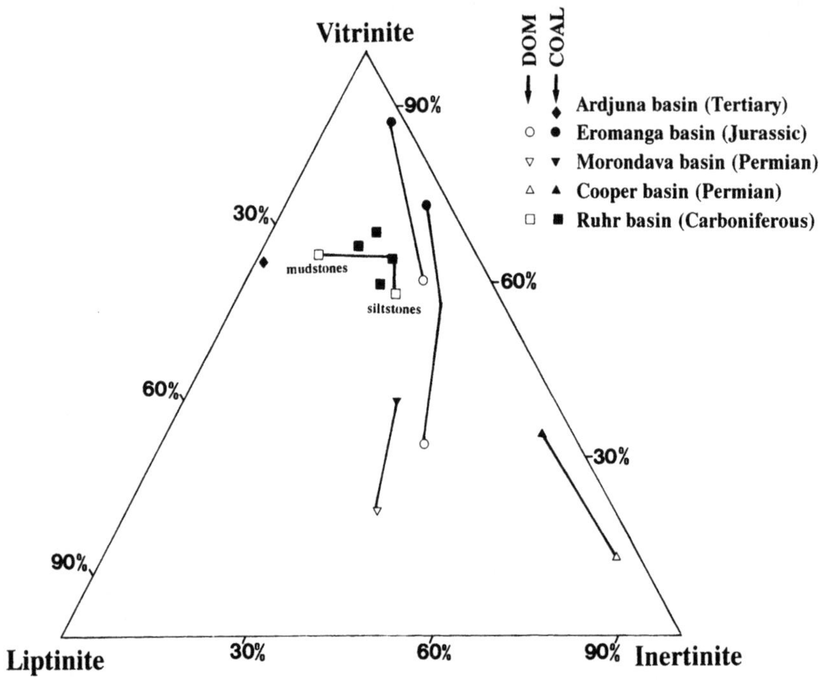

Fig. 5.20. Average vitrinite, inertinite, and liptinite percentages of coals and corresponding kerogen in clastic rocks in different coal-bearing basins. *DOM*, dispersed organic matter

matter) range from 0% to 60% in mudstones and siltstones and from 0% to 20% in sandstones (Scheidt 1988).

In summary, fluvial and deltaic sequences may be rich in organic matter which is mainly of higher land plant origin and often more gas prone than oil prone. However, there are exceptions to this rule, for example, in Indonesia, Australia, and the North Sea. On the other hand, many large gas accumulations are genetically linked to coals, and methane can even be commercially produced from coal. In coal-bearing basins mudstones and siltstones often contain elevated percentages of organic carbon which significantly adds to the hydrocarbon generation potential of the coals.

5.11 Source Rocks and Tectonics of Petroleum Basins

The relationship between source bed occurrence and basin tectonics is difficult to ascertain. This limitation results from several related deficiencies. Often the unequivocal identity, such as through geochemical characterization or correlation with accumulated hydrocarbons, of the source rock is unavailable. Further, although a petroleum basin may be classifiable on a tectonic basis (often based on a present-time view), these tectonic conditions may differ from the tectonic framework which existed during the time pertinent to source rock deposition. As noted by Klemme (1980, p. 261), "crustal mobility produces changes in the tectonics and fundamental character in basin type as individual basins develop." Thus, as for other stratigraphic units, an understanding of the relationship between source rock formation and tectonic controls requires a focused coeval analysis of both. Most analyses of source rocks vs. tectonics have been based on the indirect approach of evaluating *petroleum occurrence* in terms of basin tectonics with the implicit assumption that in petroliferous basins source rocks must also occur, and, further, that the magnitude of petroleum production and estimated reserves is a measure of relative source rock quality. Studies by Klemme (1980) which involve the tectonic classification of petroleum basins and generalizations concerning their characterizations, especially those related to petroleum potential, provide a basis for this approach.

Klemme (1980, p. 187) notes that more than 600 basins (and subbasins) are known worldwide. Of these only about one-quarter have petroleum production in some portion or in some cases almost all of the basin, i.e., three-quarters are nonproductive, although about one-third of these have never been tested. Focusing only on the productive basins, Klemme has classified them into eight types (with several important subtypes). The classification is based on their "architectural characteristics when related to the earth's crust, tectonic setting, and basin evolution (primarily in the framework of plate tectonics)." Klemme's eightfold classification, along with the comparative classifications and terminology of Bally (1975) and Huff (1979) as taken from his Fig. 2 (1980, pp. 188–189), are presented in Fig. 5.21. Further, Klemme (Fig. 3, pp. 190–191, and Fig. 23, p. 193) has summarized a large number of the characteristics of the eight petroleum basin types. From this data source several selected characteristics which may bear directly on source rock aspects of the various basin types are compiled in Fig. 5.21.

The interpretation of these characteristics in terms of the control of basin tectonics on the depositional history (accumulation), occurrence and distribution, and properties (quality) of source rocks is fraught with uncertainties. In particular, with few exceptions Klemme did not specifically identify probable source rock formations within the basins. Clearly, the occurrence of source rocks is implicit in petroliferous basins, but source assessment in relation to basin tectonics must be surmised largely by indirect reasoning, i.e., estimates of petroleum magnitude. However, as well understood, basin characteristics such as recovery, number of giant oil fields, and percentage of world reserves (see Table 5.12) are governed not only by source rock quality. Also of major importance are the collection of processes and factors which govern the whole dynamics of petroleum formation, including principally, generation (thermal maturation), expulsion from source (primary migration), secondary migration (carrier-bed transport), availability and size of traps, basin and trap seals, and post-accumulation dismigration and alteration. Similarly, and especially for older basins, for example, Paleozoic, postaccumulation tectonics (deformation) may have destroyed much petroleum. Thus, because of inferior preservation, present estimates of petroleum may be considerably less than that which existed during earlier episodes of basin history and may result in an erroneous interpretation of source rock aspects of the particular basin.

KLEMME 1980				BALLY 1975			HUFF 1979	
Intracontinental	Cratonic		Type 1 Interior Simple	Rigid Lihosphere	Consolidated Mega-Suture	Cratonic Basins	Cratonic Foreland Areas	Cratonic Sag Basins
Intracontinental	Cratonic		Type 2 Composite a) large, b) small, 2a Complex	Rigid Lihosphere	Consolidated Mega-Suture	Cratonic Foredeep	Cratonic Foreland Areas	Foreland Basins
Intracontinental / Extracontinental	Cratonic		Type 3 Rift	Rigid Lihosphere	Exten-sion	Rifted Grabens	Divergent Area	Rifts and Grabens
Extracontinental	Cratonic / Borderland		Type 4 Downwarp into Small Ocean Basin: a) closed, b) trough only, c) open	?	?	(in part) Atlantic-type Fordedeep (Episutural Basins)	?	(in part) Foreland Divergent Margin Back-arc
Extracontinental	Borderland		Type 5 Pull-Apart a) parallel, b) traverse	Rigid L.	Exten-sion	Atlantic-type Passive Margins	Divergent Area	Divergent Margin (transverse)
Extracontinental	Borderland	Intermontane	Type 6 Subduction a) fore-arc, b) back-arc, c) non-arc	Episutural	Subduction	Fore Arc Back-deep and Marginal California-type	Convergent Area	Fore-arc Back-arc Strike-slip
Extracontinental	Borderland	Intermontane	Type 7 Median	Episutural	Subduction	Pannonian-type	Convergent Area	
?	Borderland		Type 8 Deltas					Deltas and Fans

Fig. 5.21. Tectonic classifications of sedimentary basins. (Klemme 1980)

Despite these qualifications the following generalizations regarding the controls and relationships between basin tectonics and the occurrence and nature of petroleum source rocks seem justified (see Table 5.12):

1. Source rocks occur in all basin types; however, recognizing that two-thirds of all basins are unproductive (and/or untested), many basins of all types may be totally devoid of source rocks.
2. Based on recovery, percentage of giant fields, percentage of world basin area, and percentage of world reserves, source rock potential of the different basin types seems quite variable.
   - Source rock potential of Interior (cratonic) and Pull-Apart basins is low.
   - Source rock potential of Composite basins is average or intermediate.
   - Source rock potential of Rift, Downwarp, Subduction, Median and Delta basin types is high or substantial. Note that in all these basin types, recovery is high, giants are common (except for Delta type), and the ratio of percent world reserves to percent world basin area is consistently greater than for Interior and Pull-Apart basins.
3. Following the conventional view that high-gravity, low-sulfur oils are derived primarily from siliciclastic source facies, and lower-gravity high-sulfur oils from carbonate source facies, we can surmise the following from the data summarized in Fig. 5.21.
   - Although carbonate lithofacies are an important component of basin fill in Interior basins, siliciclastic facies are the prevalent source facies,
   - Both siliciclastic and carbonate facies are important sources in Composite basins.
   - Siliciclastic source facies are dominant in Rift basins.
   - Carbonate source facies are dominant in Downwarp basins.
   - Although clastic facies are predominant in Subduction and Median basins, the oils are variable in gravity and sulfur properties, showing that special conditions, for example, Monterey Formation of the Los Angeles Basin, may result in sulfur-rich kerogen in siliciclastic sediments.
   - As expected, Delta basins yield primarily high-gravity, low-sulfur oils presumably derived from their overwhelming proportions of siliciclastic source facies.
4. The high gas content of Delta basins indicates the importance of source facies characterized by the predominance of terrestrial (gas prone) organic matter. Whereas the high gas content of Composite basins may also be related in part to significant amounts of gas prone source rocks, the large volumes of gas which have accumulated in Downwarp basins may be primarily related to maturation history (i.e., thermogenic gas) and not be indicative of a terrestrially dominated source facies.

5.12 Conclusions

In petroleum exploration the knowledge of petroleum source rocks and their specific characteristics is essential because the type of their incorporated or-

Table 5.12. Basin types and selected petroleum related characteristics. (Compiled with minor modification from Klemme 1980)

Basin type	1: Interior simple	2: Composite	3: Rift	4: Downwarp (closed–open)	5: Pull-apart	6: Subduction	7: Median	8: Delta
Recovery	Low	Av	Good	High	Low(?)	High (variable)	Variable	High
Giants, % productive basins with giants	12	50	50	68	15	65	44	Few
Basin lithology (clastic vs carbonate)[a]								
% Reservoirs	60/40	75/25	60/40	35/65–50/50	70/30	90/10	90/10	100/0
Basin fill	Equal	Equal+ evaporite	>Clastic	>Carb+ evaporites	>Clastic	>>Clastic	>>Clastic	100% Clastic
Temperature	Cool	Cool	Normal-high	Normal-high	Cool	High-normal	Normal-high	Normal-low
Age[b]	Paleozoic	Paleozoic, Mesozoic	Up. Paleozoic. Mesozoic, Tertiary	Up. Paleozoic, Mesozoic, Tertiary	Mesozoic, Tertiary	Up. Mesozoic, Tertiary	Up. Mesozoic, Tertiary	Tertiary, Quaternary
Petroleum type								
Gravity	High g	High g/high g[c]	High g	Intermediate g		Variable and intermediate g		High g
Sulfur	Low s	Low s/high s[c]	Low s	High s		High s		Low s
Gas		High/high[c]		High				High
% World basin area	18.2	27.3	5.4	17.5	18.2	7.1	3.7	2.6
% World reserves	1.5	25	10	47	0.5	7.5	2.5	6

[a]From "5" to "8", increasingly clastic.
[b]From "2" to "6", increasingly young.
[c]Clastic res/carbon res.

ganic material greatly determines the quality of the petroleum. For example, sulfur-rich kerogen which is typical for iron-depleted source rocks produces sulfur-rich oil, and coaly organic matter tends to generate more gas (higher gas/oil ratios) than algal-derived kerogen. However, not only the type of oil or gas produced but also the total mass and volume are greatly governed by source rock characteristics. Masses and volumes greatly depend on the thickness of the source rocks, chemical nature of the organic matter, total amount of organic matter, areal extent, and expulsion characteristics and drainage volume. Therefore information on the distribution, thickness, and facies variation of source rocks in sedimentary basins is highly desirable. This information is also necessary in any numerical basin modeling, where it is used in combination with the burial and temperature history and kinetic data to calculate the petroleum generation history in sedimentary basins.

Depositional environments of petroleum source rocks have commonly been categorized as either "nonmarine" or "marine." Considering the broad variety of depositional settings and conditions, such a classification is too simple to provide adequate insight into source rock deposition. As an alternative we have attempted to treat source rocks in a framework similar to the depositional categories as currently discussed by sedimentologists and stratigraphers.

The depositional settings for source rocks greatly affect the source rock properties. For example, source rocks deposited in anoxic lakes or anoxic shallow marine basins tend to contain a very hydrogen-rich kerogen derived from plankton (type I or II; Table 5.13), whereas those deposited in fluvial, deltaic, and deep-marine basins tend to contain predominantly hydrogen-lean kerogen (type III) derived from higher land plants. Undoubtedly, marine source rocks provided most giant oil fields of the world (such as those of the Arabian peninsula; Table 5.13). Some of these oil shales are of great lateral extent, such as the Posidonia Shale or the Kimmeridge Clay in northern and central Europe.

Commercial oil fields also result from source rocks deposited in lakes and the largest oil field related to such lacustrine source rocks is probably the Daqing field in the Songliao basin of China (Kulke 1994). Other examples exist along the coasts of western Africa and Brazil, where hydrogen-rich, alginitic kerogen is present in Lower Cretaceous deposits which witness the early rifting of the Atlantic Ocean.

The most giant gas accumulations in the world occur in northwestern Siberia and originate from coaly (type III) source rocks situated at shallow depth (Lopatin et al. 1995; Table 5.13). The large gas fields of the eastern Netherlands and northern Germany are also derived from coaly organic matter, but these source rock sequences are situated at great depth. In addition to methane, the coals also generated significant amounts of molecular nitrogen gas in this area (Littke et al. 1995).

The concentration of organic matter both in progradational submarine fans and in evaporitic environments is often smaller than in the above discussed settings whereas the quality with respect to hydrocarbon generation may be good or at least moderate. Evaporitic sequences such as the Staßfurth-Karbonat of the Late Permian Zechstein formation in Germany can act as effective

Table 5.13. Characteristic features of different settings for petroleum source rock deposition

Settings	Contribution to world petroleum	Predominant source rock types	Lateral extent	Predominant organic particles	Characteristic kerogen types
Marine silled basins, anoxic continental shelves	+++	Marlstones, shales, siltstones, carbonates	**	Alginite, bituminite	II, II-III
Anoxic lakes	++	Shales, marlstones	*	Alginite	I, II
Fluviodeltaic (coal-bearing) basins	+	Siltstones, shales, coals	*	Vitrinite	III
Progradational submarine fans	+	Shales, siltstones, marlstones, turbidites	**	Often vitrinite, also inertinite, liptinite	III
Evaporitic environments	0	Carbonates, sulfates, interlayered, thin shales, salt	*	Alginite	II or III
Deep marine upwelling areas	0	Siliceous shales, shales, marlstones	*	Predominant unstructured (marine derived), alginite	II-III

+++, Extremely important; ++, very important: +, important especially for gas generation; 0, less important or unknown; **, extensive to very extensive; *, small to extensive.

source rocks for petroleum. One advantage of this type of source rock is the intimate connection with rather impermeable salt cap rocks.

However, many exceptions exist for the general pattern summarized in Table 5.13. For example, coaly organic matter is regarded as the source of some important oil (rather than gas) accumulations (e.g., Gippsland basin, Australia). The many factors controlling petroleum generation and expulsion from source rocks are the principal reason why any classification of source rocks remains unsatisfactory. Therefore source rocks must be specifically mapped and studied for each individual basin and petroleum play. The results of these specific investigations are necessary ingredients of any sophisticated petroleum exploration and basin modeling.

References

Arrhenius GOS (1952) Sediment cores from the east Pacific. Rep Swed Deep Sea Exped 1947–1948, vol 5. Eleander, Göteborg, 288 pp

Baker DR (1962) Organic geochemistry of Cherokee group in southeastern Kansas and northern Oklahoma. Am Assoc Petr Geol Bull 46: 1621–1642

Bailey GW (1991) Organic carbon flux and development of oxygen deficiency on the northern Benguela continental shelf south of 22°S: spatial and temporal variability. In: Tyson RV, Pearson TH (eds) Modern and ancient continental shelf anoxia. Geol Soc Spec Publ 58: 171–183

Bally AW (1975) A geodynamic scenario for hydrocarbon occurrences. Proc 9th World Petr Congr, vol 2. Applied Science Publ, Essex, pp 33–44

Barbé A, Grimalt JO, Pueyo JJ, Albaigés J (1990) Characterization of model evaporitic environments through the study of lipid components. In: Durand B, Béhar F (eds) Advances in organic geochemistry 1989. Org Geochem 16: 815–828

Baskin DK, Peters KE (1992) Early generation characteristics of a sulfur-rich Monterey kerogen. Am Assoc Petrol Geol Bull 76: 1–13

Baturin GN (1983) Some unique sedimentological and geochemical features of deposits in coastal upwelling regions. In: Thiede J, Suess E (eds) Coastal upwelling. Part B: Sedimentary records of ancient coastal upwelling. Plenum Press, New York, pp 11–28

Baudin F, Herbin J-P, Bassoulet J-P, Dercourt J, Lachkar G, Manivit H, Renard M (1990) Distribution of organic matter during the Toarcian in the Mediterranean Tethys and Middle East. In: Huc AY (ed) Deposition of organic facies. Am Assoc Petrol Geol Stud Geol 30: 73–92

Berger WH (1989) Global maps of ocean productivity. In: Berger WH, Smetacek VS, Wefer G (eds) Productivity of the ocean: present and past. Dahlem Worksh Rep, Life Sci Res Rep 44. John Wiley and Sons, Chichester, pp 429–456

Berger WH, Smetacek VS, Wefer G (eds) (1989a) Productivity of the ocean: present and past. Dahlem Worksh Rep, Life Sci Res Rep 44. John Wiley and Sons, Chichester

Berger WH, Smetacek VS, Wefer G (1989b) Ocean productivity and paleoproductivity - an overview. In: Berger WH, Smetacek VS, Wefer G (eds) Productivity of the ocean: present and past. Dahlem Worksh Rep, Life Sci Res Rep 44. John Wiley and Sons, Chichester, pp 1–34

Berner RA (1984) Sedimentary pyrite formation: an update. Geochim Cosmochim Acta 48: 605–615

Betzer PR, Showers WJ, Laws EA, Winn CD, Ditullo GR, Kroopnick PM (1984) Primary productivity and particle fluxes on a transect of the equator at 153 #W in the Pacific Ocean. Deep-Sea Res 31: 1–11

Bishop JKB (1989) Regional extremes in particulate matter composition and flux: effects on the chemistry of the ocean interior. In: Berger WH, Smetacek VS, Wefer G (eds) Productivity of the ocean: present and past. Dahlem Worksh Rep, Life Sci Res Rep 44. John Wiley and Sons, Chichester, pp 117–138

Blanc-Valleron M-M, Gely J-P, Schuler M, Dany F, Ansart M (1991) La matière organique associée aux évaporites de la base du sel IV (Oligocène inférieur) du bassin de Mulhouse (Alsace, France). Bull Soc Geol Fr 162: 113–122

Boudou JR, Pelet R, Letolle R (1984) A model of the diagenetic evolution of coaly sedimentary organic matter. Geochim Cosmochim Acta 48: 1357–1362

Bralower TJ, Thierstein HR (1987) Organic carbon and metal accumulation in Holocene and mid-Cretaceous marine sediments: paleoceanographic significance. In: Brooks J, Fleet AJ (eds) Marine petroleum source rocks. Geol Soc Spec Publ 26: 345–369

Brumsack H-J (1980) Geochemistry of Cretaceous black shales from the Atlantic Ocean (DSDP Legs 11, 14, 36, and 41). Chem Geol 31: 1–25

Calvert SE (1987) Oceanographic controls on the accumulation of organic matter in marine sediments. In: Brooks J, Fleet AJ (eds) Marine petroleum source rocks. Blackwell Scientific, London, pp 137–151

Chandra D, Taylor GH (1982) Gondwana coals. In: Stach E, Mackowsky MTh, Teichmüller M et al. (eds) Stach's textbook of coal petrology. Gebrüder Bornträger, Berlin, pp 177–197

Cohen Y, Krumbein WE, Shilo M (1977) Solar Lake (Sinai). II. Distribution of photosynthetic microorganisms and primary production. Limnol Oceanogr 21: 609–620

Collins MJ, Bishop AN, Farrimond P (1995) Sorption by mineral surfaces: rebirth of the classical condensation pathway for kerogen formation? Geochim Cosmochim Acta 59: 2387–2391

Crisp PT, Ellis J, Hutton AC, Korth J, Martin FA, Saxby JD (1987) Australian oil shales: a compendium of geological and chemical data. Departments of Geology and Chemistry, University of Wollongong, 109 pp

Curtis CD (1980) Diagenetic alteration in black shales. J Geol Soc Lond 137: 189–194

Dean WE, Arthur MA (1989) Iron-sulfur-carbon relationships in organic-carbon-rich sequences. I. Cretaceous Western Interior seaway. Am J Sci 289: 708–743

Degens ET, Ittekott V (1987) The carbon cycle-tracking the path of organic particles from sea to sediment. In: Brooks J, Fleet AJ (eds) Marine petroleum source rocks. Geol Soc Spec Publ 26: 121–136
Degens ET, Mopper K (1976) Factors controlling the distribution and early diagenesis of organic material in marine sediments. In: Riley JP, Chester R (eds) Chemical Oceanography, vol 6. Academic Press, London, pp 60–114
Degens ET, Stoffers P, Golubic S, Dickmann MD (1978) Varve chronology: estimated rates of sedimentation in the Black Sea deep basin. In: Ross DA, Neprochnor YP et al. (eds) Initial Reports Deep Sea Drilling Project 42. US Government Printing Office, Washington, pp 499–508
Degens ET, Emeis K-C, Mycke B, Wiesner MG (1986) Turbidites: the principle mechanism yielding black shales in the early deep Atlantic Ocean. Geol Soc Spec Publ 21: 361–376
Demaison GJ (1991) Anoxia vs productivity: what controls the formation of organic-carbon-rich sediments and sedimentary rocks? Discussion. Am Assoc Petrol Geol Bull 75: 499
Demaison GJ, Moore GT (1980) Anoxic environments and oil source bed genesis. Am Assoc Petrol Geol Bull 64: 1179–1209
Demaison GJ, Holck AJJ, Jones RW, Moore GT (1984) Predictive source bed stratigraphy: a guide to regional petroleum occurrence. Proc 11th World Petrol Congr, London, vol 2, pp 17–30
Deuser WG (1974) Evolution of anoxic conditions in Black Sea during Holocene. In: Degens ET, Ross DA (eds) The Black Sea – geology, chemistry, and biology. Am Assoc Petrol Geol Mem 20: 133–136
Diessel CFK (1987) On the correlation between coal facies and depositional environments. Proc 20th Symp Adv in the study of the Sydney Basin, Newcastle, pp 19–22
Dinur D, Spiro B, Aizenshtat Z (1980) The distribution and isotopic composition of sulfur in organic-rich sedimentary rocks. Chem Geol 31: 37–51
Durand B (ed) (1980) Kerogen, insoluble organic matter from sedimentary rocks. Editions Technip, Paris, 519 pp
Durand B, Paratte M (1983) Oil potential of coals. In: Brooks J (ed) Petroleum geochemistry and exploration of Europe. Blackwell Scientific, Oxford, pp 285–292
Durand B, Espitalié J, Nicaise G, Combaz A (1972) Études de la matière organique insoluble (kérogène) des argiles du Toarcian. Analyse élémentaire, études en microscopie et diffraction électroniques. Rev Inst Fr Pétr 27: 865–884
Emeis K-C, Morse JW (1990) Organic carbon, reduced sulfur and iron relationships in sediments of the Peru margin, sites 680 and 688. In: Suess E, von Huene R et al. (eds) Proc of the Ocean Drilling Program, Sci Results 112, pp 441–454
Espitalié J, Laporte JL, Madec M, Marquis F, Leplat P, Paulet J, Boutefeu A (1977) Méthode rapide de caractérisation des roches mères, de leur potentiel pétrolier et de leur degré d' évolution. Rev Inst Fr Pétr 32: 23–42
Espitalié J, Marquis F, Sage L, Barsony I (1987) Géochemie organique du basin de Paris. Rev Inst Fr Pétr 42: 271–302
Ferguson J (1987) The significance of carbonate ooids in petroleum source rock studies. In: Brooks J, Fleet AJ (eds) Marine petroleum source rocks. Geol Soc Spec Publ 26: 207–215
Ferguson J, Ibe AC (1982) Some aspects of the occurrence of proto-kerogen in recent ooids. J Petrol Geol 4: 267–285
Fleet AJ, Brooks J (1987) Introduction. In: Brooks J, Fleet AJ (eds) Marine petroleum source rocks. Geol Soc Spec Publ 26: 1–11
Franzen JL, Michaelis W (eds) (1988) Der eozäne Messelsee – Eocene Lake Messel. Cour Forsch-Inst Senckenberg 107: 452 pp
Galloway WE, Hobday DK (1983) Terrigenous clastic depositional systems. Applications to petroleum, coal and uranium exploration. Springer, Berlin Heidelberg New York, 423 pp
Glenn CR, Arthur MA (1984) Sedimentary and geochemical indicators of productivity and oxygen contents in modern and ancient basins: the Holocene Black Sea as the "type" anoxic basin. Chem Geol 48: 325–354
Hallam A (1981) Facies interpretation and the stratgraphic record. WH Freemand, Oxford, 291 pp

ten Haven HL, Rullkötter J (1991) Preliminary lipid analysis of sediments recovered during Leg 117. In: Prell WL, Niitsuma N et al. (eds) Proc of the Ocean Drilling Program, Sci Results 117, pp 561–570

ten Haven HL, Baas M, de Leeuw JW, Schenck PA (1987) Late Quaternary Mediterranean sapropels. I. On the origin of organic matter in sapropel S7. Mar Geol 75: 137–156

ten Haven HL, de Leeuw JW, Rullkötter J, Sinninghe-Damsté JS (1988) Restricted utility of the pristane/phytane ratio as palaeoenvironmental indicator. Nature 330: 641–643

ten Haven HL, Littke R, Rullkötter J, Stein R, Welte DH (1990) Accumulation rates and composition of organic matter in Late Cenozoic sediments underlying the active upwelling area off Peru. In: Suess E, von Huene R et al. (eds) Proc of the Ocean Drilling Program, Sci Results 112, pp 591–605

Heath GR, Moore TC, Dauphin JP (1977) Organic carbon in deep-sea sediments. In: Anderson NR, Malahoff A (eds) The fate of fossil fuel CO_2 in the oceans. Plenum Press, New York, pp 605–625

Heckel PH (1991) Thin widespread Pennsylvanian black shales of Midcontinent North America: a record of a cyclic succession of widespread pycnoclines in a fluctuating epeiric sea. In: Tyson RV, Pearson TH (eds) Modern and ancient continental shelf anoxia. Geol Soc Spec Publ 58: 259–274

Herbin JP, Montadert LO, Müller C, Gomez R, Thurow J, Wiedmann J (1986) Organic-rich sedimentation at the Cenomanian/Turonian boundary in oceanic and coastal basins in the North Atlantic and Tethys. In: Summerhayes CP, Shackleton NJ (eds) North Atlantic paleoceanography. Geol Soc Spec Publ 21: 389–422

Hite RJ, Anders DE (1991) Petroleum and evaporites. In: Melvin JL (ed) Evaporites, petroleum and mineral resources. Dev Sedimentol 50: 349–410

Hofmann P (1992) Sedimentary facies, organic facies and hydrocarbon generation in evaporitic sediments of the Mulhouse basin, France. Ber Forschungszentr Jülich 2664: 1–288

Horsfield B, Yordy KL, Crelling JC (1988) Determining the petroleum-generating potential of coal using organic geochemistry and organic petrology. In: Mattavelli L, Novelli L (eds) Advances in organic geochemistry 1987. Org Geochem 13: 121–129

Horsfield B, Heckers J, Leythaeuser D, Littke R, Mann U (1991) A study of the Holzener Aspaltkalk, northern Germany: observations regarding the distribution, composition and origin of organic matter in an exhumed petroleum reservoir. Mar Petrol Geol 8: 198–211

Horsfield B, Bharati S, Larter SR, Leistner F, Littke R, Schenk HJ, Dypvik H (1992) On the atypical petroleum-generating characteristics of alginite in the Cambrian Alum Shale. In: Schidlowski M, Golubic S, Kimberley M, McKirdy DM, Trudinger PA (eds) Early organic evolution. Implications for mineral and energy resources. Springer, Berlin Heidelberg New York, pp 257–266

Horsfield B, Curry DJ, Bohacs K, Littke R, Rullkötter J, Schenk HJ, Radke M, Schaefer RG, Caroll AR, Isaksen G, Witte EG (1994) Organic geochemistry of freshwater and alkaline lacustrine sediments in the Green River Formation of the Washakie basin, Wyoming. Org Geochem 22: 415–440

Huc AY (1977) Contribution de la geochimie organique a une esquisse palaeoécologique des schistes bitumineux de Toarcien de l'est du bassin de Paris. Étude de la matière organique insoluble (kérogène). Rev Inst Fr Pétr 32: 703–718

Huc AY (1988a) Aspects of depositional processes of organic matter in sedimentary basins. In: Mattavelli L, Novelli L (eds) Advances in organic geochemistry 1987. Org Geochem 13: 263–272

Huc AY (1988b) Sedimentology of organic matter. In: Frimmel FH, Christman RF (eds) Humic substances and their role in the environment. John Wiley and Sons, Chichester, pp 215–243

Huc AY, Durand B, Roucachet J, Vandenbroucke M, Pittion JC (1986) Comparison of three series of organic matter of continental origin. In: Leythaeuser D, Rullkötter J (eds) Advances in organic geochemistry 1985. Org Geochem 10: 65–73

Huff KF (1979) Frontiers of world exploration. Oil Gas J 76: 214

Hvoslef S, Larter SR, Leythaeuser D (1988) Aspects of generation and migration of hydrocarbons from coal-bearing strata of the Hitra Formation, Haltenbanken area, offshore

Norway. In: Mattavelli L, Novelli L (eds) Advances in organic geochemistry 1987. Org Geochem 13: 525–536

Ibach LE (1982) Relationships between sedimentation rate and total organic carbon content in ancient marine sediments. Am Assoc Petrol Geol Bull 66: 170–188

Jankowski B (1981) Die Geschichte der Sedimentation in Nördlinger Ries und Randecker Maar. Boch Geol Geotech Arb 6: 1–315

Jankowski B, Littke R (1986) Das organische Material der Ölschiefer von Messel. Geowiss Unserer Zeit 4: 73–80

Jendrzejewski L (1995) Organische Geochemie der höheren Unterkreide Nordwestdeutschlands: Ablagerungsmilieu und Zyklik. Ber Forschungszentrum Jülich 3134, 211pp, ISSN 0944–2952

Jenkyns HC (1985) The early Toarcian and Cenomanian–Turonian anoxic events in Europe – comparisons and contrasts. Geol Rundsch 74: 505–518

Jones RW (1983) Organic matter characteristics near the shelf-slope boundary. SEPM Spec Publ 33: 391–405

Kenig F, Huc AY (1990) Incorporation of sulfur into recent organic matter in a carbonate environment (Abu Dhabi, United Arab Emirates). In: Orr WL, White CM (eds) Geochemistry of sulfur in fossil fuels. ACS Symp Ser 429: 170–185

Kenig F, Huc AY, Purser BH, Oudin J-L (1990) Sedimentation, distribution and diagenesis of organic matter in a recent carbonate environment, Abu Dhabi, UAE. In: Durand B, Béhar F (eds) Advances in organic geochemistry 1989. Org Geochem 16: 735–747

Kennicutt MC II, De Freitas DA, Joyce JE, Brooks JM (1986) Nonvolatile organic matter in sediments from sites 614 to 623. Deep Sea Drilling Project Leg 96. In: Bouma AH, Coleman JM, Meyer AW et al. (eds) Init Rep DSDP 96, pp 747–756

Khavari-Khorasani G (1987) Oil-prone coals of the Walloon coal measures, Surat Basin/Australia. In: Scott AL (ed) Coal and coal-bearing strata: recent advances. Geol Soc Spec Publ 32: 303–310

Klemme HD (1980) Petroleum basins – classifications and characteristics. J Petrol Geol 3: 187–207

Koblents-Mishke OI (1977) Primary production. In: Vinogradow ME (ed) Oceanology biology of the ocean, vol 1. Nauka, Moscow, p 62 (in Russian)

Kruijs E, Barron EJ (1990) Climate model prediction of paleoproductivity and potential source-rock distribution. In: Huc AY (ed) Deposition of organic facies. Am Assoc Petrol Geol Stud Geol 30: 195–216

Kulke H (1994) Eastern China. In: Kulke H (ed) Regional petroleum geology of the world, part I. Europe and Asia. Gebrüder Bornträger, Stuttgart, pp 819–868

Küspert W (1982) Environmental changes during oil shale deposition as deduced from stable isotope ratios. In: Einsele G, Seilacher A (eds) Cyclic and event stratification. Springer, Berlin Heidelberg New York, pp 482–501

Lallier-Vergès E, Bertrand P, Huc AY, Bückel D, Tremblay P (1993) Control of the productivity and the sulphate reduction on the fossilisation of organic matter in marine anoxic sediments: Kimmeridgian shales from Dorset Formation (UK). Mar Petrol Geol 10: 600–605

Leventhal JS (1983) An interpretation of the carbon and sulphur relationships in the Black Sea sediments as indicators of the environment of deposition. Geochim Cosmochim Acta 47: 133–137

de Leeuw JW, Sinninghe Damsté JS (1990) Organic sulfur compounds and other biomarkers as indicators of palaeosalinity. In: Orr WL, White CM (eds) Geochemistry of sulfur in fossil fuels. ACS Symp Ser 429: 417–443

Leythaeuser D, Schaefer RG, Radke M (1988) Geochemical effects of primary migration of petroleum in Kimmeridge source rocks from Brae field area, North Sea. I. Gross composition of C_{15+} -soluble organic matter and molecular composition of C_{15+} -saturated hydrocarbons. Geochim Cosmochim Acta 52: 701–713

Leythaeuser D, Borromeo O, Mosca F, di Primio R, Radke M, Schaefer RG (1995) Pressure solution in carbonate source rocks and its control on petroleum generation and migration. Mar Petrol Geol 12: 717–733

Littke R (1985) Flözaufbau in den Dorstener, Horster und Essener Schichten der Bohrung Wulfener Heide 1 (nördliches Ruhrgebiet). Fortschr Geol Rheinl Westfalen 33: 129–159

Littke R (1987) Petrology and genesis of Upper Carboniferous coal seams from the Ruhr region, western Germany. Int J Coal Geol 7: 147–184
Littke R (1993) Deposition, diagenesis, and weathering of organic matter-rich sediments. Springer, Berlin Heidelberg New York, 216 pp
Littke R, ten Haven HL (1989) Palaeoecologic trends and petroleum potential of Upper Carboniferous coal seams of western Germany as revealed by their petrographic and organic geochemical characteristics. Int J Coal Geol 13: 529–574
Littke R, Leythaeuser D (1993) Migration of oil and gas in coals. In: Law BE, Rice DD (eds) Hydrocarbons from coal. Am Assoc Petrol Geol Stud Geol 38: 219–236
Littke R, Rullkötter J (1987) Mikroskopische und makroskopische Unterschiede zwischen Profilen unreifen und reifen Posidonienschiefers aus der Hilsmulde. Facies 17: 171–180
Littke R, Sachsenhofer RF (1994) Organic petrology of deep sea sediments: a compilation of results from the Ocean Drilling Program and the Deep Sea Drilling Project. Energy and Fuels, 8: 1498–1512
Littke R, Welte DH (1992) Hydrocarbon source rocks. In: Brown GC, Hawkesworth, CJ, Wilson C (eds) Understanding the earth. Cambridge University Press, Cambridge, pp 364–374
Littke R, Baker DR, Leythaeuser D (1988) Microscopic and sedimentologic evidence for the generation and migration of hydrocarbons in Toarcian source rocks of different maturities. In: Mattavelli L, Novelli L (eds) Advances in organic geochemistry 1987. Org Geochem 13: 549–559
Littke R, Horsfield B, Leythaeuser D (1989) Hydrocarbon distribution in coals and dispersed organic matter of different maceral composition and maturities. Geol Rundsch 78: 391–410
Littke R, Rotzal H, Leythaeuser D, Baker DR (1991a) Lower Toarcian Posidonia Shale in southern Germany (Schwäbische Alb) - organic facies, depositional environment, and maturity. Erdöl Kohle, Erdgas, Petrochem/Hydrocarbon Technol 44: 407–414
Littke R, Baker DR, Leythaeuser D, Rullkötter J (1991b) Keys to the depositional history of the Posidonia Shale (Toarcian) in the Hils syncline, northern Germany. In: Tyson RV, Pearson TH (eds) Modern and ancient continental shelf anoxia. Geol Soc Spec Publ 58: 311–334
Littke R, Rullkötter J, Schaefer RG (1991c) Organic and carbonate carbon accumulation on Broken Ridge and Ninetyeast Ridge, central Indian Ocean. In: Weissel J, Peirce J et al. (eds) Proc of the Ocean Drilling Program, Sci Results 121, College Station, Texas (Ocean Drilling Program), pp 467–487
Littke R, Krooss BM, Idiz E, Frielingsdorf J (1995) Molecular nitrogen in natural gas accumulations: generation from sedimentary organic matter at high temperatures. Am Assoc Petrol Geol Bull 79: 410–430
Lopatin NV, Cramer B, Littke R, Poelchau HS, Schaefer RG, Welte DH (1995) West Siberian gas: geochemical features, hydrocarbon generation modelling and origin of gas from low maturity sediments. In: Grimalt JO, Dorronsoro C (eds) Organic geochemistry: developments and applications to energy, climate, environment, and human history. AIGOA, Donostia-San Sebastian, Spain, pp 869–871
Lückge A, Böttcher P, Littke R (1996) Sedimentation und Frühdiagenese von quartären und tertiären Auftriebssedimenten vor Peru: ein Modell für die Bildung von Erdölmuttergesteinen? Zentralbl Geol Paläont, Teil I, 1995, pp 229–247
Lutz M, Kaasschieter JPH, van Wijke DH (1975) Geological factors controlling Rotliegend gas accumulations in the Mid-European basin. Proc 9th World Petroleum Congr, Applied Science Publ II, pp 93–103
Lyons WB, Hines ME, Gaudette HE (1984) Major and minor pore water geochemistry of modern marine sabkhas: the influence of cyanobacterial mats. In: Yehuda C, Castenholz RW, Halvorson HO (eds) Microbial mats: stromatolites. AR Liss Inc, New York, pp 411–423
Marzi R, Rullkötter J (1986) Organic matter accumulation and migrated hydrocarbons in deep sea sediments of the Mississippi fan and adjacent intraslope basins, northern Gulf of Mexico. Mitt Geol-Paläont Inst Univ Hamburg 60: 359–379
Masters JA (ed) (1984) Elmworth, case study of a deep basin gas field. Am Assoc Petrol Geol Mem 38: 316 pp
Mello MR, Telnaes N, Gaglionone PC, Chicarelli MI, Brassell SC, Maxwell JR (1988) Organic geochemical characterisation of depositional palaeoenvironments of source rocks and oils

in Brazilian margin basins. In: Mattavelli L, Novelli L (eds) Advances in organic geochemistry 1987. Org Geochem 13: 31–46

Miller RG (1989) Prediction of ancient coastal upwelling and related source rocks from palaeo-atmospheric pressure maps. Mar Petrol Geol 6: 277–283

Mix AC (1989) Pleistocene paleoproductivity: evidence from organic carbon and forminiferal species. In: Berger WH, Smetacek VS, Wefer G (eds) Productivity of the ocean: present and past. Dahlem Worksh Rep, Life Sci Res Rep 44. John Wiley and Sons, Chichester, pp 313–340

Moldowan JM, Sundararaman P, Schoell M (1986) Sensitivity of biomarker properties to depositional environment and/or source input in the Lower Toarcian of SW-Germany. In: Leythaeuser D, Rullkötter J (eds) Advances in organic geochemistry 1985. Org Geochem 10: 915–926

Morris KA (1980) Comparison of major sequences of organic-rich mud deposits in the British Jurassic. J Geol Soc Lond 137: 157–170

Müller G, Blaschke R (1969) Zur Entstehung des Posidonienschiefers (Lias epsilon). Naturwissenschaften 12: 635–636

Müller PJ, Suess E (1979) Productivity, sedimentation rate and sedimentary organic matter in the oceans-organic carbon preservation. Deep-Sea Res 27A,222: 1347–1362

O'Brian NR (1990) Significance of lamination in Toarcian (Lower Jurassic) shales from Yorkshire, Great Britain. Sediment Geol 67: 25–34

Pedersen TF, Calvert SE (1990) Anoxia versus productivity: what controls the formation of organic-carbon-rich sediments and sedimentary rocks? Am Assoc Petrol Geol Bull 74: 454–466

Pedersen TF, Calvert SE (1991) Anoxia vs productivity: what controls the formation of organic-carbon-rich sediments and sedimentary rocks?: reply. Am Assoc Petrol Geol Bull 75: 500–501

Peirce J, Weissel J et al. (eds) (1989) Proceedings of the Ocean Drilling Program, Initial Reports, 121. College Station, Texas (Ocean Drilling Program), 1015 pp

Pelet R (1983) Preservation and alteration of present-day sedimentary organic matter. In: Bjøroy M (ed) Advances in organic geochemistry. John Wiley and Sons, Chichester, pp 241–250

Peters KE, Kontorovich AEh, Huizinga BJ, Molodowan JM, Lee CY (1994) Multiple oil families in the West Siberian Basin. Am Assoc Petrol Geol Bull 78: 893–909

Piasecki S, Stemmerik L (1991) Late Permian anoxia in central East Greenland. In: Tyson RV, Pearson TH (eds) Modern and ancient continental shelf anoxia. Geol Soc Spec Publ 58: 275–290

Platt T, Subba Rao DV (1975) Primary production of marine microphytes. Photosynthesis and productivity in different environments. In: International Biological Programme, vol 3. Cambridge University Press, Cambridge, pp 249–279

Prell WL, Niitsuma N et al. (eds) (1989) Proceedings of the Ocean Drilling Program, Initial Reports 117. College Station, Texas (Ocean Drilling Program), 1236 pp

Ramanampisoa L, Radke M, Schaefer RG, Littke R, Rullkötter J, Horsfield B (1990) Organic-geochemical characterisation of sediments from the Sakoa coalfield, Madagascar. In: Durand B, Béhar F (eds) Advances in organic geochemistry 1989. Org Geochem 16: 235–243

Reimers CE, Suess E (1983) Spatial and temporal patterns of organic matter accumulation on the Peru continental margin. In: Thiede J, Suess E (eds) Coastal upwelling. Part B: Sedimentary records of ancient coastal upwelling. Plenum Press, New York, pp 311–346

Rice DD, Clayton JL, Pawlewicz MJ (1989) Characterization of coal-derived hydrocarbons and source-rock potential of coal beds, San Juan Basin, New Mexico and Colorado, USA. In: Lyons PC, Alpern B (eds) Coal: classification, coalification, mineralogy, trace-element chemistry, and oil and gas potential. Int J Coal Geol 13: 597–626

Riley GA (1970) Particulate organic matter in seawater. Adv Mar Biol 8: 1–118

Risk MJ, Rhodes EG (1985) From mangroves to petroleum precursors. An example from tropical northeast Australia. Am Assoc Petrol Geol Bull 69: 1230–1240

Romankevitch EA (1984) Geochemistry of organic matter in the ocean. Springer, Berlin Heidelberg New York, 334 pp

Ronov AB (1958) Organic carbon in sedimentary rocks (in relation to presence of petroleum). Transl Geochem 5: 510–536

Rullkötter J, Aizenshtat Z, Spiro B (1984) Biological markers in bitumens and pyrolysates of Upper Cretaceous bituminous chalks from the Ghareb formation (Israel). Geochim Cosmochim Acta 48: 151–157

Rullkötter J, Littke R, Hagedorn-Götz I, Jankowski B (1988) Vorläufige Ergebnisse der organisch-geochemischen und organisch-petrographischen Untersuchungen an Kernproben des Messeler Ölschiefers. In: Franzen JL, Michaelis W (eds) Der eozäne Messelsee - Eocene Lake Messel. Cour Forsch-Inst Senckenberg 107: 37–52

Rullkötter J, Littke R, Schaefer RG (1990) Characterization of organic matter in sulfur-rich lacustrine sediments of Miocene age (Nördlinger Ries, southern Germany). In: Orr WL, White CM (eds) Geochemistry of sulfur in fossil fuels. Am Chem Soc Symp Ser 429: 149–169

Sarnthein M, Winn K, Zahn R (1987) Paleoproductivity of oceanic upwelling and the effect of atmospheric CO_2 and climatic change during deglaciation times. In: Berger WH, Labeyrie LD (eds) Abrupt climatic change. Reidel, Dordrecht, pp 311–337

Sarnthein M, Winn K, Duplessy JC, Fontugne MR (1988) Global variations of surface water productivity in low- and mid-latitudes: influence on CO_2 reservoirs of the deep ocean and atmosphere during the last 21,000 years. Paleoceanography 3: 361–399

Sassen R, Moore CH, Meendsen FC (1987) Distribution of hydrocarbon source potential in the Jurassic Smackover formation. Org Geochem 11: 379–383

Savrda CE, Bottjer DJ (1991) Oxygen-related biofacies in marine strata: an overview and update. In: Tyson RV, Pearson TH (eds) Modern and ancient continental shelf anoxia. Geol Soc Spec Publ 58: 201–220

Scheidt G (1988) Ausbildung und Verteilung des dispersen organischen Materials im Ruhrkarbon. Boch Geol Geotechn Arb 28: 210 pp

Scheidt G, Littke R (1989) Comparative organic petrology of interlayered sandstones, siltstones, mudstones and coals in the Upper Carboniferous Ruhr basin, northwest Germany, and their thermal history and methane generation. Geol Rundsch 78: 375–390

Scheihing MH, Pfefferkorn HW (1984) The taphonomy of landplants in the Orinoco delta: a model for the incorporation of plant parts in clastic sediments of Late Carboniferous age of Euramerica. Rev Palaeobot Palynol 41: 205–240

Schlanger SO, Jenkyns HC (1976) Cretaceous oceanic anoxic events: causes and consequences. Geol Mijnbouw 55: 179–184

Schlanger SO, Arthur MA, Jenkyns HC, Scholle PA (1987) The Cenomanian-Turonian oceanic anoxic events. I. Stratigraphy and distribution of organic carbon-rich beds and the marine $\delta^{13}C$ excursion. In: Brooks J, Fleet AJ (eds) Marine petroleum source rocks. Geol Soc Spec Publ 26: 371–400

Shanmugam G (1985) Significance of coniferous rain forests and related organic matter in generating commercial quantities of oil, Gippsland Basin, Australia. Am Assoc Petrol Geol Bull 69: 1241–1254

Shibaoka M, Smyth M (1975) Coal petrology and the formation of coal seams in some Australian sedimentary basins. Econ Geol 70: 1463–1473

Shimkus KM, Trimonis ES (1974) Modern sedimentation in Black Sea. In: Degens ET, Ross DA (eds) The Black Sea - geology, chemistry, and biology. Am Assoc Petrol Geol Mem 20: 249–278

Smith AHV (1962) The paleocology of Carboniferous peats based on the microspores and petrography of bituminous coals. Proc Yorkshire Geol Soc 33: 423–478

Smyth M (1989) Organic petrology and clastic depositional environments with special reference to Australian coal basins. Int J Coal Geol 17: 635–656

Spaulding S (1991) Neogene nannofossil biostratigraphy of sites 723 through 730, Oman continental margin, northwestern Arabian Sea. In: Prell WL, Niitsuma N et al. (eds) Proceedings of the Ocean Drilling Program, Sci Results, 117, pp 5–36

Spiro B, Dinur D, Aizenshtat Z (1983) Evaluation of source, environment of deposition and diagenesis of some Israeli "Oil Shales", n-alkanes, fatty acids, tetrapyrroles and kerogen. Chem Geol 39: 189–214

Stach E (1982) The macerals of coal. In: Stach E, Mackowsky MTh, Teichmüller M et al. (eds) Stach's textbook of coal petrology. Gebrüder Bornträger, Berlin, pp 87–139

Stach E, Mackowsky MTh, Teichmüller M, Taylor GH, Chandra D, Teichmüller R (1982) Stach's textbook of coal petrology. Gebrüder Bornträger, Stuttgart, 535 pp

Stasiuk LD (1994) Oil-prone alginite macerales from organic-rich Mesozoic and Palaeozoic strata, Saskatchewan, Canada. Mar Petrol Geol 11: 208–218
Stein R (1986a) Surface-water paleo-productivity as inferred from sediments deposited in oxic and anoxic deep-water environments of the Mesozoic Atlantic Ocean. Mitt Geol-Paläont Inst Univ Hamburg 60: 55–70
Stein R (1986b) Organic carbon and sedimentation rate – further evidence for anoxic deep-water conditions in the Cenomanian/Turonian Atlantic Ocean. Mar Geol 72: 199–209
Stein R (1990) Organic carbon content/sedimentation rate relationship and its paleoenvironmental significance for marine sediments. Geo-Mar Lett 10: 37–40
Stein R (1991) Accumulation of organic carbon in marine sediments. Lect Notes Earth Sci 34: 1–217
Stein R, Littke R (1990) Organic-carbon-rich sediments and palaeoenvironment: results from Baffin Bay (ODP-Leg 105) and the upwelling area off northwest Africa (ODP-Leg 108). In: Huc AY (ed) Deposition of organic facies. Am Assoc Petrol Geol Stud Geol 30: 41–56
Stein R, Rullkötter J, Welte DH (1986) Accumulation of organic-carbon-rich sediments in the Late Jurassic and Cretaceous Atlantic Ocean – a synthesis. Chem Geol 56: 1–32
Stein R, ten Haven HL, Littke R, Rullkötter J, Welte DH (1989) Accumulation of marine and terrigenous organic carbon at upwelling site 658 and nonupwelling sites 657 and 659: implications for the reconstruction of paleoenvironments in the eastern subtropical Atlantic through Late Cenozoic times. In: Ruddiman W, Sarnthein M et al. (eds) Proceedings of the Ocean Drilling Program, Sci Results 108, pp 361–385
Styan WB, Bustin RM (1983) Petrography of some Fraser river delta peat deposits: coal maceral and microlithotype precursors in temperate-climate peats. Int J Coal Geol 2: 321–370
Suess E, Thiede J (eds) (1983) Coastal upwelling. Part A: Responses of the sedimentary regime to present coastal upwelling. Plenum Press, New York, 610 pp
Suess E, von Huene R et al. (eds) (1988) Proceedings of the Ocean Drilling Program, Initial Rep 112, 1015 pp
Summerhayes CP (1981) Organic facies of middle Cretaceous black shales in deep north Atlantic. Am Assoc Petrol Geol Bull 65: 2364–2380
Summerhayes CP (1983) Sedimentation of organic matter in upwelling regimes. In: Thiede J, Suess E (eds) Coastal upwelling. Part B: Sedimentary records of ancient coastal upwelling. Plenum Press, New York, pp 29–72
Summerhayes CP, Prell WL, Emeis KC (eds) (1992) Upwelling systems: evolution since the early Miocene. Geol Soc Spec Publ 64: 519 pp
Swanson VE, Palacas JG (1965) Humate in coastal sands of northwest Florida. US Geol Surv Bull: 1214-B
Tannenbaum E, Aizenshtat Z (1985) Formation of immature asphalt from organic-rich carbonate rocks. I Geochemical correlation. Org Geochem 8: 181–192
Teichmüller M (1982) Origin of the petrographic constituents of coals. In: Stach E, Mackowsky MTh, Teichmüller M et al. (eds) Stach's textbook of coal petrology. Gebrüder Bornträger, Stuttgart, pp 219–294
Teichmüller M, Ottenjann K (1977) Liptinite und lipoide Stoffe in einem Erdölmuttergestein. Erdöl Kohle 30: 387–398
Thiede J, Suess E (eds) (1983) Coastal upwelling. Part B: Sedimentary records of ancient coastal upwelling. Plenum Press, New York, 610 pp
Thierstein HR (1989) Inventory of paleoproductivity records. The mid-Cretaceous enigma. In: Berger WH, Smetacek VS, Wefer G (eds) Productivity of the ocean: present and past. Dahlem Worksh Rep, Life Sci Res Rep 44. John Wiley and Sons, Chichester, pp 355–375
Thomas BM (1982) Land plant source rocks for oil and their significance in Australian basins. Aust Petrol Expl Assoc J 22: 166–178
Thompson S, Cooper BS, Morley RJ, Barnard P (1985) Oil-generating coals. In: Thomas BM, Doré AG, Eggen SS et al. (eds) Petroleum geochemistry in exploration of the Norwegian Shelf. Graham and Trotman, London, pp 59–73
Thunell RC, Williams DF, Belyea PR (1984) Anoxic events in the Mediterranean sea in relation to the evolution of late Neogene climates. Mar Geol 59: 105–134

Tissot BP, Welte DH (1984) Petroleum formation and occurrence, 2nd edn. Springer, Berlin Heidelberg New York, 699 pp
Tourtelot HA (1979) Black shale - its deposition and diagenesis. Clays Clay Miner 27: 313–321
Trask PD, Wu CC (1930) Does petroleum form in sediments at time of deposition? Am Assoc Petrol Geol Bull 14: 1451–1463
Tyson RV (1987) The genesis and palynofacies characteristics of marine petroleum source rocks. In: Brooks J, Fleet AJ (eds) Marine petroleum source rocks. Geol Soc Spec Publ 26: 47–67
Tyson RV, Pearson TH (eds) (1991) Modern and ancient continental shelf anoxia. Geol Soc Spec Publ 58: 470 pp
Urlichs M (1977) The Lower Jurassic in southwestern Germany. Stuttg Beitr Naturk Ser B 24: 1–30
van der Zwaan GJ, Jorissen FJ (1991) Biofacial pattern in river-induced shelf anoxia. In: Tyson RV, Pearson TH (eds) Modern and ancient continental shelf anoxia. Geol Soc Spec Publ 58: 65–82
van Krevelen DW (1961) Coal typology - chemistry-physics-constitution. Elsevier, Amsterdam, 513 pp
Vetö I, Hetényi M, Demény A, Hertelendi E (1994) Hydrogen index as reflecting intensity of sulphidic diagenesis in non-bioturbated, shaly sediments. Org Geochem 22: 299–310
von Engelhardt W (1973) Sedimentpetrologie, Teil III: Die Bildung von Sedimenten und Sedimentgesteinen. Schweizerbarth, Stuttgart, 378 pp
Wefer G, Heinze P, Suess E (1990) Stratigraphy and sedimentation rates from oxygen isotope composition, organic carbon content and grain-size distribution at the Peru upwelling region: holes 680B and 686B. In: Suess E, von Huene R et al. (eds) Proceedings of the Ocean Drilling Program, Sci Results 112, pp 355–368
Wehner H, Gerling P, Hiltmann W, Kockel F (1989) Erdöl-Charakteristik und Öl-Muttergestein-Korrelation im Niedersächsischen Becken. Nachr Dtsch Geol Ges 41: 77–78
Weissel J, Pierce J et al. (eds)(1991) Proceedings of the Ocean Drilling Program, Sci Results 121, 990 pp
Welte DH (1979) Organisch-geochemische Untersuchungen zur Bildung von Erdöl-Kohlenwasserstoffen an Gesteinen des mittleren Oberrhein-Grabens. Fortschr Geol Rheinl Westfalen 27: 51–74
Welte DH, Waples DW (1973) Über die Bevorzugung gradzahliger n-Alkane in Sedimentgesteinen. Naturwissenschaften 60: 516
Wenger LM (1987) Variations in organic geochemistry of anoxic-oxic black shale-carbonate sequences in the Pennsylvanian of the Midcontinent, USA. PhD Diss, Rice University, Houston, 628 pp
Wenger LM, Baker DR (1986) Variations in organic geochemistry of anoxic-oxic black shale-carbonate sequences in the Pennsylvanian of the Midcontinent, USA. In: Leythaeuser D, Rullkötter J (eds) Advances in organic geochemistry 1985. Org Geochem 10: 85–95
Zimmerman HB, Boersma A, McCoy FW (1987) Carbonaceous sediments and palaeoenvironment of the Cretaceous South Atlantic Ocean. In: Brooks J, Fleet AJ (eds) Marine petroleum source rocks. Geol Soc Spec Publ 26: 271–286

Chapter 6
The Bulk Composition of First-Formed Petroleum in Source Rocks

Chapter 6: Overview and Insights

On its way from a source rock to a trap petroleum is subjected to many fractionation and alteration phenomena. Hence a comparison is essential between the original product, first formed in a source kitchen, and the kind of petroleum ultimately emplaced and surviving in a reservoir. This chapter reviews two approaches to performing this comparison, the analysis of rock and oil samples from selected geological case histories, and it describes in detail recent laboratory experiments focusing on open- and closed-system pyrolysis and accompanying quantitative measurements on a molecular level.

The molecular and structural relationship between kerogen and newly formed petroleum is still only poorly understood. To help overcome this, a bulk geochemical correlation has been established between petroleums and kerogens in terms of its chemical-structural entities, making extensive use of pyrolysis. Along these lines a kerogen decomposition model has been outlined (K→K′+B; B→B′+O) whereby the rate-controlling step in pyrolysis is the bitumen to oil reaction (B→B′+O), whereas for geological heating conditions, the rate-controlling step would be the kerogen to bitumen reaction (K→K′+B). Alternatively, the reactive low molecular weight breakdown entities of kerogen may form a polar bitumen by condensation reactions and aggregation during the high rates of product generation in pyrolysis, while there are far fewer chances for condensation reactions and aggregation during slow maturation in sedimentary basins and with slow product generation rates and low concentrations of reaction partners. From this concept it is understandable why the chemical structural pattern of a given kerogen is very similar in the resulting pyrolysis product and the naturally formed petroleum. The types of moieties seen in pyrolysates and petroleum are qualitatively similar. However, it is not as easy to predict the quantitative relationships among the various molecular weight fractions of a petroleum, i.e. the gas-oil ratio and the heavy end with the asphaltene fraction. A way out of this dilemma may be detailed analysis of the resolved components in gas chromatograms of the kerogen pyrolysates. From this information gas-oil ratios can be calculated.

Recognition of organofacies and establishing better links between the source kerogen and a specific petroleum and vice versa are other topics of interest. Here preliminary findings suggest that macromolecular organic matter in petroleum accumulation, seeps, and stains might be a source of data to determine kerogen type/source facies and gross compositional characteristics of first-formed petroleum.

The Bulk Composition of First-Formed Petroleum in Source Rocks

B. Horsfield[1]

6.1 Introduction

Petroleum consists of an exceedingly complex mixture of hydrocarbons and non-hydrocarbons, extending from methane to macromolecular aggregates. The relative proportions of these components are quite variable and depend initially on the nature of the kerogen in the parent source rock and its level of maturity at the time of expulsion, and subsequently upon the pressure and temperature conditions of the source-carrier-reservoir system during expulsion, migration and accumulation. Under subsurface reservoir conditions the petroleum may be in a single vapour phase, in which case it is termed a gas accumulation, a single liquid phase, as exemplified by crude oils which are undersaturated with respect to gas, or a two-phase system comprised of a gas cap in equilibrium with an underlying crude oil accumulation. When produced at the surface, the petroleum changes from its sub-surface state as gaseous and liquid components segregate. For instance, the C_1–C_4 components that are dissolved in undersaturated petroleums under sub-surface conditions exsolve to give gas, while the major remaining liquid portion shrinks in volume and is termed crude oil. Similarly, as natural gases are produced dissolved C_{5+} components condense out as liquids or solids, giving rise to "condensate".

Most natural gases consist of methane and subordinate amounts of ethane, propane and butanes. Carbon dioxide and nitrogen are also often present and can actually exceed hydrocarbon gas contents where contact metamorphism of carbonates or very high levels of coalification are regionally important. In carbonate reservoirs hydrogen sulphide may also occur as a result of thermochemical sulphate reduction reactions involving mineral sulphates and hydrocarbon gases. Condensates and crude oils are highly complex, consisting of aliphatic, alicyclic and aromatic hydrocarbons, and heteroelement-containing (N,S,O) compounds. The distribution of these molecular types varies according to average molecular size, as illustrated by the example shown in Fig. 6.1 (after Hunt 1979).

[1]Institut für Erdöl und Organische Geochemie (ICG-4), Forschungszentrum Jülich GmbH, 52425 Jülich, Germany

Welte et al. (eds)
Petroleum and Basin Evolution
© Springer-Verlag Berlin Heidelberg 1997

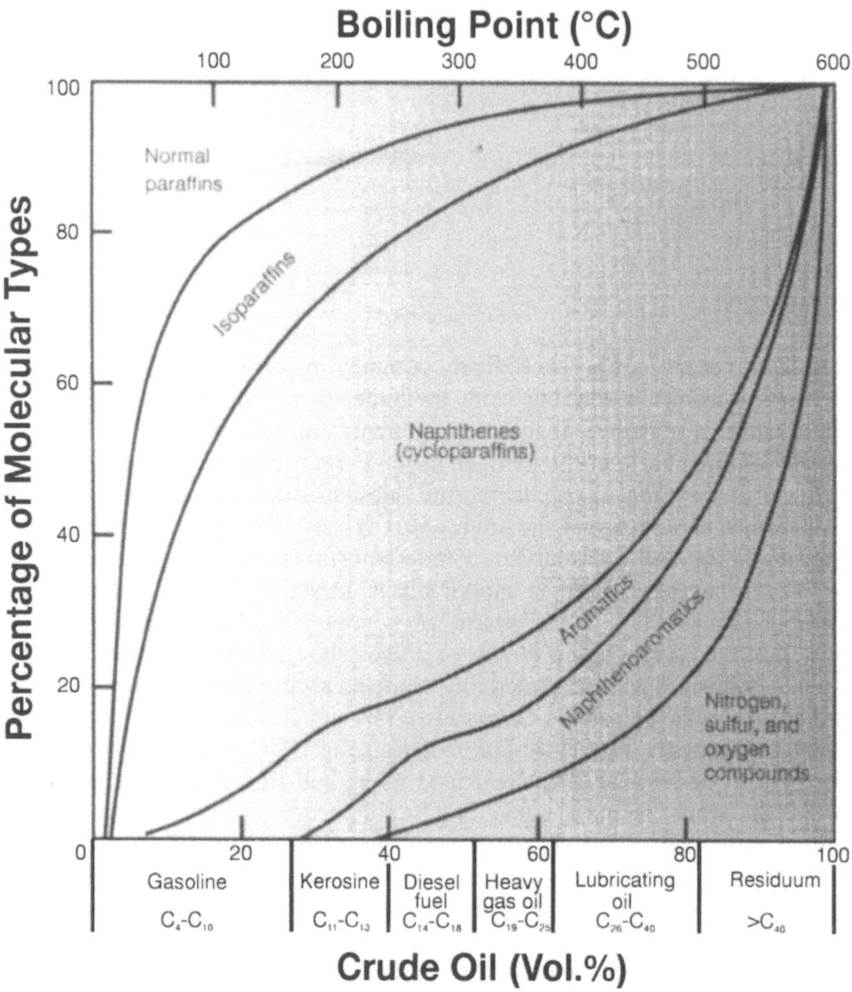

Fig. 6.1. Chemical composition of a crude oil in terms of molecular type. (Hunt 1979). From: PETROLEUM GEOCHEMISTRY AND GEOLOGY by Hunt. Copyright © by W. H. Freeman and Company. Used with permission.

Until the middle 1980s and early 1990s, predicting the occurrence of gas versus crude oil using organic geochemistry relied only minimally on the actual composition of petroleum. For exploration purposes, especially in virgin territory, crude oil was considered mainly as a single entity and gas as another, and the presence of one or the other was predicted using an overriding emphasis on measured or anticipated source rock quality and maturity. These concepts, established 30 years ago and more (Forsman and Hunt 1958; Philippi 1965; Landes 1967; McIver 1967; Tissot et al. 1974), were that hydrogen-rich kerogen disproportionates when mature to form liquid-rich, gas-poor petroleums and a carbon-enriched residue, whereas mature hydrogen-poor kerogens, on the other hand, have a higher probability of generating gas. This is because hydrogen-rich kerogens largely consist of algal-derived aliphatic cell membranes and lipid components (Cane and Albion 1973; Philp and Calvin 1976; Hutton et al. 1980; Largeau et al. 1984; Tegelaar et al. 1989a) whereas originally hydrogen-poor kerogens often contain high proportions of altered lignocellulosic materials whose aliphatic constituents consist of alicyclic moieties and short alkyl chains (Cooper and Murchison 1969; Mycke and Michaelis 1986). At higher levels of thermal stress long alkyl chains in unmigrated bitumen are cracked to short ones, and naphthenoaromatic hydrocarbons aromatise and condense, with the result that overmature sediments are characterised by an increasing tendency to form gas. Predicting petroleum composition in terms of oil versus gas therefore relied on the typing of kerogens (mainly according to organic hydrogen availability using elemental H/C ratios and hydrogen indices), defining where these kerogens occurred in the sedimentary record (organic facies concepts), establishing maturity levels using parameters based on rock bitumen and kerogen, and in later years deducing maturation histories.

Wherever petroleum composition was considered in geochemistry, usually later in the exploration and development cycle, emphasis focussed on trace components in the high boiling range fractions of crude oils, condensates and rock bitumens rather than on its major constituents. This was because the high information content contained in biological marker molecules of source rocks and crude oils could be readily applied to another set of major research thrusts of the time, these being to determine marine versus non-marine source contributions, compile indices of maturity and infer migration pathways based on oil-oil and oil-source correlations (see Mackenzie 1984; Radke 1987; Peters and Moldowan 1993). These research avenues still exist today, and are important in helping identify petroleum systems worldwide (Magoon and Dow 1994), in providing molecular maturity parameters for basin modelling applications (Doré and et al. 1993; Radke et al. this book) and for determining the filling histories of petroleum reservoirs using detailed core analyses (Larter et al. 1990; Hall et al. 1994).

Up until about 10 years ago the major constituents of petroleum were largely ignored because they were not considered important for achieving exploration-orientated research goals. However, the following observations now strongly suggest that bulk petroleum compositions are important:

1. The first concerns predicting the petroleum generating potential of source rocks. It is now clear that the total organic hydrogen budget is inadequate for assessing oil or gas generating potential. This has been demonstrated by the liquid hydrocarbon-prone nature of type III source rocks in southeast Asia and Australia (Durand and Paratte 1983; Smith and Cook 1984; Thompson et al. 1985; Horsfield et al. 1988) and proposed for the gas-condensate-prone type II kerogen from the Alum Shale of Scandinavia (Horsfield et al. 1992). Predicting the actual composition of petroleums first generated in source rocks must therefore extend beyond elemental typing schemes to molecular considerations, as further discussed below.

2. The second observation is that the occurrence of gas versus oil in petroleum reservoirs is strongly dependent on secondary migration phenomena, and that source rock influences can be vastly overshadowed. This is based on the application of reservoir engineering principles to carrier-reservoir systems (Schowalter 1979; Heum et al. 1986; England et al. 1987; England and Mackenzie 1989). Petroleum density and viscosity are controlled directly by the relative proportions of low-density, low-viscosity consitituents in the C_1–C_4 range, on the one hand, and high density, high viscosity constituents, especially colloidal components such as asphaltenes, on the other. Thus a fall in confining pressure below the bubble point pressure during secondary migration leads to the progressive exsolution of free gas from the petroleum (Fig. 6.2), and a rapid increase in viscosity and shrinkage of the remaining liquid phase with continuing pressure decline (Neumann et al. 1981). High-GOR petroleums generally begin to exsolve a separate gas phase at higher pressures and temperatures than do low-GOR petroleums. Density is also controlled by the gas-oil ratio and the relative abundances of hydrocarbons and non-hydrocarbons in the C_{15+} fraction (Hunt 1979; Connan and Coustau 1987). Additionally, the absolute value of viscosity for the liquid phase, and hence Darcy flow behaviour, is dependent on its composition with asphaltic crude oils generally being more viscous than low-wax paraffinic oils, for example (Neumann et al. 1981). Because viscosity increases with the molar mass for *n*-alkane homologues, high-wax paraffinic petroleum is more viscous than the low-wax paraffinic type. However, this relationship is not always straightforward (Hernandez et al. 1983) because the physical interactions of monomolecular and macromolecular components in highly complex mixtures are not well understood (Mitchell and Speight 1973). It is consequent that the *quantitative* determination of bulk composition for petroleums first-formed in source rocks is a requirement for modelling the migration of petroleum charges from source rock kitchens to traps using a petroleum engineering approach. This is illustrated in Fig. 6.2 by the example of hypothetical petroleum charges with differing original GORs which migrate from a depth of 4500 m to a trap at 2000 m. Using the Prospect Analyzer computer program, with input of the petroleum properties published by England et al. (1987), the temperature profile from Dahl and Augustson (1993) and an average pressure gradient 13 mpa/km, it can be seen that the absolute volumes of gas-saturated oil and oil-saturated gas at the prospect are highly dependent on the original petroleum composition.

3. Related to the above is the finding by Thompson (1988), supported by Dzou and Hughes (1993), that petroleums in any given region can be considered to have one original composition, but that this is strongly modified as a result of evaporative fractionation during reservoir leakage. Original oils exhibit varying degrees of light end loss, whereas the lost material form migratory gas/condensates, some of which can accumulate in shallower traps. The losses are accompanied by fractionation so that the remaining residual oils show changes in the intermolecular ratios termed aromaticity, normality and paraffinicity, their extent being proportional to the degree of light end loss. Further modifications can occur within petroleum accumulations during in-

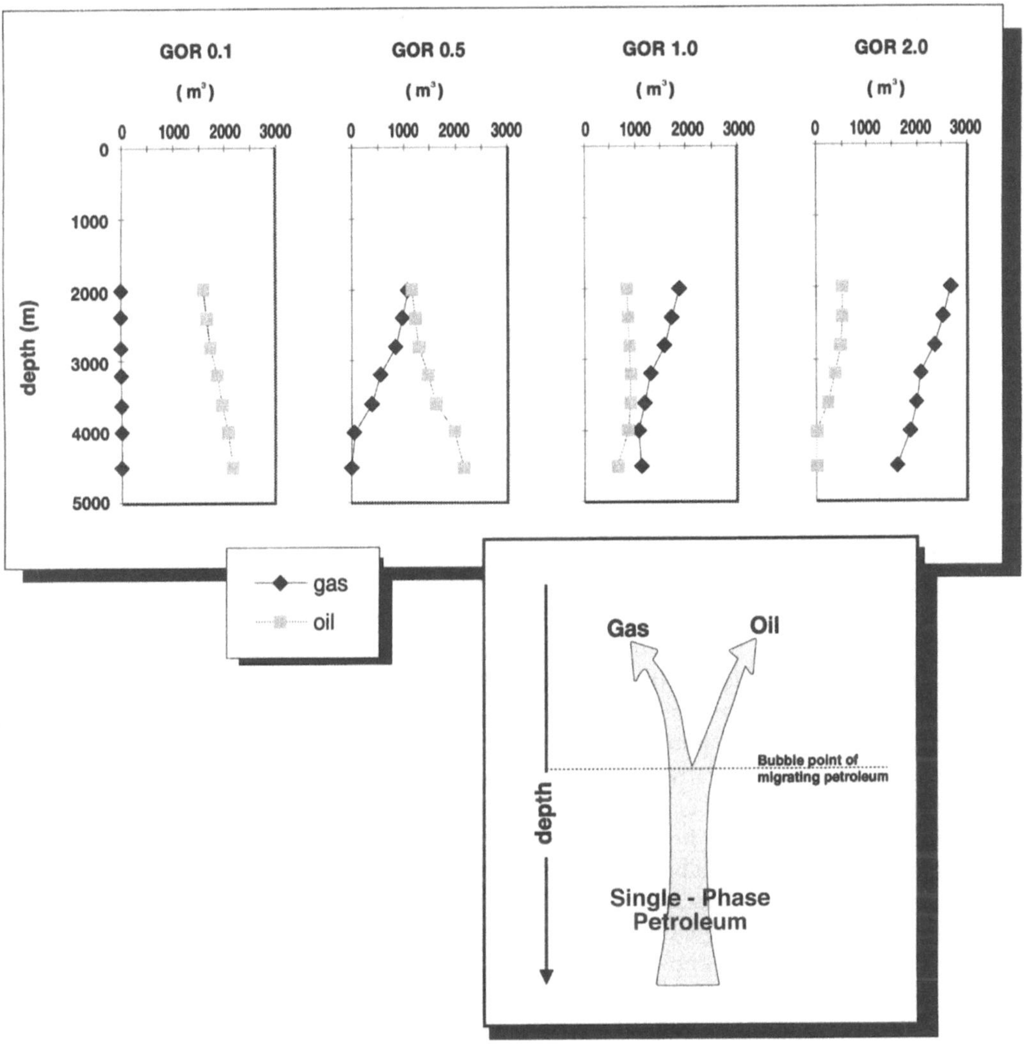

Fig. 6.2. Changes in the volumes of gas and oil phases as they migrate from 4500 m to shallower depths. Scenarios for different starting GORs are shown

version when gas cap expansion displaces oil out of the trap, by preferential leakage of light components through seals, or via microbial biodegradation and water washing (Bailey et al. 1974; Krooss et al. 1992). Knowing the composition of the initially reservoired petroleum is required in order to recognise the full extent of evaporative fractionation (Thompson, personal communication) and then to infer migration pathways and predict the existence of shallower secondary targets.

The first step in unravelling all types of fractionation phenomena is to determine the bulk composition of the petroleum that is first-formed in the source rock because all subsequent processes simply act upon and modify this original composition. There are two basic ways of doing this: by the analysis of rock and oil samples from selected geological study areas and by laboratory experiment. Cuttings gas and extractable hydrocarbon data from source rocks at different maturity levels may give some clues as to the in situ composition of the generated petroleum but sampling and migration losses are regarded as being high, making absolute quantitation difficult (Price 1989). Additionally absolute gas-oil ratios of generated and expelled petroleums are not actually known with any certainly because of the impracticality of sampling deep (ca. 5–8 km) source rock kitchens and in view of the difficulty in distinguishing the respective influences of co-occurring generation and expulsion. It is therefore left mainly to indirect methods such as the laboratory pyrolysis of kerogen, with their high precision but inherent uncertainties as to direct geological applicability, to infer gas versus oil generating potential or define likely compositional attributes of petroleums in source rock kitchens. This chapter reviews both approaches, but concentrates on laboratory experiments which have proved more valuable in recent years.

6.2 The Direct Analysis of First-Formed Petroleum

Determining the bulk composition of first-formed petroleum using direct methods is beset with problems. The first problem is geological, and that is that exploration wells are drilled on structural highs where sediments with significant source rock potential are often only immature or early mature. The deep reaches of generative basins where source rocks are at peak maturity and petroleum generation is actually going on are hardly ever drilled. Where mature source rocks are near the surface because of post-generation uplift, diffusive losses of especially gaseous components must be expected. For example, very little gas was detected in near-surface subcrops of the Posidonia Shale in the Hils Syncline, Germany, despite its broad maturity range (Rullkötter et al. 1988), and coals are known to lose a high proportion of their originally generated gas during basin uplift (Rice 1993). Additionally, and very importantly, present-day concentrations of hydrocarbons and related compounds in source rocks are the result of two processes, namely build-up by generation and loss by expulsion. If expulsion from *de facto* source rocks is very efficient, as examples from northern Europe, the Middle East and Alaska suggest (Cooles et al. 1986; Larter 1988; Rullkötter et al. 1988), non-migrated residues of light or heavy fractions cannot be easily related to the process of generation.

Nonetheless, it is also clear that field data must be used for calibrating and validating geochemical predictions made by laboratory experiment or basin modelling, in which case it becomes necessary to consider the quality and comprehensiveness of these geochemical field data. In the case of C_{15+} yields, allowance must be made for the different types of extraction solvents and procedures that are used as this most strongly affects the yields of non-hydrocarbons and makes direct comparisons between published datasets difficult. C_5–C_{14} compounds make up 50% of reservoired petroleums and probably constitute a significant proportion of the dispersed organic matter in source rocks, yet these abundances are hardly ever measured. C_1–C_4 data from headspace and cuttings analyses are often used only qualitatively, in part because of significant losses during passage to the shale shaker (where cuttings are collected) and because collection procedures are not standardised, resulting in further variable losses. Price (1989) has indicated that gas losses are likely to be very large as a result of the significant drop in pressure as the samples come to the surface from great depth. He concluded further that the important role of gas as a migration agent has gone unnoticed mainly because these sampling problems exist. On a more positive note, Snowdon and McCrossan (1973) showed that once the cuttings were at the surface, gas loss was not significant even up to 3 h after the sample reached the shale shaker, meaning that a relatively large time window exists for the careful selection and collection of samples without any detrimental impact on data quality. Finally, a major hurdle as far as statistically meaningful data are concerned is a paucity of published datasets that include both light and heavy components, and the qualitative way in which headspace and cuttings gas data are reported (e.g. volume gas/unit volume of wet cuttings). For instance, although geochemical studies of the Brae Field area (Mackenzie et al. 1988), Dampier Sub-Basin (Powell 1975) and the Douala Basin (Albrecht et al. 1976) actually penetrated deep generative source rocks, no data on gaseous or light hydrocarbons were presented. Similarly, no heavy hydrocarbon data accompanied the light hydrocarbon data presented by Snowdon and Roy (1975).

While it is difficult to measure the relative abundances of light and heavy hydrocarbons generated in source rocks during catagenesis, some recent results serve to stress that direct measurements of gases and light hydrocarbons can shed some useful light on this matter. Changes in gas-oil ratio were summarised as a function of maturity for the Kimmeridge Clay (Upper Jurassic) by Cornford (1993) based on a dataset that included cuttings gas, light hydrocarbon and C_{15+} extract yields. His results indicated that gas-oil ratios during early maturity (<130 °C) were of the order of 0.3 kg/kg (ca. 1500 scf/bbl), climbing to 0.8 kg/kg (ca. 4000 scf/bbl) at peak maturity (130–140 °C) up to and beyond 2.0 kg/kg (ca. 10000 scf/bbl) where late mature (>140 °C). While these results reflect the combined influences in generation and expulsion, the value for peak maturity closely corresponds to that composition from which reservoired crude oils and condensates in a part of the North Sea were derived, according to trends of gas-oil ratio and condensate-gas ratio (England and Mackenzie 1989; Fig. 6.3). Additionally, a gas-oil ratio of approximately 0.2 kg/kg has been reported for early mature Upper Miocene sediments of the Los Angeles Basin using cuttings gas and thermovaporisation data (Philippi 1975), corroborating that gas-oil ratios in source rocks are initially low. Using thermovaporisation–gas chromatography (GC) alone, Erdmann (1995) showed that the Draupne and Heather Formations (Upper Jurassic) of the Norwegian North Sea contain variably high proportions of light hydrocarbons (C_5–C_{10}), but gas-oil ratios (C_1–C_5/C_{6+}) are uniformly less than 0.1 kg/kg throughout the entire oil window, probably pointing to losses during sample collection. Using the same thermovaporisation-GC technique, Muscio et al. (1994) found that

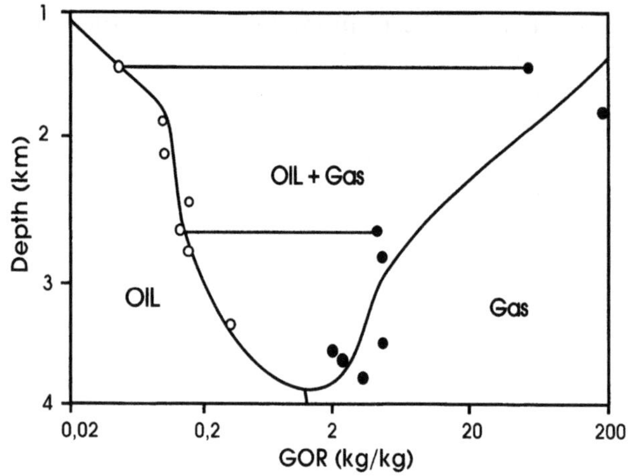

Fig. 6.3. Surface gas-oil ratios of oils and gases from offshore Norway as a function of depth. *Open circles*, oil accumulation; *closed circles*, gas accumulation; *horizontal lines* connect co-existing gas caps and their underlying oil leg. (After England and Mackenzie 1989)

immature through early mature Bakken Shale (Williston Basin, USA) contained both light and heavy aliphatic, alicyclic and aromatic hydrocarbons, whose gas-oil ratio was 0.18 kg/kg, again confirming the above results. Additionally, a strong positive correlation with TOC attested to the indigenous nature of the gas (Fig. 6.4). Interestingly, average gas yields were on the order

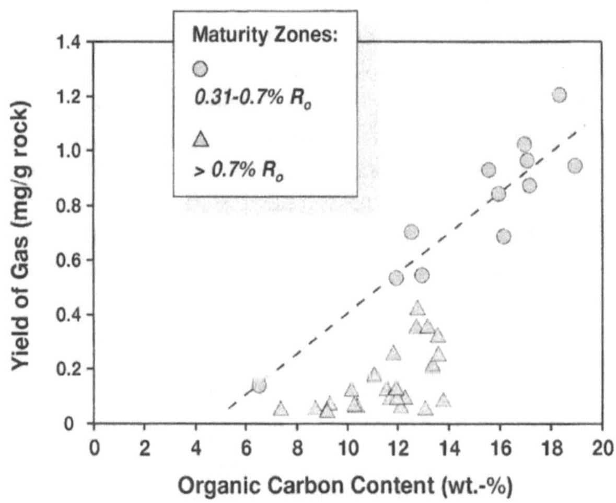

Fig. 6.4. Relationship between organic richness (TOC) and gas yield (from thermovaporisation) for the Bakken Shale. (Muscio et al. 1994). (Reprinted from Muscio et al., Occurrence of thermogenic gas in the immature zone - implications from the Bakken in-source reservoir system, 1994, pp 461-476, with kind permission from Elsevier Science Ltd, The Boulevard, Langford Lane, Kidlington OX5 1GB, UK)

of 6 mg/g TOC, which should amount to about 20% of the the total gas generating capacity according to pyrolysis-GC. The result signified not only that a high proportion of the kerogen's total gas-generating potential had been realised early during maturation, but also that the generated gas had been retained in the rock during recovery. The correlation with TOC indicated that the gas was adsorbed on or absorbed within the kerogen, whose open structure partly consists of aromatic or readily aromatisable moieties. Absorption is feasible according to the results of Young and McIver (1977) who showed that light hydrocarbon retention is directly related to the presence of solid organic matter. The higher GOR values noted for higher maturity levels by Cornford (1993) could not be corroborated in the case of the Bakken Shale because of losses resulting from migration or during sample recovery, possibly in part due to changes in kerogen structure and hence absorptive capacity with increasing maturation (Fig. 6.4).

High C_1–C_{10} contents were detected in the thermovaporisation products of the alginitic Alum Shale of Scandinavia over a wide range of H/C ratios (Horsfield et al. 1992), and gas-oil ratios fell in the range 0.3–0.6 kg/kg. Simulated maturation experiments verified that gases were amongst the main generation products. A common denominator with the Bakken Shale is that this kerogen has a high proportion of thermally labile aromatic and/or alicyclic groups. Difficult to reconcile are unpublished light hydrocarbon data (headspace and cuttings gas) and solvent extract yields (C_{15+} range) for early mature (R_o=0.55–0.70%) Jurassic-Cretaceous source rocks from two wells on the northern slope of Alaska (Fig. 6.5). High values of gas-oil ratio, approximated by C_1 -C_4 and C_{15+} concentrations and represented by two trends (20 and 46 kg/kg), were stratigraphically distinct in both wells, and possibly related to organic matter type or petrophysical properties. The observed correlation of gas and C_{15+} extract yields appears to rule out gas contamination ("trip gas") from the mudstream. The range of values is actually more typical of condensate dissolved in gas than gas dissolved in oil, according to the results of England and Mackenzie (1989) summarised in Fig. 6.3. It is not known whether this composition truly represents an early generation product (see Muscio et al. 1994), or whether the high-GOR components seen in the rock represent the oil-saturated gaseous residuum after preferential expulsion of a low-GOR liquid phase (see Fig. 6.3). Nevertheless, the example serves to illustrate that yields of volatile hydrocarbons may provide a source of important information for studying geochemical processes.

The results discussed above point to a substantial proportion of a kerogen's total gas generating potential being realised early on in catagenesis, but in amounts that are usually subordinate to those of heavier components. Gas-oil ratios are therefore generally low, falling in the range 0.2–0.3 kg/kg, and possibly as low as 0.1 kg/kg or less, though in this case material losses can be suspected too. It is not known how common it is to find occurrences of first-formed petroleums with higher gas-oil ratios, but they undoubtedly exist, not only at high levels of thermal maturation but also at low levels of thermal stress. The roles played by kerogen type – and this revolves around the response of its molecular composition to increasing thermal stress – have yet to

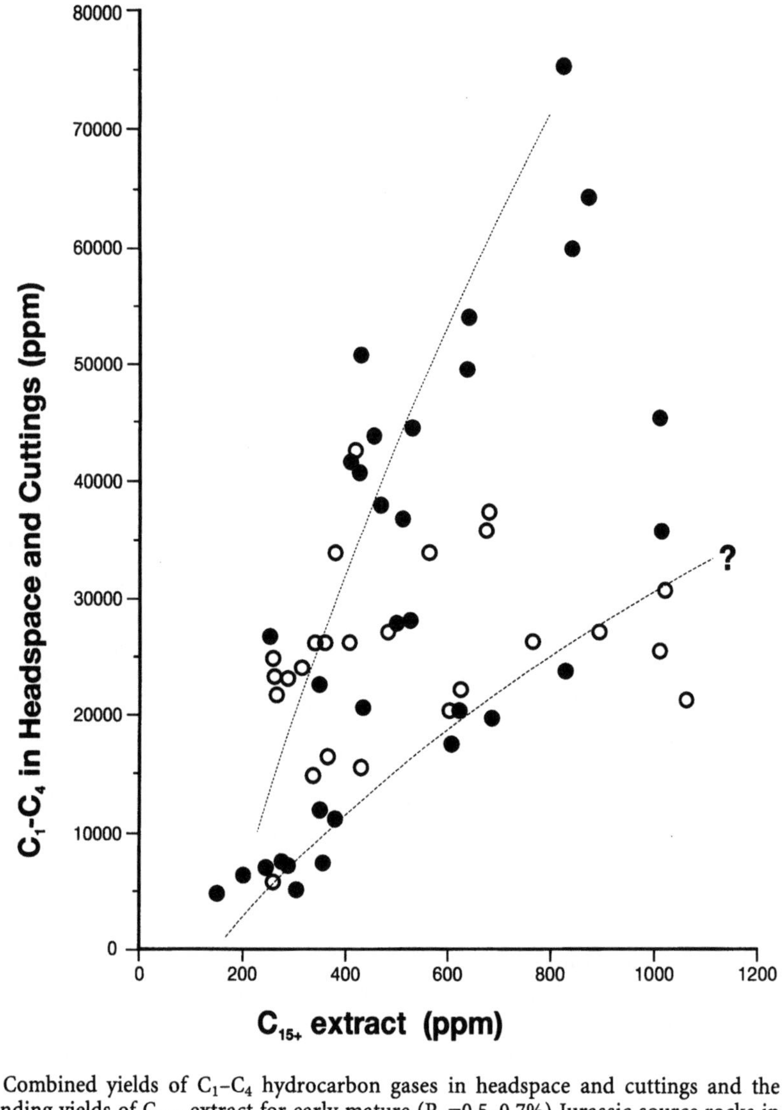

Fig. 6.5. Combined yields of C_1–C_4 hydrocarbon gases in headspace and cuttings and the corresponding yields of C_{15+} extract for early mature (R_o=0.5–0.7%) Jurassic source rocks in two wells (*open and closed symbols*) from Alaska. The steeper trend corresponds to approximately 46 kg/kg and the other to 20 kg/kg

be demonstrated satisfactorily using a statistically meaningful dataset. Nevertheless, and in accordance with the views of Price (1989), the significant generation of gas during early maturation and the role of gas in petroleum expulsion are worthy of further detailed investigation.

6.3 Kerogen Composition

Analysing kerogens is an alternative and indirect way of inferring the composition of petroleum generated in source rocks and is not affected by the losses (expulsion) and gains (staining, contamination) that affect petroleum-like compounds. Kerogen studies can be based on laboratory experiments or regional geochemical considerations. To draw a comparison between their respective compositions data must be gathered that are common to both kerogen and petroleum. However, the molecular relationship between kerogen and petroleum is poorly understood because each is routinely examined using mutually exclusive frames of reference. Crude oil and solvent extract geochemistry has centred on the trace biomarker components that form the basis of oil correlations, molecular maturation parameters and palaeoenvironmental reconstructions, and very little regard has been afforded to bulk constituents. Kerogen geochemistry, on the other hand, has been overwhelmingly dominated by elemental measurements and microscopic distributions, with molecular considerations again often being largely focussed on biomarker moieties. A bulk geochemical correlation in terms of alkyl and aromatic entities in both petroleums and kerogens has been largely ignored. This section first briefly illustrates the deficiency of elemental kerogen typing and then introduces some bulk chemical features that are common to kerogens and petroleums that may be used to bridge the gap.

6.3.1 The Typing of Kerogens by Elemental Composition

"Kerogen typing" simply consists of classifying the bulk indigenous fraction of sedimentary organic matter according to practical and useful criteria. It can be performed in various ways, depending on the application. Microscopy used to feature strongly in this regard, but since the mid-seventies petroleum geochemists have almost exclusively relied upon the element-based (H/C versus O/C) classification scheme which discriminates kerogen types I, II and III (Tissot et al. 1974). The development of the Rock-Eval ensured that these kerogen types could be identified quickly and cheaply using whole rock samples (Espitalié et al. 1977). The effectiveness of any kerogen typing scheme can be judged by how it is used directly in problem-solving, or in this case how pertinent it is to the demands of quantitative analysis and basin modelling. The elemental scheme for kerogen classification discriminates solely according to hydrogen availability, which can be said to equate to genetic potential (total pyrolysate yield per gramme TOC). This scheme is therefore well suited for predicting the masses of petroleum generated in and thence expelled from source rock kitchens. Unfortunately, the explorationist has no means for predicting petroleum compositions (e.g. GOR) or maturation characteristics because the types of petroleum generated by kerogens and the maturation levels needed to bring about generation may be substantially variable within each of the classical kerogen types (Horsfield 1984; Larter 1984; Tegelaar and

Noble 1994). Choosing both kerogen quality parameters and kinetic parameters is therefore problematical where phase and volume and kinetic modelling must be performed without direct access to calibration samples. For this reason a classification of organic matter based on not only genetic potential but also maturation characteristics and kerogen quality is highly desirable for geochemical modelling (Horsfield 1989). Both these additional attributes can be determined by molecular typing using pyrolysis methods. Furthermore, because molecular similarities exist between genetically related asphaltenes, heterocompounds and kerogens (Behar and Pelet 1985; Solli and Leplat 1986; Muscio et al. 1991), molecular kerogen typing criteria can be extended to the macromolecular fractions of crude oils and rock bitumen. In this way reservoir and source facies could be classified using a single frame of reference for applications in basin modelling (Horsfield et al. 1993a).

The use of molecular typing can be illustrated using the example of the Ardjuna Basin in Indonesia (see Gordon 1985; Horsfield et al. 1988; Noble et al. 1991 for details). The Ardjuna Basin is a major petroleum province in northwestern Java, Indonesia. Non-marine crude oil is produced from Oligocene through Middle Miocene clastic and carbonate reservoirs. Although a great variety of compositions are attributable to alteration phenomena such as evaporation-fractionation and biodegradation, primary petroleums are of the high-wax type. Based on geological and geochemical considerations the primary source of all petroleum discovered to date in the basin is the deltaic member of the Oligocene Talang Akar Formation. Within the Talang Akar Formation there are three fine-grained lithofacies, these being lower delta plain coals and interdistributory bay black shales each containing kerogen types II through III, and delta plain grey shales whose kerogen has a type III composition. On the logical premise that the immature source of the high-wax oils must contain long-chain *n*-alkyl structures, the grey shale lithofacies could be ruled out as a potential source as these substituents were absent from kerogen concentrates. These components were present in the pyrolysates of the other two lithofacies, pointing to high-wax oil-generating potential. There was, however, no relationship between the abundance of *n*-alkyl substituents and kerogen type assignment (II or III). Indeed, within the coal source facies the best correlation found was that with the proportion of matrix liptinite (Fig. 6.6A). The hydrogen index was correlated with the contributions of amber and microscopic resinite particles rather than being equatable with generative yields of *n*-alkane precursors (Fig. 6.6B). In other words, good high-wax oil-generating potential did not coincide with high hydrogen index values at any given maturity level. Instead, elemental composition was being overwhelmingly influenced by resin-related macerals whose maturation products are known to contribute to petroleums (Grantham et al. 1983; Alexander et al. 1987; van Aarssen et al. 1992), but whose fate in quantitative terms is poorly understood and is certainly subordinate in this particular case. The assertion from marine source rock studies that Hydrogen Index is a measure of oil generating potential (Tissot et al. 1974) cannot therefore be automatically extended to non-marine source sequences because resinite is a commonly occurring maceral in Tertiary fluviodeltaic systems.

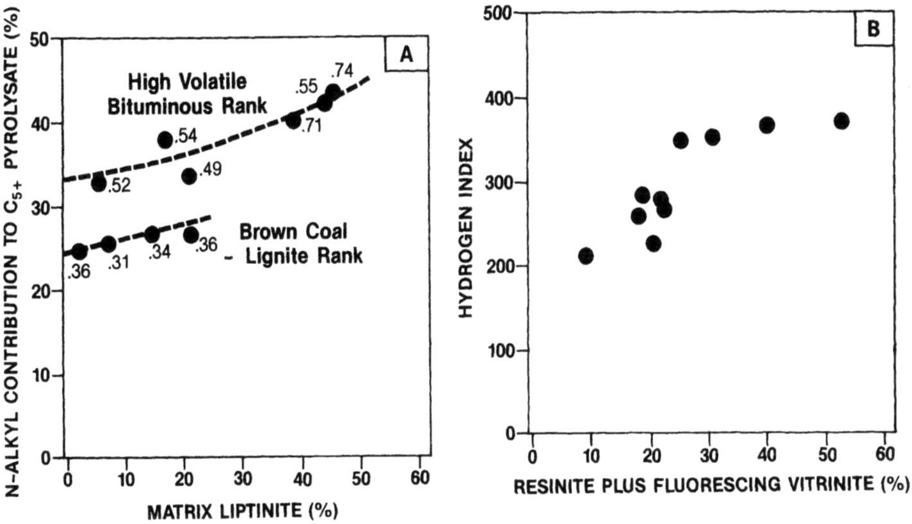

Fig. 6.6A,B. Organic petrographic and pyrolysis data for Talang Akar coals (Horsfield et al. 1988). **A** Relative abundance of the coal maceral matrix liptinite versus relative abundance of *n*-alkyl moieties in C_{5+} pyrolysates. **B** Relative abundance of resin-derived coal macerals versus hydrogen index and isotopic composition. (Reprinted from Horsfield et al., Determining the petroleum-generating potential of coal using organic geochemistry and organic petrology, 1988, pp 121-131, with kind permission from Elsevier Science Ltd, The Boulevard, Langford Lane, Kidlington OX5 1GB, UK)

6.3.2
Kerogen Composition and Structure – A Brief Overview

The molecular composition and structure of kerogen is complex, and determined by its biological precursors and the modifications brought about during diagenesis and catagenesis. Kerogen can form via two pathways, one being the random "repolymerisation" or "condensation" of the microbial breakdown products of proteins, polysaccharides and lignins, namely amino acids, sugars and phenols (Nissenbaum and Kaplan 1972; Stevenson 1974), and the other from the selective preservation of resistant, often morphologically structured biopolymeric materials such as spores, pollen and degraded cellular debris (Philp and Calvin 1976; Stach et al. 1982; Largeau et al. 1984; Tegelaar et al. 1989a). The two pathways are shown in Fig. 6.7 (after Rullkötter and Michaelis 1990).

The proportion of kerogen resulting from either diagenetic reactions or via direct preservation probably depends on the nature of the starting material and the depositional environment in question (as reviewed by Horsfield and Rullkötter 1994). Thus, a portion of the polymethylene components in aquatic autochthonous kerogen particles may originate by the random polymerisation of algal-derived polyunsaturated lipids (Knights et al. 1970; Cane and Albion 1973; Saxby 1981) possibly via sulphur atoms (Sinninghe Damsté and De Leeuw 1990; Adam et al. 1993) whereas the rest could simply consist of preserved aliphatic cell wall material. Similarly, the polymethylene components in

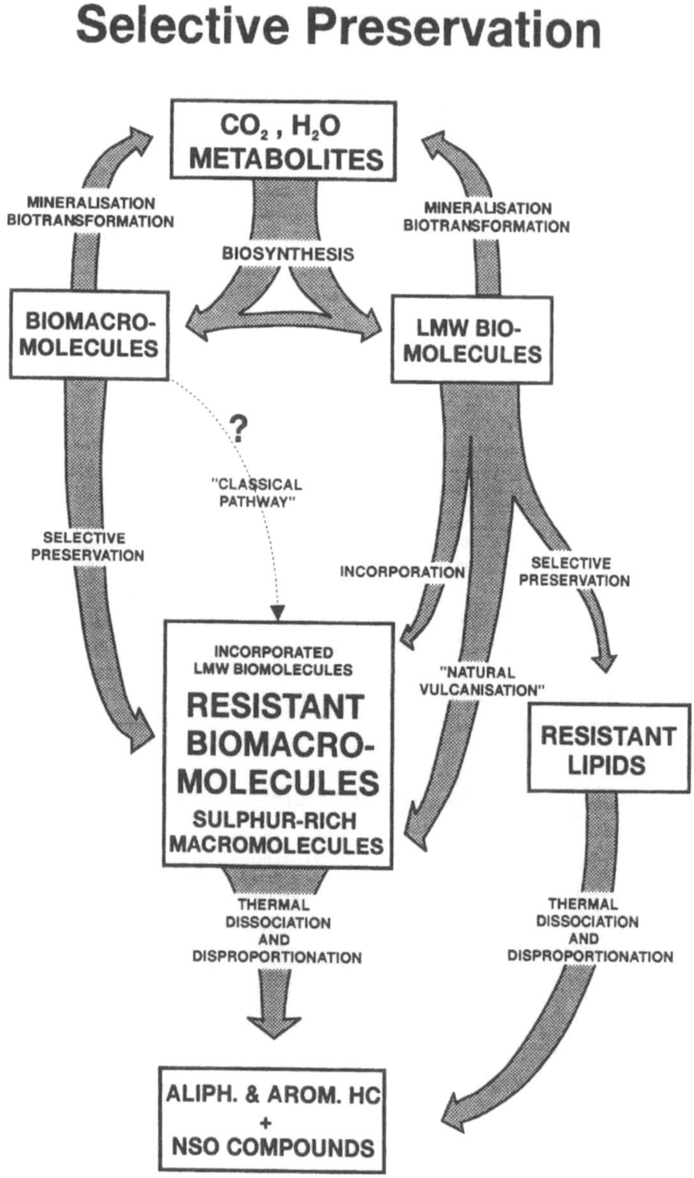

Fig. 6.7. The two paths for kerogen formation: selective preservation and neo-condensation. (Reprinted from Rullkötter and Michaelis, The structure of kerogen and related materials. A review of recent progress and future trends, 1990, pp 829-852, with kind permission from Elsevier Science Ltd, The Boulevard, Langford Lane, Kidlington OX5 1GB, UK)

allochthonous kerogen particles may originate by the random grafting (initially esterification) of wax acids and wax alcohols from higher plant cuticles to a polycondensed nucleus (Larter et al. 1983), or they may represent preserved

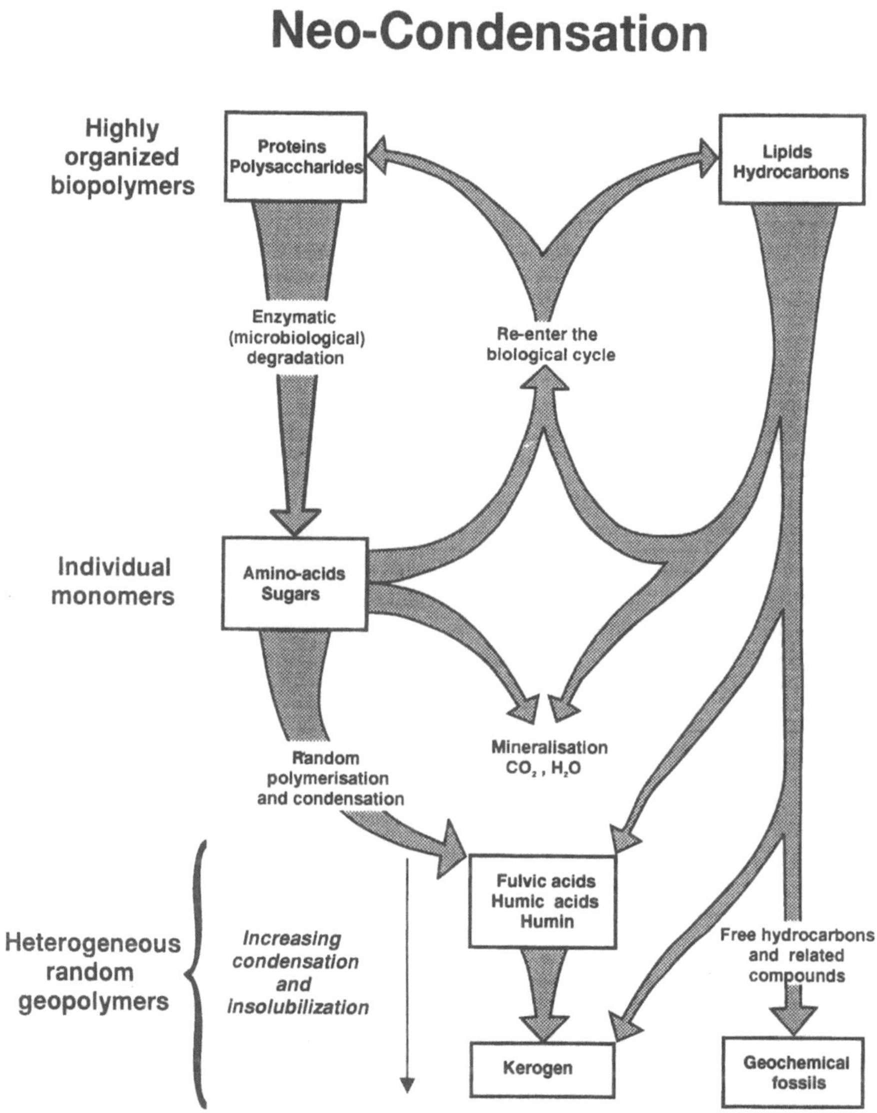

Fig. 6.7 (*Contd.*)

cuticular aliphatic biopolymeric materials. In the case of vitrinite, whose petrological habits indicate a variety of possible origins (Stach et al. 1982), alkylphenolic and methoxyphenolic moieties may enter the kerogen via random repolymerisation of lignocellulosic degradation products or the preservation of lignin and similar materials (Mycke and Michaelis 1986; de Leeuw and Largeau 1993). Because they may form via one or both of these processes, even monomaceralic kerogens such as vitrites and torbanites, occurring in banded coals and as oil shales respectively, may display structural and compositional heterogeneity at the molecular-macromolecular level. This complexity is compounded many-fold in the case of the complex polymaceralic assemblages that occur in most sediments because of the large number of particle types, each with its own structural and compositional heterogeneity.

It follows that average molecular structural models, derived via combinations of elemental analysis, infrared, ^{13}C nuclear magnetic resonance (^{13}C-NMR) and X-ray spectroscopy, solvent extract data and the results of oxidation and pyrolysis experiments (Burlingame et al. 1969; Yen 1972; Béhar and Vandenbroucke 1987), are actually meaningless in a true structural sense. The models nevertheless give a valuable conceptual and statistical insight into the chemical building blocks of kerogen. Figure 6.8, taken from Béhar and Vandenbroucke (1987) illustrates the differences between selected immature kerogens. Type I kerogen from the Uinta Basin is rich in aliphatic chains, and some ether groupings are present, whereas type II kerogen from the Toarcian Shale consists of mainly alicyclic structures, shorter *n*-alkyl chain lengths and abundant ester/amide linkages, and type III kerogen from the Mahakam Delta consists of polycyclic structures containing oxygenated functional groups and long paraffinic substituents.

The cleavage of major substituents from the kerogen structure according to bond strength and the concomitant structural rearrangement of the remaining macromolecular structure leads to the formation of a mobile petroleum and an immobile residue. Knowing the chemical bond strengths of alkyl chains in a variety of likely structural configurations is important as this may serve to identify points of relative weakness (Claxton et al. 1993). Chain length and boiling point distributions of the products govern the physical state of the petroleum. Therefore, far from being of only academic interest with little practical value, identifying the major building blocks using molecular kerogen typing is actually the key link to predicting bulk petroleum compositions.

6.4 Choice of Pyrolysis

A barrage of analytical techniques can be applied to characterise kerogen on a molecular basis. Organic petrology, while essential for recognising organic matter assemblages in organo- and palynofacies studies, is not one of these techniques because morphologically defined classes of organic particles can be chemically quite diverse. Rullkötter and Michaelis (1990) listed a total of 22 oxidation and reduction methods of variable specificity that have been applied to macromolecular sedimentary organic matter in addition to the more rapid microscopic, spectroscopic (^{13}C-NMR and infrared) and pyrolytic approaches. They also illustrated how *inter alia* the specific degradation of ester, alkyl-aromatic and sulphur linkages using hydrolysis, RuO_4 oxidations, and Raney Nickel hydrogenation has provided detailed information on some of the major building blocks of marine and non-marine kerogens, humic substances and asphaltenes, and concluded that specificity was the crucial step for kerogen characterisation. While this may be true as long as the physicochemical complexity of kerogen is treated statistically, it is unclear how such detailed

1a

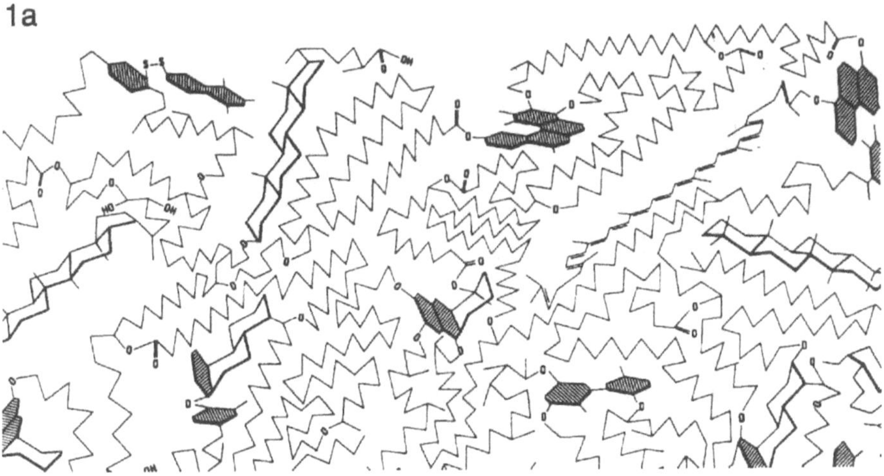

2a

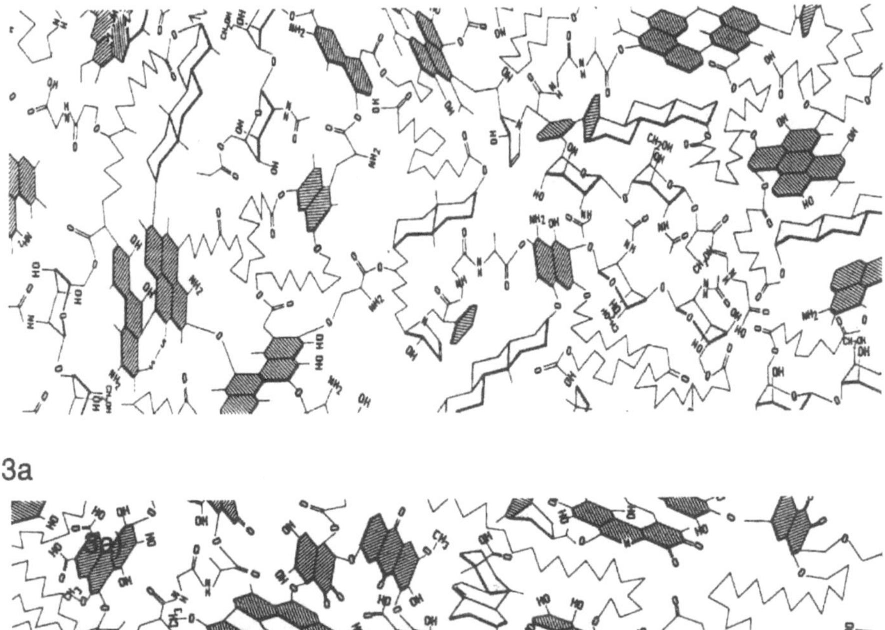

3a

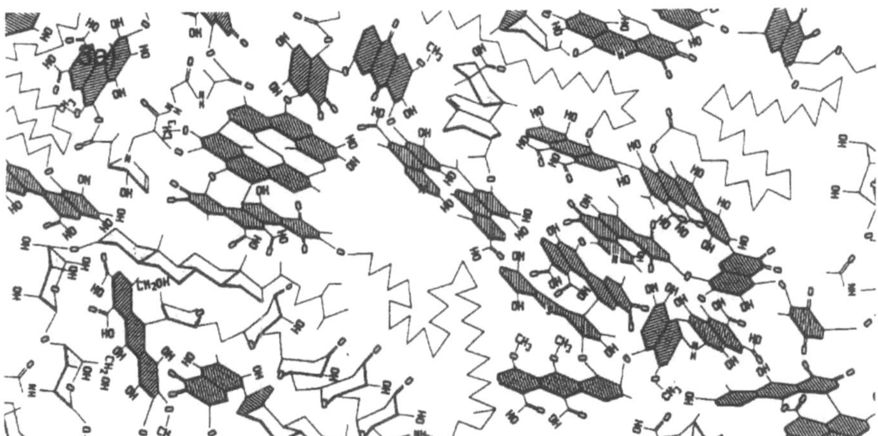

Fig. 6.8. Conceptual models of kerogen structure (after Béhar and Vandenbroucke 1987). *1a, 2a, 3a,* the structures of selected type I, II and III kerogens at the immature stage

information might be used to predict the bulk composition of crude oils. Similarly, solid state ^{13}C-NMR and infrared spectroscopy have been used to quantify aliphatic carbon, aromatic carbon, carboxyl groups and chain length distributions in kerogens (Rouxhet and Robin 1978; Wilson and Vassallo 1985; Schenk et al. 1986, 1993; Ganz and Kalkreuth 1987), but are not readily transposable into bulk properties of generated petroleums such as gas-oil ratio, aromaticity, wax content and sulphur content. Also, the parameters refer to the kerogen in total, i.e. both dead carbon and live carbon (Gransch and Eisma 1970; Cooles et al. 1986) and not only the part that is evolved as petroleum during catagenesis. Thus, while playing a key part of kerogen and coal characterisation, ^{13}C-NMR and infrared spectroscopy alone appear severely limited for the work under discussion here.

Pyrolysis lends itself well to molecular kerogen typing. The term pyrolysis has been defined as "a chemical degradation reaction that is induced by thermal energy alone" (Ericsson and Lattimer 1988). High-temperature pyrolysis for short periods of time (e.g. 600 °C/1 s) in a flow of inert carrier gas is a way of rapidly cracking kerogens into primary smaller molecular fragments for on-line bulk compositional determinations, whereas low-temperature pyrolysis for long periods of time (e.g. 300 °C/5 days) in a closed system more closely simulates catagenesis because degradation proceeds according to bond strength under conditions that favour secondary reactions (see Philp 1982; Horsfield 1984; Larter 1984 for reviews). The advantage of pyrolysis, especially when performed on-line with GC or GC–mass spectrometry (Giraud 1970; Horsfield et al. 1983), is that the volatile products are formed according to the same principles as in natural catagenesis, i.e. thermal cleavage reactions. Therefore if the reactions induced by geological heating rates (10^{-10}–10^{-12} K/min) are closely similar to those in the laboratory (10^{-1}–10^{-5} K/min) there is a high likelihood that pyrolysis products can be used to predict the compositions of petroleums generated in nature. The similarities and dissimilarities between natural petroleums and pyrolysates are discussed in some detail below to address this point. Additional arguments are presented in Chapter 4 of this book.

6.4.1
The Concept of Structural Moieties

Kerogens degrade upon pyrolysis to yield many compound types including hydrocarbons, ketones, alcohols, nitriles and thiols, as represented by cyclic and acyclic, saturated and unsaturated carbon skeletons (Rovere et al. 1983; Wilson et al. 1983). Of these the most commonly occurring major identifiable components seen by pyrolysis-GC are doublets of normal alk-1-enes and alkanes, alkylphenols, alkylbenzenes, alkylnaphthalenes and alkylthiophenes (Fig. 6.9). The combined sum of major and minor resolved components makes up on average only 20–50% of the GC-amenable C_{6+} pyrolysate, the majority being present in the form of an unresolved complex mixture or "hump". Additionally, high proportions of polar compounds are generated during the pyrolysis of all kerogen types (Urov 1980; Castelli et al. 1990; Landais et al. 1991) and tarry residues are deposited in the GC interface. These factors,

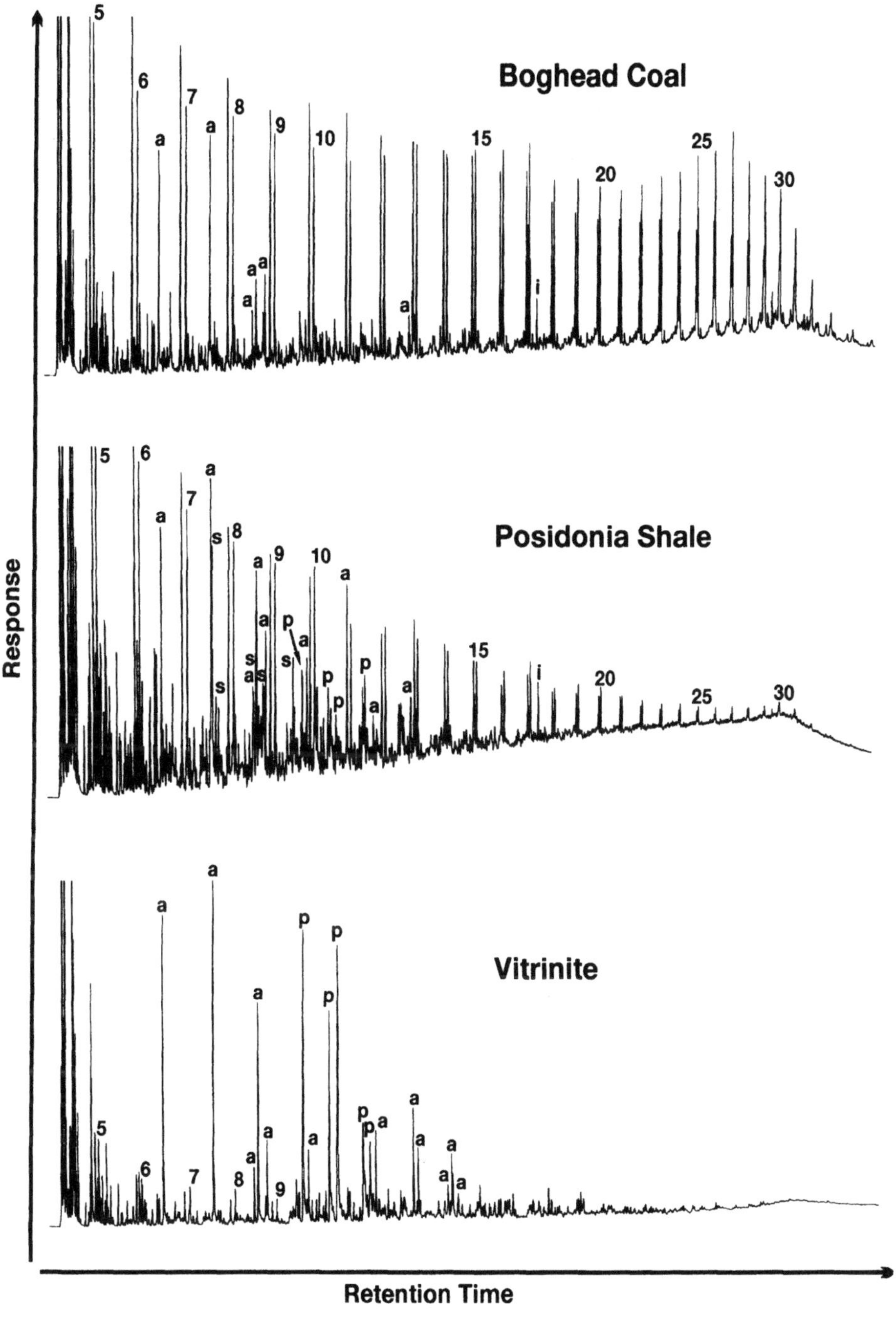

Fig. 6.9. Open-system pyrolysis-gas chromatograms of three immature kerogens – Boghead Coal, Posidonia Shale and Vitrinite – reveal the major resolved components seen in geological samples. *n*-Alkene/-ane doublets extending to variable chain lengths (numbered), isoprenoid alkanes (i), aromatic hydrocarbons (*a*), phenols (*p*) and the thiophenes (*s*) are marked

coupled with the fact that the method takes no account of the inert kerogen fraction, question whether the relative abundances of the major pyrolysis products are representative of the kerogen as a whole, and/or whether the products are representative of the volatile fraction that in nature evolves as petroleum. In this discussion the latter item is particularly important.

Horsfield (1989) determined aromaticities for a large suite of coal macerals and kerogens using the relative proportions of major aromatic (hydrocarbon and alkylphenol) and aliphatic (*n*-alkenes and *n*-alkanes) compounds in high-temperature pyrolysates, and found that these values, when plotted against atomic H/C ratio, fall on exactly the same trend as the f_a-H/C trend from published ^{13}C-NMR literature. The remarkable overlap (Fig. 6.10) provided firm evidence that structural information is contained within the readily identifiable and major components of pyrolysis products. The agreement between ^{13}C-NMR and pyrolysis-GC was particularly good for kerogens with H/C ratios less than 1.1, as was illustrated by the excellent agreement for the vitrinite and sporinite samples (see inset) that were common to both datasets. The divergence in trends above H/C ratios of 1.1 is probably linked to ^{13}C-

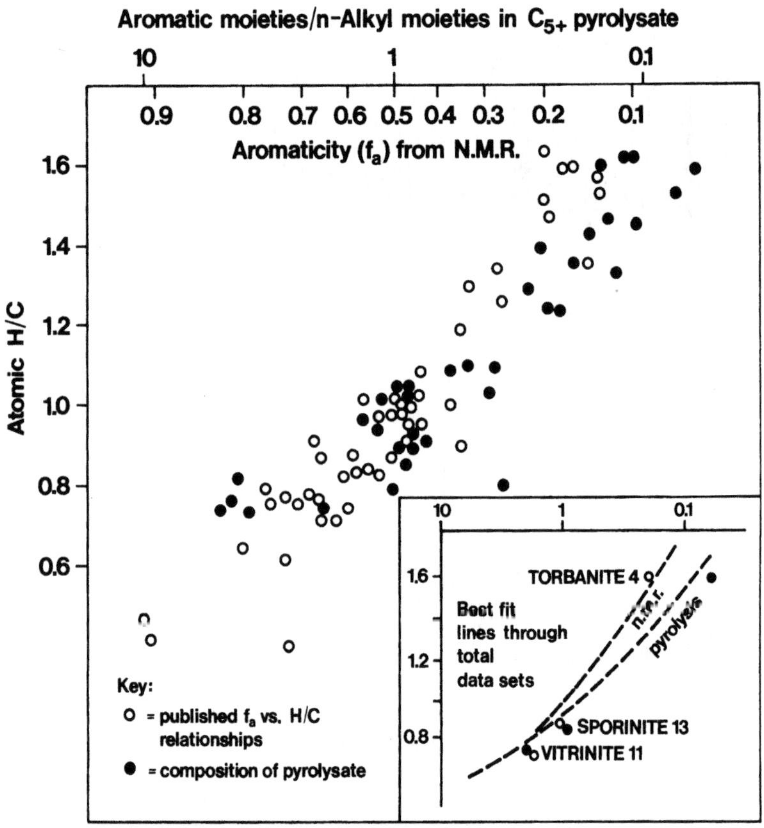

Fig. 6.10. The relationship between kerogen aromaticity and atomic H/C ratio according to ^{13}C-NMR and pyrolysis-GC (Horsfield 1989)

NMR being unable to differentiate aromatic from olefinic unsaturated species (Derenne et al. 1987). As reviewed by Larter and Horsfield (1993), other bulk compositional features of kerogens can be predicted using pyrolysis. The alkylphenol content of Carboniferous coal pyrolysates in United States is directly proportional to the hydroxyl oxygen content determined by wet chemical methods (Yarzab et al. 1979; Senftle et al. 1986), pointing to phenolic structures in pyrolysates being proportionally representative of oxygen-substituted benzenoid structures in kerogens. Similarly the relative abundances of thiophenes versus aromatic plus aliphatic hydrocarbons were proportional to the atomic S/C ratio (Eglinton et al. 1990a), suggesting that pyrolytic sulphur species are proportionally representative of kerogen-bound sulphur species.

It is deduced that major resolved species in high-temperature pyrolysates can actually give compositional information on the kerogen as a whole rather than on atypical part-structures, at least as far as the lignite through medium volatile bituminous coal rank range is concerned (R_o=0.27–1.04%). On the premise that kerogen composition directly controls the types and yields of volatile products generated during natural maturation, it can be further concluded that the abundances and distributions of resolved pyrolysis products give clues as to the bulk compositions of natural petroleums, such as paraffinicity and aromaticity.

6.4.2
Simulating Catagenesis

Low-temperature pyrolysis is employed to try to replicate the composition of products generated in nature. Partial rather than exhaustive fragmentation under closed-system conditions, sometimes in the presence of water or under high pressure, is used for this purpose. Although secondary reactions such as transalkylation should be more favoured under closed-system conditions (Weres et al. 1988), and indeed have been proposed as being important in the case of alkylcycloalkanes (Williams et al. 1988), there is no evidence that bimolecular reactions are quantitatively significant as far as aliphatic hydrocarbons, for instance, are concerned. Thus, long alkyl chains are not formed from kerogens whose in situ *n*-alkyl chain length is short. This is a very important consideration when predicting petroleum compositions. However, free radicals on first-formed volatile products readily abstract hydrogen from the kerogen structure so that *n*-alkenes are not a major product of the closed-system pyrolysis of kerogens. Closed-system pyrolysates therefore closely resemble natural crude oils in this regard.

A direct comparison of open versus closed-system pyrolysis is not easy to make because of large analytical configurational variability (macro- and microscale, on-line and off-line, etc). Similarly, there has been a great deal of controversy regarding the roles played by water and pressure during closed-system heating experiments. For these reasons it has always been very difficult to decifer apparently contradictory laboratory findings, let alone apply the results to natural systems. Schenk and Horsfield (1993) were the first to make a direct comparison of open- versus closed-system (<10 MPa pressure) pyrolysis, independently of extraneous variables, by means of the microscale sealed vessel pyrolysis capability (Horsfield et al. 1989). Using programmed heating of Posidonia Shale, they showed that the kinetic parameters for open- and closed-system pyrolysis were closely similar (both pre-exponential factors and activation energy distributions). The result pointed to pyrolysis scission reactions having the same heating rate dependence under both laboratory and geological heating conditions. However, the 40% lower yield for the closed-system experiment pointed to enhanced tar and coke formation from secondary radical reactions under the confined system that had been employed, signalling that petroleum yield predictions depend on system configuration. The roles played by pressure and water on the thermal cracking of sedimentary organic materials has recently been comprehensively addressed by Michels et al. (1995 and references therein). Using gold vessels (260–365 °C, 200–1300 bars), it was shown that the nature of the pressuring medium plays a crucial role in the results obtained. Pressure was shown to exert only a minor influence in the case where water was essentially absent but a large influence in its presence. Water is an additional important hydrogen source for hydrocarbon formation during hydrous pyrolysis whereas only organic hydrogen is important in anhydrous systems, and it is the actual contact between these potential hydrogen donors and acceptors which basically determines reaction pathways in nature. The predominant mechanism is not known with certainty. However, because the matrix porosity and permeability of mature organic-rich sediments is low, thus reducing the chance of organic matter-water interactions, and that the solubility of bitumen in water in the temperature range 60–150 °C is much lower than that typical used for laboratory pyrolysis, it can be deduced that hydrous pyrolysis overemphasises the role played by water.

In conclusion, two approaches can be used for predicting petroleum composition. *High-temperature, open-system pyrolysis* gives information on moieties in kerogens and thence gives a measure of the likely products generated in nature, for example, paraffinic, aromatic etc, assuming that the liabilities of these moieties in nature can be related to, and preferably be the same as, those displayed during laboratory pyrolysis. *Low-temperature, closed-system pyrolysis* attempts to simulate natural reactions by utilising milder heating conditions and allowing secondary reactions to take place. Not only should the labilities of these moieties in nature preferably be the same as during laboratory pyrolysis, but also finer details such as isomeric distribution patterns should be replicated.

6.5 Pyrolysates and Petroleum

This section describes the major classes of compounds seen in kerogen pyrolysates and further examines whether these same products are seen in natural petroleums.

6.5.1 Aliphatic Hydrocarbons

Linear aliphatic structures in kerogens crack to give *n*-alkene and *n*-alkane doublets during analytical pyrolysis, and *n*-alkanes under low-temperature closed-system conditions. The relative abundance of each homologue is controlled to variable degrees by the chain length of the precursor moiety in the kerogen and the secondary reactions that occur during pyrolytic cleavage. Bimolecular combination reactions are not prevalent, and therefore *n*-alkyl chain lengths for pyrolysates represent minimum values of those present in the kerogen.

There is strong evidence from the empirical relationship between the chain length distribution of *n*-alkenes and *n*-alkanes in high-temperature pyrolysates and the distributions of *n*-alkanes in genetically related crude oils to suggest that *n*-alkane forming scission reactions are essentially the same as those in nature. Horsfield (1989) showed that the source kerogens of high-wax oils have, not entirely surprisingly, a high proportion of long-chain *n*-alkenes and -anes in their pyrolysates (e.g. Talang Akar Formation of Java and the Green River Shale of Utah) whereas major sources of gas (e.g. vitrinites and sporinites in coals of the southern North Sea) are characterised by very short average *n*-alkyl chain length. Intermediate chain lengths were found to characterise anoxic marine shales such as the Toarcian Shales (France), Woodford Shale (Oklahoma) and Glenwood Shale (Iowa) which generate low-wax crude oils in nature. Also, kerogens that yielded pyrolysates rich in *n*-alkenes and -anes relative to other resolved components were found to generate paraffinic crude oils in nature. Further evidence that pyrolytic scission reactions are essentially the same as in nature comes from the fact that subtle features of high-temperature pyrolysates have in some instances also been documented for naturally formed crude oils. For example, *n*-alkanes in Ordovician oils from the Michigan Basin display a strong odd carbon predominance in the C_9–C_{19} range (Martin et al. 1963; Reed et al. 1986; Williams et al. 1988), and a similar periodicity has been documented for *n*-alkenes, *n*-alkanes and *n*-alkylcyclohexanes in their source rock pyrolysates (Klesment 1974; Fowler 1992). Similarly, *n*-alkanes in fluviodeltaic/lacustrine-associated waxy crude oils often display an odd predominance in *n*-alkanes in the C_{25}-C_{35} region (Sutton 1979; Thomas 1982; Kelley et al. 1985), a feature that has been documented in pyrolysates of algal kerogens (Goth et al. 1988; Horsfield et al. 1994), terrigenous organic matter (Van de Meent et al. 1980; Dong et al. 1986) and biopolymers from lower and higher plants and bacteria (Chalansonnet et al. 1988; Fukushima et al. 1989; van Bergen et al. 1994). Interestingly, the low-temperature closed-system pyrolysis of a cutan from *Agave americana* yielded *n*-alkanes and a series of iso- and anteisoalkanes with distributions that were identical to that of an Indonesian crude oil (Tegelaar et al. 1989b), and the low-temperature pyrolysates of Talang Akar coals were closely similar to waxy crude oils generated by mature equivalents (Noble et al. 1991).

These results confirm *qualitatively* that *n*-alkyl moieties respond similarly to thermal stress under geological and laboratory heating conditions, and

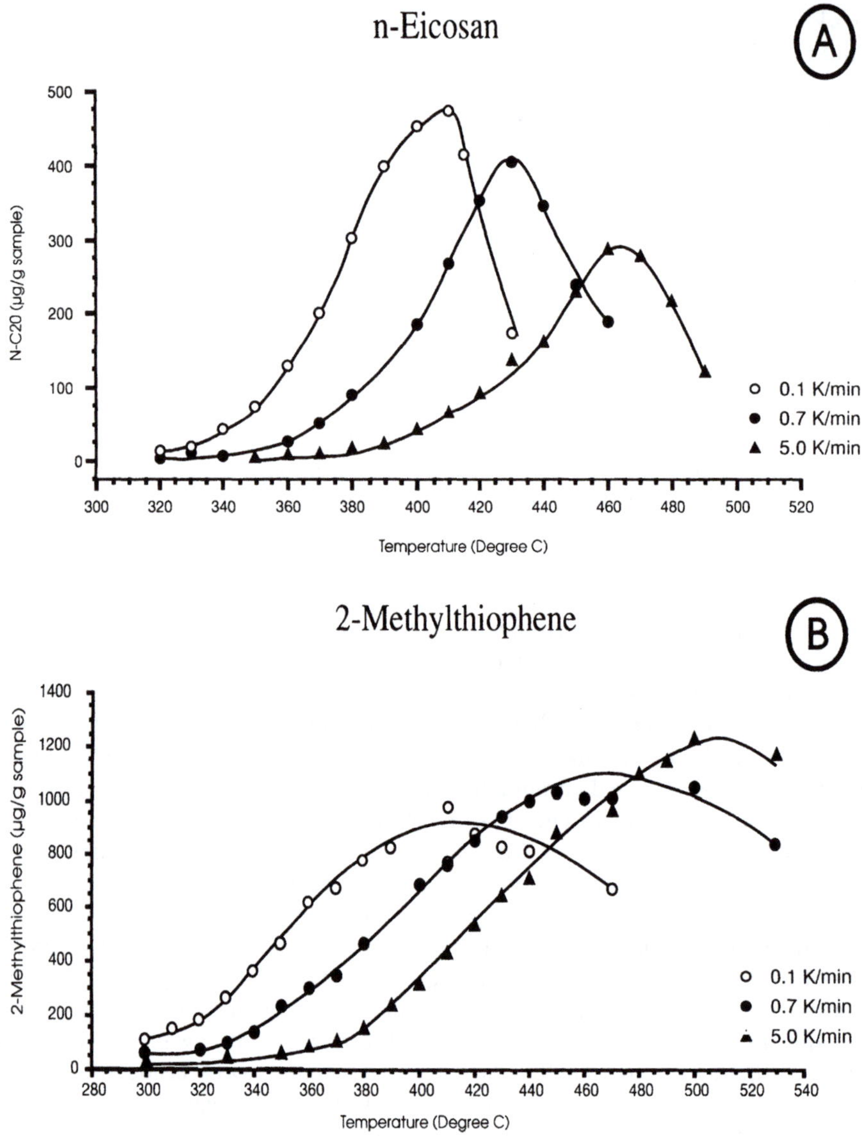

Fig. 6.11A,B. Changes in the yields of individual pyrolysate components as a function of heating rate: *n*-eicosane increases whereas 2-methylthiophene decreases as heating rates are lowered

therefore that the formation of these components of natural petroleums can be successfully predicted. *Quantitative* predictions of paraffin content must consider that component yields are heating rate-dependent. As an example, the maximum yield of *n*-C_{20} from Posidonia Shale kerogen pyrolysis [microscale sealed vessel (MSSV) conditions] increases with decreasing heating rate (Fig. 6.11A). This topic is discussed further in Section 6.5.4.

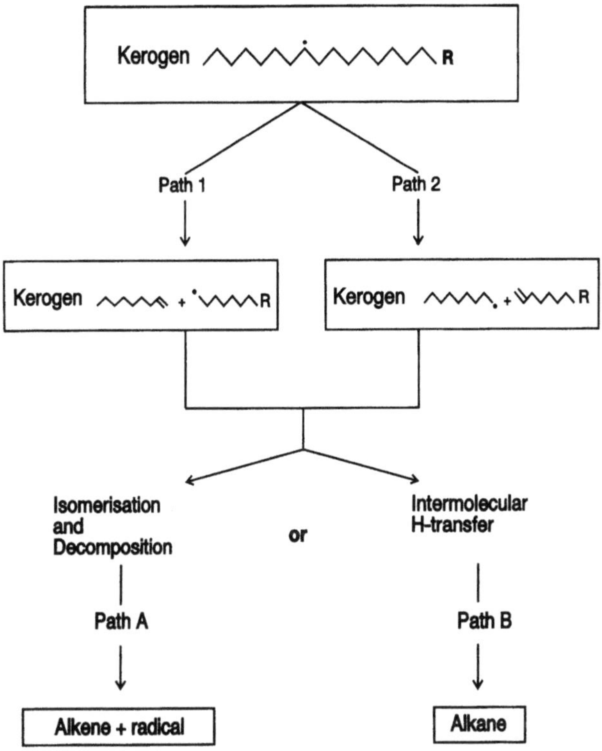

Fig. 6.12. Generalised reaction pathway for the formation of *n*-alkenes and *n*-alkanes in open-system pyrolysates. (Kiran and Gillham 1976)

The degradation of kerogen proceeds via free radical intermediates. The generalised model shown in Fig. 6.12 serves as an example of how *n*-alkenes and *n*-alkanes products are formed from a polymethylene precursor in this way (Kiran and Gillham 1976). No single mechanism can be assigned to all kerogens because of their different origins and precursors. Klesment (1974) and Van de Meent et al. (1980) ascribed the even carbon preference in *n*-alkenes and the odd carbon preference in *n*-alkanes of pyrolysates from Kukersite and terrigenous organic matter as originating from esterified even carbon numbered long-chain alcohols and acids that had undergone dehydration and decarboxylation, respectively. Similarly, model experiments (Larter et al. 1983) verified that alkyl chain lengths may indeed be closely similar to those of lipids incorporated into macromolecular organic matter by neocondensation reactions. However, while ester-bound moieties can indeed contribute appreciably to some kerogen pyrolysates (Brooks and Smith 1969; Douglas et al. 1977), periodicity in pyrolysis products may also be related to the specific cleavage of carbon-carbon and ether bonds in biopolymeric algans and cutans. The marked periodicity seen for highly mature Proterozoic and Lower Palaeozoic kerogen pyrolysates (Dungworth and Schwartz 1972; McKirdy et al. 1980; Jackson et al. 1984) is further evidence that such features can arise from carbon-carbon or ether-bound substituents rather than ester-bound moieties.

Other algal kerogen pyrolysates resemble those of polyethylene in that they display a dominance of the C_6, C_{10} and C_{14} *n*-alk-1-enes. This feature, documented for numerous torbanites, has been attributed to intramolecular radical transfer processes occurring after the random scission of *n*-alkyl chains has taken place (Larter 1978; Hall and Douglas 1983). It is clear therefore that not all subtle features in *n*-alkene/-ane distributions may be related to actual chain length distributions in the kerogen.

The thermal lability of *n*-alkyl moieties is highly variable. Those in very aliphatic kerogens (e.g. Glenwood Shale in the Michigan Basin, Green River Shale in the Uinta Basin) are frequently very stable, whereas those in kerogens of mixed origin such as occur in coals (e.g. Talang Akar Formation in the Ardjuna Basin) are significantly less so. While long-chain alkanes are produced at lower temperatures than are short-chain homologues in some cases, this is not always the case (Leventhal 1976; Horsfield et al. 1989; Düppenbecker and Horsfield 1990; Gray et al. 1991; Muscio et al. 1991). Kerogens displaying great differences in chain length distribution with increasing maturity, crudely simulated using two-step pyrolysis with an equivalent vitrinite reflectance cut-off of 1.5% (see Asakawa and Takeda 1988), have been said to have *heterogeneous* distributions of *n*-alkyl precursors whereas those displaying little or no difference have been termed *homogeneous* (Horsfield 1989). Homogeneous kerogens are thermally stable, retaining compositional characteristics to high levels of thermal stress (Fig. 6.13), and probably derive from selectively preserved aliphatic biopolymers (Horsfield 1989; Tegelaar and Noble 1994). Heterogeneous kerogens, on the other hand, lose long-chain *n*-alkyl components early during maturation, become enriched in short-chain components with increasing maturation (Fig. 6.13) and appear likely to have originated via the diagenetic grafting of lipids onto a more functionalised and condensed nucleus (Horsfield, *ibid*).

Acyclic isoprenoid, steroid and terpenoid hydrocarbons have been identified in both high- and low-temperature pyrolysates of kerogens and coals (e.g. Philp and Gilbert 1985). The major isoprenoids in laboratory pyrolysates are C_{19} isomers, followed in order of decreasing abundance by components with 14, 18 and 20 carbon atoms. The unsaturated components prist-1-ene and prist-2-ene are dominant in open-system pyrolysis products of kerogens and whole rocks, whereas pristane is present in closed-system products. Macromolecularly bound tocopherols are considered to be the major precursor (Goossens et al. 1984). The C_{14} alkene and C_{18} alkane are common constituents of analytical kerogen pyrolysates. No single phytene isomer dominates open system pyrolysates, but phytane is frequently found in closed-system experiments (Eglinton 1988). Isoprenoid moieties are particularly abundant in immature coals (see Fig. 6.12) and type II kerogens (Van Graas et al. 1981) and even more abundant in associated asphaltene fractions (Muscio et al. 1991). The occurrence of isoprenoid structures in kerogens might be controlled initially by the redox environment during early diagenesis (Curry and Simpler 1988). Isoprenoid and other biomarker structures are always thermally labile in laboratory experiments. They decrease with increasing levels of natural maturation and are absent in naturally matured samples above ca. 1% Ro (Larter

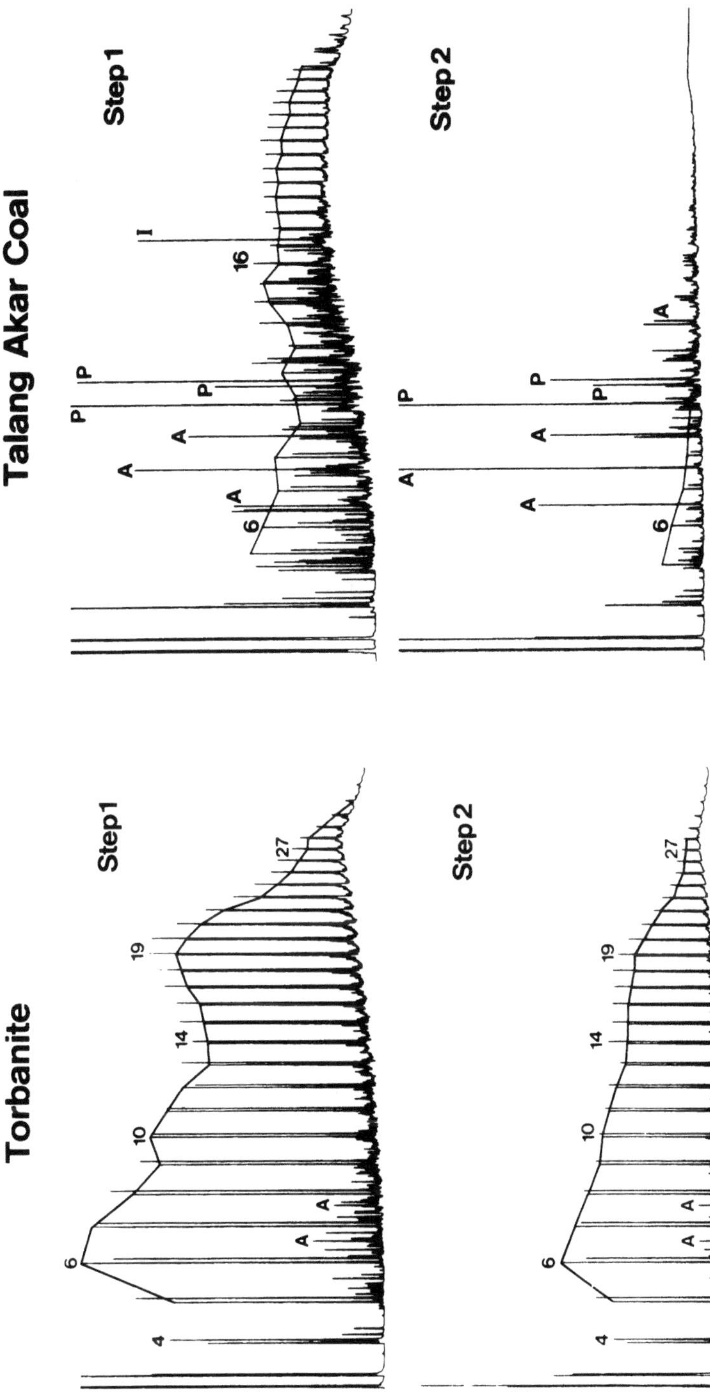

Fig. 6.13. Two-step pyrolysis-GC of Torbanite and Talang Akar Coal. *Step 1*, 320–450 °C; *step 2*, 450–600 °C. *n*-Alkyl precursors in torbanite have a "homogeneous" distribution whereas those in the coal are "heterogeneous" (Horsfield 1989)

et al. 1983; Curry and Simpler 1988; Øygard et al. 1988; Muscio et al. 1991; Horsfield et al. 1992).

Alkylcycloalkane moieties in kerogens may be released as such during pyrolysis or retained in condensed and aromatised pyrolysis residues. Where present in high-temperature pyrolysates they are usually only minor constituents. For instance, they make up only a small proportion of high-temperature pyrolysates from algal kerogens. In the case of *Gloeocapsomorpha*-containing kerogens they display a strong odd carbon predominance in the C_9–C_{19} range similar to the *n*-alkanes and cycloalkanes in related natural petroleums (Klesment 1974; Reed et al. 1986; Williams et al. 1988). In the case of the Alum Shale and Bakken Shale it has been noted that alkylcycloalkanes made up a higher proportion of low-temperature closed-system pyrolysates than they do of high-temperature open-system pyrolysates (Horsfield et al. 1992; Muscio et al. 1994). This is in agreement with the results of oil shale pyrolysis where high-temperature products contain the highest yields of aromatics and alkenes and lowest yields of alkanes and char (Burnham and Happe 1984). It is also noteworthy that alicyclic moieties in some algal lacustrine kerogens form involatile aromatised residues on pyrolysis (Schenk et al. 1993). Fossil resins are primarily alicyclic. Those from gymnosperms are diterpenoid whilst those from angiosperms (e.g. dammar resin) are sesquiterpenoid (Alexander et al. 1987; van Aarssen et al. 1990, 1991, 1992). Interestingly the closed-system artificial maturation products of dammar resin yielded an abundance of 1,6 dimethylnaphthalene, 1,6 dimethyl, 4-isopropyl-naphthalene, 1,6 dimethyl, 4-isopropyl-decahydronaphthalene and bicadinanes, these also being prominent components of Southeast Asian crude oils (van Aarssen et al. 1991). The latter observation indicates that reaction mechanisms involved in the catagenesis of fossil resins are similar to those performed in the laboratory.

6.5.2 Aromatic Compounds

The major resolved aromatic pyrolysis products in all kerogen types are the C_0–C_2 alkylbenzenes and alkylphenols. In most type I kerogens the contributions are very small compared to the *n*-alkenes and -anes. Phenolic building blocks constitute an even smaller part of type I kerogens, with proportions purportedly increasing in response to increasing salinity (Derenne et al. 1992), the degree of their cross-linking and hydroxylation/oxidation during early diagenesis (Horsfield et al. 1994; Curry and Horsfield, unpublished data) and the carbon skeleton (polymethylene, polyterpenoid or polyisoprenoid) of polyunsaturated lipids entering into neocondensation reactions (Cane and Albion 1971; Larter 1978; Metzger et al. 1985; Derenne et al. 1988, 1989). Type II kerogens contain higher proportions of aromatics than do type I kerogens (Fig. 6.9), with aromatic hydrocarbons being major components and phenols present in only very low abundance. Toluene and the xylenes are prominent in the vast bulk of cases, with amounts equalling or exceeding those of *n*-hydrocarbons in the same boiling range. The Alum Shale generates exceptionally high yields of aromatics because of its unusual molecular structure (Dahl et al.

1988; Lewan and Buchardt 1989; Horsfield et al. 1992). Also unusual are the Bakken and Duvernay Shales (Devonian) of the Williston and Western Canada Basins (Gray et al. 1991; Requejo et al. 1992; Muscio et al. 1994) and the Womble Shale (Ordovician) of Oklahoma (Douglas et al. 1991). Petrographically consisting of bituminite and alginite, these kerogens yield abundant 1,2,3,4-tetramethylbenzene by β-cleavage of aromatic carotenoid residues of photosynthetic sulphur-oxidising bacteria (Radke 1987; Requejo et al. 1992). Additionally, type II-S kerogen yields 1,2,3-trimethylbenzene when pyrolysed, possibly from aromatised β-β-carotene (Hartgers et al. 1994). At low levels of naturally and artificially induced thermal stress, increasing maturation generally causes a fall in the abundance of aromatic moieties in pyrolysates (Van Graas et al. 1981; Bjorøy et al. 1988; Boreham and Powell 1991; Muscio et al. 1994), ostensibly because aromatic-rich bitumen is expelled or the aromatic systems become non-volatilisable as the case may be. While type II kerogens generate higher absolute quantities of aromatic hydrocarbons on pyrolysis, type III vitrinitic kerogens yield the highest relative amounts so that their pyrolysates are dominated by alkylbenzenes, alkylnaphthalenes and alkylphenols (Larter and Senftle 1985). These products originate from the pyrolytic degradation of lignin, sporopollenin and polycarboxylic acids (Stout and Boon 1994). Coals of increasing rank yield pyrolysates that are progressively enriched in total and low molecular weight aromatic compounds (Romovacek and Kubat 1968; McHugh et al. 1976), reflecting the aromatisation of the coal macromolecule. Contemporaneously there is a decrease in phenolic structures associated with the loss of oxygenated species during catagenesis so that phenolic oxygen is essentially absent in coals of low volatile bituminous rank (Van Graas et al. 1981; Senftle et al. 1986).

The fate of aromatic and naphthenoaromatic moieties during laboratory pyrolysis and natural maturation is one of either cracking to form a volatile component or condensation and its retention in the macromolecular structure. In the laboratory, cracking reactions via β-cleavage relative to the aromatic ring are certainly favoured, resulting in a high abundance of mono- and diaromatic compounds, whereas their natural fate is not known with certainty. The problem is not quantitatively significant for most type I kerogens because aromatic component abundances are low. In the case of type II kerogens, there is an empirical relationship between pyrolysate- and petroleum compositions, with aromatic components making a pronounced contribution in both cases. Thus, the paraffinic-naphthenic and paraffinic-aromatic oil families in nature (Tissot and Welte 1978) could be related to the paraffinic-naphthenic-aromatic pyrolysate family of kerogen pyrolysates by Horsfield (1989). Perhaps coincidentally, toluene and meta-xylene are the most abundant single aromatic compounds in both this oil type and pyrolysates of parent kerogens. The remaining aromatic hydrocarbons do not occur as major distinct peaks in gas chromatograms of either unaltered "marine" crude oils (e.g. Illich et al. 1977; Bockmeulen et al. 1983; Thompson 1988; Williams et al. 1988) or their source rock pyrolysates. In the case of crude oils they occur as di- through polyaromatics in the gas oil, lubricating oil and residuum fractions (Hunt 1979; Radke 1987). Aromatics therefore seem to be generated from the non-condensed aromatic structure of type II kerogen (Behar and Vandenbroucke 1987) in the form of labile moieties during both laboratory and natural thermal stress. Nevertheless, partition between petroleum and involatile polyaromatic ring systems can be anticipated according to Patience et al. (1992) who used ^{13}C-NMR to show that new aromatic systems develop in many marine source rocks, liberating hydrogen-rich molecules. One extreme example of this is the Bakken Shale based on mass balance calculations utilising naturally matured samples (Fig. 6.14A; Muscio and Horsfield 1996). Another is the Alum Shale, which forms dead carbon during simulated maturation experiments and possibly in sedimentary basins (Fig. 6.14B).

The fate of aromatics is of particular importance when evaluating the petroleum generating potential of type III kerogens and coals because many yield high abundances of alkylbenzenes, alkylnaphthalenes and alkylphenols as well as long-chain hydrocarbons on pyrolysis. The subject of whether humic coals are a source of liquid hydrocarbons is a complex one, and the reader is referred to Mukhopadhyay et al. (1991) and Law and Rice (1993) for updates on the different points of view. Of relevance to the present discussion is whether "liquid petroleum" is defined according to solvent extract or thermal extract yield. Solvent extracts are very aromatic and asphaltic and to a large degree involatile whereas thermal extracts are relatively paraffinic; solvent extract yields are five to ten times higher than thermal extract yields (see data in Durand et al. 1977; Boudou et al. 1984; Hvoslef et al. 1988; Littke and Leythaeuser 1993). Therefore estimating the composition and yield of petroleum strongly depends on how the term petroleum is defined. Pyrolysis cannot directly determine the potential of a coal to generate aromatic-asphaltic petroleum (solvent extract definition) because these materials are involatile and are cracked to secondary products. Therefore for coals that have not been solvent extracted, the low molecular weight alkylaromatic hydrocarbons and alkylphenols seen in pyrolysates could originate from either heavy bitumen or kerogen. The presence of low molecular weight alkylaromatic hydrocarbons and phenolic compounds in coal pyrolysates might therefore indicate that aromatic-rich petroleum could be generated in nature, but cannot discriminate what molecular weight it has. Whether aromatic petroleum, irrespective of molecular weight, is generated in nature depends upon (1) whether aromatic moieties are distributed between high-temperature pyrolysate and char in the same way as aromatic moieties are distributed between catagenetic products and fixed carbon in nature, and (2) how liquid hydrocarbons are expelled from coal, it being well-known that liquid petroleums associated with coal are of the high wax type (Kent 1954; Durand and Paratte 1983). Based on extensive literature review, the formation of "dead" carbon during natural maturation from carbon that is "live" in immature coals seems now to be the major reaction pathway, i.e. aromatic moieties are mainly incorporated into polyaromatic structures rather than being released mainly as volatile products (Levine 1993; also documented by Patience et al. 1992 for type II kerogen).

For predicting petroleum compositions, the pragmatic solution is to accept that the aromatic hydrocarbons and phenols in coal pyrolysates are either pointing to the presence of a non-migratable aromatic-asphaltic bitumen or to

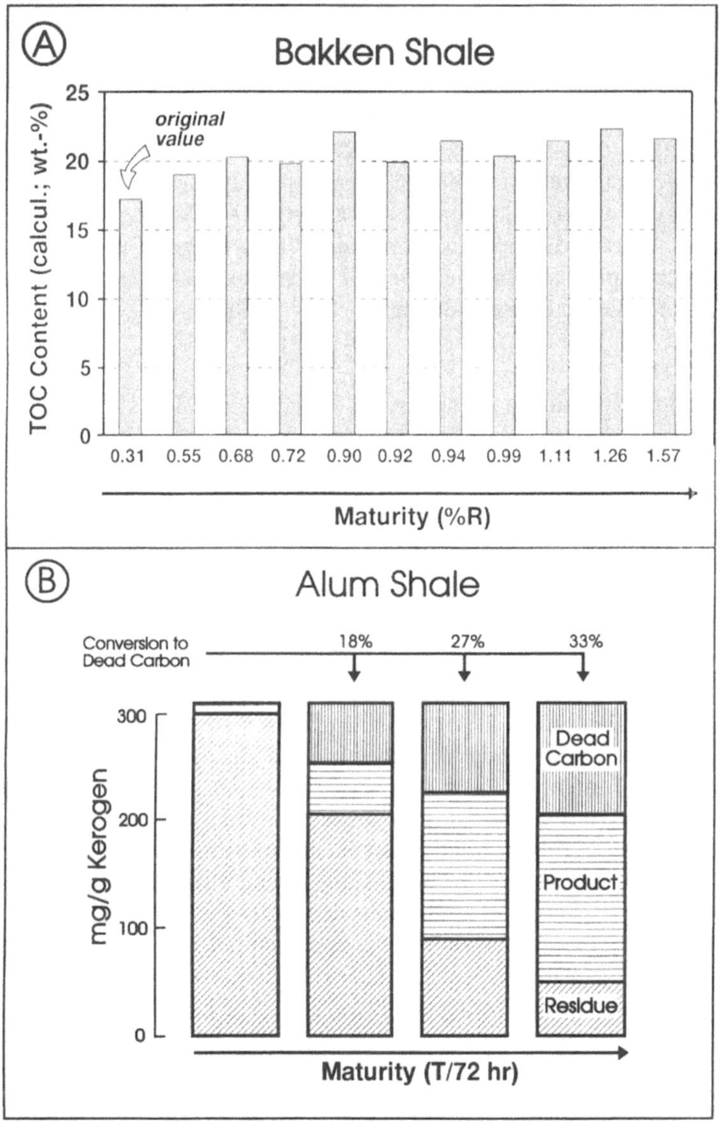

Fig. 6.14A,B. The postulated neo-formation of inert carbon. **A** During natural catagenesis of the Bakken Shale (Muscio and Horsfield 1996). **B** During the MSSV pyrolysis of Alum Shale kerogen (Horsfield et al. 1992). (Reprinted with kind permission from Muscio GPA and Horsfield B, Enhanced formation of inert carbon during the natural maturation of the marine source rock; Bakken Shale, pp 10-16, 1996. American Chemical Society)

a moiety that undergoes condensation reactions during natural maturation and is retained within the kerogen. In either circumstance, phenol is classed as an immobile constituent. This approach was adopted by Horsfield (1989) when classifying vitrinites and sporinites as gas-prone based purely on average *n*-alkyl chain length of the high-temperature pyrolysate and their gas-generating potential in nature. It is nevertheless interesting to note that alkylphenols, monoaromatic- and diaromatic hydrocarbons are present in the thermal extracts (300 °C) of high volatile through medium volatile bituminous coals from Germany (0.9–1.5% Ro; Horsfield and Idiz, unpublished results), and that humic coals generate phenol as a result of low-temperature closed-system pyrolysis under both hydrous and anhydrous conditions (Teerman and Hwang 1991; Horsfield, unpublished results). This suggests that at least some light aromatic products can be expected to be generated and possibly expelled from the thermal maturation of coals.

6.5.3 Sulphur-Containing Compounds

The major resolved sulphur compounds in kerogen pyrolysates are H_2S, COS and SO_2, which make up the gas peak in gas chromatograms, and the alkylthiophenes and alkylated benzothiophenes which dominate the C_{6+} fraction; alkylated dibenzothiophenes and alkylthiolanes are also present but in smaller abundances (Sinninghe-Damste et al. 1989). The carbon skeletons of the alkylthiophenes possess straight-chain, branched-chain, hopanoid and steroid configurations and are formed during early diagenesis by intermolecular sulphur incorporation reactions involving functionalised lipids and hydrogen sulphide (Brassell et al. 1986; Sinninghe Damsté and de Leeuw 1990). The yield of sulphur compounds from pyrolysis is directly proportional to the organic sulphur content of kerogens (see above, Sect. 6.4.1). Thus, high-sulphur kerogens from clay-poor depositional settings yield abundant sulphur compounds on pyrolysis whereas kerogens deposited in freshwater or in the presence of excess reactive iron yield only small quantities (Eglinton et al. 1990b). Sulphur-rich kerogens are usually also hydrogen-rich and classified as type II-S (Orr 1986), though instances have also been reported of sulphur-rich type I and type III kerogens (Sinninghe Damsté et al. 1992, 1993; Radke and Willsch 1993). Sulphur-containing products from type I kerogens are dominated by the 2-alkylthiophenes and those from type II-S kerogens by the 2,5 dialkylthiophenes whereas those from coals and type III kerogens consist dominantly of branched isomers such as 2,4 and 3,4 dialkylthiophenes (Eglinton et al. 1992). Maturation may also play an influence, as exemplified by the decreasing ratio of 2-methylthiophene/2,5 dimethylthiophene during the MSSV pyrolysis of Posidonia Shale (transformation ratios >40%). Changes in the relative abundances of 2,3 dimethylthiophene, 2-ethyl,5-methylthiophene and 2,3,5 trimethylthiophene have been documented for the Posidonia Shale (R_o = 0.48–1.45%; Muscio et al. 1991). Maturation sequences from the Monterey Formation, Kimmeridge Clay and Mahakam Delta, supplemented by simulated maturation results on corresponding immature samples, showed decreasing thiophene ratios (2,3 dimethylthiophene/(*o*-xylene + *n*-non-1-ene)) and increasing proportions of branched versus linear isomers (Eglinton et al. 1990b). An increase in the alkylbenzothiophene /alkylthiophene ratio was also reported, though, quantitatively, the formation of benzothiophenes is minor compared to the overall fall in sulphur content.

Sulphur makes up between 0.05 and 10% of most crude oils (Tissot and Welte 1978; Orr and Sinninghe Damsté 1990). High-sulphur petroleums (>0.4%) are low in gasoline components, have low API gravities and high viscosities (Baskin and Peters 1992). According to Ho et al. (1974) it is present in intermediate as well as heavy distillation fractions, consists of sulphides, alkylthiophenes, benzothiophenes and dibenzothiophenes and high molecular weight asphaltenes, respectively. Within the asphaltenes, the sulphur is mainly thiophenic according to the XANES X-ray method, though numerous other sulphur species have been determined using degradative techniques (Wilhelms 1992; Kasrai et al. 1994). Immature oils from carbonate-evaporite and siliceous sources have been documented as containing non-polar macromolecular organic matter consisting of hydrocarbon building blocks cross-linked by polysulphide or disulphide bridges (Adam et al. 1993). Sulphides with alkyl chains closely similar to those of corresponding free *n*-alkanes have been detected in low maturity high-sulphur crude oils (Payzant et al. 1989), but maturation brings about an increase in the proportion of thiophenic carbon (Ho et al. 1974) and a lowering of sulphur contents by disproportionation into H_2S and pyrobitumen (Orr and Sinninghe Damsté 1990). The sulphur content of crude oil is initially determined by that of the source kerogen (Gransch and Posthuma 1974), with clay-poor source rocks having the highest organic sulphur contents, after which the influence of maturity causes sulphur contents to fall. Sulphur-rich kerogens are generally considered to be thermally labile and to generate sulphur-rich crude oils at relatively low levels of thermal stress as a result of the cleavage of C-S (Orr 1986).

High-temperature pyrolysis cannot directly determine whether a kerogen can generate low-API aromatic-asphaltic petroleums because these materials are thermally involatile and only their secondary cracking products can be seen by on-line techniques. High yields of low molecular weight alkylthiophenes signify only that a kerogen is rich in organic sulphur, and it is left to known correlations to infer bulk composition of the oil formed from that kerogen during maturation (di Primio and Horsfield 1996; see below, "Asphaltic Oils"). It is nevertheless enigmatic that low molecular weight alkylthiophenes are the most abundant volatile sulphur-containing products, even under very mild conditions (e.g. 240 °C/3 months; Kohnen and Horsfield, unpublished data). Considering the low maturity at which high-sulphur kerogen is reputed to generate petroleum (Tissot et al. 1987; Baskin and Peters 1992), it is surprising that the oils they generate do not contain these compounds in high relative abundance within the gasoline range. One explanation is that sulphur-compound abundances are dependent upon heating rate. Sealed vessel heating experiments (MSSV, Posidonia Shale kerogen) reveal that the sum of low molecular weight alkylthiophene abundances behave oppositely to *n*-alkanes in that they decrease with decreasing heating rate, as exemplified by the case of 2-methylthiophene (see Fig. 6.11). This signifies that the natural evolution of sulphur-containing compounds is difficult to predict using laboratory pyrolysis.

6.5.4 "Unresolved" Compounds

Readily resolvable pyrolysate components give useful information on the kerogen structure as a whole. It must follow that similar structural information is also contained within the remaining pyrolysate, possibly as oligomeric or functionalised components within GC humps and in tarry pre-column residues. Shown in Fig. 6.15 are C_{15+} liquid chromatographic data for pyrolysates and several crude oils that are worldwide representative. This shows that the bulk composition of unaltered petroleum, a product of kerogen degradation under geological conditions, is fundamentally different to that of kerogen pyrolysates, produced under laboratory conditions, irrespective of crude oil class or kerogen type (Larter and Horsfield, 1993). Importantly, the difference is also irrespective of pyrol-ysis conditions – Rock-Eval-style, flash pyrolysis, autoclave/hydrous and MSSV (Alum Shale only) conditions are displayed. Crude oils are undoubtedly seen to be hydrocarbon-rich systems, whereas pyrolysates from either hydrous or anhydrous systems are without exception rich in polar and aromatic components. Included, surprisingly, are torbanites and other type I kerogens which, by spectroscopic and elemental considerations, are predominantly aliphatic. The predominantly polar nature of pyrolysates can also be seen in the results of Urov (1980) and Castelli et al. (1990). Additionally, the laboratory pyrolysis of oil-prone type II source rocks, such as the Kimmeridge Clay, generates proportionately much more aromatic fraction

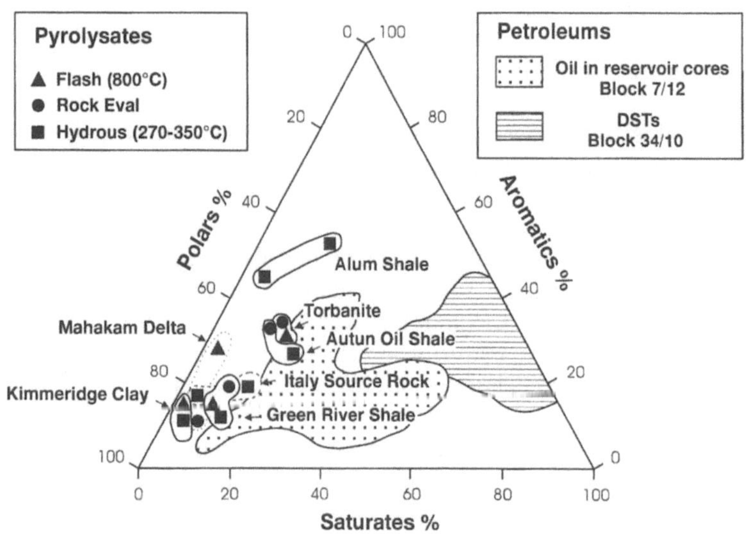

Fig. 6.15. The bulk chemical composition of kerogen pyrolysates (origin stated within the ternary diagram), petroleum from drill-stem tests (34/10 area, North Sea) and reservoir bitumens (7/12 area, North Sea). All kerogen pyrolysates are richer in polar components, irrespective of their origins or the pyrolysis method used

than is seen in naturally occurring solvent extracts (cf. Cornford et al. 1983; Rowland et al. 1986).

One plausible explanation for the compositional discrepancy is that migration in the subsurface tends to concentrate hydrocarbons in the reservoirs and polar components in the source rocks (Tissot and Welte 1978). In other words, products generated in the laboratory and during natural petroleum generation might be compositionally similar, but fractionation could lead to crude oils becoming enriched in hydrocarbons. There is no doubt that this is a contributory factor. However, the generally high expulsion efficiencies associated with mature oil-prone source rocks (Cooles et al. 1986; Larter 1988) and the hydrocarbon-dominated nature of most mature source rock extracts (Philippi 1965; Powell 1975) demonstrate that kerogen decomposition in the subsurface, not simply migration phenomena, produces a hydrocarbon-dominated C_{15+} fraction. It can therefore be deduced that laboratory pyrolysis generates products which *in quantitative terms* are quite unlike those derived from natural maturation.

The Bakken Shale is an ideal material for studying the nature of polar pyrolysate components because both flash pyrolysis and MSSV pyrolysis of its kerogen yield 1,2,3,4 tetramethylbenzene as a distinctive and prominent component (Muscio et al. 1994). Using this compound as a tracer it can be shown that macromolecular polar materials in pyrolysates contain the same precursor moieties that occur in the parent kerogen. Specifically, the open-system pyrolysis of a bitumen, collected from a MSSV experiment (330 °C/2 days) produced a series of *n*-alkenes and -anes plus the notable presence of 1,2,3,4 tetramethyl benzene from its macromolecular components (Fig. 6.16). These results clearly demonstrate that structural information contained in the resolved alkenes/-anes and aromatic hydrocarbons of pyrolysates is also present in the higher molecular weight bitumen fractions. This observation, also reported for coal tars (Koplick et al. 1983; Nelson 1987), is consistent with the macromolecular part of the polar fraction consisting of oligomeric fragments of the kerogen or aggregates of small molecules.

6.5.5 Model of Kerogen Decomposition

A conceptual model which accounts for the product-precursor relationships discussed above for laboratory and geological heating rates is illustrated in Fig. 6.17 (Larter and Horsfield, 1993). Kerogen is considered to be a complex macromolecular structure capable of releasing aliphatic, aromatic and polar moieties, the main difference between the volatile (reactive) and involatile (inert) kerogen is that one is the source and the other the sink of hydrogen. During natural or laboratory degradation, thermal reactions decompose the macromolecule into lower molecular weight fragments. The high laboratory pyrolysis heating rate produces mainly compounds which elute in the polar liquid chromatographic fraction, though some hydrocarbons and other monomers are also present. Conversely, the low geological heating rate produces proportionately fewer polar compounds. Pyrolysate constituents that are

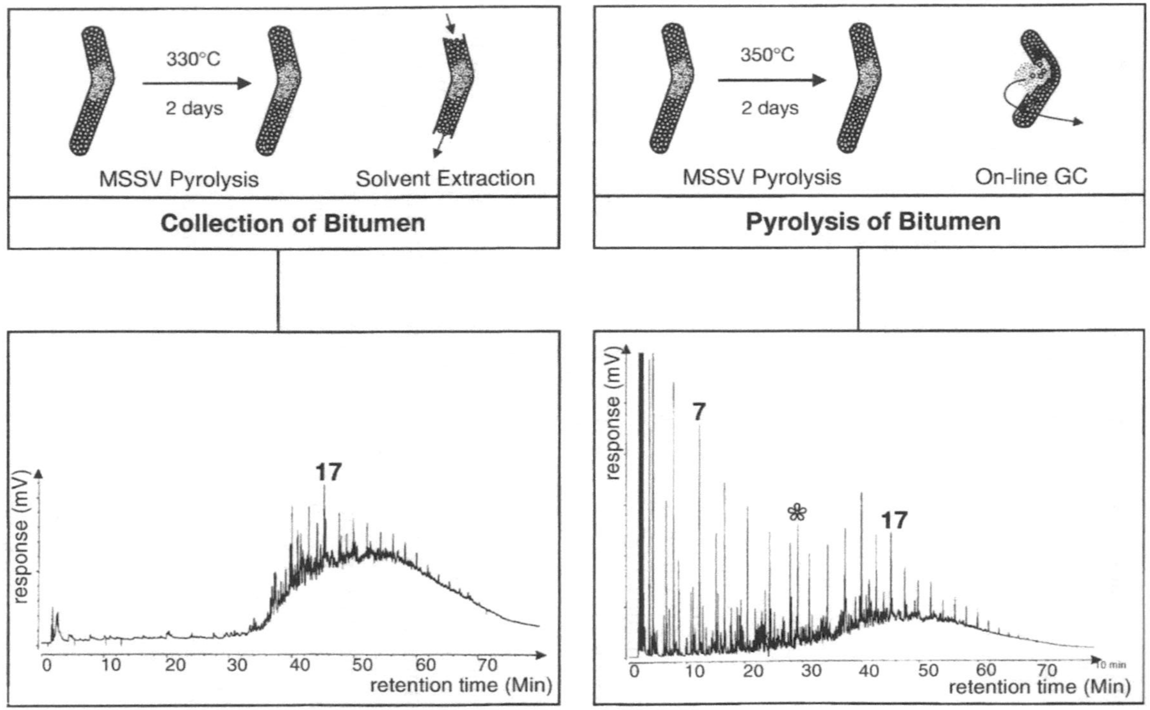

Fig. 6.16. Experiments reveal that the GC-unamenable polar bitumen generated on pyrolysis contains the same types of moieties as occur in the GC-amenable fraction. Here, bitumen collected from a 330 °C/3-day MSSV pyrolysis experiment was pyrolysed and yielded *inter alia* 1,2,3,4 tetramethylbenzene, a major pyrolysate component of Bakken Shale kerogen

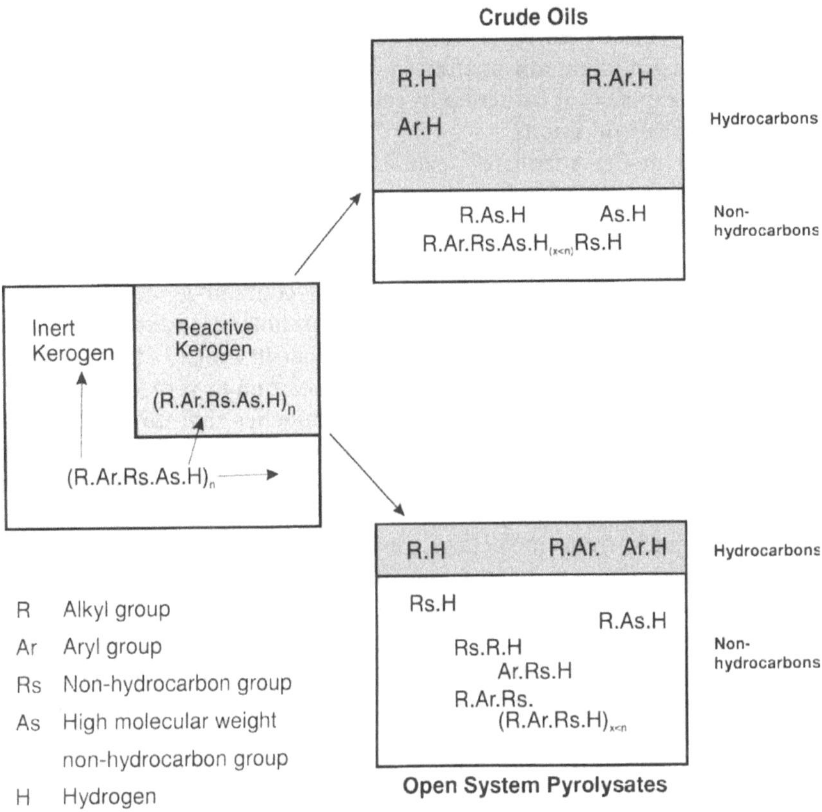

Fig. 6.17. A schematic kerogen decomposition model suggesting a conceptual explanation of the utility of open-system pyrolysis as a proportionally representative structural tool for characterising kerogens. (Larter and Horsfield 1993)

resolved by GC and which actually appear prominent, such as *n*-hydrocarbons (R.H) and aromatic hydrocarbons (Ar.H, R.Ar.H) do not represent quantitatively the whole kerogen but provide nevertheless an accurate, proportional account of the distributions of the major structural moieties in the kerogen itself (Horsfield 1989; Larter and Horsfield 1993). The so-called polar compounds may be oligomeric and/or highly functionalised. They too inherit the distributions of the major structural moieties from the kerogen. Thus, the fate of aliphatic moieties in kerogen could, in the case of geological heating rates, be mainly as hydrocarbons with subsidiary proportions of polar compounds or, in the case of laboratory pyrol-ysis, mainly as alkyl substituents on polar compounds with subsidiary proportions of hydrocarbons. The closely similar *n*-alkyl chain length distributions in genetically related kerogens, asphaltenes and resin fractions (Solli and Leplat 1986; Horsfield et al. 1991; Muscio et al. 1991) corroborate that the model indeed applies to aliphatic moieties in a variety of macromolecular sedimentary organic materials. In this regard, *n*-alkane yields increase as a function of decreasing heating rate, according to heating experiments on the Posidonia Shale (Fig. 6.11A) though, as noted previously for the Green River Shale (Burnham et al. 1982), homologue distributions remain essentially unaffected. This again suggests that hydrocarbons form at the expense of structurally related alkyl-substituted polar moieties under geological conditions.

While the model adequately explains the relationship between kerogen structure and pyrolysis products in general, and also the fate of alkyl moieties under laboratory versus geological heating rates, it takes no account of how aromatic or sulphur-containing moieties behave under geological conditions. As pointed out earlier, aromatic and sulphur-containing moieties are more likely to enter into aromatisation and condensation reactions, a point that is supported by the fall in 2-methylthiophene concentrations as heating rates are reduced (Fig. 6.11B). In this case, linear extrapolation to very low heating rates would result in sulphur being contained in moieties that do not generate 2-methylthiophene on pyrolysis. Some proportion may actually be incorporated into "dead" rather than "live" carbon rather than from oligomers to monomers as is the case with alkyl moieties.

Stepping back from molecular considerations, three basic components, namely kerogen (K), a polar bitumen (B) and a hydrocarbon-rich oil (O), can be considered (Louis and Tissot 1967; Braun and Rothman 1975; Lewan 1983). If it is assumed for simplicity that two sequential bulk reactions are active (see Larter and Horsfield 1993), $K \rightarrow K' + B$ (kerogen-to-bitumen), followed by $B \rightarrow B' + O$ (bitumen-to-oil), the rate-controlling step under laboratory conditions where bitumen yields are high would be the bitumen-to-oil reaction, whereas the rate-controlling step under geological heating conditions where hydrocarbon yields are high would be the kerogen-to-bitumen reaction. This phenomenon would be manifested on an Arrhenius diagram by a crossover of rate curves (Braun and Rothman 1975). The divergence of the curves ought nevertheless to be rather small because the kinetic parameters determined for the $K \rightarrow K' + B$ reaction by open-system pyrolysis adequately predict the conversion of kerogen to petroleum (i.e. $K \rightarrow K' + B$, and $B \rightarrow B' + O$) in the subsurface (Ungerer and Pelet 1987; Burnham and Braun 1990).

According to the traditional scheme above, which corresponds to the depolymerisation model of Ungerer (1990), heavy bitumen is envisaged as a reaction intermediate. Significant is the observation that tar is effectively removed from the site of pyrolysis when low generation rates (e.g. by heating at 350 °C for 2 days) and high carrier gas velocities are employed, suggesting that first-formed products are small and highly volatile rather than being large and involatile. Accordingly, it can deduced that polar bitumen may form by the condensation or aggregation of these small reactive species where concentrations are high enough. High rates of product generation, as in the case of most pyrolysis experiments, would favour bitumen formation. Conversely, under low product generation rates such as in naturally subsiding sedimentary basins, condensation and aggregation should be less probable, and if the first-formed species is a small radical, hydrogen abstraction to form a hydrocarbon should be favoured. The component reactions given above might then be modified so that K→R, is followed by R+R→B or R+H→O, where R denotes the first-formed product.

6.6 Predicting Petroleum Compositions

6.6.1 Qualitative Versus Quantitative Predictions

By far the easiest means of predicting gas-oil ratio by pyrolysis is to calculate the ratio of light and heavy pyrolysis products, on the assumption that the kerogen forms gas and oil in the same proportions in nature (Larter et al. 1977; Bailey 1981; Dembicki et al. 1983; Horsfield et al. 1983). Espitalié et al. (1988) have used C_1, C_{2-4}, C_{5-14} and C_{15+} bulk chromatogram splits for petroleum composition kinetic modelling, and Mackenzie and Quigley (1988) used a simple C_1–C_5 versus C_{6+} ratio as input for modelling the secondary migration of first-formed petroleums. Such bulk chromatogram splits are made by integrating all products above baseline in pyrolysis-GC or MSSV chromatograms

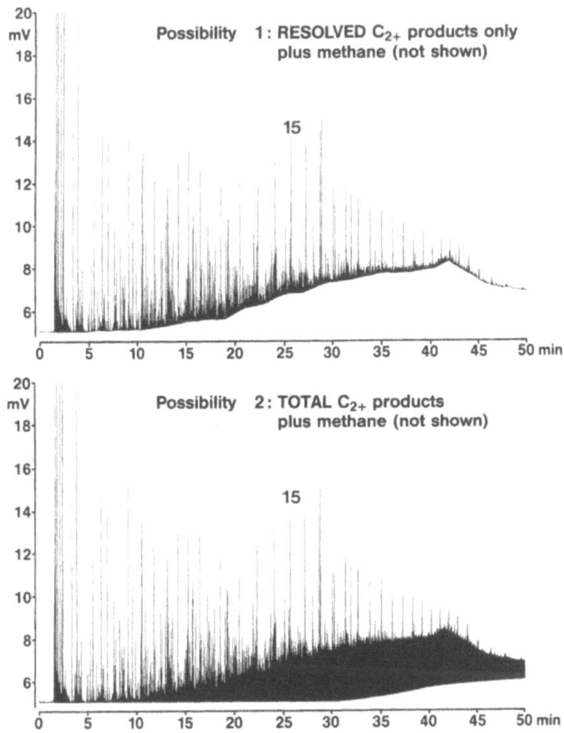

Fig. 6.18. The gas-oil ratio of pyrolysates can be calculated using either resolved products from a skim integration (upper) or from total products above baseline (lower)

(Fig. 6.18 lower part). The oil fraction consists largely of unresolved peaks in the form of a hump above which can be discerned resolved components dominated by *n*-alkenes, *n*-alkanes, alkylaromatic hydrocarbons and, in the case of coals, alkylphenols. The unresolved components make up between 40 and 60% of the C_{6+} chromatogram for types II and III kerogens heated up to transformation ratios of 0.70.

The analysis of a large sample set (see Horsfield 1989) reveals that gas-oil ratios from all organic matter types (I, II and III) and origins (marine, lacustrine etc) are uniformly very low and fall in the narrow range <0.4 kg/kg (Fig. 6.19B). The average value (ca. 0.2 kg/kg) is in good agreement with previously published values obtained by pyrolysis (Mann et al. 1987; England and Mackenzie 1989). Because this value also corresponds to that of solution gas-oil ratios in undersaturated marine-source rock-derived petroleums (also see Bailey et al. 1974; Glasø 1980), it has been argued that this pyrolysis method is indeed a good way of assessing the gas-oil ratio of petroleums generated in marine source rocks, and that the broad range of GOR seen in most petroleum provinces is linked to fractionations occurring during expulsion, secondary migration and dysmigration. Accepting for the moment that these arguments are valid, it can be seen from Fig. 6.19B that all first-formed petroleums should have very low gas-oil ratios, no matter what type of source rock is involved, be they marine, paralic or non-marine. One specific con-

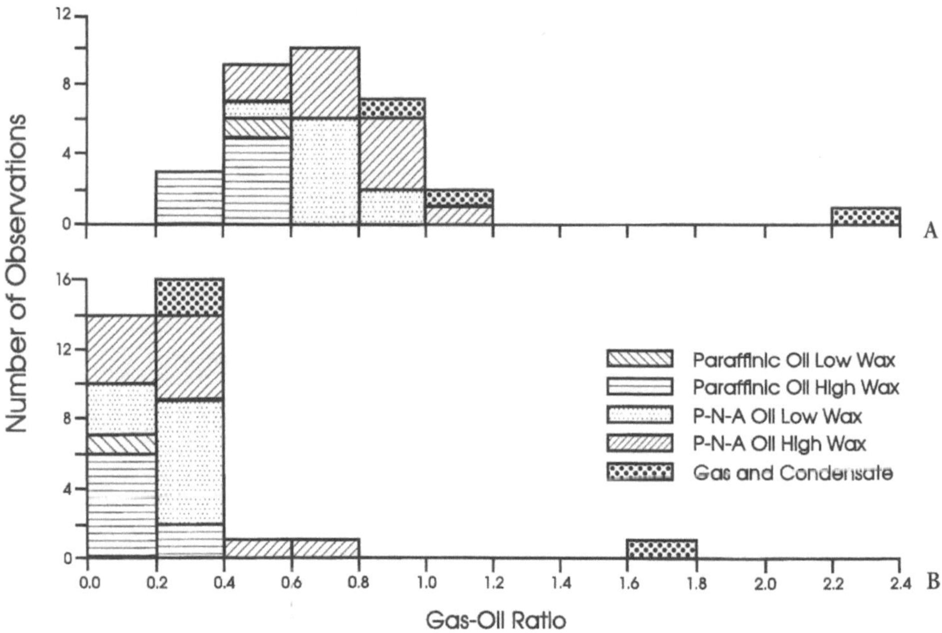

Fig. 6.19A,B. Gas-oil ratios (C_1-C_5/C_{6+}) of pyrolysates from a variety of kerogen types. **A** Based on resolved components only. **B** Based on total products, largely unresolved. (Horsfield 1989)

sequence of this would be that all humic coals are oil-prone, and that the gas-prone nature of petroleums associated with coals is due to the secondary cracking of oil that could not cannot migrate from micropores (see Levine 1993). While this mechanism indeed operates to some degree - bituminous coal extracts are extremely rich in involatile polar components - this does not mean that C_{6+} pyrolysate is a measure of these component yields. Indeed, coming back to the point made earlier, it is conceptually erroneous to calculate gas-oil ratio from the quotient of C_1–C_5/C_{6+} products. The distribution of what can be termed gas versus oil in the laboratory K→B reaction is not necessarily the same as in the B→O reaction (Horsfield and Düppenbecker 1991). In particular, "unresolved" components equating to "oil" (C_{6+}) in the bitumen B are likely to yield both oil and gas in nature, as inferred by the heating experiments on Bakken Shale kerogen (Fig. 6.16).

An alternative way of directly calculating GOR from on-line pyrolysate data is to utilise only the resolved components (Fig. 6.18 upper part) because, as described above, their chain length distribution and paraffinicity are empirically related to those of genetically related natural petroleums. Two points then emerge. First, GORs are higher by an average factor of 3 (Fig. 6.19A), and fall mainly in the range 0.2–1.2 kg/kg. Such GORs give more conservative estimates of relative oil yield and therefore of undersaturated petroleum occurrence than do those from total pyrolysates (Düppenbecker and Horsfield 1990). Secondly, a relationship between GOR and source rock type begins to emerge, with lacustrine petroleums predicted to have the lowest GORs, marine petroleums intermediate values and deltaic petroleums the highest. Such a result is in line with organic facies concepts where petroleum composition and sedimentary facies are closely linked (Jones 1987). It was this consideration that led Düppenbecker and Horsfield (1990) to measure GORs by the skim method, and these were used when modelling petroleum generation in the Posidonia Shale of northwestern Germany instead of the total pyrolysate approach (Düppenbecker and Welte 1991). Transformation ratios were predicted by a kinetic model, and the GOR of the generated petroleum calculated using the laboratory-determined (MSSV pyrolysis) relationship between GOR and transformation ratio. Even these GOR values might be considered minima because the aromatic content of petroleums are overemphasised by pyrolysis predictions (see earlier). To some degree this effect is counterbalanced because on-line pyrolysis takes no account of the macromolecular oil fraction, but this effect is relatively minor because asphaltenes make up only a minor proportion of most petroleums (Cornford 1993; Wilhelms and Larter 1994).

In summary, it is presently unrealistic to imagine that bulk petroleum compositions can be determined quantitatively and directly using pyrolysis because (1) the nature of the unresolved "hump" in pyrolysates remains problematic and (2) the natural fate of aromatic and sulphur-containing moieties is poorly understood, with the result that heavy oil generation and the petroleum potential of coal are especially difficult to evaluate. The similarity between the solution GORs of natural petroleums and marine kerogen pyrolysate GORs appears to be coincidence. GOR based on resolved components might be a temporary "prediction engineering" solution to this problem.

6.6.2 Organofacies Based on Petroleum Composition

Now that some of the deficiencies of quantitative pyrolysis data have been explored, this section reviews how qualitative approaches may be used to infer the link between bulk petroleum compositions, kerogen compositions and depositional environment.

In recent years "determining organic facies" has become synonymous with evaluating the composition and concentration of organic matter within a given sediment as it is, or was, in its immature stage. The term "organic facies" has been used in an alarming variety of contexts and connotations (Jones 1987), but these boil down to two basic usages. One seeks to relate variations in organic matter compositions and concentrations to specific biological inputs and depositional processes, and thence recognise "anoxic and oxic facies", "marine and deltaic facies", and the like. This involves the integration of detailed geological information with, *inter alia*, organic petrographic and biomarker data. Details can be found in treatises by Reading et al. (1978) and Walker (1979) on sedimentology, Peters and Moldowan (1993) on biomarkers and Chapter 5 (this Vol.) on organic sedimentation. The second usage focussed on here, is to apply the information in a practical way. Because dramatic chemical and physical transformations and mass transfer processes ensue under "mild" temperature conditions (<200 °C; see metamorphic processes), it is logical that organofacies should be defined according to criteria that allow the transformation of the static kerogen phase into mobile products, such as petroleum, to be defined qualitatively and quantitatively.

Whereas Jones (1987) defined and mapped organofacies according to petroleum *yield* potential, this chapter deals with predicting the *composition* of petroleums using the pyrolysate classification scheme of Horsfield (1989; see also Cooper and Barnard 1984; Horsfield 1984; Larter and Senftle 1985; Orr 1986). Gas/condensate-, paraffinic-naphthenic-aromatic crude oil (high- and low-wax varieties) and paraffinic crude oil generating (high- and low-wax varieties) potentials have been recognised (Fig. 6.20) and are described below. Similarities between these divisions and those of naturally occurring crude oils, as described by Tissot and Welte (1978) are pointed out where appropriate. The divisions of Jones (1987) are also included where feasible.

Low-Wax, Paraffinic-Naphthenic-Aromatic Crude Oil Generating Facies

Woodford Shale and Toarcian Shale pyrolysates define the low-wax, mixed-base crude oil generating facies. In nature the petroleums they generate belong to the paraffinic-naphthenic and aromatic-intermediate classes. The low-wax, mixed-base crude oil generating facies consists of black laminated shales and carbonates with a high petroleum-generating potential (Jones' facies B). The organic matter is derived mainly from algal and bacterial inputs though terrigenous macerals such as sporinite and inertinite are common. This facies most likely forms in silled basins and on stagnant shelves where reducing

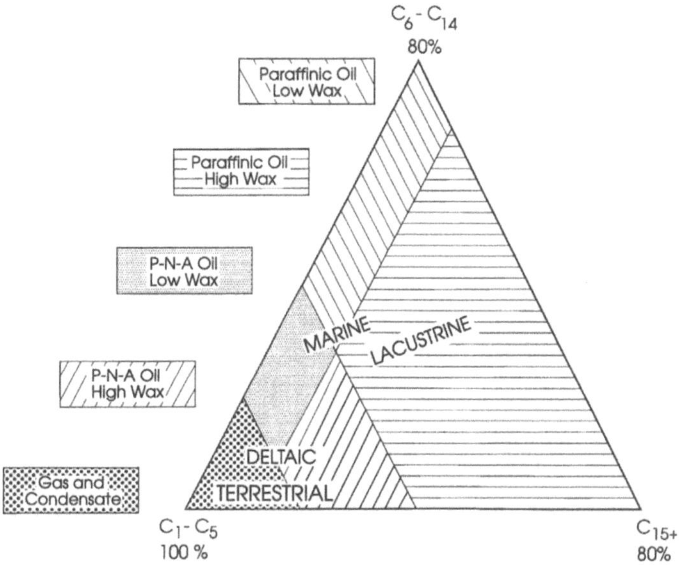

Fig. 6.20. The empirical relationship between petroleum types, source depositional environment and pyrolysate composition for clastic petroleum systems. (Newly calibrated after Horsfield 1989)

conditions exist at or above the sediment-water interface. Organic productivity related to nutrient availability, residence time in the water column and the degree of water column stratification are important controlling factors. The Woodford Shale (Devonian) of the United States, the Kimmeridge Clay (Jurassic) of the United Kingdom, the Fischschiefer (Cretaceous) of Germany, and the Posidonia Shale/early Toarcian Shales (Jurassic) of Germany and France are examples of this facies (Thomas et al. 1985; Horsfield 1989; Lafargue and Béhar 1989; Muscio et al. 1991; Ropertz 1994), as are the Pennsylvanian Black Shales of Kansas (Baker and Horsfield, unpublished data). Chain length distributions have been noted to remain essentially constant throughout the oil window (R_o <1.0%) in the case of the Posidonia Shale, signifying that the facies can be recognised by comparison of immature and mature source rocks alike.

The pyrolysates of kerogens and asphaltenes from Triassic source rocks (Italy) indicate that they should generate low-wax mixed-base crude oil in nature, but the natural petroleums are actually of the heavy, sulphur-rich and asphaltic type (di Primio and Horsfield, 1995). This raises the issue of how to predict the occurrence and composition of crude oils that are rich in non-volatile components, and is discussed below (see "Asphaltic Oils").

Low-Wax, Paraffinic Crude Oil Generating Facies

The low-wax, paraffinic oil category is named after the composition of Ordovician oils in the Michigan Basin (paraffinic class of Tissot and Welte 1978) and defined by the kerogen pyrolysate composition of the marine Glenwood and Trenton Formations. Algal kerogens derived from *Tasmanites* and *Gloeocapsomorpha* (and associated biomass) have an exceedingly high potential for generating low-wax paraffinic oil (Fowler and Douglas 1984; Hoffmann et al. 1987; Jacobson et al. 1988). Deposited in the marine system under conditions of persistent anoxia (Jones 1987), this facies is relatively rare and grades into, and intercalates with the more common low-wax, mixed-base crude oil generating facies.

High-Wax, Paraffinic Crude Oil Generating Facies

The "high-wax paraffinic" field was named after the composition of Uinta Basin oils (paraffinic class of Tissot and Welte 1978) and defined using pyrolysates of the lacustrine Green River Shale. The high-wax, paraffinic crude oil generating facies is a well-documented source of high-wax oil in the Tertiary freshwater and saline lacustrine deposits of Asia, Australasia and the United States (Hu 1985; Hong et al. 1986; Moore et al. 1986; Powell 1986). This facies grades into, and may intercalate with, the gas/condensate- and high-wax, mixed-base crude oil generating facies described above. The kerogen consists of the remains of algae (*Botryococcus*) and bacteria and is a prolific petroleum source (Jones' facies A). A stable stratified water column, as controlled by salinity and temperature gradients and therefore by climate and topography, is required to form this facies.

Lacustrine sediments may be deposited under oxic through anoxic conditions and within saline through freshwater environments. The Laney Shale member of the Green River Shale Formation (Washakie Basin) was deposited in a hydrologically closed, alkaline lake under an arid climate and was characterised by high concentrations (6–20% TOC) of alginite-rich organic matter. The Luman Tongue member, on the other hand, was deposited in a hydrologically open freshwater lake under a humid climate and consisted of relatively organic-poor (1–5% TOC) deep-lake mudstones and coaly lake-margin sediments (see Horsfield et al. 1993b, 1994; Schenk et al. 1993). Three major sequences occur in the Laney, corresponding to early lake deepening, maximum lake extension and a late rejuvenation phase that followed a period of desiccation. The highest wax contents are associated with the first and third of these, whereas the middle sequence (all are immature; R_o<0.4%) has a pyrolysate wax content approaching that of the low-wax, mixed-base crude oil generating facies (Fig. 6.21). Interestingly, the deep-lake sediments of the freshwater Luman Tongue member fall within the same trend as the alkaline Laney Shale member. It can be concluded from this that alkalinity does not exert a major influence on the chain length distribution of *n*-alkane precursors in lake sediments.

Chain length distribution has been linked to redox conditions in the case of relatively organic-poor (0.9–2.2% TOC) Cretaceous lacustrine sediments from the eastern Alps (Sachsenhofer et al. 1996). Short-chain lengths characterised the pyrolysates of organic-poor hydrogen-sparse bioturbated marls and re-

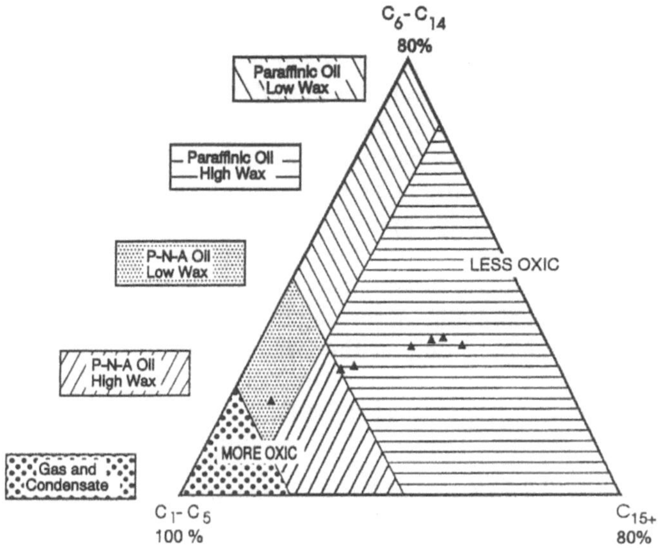

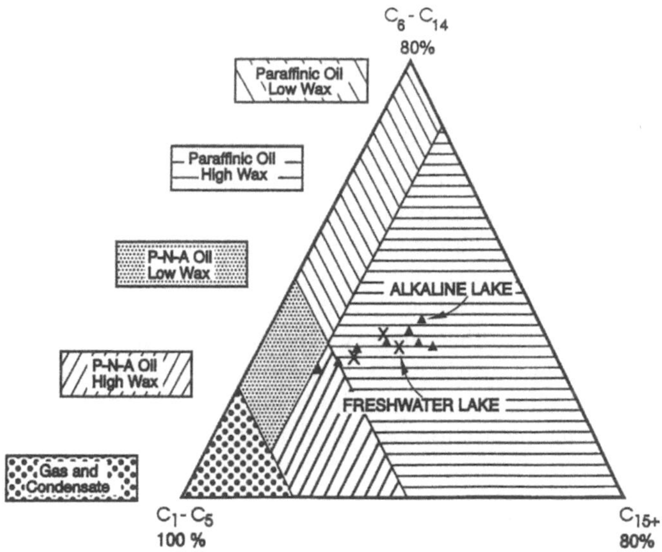

Fig. 6.21. Inferred petroleum compositions in lacustrine environments. Oxicity has a strong bearing on source rock composition and hence likely petroleum composition in the Styrian Basin (Sachsenhofer et al. 1996). Alkalinity has no measurable effect in the Washakie Basin (Horsfield et al. 1994): crosses, freshwater lake; triangles, alkaline lake

worked bioclasts (coquinas) deposited under oxic conditions. In contrast, higher wax contents are associated with shales richer in hydrogen-rich organic matter, and which had been deposited under more anoxic conditions. The extent of the change in chain length distribution is shown in Fig. 6.21. In other depositional systems of the Alpine region, such as the Ottnangian and Karpathian of the Styrian Basin (Sachsenhofer 1993), this trend has been related to a change from a non-marine environment, with highest wax contents, to anoxic marine deposition for pyrolysates with the lowest wax contents.

The chain length distribution of Green River Formation kerogen pyrolysates for the Uinta Basin become shorter with increasing levels of thermal maturation (Gray et al. 1991). According to MSSV pyrolysis experiments, this appears to be pronounced only at the earliest levels of conversion; thereafter chain length distributions remain constant (Düppenbecker and Horsfield 1990; Horsfield and Düppenbecker 1991).

High-Wax, Paraffinic-Naphthenic-Aromatic Crude Oil Generating Facies

Coal of the deltaic member of the Talang Akar Formation is the source of waxy oil, gas and condensate in the Ardjuna Basin and was used to define the boundaries of the high-wax, paraffinic-naphthenic-aromatic crude oil generating facies. This sub-class of high-wax oils, not recognised in the classification scheme of Tissot and Welte (1978), usually occurs in lower delta plain and inner shelf environments and commonly grades into and intercalates with the high-wax paraffinic crude oil facies described above. The organic matter often has a variable type II/III to type III composition and therefore falls in facies BC and C of Jones (1987). The kerogen consists of vitrinite, cutinite, liptodetrinite and resinite in variable proportions, and is derived in part from lignocellulosic materials, preserved and partially degraded cuticular tissue and plant resins, which give phenolic compounds, long-chain *n*-alkanes and naphthenoaromatic compounds on analytical pyrolysis. Maceral abundances are thought to be strongly influenced by the actions of meandering river channels and tidal currents which erode peat deposits and segregate the macerals (Thompson et al. 1985). Many Tertiary coals from Southeast Asia and Australia fall in this category (Thomas 1982; Gordon 1985; Kelley et al. 1985). Chain length distributions may be maturity-sensitive according to two-step pyrolysis data, with long chains being released in preference to short ones with increasing maturation (Horsfield 1989).

Gas/Condensate-Generating Facies

The gas/condensate-generating facies is defined by vitrinite and sporinite pyrolysates because Carboniferous coals rich in these macerals/maceral groups are considered to be a prolific source of gas in northwestern Europe. Organic-rich sediments of the facies display a pyrolysate chain length distribution which is rich in low molecular weight components (see Fig. 6.20). They often consist of coals, shales and siltstones which were deposited in continental/

deltaic settings such as in swamps on lake margins or on the delta plain. When total petroleum-generating potential is low (facies C and CD of Jones 1987), telocollinite, sporinite and inertinite are the predominant macerals. This is typified by the Carboniferous of northwestern Europe (e.g. Scheidt and Littke 1989). When total petroleum-generating potential is high (facies B and BC), desmocollinite and resinite are usually the major macerals present. This is a common feature of Southeast Asian coals.

The analysis of Westphalian coals (Krooss et al. 1995) revealed that the yield of pyrolysis gases falls by almost an order of magnitude in going from high volatile bituminous to meta-anthracite rank. This is exemplified by methane yields which fell from 33–50 mg/g TOC at low rank to <5 mg/g TOC at high rank. Primary gas potential may be estimated from open-system pyrolysis whereas primary and secondary gas yields can be determined from closed-system pyrolysis. The most significant differences in gas yield between the open and closed methods occur for those coals containing labile alkylaromatic and/or aliphatic moieties, this corresponding to the R_o range <1.4% where the structure is open through "liquid", according to Hirsch (1954). In the open-system, pyrolysis products are swept from the heated zone and to a large degree reflect the types of structures in the parent material. In the closed-system, first-formed products remain in the heated zone and react further to form secondary products. Thus, as an example, alkylaromatic moieties cracked from the coal molecule during open-system pyrolysis may be further transformed to yield hydrocarbon gases (from methylene and methyl substituents) and a polyaromatic residue upon further thermal degradation under closed-system conditions. The yield of methane from the coal is between five and ten times more than from type II marine kerogen of equivalent rank whereas the yields of wet gas are about the same.

Surprisingly, the organic-rich alginitic type II marine shales of the Bakken and Alum Shales also belong to this facies. In both cases the confirmation of light hydrocarbon-generating capacity was afforded by thermovaporisation–GC. In the case of the Alum Shale, aliphatic carbon was abundant in the kerogen structure but was not released as such during pyrolysis, there being a predominance of C_1–C_9 aliphatics and mono- plus dialkylaromatics. MSSV pyrolysis experiments also revealed the contemporaneous formation of polyaromatic "dead carbon". While radiation damage might account for cross-linking and thence aromatisation/condensation on pyrolysis (Dahl et al. 1988; Lewan and Buchardt 1989) primary biological structures, possibly carotenoid-like, represent a viable alternative (Horsfield et al. 1992). The case of the immature Bakken Shale is similar in that alkylaromatic hydrocarbons and enhanced "dead carbon" formation are associated with pyrolysis (Muscio and Horsfield 1996). Up to 30% of originally volatilisable carbon may be transformed into involatile aromatic structures during natural maturation according to this experimental evidence (Fig. 6.14). Mass balance models which assume that involatile carbon remains constant through maturation (Larter 1985; Cooles et al. 1986) will therefore give overestimates petroleum yields in these cases.

Facies Recognition from Asphaltene Pyrolysis

Pyrolysis data for the kerogen, asphaltene and heterocomponent fractions of the Posidonia Shale is shown in Figs. 6.22 and 6.23. The pyrolysates' *n*-alkyl chain length distribution for the asphaltenes and heterocompounds are very similar to each other and, as noted for the kerogen fraction, fall in a restricted part of the ternary plot despite the large maturity span (Fig. 6.22). While the points actually fall in a crescent-shaped evolution path, the data points covering petroleum generation ($R_m = 0.53–0.88\%$) fall in an extremely restricted range. This range is within the low-wax paraffinic-naphthenic-aromatic oil category, directly adjacent to the tight cluster of kerogen data points.

It has been shown by other workers that, despite apparent differences in average molecular weight, the asphaltenes in crude oils show a strong compositional resemblance to those in their source rocks and thence to the parent kerogen (Pelet et al. 1986; Jones et al. 1988). From this it can be deduced that the pyrolysate chain length distribution of asphaltenes in Posidonia Shale derived crude oils will be the same as that of the asphaltenes in the source rock bitumen at the time of expulsion. Because the data scatter for the oil-generating maturity range is extremely narrow and closely adjacent to the equivalent kerogen data points (see Fig. 6.22), it is concluded that organic matter quality can be determined either from kerogens or from genetically related macromolecular organic matter in asphaltene and heterocomponent fractions of rock bitumens and crude oils. Interestingly, pyrolysate gas-oil ratios for the kerogen and asphaltene fractions are essentially identical (Fig. 6.23) for maturities up to 0.88% R_m. This further attests to the structural similarity between kerogens and asphaltenes, and extends kerogen typing and petroleum prediction concepts from source to reservoir lithologies.

The Holzener Asphaltkalk of northern Germany consists of a series of Kimmeridgian-Portlandian bitumen-impregnated limestones which probably represent an exhumed and highly altered multi-pay oil accumulation (Herrmann 1971). The source of this deposit and the relative mass of unaltered petroleum originally present could be inferred from pyrolysis data using Holzener Asphaltkalk asphaltenes, Posidonia Shale asphaltenes and Posidonia Shale kerogens and the concepts described above.

Carbon isotope data and regional geochemical considerations point towards the original petroleum as having been derived from the Posidonia Shale source rock (Horsfield et al. 1992). MSSV pyrolysis at 330 °C/72 h and 350 °C/72 h revealed that Posidonia Shale kerogens and asphaltenes plot close together on a ternary diagram. Simulation experiments confirm that the data scatter is very small in the case of mature (>20% conversion) samples, and that data points are close to but distinct from those for Holzener Asphaltkalk asphaltenes (Fig. 6.24). The kerogen-based calibration points in Fig. 6.24 indicate that an active algal lacustrine source rock may have co-sourced the Holzener Asphaltkalk petroleum.

The original form of the accumulation was significantly different to that of today because most of the original reservoired oil was lost, preferentially from pores >25 nm, during post-Cretaceous uplift. Assuming the original petroleum

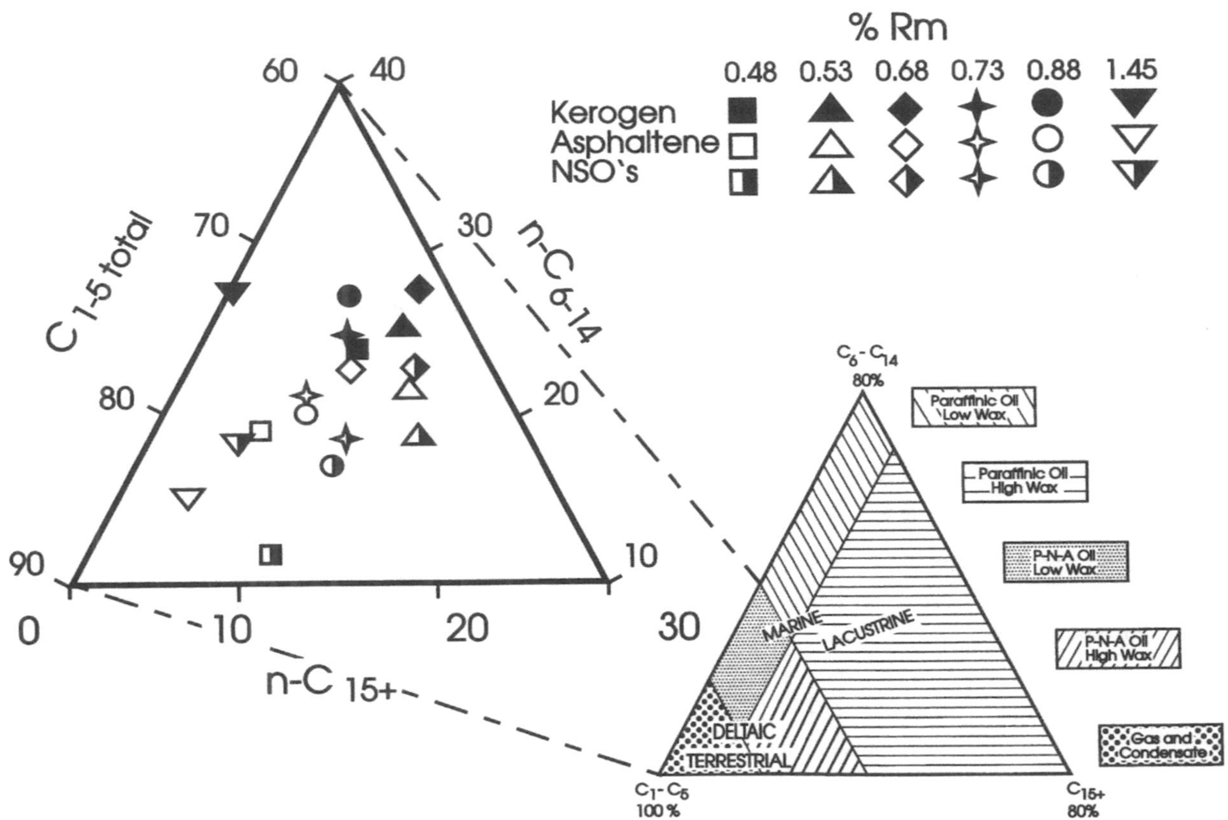

Fig. 6.22. Pyrolysis-GC of macromolecular organic matter in the Posidonia Shale reveals the relative distribution of C_1–C_5 total, C_6–C_{14} *n*-alkenes/-anes and *n*-C_{15+} *n*-alkenes/-anes. (Muscio et al. 1991)

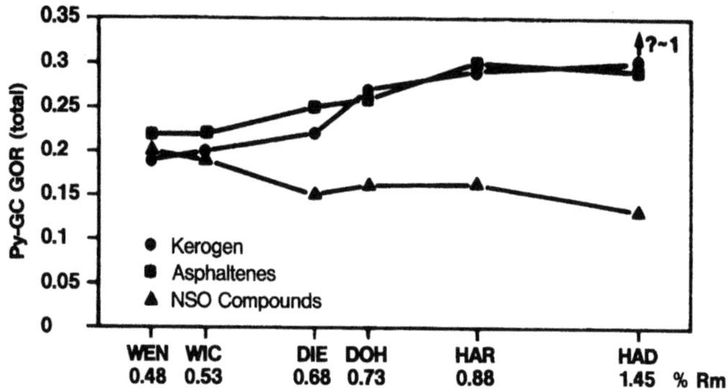

Fig. 6.23. Changes in the gas-oil ratios (C_1-C_5/C_{6+} total components) of pyrolysates from kerogens, asphaltenes and resins from the Posidonia Shale at different levels of maturity. (Reprinted from Basin Modelling: advances and applications, Horsfield et al., Kerogen typing concepts designed for the quantitative geochemical evaluation of petroleum potential, 1993, pp 243-249, with kind permission from Elsevier Science Publishers – NL Sara Burgerhartstraat 25, 1055 KV Amsterdam, The Netherlands)

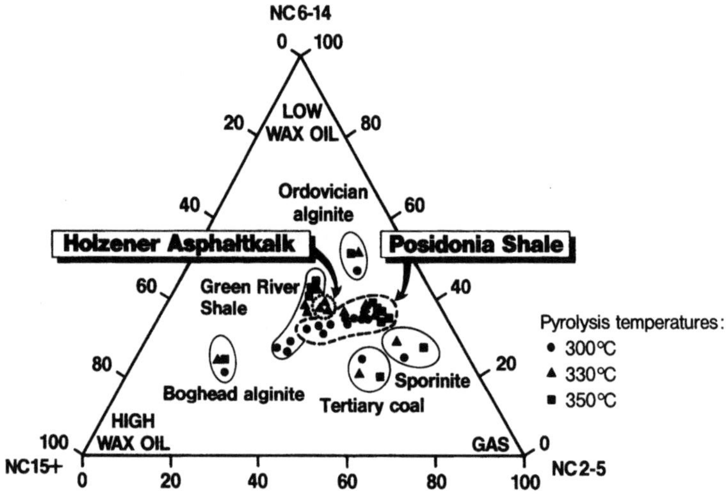

Fig. 6.24. Chain length distribution of *n*-alkanes in MSSV pyrolysates of immature Posidonia Shale kerogen and Holzener Asphaltkalk asphaltenes (Horsfield et al. 1993). (Reprinted from Basin Modelling: advances and applications, Horsfield et al., Kerogen typing concepts designed for the quantitative geochemical evaluation of petroleum potential, 1993, pp 243-249, with kind permission from Elsevier Science Publishers – NL Sara Burgerhartstraat 25, 1055 KV Amsterdam, The Netherlands)

had a GOR close to 0.2 kg/kg based on pyrolysis-GC data for Holzener Asphaltkalk asphaltenes (Heckers, unpublished results) and Posidonia Shale kerogens and asphaltenes, and equating buoyancy and capillary pressure, an original oil column of 100 m was estimated (Horsfield et al. 1991). Petroleum density, water density and interfacial tension values were calculated using a maximum temperature of 100 °C and a hydrostatic pressure of up to 15 MPa. The amount of petroleum lost during alteration was calculated assuming that the gross C_{15+} composition of Posidonia Shale derived crude oils (Lower Saxony Basin) and the Holzener Asphaltkalk were representative of the unaltered starting material and altered end-product, respectively, that no polar components were added during biodegradation, and that the total amount of C_{1+} original petroleum could be inferred from the ratio of light to heavy products (C_{15-}/C_{15+}) from pyrolysis. The mass balance calculation revealed that a 50–90% mass loss of petroleum could be attributed to alteration phenomena. GOR values of 0.2 kg/kg and lower were considered. Figure 6.19 indicates this GOR range is valid, even allowing for an algal lacustrine co-source.

These preliminary findings pointed to macromolecular organic matter in petroleum accumulations, seeps and stains as being a source of data from which the basin modeller might determine, on the one hand, kerogen type/ source facies and, on the other, gross compositional characteristics of the original unaltered petroleum expelled from the source rock.

Asphaltic Oils

Carbonate-evaporitic and carbonate-siliceous rocks are the sources of large quantities of immature to marginally mature, non-biodegraded heavy oil deposits that occur as liquids or semisolids in porous and fractured media (Powell 1984; Zumberge 1984; Tannenbaum and Aizenshtat 1985). These non-biodegraded heavy oils, found, for example, in the southern United States, Venezuela and Italy (Gransch and Posthuma 1974; Baskin and Peters 1992; Matavelli et al. 1993), are considered to be the products of early petroleum generation of oil-prone, sulphur-rich kerogens (Orr 1986). Despite the fact that in reality they are seen to have generated low API gravity, high-sulphur heavy crude oils, most sulphur-rich kerogen pyrolysates have a *n*-alkyl chain length distribution that indicates they would generate low-wax mixed-base crude oils in nature (see earlier). Examples are the Maastrichtian Limestone of Jordan and Mesozoic source rocks from Italy (di Primio and Horsfield 1996). The explanation for this discrepancy is that thermally labile (poly)sulphide bridges in the kerogen structure rupture very easily to produce low API gravity asphaltene-rich petroleum (Orr 1986; Baskin and Peters 1992), whereas low-sulphur kerogens from argillaceous source rocks with the same *n*-alkyl chain distribution crack at higher maturities to give a higher API gravity product that is low in asphaltenes.

In order to discriminate these two petroleum types an additional ternary diagram has been developed (di Primio and Horsfield 1996), based on the concept that thiophenes generated on pyrolysis are representative of the total pool of organically bound sulphur (Eglinton et al. 1990b), and that high-sulphur crude oils are generated from high-sulphur kerogens (Gransch and Posthuma 1974). It is based on toluene, 2,5 dimethylthiophene and the sum of one short-chain and one long-chain *n*-alkene (n-$C_{9:1}$+n-$C_{25:1}$). The extent to which these compounds are representative of the aromatic, sulphur and *n*-alkyl moieties in kerogens is discussed above (Sect. 6.4.1). Figure 6.25 shows the new ternary diagram in conjunction with the original. The samples shown in

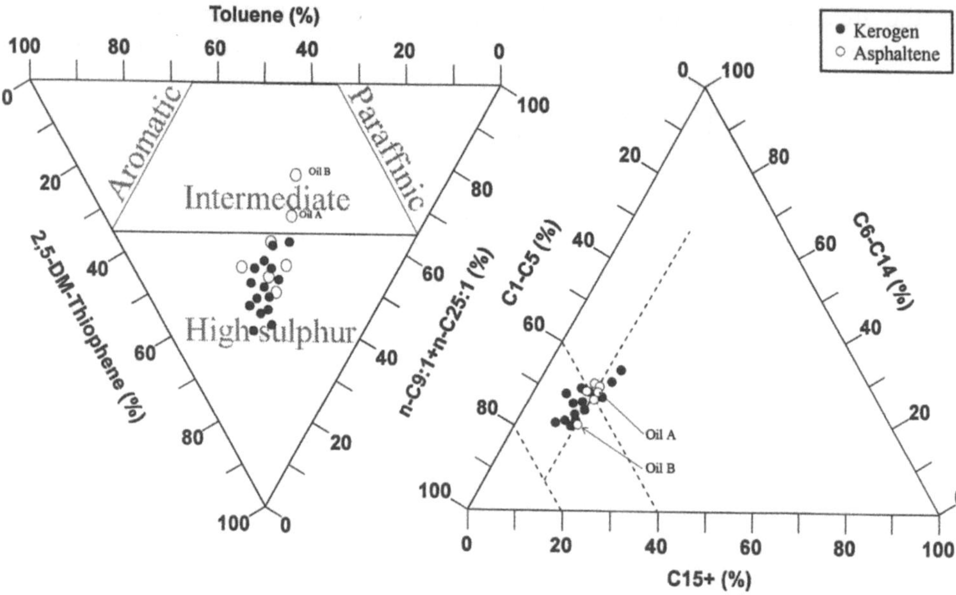

Fig. 6.25. The chain length distribution diagram (*right*), based on clastic petroleum system correlations (Horsfield 1989), cannot distinguish between sulphur-rich heavy oil potential and normal marine oil (paraffinic-naphthenic-aromatic) potential. Utilisation of sulphur-compound abundance in pyrolysates allows high-sulphur from low-sulphur petroleum potential to be recognised. (di Primio and Horsfield 1996)

these ternary plots include high-wax, intermediate, aromatic and high-sulphur kerogens. The distinction of high-sulphur from low-sulphur kerogens is impossible in the diagram of Horsfield (1989), whereas in the new ternary plot the proportion of 2,5 dimethylthiophene allows a clear differentiation of the extremely sulphur-rich kerogens from Italy and from the Jordan oil shale on the one hand, and even between relatively low-sulphur kerogens like the Toarcian of the Paris Basin and Posidonia Shale kerogens from Germany on the other. Additionally the distinction between kerogens which generate high-wax or aromatic, generally gas-rich, oils is possible.

Predicting the early generation of heavy oils from sulphur-rich kerogens by open-system pyrolysis is complicated by the fact that these methods cannot detect involatile compounds which make up a major portion of heavy oils. A combination of multistep open-system pyrolysis and the results from kinetic analysis of oil generation using a kerogen sample allows an insight into the activation energy-dependant release of compounds during petroleum generation. Figure 6.26 exemplifies that at low energies sulphur-rich products (mainly thiophenes) dominate in the pyrolysates. This fact can be attributed to the preferential cleavage of S-C bonds, although the presence of the thiophenes indicates that much of the sulphur is present in stable rings as well as in labile S-C bonds. These first pyrolysates indicate that in nature sulphur-rich petroleum is generated at this stage. The molecular composition of these oils differs

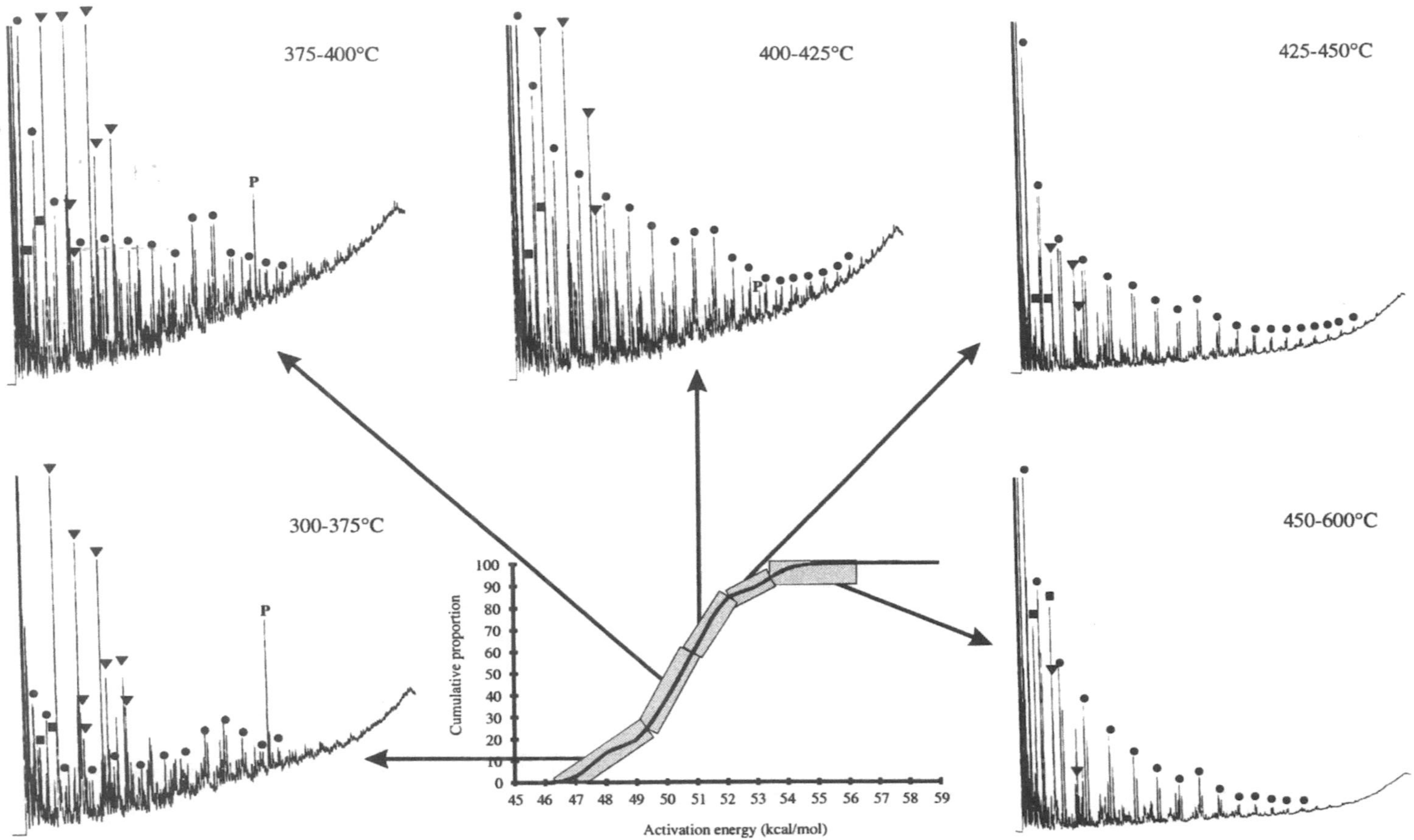

Fig. 6.26. Five-step pyrolysis-GC of an Italian source rock kerogen showing the abundances of thiophenes (*triangles*), aromatic hydrocarbons (*squares*) and *n*-alkenes/-anes (*circles*). The equivalent stages on the activation energy distribution from bulk-flow pyrolysis are indicated. (di Primio and Horsfield 1996)

from that of the pyrolysis products, since the major proportion of high molecular weight compounds of aromatic/asphaltic oils are thermally involatile and therefore cracked to secondary products during pyrolysis. High yields of low molecular weight alkylthiophenes signify only that the kerogen is rich in sulphur, and it is left to known correlations to infer bulk oil composition. At higher energies, in which the dominant activation energy is included, the pyrolysates contain higher proportions of *n*-alkanes/alkenes with a fixed chain length distribution and lower sulphur contents, which in nature would be manifested as higher API oils.

6.7 Concluding Remarks

Despite conceptual advances in recent years, there remain important gaps in our knowledge of the nature of first-formed petroleum in source rocks. This is partly because it is largely impossible directly to observe source rock "kitchens". We must simply rely to a large degree on highly precise but indirect methods such as pyrolysis, whose results must be extrapolated over nine orders of magnitude of heating rate. When doing so, it appears valid to consider "total open-system pyrolysate" as being equivalent to "total petroleum" for most oil source rock systems because mass balance models (yields) and kinetic predictions (rates) work well under the flexible boundary conditions employed. Exceptions to this are mentioned again shortly, below. Water can be viewed as a hydrogen donor, but recent experiments suggest this not to be important in sedimentary basins (Michels et al. 1995). As far as compositional considerations are concerned, we can feel comfortable with some of the qualitative relationships that exist between pyrolysates and petroleums. For instance, normal paraffin chain length distributions form the basis for linking petroleum composition, kerogen composition and sedimentary facies in a predictive organic facies model. However, there are major problems when it comes to using quantitative pyrolysis data in geochemical models. Gas-oil ratio predictions are ubiquitously low because of differences in reaction mechanisms for slow versus fast heating rates. The natural fate of kerogen moieties which yield aromatic hydrocarbons and sulphur-containing compounds on pyrolysis remains uncertain. In the laboratory, both originate via thermal cracking reactions. In nature, condensation and aromatisation reactions are more prevalent, so that these moieties are retained in the residual kerogen rather than being evolved as petroleum. The outcome is two-fold. Predicted petroleum yields are too high and predicted compositions are incorrect. As far as humic coals are concerned, oil potential may be overestimated. The same may be true for some hydrogen-rich kerogens (exemplified by the Alum Shale and Bakken Shale). Studying the relative rates of aromatisation/condensation and cracking reactions under laboratory and natural burial conditions will be important in the future because of its application in natural gas research.

Acknowledgments. I would like to thank my contemporaries at Newcastle University, Conoco Inc, Atlantic Richfield Company and KFA Jülich, whose input, both direct and indirect, helped me formulate the ideas expressed in this publication. Grateful thanks are also extended to a great many friends and colleagues in industry and academia whose encouragement and/or sponsorship has allowed this research theme to be pursued.

References

Adam PC, Schmid B, Mycke B, Strazielle C, Connan J, Huc A, Riva A, Albrecht P (1993) Structural investigation of non-polar sulfur cross-linked macromolecules in petroleum. Geochim Cosmochim Acta 57: 3395–3419

Albrecht P, Vandenbroucke M, Mandengué M (1976) Geochemical studies on the organic matter from the Douala basin (Cameroon). 1. Evolution of the extractable organic matter and the formation of petroleum. Geochim Cosmochim Acta 40: 791–799

Alexander R, Noble RA, Kagi RI (1987) Fossil resin biomarkers and their application in oil to source-rock correlation, Gippsland basin, Australia. APEA J 27: 63–72

Asakawa T, Takeda N (1988) Study of petroleum generation by pyrolysis 1. Pyrolysis experiments by rock eval and assumption of molecular structural change of kerogen using 13-CNMR. Appl Geochem 3: 441–453

Bailey NJL (1981) Hydrocarbon potential of organic matter. In: Brooks J (ed) Organic maturation studies and fossil fuel exploration. Academic Press, New York, pp 283–301

Bailey NJL, Evans CR, Milner CWD (1974) Applying petroleum geochemistry to search for oil: examples from Western Canada basin. Bull Am Assoc Petol Geol 58: 2284–2294

Baskin DK, Peters KE (1992) Early generation characteristics of a sulfur-rich monterey kerogen. Bull Am Assoc Petrol Geol 76(1): 1–13

Béhar F, Pelet R (1985) Pyrolysis-gas chromatography applied to organic geochemistry – structural similarities between kerogens and asphaltenes from related rock extracts and oils. J Anal Appl Pyrolysis 8: 173–187

Béhar F, Vandenbroucke M (1987) Chemical modelling of kerogens. Org Geochem 11(1): 15–24

van Bergen PF, Collinson ME, Sinninghe Damste JS, de Leeuw JW (1994) Chemical and microscopical characterization of inner seed coats of fossil water plants. Geochim Cosmochim Acta 58: 231–239

Bjorøy M, Williams JA, Dolcater DL, Winters JC (1988) Variation in hydrocarbon distribution in artificially matured oils. Org Geochem 13(4–6): 901–913

Bockmeulen H, Barker C, Dickey PA (1983) Geology and geochemistry of crude oils, Bolivar coastal fields, Venezuela. Bull Am Assoc Petrol Geol 67(2): 242–270

Boreham CJ, Powell TG (1991) Variation in pyrolysate composition of sediments from the Jurassic Walloon coal measures, eastern Australia as a function of thermal maturation. Org Geochem 17(6): 723–733

Boudou J-P, Durand B, Ouding J-L (1984) Diagenetic trends of a tertiary low-rank coal series. Geochim Cosmochim Acta 48: 2005–2010

Brassell SC, Lewis CA, de Leeuw JW, de Lange F, Sinninghe Damsté JS (1986) Isoprenoid thiophenes: novel diagenetic products in sediments? Nature 320: 160–162

Braun RL, Rothman AJ (1975) Oil shale pyrolysis: kinetics and mechanism of oil production. Fuel 54: 129–131

Brooks JD, Smith JW (1969) The diagenesis of plant lipids during the formation of coal, petroleum and natural gas. Geochim Cosmochim Acta 33: 1183–1194

Burlingame AL, Haug PA, Schnoes HK, Simoneit BR (1969) Fatty acids derived from the Green River formation oil shale by extractions and oxidations-a review. In: Schenck PA, Havenaar I (eds) Advances in organic geochemistry 1968. Pergamon Press, Oxford, pp 85–129

Burnham AK, Braun RL (1990) Development of a detailed model of petroleum formation, destruction, and expulsion from lacustrine and marine source rocks. Org Geochem 16(1–3): 27–39

Burnham AK, Happe JA (1984) On the mechanism of kerogen pyrolysis. Fuel 63(10): 1353–1356

Burnham AK, Clarkson JE, Singleton MF, Wong CM, Crawford RW (1982) Biological markers from Green River kerogen decomposition. Geochim Cosmochim Acta 46: 1243–1251

Cane RF, Albion PR (1971) The phytochemical history of torbanites. J Proc R Soc NSW 104: 31–37

Cane RF, Albion PR (1973) The organic geochemistry of torbanite precursors. Geochim Cosmochim Acta 37: 1543–1549

Castelli A, Chiaramonte MA, Beltrame PL, Carniti P, Del Bianco A, Stroppa F (1990) Thermal degradation of kerogen by hydrous pyrolysis. A kinetic study. Org Geochem 16(1–3): 75–82

Chalansonnet S, Largeau C, Casadevall E, Berkaloff C, Peniguel G, Couderc R (1988) Cyanobacterial resistant biopolymers. Geochemical implications of the properties of Schizothrix sp. resistant material. Org Geochem 13(4–6): 1003–1010

Claxton MJ, Patience RL, Park PJD (1993) Molecular modelling of bond energies in potential kerogen sub-units. In: Oygard K (ed) Organic geochemistry. Falch Hurtigtrykk, Oslo, pp 198–201

Connan J, Coustau H (1987) Influence of the geological and geochemical characteristics of heavy oils on their recovery. In: Meyer RF (ed) Exploration for heavy crude oil and natural bitumen. AAPG studies in geology, vol 25. AAPG, Tulsa, pp 261–179

Cooles GP, Mackenzie AS, Quigley TM (1986) Calculation of petroleum masses generated and expelled from source rocks. In: Leythaeuser D, Rullkötter J (eds) Advances in organic geochemistry 1985. Organic geochemistry, vol 10. Pergamon Press, Oxford, pp 235–245

Cooper BS, Barnard PC (1984) Source rocks and oils of the central and northern North Sea. In: Demaison G, Murris RJ (eds) Petroleum geochemistry and basin evaluation. AAPG Mem 35: 303–314

Cooper BS, Murchison DG (1969) Organic geochemistry of coal. In: Eglinton G, Murphy MTJ (eds) Organic geochemistry. Springer, Berlin Heidelberg New York, pp 699–726

Cornford C (1993) Hydrocarbon flux efficiences in the central Graben of the North Sea. Geofluids '93 Extended Abstr: 76–81

Cornford C, Morrow JA, Turrington A, Miles JA, Brooks J (1983) Some geological controls on oil composition in the UK North Sea. In: Brooks J (ed) Petroleum geochemistry and exploration of Europe. Geological Society of London Spec Publ 12. Academic Press, London, pp 175–194

Curry DJ, Simpler TK (1988) Isoprenoid constituents in kerogens as a function of depositional enviroment and catagenesis. In: Mattavelli L, Novelli L (eds) Advances in organic geochemistry 1987. Pergamon Journals, Oxford, pp 995–1001

Dahl B, Augustson JH (1993) The influence of tertiary and quaternary sedimentation and erosion on hydrocarbon generation in Norwegian offshore basins. In: Doré AG, Augustson JH, Hermanrud C, Stewart DJ, Sylta Ø (eds) Basin modelling: advances and applications, NPF Spec Pub, vol 3. Elsevier, Amsterdam, pp 419–431

Dahl J, Hallberg R, Kaplan IR (1988) The effects of radioactive decay of uranium on elemental and isotopic ratios of alum shale kerogen. Appl Geochem 3: 583–589

de Leeuw JW, Largeau C (1993) A review of macromolecular organic compounds that comprise living organisms and their role in kerogen, coal and petroleum formation. In: Engel MH, Macko SA (eds) Organic geochemistry, principles and applications. Plenum Press, New York, pp 23–72

Dembicki H, Horsfield B, Ho T (1983) Source rock evaluation by pyrolysis GC. Bull Am Assoc Petrol Geol 67: 1094–1103

Derenne S, Largeau C, Casadevall E, Laupretre F (1987) Structural analysis of two torbanites at different evolutionary stages. Investigation of the quantitative reliability of fa determination by 13C CP/MAS n.m.r. Fuel 66: 1084–1090

Derenne S, Largeau C, Casadevall E, Connan J (1988) Mechanism of formation and chemical structure of coorongite. I. Structure and origin of the labile fraction. Fate of botryococennes during early diagenesis. Org Geochem 13(4–6): 965–971

Derenne S, Largeau C, Casadevall E, Berkaloff C (1989) Occurrence of a resistent biopolymer in the L race of *Botryococcus braunii*. Phytochemistry 28: 1137–1142

Derenne S, Metzger P, Largeau C, van Bergen PF, Gatellier JP, Sinninghe Damsté JS, de Leeuw JW, Berkaloff C (1992) Similar morphological and chemical variations of *Gloeocapsomorpha prisca* in Ordovician sediments and cultured *Botryococcus braunii* as a response to changes in salinity. In: Eckardt C, Maxwell JR, Larter SR, Manning DAC (eds) Advances in organic geochemistry 1991. Org Geochem 19(4–6): 299–313

di Primio R, Horsfield B (1996) Predicting the generation of heavy oils in carbonate/evaporitic environments using pyrolysis methods. Org Geochem 24, 10\11 (in press)

Dong JZ, Katoh T, Itoh H, Ouchi K (1986) n-Paraffins obtained by extraction and mild stepwise hydrogenation of Wandoan coal. Fuel 65(8): 1073–1078

Doré AG, Augustson JH, Hermanrud C, Stewart DJ, Sylta Ø (eds) (1993) Basin modelling: advances and applications. Norwegian Petroleum Society Spec Publ, 3rd edn. Elsevier, Amsterdam

Douglas AG, Coates RC, Bowler BFJ, Hall K (1977) Alkanes from pyrolysis of recent sediments. In: Gomez-Angulo JA, Campos R (eds) Advances in organic geochemistry 1975. Enadimsa, Madrid, pp 357–374

Douglas AG, Sinninghe Damsté JS, Fowler MG, Eglinton TI, de Leeuw JW (1991) Unique distribution of hydrocarbons and sulphur compounds released by flash pyrolysis from the fossilised alga *Gloecapsomorpha prisca*, a major constituent in one of four Ordovician kerogens. Geochim Cosmochim Acta 55: 275–291

Düppenbecker SJ, Horsfield B (1990) Compositional information for kinetic modelling and petroleum type prediction. Org Geochem 16: 259–266

Düppenbecker SJ, Welte DH (1991) Petroleum expulsion from source rocks – insights from geology, geochemistry and computerized basin modelling. Proc 13th World Petrol Congr 3: 1–13

Dungworth G, Schwartz AW (1972) Kerogen isolated from the precambrian of South Africa and Australia. In: Gaertner HRV, Wehner H (eds) Advances in organic geochemistry 1971. Pergamon Press, Oxford, pp 699–706

Durand B, Paratte M (1983) Oil potential of coals: a geochemical approach. In: Brooks J (ed) Petroleum geochemistry and exploration of Europe. Blackwell, Oxford, pp 285–292

Durand B, Nicaise G, Roucache J, Vandenbroucke M, Hagemann HW (1977) Etude geochimique d'une serie de charbons. In: Campos R, Goni J (eds) Advances in organic geochemistry 1975. Enadimsa, Madrid, pp 601–632

Dzou LIP, Hughes WB (1993) Geochemistry of oils and condensates, K field, offshore Taiwan: a case study in migration fractionation. Org Geochem 20(4): 437–462

Eglinton TI (1988) An investigation of kerogens using pyrolysis methods. PhD Diss, Univeristy of Newcastle upon Tyne

Eglinton TI, Sinninghe-Damsté JS, Kohnen MEL, de Leeuw JW (1990a) Rapid estimation of the organic sulfur content of kerogens, coals and asphaltenes by pyrolysis-gas chromatography. Fuel 69(11): 1394–1404

Eglinton TI, Sinninghe-Damsté JS, Kohnen MEL, De Leeuw JW, Larter SR, Patience RL (1990b) Analysis of maturity-related changes in the organic sulfur composition of kerogens by flash pyrolysis-gas chromatography. In: Orr WL, White CM (eds) ACS Symposium series 429: geochemistry of sulfur in fossil fuels. Am Chem Soc, Washington, pp 529–565

Eglinton TI, Sinninghe Damsté JS, Pool W, De Leeuw JW, Eijkel G, Boon JJ (1992) Organic sulphur in macromolecular sedimentary organic matter. II. Analysis of distributions of sulphur-containing pyrolysis products using multivariate techniques. Geochim Cosmochim Acta 56: 1545–1560

England WA, Mackenzie AS (1989) Some aspects of the organic geochemistry of petroleum fluids. Geol Rundsch 78: 291–304

England WA, Mackenzie AS, Mann DM, Quigley TM (1987) The movement and entrapement of petroleum fluids in the subsurface. J Geol Soc 144: 327–347

Erdmann M (1995) PhD Thesis, RWTH Aachen (in preparation)

Ericsson I, Lattimer RP (1988) Pyrolysis nomenclature. J Anal Appl Pyrolysis 14: 219–221

Espitalié J, Laporte JL, Madec M, Marquis F, Leplat P, Paulet J, Boutefeu A (1977) Méthode rapide de caractérisation des roches mères de leur potentiel pétrolier et de leur degré d'évolution. Rev Inst Fr Pétr 32: 23–42

Espitalié J, Ungerer P, Irwin I, Marquis F (1988) Primary cracking of kerogens. Experimenting and modelling C1, C2–C5, C6–C15 and C15+ classes of hydrocarbons formed. In: Matavelli L, Novelli L (eds) Advances in organic geochemistry 1987. Pergamon Journals, Oxford, pp 893–900

Forsman JP, Hunt JM (1958) Insoluble organic matter (kerogen) in sedimentary rocks of marine origin. In: Weeks LG (ed) Habitat of oil: a symposium. American Association of Petroleum Geologists, Tulsa, pp 747–778

Fowler MF (1992) The influence of *Gloeocapsomorpha prisca* on the organic geochemistry of oils and organic-rich rocks of late ordovician age from Canada. In: Schidlowski M, Golubic S, Kimberley MM, McKirdy DM, Trudinger PA (eds) Early organic evolution. Springer, Berlin Heidelberg New York, pp 336–356

Fowler MG, Douglas AG (1984) Distribution and structure of hydrocarbons in four organic-rich Ordovician rocks. Org Geochem 6: 105–114

Fukushima K, Morinaga S, Uzaki M, Ochiai M (1989) Hydrocarbons generated by pyrolysis of insoluble kerogen-like materials isolated from microbially degraded plant residues. Chem Geol 76: 131–141

Ganz H, Kalkreuth W (1987) Application of infrared spectroscopy to the classification of kerogen types and the evaluation of source rock and oil shale potentials. Fuel 66: 708–711

Giraud A (1970) Application of pyrolysis and pyrolysis gas chromatography to the geochemical characterisation of kerogen in sedimentary rocks. Bull Am Assoc Petrol Geol 54: 439–455

Glasø O (1980) Generalized pressure-volume-temperature correlations. J Petrol Technol: 785–795

Goossens H, de Leeuw JW, Schenck PA, Brassell SC (1984) Tocopherols as likely precursors of pristane in ancient sediments. Nature 312: 440–442

Gordon TL (1985) Talang Akar coals - Ardjuna subbasin oil source. Proc of the Indonesian Petroleum Association, 14th Annu Convention, pp 91–120

Goth K, De Leeuw JW, Püttmann W, Tegelaar EW (1988) Origin of messel oil shale. Nature 336: 759–761

Gransch JA, Eisma EE (1970) Characterisation of the insoluble organic matter of sediments by pyrolysis. In: Hobson GG, Spears GC (eds) Advances in organic geochemistry. Pergamon Press, New York, pp 407–426

Gransch JA, Posthuma J (1974) On the origin of sulphur in crudes. In: Tissot BP, Bienner F (eds) Advances in organic geochemistry 1973. Editions Technip, Paris, pp 727–739

Grantham PJ, Posthuma J, Baak A (1983) Triterpanes in a number of Far-Eastern crude oils. In: Bjorøy M (ed) Advances in organic geochemistry 1981. John Wiley, Chichester, pp 675–683

Gray NR, Lancaster CJ, Gethner J (1991) Chemometric analysis of pyrolysate compositions: a model for predicting the organic matter type of source rocks using pyrolysis-gas chromatography. J Anal Appl Pyrolysis 20: 87–106

Hall PB, Douglas AG (1983) The distribution of cyclic alkanes in two lacustrine deposits. In: Bjorøy M (ed) Advances in organic geochemistry 1981. Wiley, Chichester, pp 576–587

Hall PB, Stoddart D, Bjoroy M, Larter SR, Brasher JE (1994) Detection of petroleum heterogeneity in eldfisk and satellite fields using thermal extraction, pyrolysis-GC, GC-MS and isotope techniques. Adv Org Geochem 22: 383–402

Hartgers WA, Sinninghe Damsté JS, de Leeuw JW (1994) Geochemical significance of alkylbenzene distributions in flash pyrolysates of kerogens, coals and asphaltenes. Geochim Cosmochim Acta 58: 1759–1775

Hernandez ME, Vives MT, Pasquali J (1983) Relationships among viscosity, composition, and temperature for two groups of heavy crudes from the eastern Venezuelan basin. Org Geochem 4(3/4): 173–178

Herrmann A (1971) Die Asphaltkalk-Lagerstätte bei Holzen/Ith auf der Südwestflanke der Hils-Mulde. Beih Geol Jahrb: 125

Heum OR, Dalland A, Meisingset KK (1986) Habitat of hydrocarbons at Haltenbanken (PVT-modelling as a predictive tool in hydrocarbon exploration). Norwegian Petroleum Society, Graham and Trotman, London, pp 259–274

Hirsch PB (1954) X-ray scattering from coals. Proc R Soc A226: 143–169

Ho TY, Rogers MA, Drushel HV, Koons CB (1974) Evolution of sulfur compounds in crude oils. Bull Am Assoc Petrol Geol 50(11): 2338–2348

Hoffmann CF, Foster CB, Powell TG, Summons RE (1987) Hydrocarbon biomarkers from ordovician sediments and the fossil alga *Gloeocapsomorpha prisca* Zalessky 1917. Geochim Cosmochim Acta 51: 2681–2697

Hong Z, Li H, Rullkötter J, Mackenzie AS (1986) Geological application of sterane and triterpane biological marker compounds in the Linyi basin. Org Geochem 10: 433–439

Horsfield B (1984) Pyrolysis studies and petroleum exploration. In: Brooks J, Welte DH (eds) Advances in petroleum geochemistry, vol 1. Academic Press, London, pp 47–298

Horsfield B (1989) Practical criteria for classifying kerogens: some observations from pyrolysis-gas chromatography. Geochim Cosmochim Acta 53: 891–901

Horsfield B (1990) The rapid characterization of kerogens according to source quality, compositional heterogeneity and thermal lability. Rev Palaeobot Palynol 65: 357–365

Horsfield B, Düppenbecker SJ (1991) The decomposition of Posidonia shale and Green River shale kerogens using microscale sealed vessel (MSSV) pyrolysis. J Anal Appl Pyrolysis 20: 107–123

Horsfield B, Rullkötter J (1994) Diagenesis, catagenesis and metagenesis of organic matter. In: Magoon LB, Dow WG (eds) The petroleum system – from source to trap. American Association of Petroleum Geologists, Tulsa, pp 189–199

Horsfield B, Dembicki H, Ho TTY (1983) Some potential applications of pyrolysis to basin studies. J Geol Soc 140: 431–443

Horsfield B, Yordy KL, Crelling JC (1988) Determining the petroleum generating potential of coal using organic geochemistry and organic petrology. In: Matavelli L, Novelli L (eds) Advances in organic geochemistry 1987. Pergamon Press, Oxford, pp 121–131

Horsfield B, Disko U, Leistner F (1989) The microscale simulation of maturation: outline of a new technique and its potential applications. Geol Rundsch 78: 361–373

Horsfield B, Heckers J, Leythaeuser D, Littke R, Mann U (1991) A study of the Holzener Asphaltkalk, northern Germany: observations regarding the distribution, composition and origin of organic matter in an exhumed petroleum reservoir. Mar Petrol Geol 8: 198–211

Horsfield B, Bharati S, Larter SR, Leistner F, Littke R, Mann U, Schenk HJ, Dypvik H (1992) On the atypical petroleum generating characteristics of alginite in the Cambrian alum shale. In: Schidlowski MEA (ed) Early organic evolution:implications for mineral and energy resources. Springer, Berlin Heidelberg New York, pp 257–266

Horsfield B, Düppenbecker SJ, Schenk HJ, Schaefer RG (1993a) Kerogen typing concepts designed for the quantitative geochemical evaluation of petroleum potential. In: Doré AG, Augustson JH, Hermanrud C, Stewart DJ, Sylta Ø (eds) Basin modelling: advances and applications. NPF Spec Publ, vol 3. Elsevier, Amsterdam, pp 243–249

Horsfield B, Littke R, Schenk HJ, Mann U, Curry DJ (1993b) Organic geochemistry of freshwater and alkaline lacustrine environments, Green River formation, Wyoming. Part II. Kerogen characterisation and source rock evaluation. In: Øygard K (ed) Organic geochemistry. Poster sessions from 16th Int Meet on Organic geochemistry, Stavanger 1993. Falch Hurtigtrykk, Oslo, pp 321–324

Horsfield B, Curry DJ, Bohacs K, Littke R, Rullkötter J, Schenk HJ, Radke M, Schaefer RG, Carroll AR, Isaksen G, Witte EG (1994) Organic geochemistry of freshwater and alkaline lacustrine sediments in the Green River formation of the Washakie basin, Wyoming,USA. In: Telnaes N, van Graas G, Oygard K (eds) Advances in organic geochemistry 1993. Organic geochemistry, vol 22. Elsevier Science, Oxford, pp 415–440

Hu C (1985) Geologic characteristics and oil exploration of small depressions in eastern China. Geology 13: 303–306

Hunt JM (1979) Petroleum geochemistry and geology. Freeman, San Francisco, 617 pp

Hutton AC, Kanstler AJ, Cook AC, McKirdy DM (1980) Organic matter in oil shales. J Aust Petrol Expl Assoc 20: 68–86

Hvoslef S, Larter SR, Leythaeuser D (1988) Aspects of generation and migration of hydrocarbons from coal-bearing strata of the Hitra formation, Haltenbanken area, offshore Norway. Org Geochem 13(1–3): 525–536

Illich HA, Haney FR, Jackson TJ (1977) Hydrocarbon geochemistry of oils from Maranon basin, Peru. Bull Am Assoc Petrol Geol 61(12): 2103–2114

Jackson KS, McKirdy DM, Deckelmann JA (1984) Hydrocarbon generation in the Amadeus basin, central Australia. APEA J 24: 43–65

Jacobson SR, Hatch JR, Teerman SC, Askin RA (1988) Middle Ordovician organic matter assemblages and their effect on Ordovician-derived oils. Bull Am Assoc Petrol Geol 72: 1090–1100

Jones DM, Douglas AG, Connan J (1988) Hydrous pyrolysis of asphaltenes and polar fractions of biodegraded oils. Org Geochem 13(4–6): 981–993

Jones RW (1987) Organic facies. In: Brooks J, Welte DH (eds) Advances in petroleum geochemistry, vol 2. Academic Press, London, pp 1–90

Kasrai M, Bancroft GM, Brunner RW, Jonasson RG, Brown JR, Tan KH, Feng X (1994) Sulphur speciation in bitumens and asphaltenes by X-ray absorption fine structure spectroscopy. Geochim Cosmochim Acta 58: 2865–2872

Kelley PA, Bissada KK, Burda BH, Elrod LW, Pfeiffer RN (1985) Petroleum generation potential of coals and organic-rich deposits: significance in teriary coal-rich basins. Proc Ind Petrol Assoc 14: 3–21

Kent PE (1954) Oil occurrence in coal measures of England. Bull Am Assoc Petrol Geol 38: 1699–1712

Kiran P, Gillham JG (1976) Pyrolysis - molecular weight chromatography: a new on-line system for analysis of polymers. II. Thermal decomposition of polyolefins - polyethylene, polypropylene, polyisobutylene. J Appl Polymer Sci 20: 2045

Klesment I (1974) Applications of chromatographic methods in biogeochemical investigations. Determination of the structures of sapropelites by thermal decomposition. J Chromatogr 91: 705–713

Knights BA, Brown AC, Conway E, Middleditch BS (1970) Hydrocarbons from the green form of the freshwater alga *Botryococcus braunii*. Phytochemistry 9: 1317–1324

Koplick AJ, Wailes PC, Galbraith MN, Vit I (1983) Constitution of tars from the flash pyrolysis of Australian coals: 2. Structural study of Millmerran coal-tar resins by hydrogenolysis. Fuel 62(10): 1167–1176

Krooss BM, Leythaeuser D, Schaefer RG (1992) The quantification of diffusive losses through cap rocks of natural gas reservoirs - a reevaluation. AAPG Bull 76(3): 403–406

Krooss BM, Idiz EF, Horsfield B (1995) The generation of nitrogen and methane from coal as revealed by pyrolysis experiments: implications for natural gas compositions. AAPG Annu Convention, Houston (Abstr)

Lafargue E, Behar F (1989) Application of a new preparative pyrolysis technique for the determination of source-rock types and oil/source-rock correlations. Geochim Cosmochim Acta 53: 2973–2983

Landais P, Zaugg P, Monin JC, Kister J, Muller JF (1991) Experimental simulation of the natural coalification of coal maceral concentrates. Bull Soc Geol Fr 2: 211–217

Landes KK (1967) Eometamorphism and oil and gas in time and space, part 1. AAPG Bull 51: 828–841

Largeau C, Casadevall E, Kadouri A, Metzger P (1984) Formation of *Botryococcus*-derived kerogens-comparative study of immature torbanites and of the extant alga *Botryococcus braunii*. Org Geochem 6: 327–332

Larter SR (1978) A geochemical study of kerogen and related materials. PhD Diss, Univeristy of Newcastle upon Tyne

Larter SR (1984) Application of analytical pyrolysis techniques to kerogen characterization and fossil fuel exploration/exploitation. In: Voorhees K (ed) Analytical pyrolysis, methods and application. Butterworth, London, pp 212–275

Larter SR (1985) Integrated kerogen typing in the recognition and quantitative assesment of petroleum source rocks. In: Petroleum geochemistry in exploration of the Norwegian Shelf. Norwegian Petroleum Society, Graham & Trotman, pp 269–285

Larter SR (1988) Some pragmatic perspectives in source rock geochemistry. Mar Petrol Geol 5: 194–204

Larter SR, Horsfield B (1993) Determination of structural components of kerogens using analytical pyrolysis methods. In: Engel M, Macko S (eds) Organic geochemistry. Plenum Press, New York, pp 271–287

Larter SR, Senftle JT (1985) Improved kerogen typing for petroleum source rock analysis. Nature 318: 277–280

Larter SR, Horsfield B, Douglas AG (1977) Pyrolysis as a possible means of determining the petroleum generating potential of sedimentary organic matter. In: Jones CER, Cramers CA (eds) Analytical pyrolysis. Proc 3rd Int Symp Analytical Pyrolysis 1975, Amsterdam, pp 189–202

Larter SR, Solli H, Douglas AG (1983) Phytol containing melanoidins and their bearing on the fate of isoprenoid structures in sediments. In: Bjoroy M, Albrecht P, Cornford C et al. (eds) Advances in organic geochemistry 1981. Wiley, Chichester

Larter SR, Bjørlykke KO, Karlsen DA, Nedkvitne T, Eglinton TI, Johansen PE, Leythaeuser D, Mason PC, Mitchell AW, Newcombe GA (1990) Determination of petroleum accumulation histories: examples from the Ula field, Central Graben, Norwegian North Sea. In: Buller A (ed) North Sea oil and gas reservoirs, vol 2. Graham & Trotman, London, pp 319–330

Law BE, Rice DD (eds) (1993) Hydrocarbons from coal. AAPG studies in geology 38. American Association Petroleum Geologists, Tulsa, 400 pp

Leventhal JS (1976) Stepwise pyrolysis-gas chromatography of kerogen sedimentary rocks. Chem Geol 18: 5–20

Levine JR (1993) Coalification: the evolution of coal as a source rock and reservoir rock for oil and gas. In: Law BE, Rice DD (eds) Hydrocarbons from coal, AAPG studies in geology 38. American Association Petroleum Geologists, Tulsa, pp 39–77

Lewan MD (1983) Effects of thermal maturation on stable organic carbon isotopes as determined by hydrons pyrolysis of Woodford Shale. Geochim Cosmochim Acta 47: 1471–1479

Lewan M, Buchardt B (1989) Irradiation of organic matter by uranium decay in the Alum Shale, Sweden. Geochim Cosmochim Acta 53: 1307–1322

Littke R, Leythaeuser D (1993) Migration of oil and gas in coals. In: Law BE, Rice DD (eds) Hydrocarbons from coal. AAPG studies in geology 38. American Association Petroleum Geologists, Tulsa, pp 219–236

Louis MC, Tissot BP (1967) Influence de la temperature et de la pression sur la formation des hydrocarbures dans les argiles à kerogen. In: Proc 7th World Petroleum Congr, Mexico. Elsevier, London, pp 47–60

Mackenzie AS (1984) Applications of biological markers in petroleum geochemistry. In: Brooks J, Welte DH (eds) Advances in petroleum geochemistry, vol 1. Academic Press, London, pp 115–214

Mackenzie AS, Quigley TM (1988) Principles of geochemical prospect appraisal. Bull Am Assoc Petrol Geol 72: 399–415

Mackenzie AS, Leythaeuser D, Altebäumer FJ, Disko U, Rullkötter J (1988) Molecular measurements of maturity for Lias δ shales in N.W. Germany. Geochim Cosmochim Acta 52: 1145–1154

Magoon LB, Dow WG (eds) (1994) The petroleum system from source to trap, AAPG Mem 60. American Association Petroleum Geologists, Tulsa, 655 pp

Mann AL, Goodwin NS, Lowe S (1987) Geochemical characterisation of lacustrine source rocks: a combined palynological/molecular study of a tertiary sequence from offshore China. Proc Ind Petrol Assoc 16: 241–258

Martin RL, Winters JC, Williams JA (1963) Distributions of n-paraffins in crude oils and their implications to origin of petroleum. Nature 199: 110–113

Mattavelli L, Pieri M, Groppi G (1993) Petroleum exploration in Italy: a review. Mar Petrol Geol 10: 410–425

McHugh DJ, Saxby JD, Tardif JW (1976) Pyrolysis hydrogenation gas chromatography of carbonaceous material from Australian sediments. Part 1: Some Australian coals. Chem Geol 17: 243–259

McIver RD (1967) Composition of kerogen-clue to its role in the origin of petroleum. Proc 7th World Petrol Congr, Mexico City, vol 2, p 26

McKirdy DM, McHugh DJ, Tardif JW (1980) Comparitive analysis of stromatolitic and other microbial kerogens by pyrolysis-hydrogenation gas chromatography (PHGC). In: Trudinger PA, Walter MR, Ralph BJ (eds) Biogeochemistry of ancient and modern environments. Australian Academy of Sciences and Springer, Berlin Heidelberg New York, pp 187–200

Metzger P, Berkaloff C, Casadevall E, Coute A (1985) Alkadiene- and botryococcene-producing races of wild strains of *Botryococcus braunii*. Phytochemistry 24: 2305–2312

Michels R, Landais P, Torkelson BE, Philp RP (1995) Effects of effluents and water pressure on oil generation during confined pyrolysis and high-pressure hydrous pyrolysis. Geochim Cosmochim Acta 59

Mitchell DL, Speight JG (1973) The solubility of asphaltenes in hydrocarbon solvents. Fuel 52: 149–152

Moore PS, Hobday DK, Mai H, Sun ZC (1986) Comparison of selected non-marine petroleum-bearing basins in Australia and China. APEA J 26: 285–309

Mukhopadhyay PK, Fowler MG, Dow WG (1991) Selected papers from the symposium on coal and terrestrial organic matter as a source rock for petroleum. 199th ACS Meet, Boston, 23–24 April 1990. Org Geochem 17(6): 671–872

Muscio GPA, Horsfield B (1996) Enhanced formation of inert carbon during the natural maturation of a marine source rock; Bakken Shale, Williston basin. Energy Fuels 10: 10–16

Muscio GPA, Horsfield B, Welte DH (1991) Compositional changes in the macromolecular organic matter (kerogens, asphaltenes and resins) of a naturally matured source rock sequence from northern Germany as revealed by pyrolysis methods. In: Manning D (ed) Organic geochemistry advances and applications in energy and the natural environment. Manchester University Press, Manchester, pp 447–449

Muscio GPA, Horsfield B, Welte DH (1994) Occurrence of thermogenic gas in the immature zone – implications from the Bakken in-source reservoir system. Org Geochem 22(3–5): 461–476

Mycke B, Michaelis W (1986) Molecular fossils from chemical degradation of macromolecular organic matter. Org Geochem 10: 847–858

Nelson PF (1987) Chemically-bound n-alkyl groups in coal. Fuel 66(9): 1264–1268

Neumann HJ, Paczynska-Lahme B, Severin D (1981) Geology of petroleum, 5. Composition and properties of petroleum. Wiley, Chichester, 137 pp

Nissenbaum A, Kaplan IR (1972) Chemical and isotopic evidence for the in situ origin of marine humic substances. Limnol Oceanogr 17: 570–582

Noble RA, Wu CH, Atkinson CD (1991) Petroleum generation and migration from Talang Akar coals and shales offshore N.W. Java, Indonesia. Org Geochem 17(3): 363–374

Orr WL (1986) Kerogen/asphaltene/sulfur relationships in sulfur-rich Monterey oils. In: Leythaeuser D, Rullkötter J (eds) Advances in organic geochemistry 1985. Pergamon Journals, Oxford, pp 499–516

Orr WL, Sinninghe Damsté JS (1990) Geochemistry of sulfur in petroleum systems. In: Orr WL, White CM (eds) ACS symposium series 429: geochemistry of sulfur in fossil fuels. Am Chem Soc, Washington, DC, pp 2–29

Øygard K, Larter SR, Senftle JS (1988) The control of maturity and kerogen type on quantitative analytical pyrolysis data. In: Mattavelli L, Novelli L (eds) Advances in organic geochemistry 1987. Organic geochemistry. Pergamon Journals, Oxford, pp 1153–1162

Patience RL, Mann AL, Poplett IJF (1992) Determination of molecular structure of kerogens using ^{13}C NMR spectroscopy. II. The effects of thermal maturation on kerogens from marine sediments. Geochim Cosmochim Acta 56: 2725–2742

Payzant JD, McIntyre DD, Mojelsky TW, Torres M, Montgomery DS, Strausz OP (1989) The identification of homologous series of thiolanes and thianes possessing a linear carbon framework from petroleums and their interconversion under simulated geological conditions. Org Geochem 14(4): 461–473

Pelet R, Behar F, Monin JC (1986) Resins and asphaltenes in the generation and migration of petroleum. Org Geochem 10: 481–498

Peters KE, Moldowan J (1993) The biomarker guide. Prentice Hall, New York, 363 pp

Philippi GT (1965) On the depth, time and mechanism of petroleum generation. Geochim Cosmochim Acta 29: 1021–1049

Philippi GT (1975) The deep subsurface temperature controlled origin of the gaseous and gasoline-range hydrocarbons of petroleum. Geochim Cosmochim Acta 39: 1353–1373

Philp RP (1982) Application of pyrolysis-GC and pyrolysis-GC-MS to fossil fuel research. Trends Anal Chem 10: 237–241

Philp RP, Calvin M (1976) Possible origin for insoluble kerogen, debris in sediments from insoluble cell-wall materials of algae and bacteria. Nature 262: 134–136

Philp RP, Gilbert TD (1985) Source rock and asphaltene biomarker characterization by pyrolysis gas chromatography-mass spectrometry-multiple ion detection. Geochim Cosmochim Acta 49: 1421–1432

Powell TG (1975) Geochemical studies related to the occurrence of oil and gas in the Dampier sub-basin, western Australia. J Geochem Expl 4: 441–466

Powell TG (1984) Some aspects of the hydrocarbon geochemistry of a middle Devonian barrier reef complex, western Canada. In: Palacas JG (ed) Petroleum geochemistry and source rock potential of carbonate rocks. AAPG Stud Geol 18: 45–62

Powell TG (1986) Petroleum geochemistry and depositional setting of lacustrine source rocks. Mar Petrol Geol 3: 200–219

Price LC (1989) Primary petroleum migration from shales with oxygen-rich organic matter. J Petrol Geol 12: 289–324

Radke M (1987) Organic geochemistry of aromatic hydrocarbons. In: Brooks J, Welte D (eds) Advances in petroleum geochemistry, vol 2. Academic Press, London, pp 141–207

Radke M, Willsch H (1993) Generation of alkylbenzenes and benzo(b)thiophenes by artificial thermal maturation of sulfur-rich coal. Fuel 72(8): 1103–1108

Reading HG, Collinson JD, Edwards MB, Elliott T, Jenkyns HC, Johnson HD, Mitchell AHG, Rupke NA, Sellwood BW, Till R (1978) Sedimentary environments and facies. Elsevier, New York

Reed JD, Illich HA, Horsfield B (1986) Biochemical evolutionary significance of Ordovician oils and their sources. In: Leythaeuser D, Rullkötter J (eds) Advances in organic geochemistry 1985. Pergamon Press, Oxford, pp 347–358

Rice DD (1993) Composition and origins of coalbed gas. In: Law BE, Rice DD (eds) Hydrocarbons from coal, AAPG studies in geology, vol 38. American Association Petroleum Geologists, Tulsa, pp 159–184

Romovacek J, Kubat J (1968) Characterisation of coal substance by pyrolysis-gas chromatography. Anal Chem 40: 1119–1127

Ropertz B (1994) Wege der primären Migration:eine Untersuchung über Porennetze, Klüfte und Kerogennetzwerke als Leiterbahnen für den Kohlenwasserstoff-Transport. PhD Diss, RWTH Aachen (unpublished)

Rouxhet PG, Robin PL (1978) Infrared study of the evolution of kerogens of different origins during catagenesis and pyrolysis. Fuel 57: 533–540

Rovere CE, Crisp PT, Ellis J, Bolton PD (1983) Chemical characterization of shale oil from Condor, Australia. Fuel 62: 1274–1282

Rowland SJ, Aareskjold K, Xuemin G, Douglas AG (1986) Hydrous pyrolysis of sediments: composition and proportions of aromatic hydrocarbons in pyrolysates. In: Leythaeuser D, Rullkötter J (eds) Advances in organic geochemistry 1985. Organic geochemistry, vol 10. Pergamon, Oxford, pp 1033–1040

Rullkötter J, Michaelis W (1990) The structure of kerogen and related materials. A review of recent progress and future trends. In: Durand B, Behar F (eds) Advances in organic geochemistry 1989. Part II: Molecular geochemistry. Pergamon Press, Oxford, pp 829–852

Rullkötter J, Leythaeuser D, Horsfield B, Littke R, Mann U, Müller PJ, Radke M, Schaefer RG, Schenk H-J, Schwochau K, Witte EG, Welte DH (1988) Organic matter maturation under influence of a deep intrusive heat source: a natural experiment for quantification of hydrocarbon generation and expulsion from a petroleum source rock (Toarcien shale, northern Germany). In: Mattavelli L, Novelli L (eds) Advances in organic geochemistry 1987. Organic geochemistry. Pergamon Journals, Oxford, pp 847–856

Sachsenhofer RF (1993) Petroleum generation and migration in the Styrian basin (Pannonian basin system, Austria): an integrated geochemical and numerical modelling study. Mar Petrol Geol 11: 684–701

Sachsenhofer RF, Curry DJ, Horsfield B, Rantitsch G, Wilkes H (1996) Characterization of organic matter in late Cretaceous black shales of the Eastern alps (Kainach Gosau-Group, Austria). Org Geochem (in press)

Saxby JD (1981) Kerogen genesis and structure – similarities to rubber. Fuel 60: 994–996

Scheidt G, Littke R (1989) Organic petrology of clastic sediments and coals from western Germany. Geol Rundsch 78: 375–390

Schenk HJ, Horsfield B (1993) Kinetics of petroleum generaion by programmed-temperature closed- versus open-system pyrolysis. Geochim Cosmochim Acta 57: 623–630

Schenk HJ, Witte EG, Müller PJ, Schwochau K (1986) Infrared estimates of aliphatic kerogen carbon in sedimentary rocks. Org Geochem 10: 1099–1104

Schenk HJ, Horsfield B, Witte EG (1993) Organic geochemistry of freshwater and alkaline lacustrine environments, Green River formation, Wyoming. Part III. Comparative characterisation of selected shale samples by various spectroscopic and pyrolysis techniques including kinetic measurements. In: Øygard K (ed) Organic geochemistry. Poster sessions from the16th Int Meet on Organic geochemistry, Stavanger 1993. Falch Hurtigtrykk, Oslo, pp 324–327

Schowalter TT (1979) Mechanics of secondary hydrocarbon migration and entrapment. Bull Am Assoc Petrol Geol 63: 723–760

Senftle JT, Larter SR, Bromley BW, Brown JH (1986) Quantitative chemical characterization of vitrinite concentrates using pyrolysis-gas chromatography. Rank variation of pyrolysis products. Org Geochem 9(6): 345–350

Sinninghe Damsté JS, De Leeuw JW (1990) Analysis, structure and geochemical significance of organically-bound sulphur in the geosphere: state of the art and future research. Org Geochem 16(4–6): 1077–1101

Sinninghe-Damsté JS, Eglinton TI, de Leeuw JW, Schenck PA (1989) Organic sulphur in macromolecular sedimentary organic matter. 1.Structure and origin of sulphur containing moieties in kerogen, asphaltenes and coal as revealed by flash pyrolysis. Geochim Cosmochim Acta 53: 873–889

Sinninghe Damsté JS, de la Heras FXC, de Leeuw JW (1992) Molecular analysis of sulphur-rich brown coals by flash pyrolysis-gas chromatography-mass spectrometry. J Chromatogr 607: 361–376

Sinninghe Damsté JS, de las Heras FXC, van Bergen PF, de Leeuw JW (1993) Characterization of tertiary catalan lacustrine oil shales: discovery of extremely organic sulphur-rich type I kerogens. Geochim Cosmochim Acta 57: 389–415

Smith GC, Cook AC (1984) Petroleum occurence in the Gippsland basin and its relationship to rank and organic matter type. APEA J 24: 196–216

Snowdon LR, McCrossan RG (1973) Identification of petroleum source rocks using hydrocarbon gas and organic carbon content. Geol Surv Can 72-36: 1–12

Snowdon LR, Roy KJ (1975) Regional organic metamorphism in the mesozoic strata of the Sverdrup basin. Bull Can Petrol Geol 23: 131–148

Solli H, Leplat P (1986) Pyrolysis-gas chromatography of asphaltenes and kerogens from source rocks and coals – a comparative structural study. In: Leythaeuser D, Rullkötter J (eds) Advances in organic geochemistry 1987. Organic geochemistry, vol 10. Blackwell, Oxford, pp 313–329

Stach E, Mackowsky M-T, Teichmüller M, Teichmüller R, Taylor GH, Chandra D, Murchinson DG, Zierke F (eds) (1982) Stach's textbook of coal petrology. Gebrüder Bornträger, Berlin

Stevenson FJ (1974) Non biological transformations of amino acids in soils and sediments. In: Tissot B, Bienner F (eds) Advances in organic geochemistry 1973. Editions Technip, Paris, pp 701–714

Stout SA, Boon JJ (1994) Structural characterization of the organic polymers comprising a lignite's matrix and megafossils. Org Geochem 21(8/9): 953–970

Sutton C (1979) Depositional environments and their relation to chemical composition of Java Sea crude oils. UN/ESCAP, CCOP Tech Publ, Bangkok, pp 163–179

Tannenbaum E, Aizenshtat Z (1985) Formation of immature asphalt from organic rich carbonate rocks-I: geochemical correlation. Org Geochem 8(2): 181–192

Teerman SC, Hwang RJ (1991) Evaluation of the liquid hydrocarbon potential of coal by artificial maturation techniques. Org Geochem 17(6): 749–764

Tegelaar EW, Noble RA (1994) Kinetics of hydrocarbon generation as a function of the molecular structure of kerogen as revealed by pyrolysis-gas chromatography. In: Telnaes N, van Graas G, Oygard K (eds) Advances in organic geochemistry 1993. Pergamon Press, Oxford, pp 543–574

Tegelaar EW, de Leeuw JW, Derenne S, Largeau C (1989a) A reappraisal of kerogen formation. Geochim Cosmochim Acta 53: 3103–3106

Tegelaar EW, Matthezing RM, Jansen BH, Horsfield B, de Leeuw JW (1989b) Possible origin of n-alkanes in high-wax crude oils. Nature 342: 529–531

Thomas BM (1982) Land-plant source rocks for oil and their significance in Australian basins. APEA J 22: 164–178

Thomas BM, Moller-Pedersen P, Whitaker MF, Shaw ND (1985) Organic facies and hydrocarbon distributions in the Norwegian North Sea. In: Thomas BM (ed) Petroleum geochemistry in exploration of the Norwegian shelf. Graham & Trotman, London, pp 3–26

Thompson KFM (1988) Gas-condensate migration and oil fractioning in deltaic systems. Mar Petrol Geol 5: 237–246

Thompson S, Cooper BS, Morley RJ, Barnard PC (1985) Oil-generating coals. Norwegian Petroleum Society, Graham & Trotman, London, pp 59–73

Tissot BP, Welte DH (1978) Petroleum formation and occurrence. Springer, Berlin Heidelberg New York

Tissot BP, Durand B, Espitalie J, Combaz A (1974) Influence of nature and diagenesis of organic matter in formation of petroleum. AAPG Bull 58: 499–506

Tissot BP, Pelet R, Ungerer P (1987) Thermal history of sedimentary basins, maturation indices, and kinetics of oil and gas generation. Bull Am Assoc Petrol Geol 71: 1445–1466

Ungerer P (1990) State of the art of research in kinetic modelling of oil formation and expulsion. In: Durand B, Behar F (eds) Advances in organic geochemistry 1989. Org Geochem 16: 1–25

Ungerer P, Pelet R (1987) Extrapolation of the kinetics of oil and gas formation from laboratory experiments to sedimentary basins. Nature 327: 52–54

Urov KE (1980) Thermal decomposition of kerogens – mechanism and analytical application. J Anal Appl Pyrolysis 1: 323–338

van Aarssen BGK, Cox HC, Hoogendoorn P, De Leeuw JW (1990) A cadinene biopolymer, in fossil and extant dammar resins as a source for cadinanes and bicadinanes in crude oils from South East Asia. Geochim Cosmochim Acta 54: 3021–3031

van Aarssen BGK, de Leeuw JW, Horsfield B (1991) A comparative study of three different pyrolysis methods used to characterise a biopolymer isolated from fossil and extant dammar resins. J Anal Appl Pyrolysis 20: 125–139

van Aarssen BGK, Hessels JKC, Abbink OA, de Leeuw JW (1992) The occurrence of polycyclic sesqui-, tri- and oligoterpenoids derived from a resinous polymeric cadinene in crude oils from southeast Asia. Geochim Cosmochim Acta 56: 1231–1246

van de Meent D, Brown SC, Philp RP, Simoneit BRT (1980) Pyrolysis-high resolution gas chromatography and pyrolysis gas chromatography-mass spectrometry of kerogens and kerogen precursors. Geochim Cosmochim Acta 44: 999–1013

van Graas G, De Leeuw JW, Schenck PA, Haverkamp J (1981) Kerogen of toarcian shales of the Paris basin. A study of its maturation by flash pyrolysis techniques. Geochim Cosmochim Acta 45: 2465–2474

Walker RG (1979) Facies and facies models. General introduction. In: Walker RG (ed) Facies models, geoscience Canada reprint series edn, vol 1. Geosci Can, Ottawa (N): 1–8

Weres O, Newton AS, Tsao L (1988) Hydrous pyrolysis of alkanes, alkenes, alcohols and ethers. Org Geochem 12(5): 433–444

Wilhelms A (1992) An investigation into the factors influencing tar mat formation in petroleum reservoirs. PhD Diss, May 1992, University of Oslo, Norway

Wilhelms A, Larter SR (1994) On the origin of tar mats in petroleum reservoirs: part I. Introduction and case studies. Mar Petrol Geol 11: 418–441

Williams JA, Dolcater DL, Torkelson BE, Winters JC (1988) Anomalous concentrations of specific alkylaromatic and alkylcycloparaffin components in West Texas and Michigan crude oils. Org Geochem 13(1–3): 47–59

Wilson MA, Vassallo AM (1985) Developments in high-resolution solid-state 13C NMR spectroscopy of coals. Org Geochem 8(5): 299–312

Wilson MA, Philp RP, Gillam AH, Gilbert TD, Tate KR (1983) Comparison of the structures of humic substances from aquatic and terrestrial sources by pyrolysis gas chromatography-mass spectrometry. Geochim Cosmochim Acta 47: 497–502

Yarzab RF, Abdel-Basset Z, Given PH (1979) Hydroxyl contents of coals. Geochim Cosmochim Acta 43: 281–287

Yen TF (1972) Present status of the structure of petroleum heavy ends and its significance to various technical applications. Am Chem Soc Div Petrol Chem Prepr 17: F102–114

Young A, McIver RD (1977) Distribution of hydrocarbons between oils and associated fine-grained sedimentary rocks – physical chemistry applied to petroleum geochemistry II. AAPG Bull 61(9): 1407–1436

Zumberge JE (1984) Source rocks of the La Luna formation (upper Cretaceous, in the Middle Magdalena Valley, Columbia. In: Palacas JG (ed) Petroleum geochemistry and source rock potential of carbonate rocks. AAPG studies in geology, vol 18. American Association of Petroleum Geologists, Tulsa, pp 127–134

Chapter 7
Petroleum Migration: Mechanisms, Pathways, Efficiencies and Numerical Simulations

Chapter 7: Overview and Insights

During the past decade considerable progress has been made in unraveling migration effects and processes. This has been due mainly to thoroughly selecting and studying proper geological settings relevant to petroleum migration and by ingenious methods for investigating the geochemical effects of migration on hydrocarbon and nonhydrocarbon mixtures. The range of geological settings investigated reached from the microscopic to the regional scale, from typical pore size distributions of source rocks to macroscopic fractures and faults, and finally to specific case histories.

The strict comparison of migration phenomena between thick and thin source beds, between hydrocarbon depleted and enriched rock segments, between siliciclastic and calcareous source rocks has revealed very valuable information confirming and refining the concept of a hydrocarbon phase movement. Mass balance studies down to the molecular level suggest for rich source rocks high expulsion efficiencies of more than 80%. All this information together with aspects derived from petroleum engineering has been incorporated into a comprehensive approach on numerical simulation of petroleum migration.

An implementation of numerical modeling of petroleum migration as an integrated part of basin modeling has been achieved only recently. The main reason for this time delay has been the fact that existing equations and models for fluid flow in porous media, such as those applied in reservoir simulation for many years, offered no easy connection to current basin modeling systems. Meanwhile the proper geological and mathematical simulation tools have been developed, connecting the modeling of polyphasic fluid flow and basin modeling. The numerical simulation of petroleum migration presented here incorporates fluid flow of the three phases water, oil, and gas and compositional changes in each phase. This compositional concept is also applied in the framework of a multiple source rock concept which allows tracing of petroleum from different sources throughout the entire migration history. The fluid flow simulation model includes transport in separate phases and also transport by diffusion. The petroleum migration models differ from "classical" fluid flow problems in applied mechanics – and these include reservoir modeling – mainly due to the fact that two additional processes are integrated. Firstly, the thermally induced generation of oil and gas introduces an additional fluid source into the model, and secondly deposition, compaction, and other geological processes cause pressure build-up, which acts as a driving force for fluid movement. The effects of these processes are included in our migration concept and are implemented in the flow equations.

Modeling of petroleum migration as understood here is not only a description of fluid velocities and flow paths but also takes into account the transient changes induced by ongoing geological processes throughout the matrix of sedimentary rocks. Some of these processes and their implementation in the numerical analysis are discussed in more detail. Those include: fault modeling, salt doming, rock fracturing, cementation of pores, solution, and exsolution of light hydrocarbons in the aqueous pore fluid. The incorporation of such processes in migration modeling allows a fairly realistic simulation of different geological scenarios in context with the assessment of the formation of petroleum accumulations.

Petroleum Migration: Mechanisms, Pathways, Efficiencies and Numerical Simulations

U. Mann[1], T. Hantschel[2], R.G. Schaefer[1], B. Krooss[1], D. Leythaeuser[3], R. Littke[1], and R.F. Sachsenhofer[4]

7.1 Introduction

Movement of petroleum from the source via carrier bed to the reservoir rocks is called migration. This is divided into *primary migration,* defined as the movement of oil and gas through and out of the fine-grained source rocks, and *secondary migration*, the movement through wider pores in carrier and reservoir rocks to the trap. (Not discussed here are the aspects of petroleum entrapment, redistribution in traps and petroleum loss from traps - generally called "dismigration" - or the various types of traps.) In addition to petroleum generation, petroleum migration is the principal process in the formation of explorable petroleum accumulations (which, however, represent only a special case of petroleum migration in which an otherwise diffuse fluid flow field is highly focussed towards a reservoir structure).

An understanding how and why petroleum migration is sometimes focussed to form a petroleum accumulation would require a four-dimensional (x, y, z and time-lapse) reconstruction of the characteristics of migration pathways - i.e. its rock and fluid properties - as a function of the palaeohistory of a sedimentary basin. Although such a complete reconstruction represents an idealisation that will never be reached, many significant recent developments have elucidated the basic problems of petroleum migration. These concern the mechanisms, timing, pathways and efficiency of petroleum migration. In total, this has been obtained only by the consecutive integration of:

- sedimentology and petrophysics in order to identify and quantify the capacity of a pathway,
- organic geochemistry in order to establish migration mechanism, stage and efficiency, and
- numerical simulation in order to reconstruct the petroleum migration process within the geological framework.

[1]Institut für Erdöl und Organische Geochemie (ICG-4), Forschungszentrum Jülich GmbH, 52425 Jülich, Germany
[2]IES Gesellschaft für Integrierte Explorationssysteme mbH, Bastionstr. 11-19, 52428 Jülich, Germany
[3]Geologisches Institut, Universität zu Köln, Zülpicher Str. 49a, 50674 Köln, Germany
[4]Institut für Geowissenschaften, Montan-Universität Leoben, 8700 Leoben, Austria

Welte et al. (eds)
Petroleum and Basin Evolution
© Springer-Verlag Berlin Heidelberg 1997

This chapter therefore addresses petroleum transport, the individual mechanisms of migration, the characteristic differences of individual migration pathways, and the timing and efficiency of a petroleum migration process.

7.2 Migration Mechanisms

7.2.1 Primary Migration Mechanisms

The most important form of primary migration during the main phase of oil and gas formation is discrete hydrocarbon phase movement (Dickey 1975; Hunt 1979; Momper 1978; Tissot and Welte 1984; Welte 1987; Durand 1988; Ungerer 1990; Mann 1994). However, as true hydrocarbon flow cannot take place in micro- and mesopores of a source rock, and petroleum must yet reach the macropores, diffusion directly after petroleum generation is the most likely initial mechanism until a pressure-driven flow takes over (Fig. 7.1).

Pressure-Driven Fluid Flow

Obtaining petroleum flow within macropores or fractures of a source rock requires a pressure gradient. Compaction probably represents the most common case to provide the necessary driving force for expulsion (Meissner 1978; Hunt 1979; du Rouchet 1981). The weight of the overburden is partly transferred from the solid grains to the pore fluids. If burial is rapid, pore waters cannot escape from the tight lithologies, such as source rocks, and undercompaction and overpressuring occur.

The volume increase in organic matter due to *solid to liquid* and *solid to gas conversion* during hydrocarbon generation provides internal pressure to the source rock (Snarsky 1962; Sokolov et al. 1964; Hedberg 1974; Momper 1978). Ungerer et al. (1983) and Goff (1983) estimated that this volume increase is as high as 10–20% and is most effective if gas is generated, as in the Tertiary of the northern Gulf of Mexico (Hunt et al. 1994; Yükler and Dow 1990).

In the deeper parts of sedimentary basins where oil is generated, *seal-bounded fluid compartments* may exist which are characterised by different internal pressures and which control flow (Vandenbroucke et al. 1983; Demaison 1984). Thus, deep petroleum systems do not exhibit a single basin wide compartment such as most shallow systems but consist of a series of individual fluid compartments that are not in hydraulic pressure communication with each other or with the overlying hydrodynamic regime (Hunt 1990, 1991). Oil and gas are able to migrate downwards, but only from a source rock to an underlying permeable rock which releases pressure. The overpressure must be sufficiently high that downward flow becomes possible. The gradient required to overcome buoyancy is such that it is unlikely that petroleum is able to migrate more than 300–500 m downwards (Mann and Mackenzie 1990).

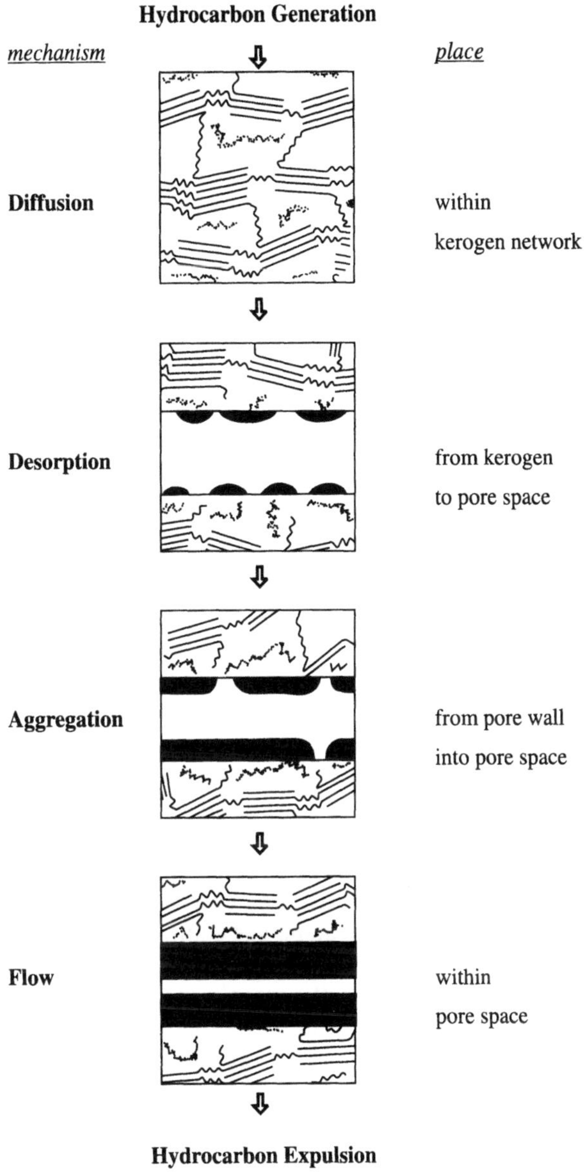

Fig. 7.1. Mechanisms and place of mechanisms acting during primary migration (Mann 1994)

If oil droplets block the pore openings in a source rock, the water trapped behind the oil expands as the temperature rises. The increased pressure caused by the thermal expansion of water has been termed *aquathermal pressuring* by Barker (1972). This mechanism should be responsible for abnormal pore pressures primarily in shallow, impermeable sediments, particularly in areas of high geothermal gradients (Daines 1982).

Clay mineral dehydration is thought to be an additional driving force in primary migration. The existence of carbonate and evaporite source rocks indicates, however, that water from clay mineral dehydration is certainly not a requirement for primary migration. Furthermore, most water of mineral dehydration leaves a siliciclastic source rock prior to petroleum expulsion (Burst 1969; Perry and Hower 1977).

Other Mechanisms

Diffusion of petroleum compounds through and in water alone seems several orders of magnitude too small to fill a petroleum reservoir (Thomas and Clouse 1990). For the same reason the postulation by Stainforth and Reinders (1990) that transport of organic molecules by active *diffusion through organic matter networks* represents the rate-limiting petroleum migration process may only hold for short migration distances (adjacent source and reservoir rocks and/or relatively poor source rocks). Short migration paths for diffusion, meaning between tens to hundred of feet, are required to account for a commercial oil accumulation in a reasonable time scale such as 10 Ma (Hunt 1979). Fluid movement can augment diffusion processes. More than one migration mechanism acting in parallel increases the distance of migration.

During primary migration, hydrocarbon transport by *solution in water* is limited to the most soluble light hydrocarbons such as methane, ethane, benzene and other low-boiling aromatics (McAuliffe 1966, 1979; Price 1976). The fact that aromatics show a much higher solubility than aliphatic hydrocarbons can be used to show whether solution in water may have contributed to primary migration.

Hydrocarbon transport in *gaseous solution* has been proposed by Sokolov (1964). Price (1989) and Leythaeuser and Poelchau (1991) re-evaluated specifically transport in methane gas solution. This mechanism may overcome bitumen saturation that is too low for oil phase flow, especially in source rocks with oxygen-rich type III organic matter. Further, the experimental data by Bray and Foster (1980) support a migration process where carbon dioxide and hydrocarbon gases mobilise the liquid hydrocarbons. Carbon dioxide from great depth may also support primary migration of hydrocarbons (Kvenvolden and Claypool 1980).

7.2.2 Secondary Migration Mechanisms

When petroleum has been expelled from a dense, fine-grained, low-permeability source bed as a discrete hydrocarbon phase and enters the larger pores of the carrier rock, larger globules of oil and gas form immediately. Hydrocarbons are moved by vertically directed buoyancy forces. The intensity of the forces depends on the density difference between water and petroleum. The buoyancy forces must overcome capillary resistance which is controlled by variations in the diameter of pores and pore throats. Therefore hydrocarbons migrate along a tortuous pathway, always following the path of least resistance, i.e. through that part of the rock with the largest interconnected pore throats. Thus a continuous migration of oil stringers occurs as long as hydrocarbons are expelled from the source rock, and as long as oil droplets coalesce to form new stringers. These principles have been elaborated by Gussow (1954); Hobson (1954); Poulet (1968); Berg (1975); Hobson and Tiratsoo (1975) and Schowalter (1979). England et al. (1987) calculated that petroleum must fill about 50% of the pore volume of a carrier bed or between 1 and 10% of the total cross section of the carrier bed before an interconnected pathway for hydrocarbon migration is established.

The pattern of migration pathways under hydrostatic conditions was elaborated by Pratsch (1982). Using empirical data from many sedimentary basins, he concluded that secondary migration starts at the depocentres of a basin (where the most mature source rocks can be assumed) and continues along migration pathways which follow the geometry of the individual sedimentary basin to the basin edges (Fig. 7.2). This allows differentiation between those parts of a basin where petroleum migration is focussed, and where it is dispersed. However, these migration patterns do not take into account the lateral discontinuity of source and carrier rocks or a specific tectonic situation.

Secondary migration occurs under varying pT conditions. Underway, a change in conditions may cause precipitation of asphaltenes. Bitumens with a composition rich in heavier and longer chain compounds are retarded by the carrier, and a fraction with lighter and shorter chain compounds move on to the reservoir. Compositional evidence for such processes is provided by case histories reported for example by Wilhelms and Larter (1994a,b).

According to the individual geological situation, path length and flux rates for secondary migration vary. Observed rates of oil migration in experiments (around 10–50 ms^{-1}) were much higher than parametric estimates of basin-scale oil migration (around 10^{-9} ms^{-1}; England et al. 1987) although this corresponds well to flux rates of several meters per 100 years for a 100 mD carrier bed (1 darcy = 10^{-12} m^2) as shown by experimental results (Thomas and Clouse 1995). Core flood experiments by Selle et al. (1993) have investigated secondary migration through real rocks at a more realistic (for the North Sea) dip angle of 7.5°. These experiments produced geologically plausible, but still high, migration rates (10^{-2} ms^{-1}) and showed the influence of capillary forces. On the other hand, in a high permeability rock, with low threshold capillary pressures (e.g. Bentheimer Sandstone with 2260 mD) Selle et al. (1993) found secondary migration to be dominated by buoyancy forces.

It appears evident that the limitations for secondary migration are not the flux rates, but the time necessary to create the minimum oil saturations within the carrier rock. An understanding of secondary migration cannot assume unlimited flow capacity due to an infinite source. Thus, for actual source rock release rates, it would take more time to develop the critical saturation than for the secondary migration process along the carrier bed. The rate-limiting step in charging a reservoir is not secondary migration itself but rather the rate of petroleum release from the source rock.

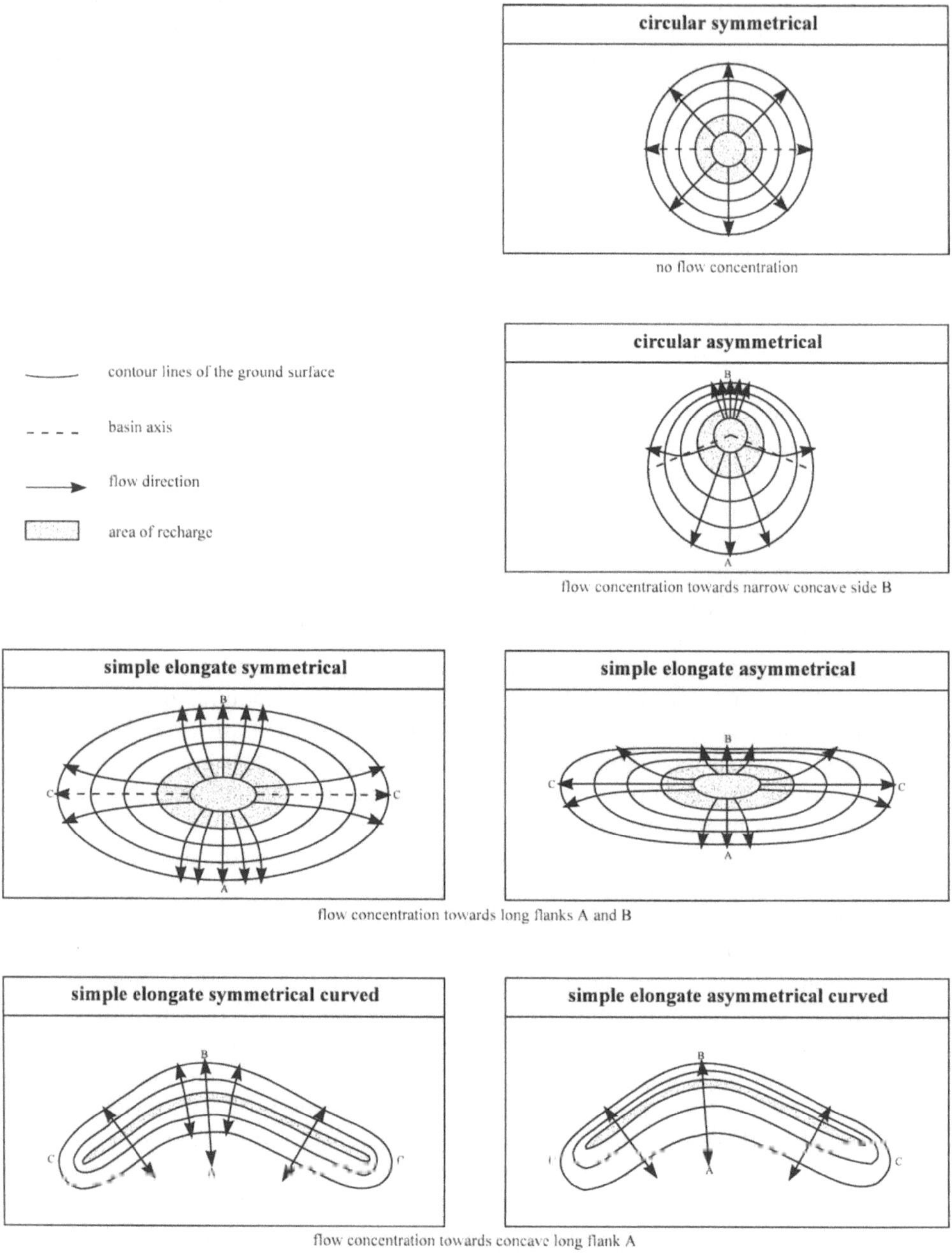

Fig. 7.2. Migration pathways according to the geometry of the sedimentary basin under hydrostatic conditions. (Pratsch 1982)

In summary, mechanisms and parameters controlling secondary migration seem to be much better understood than for primary migration. Only few parameters – buoyancy, capillary resistance, hydrodynamics and the specific migration system – determine the characteristics of the migration process after expulsion from the source bed.

7.3 Migration Pathways

Recent drilling into vertical fault zones in the Gulf of Mexico (Durham 1994; Anderson et al. 1994) brought attention to a new type of possible future exploration target: active migration pathways (Fig. 7.3). Active migration pathways demonstrate the essential conditions for petroleum migration to result in an exploitable accumulation: high and concentrated flow rates. Further, it shows that there are still other types of exploration targets than a conventional reservoir. Direct evidence for active and ancient migration pathways as well as potential migration pathways of different scales are evaluated in the following two sections.

7.3.1 Potential Migration Pathways

Potential Migration Pathways at the Basin Scale

Start and End Points of the Migration Pathway: The Role of Oil-Source Rock Correlation

At the scale of a sedimentary basin, a petroleum migration pathway represents the interconnnecting part within a petroleum system (Magoon and Dow 1994). Start and end points are defined by a genetic oil-source rock relationship. Delineation of migration patterns of genetically related petroleums allows inference not only of the depositional environments and organofacies of potential source rocks (see Chap. 6) but also helps to identify the distal and proximal limits of the migration pathway. To emphasise the importance of establishing a proven genetic relationship within a petroleum system Table 7.1 gives a very brief summary of the attributes of oils from carbonate source lithologies versus those from marine shales and from deltaic source rocks. The techniques for oil-oil and oil-source rock correlation include elemental analysis, gas chromatography, combined gas chromatography–mass spectrometry, and isotope mass spectrometry. These rely primarily on analyses of biological marker ("chemofossil") molecules (Peters and Moldowan 1993) and on analyses of isotope ratios (Fuex 1977; Stahl 1977; Schoell 1984a,b; Chung et al. 1994). For more detailed aspects of the oil-source relationship, the reader is referred to the respective text books (e.g. Peters and Moldowan 1993).

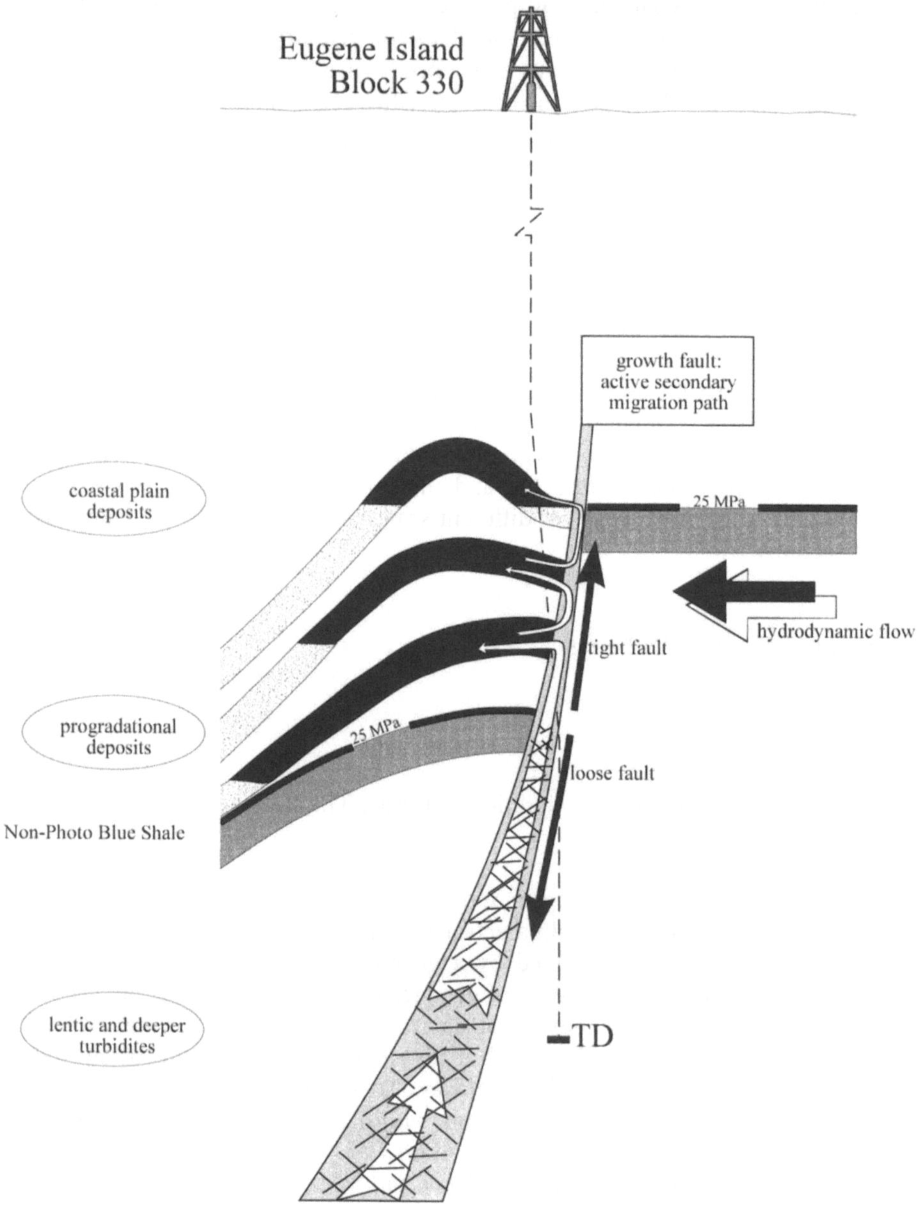

Fig. 7.3. Drilling into an active secondary migration pathway. Sketch of hydrostratigraphic model for migration and seal conditions in growth-faulted reservoir and how secondary migration of hydrocarbons occurs through the fault zone at Eugene Island Block 330, Gulf of Mexico. (Durham 1994, courtesy of Flemings, Penn State University)

Table 7.1. Summary of some established oil-oil and oil-source rock correlation parameters

Parameter	Carbonate SR (marine organofacies)	Marine Shale (marine organofacies)	Deltaic SR (terrestrial organofacies)
Basic parameters:			
n-Alkane predominance	Even	Odd	None
Sulphur content	High (1–5%)	Variable (<2%)	Low (<0.3)
SAT/ARO	Low-medium	Medium-high	High
Ni, Vcontent	High	Low-high (1–2)	Low or absent
Ni/V-ratio	Low (≤1)	Low (≤1)	High (>2) only for lacustrine sources
Pristane/phytane	Low (<1)	High (>1)	Very high (>3)
Pristane/n-C_{17}	High (>0.6)	Low (<0.6)	High (>0.6)
Phytane/n-C_{18}	High (>0.3)	Low (<0.3)	Low
Biomarkers			
General			
Sterane predominance (also diasteranes and monoaromatic steroids)	$C_{27}>C_{29}$	High C_{28}, $C_{27}<C_{29}$	High C_{29}
$C_{30}/(C_{27}–C_{30})$ steranes	High	Low	Low
Diasterane/regular sterane	Low-moderate	Moderate-high	Medium-high
Sterane/17α(H)-hopane	Low	High	Low
Oleanane index	None	None	Since Lower Cretaceous
Specific environment			
γ-Cerane index	–	For high salinity	–
4-Methylsteranes	–	For dinoflagellates	–
C_{35}-Homohopane index	–	Degree oxic environment	–
Carbon isotopes			
Type curves (sat.→ oil → extract→ kerogen→ aro. → NSO→ asphaltenes)	Heavier	Heavier	Heavier
Isotope profile	Light-heavy	Light-heavy-light	Heavy-light

Migration Pathways and Basin Type

Several authors (Bally 1975; Stonely 1981; Green 1983; Kingston et al. 1983) have attempted to explain the relationships between the geodynamic (tectonic) situation of a sedimentary basin and hydrocarbon flow. The evaluations have in common that three principal migration systems can be differentiated according to three major types of sedimentary basins.

The *fault migration system* (Fig. 7.4 top) is associated with continental rift basins (pull-apart basins and back-arc basins). High tectonic subsidence rates and high heat flow cause the rapid deposition and maturation of source rocks. Petroleum migrates predominantly in vertical direction along a network of frequently reactivated faults. The presence of permeable fractures in source rocks supports the vertical petroleum transport but also encourages dismigration and leakage. Temporarily closed faults do not only stop migration but provide individual tectonic compartments with high overpressures.

Long-range migration systems (Fig. 7.4 middle) are found primarily in older cratonic basins. These basins have been very stable during relatively long time periods with low subsidence rates and low heat flows. The number of source rock layers is generally very limited and often confined to one single stratigraphic unit. Maturation levels sufficient for petroleum generation are often reached only in the basin centre, whereas suitable traps for petroleum accumulation occur at the basin edges. Shallow platforms built-up in the course of these long time periods which include many hiatuses due to intervals of nondeposition and erosion. These platforms provide relatively uniform carrier and reservoir rocks. This fact and the repeated occurrence of seals favour short, vertical, primary migration pathways and long-range lateral (secondary) migration routes. The long pathways provide large drainage areas. But the chance for loss of relatively large proportions of petroleum as residual oils, tar mats and tar sands is also substantial. Where the platform formations are lithologically discontinuous, stratigraphic traps are common. Due to the shallow overburden the chances for the formation of overpressure zones are low.

A widespread scatter of reservoired oil and gas within a sedimentary basin is the result of dispersive migration systems. *Dispersive migration systems* (Fig. 7.4 bottom) are predominantly encountered in basins situated in orogenic belts which are areas of compression with active orogenesis. Orogenic belts possess highly variable but often low geothermal gradients. Subsidence is caused not only by gravity but also by tectonic movements. The dispersion of oil accumulations is due to a migration system formed by rapid and heterogeneous sedimentation periods with resulting irregular sedimentation. This geometry is not favourable for the formation of permeable, long-distance migration pathways. Relatively short lateral and vertical migration paths contribute to varying degrees. In addition, there is a high risk for dismigration due to the ongoing orogeny and activation of faults. Therefore leakage, surface shows and biodegraded oils occur frequently. One attraction of orogenic belts is the existence of anticlinal trap situations. The intensity of the tectonic stress and the resulting number of folds and faults per area determine the degree of dispersion within such a migration system. Comparable heterogeneous sedimentation patterns with the same implication for dispersive hydrocarbon migration as in orogenic belts are also known from deltaic plays.

The Timing Between Individual Migration-Related Processes

The overall possibility and efficiency for petroleum transport is related to the timing of several individual processes during the geological evolution of a sedimentary basin. Which processes are interrelated to petroleum migration is best demonstrated by an events chart (Magoon and Dow 1994), a summary of the evolution of a petroleum system. This chart relates the timing between the following essential elements and processes: deposition of source, reservoir, seal and overburden rock, trap formation, generation, migration and accumulation,

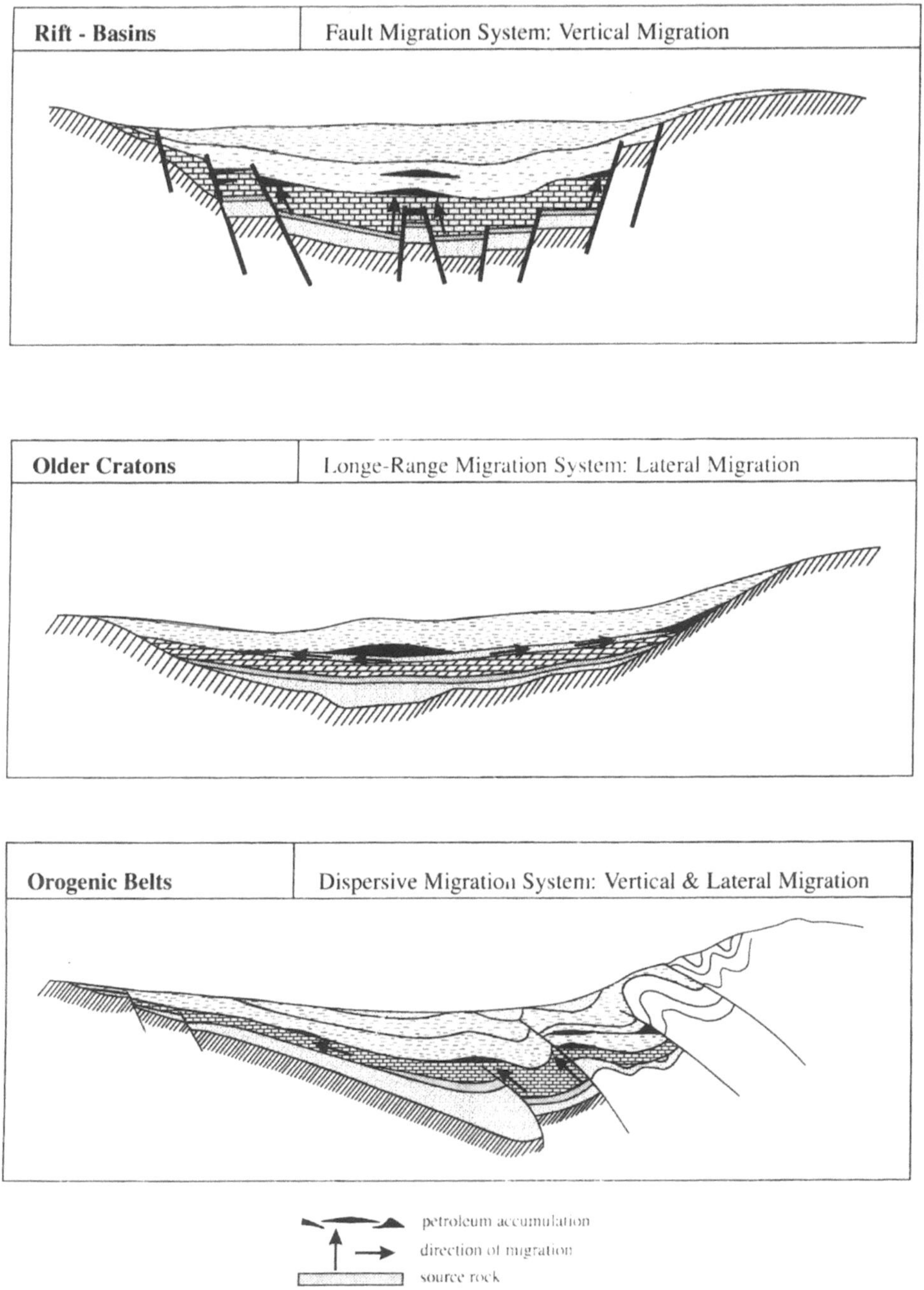

Fig. 7.4. The three principal migration systems (fictitious case histories): fault migration system, long-range migration system, dispersive migration system

and preservation. Furthermore, it makes obvious the most critical moment of the total system (Fig. 7.5). Although the events chart summarises in an easy way the individual characteristics of a petroleum system, it cannot serve for the respective argumentation. This can only be provided by a numerical simulation within the respective geological framework (see Sect. 7.5).

Potential Pathways at the Macro- and Microscopic Scale

Sedimentation is the primary controlling process which establishes the pathway for subsequent fluid transport. If no changes other than compaction occur during burial of the sediment, petroleum migration is governed directly by the primary interparticle pore space. Compaction reduces porosity but does not eliminate the primary heterogeneity or homogeneity of a sedimentary rock. Thus, chemical diagenesis and fracturing are the important secondary controlling processes. Results of these processes have significant impacts on the timing and the direction of petroleum migration. Diagenesis is especially important for carbonate, evaporite and fine-grained siliceous rocks.

Lithology and Lithofacies Variation

Macro- and microscopic petrography permit evaluation of the specific sedimentological properties of a migration pathway. Areas of higher and lower

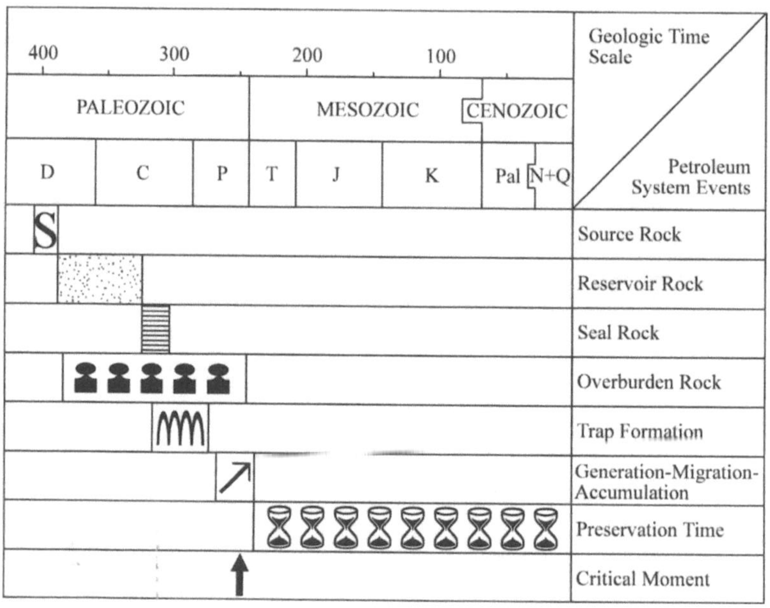

Fig. 7.5. Events chart showing the relationship between the essential elements and processes as well as the preservation time and critical moment for a fictitious petroleum system. (Magoon and Dow 1994)

porosity can be differentiated and mapped in carrier and reservoir rocks by conventional methods. Source rocks often convey the macroscopic impression of being homogeneous, dense and impermeable except for an occasional fracture or other discontinuities. Scanning electron micrographs, however, reveal a remarkably open fabric consisting of a network of regular and irregular macropores and vugs enclosed by detrital and secondary minerals which are often corroded. Thus, even in tight source rocks detailed optical investigations by normal light microscopy (e.g. Littke et al. 1988), incident-light fluorescence microscopy (e.g. Soedder 1990), and by scanning electron microscopy (e.g. Lindgreen 1987) are effective procedures for direct observation into the microfabric of a potential migration (expulsion) pathway, i.e. the most permeable part of a source rock. These methods reveal features and minerals which contribute to porosity and permeability.

The diagenetic evolution of a sedimentary rock unit must be retraced to check its local potential as migration pathway. This is performed by establishing a correct and consistent sequence of cement generations and dissolution phases. Many diagenetic events occur, for example, in the shallow marine, arenitic sandstones of the Gifhorn Trough of northern Germany (Fig. 7.6, Schwarzkopf 1990): (a) Fe-calcite cement formation, (b) Fe-dolomite formation

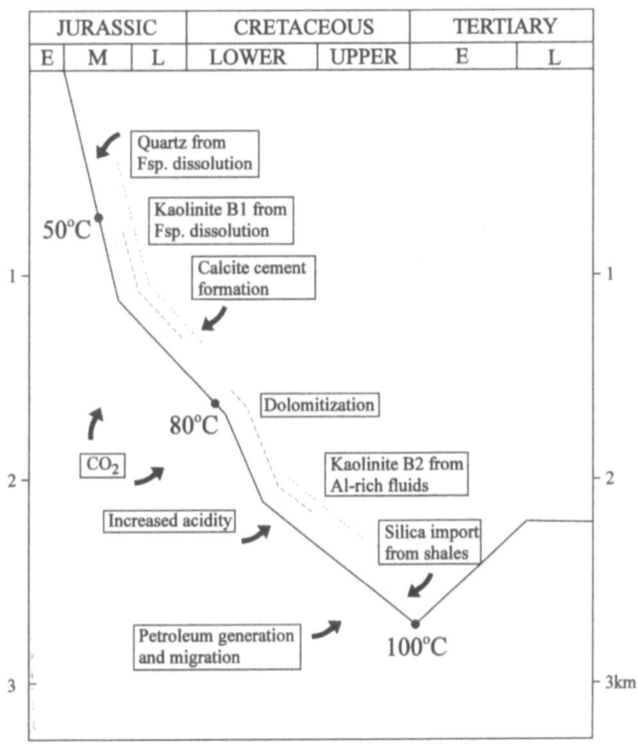

Fig. 7.6. Synthesis for burial, maturation and diagenesis. (Schwarzkopf 1990)

– dolomitisation of early formed Fe-calcite, (c) late kaolinite and secondary porosity formation, and (d) late quartz overgrowths. Based on his analyses Schwarzkopf (1990) suggested the following history of petroleum migration: early generated condensate was trapped during carbonate precipitation, prior to the main phase of petroleum migration. These early formed carbonate-cemented horizons compartmentalised the reservoir, controlled migration routes and promoted the formation of diagenetic traps.

Further, the origin of fractures must be assessed to evaluate whether they were already present at the time of petroleum migration or not. This is generally performed by microscopic observations, supported by luminescence investigations (Fig. 7.7), (trace) element analyses and by analyses of stable carbon and oxygen isotopes. For instance, the carbon isotopic composition can be used to identify the origin of carbonate cements. On the basis of the oxygen isotopic composition one can deduce the precipitation temperature of the cement and the isotopic composition of the pore water at the time of precipitation (Irwin and Hurst 1983). Cemented intersections of partly mineral-filled fractures allow the relative dating of individual fracture generations and the delineation of respective stress directions and the respective tectonic event (Kulander et al. 1990). In this way the open stage of fractures can be determined and related to the time stage of petroleum migration.

Secondary Porosity

Views on the evolution of sedimentary rocks during burial diagenesis have been proposed by Hower et al. (1976), Foscolos and Powell (1980) and Curtis (1983). In most petroleum systems new migration pathways by secondary porosity are developed during the catagenetic stage of about 65–95 °C. Potential candidates for mineral dissolution are calcite, potassium-feldspar, kaolinite and possibly other clays. They may provide secondary porosity which may develop into source rocks as well as reservoir rocks.

Concentrated flow through the more permeable horizons of mudrocks may also create secondary migration pathways (Curtis 1983). Intensified flow of undersaturated formation waters against local permeability barriers may cause dissolution of mineral phases such as carbonate and redistribution to other locations (Mann et al. 1990, 1991).

Pressure solution seams

Pressure solution seams also provide primary migration pathways (Leythaeuser et al. 1995; Mann 1996). However, expulsion from source rocks via solution seams and stylolites may function in a somewhat different way than that via other pathways. Due to the restricted pathway and the short time interval during which the actual migration process takes place Sassen et al. (1987b) coined the term "explosive migration". Oil stainings and hydrocarbon inclusions provide convincing evidence for active or palaeomigration pathways

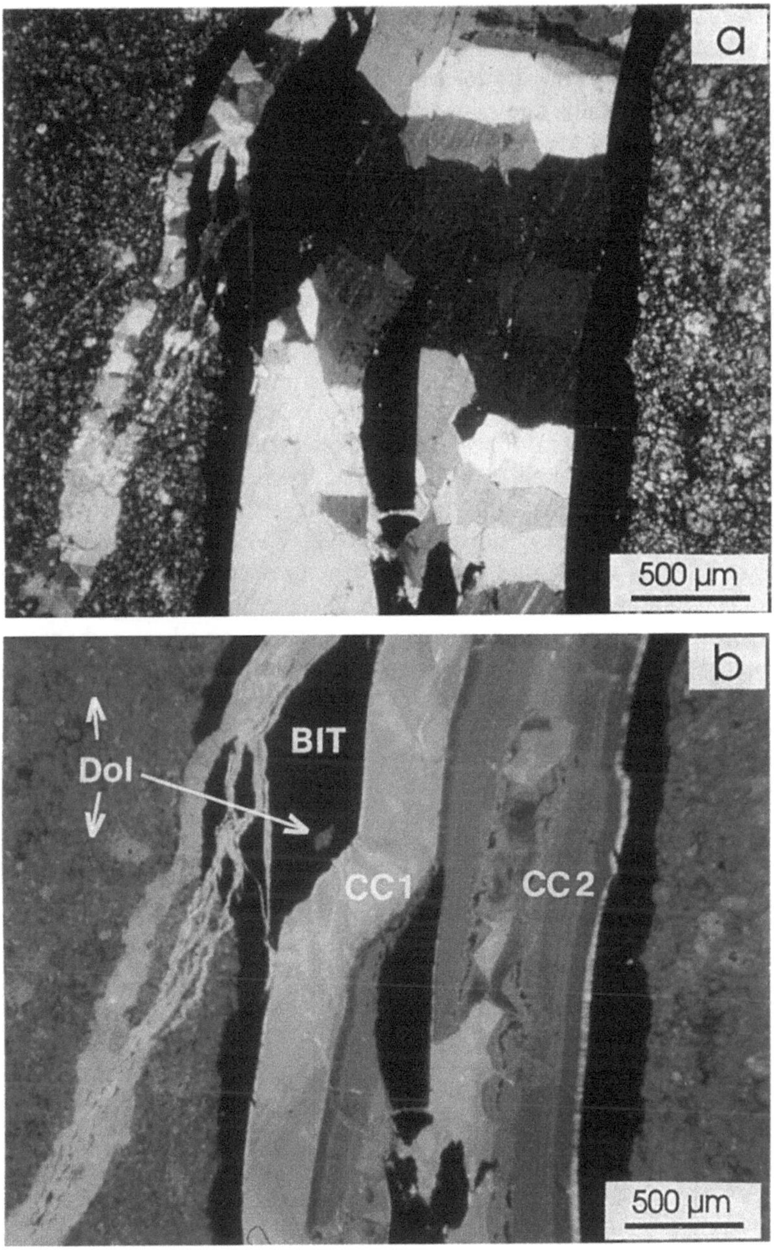

Fig. 7.7a,b. Vertical fracture from Lower Toarcian Posidonia shale at 0.73%R_r. **a** Normal light, crossed nicols. Fracture fill bears calcite and solid bitumen. **b** Luminescence excitation. The following phases can be identified according to their relative chronological sequence: solid bitumen (*BIT*); light-orange luminescing calcite (*CC1*, Fe^{2+}/Mn^{2+} ratio <0.3) and weakly dark-orange luminescing calcite (*CC2*, Fe^{2+}/Mn^{2+} ratio >5.7). At this position, the adjacent Posidonia shale is heavily dolomitisized (*Dol*). (Jochum 1993)

via solution seams in carbonate, evaporite and siliceous source rocks from the Palaeozoic to the Tertiary (Grabowski 1984; Palacas 1984; Mann 1994, 1996).

Reactions of "clay volatiles" with quartz or carbonate may also be able to create secondary porosity for potential migration pathways. According to results of Heller-Kallai and co-workers (1987, 1988, 1989; Miloslavski et al. 1991; Heller-Kallai and Miloslavski 1992), these volatiles form by thermal dissociation of clay minerals at individual temperature stages.

Tectonic Fractures and Faults

Fractures have been reported as migration conduits not only for hydrocarbons which have left the source rock but especially also for petroleum expulsion from source rocks (Meissner 1978; Littke and Rullkötter 1987; Talukdar et al. 1987; Leythaeuser et al. 1988c; Mann 1990; Mann et al. 1991). Fractures may open and close several times due to stress cycling effects, also called "seismic pumping" (Sibson et al. 1975). Fault/fracture meshes acting as conduits within an extensional stress regime tend to form either as subvertical extensional "chimneys", or "fuzzy" normal faults (Fig. 7.8). In the Monterey shales they have acted as major fluid expulsion structures, transporting large volumes of aqueous and hydrocarbon fluids across the bedding anisotropy, the relict mesh structure being preserved through cementation by silica, carbonate, and bitumen (Redwine 1981; Sibson 1994). In total, faults, extension fractures, and stylolites can substantially affect permeability of the rock mass.

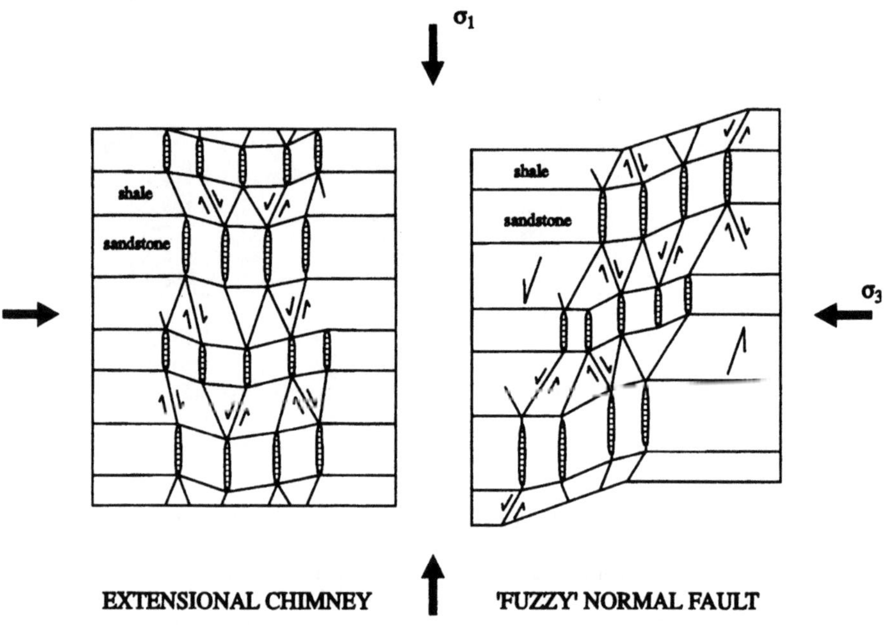

Fig. 7.8. Fault fracture meshes. (Sibson 1994)

Information about the presence, number, width, orientation and distribution of fractures can be obtained efficiently from well logs. For the geological past information about fracturing events can in most cases be deduced only from the numerical simulation of the evolution of rock properties and pressure build up. An example for the build up of overpressure as a function of permeability and sedimentation rate is given in Fig. 7.9.

Expulsion Fractures

Because of the low permeability of source rocks, a number of scientists have suggested that oil and gas generation may cause overpressure, and that this may induce microfractures in source rocks (Snarsky 1962; Tissot and Pelet 1971; Momper 1978; Palciauskas and Domenico 1980; du Rouchet 1981; Talukdar et al. 1987; Littke et al. 1988; Özkaya 1988; Lehner 1991; Düppenbecker and Welte 1992). This lithostatic failure due to high natural pore pressure in overpressured lithologies has been termed "expulsion fracturing" (Mann 1994). This term is preferred to differentiate it from the technical term "hydraulic fracturing". However, microfractures due to expulsion fracturing can seldom be confirmed by direct observations because in most clay-shale source rocks they probably completely heal. Sandvik and Mercer (1990) argue that due to

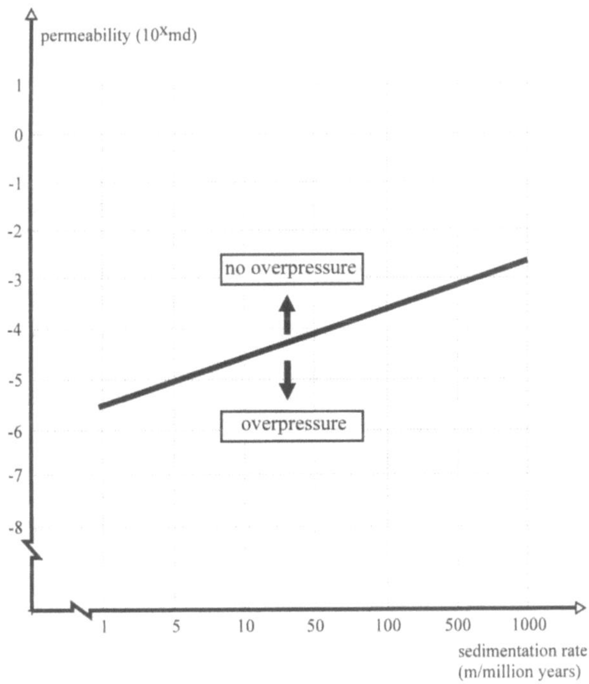

Fig. 7.9. Relationship between sedimentation rate and permeability in order to build-up overpressure

their organic matter content and small grain size source rocks are generally highly ductile and therefore require a high excess pressure for fracturing. After careful investigations most expulsion fractures turn out to be tectonic-induced fractures. Because there is little petrographic evidence, many petroleum geologists are still sceptical that expulsion fractures act as migration pathways.

The Tool for Combining Migration Pathways from the Core and Basin Scale: Sequences in Seismic Stratigraphy

The findings about potential migration pathways from the core and basin scale must be integrated into each other. This means a lateral extrapolation of the facies from the core to the basin scale. System tracts from sequence stratigraphy (see Chap. 1; Van Wagoner et al. 1990) provide a very high degree of facies predictability in the subsurface. This predictability can be used for extrapolations to the basin scale and applied within a numerical simulation of petroleum migration to trace facies changes of carrier beds and subsequent migration routes. During such an extrapolation correct up-scaling procedures are important and must be developed without neglecting the anisotropy of a sedimentary basin fill.

7.3.2 Evidence for Migration Pathways

Petrophysical Evidence for Migration Pathways: Transport-Related Properties

High permeability provides the first direct evidence for a migration pathway. After potential migration pathways have been qualitatively identified, effective from less affected or noneffective pathways need to be distinguished. These distinctions are generally based on direct and indirect permeability data similar to the approach used by reservoir engineers. The following paragraphs present both the principle objectives and limitations of such an approach.

The intrinsic permeability of rocks in the subsurface varies over more than 13 orders of magnitude from 10^{-10} m^2 to about 10^{-23} m^2 (Brace 1980; Bredehoeft and Norton 1990). These values vary according to lithology, compaction and cementation/dissolution processes (Table 7.2). Pore size and fracture width represent the main controlling agent for petroleum flow as permeability is proportional to several powers of pore size or fracture width. It is known that very few large pore channels need be present to have a dramatic effect on permeability (e.g. Khanin 1968). Furthermore, different types of migration mechanisms are involved at different pore size levels. Thus, it is essential to determine the relative amounts of pore size classes present in a representative rock sample from a migration pathway, especially to differentiate between the low permeability domain of the source rock (dominated by diffusion), the higher permeability domain of the source rock (dominated by

Table 7.2. Permeability ranges for sedimentary, igneous and metamorphic rocks

Lithology	Permeability (mD)	Permeability (D)	Permeability (m^2)
Karst limestone	10^2–10^6	10^{-1}–10^3	10^{-13}–10^{-9}
Fractured igneous and metamorphic rocks	1–10^4	10^{-3}–10	10^{-15}–10^{-11}
Limestone, dolomite	10^{-7}–10^{-4}	10^{-4}–10^{-1}	10^{-16}–10^{-13}
Sandstone	10^{-8}–10^{-4}	10^{-5}–10^{-1}	10^{-17}–10^{-13}
Siltstone	10^{-9}–10^{-6}	10^{-6}–10^{-3}	10^{-18}–10^{-15}
Shale	10^{-14}–10^{-8}	10^{-11}–10^{-5}	10^{-23}–10^{-17}
Unfractured igneous and metamorphic rocks	10^{-14}–10^{-7}	10^{-11}–10^{-4}	10^{-23}–10^{-16}

Darcy flow), and the high permeability domain of the carrier rock (also dominated by Darcy flow).

A classification of pores as a function of their size from rough pores to ultramicropores is provided in Table 7.3 (according to the IUPAC classification) by Singh et al. (1985), the supplement of ultramicropores by Dubinin (1975) and the classification of macropores by Setzer (1977). A quantification is possible by a combination of microscopy, mercury porosimetry, gas adsorption and other measurements (Table 7.4).

However, quantification of transport-related properties of rocks and fluids requires the evaluation of the relative importance of three competitive forces: viscosity, capillarity and gravity. Their importance varies according to the scale of observation, for example, whether migration takes place at the source bed where capillary forces prevail, or by transport through more permeable rocks where gravity and viscosity forces predominate.

Low Permeability Domain of Source Rocks

Transport processes in the low permeability domain of sedimentary rocks still involve a considerable number of unresolved problems. This field of research is receiving increased attention because it bears relevance for both primary migration in source rocks and the sealing efficiency of cap rocks. For the present

Table 7.3. Pore classification according to size (Singh et al. 1985; Dubinin 1975; Setzer 1977)

Type of pore	Size (diameter)
Rough pore	$d > 2$ mm
Macrocapillary	2 mm $> d >$ 50 μm
Capillary	50 μm $> d$ 2 μm
Microcapillary	2 μm $> d$
Macropore	$d \geq 50$ nm
Mesopore	50 nm $\geq d \geq$ 2 nm
Micropore	2 nm $\geq d \geq$ 0.8 nm
Submicropore	0.8 nm $\geq d$
Ultramicropore	≤ typical molecule diameter of the adsorptive: ≅ 0.6 nm

Table 7.4. Selected techniques for quantification of pores of various size

Method	Pore size	Advantage	Disadvantage
Total porosity			
Calculation of apparent density	All	Easy calculation	
Image analysis	d ≥ 20 nm	Pores of irregular shape measurable	Only area information, subjective measurements
Quantitative image analysis	d ≥ 20 nm	Optical check, even by TEM	Assumes spherical pores, limited to large pores
Small angle X-ray scattering	1 nm ≤ d ≤ 100 nm	Short-term method	Complicated evaluation program
Small angle neutron scattering	1 nm ≤ d ≤ 100 nm	Short-term method	Complicated evaluation program
Accessible porosity			
Xylene and water impregnation	d >100 nm	Simple, short	Small specimen, no pore size distributions
Rubber or liquid metal impregnation		True pore network skeleton	No measurable pore size distribution
Air/He penetration	Some nm	Short, good reproducibility	No pore size distribution
N_2 adsorption	d < 50 nm	Correlation to BET surface	Physical model, complicated evaluation program
Mercury porosimetry	3.4 nm < d < 60 μm	Most extended range of pore size, comparable data, standard use	Assumes cylindrical pores

discussion the low permeability domain is defined by permeability values of 1 μD (10^{-18} m^2) and below. The lower limit for experimental measurements of permeability ranges around 0.1 nanodarcy (10^{-22} m^2; Katsube et al. 1991; Hanebeck 1994).

Concentration-driven molecular transport is commonly described by Fick's law of diffusion:

$$J = -D \text{ grad } C$$

where C denotes concentration (mass/volume), D is the diffusion coefficient (length squared/time) and J the diffusive flux (mass/area/time). For diffusion in porous media, for example, sedimentary rocks, an effective diffusion coefficient D_{eff} is used in combination with the bulk volume concentration (C_{bulk}) of the diffusing substance (see Krooss et al. 1992a,b).

Although a general consensus exists that primary migration proceeds predominantly as a pressure-driven volume flow or phase flow, some workers (Stainforth and Reinders 1990; Thomas and Clouse 1990) have argued for molecular diffusion as an important transport mechanism during certain stages of primary migration and under specific conditions. However, due to its relatively low transport efficiency, diffusive transport in geological systems can generally be considered negligible whenever volume flow can take place. On the other hand, in situations with very low volume flow due to low permeabilities and/or absence of fluid pressure gradients, diffusion may become a relevant transport process. These situations are most likely to occur in the low-permeability domain of source rocks and other shales, carbonates and evaporites which are able to act as cap rocks above petroleum reservoirs.

Given a pressure gradient of 1 bar/100 m (10^3 Pa/m), a fluid viscosity of 1 centipoise (cP; 10^{-3} Pa s) and a permeability of 1 nD (10^{-21} m^2) the fluid flow from a source rock is 10^{-15} m^3 m^{-2} s^{-1}, corresponding to a Darcy velocity of 3 cm/Ma. In a source rock with 1% porosity this corresponds to a fluid velocity of approximately 3 m/Ma (10^{-13} m/s) within the pore system. The average diffusion distance of appreciable amounts of molecules with a diffusion coefficient of 10^{-12} m^2/s, the lower limit of experimentally measurable diffusion coefficients, can be estimated to amount to 18 m in the same period of time. Therefore diffusion is expected to play a significant role in the redistribution of hydrocarbons under these circumstances.

The above fluid velocities are in agreement with estimates by Mackenzie et al. (1987) who infer a Darcy velocity of 5×10^{-15} m/s for petroleum expulsion from a mudstone source rock sequence in the Brae oil field area of the North Sea. To compare the relevance of diffusive transport with respect to volume flow Mackenzie et al. (1987) used the Peclet number:

$$N_{Pe} = \frac{v \cdot l}{D} \quad .$$

This dimensionless number which is commonly used in fluid dynamics and heat flow theory relates the Darcy velocity (v; m/s), the diffusion coefficient (D; m^2/s) and a characteristic length (l; m). Geological applications of the Peclet number comprise the investigation of fluid and heat flow in the lithosphere (e.g. Clauser and Neugebauer 1991). Large Peclet numbers (>1) indicate convection dominated transport, whereas the diffusion controlled flow regime is characterised by low Peclet numbers (<0.1) with a transition region of $1 > N_{Pe} > 0.1$ (Clauser and Neugebauer 1991).

Mackenzie et al. (1987) conclude that in their case history diffusion is important for light hydrocarbon (LHC) molecules up to C_7 with Peclet numbers up to 0.5, whereas the transport of higher molecular weight hydrocarbons is controlled by volume flow. Considering a range of pressure gradients within shale source rocks of 1–10 bar/100 m (10^3–10^4 Pa/m) and viscosities between 1 and 0.1 cPoise (10^{-3}–10^{-4} Pa s), Darcy velocities of bitumen in source rocks can be expected to extend from 10^{-10}–10^{-15} m/s. Light hydrocarbons (LHC; C_1–C_8) diffusion coefficients in sedimentary rocks range between 10^{-10} and 10^{-12} m^2/s. Thus, using a characteristic length of 10 m for an average source rock layer the Peclet numbers for source rock systems can be estimated to range from 10^{-4} to 10^3.

This result documents that the transport regime for low permeable source rocks extends from diffusion dominated to volume flow (advection) dominated. Depending on the transport parameters of the source rock system and the fluid properties, diffusion may therefore contribute significantly to hydrocarbon transport during primary migration and expulsion. In view of these estimates the dispute about the relevance of diffusive transport in primary migration is understandable.

Diffusion of hydrocarbon gases through the water-saturated pore space must be accepted as a contributing process of primary migration (Sokolov et al. 1964; Watts 1963; Hinch 1980; Huc and Hunt 1980; Krooss and Leythaeuser 1988). Based on the calculations of Leythaeuser et al. (1982), the rate of mass transport for gas can be sufficiently high to account for commercial-sized gas fields.

Some characteristic features of transport in low permeability systems deserve more detailed discussion. For example, volume flow in a porous medium requires an interconnected pore system whereas diffusion can proceed both in fluid (gases and liquids) and solid phases. Hydrocarbon diffusion in source rocks can take place in the water- or bitumen-saturated pore space and in the solid organic matter (kerogen) which has a cross-linked polymer structure. Consequently hydrocarbons can be transported even in source rocks which possess no measurable permeability. The rate and efficiency of the diffusive transport of hydrocarbons is controlled by the mobility and the solubility of the diffusing species in the different phases and depends largely on the size and chemical structure of the molecules. Experimental work by Hanebeck (1994) has shown that due to their high aqueous solubility aromatic hydrocarbons diffuse preferentially in the water-saturated pore space of source rocks. In the absence of an interconnected aqueous phase diffusion can still proceed through the organic network. Under these conditions saturated hydrocarbons diffuse faster than aromatic compounds of the same carbon number.

Thus diffusion through organic matter networks represents an alternative concept to explain primary migration (Stainforth and Reinders 1990). In fact, the activation energies measured for the diffusion of light hydrocarbons in tight rocks are much higher than for diffusion in water (e.g. Witherspoon and Saraf 1965). Krooß (1988) can better fit and explain the data from his migration experiments with C_1–C_6 hydrocarbons by applying a model of pore diffusion plus adsorption, rather than diffusion alone. Stainforth and Reinders (1990) presented a quantitative approach for calculation of thermally activated diffusion. In addition to temperature, a major control of their migration model is represented by a characteristic length of the primary migration system which takes into account the tortuosity of the petroleum pathway. However, the migration of hydrocarbons through a three-dimensional organic matter network does not apply to source rocks with low organic matter content.

Experimental work and theoretical considerations indicate that it is questionable whether pressure-driven fluid transport in the nanodarcy permeability range and below can still be considered as “volume flow” and treated by Darcy’s law. With decreasing pore diameters fluid molecules interact increasingly with the pore walls, and ultimately transport is effected by movement of individual molecules through “molecular size” pore throats. Under these circumstances a differentiation between molecular transport and volume flow becomes irrelevant.

Higher Permeability Domain of Source Rocks

With increasing pore size and permeability, the dominance of diffusive transport gives way to a Darcy-dominated flow regime. Flow can be quantified by using Darcy's law with the most effective migration pathways of a given source rock being those of highest permeability during expulsion.

The relationship between pressure gradient (grad P) and volume flux (*J*, mass per unit cross section area per unit time) in porous media is expressed by Darcy's law:

$$J = -\frac{k}{\eta} \text{grad } p$$

where k denotes the permeability of the porous medium expressed in units of length^2 (m^2) or in Darcy (1 Darcy = 1.01×10^{-12} m^2) and η is the dynamic viscosity expressed in centipoise or Pa s (1 cP = 10^{-3} Pa s). Strictly speaking, the law pertains only to isothermal, isochemical and nonturbulent conditions. Darcy's equation describes flow by hydraulic head gradients only. However, this equation can be generalised in order also to take temperature gradients, electrical gradients and chemical gradients into account.

For practical purposes the intrinsic permeability k of a pore network can be estimated from its mean pore radius r_m (Amyx et al. 1960) and by taking into account the respective tortuosity T according to the relationship:

$$k = r_m^2/8\ T^2$$

In petroleum source rocks relatively permeable pathways (in which Darcy flow is effective) exist, and a classification of source rock expulsion efficiencies based solely on geochemical characteristics can be misleading. Burrus et al. (1993) compared organic-matter poor, kerogen type III pelagic shales from the Mahakam Delta, organic-matter rich marine kerogen type II Bakken shales, organic-matter rich Lower Toarcian shales, and less organic-matter rich Hettangian-Lotharingian marls, both from the Paris Basin to determine the relative importance of geochemical characteristics compared to transport properties during expulsion. His case histories showed that overpressures in Mahakam and Bakken shales were too low to allow the formation of expulsion fractures, and thus that expulsion is relatively insignificant compared to the two Lower Liassic sources from the Paris Basin which have a 10–100 times better permeability due to tectonic fracturing.

Following reservoir engineering principles, a critical oil saturation relative to water must be reached before oil begins to flow. Although Dickey (1975) discussed a "possible primary migration of oil from source rocks in oil phase", Durand (1983) first applied rigorously the notion of relative permeability to source rocks. By analogy with low permeability reservoirs, Durand (1983) used two examples of sedimentary basins having different sedimentation styles. One of these had a low sedimentation rate and discrete organic-matter rich layers and the other a high sedimentation rate and numerous organic-matter poor layers. As a result the relative permeability is several orders of magnitude lower in the latter basin than in the former, although Durand (1983) applied largely arbitrary values for the relative permeability versus oil saturation curves because they had been extrapolated from a production study. The principal conclusion was correct, however, according to more recent results relative permeability curves for source rocks differ considerably from conventional curves for reservoir rocks (compare Fig. 7.10). This must be considered for numerical modelling of expulsion processes. According to the calculations of Okui and Waples (1993) the maximum oil saturation should not be higher than 25% (see Fig. 7.10). However, theoretical considerations and case studies show that 15–25% is only the minimum oil saturation required for the beginning of the expulsion process (Welte 1987; Mann 1994). Measured oil saturations of the pore space around peak oil window are in the range between 40% and 60% (Mann et al. 1991; Ropertz 1994; Mann 1996). Variations seem to be effected predominantly by the range of porosity which act as a buffer and may thus even retard petroleum expulsion for a limited time.

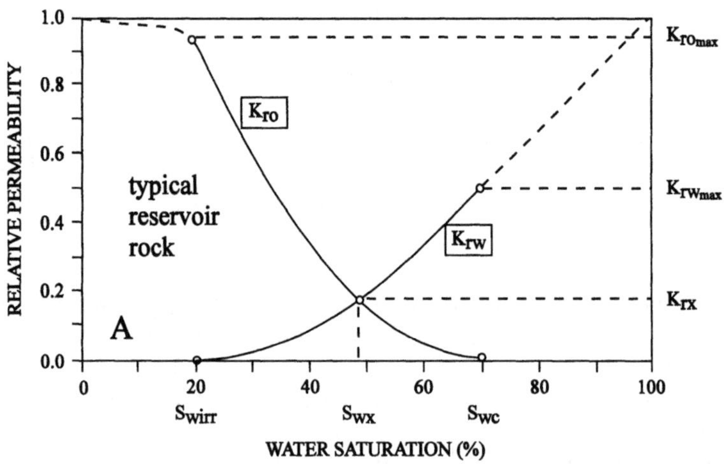

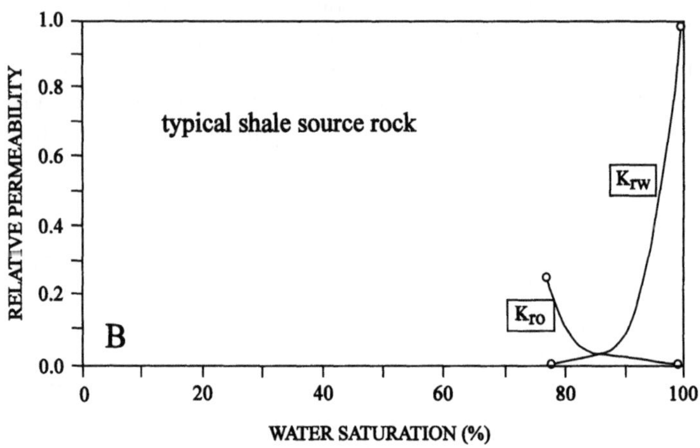

Fig. 7.10. Relative permeabiltiy curve for a typical reservoir rock (A) compared to that for a typical shale source rock (B), as proposed by Okui and Waples (1993)

Porosity, mineral matrix, wettability and adsorption all influence primary migration because interaction exists between kerogen, minerals and petroleum. Petroleum may be absorbed by kerogen, adsorbed on the kerogen or mineral surface or trapped within the kerogen network (Mann 1994). Porosity can act as a buffer and retard petroleum expulsion for a limited time (Ropertz 1989; Sweeney 1992; Mann 1996). Results from pyrolysis measurements show that a clay mineral matrix of a source rock retains hydrocarbons much longer than a carbonate matrix (Langford and Blanc-Valleron 1990; Mann 1994). If this interpretation can be transferred from laboratory to nature, it indicates that hydrocarbon migration in a clay-shale would be retarded compared to a carbonate source rock.

Better wetting conditions facilitate hydrocarbon migration within large pores because they reduce the required pressure for expulsion due to an increase in the contact angle. Wettability characteristic of individual minerals and rocks at different maturity stages (Mann 1994) clearly indicate that oil wetting in carbonate rocks is more likely than in siliciclastic rocks. Results from cryo-SEM studies of reservoir rocks by Robin et al. (1995) show that wettability of sandstones is related to the presence of kaolinite, while wettability of carbonates seems to be related to pore size distribution.

The role of absorption of generated petroleum by solid organic matter in source rocks has been investigated by Sandvik et al. (1992). He concluded that selective absorption is the cause for differences between migrated petroleum and source rock extracts. Further, absorption accounts for the long-noted correlation between total organic carbon (TOC) and extractable organic matter and may be the cause for high gas-oil ratios from source rocks with a low hydrogen index. Thus, during primary migration a residual part of bitumen must remain within the kerogen and smaller pores and experiences secondary cracking prior to expulsion; the mobile part of petroleum migrates along larger pores and fractures into the direction of the carrier bed (Lindgreen 1985; Mann 1989; Price and Clayton 1992; Ropertz 1994).

High Permeability Domain of Carrier Rocks

The main concepts to understanding two-phase fluid flow in the high permeability domain of carrier and reservoir rocks are the concepts of capillary pressure (Leverett 1941), relative permeability (Amyx et al. 1960) and wettability (Craig 1971). These fundamental concepts for secondary migration represent the effective flow behaviour as a function of fluid/fluid and fluid/rock interactions. Accordingly, rock and fluid properties are of fundamental importance during migration (see Sects. 7.4.2, 7.5).

The movement of petroleum during secondary migration, the principles of buoyant rise and capillary displacement, under hydrostatic and under hydrodynamic conditions are also described by Tissot and Welte (1980, p. 341ff, and references therein). The process of oil movement can be summarised as follows. For migration under hydrostatic conditions, buoyant forces of the petroleum phase must become high enough to overcome the capillary resistance of the carrier rock. Under hydrodynamic conditions, upward flow of water supplements buoyancy.

From theoretical considerations, England et al. (1987) concluded that petroleum flow in rocks less permeable than 10^{-15} m^2 is mainly vertical and petroleum flow in rocks greater than 10^{-15} m^2 is lateral. This general rule of thumb implies that during secondary migration buoyancy forces move petroleum stringers updip and may become intensified or reduced by the specific hydrodynamic conditions. However, many exceptions for this rule exist especially in source rocks, where also lateral migration avenues have been located (Espitalié et al. 1988; Mann 1990; Mann et al. 1990, 1991), and at source-carrier contacts, where capillary effects intensify expulsion (Mackenzie et al. 1987).

Secondary oil migration is observed to occur along restricted pathways and conduits within the carrier bed. Petroleum migrates by creating interconnected paths of petroleum-filled pore space which requires a certain percentage of pore volume to be filled. This has been shown from case histories (e.g. England et al. 1987) as well as by experimental (Catalan et al. 1992; Thomas and Clouse 1995) and numerical simulations (Hirsch and Thompson 1995). The specific petrophysical properties of the lithofacies determine the pattern of dendritic fingers and migration channels. Details of the control of rock and fluid properties on petroleum migration are treated in Sect. 7.4.2.

During secondary migration, compositional changes in migrating hydrocarbons result from phase changes (England and Mackenzie 1989; Larter and Mills 1991). Along the pathway decreasing pT conditions cause the homogeneous phase migrating fluid to separate into gas-saturated oil and condensate-saturated gas phases. These phase separations occur continuously along the migration pathway. The small volumes of the minority phase (i.e. gas exsolved from liquid or condensate exsolved from a gas phase) are unlikely to achieve the necessary saturation required to continue migrating (Amyx et al. 1960). Gas may disperse by diffusion into and through the water, and small accumulation of non-economic petroleum reservoirs may form along the migration pathway. With further migration the gas-oil ratios of the liquid phase further decrease while that of the gaseous phase increases (England et al. 1987). Hence, with increasing migration distance, the physical properties as well as the chemical composition of the liquid and the gas phase become increasingly dissimilar.

The migration and entrapment conditions along the migration pathway are different for liquids and gases, consequently, they may become physically separated. Similarly, as described by Gussow (1954) from reservoirs, gas may be kept downdip of the migration route from oil with successive microtraps that spill updip. The opposite takes place during leak differential entrapment of oil and gas where gas may be trapped updip from oil (Schowalter 1979). Such additional losses result especially if the carrier shows severe lithofacies changes. Any capillary pressure barrier not only shortens the actual length of the migration pathway but requires additional saturation and thus reduces the chance for an economic prospect.

Geochemical Evidence for Migration Pathways: Stage and Type of Processes Involved

Oil seeps at outcrops or oil-stainings at drill cores are the first direct and most obvious indications for petroleum migration pathways. Nevertheless, beside the "classical" occurrence from source to trap, there exist several specific settings where hydrocarbons and bitumens may occur (Fig. 7.11). Asphalt-like bitumen may itself serve as caprock (e.g. Levorsen 1967, p. 336). Bitumens are also found in a wide range of ore deposits, where they are generally used as evidence for an active role of hydrocarbons in ore genesis and a source of information about the geothermal and geochemical conditions of mineralisation (Gize 1993). On the other hand, late stage bitumens may be completely unrelated to the mineralisation process and simply used the same migration pathway as the ore generating pore waters (Parnell 1991). Further, Simoneit (1994) describes the paragenesis of bitumen together with hydrothermal fluid systems associated with oceanic spreading centres.

The residual soluble organic matter in the source rock and, if available, the comparison with its corresponding reservoir oil provide the key to elucidate the type of process(es) that took place during primary migration and expulsion. For instance, in terms of hydrocarbon composition on the molecular level a separate phase flow by bulk oil expulsion should exhibit no, or only little, fractionation, whereas hydrocarbon transport by solution in water or gas as well as diffusion show characteristic fractionation effects. To arrive at such conclusions soluble organic matter must be extracted from the rock material and separated by liquid and gas chromatography into compound groups, homologous series, isomers and individual molecules, thus forming the basis for mass balance calculations. In addition, concentrations of selected hydrocarbon compound groups or individual molecules with distinctive chemical and physical properties can be compared. For example, one must determine the ratio of saturated hydrocarbons to aromatics, or both to non-hydrocarbons (NSO compounds), more water-soluble aromatics versus less water-soluble ones, or short-chain to long-chain hydrocarbons.

Present knowledge of primary migration processes is derived from theoretical considerations and experiments but predominantly from selected case histories. Comparisons of live oil versus residual oil as well as source rock systems of siliciclastic, carbonate, siliceous and evaporite lithologies are presented in Section 7.3.3. The main problem of this kind of investigations is the need for well-characterised sample material, i.e. in general, conventional cores or rock samples of adequate quality.

Live Oil Versus Residual Oil

Migration pathways generally exhibit a petroleum phase which may be represented as fluid inclusions, normal oil-staining or highly viscous, perhaps even partly or completely solidified, asphalt-like bitumen. Interpretation of the individual petroleum compositions in respect to timing, direction and amount

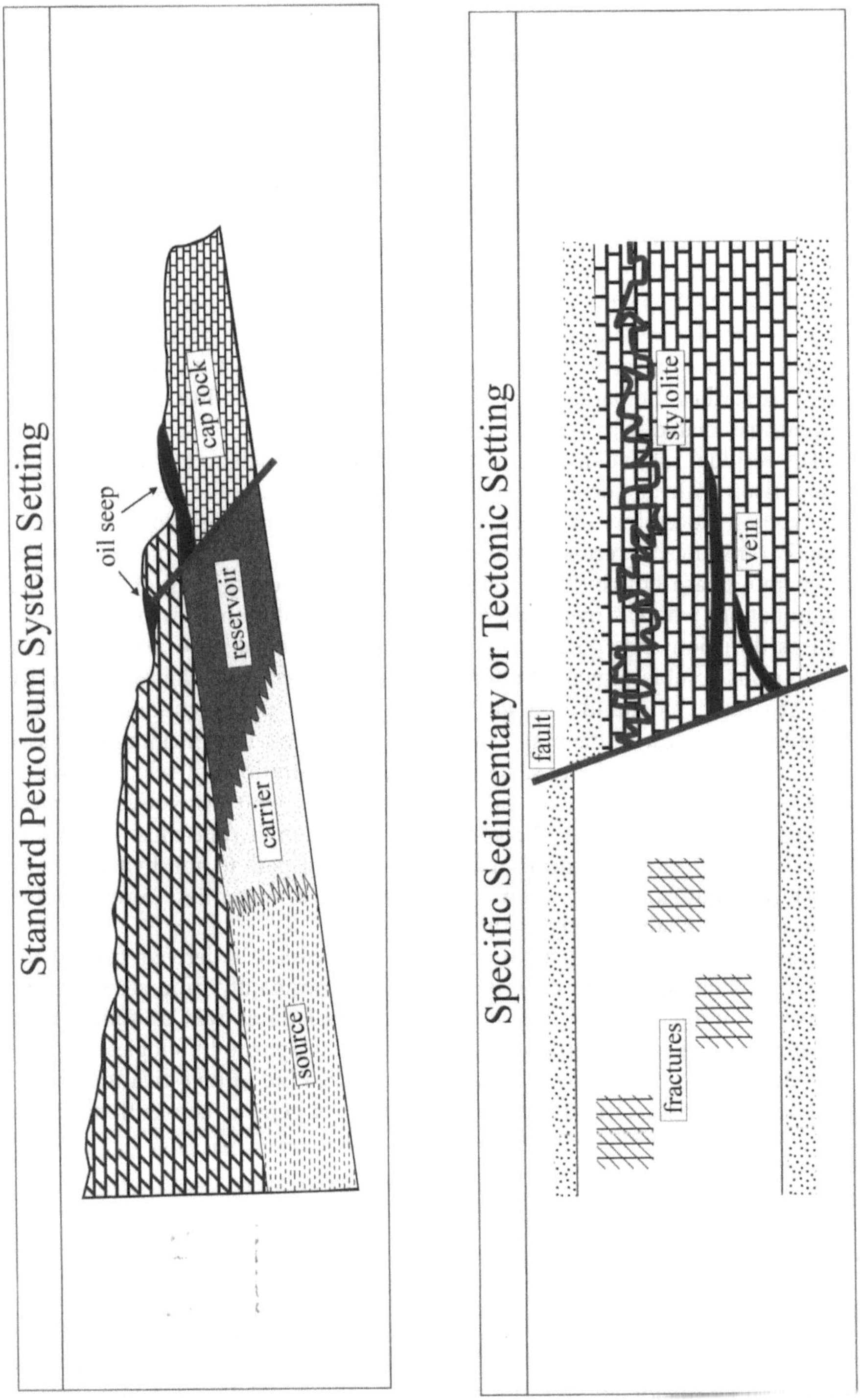
Standard Petroleum System Setting
oil seep
cap rock
reservoir
carrier
source
Specific Sedimentary or Tectonic Setting
fault
stylolite
vein
fractures

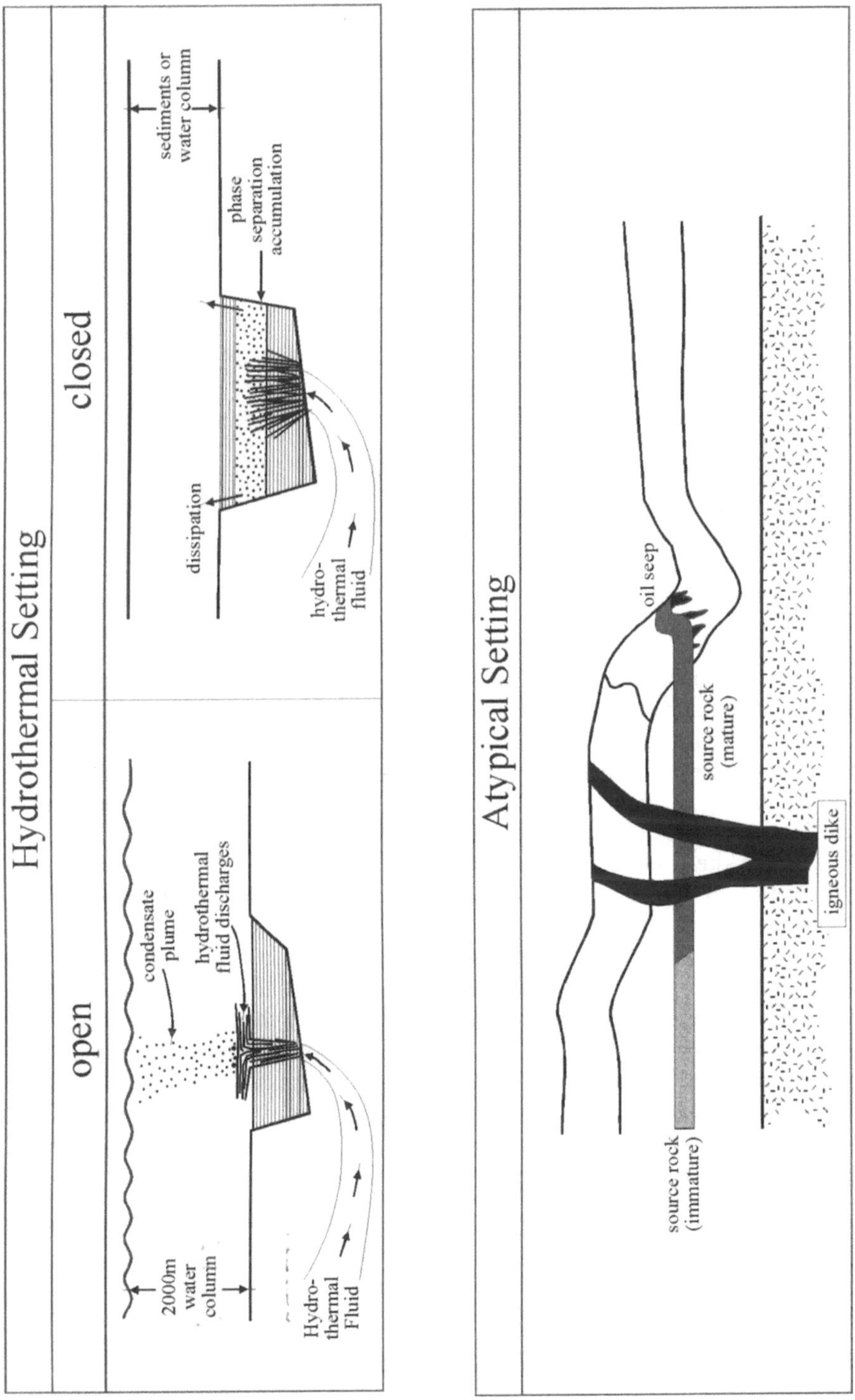

Fig. 7.11. Settings in which bitumens or paragenesis of bitumens with other phases occur

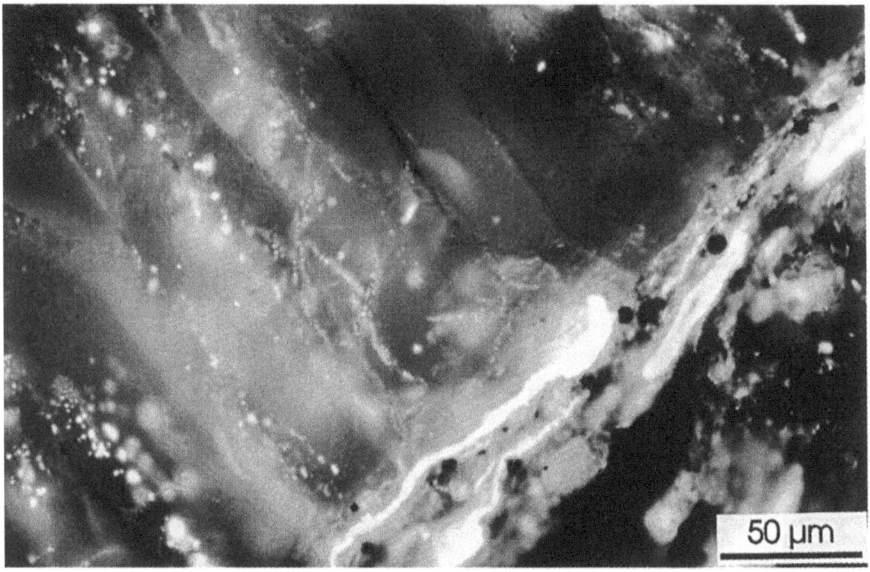

Fig. 7.12. Calcite fracture fill in Lower Toarcian Posidonia shale (0.73 %R_r). Many fluorescing petroleum inclusions are trapped in between crystal boundaries of calcite; crystal grow is vertical to bedding

plus composition of the actual migrating hydrocarbons requires an understanding of the diagenetic history and stage of the host rock.

Small petroleum inclusions in authigenic minerals may represent original samples of "pristine" petroleum (Karlsen et al. 1993), present along the migration pathway at earlier times, i.e. at the time of the diagenetic mineral formation (Fig. 7.12). The temperature of the relevant diagenetic mineral reaction (Bjørlykke et al. 1986) may be used to derive temperature estimates for the formation of the inclusion. In addition, microthermometry may allow direct measurement of inclusion formation temperatures (Roedder 1984). Roedder (1984) and Shi Jixi et al. (1988) give detailed information about the classification of organic inclusions and their relations to oil and gas occurrences. The fluorescence colours of inclusions point to the respective composition of the petroleum phase and the catagenetic stage when the expulsion took place. However, it is important to find enough inclusions for a reproducible result and to avoid inclusions with post-trapping alterations because they exhibit a re equilibrated stage and a correspondingly altered hydrocarbon composition (Fig. 7.13). Burruss et al. (1985) used the occurrence of fluorescing

Fig. 7.13a,b. Micrograph of hydrocarbon-bearing fluid inclusions from a calcite fracture fill from Lower Toarcian Posidonia shale at 0.88%R_r. **a** Primary yellowish fluorescing hydrocarbon inclusions within a calcite crystal which give homogenization temperatures of 75–15 °C. **b** Secondary yellowish and brownish fluorescing hydrocarbon inclusions along trails of with calcite resealed microcracks which give homogenization temperatures of 50–85 °C. (Jochum 1993; Jochum et al. 1995)

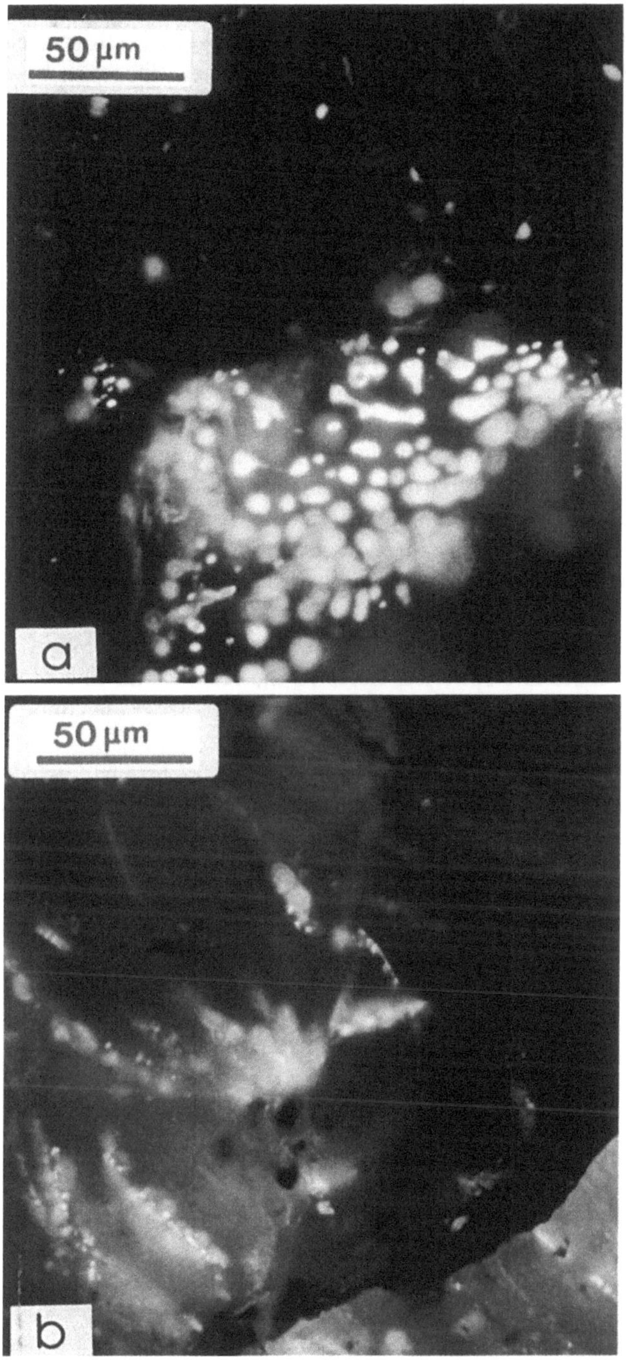
50 μm
a
50 μm
b

fluid inclusions within authigenic carbonate phases of reservoir rocks to shed light on petroleum migration in limestones from Oman and the Arabian Emirates. Karlsen et al. (1993) investigated the time of petroleum arrival in the North Sea Ula oil field. They used a new approach by concentrating the petroleum inclusions of different minerals by gravity sink float separation to obtain sufficient amounts of extract to be analyzed by conventional geochemical techniques. An example of the use of compositional relationships between hydrocarbon inclusions from fracture cement, residual oil from the open centre of the fracture, and the depleted neighbouring source rock matrix is provided by Mann (1994) for the Lower Toarcian Posidonia shale. The fracture shows a predominance of the long-chain *n*-alkanes, as typical for the composition of the residual oil, the inclusions show a predominance of the shorter-chain hydrocarbons, i.e. a composition similar to the expelled oil, whereas the rock matrix shows a composition with a maximum content in the medium chain length, i.e. the composition of the organic source material.

Core extracts and also inclusions show generally higher concentrations of polar compounds than normal crude oils or oils from production tests (e.g. Horstad et al. 1990; Karlsen and Larter 1991; Karlsen et al. 1993). During migration this is caused by preferential adsorption of polar compounds relative to saturated and aromatic hydrocarbons onto charged mineral surfaces.

Residual oil may be investigated in detail by the means of a cryo-SEM; for example, in fractures of the Posidonia shale, Mann et al. (1994a,b) found bubble-like spherical holes in the residual tar-like bitumen which was interpreted as gas-filled cavities formed by retrograde condensation during uplift of the source bed (Fig. 7.14).

In order to prove that a horizontal fracture had acted as expulsion pathway, Leythaeuser et al. (1988c) compared the hydrocarbon content of the fracture and the adjacent shale above and below in order to demonstrate the efficiency of fracture migration. A similar approach can be applied to stylolite-rich versus stylolite-poor sections in limestone source rocks.

In source rocks regular depth trends of the amount and composition of organic compounds can likewise identify drainage directions (Mann 1989). Increasing hydrocarbon depletion towards the edges of source rocks have been documented for example by Leythaeuser et al. (1984a). Examples of case histories with source rocks of different lithologies are presented in the following sections.

7.3.3 Case Studies on Primary Migration

Primary Migration in Siliciclastic Source Rocks

Until the end of the 1970s the knowledge about primary migration processes was predominantly speculative and hypothetical. Roberts and Cordell (1980), for instance, concluded that an adequate understanding of petroleum migration had not been achieved because of insufficient observations of petroleum

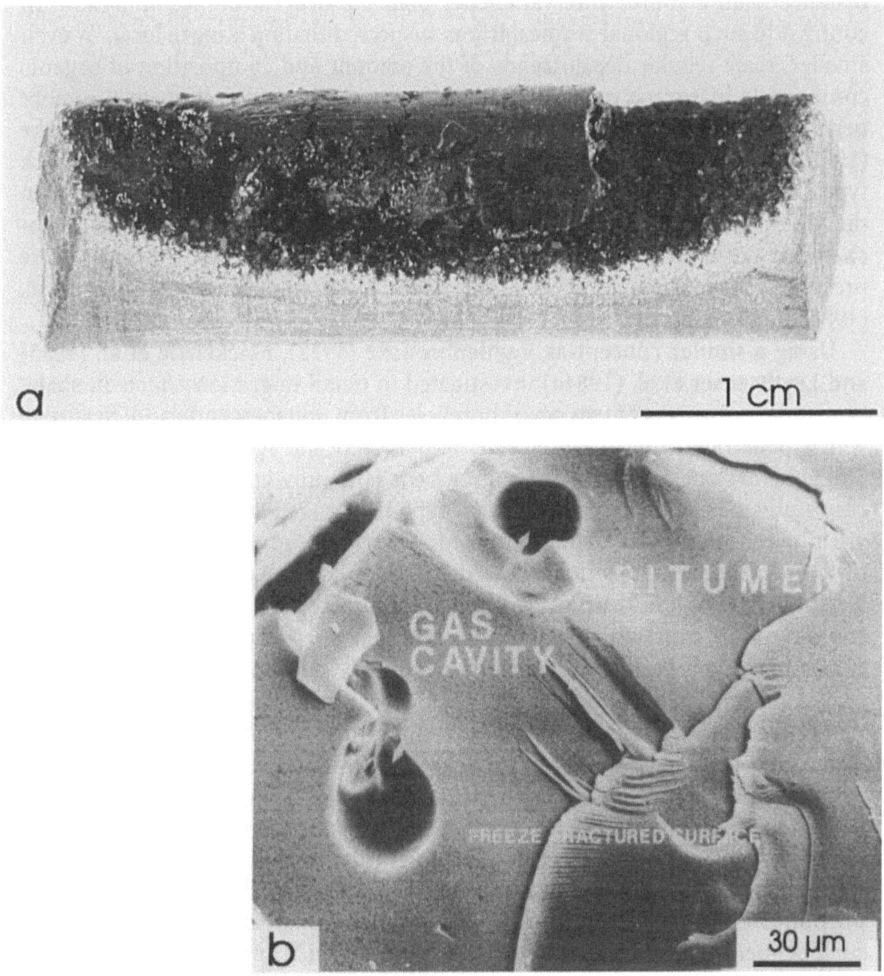

Fig. 7.14a,b. Vertical fracture from Lower Toarcian Posidonia shale at 0.88%R_r. **a** Plug with residual bitumen in the centre of a partly with calcite resealed fracture. **b** Cryo-SEM micrograph showing spherical holes in residual bitumen which are interpreted as formerly gas-filled cavities formed by retrograde condensation during uplift of the source bed. (Mann et al. 1994b)

moving from one place to another in the subsurface, although numerous attempts have been made by petroleum geoscientists to recognise petroleum migration effects and to deduce a consistent theory.

A well-documented case history based on organic geochemical methods is the Mahakam Delta area, off-shore Kalimantan, Indonesia (Durand and Oudin 1980; Vandenbroucke et al. 1983). Based on a comparison of hydrocarbon compositions between oil pools at a wide depth range in a multiple-pay field and those of the interbedded organic-rich shales and coals, migration was concluded to have occurred in a vertical direction and over a considerable distance from a source interval deeper than the deepest oil accumulation. In contrast to such regional studies, it was suspected that on a more local, or even smaller, scale regular depth trends of the amount and composition of organic compounds in source rocks may identify the drainage direction to the more permeable carrier beds or other migration pathways. Likewise, Vandenbroucke (1972) and Connan and Cassou (1980) studied drainage effects in source-rock type shales near their contacts with porous reservoir beds. Later, in line with the tremendous progress of analytical methods introduced in organic geochemical research in the 1980s, other well-documented case histories were presented by Mackenzie et al. (1983, 1987, 1988) and by Leythaeuser et al. (1984a,b, 1988a,b,c).

Using a similar concept as Vandenbroucke (1972), Mackenzie et al. (1983) and Leythaeuser et al. (1984a) investigated in detail migration effects in shale/sandstone sequences of two cored boreholes from Spitsbergen Island, Svalbard, which penetrated Lower Cretaceous and Palaeocene sequences, respectively. Since the organic matter of the source beds is mainly land-plant derived, this case study may serve as an example of type III kerogen at a maturity level with vitrinite reflectance, R_r, of about 0.8%. In terms of migration stage the patterns of hydrocarbon composition allowed the classification of most samples analyzed in one of the four categories "unmodified" shales (i.e. samples that contain all, or most, of their original bitumen due to unfavourable drainage conditions), "unmodified" sandstones with no signs of enrichment by migrated bitumen, "depleted" shales that have experienced expulsion or "enriched" sandstones and siltstones. Most interesting was the strong depletion of thin interbedded shales and of the edges of thicker source rock intervals where some samples showed a strong dependence of the degree of depletion in relation to carbon number of, for example, the *n*-alkanes. This effect is typical for diffusion-controlled and/or gas-driven migration mechanisms rather than for movement of the *n*-alkanes in a continuous oil phase. The observed migration phenomena led the authors to propose that the main phase of expulsion in such sequences is preceded by a stage during which only the edges of thick shale units and very thinly interbedded shales are depleted. This early expulsion is not a "pulsed event" but rather a slow and continuous process which is associated with pronounced compositional fractionation effects. The authors concluded also that their observations could explain the origin of accumulations of light oils and gas-condensates discovered in many low-maturity sequences bearing predominantly terrestrial-derived organic matter.

Similar effects of petroleum expulsion from two source rock intervals bearing type III and oil-prone type II kerogen, respectively, as a function of their distance from more permeable carrier beds has been documented by Leythaeuser and Schaefer (1984) and by Leythaeuser et al. (1984b) in a cored well that penetrated Upper Carboniferous strata of $R_r = 1.1\%$ in northern Germany. The authors presented material-balance type comparisons of selected samples extending from the centre portions of the two source rock units to their outer edges. Progressive depletion in petroleum compounds towards the more permeable adjacent layers could be shown clearly.

As an example of more commercial relevance, in another study Mackenzie et al. (1987, 1988) and Leythaeuser et al. (1988a,b) investigated primary migration processes in the most important source rock of one of the world's major oil provinces, the Upper Jurassic Kimmeridge Clay formation of the North Sea. As shown schematically in Fig. 7.15 this study from the Brae oilfield area in the South Viking Graben of the British continental shelf was based on cored sections of two deep exploration wells of this region. Both wells intersect oil-bearing Upper Jurassic sandstones (Brae formation) interbedded with source rocks of the Kimmeridge Clay formation. The latter has been convincingly demonstrated as the regional source rock (Reitsema et al. 1983; Cornford 1984). Present temperatures in about 4 km depth were determined to be around 100–120 °C at a pressure of about 50 MPa (500 bar), and therefore the Kimmeridge Clay formation should actively generate and expel petroleum into the perme-

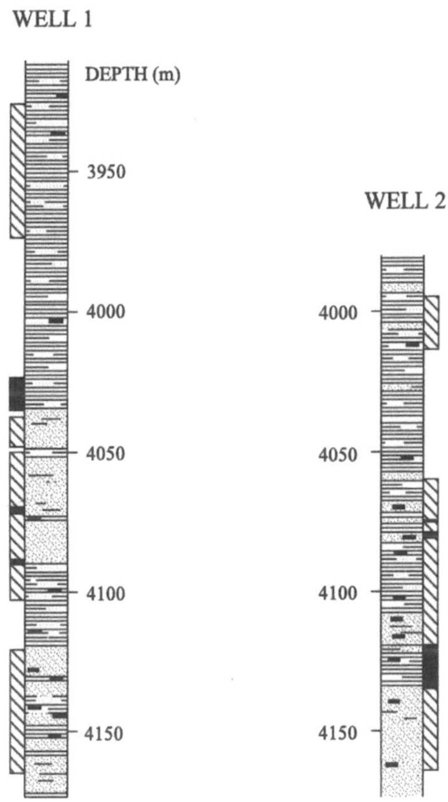

Fig. 7.15. Geological column for both wells of the Brae Oilfield area and location of cored sections studied by Mackenzie et al. (1987). Solid portions indicate zones of the Kimmeridge Clay Formation selected for detailed study. (Reprinted from Petroleum Geology of North West Europe, Mackenzie et al., The expulsion of petroleum from Kimmeridge clay source rocks in the area of the Brae oilfield, UK continental shelf, pp 865–877, 1987, with kind permission from Graham & Trotman Ltd., London SW1V 1DE, England)

able sandstone layers. The study was designed to assess the relative importance of two principal mechanisms of primary migration, i.e. pressure-driven flow of a discrete petroleum phase (see Durand et al. 1984) or diffusion of petroleum molecules through the porous rock matrix. The organic geochemical analyses were planned in such a way that the geochemical effects of petroleum expulsion should best be recognised by comparison of samples which have experienced different drainage conditions under natural conditions.

In detail, an approx. 13-m-thick interval of the Kimmeridge Clay formation overlying a sandstone sequence (well 1, core 4) and an approx. 12 m thick shale unit of same age sandwiched between two sandstone intervals (well 2, core 5/6) were chosen for this purpose, including some very thin interbedded shales. Additionally, a DST oil of well 1 [maturity in terms of vitrinite reflectance equivalent (see Radke 1987) $R_c = 0.75\%$] from the above-mentioned sandstone sequence was made available for comparison. As a first step, screening-type analyses based on Rock-Eval pyrolysis appeared suitable to determine migration effects in terms of total hydrocarbon depletion. As depicted in Fig. 7.16, the so-called transformation ratio or production index (PI; see Espitalie et al. 1977) of the Kimmeridge Clay samples showed a clear tendency to decrease

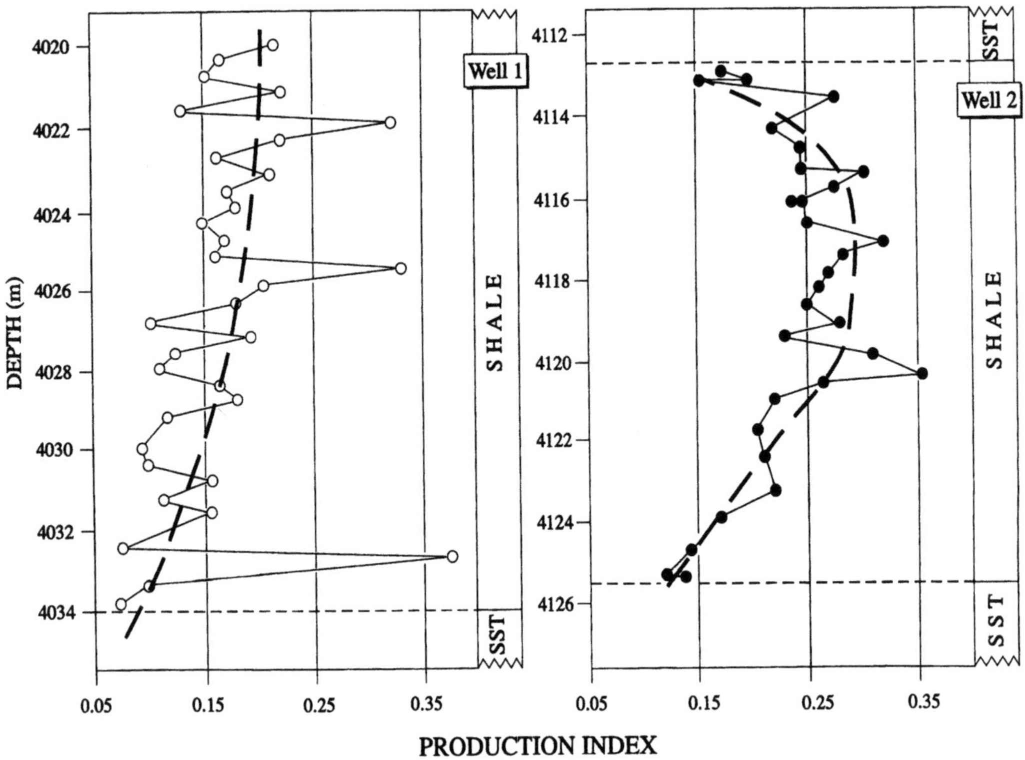

Fig. 7.16. Production index $[S_1/(S_1+S_2)]$, also called transformation ratio, calculated from Rock-Eval pyrolysis data (see Espitalie et al. 1977) versus depth for the two continuous shale source rock (Kimmeridge Clay Formation) intervals (Leythaeuser et al. 1988a). *SST*, Sandstone

towards the adjacent sandstone layers. In well 1 PI values decrease from about 0.20 at 4020 m depth to about 0.07 just above the sandstone, and in well 2 PI values decrease from about 0.30 in the centre of the shale unit to about 0.15 at the margins. Likewise, after solvent extraction of selected samples the plot of both the total soluble organic matter and the C_{15+}-hydrocarbons after liquid chromatography supported the screening data (Fig. 7.17).

The observed effects were interpreted as follows for this typical geological situation that comprises an oil-prone source rock in a favourable condition with respect to generation and expulsion. Petroleum is expelled chiefly as a petroleum-rich phase driven by excess pressures once such a phase fills most of the pores of the source rock. Therefore, the expulsion of *n*-alkanes, for instance, was nearly independent of their carbon number. There were distinct differences in composition, however, between the bitumen or "residual oil" in the source rock and the expelled reservoir oil, as is seen in Fig. 7.18. Fractionation processes have caused the relative depletion of the latter in non-hydrocarbons and the relative enrichment in saturated hydrocarbons, whereby the proportion in total aromatics was only slightly enhanced. This commonly observed discrepancy in composition of crude oils and the (residual) bitumen in their source rocks results from the chromatography-like retardation prop-

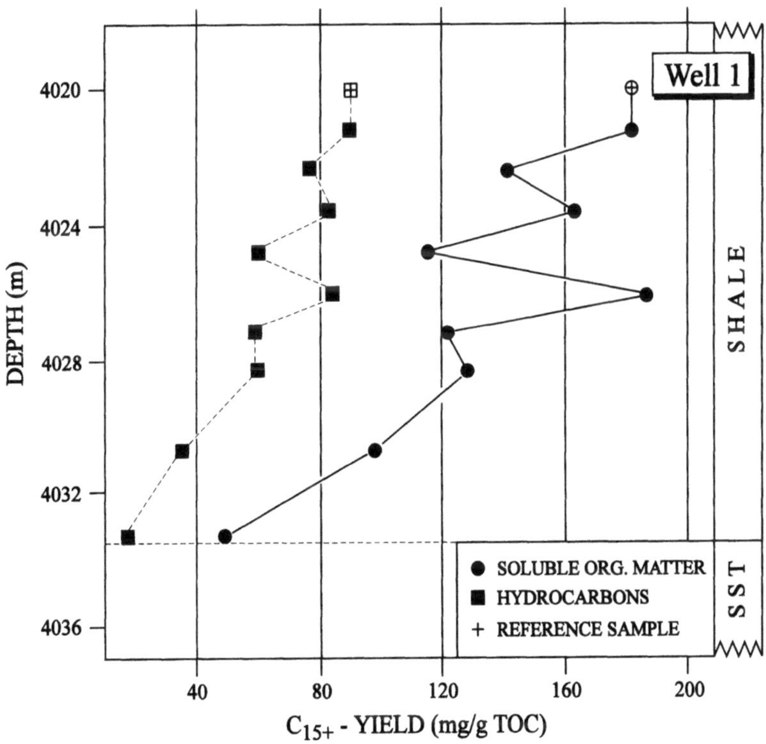

a

Fig. 7.17a,b. Yield of C_{15+}-soluble organic matter and C_{15+}-hydrocarbons versus depth for the two continuous shale source rock (Kimmeridge Clay Formation) intervals. (Leythaeuser et al. 1988a)

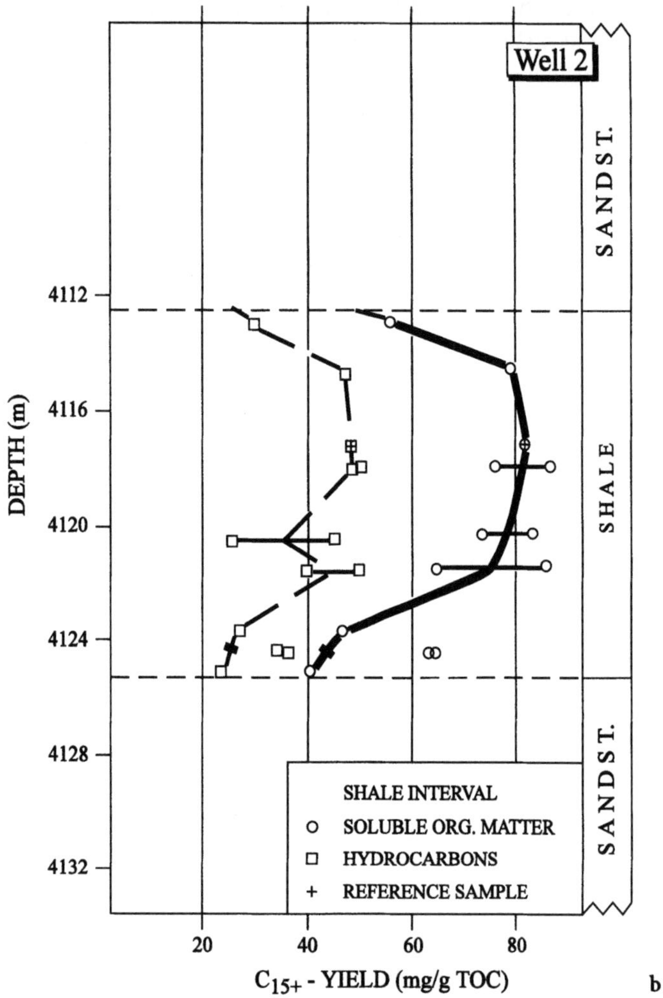

Fig. 7.17b

erties of the source rock matrix with respect to the more polar compound groups of a bitumen.

In thin interbedded shales the depletion in saturated hydrocarbons was extreme due to very favourable drainage conditions. The increasing depletion of source rock intervals towards shale/sandstone contacts ("edge effects") and the locally very high depletion of thin interbedded shales ("ideal drainage") are caused mainly by capillary forces enabling additional bulk oil expulsion as a separate phase flow. Obviously, capillary pressure differences cause oil spontaneously to drain from mudstones into adjacent sandstones, and water to replace it is imbibed by the mudstone. The authors argued that it is not necessary to invoke fracturing for expulsion to occur: the pores were sufficiently large for most petroleum flow to occur through primary porosity. This process,

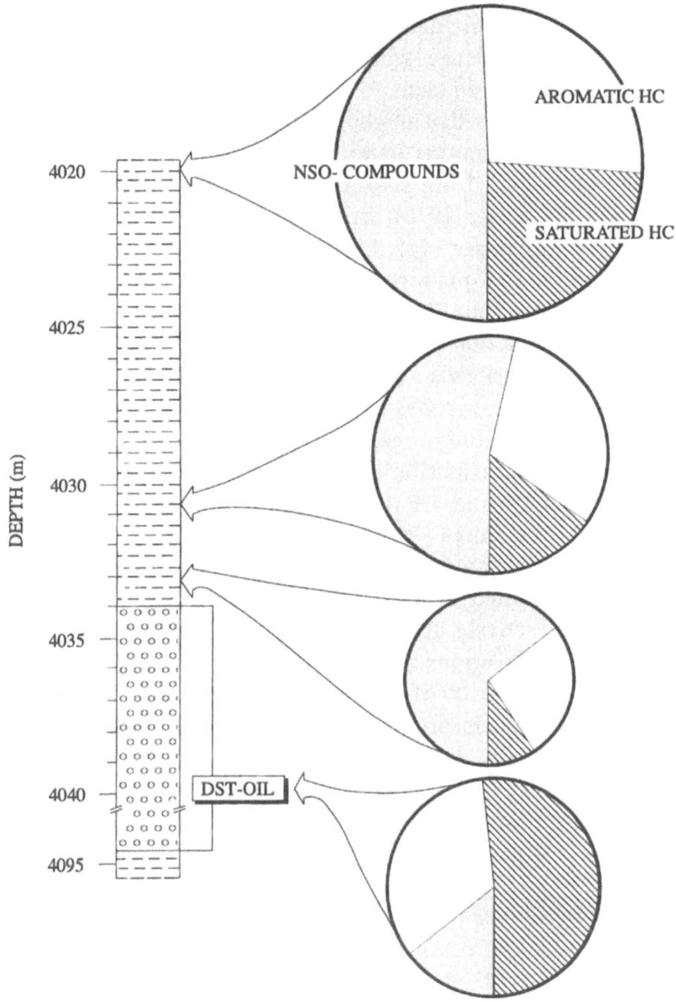

Fig. 7.18. Changes in yield and gross composition of C_{15+}-soluble organic matter for selected samples of Kimmeridge Clay Formation shales in comparison to gross composition of reservoir oil from underlying Brae sandstone unit. Diameter of circles for source rock samples is proportional to yields of C_{15+}-soluble organic matter. (Leythaeuser et al. 1987)

by which source rocks spontaneously disgorge petroleum once a sufficient saturation has been achieved, was first proposed by Hubbert (1953).

It was also concluded that molecular diffusion seeks to eliminate the gradients in concentration established by pressure-driven flow. For example, the highly depleted margins of thick mudstone intervals and the thin interbedded shales appeared less depleted in the low-molecular-weight hydrocarbon fraction because diffusion of these more mobile compounds from the centre of the mudstone and from the adjacent sandstone counteracted the loss by pressure-driven flow of a discrete petroleum phase. For this reasonable assumption, however, the available data base might be somewhat overinterpreted. There is no conclusive evidence whether the "snapshot-like" view on present-day concentrations is sufficient to characterise the combination of two dynamic migration processes that might have taken place on the geological time-scale.

Fractures, on the other hand, have been reported as migration conduits for expelled petroleum from source rocks in case studies from the Bakken Shale formation (Meissner 1978), from the Lower Toarcian in northern Germany (Littke and Rullkötter 1987; Littke et al. 1988; Leythaeuser et al. 1988c; Mann 1990), from the La Luna formation in Venezuela (Talukdar et al. 1987) and from several other locations. For instance, the study of mature, oil-prone source rocks of Toarcian age in the Hils syncline revealed that expulsion has occurred at high efficiencies primarily via a network of macro- and micro-fractures (Leythaeuser et al. 1988c). The observed effects were explained by the particular geological conditions in the study area. Overpressuring of aqueous pore fluids with intensive carbonate redistribution caused fracturing in a rapidly maturing source rock. Between 1% and 3% (by volume) of the mature source rock consisted of clearly visible fractures (Jochum 1988; Jochum et al. 1991). However, the conditions for fracturing seem to vary locally. Two later cored boreholes from the same area exhibited only very few fractures and absolutely no evidence for fractures, respectively, throughout the entire Lower Toarcian section (Heinen 1991). At these locations hydrocarbon migration was found to occur along the minus-cement porosity of cementation-rims around shell layers (Heinen 1991), or similar to other locations, via the pore network with partly secondary porosity due to calcite dissolution from coccoliths (Mann 1990). Thus, primary expulsion pathways cannot be generalised for a particular source rock even in a very restricted area. In addition to others, they form as result of locally small lithofacies changes and the respective different diagenetic overprints.

Primary Migration in Carbonate Source Rocks

Whereas the processes of primary migration and petroleum expulsion were satisfactorily explained for source rocks of siliciclastic lithologies, the corresponding effects in carbonate source rocks remained less well understood. This is remarkable in view of the fact that a major part of the known oil reserves are derived from carbonate source rocks which are mostly organic-rich, fine-grained micritic limestones. Prominent examples of carbonate source rocks which have given origin to commercial-size oil accumulations include the very prolific Upper Jurassic source rocks of the Middle East (Murris 1980; Ayres et al. 1982), the Upper Cretaceous La Luna formation of Venezuela (Zumberge 1984; Talukdar et al. 1987; James 1990), the lower member of the Jurassic Smackover formation in the Gulf of Mexico basin (Sassen 1990), the deep-water organic-rich Permian Bone Springs Limestone in the Delaware basin of west Texas and New Mexico (Barker and Halley 1986), the inter-reef algae-laminated deposits in the Devonian of the Rainbow-Zama area in western Canada (Powell 1984), and the Permian Zechstein-age "Stassfurt-Karbonat" of eastern Germany (Müller 1985).

The complexity of petroleum generation and expulsion processes in carbonates and evaporites arises from the fact that they are closely interconnected with carbonate diagenesis phenomena. The process of changing the originally deposited organic carbon contents as a consequence of carbonate redistribution during diagenesis has been demonstrated for a carbonate source rock of Triassic age from Italy by Leythaeuser et al. (1995). These authors designed a concept for a study as schematically outlined in Fig. 7.19. Since a prime objective was to examine the relationships between petroleum generation and expulsion and processes of carbonate diagenesis, selected core samples of this source rock which bear solution seams or stylolites were analyzed from three different wells which have penetrated this formation at different depths. The organic matter of this source rock has reached maturity levels (vitrinite reflectance, R_r) of 0.49%, 0.65% and 0.90% in wells B, C, and D, respectively. Because of problems commonly encountered in attempts to characterise maturity of carbonate source rocks by vitrinite reflectance measurements (see Palacas 1984), the values in Fig. 7.19 should be considered only as indications of relative differences of maturity levels between the three well sites. Accordingly, this Triassic source rock can be classified as immature to marginally mature in well B, as close to the peak of petroleum generation in well C, and as overmature in well D.

Like the Jurassic Smackover formation studied by Sassen (1990), this Triassic source rock from Italy is a fine-grained micritic lime-mudstone sequence. It was deposited in a shallow marine trough under anaerobic to dysaerobic conditions and consisted of several principal lithofacies. Leythaeuser et al. (1995) focussed their interest on those core segments which exhibited elevated organic carbon contents, i.e. laminated mudstones with parallel, usually very even laminations of light-coloured carbonate-rich, and dark-coloured clay-rich layers. Thickness of these laminations varied over a wide range from a few millimetres to a centimetre or more. The well-laminated nature and the lack of sedimentary structures suggested that this lithofacies was formed by settling of sedimentary particles from suspension through a pelagic water column under stillwater conditions. Hence the original abundance of the visible organic matter, mostly consisting of alginite and liptodetrinite structures, was controlled by sedimentological processes. The observed distribution of organic matter within the rock matrix, however, appeared to result from chemical compaction. Specifically, pressure solution processes had created local concentrations of insoluble residues (organic matter and clays) in the form of solution seams and/or stylolites. Interestingly, extensive pressure solution processes had occurred in this Triassic source rock formation at rather low maturity, i.e. prior to the organic matter maturity stage of well B (R_r = 0.49%). Due to removal of carbonate with pressure solution the organic matter particles were now in close contact with each other, thus forming a three-dimensional kerogen network. This important role of pressure solution leading to the formation of local enrichments of organic matter within carbonate sediments in the form of solution seams and stylolites, which are the predominant sites of petroleum generation and expulsion was first noticed by Sassen et al. (1987a,b) for the Smackover formation in the northern Gulf of Mexico basin area.

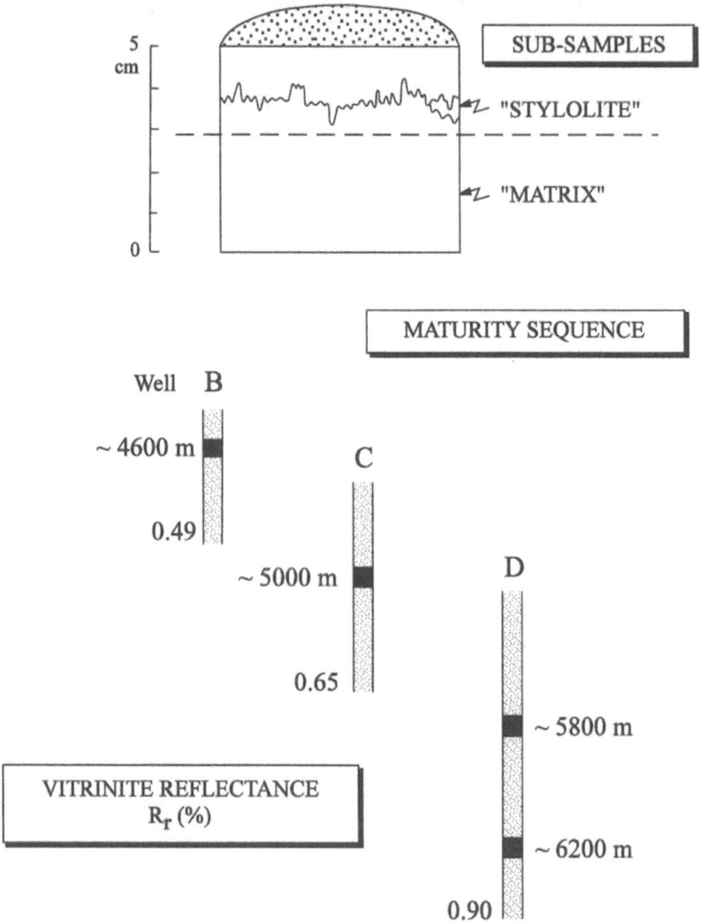

Fig. 7.19. Sampling scheme of study performed by Leythaeuser et al. (1995), based on conventional cores of a Triassic-age carbonate source rock formation from three wells representing different maturity levels. Most samples were subdivided into a solution-seam or stylolite bearing portion and a clean micritic carbonate portion (termed "matrix"). (Reprinted from Mar Petrol Geol, 12, Leythaeuser et al. Pressure solution in carbonate source rocks and its control on petroleum generation and migration, 1995, pp 717–733, with kind permission from Elsevier Science Ltd, The Boulevard, Langford Lane, Kidlington OX5 1GB, UK)

TOC contents of the samples from the Triassic source rock are low to moderate, ranging between 0.2 and 0.4% for most samples. Although these values are very low as compared to shaly source rocks, they fall within the range of TOC data reported for many carbonate source rocks (Gehmann 1962; Hunt 1979). For the well B and well C samples the TOC contents of the solution seam and the stylolite bearing samples are significantly higher than those of the adjacent carbonate matrix equivalents. To assess realistic values for the TOC contents of pure stylolites or solution seams a "corrected" TOC content can be defined under the assumption that all carbonate had been removed from these layers due to pressure solution processes:

$$TOC_{CS} = y/(1 - x)$$

where $x = (TC_S - TOC_S)/(TC_M - TOC_M)$ and $y = TOC_S - x \cdot (TOC_M)$, TC = total carbonate and organic carbon (%), TOC = total organic carbon (%), S = solution seam or stylolite bearing sample, M = adjacent matrix sample, and CS = "corrected" value for carbonate-free solution seam or stylolite. These calculated TOC_{CS} values allow a better comparison with the corresponding carbonate matrix samples to quantify the degree to which organic matter has been concentrated along solution seams or stylolites. Corrected TOC contents of the genuine solution seams and stylolites (well B) range mostly between 4 and 8% with maximum values of 11.5%, which is equivalent to TOC enrichment factors of about 10–50.

For the samples of well D, however, all stylolite bearing samples are leaner in TOC than the corresponding carbonate matrix samples. Obviously, the main hydrocarbon expulsion phase from this carbonate source rock had occurred mainly between the maturity levels represented by wells C ($R_r = 0.65\%$) and D (0.90%). This conclusion is supported by data depicted in Fig. 7.20, where both the extract yields and total hydrocarbon yields are plotted versus the TOC contents of sample pairs (solution seam or stylolite bearing samples vs. adjacent carbonate matrix samples) from the three wells. Obviously there is, as expected, a pronounced increase in extractable organic matter and hydrocarbon yields between well B and well C due to progressive petroleum generation as a result of the maturity increase from $R_r = 0.49–0.65\%$. Whereas for the well B and well C sample pairs the solution seam or the stylolite bearing sample is uniformly and significantly higher in extractable organic matter and hydrocarbon yields, this relationship is clearly reversed among the well D sample pairs. Here the stylolite bearing samples are now markedly lower in these yields, suggesting that a loss of petroleum-like material has occurred preferentially from the stylolites due to effective expulsion (Fig. 7.20).

Based on these results Leythaeuser et al. (1995) proposed a conceptual model to explain the relationships between the processes of petroleum generation/migration and those of carbonate diagenesis by pressure solution in this case history, as schematised in Fig. 7.21. Figure 7.21a depicts the initial stage of the freshly deposited carbonate sediment. Within this approximately 10-cm-thick interval (A–B) a rhythmic variation of the ratio of carbonate to noncarbonate material is assumed. The latter, consisting of organic matter and clays, makes up a higher proportion of interval C–D as compared to the intervals above and below. This difference in abundance of non-carbonate material leads to the initiation of pressure solution processes, by which carbonate is dissolved and removed from interval C–D and precipitated as cements directly above and below (Fig. 7.21b). This cementation took place during a very early stage of diagenesis, i.e. prior to significant mechanical compaction. In the more advanced stage, depicted in Fig. 7.21c, uncompacted algae are shown schematically distributed within these early formed cemented zones, whereas interval C–D has now been converted into a solution seam. The same type of

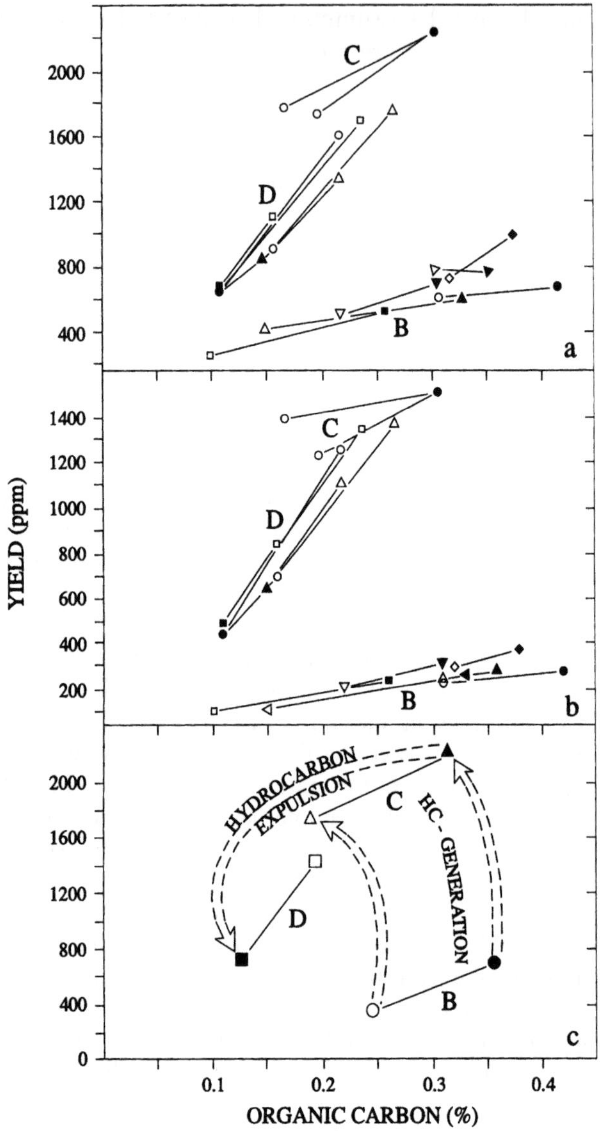

Fig. 7.20. Yields (in mg/kg of rock) of C_{15+}-soluble organic matter (**a**) and C_{15+} hydrocarbons (**b**) versus total organic carbon content for sample pairs of solution seam or stylolite bearing and adjacent carbonate matrix subsamples. **c** Mean values calculated for the data shown in a and the interpretation scheme of Leythaeuser et al. (1995). *Open symbols*, carbonate matrix samples; *filled symbols*, solution seam or stylolite bearing samples. (Reprinted from Mar petrol geol, 12, Leythaeuser et al. Pressure solution in carbonate source rocks and its control on petroleum generation and migration, 1995, pp 717–733, with kind permission from Elsevier Science Ltd, The Boulevard, Langford Lane, Kidlington OX5 1GB, UK)

algae are here compacted, and, due to removal of carbonates, brought in close contact with each other. Thus, as schematised in Fig. 7.21d, local zones of high concentration of oil-prone organic matter are formed which, upon deeper burial, allow the pore system of interval C–D to become oil-saturated and, due to effective sealing, eventually over-pressurised. Figure 7.21e finally shows the stage where the build-up of (over-)pressure has led to the formation of fractures and their filling with petroleum fluids by lateral migration along the stylolites or solution seams, respectively.

The validity of this model needs to be further verified by other case histories although the results and interpretations presented by Sassen and co-workers for the Smackover formation of the Gulf of Mexico basin (Sassen et al. 1987a,b; Sassen 1988; Sassen and Moore 1988; Sassen 1990) lead to very similar conclusions.

Primary Migration in Siliceous and Evaporite Source Rocks

Good to excellent biogenic siliceous source rocks are known from the Monterey formation from the San Joaquin basin in California (e.g. Isaacs et al. 1983). Diagenetically altered and unaltered diatoms make up silica, porcellanites (opal-A and opal-CT) and quartz cherts (e.g. Graham and Williams 1985). Such sedimentary constituents provide generally a high porosity, but at the same time a relatively low permeability. This explains why many source rocks of the Monterey formation are distinctive in yielding high S_1 signals from Rock-Eval pyrolysis (Graham and Williams 1985), thus indicating abundant generated liquid hydrocarbons, and why petroleum migration through the macropore network has not been reported from the Monterey formation so far. Primary migration in the Monterey shales has recently been evaluated by Lee (1991). He observed oil-stained solution seams and fractures and concluded that these are the pathways for primary migration. On the other hand, similar observations were made in the reservoirs of the Monterey formation, where production efficiency depends on the content of detrital material (Isaacs 1984) and the resulting degree of fracture formation.

Evaporitic rocks as such, for instance anhydrites or halites, contain only trace amounts of organic carbon (Connan et al. 1986). In this respect, such pure evaporitic lithologies do not represent typical petroleum source rocks. The petroleum potential of an evaporitic sequence arises from the fact that its sedimentary column often comprises layers of siliciclastic and/or carbonate sediments with their respective input of organic matter. Based on the particular geological and palaeoclimatic conditions, evaporite sequences may even show regular sedimentation cycles. The kerogen type in these sediments generally reflects strongly anoxic environments during deposition, and main contributions to the organic matter is due to bacterial input. Sulphur-rich type II kerogens are frequently encountered and give rise to early-mature heavy oils in these source rock systems. Generally, evaporitic source rocks are associated with close-range migration systems. They combine the petroleum source itself, the reservoir rock and the cap rock on a local scale. Accordingly, primary migration usually occurs along short migration pathways, whereby the mi-

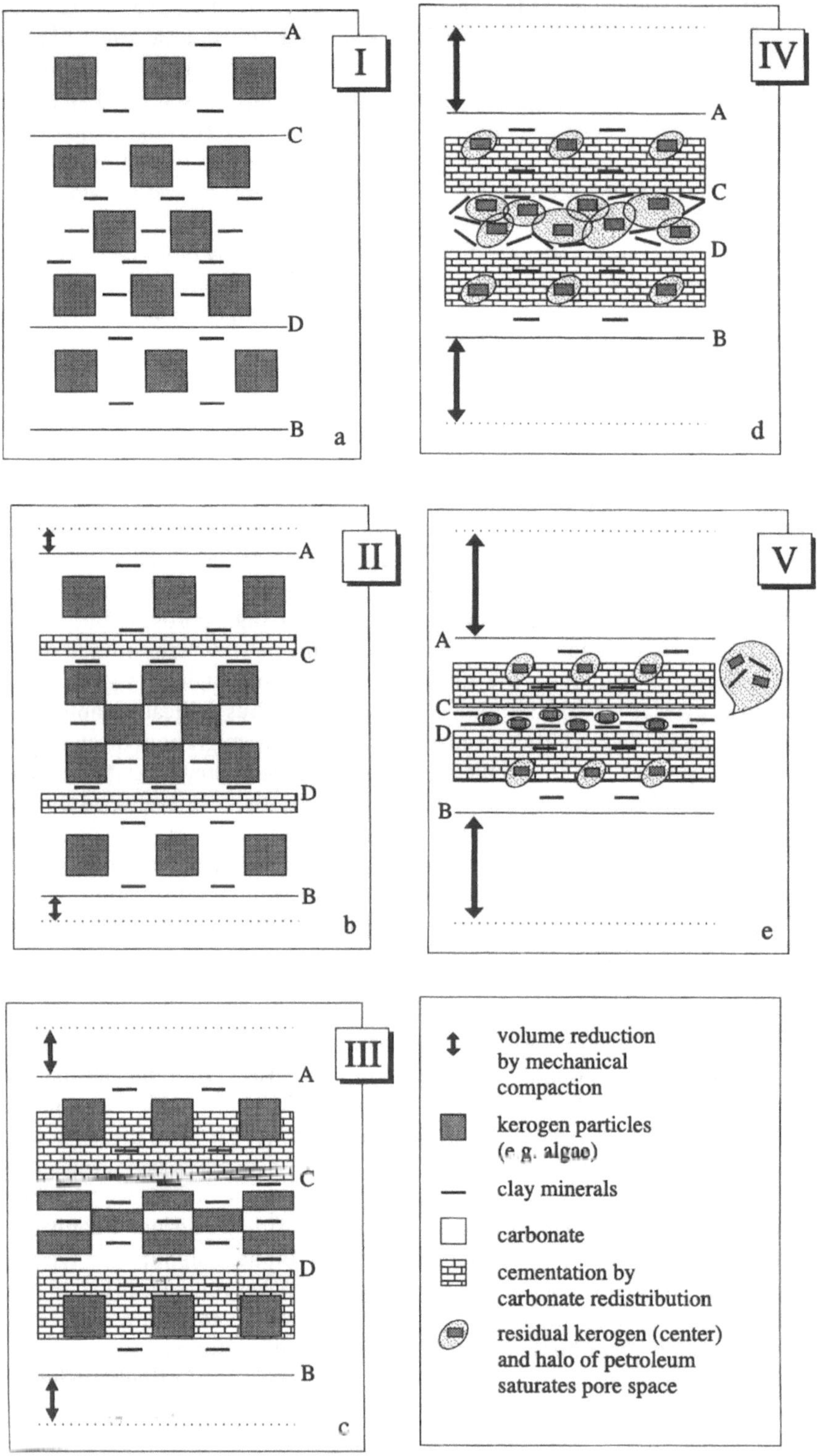

I
A
C
D
B
a
II
A
C
D
B
b
III
A
C
D
B
c
IV
A
C
D
B
d
V
A
C
D
B
e
volume reduction by mechanical compaction
kerogen particles (e.g. algae)
clay minerals
carbonate
cementation by carbonate redistribution
residual kerogen (center) and halo of petroleum saturates pore space

gration mechanism depends on the type of the source facies, i.e. its lithology and the type of organic matter.

7.4 Migration Efficiency

7.4.1 Relative and Absolute Source Rock Expulsion Efficiency

Basic Concept and Prerequisites for Mass Balances

Mass balance calculations in terms of generated and expelled petroleum from source rocks are of major importance for the assessment of basin-wide quantitations. It is generally agreed to define expulsion efficiency as the percentage of expelled petroleum in relation to total generated (and "inherited") bitumen of a particular source rock unit. According to the analytical possibilities available, expulsion efficiencies can be determined for total petroleum phases or fractions as well as for individual compounds. Well-defined samples, i.e. conventional cores or samples of adequate quality, are the absolute prerequisite for reliable results in this domain of geochemical research.

In order to determine the efficiency of primary migration processes, relative and absolute comparisons of individual source rock intervals as well as of identified migration avenues must be performed. In any case, for geochemical comparisons of source rock samples in terms of their ability to generate and expel petroleum, natural variations in source rock richness and in organic matter type must be excluded as far as possible. Ideally, to restrict the number of variables for this kind of investigations, the source rock unit should be largely homogeneous. Difficulties arise in cases where a source rock is studied at different maturity levels and hence locations. Here, particularly, the determination of expulsion efficiencies only makes sense under the assumption of equal organic matter type at the respective locations.

The basic concept of mass balance calculations for petroleum source rocks consists in comparing the present-day compositions of the organic con-

Fig. 7.21a–e. Conceptual model by Leythaeuser et al. (1995) to explain the processes of petroleum generation and migration and those of carbonate diagenesis by pressure solution. **a** Stage I: initial sediment (10-cm-thick interval) with rhythmic variation of ratio of carbonate to non-carbonate matter. **b** Stage II: initial stage of formation of solution seam by pressure solution upon burial to a few hundred metres depth. **c** Stage III: final stage of formation of solution seam by pressure solution and mechanical compaction. **d** Stage IV: optimum stage of petroleum generation due to kerogen decomposition, however, prior to primary migration. **e** Stage V: petroleum expulsion by migration in lateral direction. For sake of simplicity, fractures resulting from overpressuring are not shown in this scheme. (Reprinted from Mar Petrol Geol, 12, Leythaeuser et al. Pressure solution in carbonate source rocks and its control on petroleum generation and migration, 1995, pp 717–733, with kind permission from Elsevier Science Ltd, The Boulevard, Langford Lane, Kidlington OX5 1GB, UK)

stituents of mature source rock samples to their original immature state from the same facies. Therefore the primary requirement of a mass balance is to obtain a realistic idea of the composition of a presently mature source rock unit before the maturation process started. Usually it is attempted to find an immature equivalent of the source rock in question and use it as a reference. Very often, however, this approach encounters severe limitations as no immature equivalent of the source rock is available.

The main constituents of petroleum source rocks are the two solid phases, mineral matrix and kerogen, which together represent more than 90% of the total source rock mass. In rich source rocks the kerogen content amounts to more than 10% of the total mass, and if the term source rock is extended to coals, the solid organic matter content can even exceed the mineral content and be the predominant fraction. In contrast, the extractable organic matter (bitumen) and the low-molecular weight hydrocarbons usually represent only a minor fraction of the total mass of a source rock. In a source rock mass balance aimed at the reconstruction of petroleum generation and expulsion the four source rock components mineral matrix, kerogen, bitumen and low-molecular weight hydrocarbons must be quantified. The pore water contents of source rocks are usually not reported because standard geochemical analysis is performed with, and normalised to, dry samples. The LHC concentration in source rocks, determined by thermovapourisation techniques (Schaefer and Littke 1988; Schaefer 1992), showed some degree of variability due to the volatility of these compounds. However, as they represent only a minor fraction of the total mass, this uncertainty usually does not affect the overall results of a mass balance calculation to a significant extent.

Efforts to recognise and quantify the bulk compositional or molecular effects of petroleum expulsion by geochemical analysis have been made for more than 20 years (e.g. Vandenbroucke 1972; Connan and Cassou 1980; Durand and Oudin 1980; Vandenbroucke et al. 1983; Sajgo et al. 1983). Also, Leythaeuser et al. (1980) studied hydrocarbon generation in selected source rocks as a function of type and maturation. They presented a mass balance approach in terms of organic carbon contents, subdivided into "hydrocarbon generation potential", "carboxyl carbon" and "dead carbon", as well as the bitumen contents, with maturity increase for the three main kerogen types. However, since their data were based on a world-wide selection of source rocks, the need for well-defined case histories with a better control of the organic facies, the lithological characteristics and the geological framework became obvious.

Mass Balances Based on Petroleum Fractions

The most basic mass balance approach assumes that the immature reference sample corresponds exactly to the original state of the mature sample in terms of organic carbon content, kerogen type, and initial extract yield. A respective case history was conducted by Rullkötter et al. (1988) in a multidisciplinary approach. This study was based on four shallow boreholes drilled in the Hils syncline of northern Germany to determine quantitatively the amount of petroleum generated and expelled during maturation of a typical type II kerogen bearing source rock of Lower Toarcian age. In terms of vitrinite reflectance the cores covered a maturity range from immature (R_r = 0.48%) to overmature (R_r = 1.45%) due to the location of the Hils syncline in the vicinity of a deep igneous intrusion (Vlotho Massif). Facies variations of the Lower Toarcian in the short geographical distances of the study area were found negligible; hence this evidence of a uniform initial composition of the sedimentary organic matter allowed mass balance calculations. "Dead carbon" determinations ("inert kerogen" fraction according to Cooles et al. 1986; see below) supported this prerequisite, but were not used as a parameter in the mass balance. Only the amount of mineral matrix was taken as being constant for the different sampling sites. As shown in Fig. 7.22, about 50% of the initial kerogen was transformed into oil, gas, and inorganic compounds (CO_2, H_2O) during the vitrinite reflectance increase from 0.48% to 0.88%, and only marginally more during the maturity increase from 0.88% to 1.45%. Only a small portion of the generated material remained in the source rock even at a relatively early stage of generation (R_r = 0.68%). The expulsion efficiency of oil

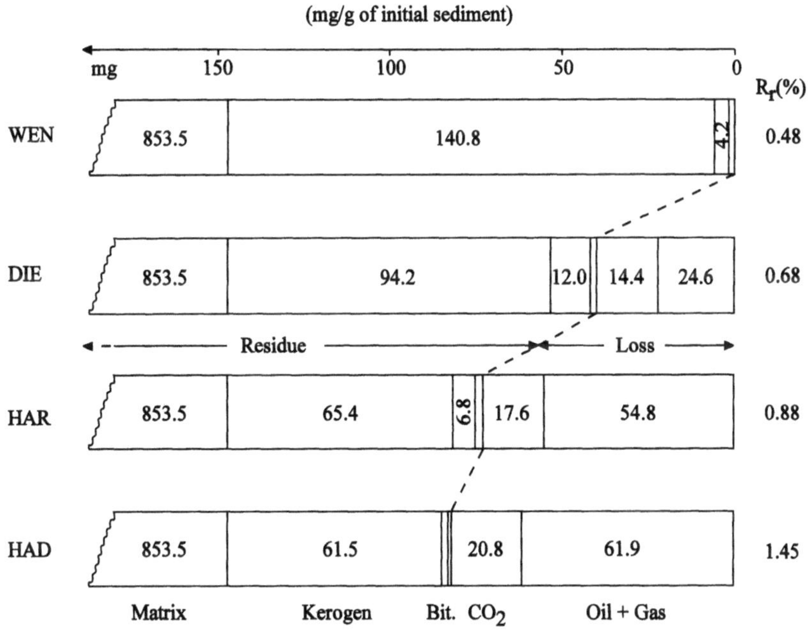

Fig. 7.22. Mass balance for petroleum generation and expulsion from the carbonate-lean upper unit of the Lias epsilon shale (Posidonia shale) in the Hils syncline, northern Germany. All yields are normalized to 1 g initial rock of immature stage and assume no loss or gain in the average amount of mineral matrix. (Reprinted from Advances in Organic Geochemistry 1987, Rullkötter et al., Organic matter maturation under the influence of a deep intrusive heat source: a natural experiment for quantitation of hydrocarbon generation and expulsion from a petroleum source rock, (Toarcian Shale, northern Germany), 1988, pp 847–856, with kind permission from Elsevier Science Ltd, The Boulevard, Langford Lane, Kidlington OX5 1GB, UK)

plus gas reached a value of 86% at the end of the main generation stage (R_r = 0.88%).

The data and conclusions of Rullkötter et al. (1988) are based on very favourable geological and sampling conditions which are rarely found in other regions of the world. Heating of an oil-prone source rock by a deep-seated igneous intrusion is a unique natural experiment for the quantitation of the effects of petroleum generation and expulsion. The important question whether their interpretations are equally valid in "normally" matured source rocks in a subsiding basin, however, could not be answered conclusively.

In many instances the original organic matter content of source rocks at the time of deposition has been subject to regional variations which could skew the overall mass balance performed with these assumptions. Therefore Cooles et al. (1986) proposed a mass balance scheme which takes the amount of "inert kerogen" as a basis of reference for the calculation of petroleum generation and expulsion efficiency. "Inert kerogen" denotes the fraction that cannot be converted into petroleum due to the limited availability of hydrogen in the organic matter. The inert kerogen content of individual samples of a given source rock is assumed to be a measure of their original organic matter content. It can be assessed from the difference between the present TOC content and the amounts of organic carbon residing in the bitumen and the "reactive kerogen" of the sample.

As a quantitative measure of the progress of petroleum generation and expulsion Cooles et al. (1986) defined the petroleum generation index (PGI) and the petroleum expulsion efficiency (PEE) as follows:

PGI = (petroleum generated + initial petroleum)/(total petroleum potential)
PEE = (petroleum expelled)/(petroleum generated + initial petroleum)

Here, the "initial petroleum" denotes the "inherited bitumen" incorporated into the kerogen at the beginning of the early-diagenetic processes after deposition of the sediment. The bitumen content of the samples is either determined by solvent extraction or by the S_1 peak of the Rock-Eval procedure whereas the reactive fraction of the kerogen corresponds to the S_2 peak. Figure 7.23 shows the reconstruction of the original composition of a presently mature source rock unit from an immature equivalent of the same source rock. Solid frames indicate the parameters which can be determined directly by the measurement of the TOC content, Rock-Eval pyrolysis, and solvent extraction.

This mass balance scheme of Cooles et al. (1986) was applied to a number of petroleum source rocks from different parts of the world, for example, the Kimmeridge Clay formation of the North Sea or the Brown Limestone of the Gulf of Suez. The petroleum expulsion efficiency was found to increase with the calculated average initial potential of the source rock and to reach values well above 80%. Expulsion efficiencies in excess of 50% were found even for source rocks with an initial petroleum potential of less than 10 kg per tonne of rock.

The approach of Cooles et al. (1986) retains the requirement of an identical kerogen type in the study area but it renders the calculations independent of

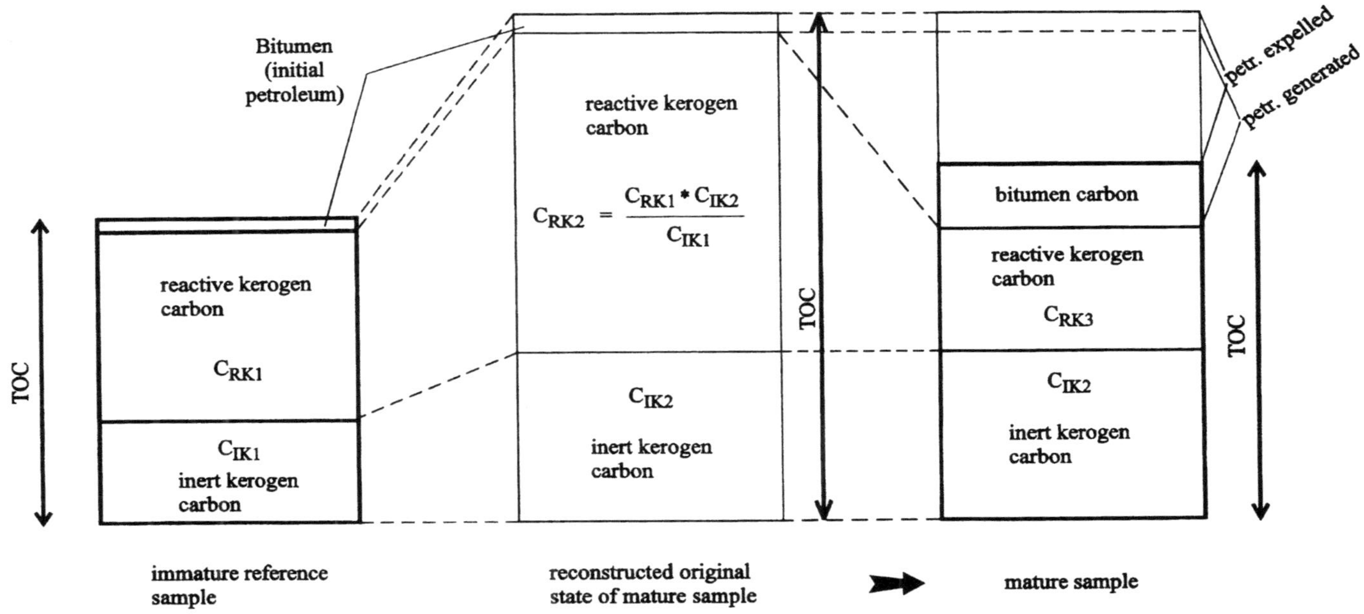

Fig. 7.23. Scheme of the reconstruction of the original composition of a presently mature source rock from an immature equivalent of the same source rock using the model of Cooles et al. (1986)

regional fluctuations in the initial content of organic matter. On the other hand, it introduces another critical assumption, namely that the S_2 peak of the Rock-Eval pyrolysis is a true quantitaive measure for the ability of source rocks to generate petroleum during natural subsurface catagenesis, i.e. that laboratory pyrolysis with the Rock-Eval instrument duplicates and continues the natural maturation process in a reproducible fashion. In particular, char formation during pyrolysis due to secondary cracking of primary cracking products would distort the results by overestimating the original inert kerogen fraction of the samples. Cooles et al. (1986) have examined this problem and conclude that this potential source of error is not significant in comparison to the errors inherent in pyrolysis and sampling techniques.

Mass Balances Based on Molecular Composition

Whereas the previous examples are mainly based on the mass balance of petroleum fractions rather than individual molecular species, other attempts focussed on molecular aspects of the residual bitumen in the source rock and the expelled and eventually accumulated petroleum to unravel the mechanisms of primary migration. For example, in a Miocene-age sequence from the Mahakam Delta region (Kalimantan, Indonesia) Vandenbroucke et al. (1983) have compared hydrocarbon compositions and concentrations of interbedded organic-matter rich shales (including coal seams as potential source rocks) in the multiple-pay Handil oilfield. Rock samples from the mature and overpressured interval of this sequence were contrasted to less mature equivalents in the hydrostatically pressured interval. The authors found depletion effects by expulsion in the order of 90% and 66% for the C_6–C_9 and C_{10}–C_{14} hydrocarbons, respectively, whilst the concentrations of C_{25}–C_{35} hydrocarbons remained constant, i.e. pronounced compositional fractionations associated with expulsion was observed in this case history. The effects caused by pressure release during sampling, however, and their possible influence on the rather volatile low and medium molecular weight hydrocarbons remained an open question. This interpretative restriction is not confined to this particular case history but pertains to all studies in which rock samples from deep wells are depressurised during the sampling procedure.

As outlined above in more detail, Mackenzie et al. (1983) and Leythaeuser et al. (1984a) investigated migration effects in Palaeocene-age type III kerogen bearing source rocks with a maturity level in terms of vitrinite reflectance of about 0.8%. Calculated relative expulsion efficiencies, defined as the concentration difference between an "unmodified" reference sample and a depleted sample, divided by the concentration of the reference sample, decrease with increasing carbon number for a thin shale sample (5 cm thickness). Nearly 90% expulsion was calculated for n-$C_{15}H_{32}$, but no significant expulsion was suspected beyond n-$C_{25}H_{52}$ (Fig. 7.24a). For thicker intervals the respective values were not so extreme; however, the tendency of decreasing expulsion efficiencies with increasing carbon number of the n-alkanes was generally observed in these type III kerogen bearing source rocks. Similar

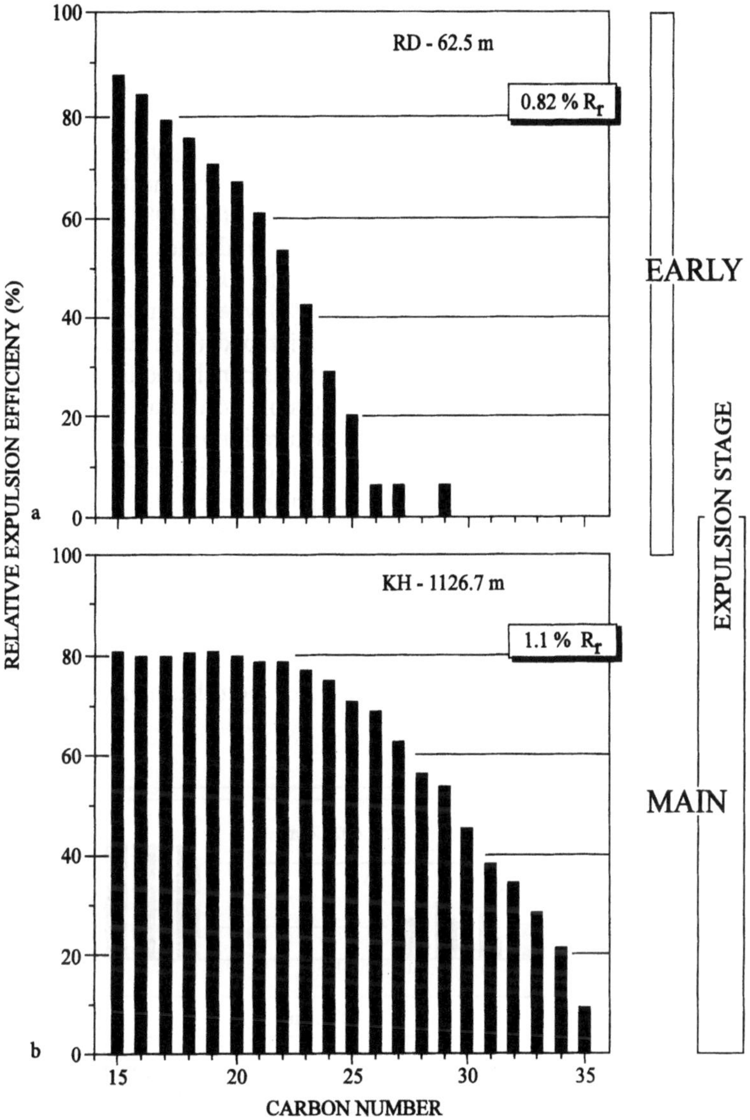

Fig. 7.24a,b. Comparison of relative expulsion efficiencies for C_{15+} *n*-alkanes as a function of carbon number for selected samples of thin interbedded shale layers at different maturity stages (R_r = 0.82 and 1.1%, respectively) for type III kerogen bearing source rocks. (Leythaeuser et al. 1987)

results were demonstrated also at higher maturity (R_r = 1.1%) for a type III kerogen bearing source rock of Carboniferous age (Leythaeuser and Schaefer 1984; Leythaeuser et al. 1984b). Relative expulsion efficiencies for a thin shale layer (Fig. 7.24b) are about 80% between *n*-$C_{15}H_{32}$ and *n*-$C_{20}H_{42}$ and decrease sharply to higher molecular weights. A comparison of these results with those of a type II kerogen bearing source rock interval from the same borehole showed that expulsion efficiencies at the edge of the source rock unit varied between ca. 80% and 95%, with no tendency to decrease with increasing carbon number.

Similar observations were made in a comprehensive study (see above) for the Upper Jurassic Kimmeridge Clay formation, the most important oil-prone source rock of the North Sea area (Mackenzie et al. 1987, 1988; Leythaeuser et al. 1988a,b). Expulsion efficiencies for the *n*-alkanes were more or less independent of their chain length although the expulsion efficiencies showed a clear tendency to increase (up to about 90%) when approaching the lithological boundary between source rock and carrier bed (Fig. 7.25) as well as in the special case of thin interbedded shales. It was concluded from this that for rich, oil-prone source rocks, petroleum expulsion occurs as a single-phase petroleum fluid but not before a minimum saturation of the shale pore system has been exceeded. For leaner quality source rocks with predominantly land-plant derived organic matter, migration mechanisms are different. Since such source rocks generate significant amounts of gas and lesser quantities of oil, the latter is probably transported in gaseous and/or hydrous solution. Fractionation effects, for example, according to chain length for the *n*-alkanes, result from differences in vapour pressure or solubility (see Leythaeuser et al. 1987).

In carbonate source rocks, solution seams and stylolites have been suspected repeatedly to serve as avenues for petroleum migration (Grabowski

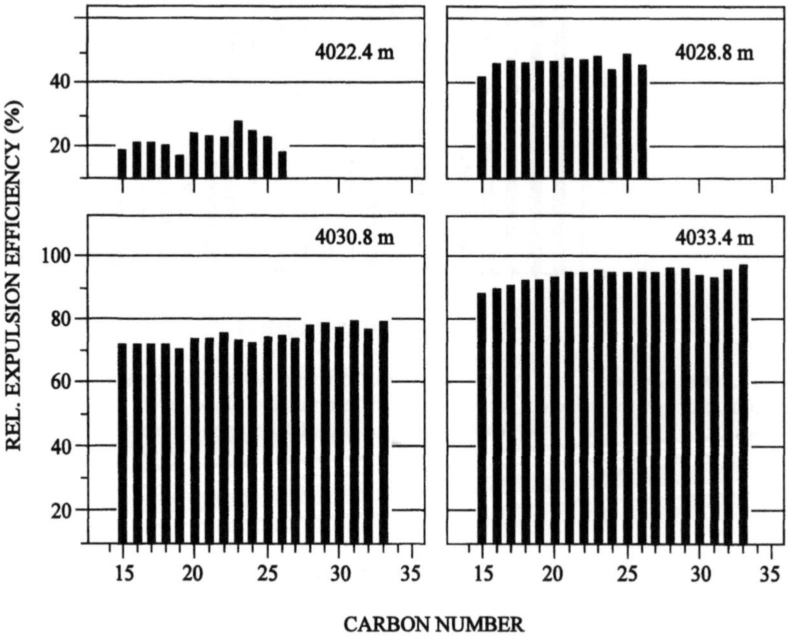

Fig. 7.25. Relative expulsion efficiencies for C_{15+} *n*-alkanes as a function of carbon number for selected samples from the Kimmeridge Clay Formation. (Leythaeuser et al. 1987)

1984; Palacas et al. 1984; Sassen et al. 1987a,b). Direct geochemical evidence for this conclusion was provided by Leythaeuser et al. (1995) for a carbonate source rock of Triassic age from northern Italy. Comparison and mass balance of hydrocarbon yields of selected samples that have experienced enhanced depletion versus those which have expelled petroleum to a lesser degree represent the key to recognise and quantify the effects of petroleum expulsion and to deduce the nature of primary migration mechanisms.

Based on the case history for this carbonate source rock outlined in detail in Section 7.3.3.2, Fig. 7.26 exemplifies this effect for stylolite bearing samples from maturity stages R_r = 0.65 and 0.90% in wells C and D, respectively. The quantities of well-defined compounds rather than total extracts or hydrocarbon fractions including long-chain *n*-alkanes, pristane, and phytane were chosen for this purpose. As is seen in Fig. 7.26a, the stylolite bearing sample from well C has much higher *n*-alkane, pristane, and phytane yields than the stylolite-bearing sample from well D. The difference in TOC content between both samples cannot account for the observed hydrocarbon concentration differences. Unless cracking reactions have played a dominant role, the opposite trend as a result of progressive generation would be expected. Therefore the yield difference is interpreted to reflect mainly the effect of petroleum expulsion from the stylolite of the well D sample. As in the examples for siliciclastic rocks discussed previously, the calculated relative expulsion efficiency denotes the ratio between the yield difference of a reference sample (well C in this case) and a depleted sample (well D) divided by the yield of the reference sample (well C).

The mass balance between both samples (Fig. 7.26b) reveals a smooth *n*-alkane distribution envelope peaking at *n*-nonadecane, and pristane and phytane in approximately equal proportions. Relative expulsion efficiency values (Fig. 7.26c) remain fairly uniform within 60–80% in the molecular range between *n*-heptadecane and *n*-pentacosane. For pristane and phytane, near-equal expulsion efficiencies are observed as for their straight-chain isomers.

For oil-prone source rocks of siliciclastic lithologies the relative expulsion efficiencies of *n*-alkanes which remain uniform with increasing carbon number have been interpreted to indicate petroleum expulsion as a separate-phase fluid (Mackenzie et al. 1987; Leythaeuser et al. 1988a). In contrast, *n*-alkanes from gas-prone type III kerogen bearing source rocks reveal regularly decreasing expulsion efficiencies of the *n*-alkanes with increasing carbon numbers, in particular at lower maturities. This compositional fractionation as a function of chain length is explained by migration of these petroleum components in gaseous solution. The trend shown in Fig. 7.26c, however, represents more analytical evidence for a continuous-phase oil migration, although the expulsion efficiencies decrease markedly both to the lower and the higher molecular weight range. A continuous-phase oil migration had been postulated previously in the literature as the predominant mechanism of petroleum expulsion from carbonate source rocks (Jones 1984).

The examples above show that expulsion efficiencies can vary greatly within an individual source rock unit: They change as a function of proximity to carrier beds and migration avenues and vary with compound type, molecular

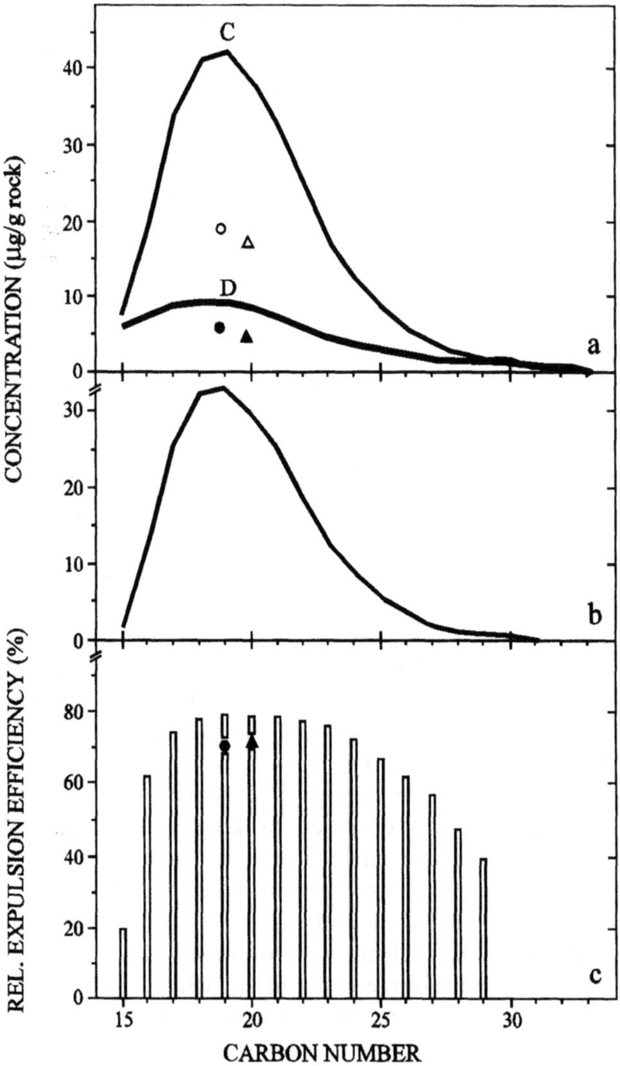

Fig. 7.26. a Concentration (in mg/kg rock) of normal and isoprenoid alkanes pristane and phytane as a function of carbon number for stylolite bearing samples from well C (vitrinite reflectance $R_r = 0.65\%$) and well D (vitrinite reflectance $R_r = 0.90\%$). **b** Concentration difference between both samples. **c** Relative expulsion efficiencies for well D sample. (Leythaeuser et al. 1995). (Reprinted from Mar Petrol Geol, 12, Leythaeuser et al. Pressure solution in carbonate source rocks and its control on petroleum generation and migration, 1995, pp 717–733, with kind permission from Elsevier Science Ltd, The Boulevard, Langford Lane, Kidlington OX5 1GB, UK)

type and/or structure. It may therefore be difficult to select appropriate mean values for the expulsion efficiencies of entire source rock sequences so as to calculate the amount of oil which migrated and possibly accumulated in the traps of a given exploration area. Previously reported values for the efficiency of petroleum expulsion from shale source rocks were estimated to be around 10–15% (Hunt 1979; Tissot and Welte 1984). However, these values result from gross budget estimates for entire shales and reservoir rock volumes of whole basins. Such general expulsion efficiencies comprise several other effects, such as the limited efficiency of secondary migration or alteration and dismigration processes.

7.4.2
Efficiency of Secondary Migration

The efficiency of secondary oil migration is governed by (a) migration distance which determines the amount of migration losses, (b) the sedimentary rock fabric and the pore size which determine the percentage of residual petroleum saturation, and (c) the composition of the migrating petroleum phase which controls via interfacial tension and contact angle the amount of residual petroleum saturation.

Migration Distance and Residual Saturation as Control of Migration Losses

Secondary hydrocarbon migration is concentrated along interconnected pathways of 1–10% of a carrier-reservoir rock system (England et al. 1987). After migration has finished, a certain amount of hydrocarbons are left behind along the pathway. The volume of these losses V_L can be estimated from the drainage volume V_D (volume of carrier-reservoir rock through which petroleum migrates), porosity P of the carrier-reservoir rock, and the residual saturation S_R of the carrier-reservoir rock according to the following relation (Mackenzie and Quigley 1988):

$$V_L = V_D \times P \times S_R$$

The distances for secondary migration are known for quite some time; e.g. Levorsen (1967) reports distances of 75 miles for the migration of oil accumulated in Pennsylvanian sands of Oklahoma, and Halbouty (1970) distances of about 100 km for the giant Athabasca tar sands in Canada. More recent calculations by Sluik and Nederlof (1984) give a range of a few metres to hundreds of kilometres, while Demaison and Huizinga (1991) speak about less than 30 km.

Although these long distances look very impressive, one must remember that long migration considerably reduces the volume of petroleum finally available for entrapment and hence limits the distance for migration to an economically exploitable petroleum accumulation. There losses must be accounted for when assessing the petroleum budget of a given region.

Mackenzie and Quigley (1988) give an example how to evaluate the residual petroleum saturation factor within a geochemical prospect appraisal. They recommend evaluating petroleum losses by using the exploration results from geologically similar regions. Based on their studies, Mackenzie and Quigley (1988) suggest a typical residual saturation factor of 0.02 m^3/m^3 for both oil

Table 7.5. Results of calculations for Malacca Strait Prospects (Mackenzie and Quigley 1988)

Prospect items	Prospect A	Prospect B
Petroleum mass expelled	5×10^{10} kg	7×10^{10} kg
Average petroleum density	700 kg/m^3	700 kg/m^3
Volumes expelled	7.1×10^7 m^3	10^8 m^3
Volumes lost	10^8 m^3	6.2×10^7 m^3
Volume of charge	2.9×10^7 m^3	3.8×10^7 m^3(240 million bbl)

Table 7.6. Necessary oil saturations for oil breakthrough. (England et al. 1987)

Sample and lithology	Porosity (%)	Saturation (%)
Yorkshire Deltaic SS 1	9.2	59.6
Yorkshire Deltaic SS 2	6.5	91.0
Yorkshire Deltaic SS 3	3.9	56.0
Millstone Grit	6.7	65.3
Costwold Oolites	15.1	47.8
Berea Sandstone	20.0	29.0
St. Bees Sandstone	18.0	24.5

and gas. However, they must admit an error bar of (0.01 m^3/m^3. The example of the calculated charge volumes from Malacca Straits, Indonesia, shows that because of the much longer migration distance in one prospect, the corresponding high losses resulted in a negative charge volume (Table 7.5).

While distances can generally be assessed relative precisely from seismic sections, more critical because of a much higher uncertainty are estimations of the residual oil saturations. Residual oil saturations in carrier rocks are controlled by the same rock and fluid properties as the required saturation to permit the beginning ("breakthrough") in oil transport, hence initial and residual saturations are of equal size. At the place of the actual migration pathway, saturation values range between 20% and more than 90% of the pore volume (see Table 7.6); however, this concerns an equivalent of only 5–10% of the total carrier rock volume which finally gives an apparent residual saturation of only about 2% (Mackenzie and Quigley 1988). On the other hand, Hirsch and Thompson (1995) claim that in a typical geological situation permeability is large enough to carry oil to the reservoir over geological time at saturations of less than 1% of the pore volume. Such different values between 1% and 90% are predominantly an effect of the sedimentary fabric, i.e. small scale heterogeneities and the pore size distribution of the carrier rock.

Rock Properties as Control of Residual Saturation

Understanding of fluid flow is generally based on the concepts of capillary pressure, relative permeability and wettability (i.e. Darcy flow equations; dimensionless capillary pressure function as introduced by Leverett 1941 and its relation to measurable bulk parameters according to Amyx et al. 1960).

However, essential to the basic understanding of residual oil saturations are the effects of the pore network within a rock.

Pore Network

A pore network is defined by the actual pores (sensu stricto) and the connecting (pore) throats. The three-dimensional, geometric relationship between pores and throats controls beside the actual fluid flow also the amount of residual oil. The three major variables are (Wardlaw 1979; Mann 1983, unpublished results): (a) the ratio of the pore radius to the throat radius (low ratios give high, ratios close to 1 give low residual saturations); (b) coordination number of throat(s) to one pore (lower coordination numbers give higher residual saturations); (c) the type and level of heterogeneity of a pore network (domains of small pore throats located within a network with large throats are easily drained, but domains of large pore throats located within a network of small throats retain oil).

England et al. (1987) carried out experiments with sandstone carriers of various porosities to investigate the conditions for an interconnected oil column through the pore network and the necessary saturation for an oil breakthrough (Table 7.6). Based on their results 50% of the pore volume must be petroleum-saturated for flow of petroleum to occur. Although their saturation values for breakthrough tend to increase from 20 to 90% with decreasing porosity (from 20 to 4%), data variation is such strong that for many carrier rocks a general mean value of 50% is too imprecise for a prospect evaluation. A 50% saturation may be necessary for lower porosity rocks with a unimodal pore size distribution (Fig. 7.27c). Other types of rocks with bi- or polymodal distribution functions bear the chance for much lower saturation values (Fig. 7.27a,b). For better approximations (if experimental data are lacking) it is therefore suggested to estimate the necessary saturation for carrier bed migration from the geometry of the pore size distribution in analogy to the geometric calculation of displacement pressure in reservoir rocks (Fig. 7.27).The inflection point of the first pore size modality with the largest pore size provides the minimum saturation and the smallest pore size of the same modality gives the maximum saturation at which fluid flow is generally obtained. In this way, carrier beds with a bi- or polymodal pore size distribution may often provide very good drainage conditions because they need comparatively few channels to the reservoir and only low saturation values. Therefore they exhibit also low residual oil saturations.

A microscopic view (cryo-SEM photomicrograph) from a meandering migration channel within a sandstone after a core-flodding experiment (water-oil-water) is given in Fig. 7.28 (Mann et al. 1994a). The residual oil phase occupies predominantly the concave (sag pores) and convex curvatures along the pore walls in contrast to the zones in between or like the smooth surfaces of well-crystallised quartz which is only sparsely covered (Fig. 7.28a). A detailed view of on one single quartz crystal provides the actual wetting conditions: quartz is wetted by a 3-to 6-μm thick brine which is coated again by a 1- to 1.5-μm-thick oil film. Regarding the pore network at a somewhat larger scale means to consider also the fabric of the sedimentary rock.

Sedimentary Rock Fabric

In typical water–wet oil/water systems, the amount of trapped oil can vary due to the fabric of the sedimentary rock. Ringrose and Corbett (1994) compared by numerical simulations the effective flow behaviour of two immiscible fluids in a variety of heterogeneous sandstones. For typical subsurface flow rates and patterns of rock heterogeneity in hydrocarbon reservoirs, capillary forces can result in significant amounts of residual oil (trapped due to the bypassing of the non-wetting phase. Ringrose and Corbett (1994) investigated the effects of uniform versus crossbedding and graded bedding, faulted layers versus parallel layering, and horizontal versus vertical flow in wavy bedded sandstones on relative oil permeability (Fig. 7.29). They found values between about 40% and 65% of residual oil trapped due to the type of heterogeneity.

The effects of small scale heterogeneities must be upscaled when large-scale flow behaviour – as in a sedimentary basin – is assessed from numerical or experimental simulations at the core scale level. Corbett et al. (1992) scale up their generalised flow models in a number of stages dictated by the respective sequence stratigraphic framework.

Other Controls of Residual Saturations

Interfacial tension between oil and water and the contact angle between the oil/water and water/pore wall contacts are parts of the Young-Laplace equation. They vary according to petroleum composition and the salts dissolved in the brine. Water-wet conditions prevail in most reservoirs, but oil-wet and mixed-wet systems also occur. The type of wetting has considerable influence on both the capillary pressure and the relative permeability (McDougall and Sorbie 1992), and thus also on residual oil saturations.

The exact composition of the migrating petroleum phase is generally unknown, and an important parameter, the gas content, can hardly be assessed. Already during expulsion of the oil phase from the source to the carrier rock, gas losses by diffusion are the rule, and significant amounts of gas losses may be involved. However, it is unclear what magnitude of losses must be taken into account. As another consequence, fluid properties such as viscosity, interfacial tension and individual wetting characteristics which together affect the relative permeability and losses during migration, are difficult to estimate.

Fig. 7.27a-c. Pore radius equivalents versus pore volume (minimum petroleum saturation S_c necessary for breakthrough are calculated according to pore volume value at first inflection point; maximum saturation S_m estimated from calculated displacement pressure). **a** bimodal, S_c = 15.2%, S_m 35.7%. **b** polymodal, S_c = 17.8%, S_m 37.0%. **c** unimodal, S_c = 53.8%, S_m 73.3%

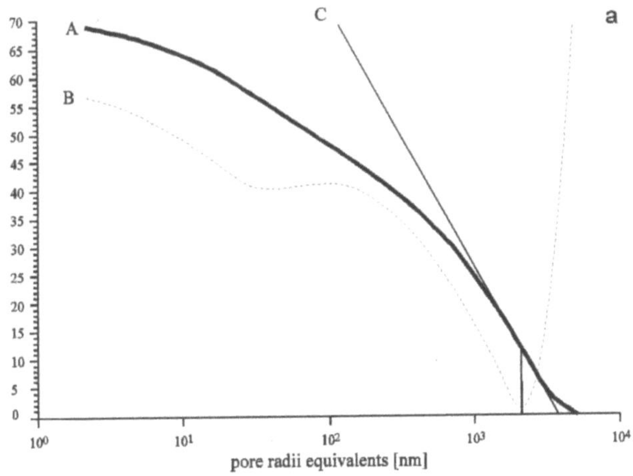
a
A
B
C
pore radii equivalents [nm]

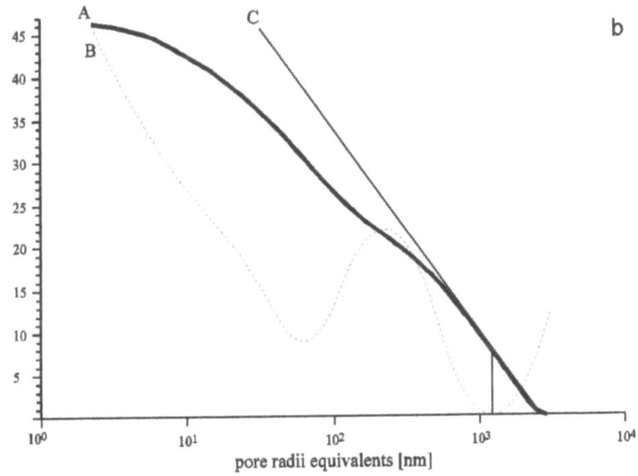
b
A
B
C
pore radii equivalents [nm]

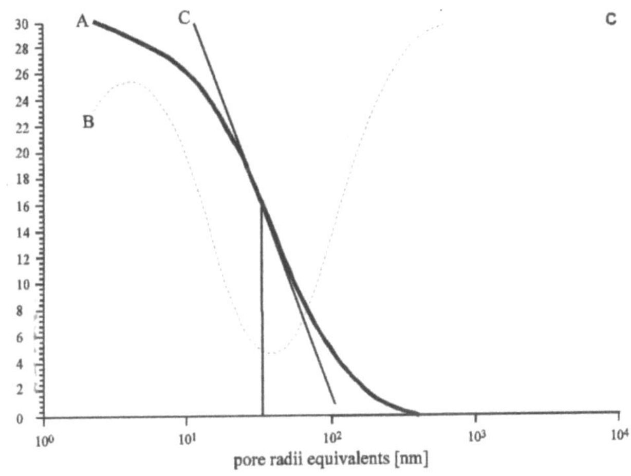
c
A
B
C
pore radii equivalents [nm]

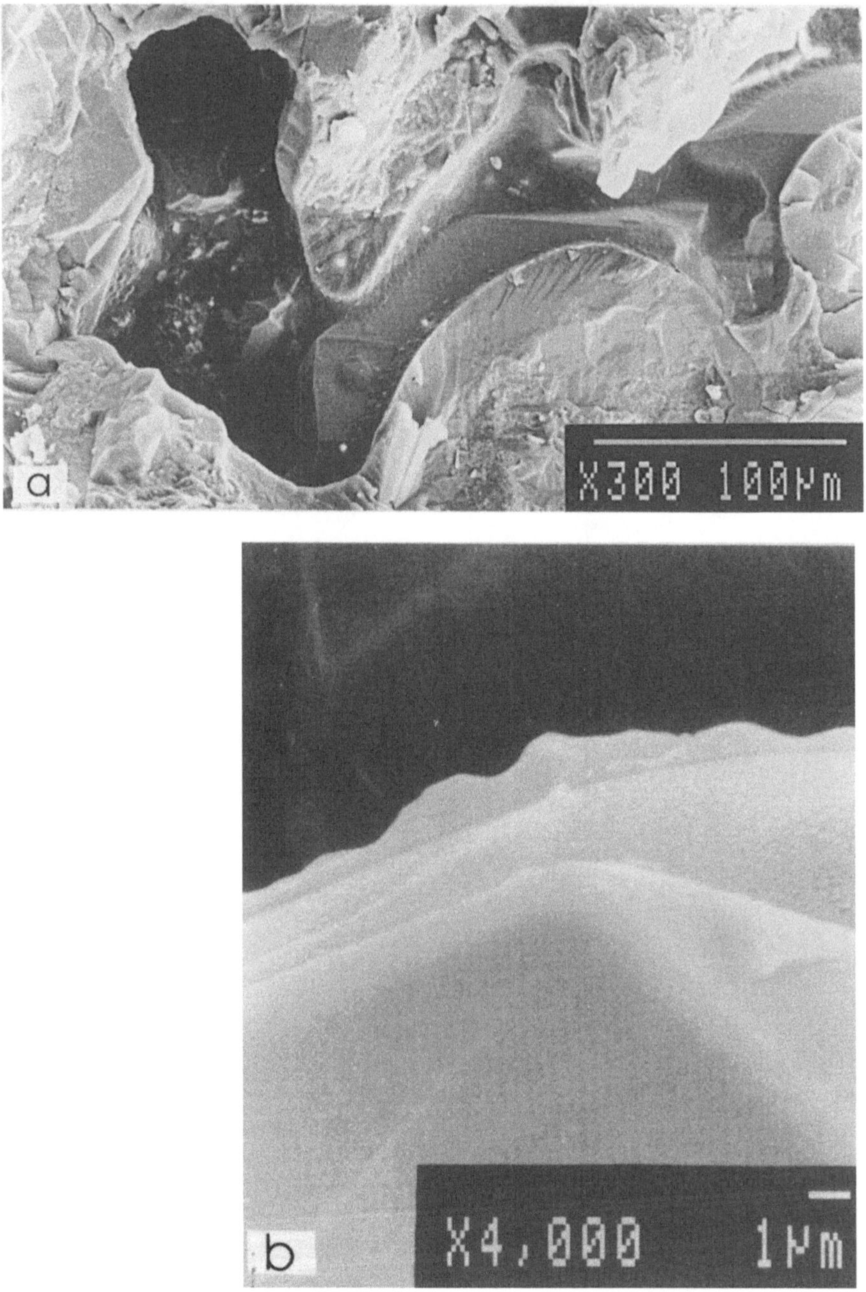

Fig. 7.28a,b. Cryo-SEM migrograph of a migration channel from a core flooding experiment (water-oil-water) with a Wealden sandstone. **a** Total view of meandering pore channel where residual oil occupies predominantly the concave (sag pores) and convex curvatures of the pore walls in contrast to the zones in between and the smooth surfaces of quartz crystalls which are only sparsly covered. **b** Detail from **a** showing wetting conditions. The quartz crystal is wetted by a 3–6 μm thick film of brine which again is coated by a 1- to 1.5-μm-thick oil film. (Mann et al. 1994a)

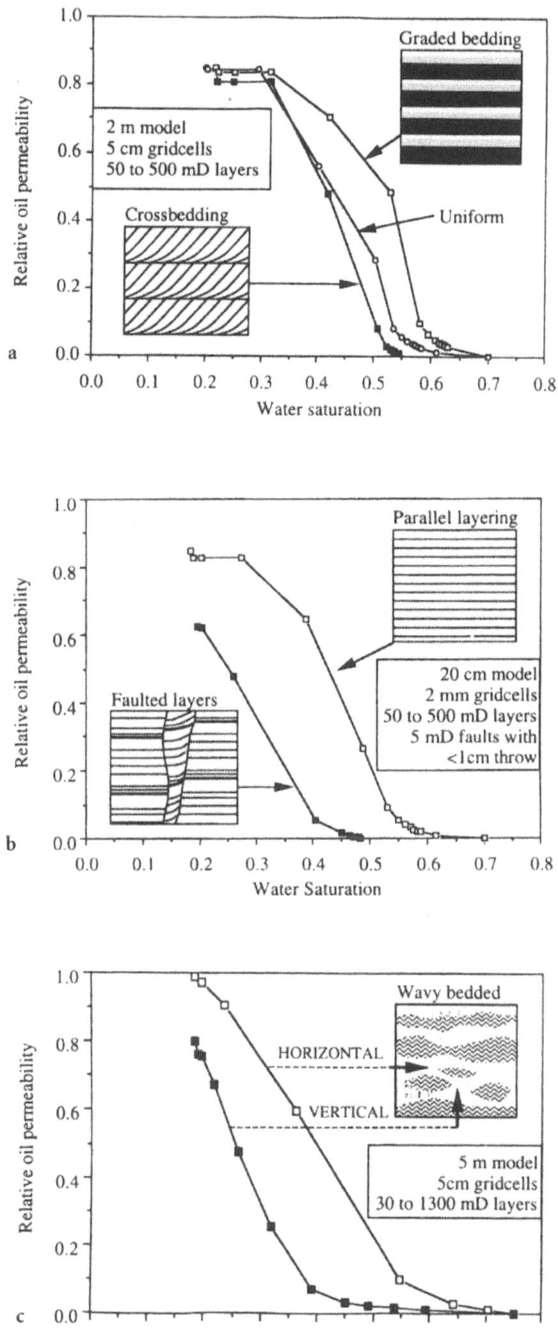

Fig. 7.29a–c. Relative oil permeabilities for different types of sedimentary fabrics of sandstones. The amount of trapped oil is given by 100% water saturation when the relative permeability approaches zero. **a** Uniform bedding versus cross-bedding and graded bedding. **b** Parallel layering versus faulted layers. **c** Horizontal versus vertical flow in wavy bedded sediments. (Ringrose and Corbett 1994, with kind permission of the Geological Society, London)

7.5 Simulation of Migration Processes: The Geological Framework

7.5.1 Prerequisite: Extension of the Conceptual Model for Migration Modelling

As described in Chapter 2, the aim of "classical" basin or thermal modelling is to determine temperature histories that enable an estimate of the timing and location of petroleum generation to be made, i.e. to interpret changing hydrocarbon generation potential values through time. These are commonly displayed as so-called "oil windows" on burial history diagrams or cross sections. Estimates of expulsion timing and efficiencies are also often made in basin modelling, but these are only of limited value as they generally use simple threshold values and are not related to the physical properties of the adjacent layers. Multi-dimensional effects such as pressures and complex geometries are also mostly neglected. The limitations of classical basin modelling are obvious, and the conclusion is clearly that more sophisticated techniques are required to investigate post-generation processes.

The modelling of petroleum migration is therefore an essential extension of thermal modelling, as the technology provides tools to analyse the processes that follow the generation of hydrocarbons in source rocks: the expulsion, migration and accumulation of petroleum. However, before these tools can be applied, the conceptual model that is used in basic thermal modelling and described before (see Chap. 2) must be extended. These extensions are of crucial importance and can be categorised *as new lithological parameters, modified lithological parameters, new modelling tools*, and – last but certainly not least – additional *user awareness* of the ramifications of each parameter.

New Lithological Parameters

Based on the assumption that pressure-controlled compaction is already being modelled with the thermal modelling tool, i.e. that compressibilities are already used to control the porosity changes, the main new lithological parameters for migration modelling are capillary pressures, migration saturation thresholds, and gas diffusion coefficients. Capillary pressures are possibly the most important new parameter, and also the most frequently underrated, as they are the controlling factors for the entire carrier system, i.e. they determine the sealing properties at unit boundaries as well as their changes through time, for example, as a function of overpressure changes. If three-phase flow is simulated, capillary pressure values are required for both oil and gas. Migration saturation threshold values are required to control expulsion (if a saturation model is used) and also to determine initial flow through the carrier system. Gas diffusion coefficients must also be established for each lithotype if gas diffusion is included in the model.

Modified Lithological Parameters

In addition to these new parameters, modified or refined lithological parameters are required, compared to those used in thermal models. These include anisotropies for thermal conductivities if the change from one to multi-dimensional models is being made. The most important refinement, however, is the definition of permeability anisotropies for the bulk units being modelled. These must include an interpretation of the properties relative to the specific processes being modelled. The standard example is an alternating sequence of sands and shales which could simply be described as a bulk sand and shale. However, the effective horizontal permeabilities that affect petroleum migration processes could well be those of the sandy units (if they are interconnected), and this value must be included in the model.

New Modelling Tools

Sedimentary sequences are rarely created in straightforward burial processes, and the limitations of simple, sequential conceptual models used for basic thermal history modelling become especially apparent when migration processes and the factors that affect them are investigated. Many of these complex geological processes occur deep within the sequence, while simultaneous burial or erosion is occurring along the surfical boundary of the system. This means that new geological modelling tools that are not commonly employed or even available in thermal history modelling are required. These include fault property handling, fracturing, cementation, intrusion, salt movement and aquifer flow modelling and are described in more detail below.

User Awareness

Classical basin or thermal modelling can be performed by petroleum geologists with little specific experience in modelling as most commercially available software packages, regardless of whether they are one- or two-dimensional, provide reasonable default values for physical parameters such as thermal conductivities, so that results are reasonably "safe", i.e. repeatable. However, in migration modelling, the results can be extremely sensitive to physical parameters (for example capillary pressures) that many geologists are initially unaware of and find difficult to quantify. The example given above in which the bulk and effective properties of a lithological unit (sand and shale) can differ according to the specific process being investigated, shows this quite clearly, and the way in which this is defined in the conceptual model of the section certainly affects the results. The user must therefore be able to combine his regional geological knowledge (for example, are the sandy units interconnected) with an awareness of the relative importance of capillary pressures and permeabilities in the carrier system. Specific technical knowledge is at least as important in migration modelling as in, for example, reservoir or structural history modelling. Users must be aware of the demands and potential pitfalls and ensure that sufficient support and training is provided by more experienced users or software vendors – a crucial factor when constructing conceptual models for petroleum migration modelling.

7.5.2 Conceptual Model: Migration System and Pathways

In addition to the geological data used to construct the general conceptual model for thermal histories, a conceptual model of the migration system must be developed that integrates the additional geological information required for the modelling of petroleum migration processes. These additional data requirements concern mainly the physical parameters of lithologies and special geological factors that affect the geometry of the migration system and its changes through geological time.

The physical parameters assigned to lithological units for thermal history modelling are insufficient to define migration systems and both new and refined parameters are required: (a) Amongst the new physical parameters that must be quantified for migration modelling, and that are not used in thermal history modelling, one of particular importance is capillary pressure. It is by far the most important single parameter in migration modelling as it actually defines the evolving carrier system, i.e. the most favourable flow path. This system is not simply a function of seal efficiencies but also of their relationship to pressures and overpressures and to the properties of the petroleum in the carrier units. Simplistic models of migration paths can lead to misinterpretations as the carrier system is in reality highly dynamic and its properties are always relative to those of adjacent units. (b) However, even the properties already employed in thermal modelling, such as permeabilities that are used to determine overpressuring, must be critically reviewed as to their role in migration processes. Bulk or average values may be completely irrelevant: for example, in a shale with interconnected sandy units the (effective) permeability of the sandy units controls flow rates within the carrier system that is initially defined by the capillary pressure relationships.

Special geological factors and processes that must be included in the conceptual model for migration modelling include, for example:

- Fracturing and fault properties, as discussed below.
- The changing geometry of the modelled sections or areas, in particular due to variations in palaeo-water depths through geological time. A failure to define this frequently neglected factor accurately through geological time can completely invalidate conclusions on migration directions and rates drawn from the modelling.
- Aquifer flow also plays an important role in some basins, both as an effective heat transfer medium that can lead to severe perturbations of the conductive temperature field, and as an additional direct force on oil and gas movements that, in extreme cases, can flush hydrocarbons from potential trap areas.

Only the consideration of these controlling parameters of the migration system and their inclusion in the original geological conceptual model can provide a sufficiently accurate representation of what is frequently described as the "migration system" in petroleum systems modelling.

7.5.3 General Numerical Model

The dominant transport mechanism of petroleum, i.e. oil and gas, strongly depends on its physical and chemical properties and the type of carrier rock. In most carrier-reservoir rock systems, separate phase hydrocarbon migration is the most important process. In this case, liquid petroleum with variable proportions of dissolved gas or gas with variable proportions of high-molecular-weight hydrocarbons is transported as separate phase in the otherwise water-saturated pore space. Buoyancy is the decisive factor controlling the direction of this transport.

Petroleum transport as a diffusion process in aqueous solutions in molecular or micellar form is of lesser importance for the migration process. Exceptions may be organic-matter-lean source rocks, in which insufficient petroleum is generated to allow separate phase transport. Certainly, diffusion is more important for low-molecular-weight hydrocarbons, i.e. gas, than for high-molecular-weight hydrocarbons due to the greater solubilities of the former. Both mechanisms - separate phase flow and the diffusion of low molecular hydrocarbons in aqueous solutions - are included in the model presented here.

The usual method for the description of separate phase flow and diffusion processes on a macroscopic or basin scale is the *continuum* approach. The object or region of interest is assumed to consist of uninterrupted volume elements and every field value (temperature, pressure, saturation, concentration) and every material parameter (permeability, thermal conductivity) is well defined in the volume element as a single, effective or average value. Furthermore, the value can vary continuously from element to element at least for parts of the system. This method requires that the size of the volume element must be small compared to the system being modelled (basin scale) but at the same time large compared to the structure elements (pores). Scale sizes used in basin modelling are shown in Table 7.7.

In the continuum approach information on the amount and composition of each fluid phase is required for each volume element (Fig. 7.30). In the model the fluid phases - a liquid aqueous phase, a liquid petroleum (oil) phase and a gaseous phase - can fill the pore space between the solid material matrix formed by lithological and organic particles. The three different fluids are treated as insoluble fluid phases in the flow module. Therefore, the presence of well-defined interfaces between the three fluid phases can be assumed. Additionally, every phase consists of several compounds (Fig. 7.30) and the fractions of all of these defined compounds are calculated and stored for each volume element. Furthermore, the methane content of the aqueous phase is important for the reservoir diffusion model, and the subdivision of the liquid

Table 7.7. Scale sizes for basin analysis

Molecular scale	10^{-9}–10^{-8} m
Pore scale	10^{-6}–10^{-3} m
Volume elements of the continuum	10^{-3}–10^{-2} m
Finite elements of the network	10^{0}–10^{2} m
Basin scale	10^{3}–10^{5} m

petroleum into compounds is crucial if different sources simultaneously produce oil with different flow behaviour. In some areas the nitrogen content in gas reservoirs must also be predicted by basin modelling (Sect. 6.6) and molecular nitrogen can be added as a special component of the gaseous phase.

The volume fractions of the phases (oil, gas, water) are represented by the saturation values, which describe the phase volume occupied by the fluid relative to the effective pore space. Mass fractions of the phases and compounds are defined as the total mass of the considered compound per mass of the rock (kg/kg rock). In general, the primary mineral mass is constant for the cells during the entire basin history, if no dissolution or cementation occurs.

The effective cell values of the thermal and hydrodynamic material properties are arithmetic or geometric average values of all compounds or phases. Densities and heat capacity values are mixed arithmetically, while thermal conductivities, viscosities and compressibilities are typical values based on geometrical averages. Effective phase values are defined with the mass fraction of the compound:

$$\text{arithmetic average}: c^{p} = \frac{1}{\sum_i m^{pi}} (m^1 c^{p1} + m^2 c^{p2} + m^3 c^{p3} + \ldots)$$

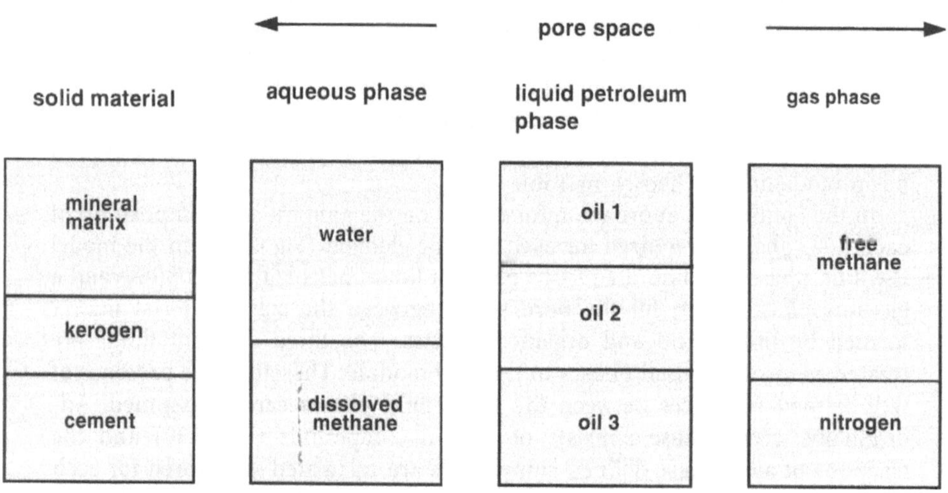

Fig. 7.30. Phases and compounds of the volume elements and cells

$$\text{geometric average}: c^{p} = \left(c^{p1^{mp1}} \cdot c^{p2^{mp2}} \cdot c^{p3^{mp3}} + \ldots\right)^{\frac{1}{\sum_i m^{pi}}}$$

with c^p = effective phase value, m^{pi} = mass fraction of the compound i (i = 1, 2...), and c^{pi} = value of the compound i (i = 1, 2...). Effective cell values are calculated with the volume fractions of the different phases:

$$\text{arithmetic average}: \; c = (1-\varphi)c^{s} + \varphi S^{a}c^{a} + \varphi S^{o}c^{o} + \varphi S^{g}c^{g}$$

$$\text{geometric average}: \; c = c^{s(1-\varphi)} \cdot c^{a^{\varphi Sa}} \cdot c^{o^{\varphi So}} + c^{g^{\varphi Sg}}$$

with c = effective cell value, ϕ = porosity, S^i = saturation of the fluid phase I (a = water, o = oil, g = gas), c^i = value of the phase i (a = water, o = oil, g = gas).

During the history of sedimentary basins that can evolve over time spans of several ten to hundreds of million years, changes in the rock and fluid compositions occur. The fluid carrier rocks experience large changes through time due to geometric (tectonic and compaction) processes or changes in their properties due to changing temperature and pressure conditions. Then the properties of the fluids themselves change, for example, due to chemical cracking of the petroleum. A dynamic analysis of the petroleum system is therefore meaningful only with an accurate and complex basin analysis, and the petroleum migration models must be embedded in the reconstruction of the thermal and pressure history of sedimentary basins. A well-calibrated thermal analysis is of immense importance for several aspects of the petroleum migration models, especially for the calculation of the petroleum generation rates and for the determination of fluid properties such as viscosities and diffusion constants. Thermal markers such as vitrinite reflectance, illite crystallinity and specific biomarker ratios are used to calibrate the geological and geothermal history. An overview of the main modules of a complete basin simulation is given in Fig. 7.31 (see Chap. 2 for the details of thermal modelling).

Primary Migration

The factors that are important for the expulsion process from source rocks differ from those that control further migration in carrier or reservoir rocks. For example, the saturation values that control initial expulsion from source rocks in separate phases are generally assumed to be higher than those used for initial movement through carrier rocks. The most common approach at present is the use of specific saturation threshold values for different source rock types. Typical bulk values for cells with sizes ranging from several tens of meters to 500 m, are 2–5%.

However, more complex expulsion models that take the various mechanisms of the petroleum generation process into account are also employed. In these models the beginning of petroleum expulsion is related to the controlling physical properties and processes, instead of assumed threshold values. The

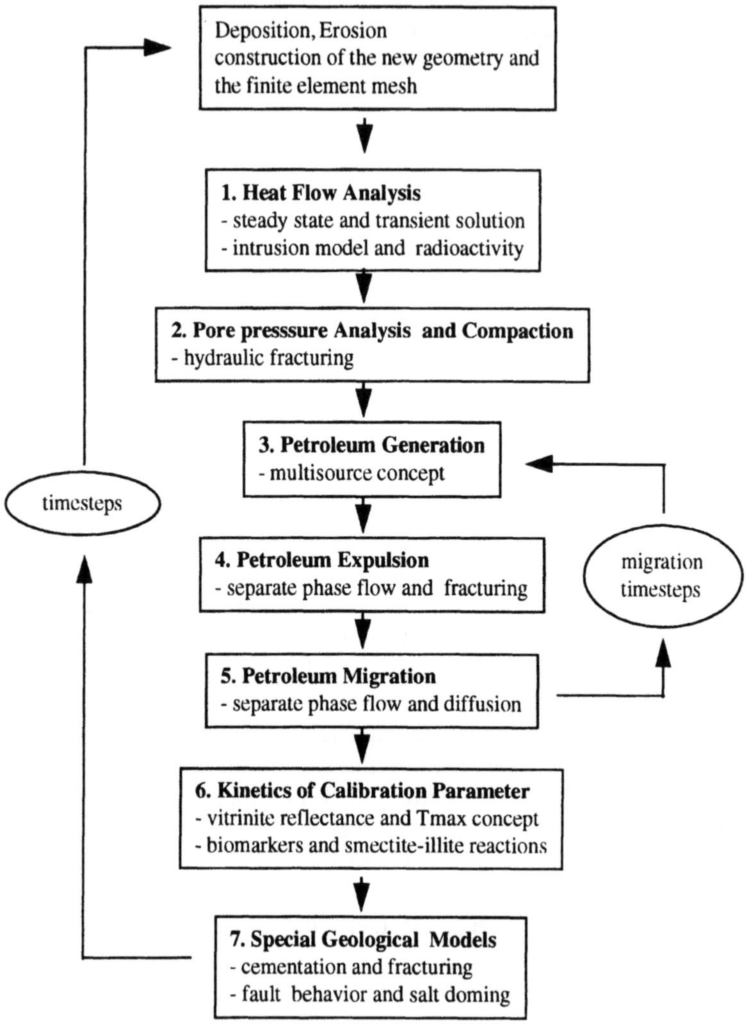

Fig. 7.31. Basin modeling processes

complex expulsion model used here in addition to the above threshold saturation concept is described by Düppenbecker (1990) and Düppenbecker and Welte (1992) and was calibrated for a specific source rock in northern Germany.

The model is based on analysis of the densities of kerogen and petroleum during kerogen cracking. Kerogen has a significantly higher density than petroleum and the densities of both vary with increasing transformation, as shown in Fig. 7.32 for a typical petroleum generation system. A potential volume increase occurs when petroleum (oil or gas) is generated in source rocks. With the assumption of isobar conditions this additional volume is proportional to the area between the two specific volume curves as shown in Fig. 7.32.

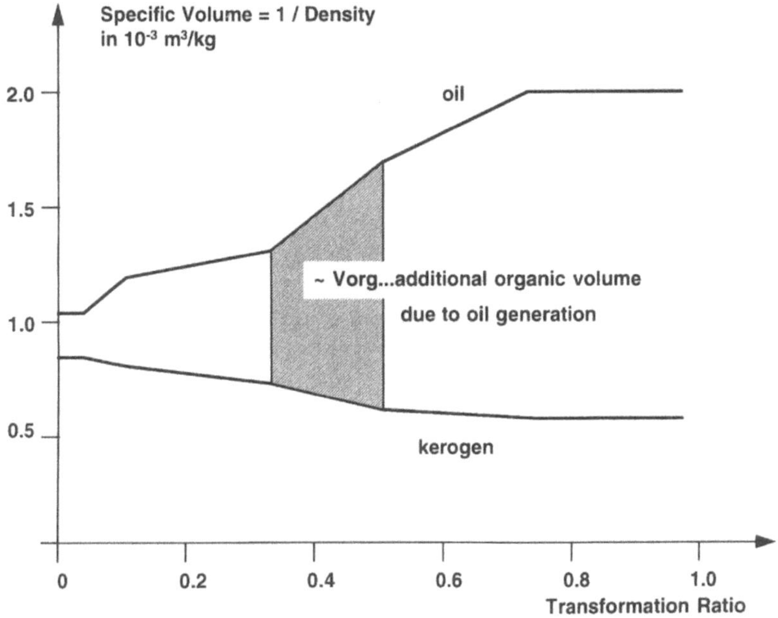

Fig. 7.32. Kerogen and petroleum specific volume changes during oil generation

Due to the limited pore volume, this potential volume increase cannot occur, and the potential increase in the volume of the material is instead transferred to a marked increase in the petroleum phase pore pressure. Several state equations (pressure-volume-temperature curves) for the water and the petroleum phase are used to describe this pressure increase. Of special significance is the role of the different pore classes (pore size distributions) and the amounts of adsorption that are considered in detail. The pore pressure increase allows the formulation of an expulsion condition. As a common concept in the continuum approach the corresponding field parameter is compared to a constant material parameter resistance, which acts against the effect. In the model presented here, two expulsion mechanisms are considered.

One process is separate phase flow through the pore system. The field parameter is the petroleum phase potential (difference of real pore pressure and hydrostatic pressure), while the capillary pressure acts as the resistant material parameter. Expulsion therefore occurs as separate phase flow through the pore system when the petroleum potential exceeds the capillary pressure of the material.

The second process is separate phase flow along fractures. Microfractures exist in most fine-grained rocks and are partly a product of the volume expansion of the organic matter during petroleum generation (Littke et al. 1988). The stress intensity factor describes the stress extension at the crack tips and is the driving force for fracture extension. This field parameter can be calculated from the pore pressure increase with the help of mechanical laws that define fracture processes. The corresponding material parameter is the fracture toughness of the rock. If the stress intensity factor exceeds the fracture toughness, crack extension occurs and petroleum can be expelled through these generated fracture paths.

Therefore the expulsion process is initiated if one of the following conditions is satisfied:

$$p^e \geq p^c \text{ or } K^I \geq K^c$$

with p^e = petroleum potential, K^i = stress intensity factor, p^c = capillary pressure, and K^c = fracture toughness.

It must be noted that a large amount of experimental data, mainly rock and fluid properties, must be available to employ this model successfully. The first calibrations and applications of the model were performed for the Posidonia Shale source rock of the Lower Saxony Basin, for which extensive laboratory data are available (e.g. Rullkötter et al. 1988; Düppenbecker 1990).

Secondary Migration

Several mechanisms such as separate phase flow, diffusion, solution and dissolution of gas in oil and water and chemical cracking simultaneously influence and define the complex process of secondary migration. Due to this multiprocessing behaviour the simulation technique must also consist of multiple models. Each model acts independently within the predefined volume elements (see Fig. 7.30) and causes changes in the composition of the phases and compounds. In this way the tracing of petroleum from different source rocks can be realised side by side within the same volume element. Mathematically the models are mainly formulated as boundary value problems of second-order differential equation systems or they are described in terms of additional operations, equations or conditions.

7.5.4 Specific Items of the Numerical Model

Separate Phase Flow

In the continuum-mechanical approach, a (hydrodynamic) potential is defined for every fluid. This potential is a measure of the energetically most favourable state or position of a fluid or its element. Conclusions about the interactions of the fluids with their environment can be obtained from the gradient of the potential. The fluid velocities and flow direction are directly derived from potential gradients. The linear form is represented with the single phase Darcy's law. A simple extension of Darcy's law to multi-phase flow can be obtained under the assumption that the flow of one phase is described as if the other phases are parts of the solid rock matrix. Then the effective permeabilities of the rock matrix must be redefined as a function of saturation. It must be noted that this approximation is not suitable if the different fluid phases have strong interactions during the flow process. Thus the adaptation of Darcy's law for multiphases has the following form:

$$v_i^f = -\frac{k_{ij} \cdot k^f}{v^f} \cdot \partial_j u^f$$

with v_i^f = velocity of the fluid phase (f: w = water, o = oil, g = gas), k_{ij} = - permeability tensor, k^f = relative permeability of the fluid phase (f: w, o, g), v^f = viscosity of the fluid phase (f: w, o, g), u^f = (hydrodynamic) fluid potential of the phase (f: w, o, g). Whereas the fluid potentials can be regarded as the driving forces of migration, the transport properties depend on the permeability and viscosity values. Fluid viscosities depend strongly on temperature. Additionally, oils of different consistency also very wide in the ranges of the viscosity (100–10 000 mPa s for high viscous oils and 1–100 mPa s for low viscous oils). Water has a viscosity of 1 mPa s (20 °C), and typical values for hydrocarbon gases are 0.01–0.1 mPa s.

Permeability values generally decreases with decreasing porosity during burial. On a permeability logarithmic scale a linear dependency on porosities is assumed. A special relationship is used for lithotypes consisting primarily of sandstones. Some typical permeability curves of the model are shown in Fig. 7.33 for the main lithology types (after Wygrala 1989).

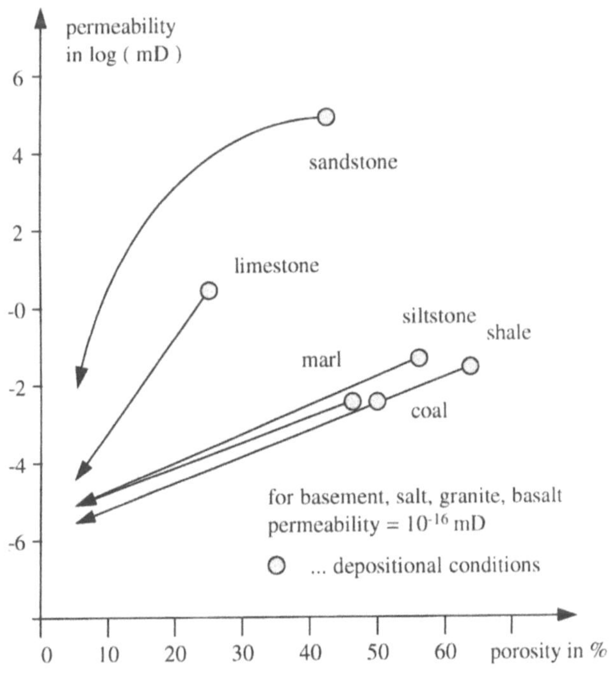

Fig. 7.33. Permeability versus porosity relationships

Some typical relative permeability curves are shown in Figs.7.34 and 7.35 (after Aziz and Settari 1985). Water is considered to be a wetting phase to both oil and gas, gas is considered to be the nonwetting phase to both water and oil, and oil is considered to be the nonwetting phase to water and the wetting phase to gas. The relative permeability functions (linear or quadratic) of water and gas are assumed to be a function of the water and the gas saturation respectively. The relative permeability of oil is assumed to be a function of both, the water and the gas saturation and it is calculated as follows.

$$k^{rw} = f(S^w) \qquad k^{ro} = f(S^w, S^g) \qquad k^{rg} = f(S^g)$$

and $\quad k^{ro} = k^{row} \cdot k^{rog}$

with k^{rw} = relative permeability of water, k^{ro} = relative permeability of oil, k^{rg} = relative permeability of gas, k^{row} = relative permeability of oil in a oil water system, k^{rog} = relative permeability of oil in a oil gas system, S^w = water saturation, and S^g = gas saturation.

Another important parameter for the flow pattern are the threshold saturation values. If the threshold saturation value for the initiation of oil mi-

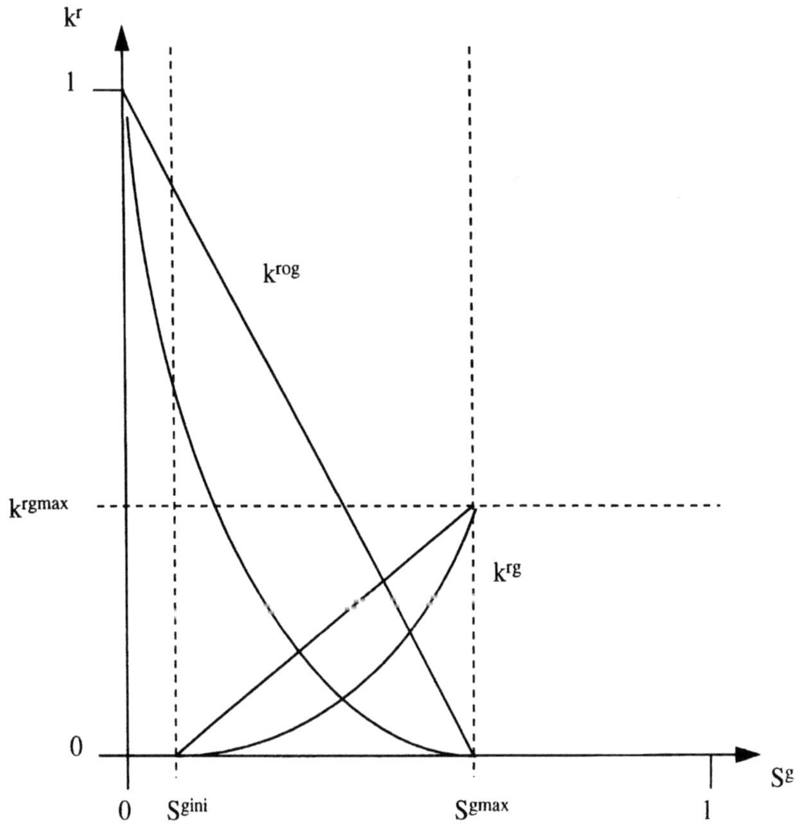

Fig. 7.34. Relative permeability function for water-oil system

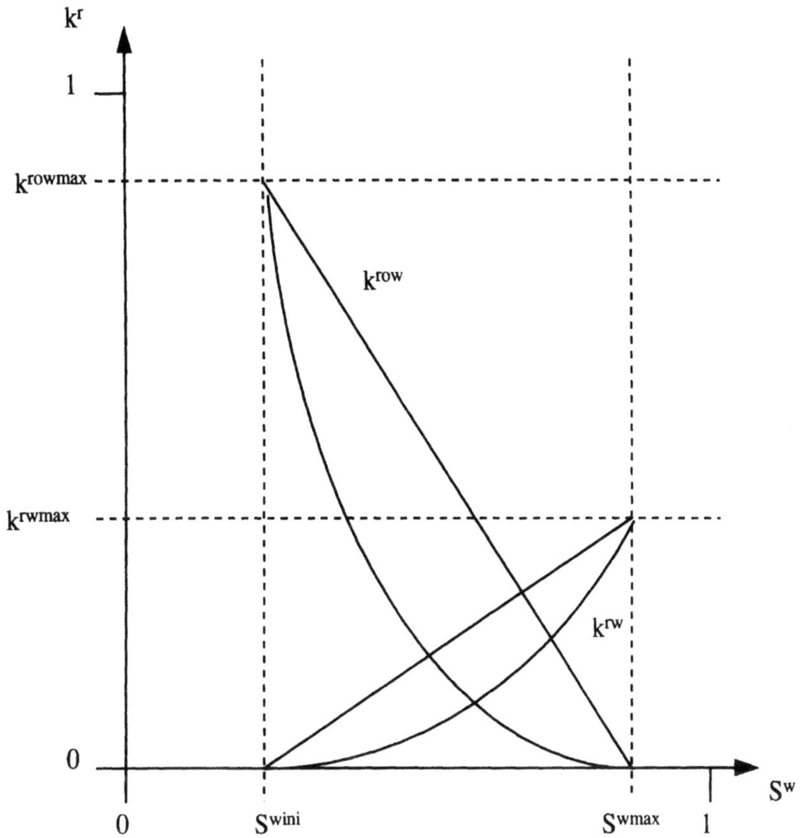

Fig. 7.35. Relative permeability function for oil-gas system

gration is relatively high (5–15%), pulsed migration or oil band movement occurs, while low threshold values result in more continuous flow patterns. In addition, this threshold value must be given for elements that are large compared to the pore size. The values are therefore bulk or average values, and they can be much lower than local or small-scale saturation values that often reach the range of 20–50%.

In the migration model initial saturation values can have a considerable effect on overall flow velocities. In the basic saturation controlled expulsion model the same saturation values are used for expulsion from the source rock and for cell saturation during migration. The cell must be filled to the saturation value before flow can continue into the next cell. After flow has occurred, the cell must be filled again before migration from the cell can continue. Saturation values of approximately 20%, which might be appropriate for thin layers (several tens of meters), therefore result in "pulsed" migration and large differences in saturation values along the migration path. Lower saturation values, for example 3–5%, which more closely approximate bulk values for thicker layers, can result in higher mean flow velocities as the cells only must wait for a smaller amount of petroleum before flow can continue.

The fluid potential (or overpressure of the fluid) is derived from the real pore pressure field under consideration of the buoyancy potential:

$$u^f = p^f - \rho^f g x_1 \qquad (f: \text{ water, oil, gas})$$

with u^f = fluid potential, p^f = real fluid pressure, x_1 = coordinate of the vertically downward directed axis, and g = acceleration constant. The difference in pore pressures of the two fluids is the capillary pressure. It is generated due to the fact that the adhesive attractive forces of two immiscible fluids are essentially greater than the cohesive repulsive forces. Capillary pressures are one of the fundamental parameters that control fluid flow in a sedimentary sequence. In a sense the capillary pressure defines the resistance in the material against migration. If a petroleum particle is to be transported in a certain direction, the capillary pressure acting in this direction must be overcome in order to displace the water phase in the new position. The capillary pressure or the corresponding potential acts against the free motion of the particles. The capillary pressures are a function of lithotypes, and in porous media the values are mainly effected by the geometry of the pore system or, more specifically, by the size of the pore throats. It ranges from 0.01 MPa (sandstone) to 2 MPa (tight shales), with an interfacial tension of 0.03 N/m. Fluid movements are therefore much more efficient in sandstones. Capillary pressures are also the controlling factor at layer boundaries with different lithotypes, for example, where a sandstone carrier is overlain by a shale seal. The differences between the capillary pressure values in adjacent layers affect petroleum flow across layer boundaries, i.e. they determine the effectiveness of seal properties.

The fluid potentials can be written in the following form:

$$u^p = u^w + (\rho^w - \rho^p) g x_1 + p^{pc} \quad (p..\text{oil, gas})$$

where p^{pc} is the capillary pressure of the oil to water or gas to water. These potential definitions produce the following petroleum driving forces:

- The petroleum moves from high pore pressure regions (produced for example by high sedimentation rates) to low pore pressure regions.
- Due to the lower density petroleum tends to migrate vertically upward.
- Petroleum migrates from high to low capillary pressure regions, obstacles such as caprocks with high capillary pressure levels inhibit migration. This can lead to changes of the flow direction and migration paths or to accumulations.

The mass balance of fluid flow in porous media requires that the average density of a volume element changes if fluids, as mass fluxes, stream through the element. The average density change can be related to porosity or saturation changes if special flow mechanisms and fluid properties are assumed. In the following model it is assumed that the solid (rock) material is ideally rigid and the fluids are immiscible and incompressible. These assumptions imply that porosity changes and compaction can only occur due to the outflow of fluids from the element. Then mass balance equations for a fluid can be formulated as follows:

$$\partial_i(\rho^f v_i^f) + S^f \partial_t(\rho^f \phi) + \phi \partial_t(\rho^f S^f) = -\phi \partial_t(\rho^f S^{add}) \quad (f: \text{ water, oil, gas})$$

with ϕ = porosity, S^f = fluid saturation, S^{add} = additional saturation due to oil and gas generation. The first two terms of the equation are well-known flow terms from general fluid mechanics, while the other two represent special basin modelling effects. These effects are compaction (porosity change) and petroleum generation (fluid sources).

The porosity change is caused by the changing overburden load potential due to subsequent sedimentation. This is described by the following compaction law:

$$\partial_t \phi = -(1 - \phi) \cdot C \cdot \partial_t(u^a - u^f)$$

with C = compressibility, u^a = potential of the overburden material, and u^f = fluid potential. The difference of the overburden and the fluid potential is the effective stress. The sedimentation rate directly affects the overburden potential. Therefore both overpressuring and compaction processes are strongly influenced by sedimentation rates. Compressibility values, used in the model, are shown in Fig. 7.36 for the main lithology types (after Wygrala 1989).

The source term for petroleum generation contains the additional or generated petroleum (oil or gas) saturation. The representative values for the petroleum kinetics are the chemical petroleum potentials (mass of hydro-

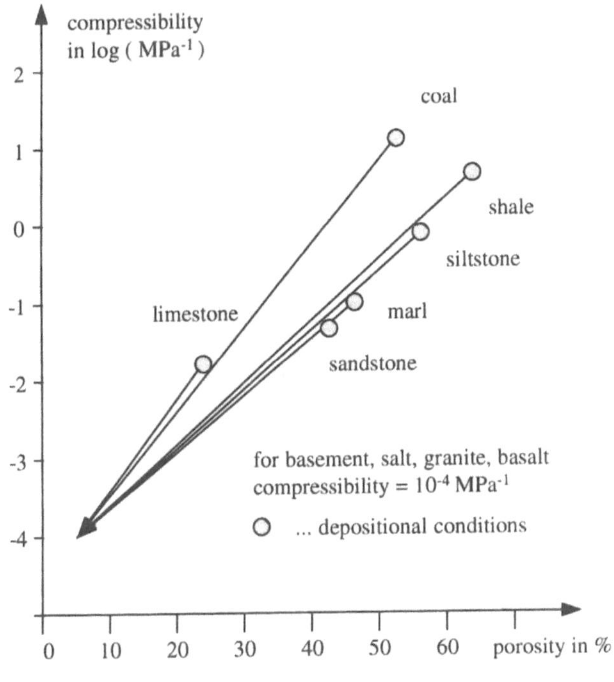

Fig. 7.36. Compressibility versus porosity relationships

carbons per TOC mass) and the relationship between the chemical potentials and the generated saturations are the following:

$$\partial_t S^{add} = m^{TOC} \cdot \frac{(1-\phi)\rho^{rock}}{\phi\rho^f} \cdot \partial_t c^f$$

with ρ^{rock} = rock density, ρ^f = oil or gas density, and c^f = chemical potential of oil or gas.

Petroleum Generation

This module describes the transformation of a part of a compound or a phase to another compound or phase by chemical reactions. A chemical reaction is characterised by a kinetic data set, which consists of a set of activation energies, frequency factors and initial masses. Every petroleum kinetics contains the definition and description of three different reaction systems:

$$\begin{array}{lll} Ker_1 \xrightarrow{k_{11}} Oil & Ker_1 \xrightarrow{k_{31}} Gas & Oil_1 \xrightarrow{k_{21}} Gas \\ Ker_1 \xrightarrow{k_{12}} Oil & Ker_1 \xrightarrow{k_{32}} Gas & Oil_1 \xrightarrow{k_{22}} Gas \\ \cdots\cdots\cdots & \cdots\cdots\cdots & \cdots\cdots\cdots \\ Ker_1 \xrightarrow{k_{1n}} Oil & Ker_1 \xrightarrow{k_{31}} Gas & Oil_1 \xrightarrow{k_{2m}} Gas \end{array}$$

The reaction parameter k that characterises the reaction velocity is a function of temperature and is described by the Arrhenius law:

$$k_i = A_i \cdot e^{-\frac{E_i}{R \cdot T}}$$

with E^i = activation energy, A^i = frequency factors, R = gas constant. Basically these equations are defined with different reaction parameters for the different kerogen types. For kerogen types I and II, the kerogen→oil→gas reaction is the main process, while for kerogen type III an important part of the gas is obtained from the kerogen→gas reaction. Figures 7.37 and 7.38 present typical kerogen type II and type III parameters (after Tissot et. al. 1987).

A simple first-order differential equation is used to describe the mass exchange during a chemical reaction. For a multibound kerogen system the formulation is the following:

$$\partial_t x_i - \; k_{11} x_1 - k_{3i} x_i \quad \text{for kerogen bonds}$$

$$\partial_t y_i = \alpha_i \left(\sum_{j=1}^{n} k_{1j} x_j\right) - k_{2i} \quad \text{for oil bonds}$$

$$\partial_t z = \left(r \sum_{j=1}^{m} k_{2j} y_j\right) + \left(\sum_{j=1}^{1} k_{3j} x_j\right) \quad \text{for gas}$$

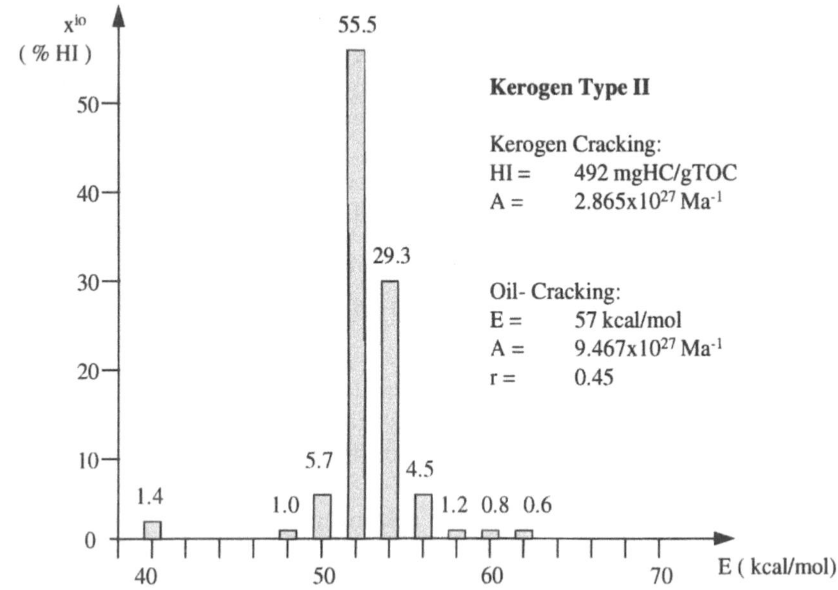

Fig. 7.37. Kinetic parameters of kerogen type II. (Tissot, in Waples et al. 1992)

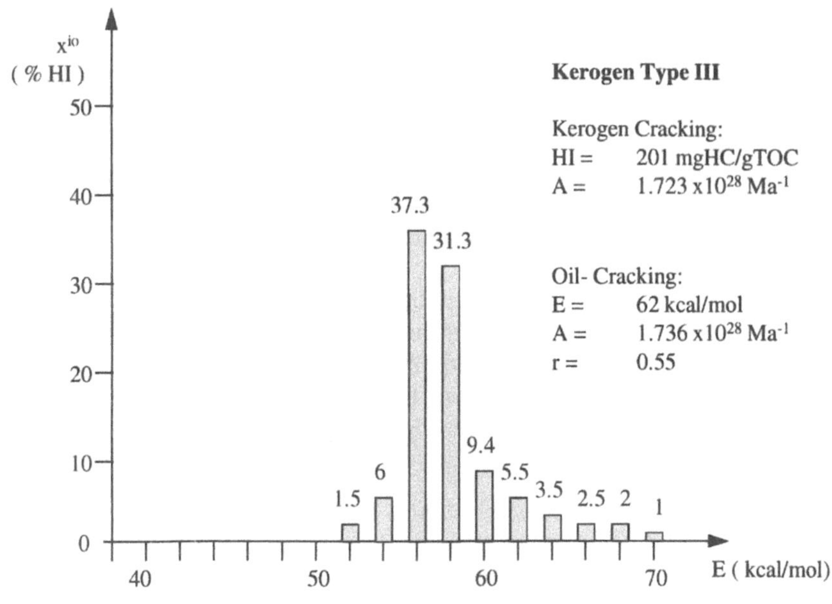

Fig. 7.38. Kinetic parameters of kerogen type III. (Tissot, in Waples et al. 1992)

with x_i = kerogen potential of the compound i, y_i = oil potential of the compound i, z = gas potential, α_i = initial oil potential distribution, k_{1i} = reaction rate of the reaction kerogen→oil, k_{2i} = reaction rate of the reaction kerogen→gas, k_{3i} = reaction rate of the reaction oil→gas. In the case of gas cracking from oil the organic mass is reduced, meaning that when 1 g oil is cracked to gas, this gas has a mass of only 0.4–0.7 g. A reduction factor r is therefore introduced in the mass balance to consider this effect. To estimate the size of the reduction factor, the following principal equation can be used:

$$\text{reaction}: \quad 2CH_2 \rightarrow C + CH_4$$

$$\text{relative molecule mass}: \quad 2 \cdot 14 \rightarrow 12 + 16$$

Thus, $r = \frac{16}{28} = 0.57$

If the presence of different oil components is considered, the distribution α_i of the mass ratios of these components must be given for the kerogen→oil reaction to decide which amounts of the different oil components are produced from the kerogen.

The petroleum kinetics described above can also be used to describe the reactions of different hydrocarbon component classes (for example C1, C2–C4, C5–C6, C7–C15, C15+). Then instead of the properties of a whole kerogen type, only the properties of the special hydrocarbon component class must be entered in the kerogen reactions. After the model has been processed with several component class kinetics, an overview over the distributions of the hydrocarbon classes can be obtained.

The kinetic approach indicates that the conversion of kerogen to oil and gas is more strongly effected by temperature (exponential influence) than by time (linear influence). For oil and gas generation calculations the use of the actual kinetic equations requires numerical integration due to the complexity of thermal histories. It should be emphasised that the extrapolation of kinetic processes from laboratory experiments into geological time is still a problem which can be controlled only by calibrating with precise temperature histories and real cases.

In a multiple source rock concept the petroleum from each different source unit is tracked or traced separately throughout its entire migration history to ensure that the properties of mixed petroleums are determined correctly. It is therefore always possible to see how much of the petroleum in a specific cell comes from which source unit (see Fig. 7.39).

Fracturing and Cementation/Dissolution

Fracturing and cementation or dissolution processes can influence the permeability values as well as the effective porosities. Although the overall process of fracturing or cementation may be based on many different mechanisms on both a microscopic and a mesoscopic scale, in the presented model very simple laws are used for the description of the effects on a macroscopic or basin scale.

The fracturing process mainly influences the permeability values and, in addition to the fracturing due to hydrocarbon generation and expulsion, hy-

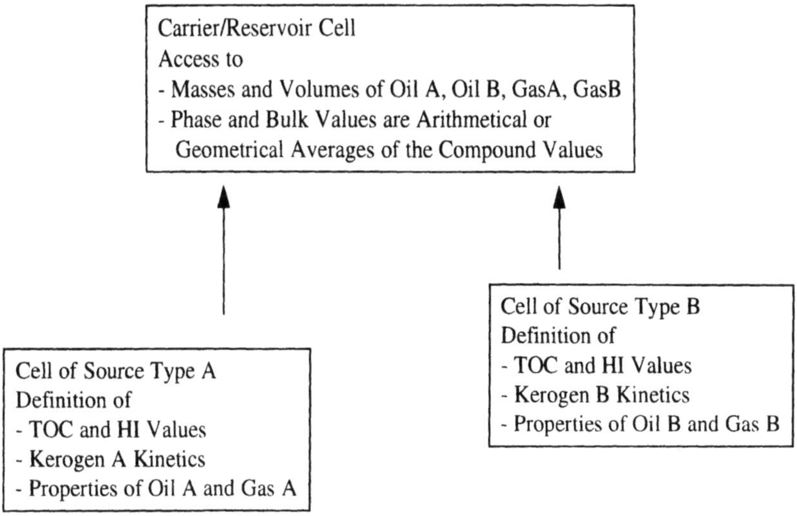

Fig. 7.39. Concept of multiple source tracing

draulic fracturing conditions can also be defined. When one fracture condition is satisfied during a time step, then the permeability value of the considered element is increased by several orders of magnitude. The following fracture conditions can be applied: (a) the excess hydraulic pressure exceeds a certain proportion of the lithostatic pressure (generally in the range of 0.6 – 0.8); (b) the real pore pressure exceeds a threshold pressure value. The increase in permeabilities is calculated on a logarithmic scale:

$$\Delta \mathrm{lgk} = \mathrm{f} \cdot \frac{\mathrm{p^e} - \mathrm{p^c}}{\mathrm{p^c}}$$

where f denotes the factor of permeability increase, p^e denotes the real (pore or over-)pressure, and p^c is the corresponding threshold pressure.

The cementation/dissolution condition is formulated in terms of a critical depth or temperature, or a specific geological age for a certain layer. If this condition is satisfied, the porosity of the volume cell is increased or decreased by the cementation volume. Because of the strong dependency of the permeability on the porosity values the permeabilities also change significantly. The bulk material parameters such as thermal conductivity, heat capacity and density are effected by the solid cement properties. Some typical cement properties are listed in Table 7.8. Especially the considerable effect on the permeability values necessitated the implementation of both processes, cementation/dissolution and fracturing in the fluid flow simulation.

Fault Model

As faults can have very different and complex effects on fluid transport processes, their effects are handled by a variety of models. They can, for example,

Table 7.8. Selected typical cement properties

Cement type	Density (kg/m^3)	Thermal conductivity (W m^{-1} K^{-1}) At 20° C	At 100° C	Heat capacity (kJ kg^{-1} K^{-1}) At 20° C	At 100° C
Silica cement	2650	7.70	6.00	0.177	0.212
Calcite cement	2721	3.30	2.70	0.199	0.232
Dolomite cement	2857	5.30	4.05	0.204	0.238
Anhydrite cement	2978	6.30	4.90	0.175	0.193
Halite cement	2150	5.70	4.85	0.206	0.214
Clay cement	2810	1.80	1.60	0.200	0.220

act as completely impermeable interfaces, or as ideal conductive migration paths that can be described as thin permeable layers. In strongly faulted sections the entire petroleum migration process can be restricted exclusively to fault systems and described by a pure pipeline model. Considerable experience is therefore required to decide which fault model is relevant for the specific geological object or basin.

In the fault model discussed here faults are handled as volume properties and can be modelled both as preferred highly permeable migration paths that conduct flow along the faults and as low-permeability zones that restrict flow across the fault planes and, for example, create pressure seals. In the numerical procedure the basin region is discretised by a network of small cells; when a fault line crosses a cell, the permeability, thermal conductivity and capillary pressure values are modified.

Permeabilities are anisotropic values in layered rocks and are typically described by two values in the main directions, namely the permeabilities along the layer and perpendicular to the layer. Mathematically the permeability values in the equations are defined by permeability tensors. In a fault element the permeability tensor is affected by the given fault transmissibility, as shown in Fig. 7.40.

Salt Tectonics

Petroleum reservoirs and salt layers and domes are often closely associated as salt can act as an ideal caprock due to its very low permeability. Moreover, salt layers affect both the temperature and pore pressure fields, due to their extremely high thermal conductivities (4.5–6 W m^{-1} K^{-1}) and their low permeabilities and compressibilities. The volumetric modelling of salt layers through geological time is therefore of great importance in basin modelling.

As a reaction to overburden load (depositional mass), salt layers tend to build domes, sometimes of up to several kilometres thickness (Fig. 7.41). During this doming the salt layer behaves as a viscous fluid. A relatively simple model that includes the description of the geometric changes on the basin scale

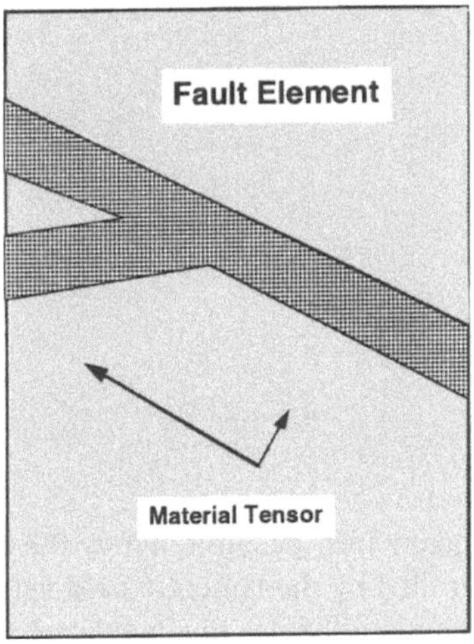

Fig. 7.40. Fault element

is employed as follows: an initial depositional model of the salt layer is reconstructed by volume balancing. Then a time period must be defined when salt doming occurs. During the doming phase the geometry of the salt layer linearly changes from the initial to the final state. This is realised by a special cell stretching procedure that takes total volume constants into account (Fig. 7.42).

Diffusion

The diffusion model is related to the transport of light hydrocarbons in aqueous solution. Although separate phase flow generally plays the dominant role, diffusion transport can be important under specific conditions. The prerequisites are: (a) the hydrocarbon components must have a high water solubility (see McAuliffe 1966), (b) the concentration of these components is very high in specific areas, and (c) special geological conditions prohibit significant separate phase flow (e.g. impermeable cap rocks). Two problems are then

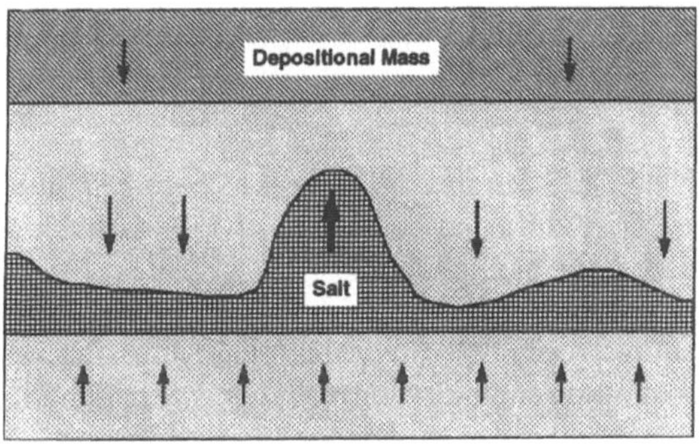

Fig. 7.41. Salt doming

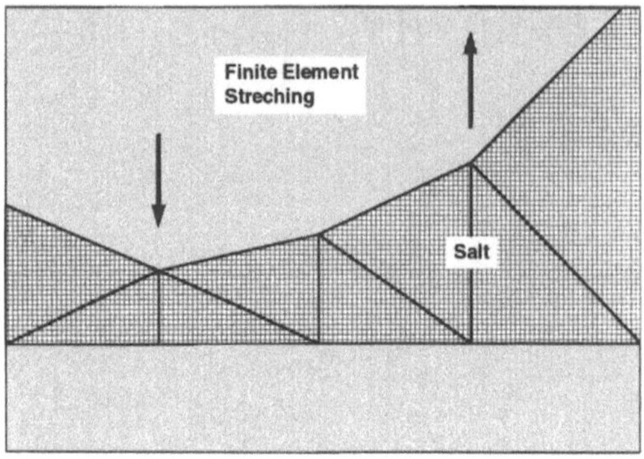

Fig. 7.42. Finite element stretching

taken into account. First, the quality of the dissolved hydrocarbons is controlled by the concrete field values (temperatures, pressures) and in the case of nonequilibrium the hydrocarbon compounds must be dissolved in or exsolved from the pore water. Secondly the diffusion fluxes are calculated and realised along hydrocarbon concentration gradients.

The solubility of light hydrocarbons in water is usually described as a function of temperature, pressure and salinity conditions as state diagrams or state laws. In Fig. 7.43 the methane solubility (in ppm) is calculated after Haas (1987). Additionally, typical pressure and temperature intervals of reservoirs at 2, 4 and 6 km depth are shown.

The solubility is the highest possible concentration under the given field parameters; the concentration depends on the available amount in the free gas phase. In every time step the free gas phases and the laws of solubility are analysed to determine the solution and dissolution process that determines the aqueous hydrocarbon concentration field. This concentration field is the basis for the calculation of the diffusion fluxes by using Fick's law applied to the porous media (see Krooß et al. 1992a,b).

$$J = -\phi \cdot D^{eff} \cdot \mathrm{grad}(C^{w})$$

with ϕ = porosity, D^{eff} = effective diffusion coefficient, c^{w} = concentration of the HC compound in water. The diffusion coefficient strongly depends on temperature. This is realised by applying the following Arrhenius-type law:

$$D(T) = D^{150} \cdot \exp\left[-\frac{E}{R} \cdot \left(\frac{1}{T} - \frac{1}{423.15}\right)\right] \quad \text{with } R = 8.314 \quad \mathrm{J/mol/K}$$

$$\text{and } E = 50\ \mathrm{kJ/mol}$$

Here D(T) denotes the diffusion coefficient at temperature T (in K), D^{150} is the effective diffusion coefficient at 150 °C, (423.15 K), E is the activation energy and R the gas constant.

A typical scenario is described in Fig. 7.44. When gas is generated in the source rock, first the water phase must be saturated with methane until the maximum solubility for methane is reached. Then methane can also form a free phase and move as separate phase transport. Therefore, in most of the section

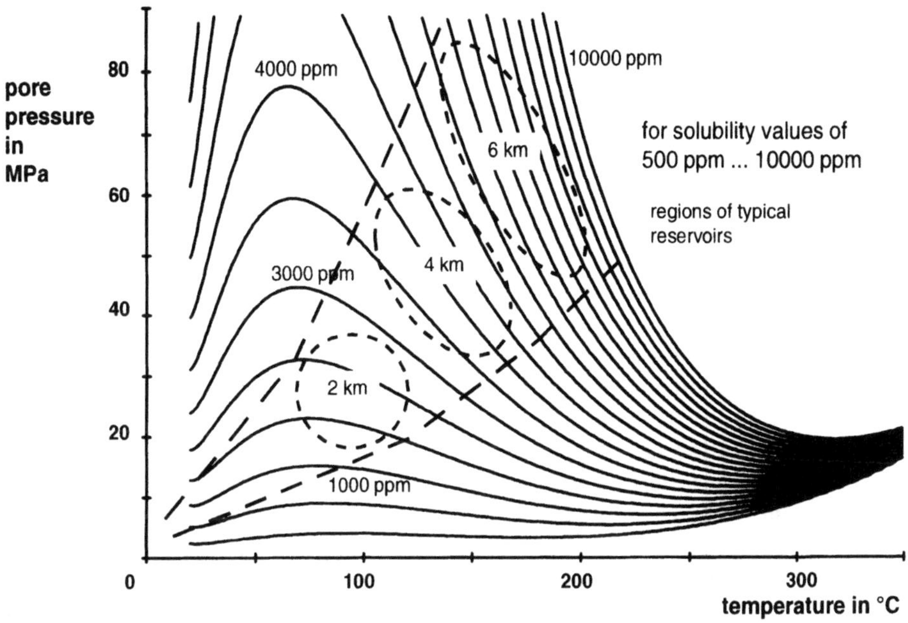

Fig. 7.43. Methane solubility in water as a function of pressure and temperature

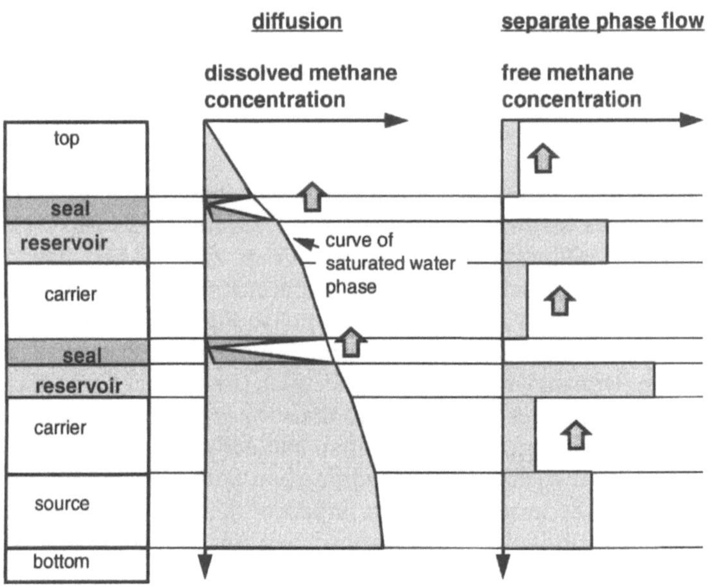

Fig. 7.44. Methane transport mechanisms

the water phase is saturated with methane and thus diffusion is a relatively small but continuous and basinwide process.

7.5.5 Case Histories

Hydrocarbon Migration in the Neogene Styrian Basin

Introduction

Basin evolution of the Styrian Basin can be subdivided into an early Miocene synrift and a middle to late Miocene postrift phase of subsidence. As presented in detail in Chapter 2, the heat flow history of the basin is governed primarily by Miocene (Karpatian to early Badenian) volcanism. Volcanic centres were characterised by extremely elevated heat flows (> 300 mW/m^2) and heat flow decreased to background values (about 120 mW/m) at a distance of about 10 km. A second volcanic phase producing basalts in Plio-/Pleistocene times had only minor influence on the regional heat flow pattern.

The basin is 100 km long, 60 km wide and more than 4 km deep and is located at the eastern margin of the Alps, forming part of the Pannonian Basin System (Fig. 7.45). Many subbasins of the Pannonian realm contain prolific hydrocarbon deposits. However, in the Styrian Basin no commercial hydrocarbon accumulations have been detected to date. Only a small subeconomic gas deposit was encountered in middle-Badenian algal reefs near the northern margin of the basin in the Ludersdorf area ("L" in Fig. 7.45).

A recent paper evaluates the hydrocarbon potential of the basin using organic geochemical and numeric modelling techniques (Sachsenhofer, 1994). For this purpose the above heat flow model, based upon extensively calibrated temperature histories along two cross sections (I, II; Fig. 7.45) and for five individual wells was established (see Chap. 2). This heat flow model and organic geochemical data were used to reconstruct hydrocarbon generation and migration along a third cross section (III; Fig. 7.46). The most important results are: (a) mature oil- and gas-prone source rocks exist in the lower Miocene section (Ottnangian and Karpatian levels); (b) hydrocarbons in close vicinity of Miocene volcanoes may be lost to the surface because no seals were deposited at the time of generation (Karpatian and early Badenian); (c) the area around cross section III has a good hydrocarbon potential, because of relatively late generation. Because of the great impact of the time factor on the hydrocarbon potential quantitative models of hydrocarbon generation and migration along cross section III are presented in this contribution.

Conceptional Geological Model

Hydrocarbon migration was studied along a 19-km-long north-south trending cross section (III) through the northern flank of the basin depocenter. In its

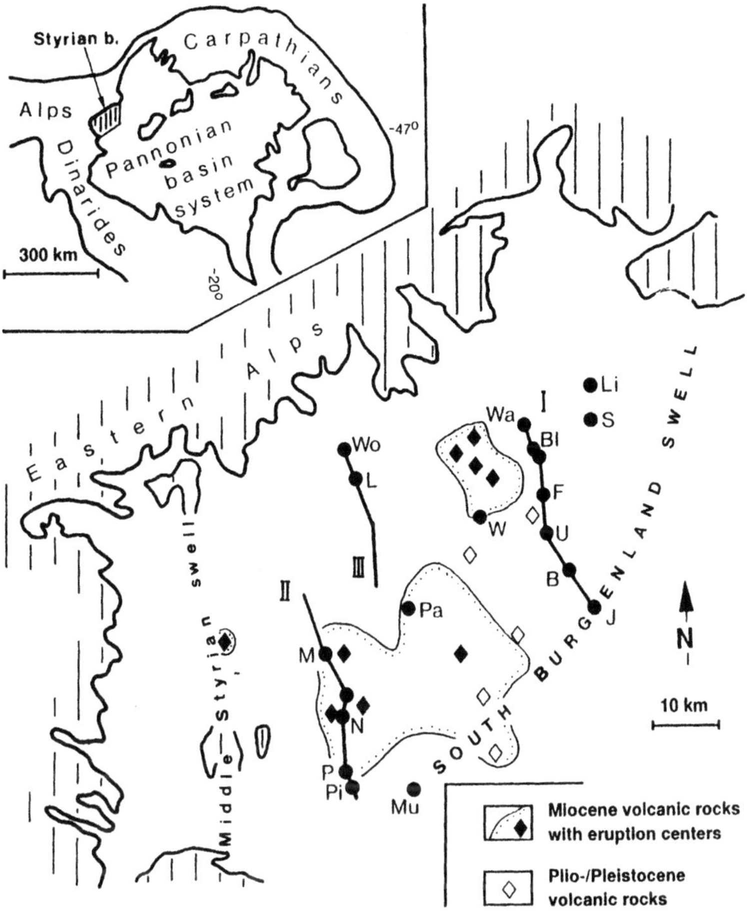

Fig. 7.45. Position of the Styrian Basin within the Alpine-Carpathian-Pannonian region and sketch map of the Styrian Basin with location of volcanic rocks (Sachsenhofer, 1994). The heat flow model for the Styrian Basin (Fig. 7.49) was calibrated along cross sections I and II. Numeric modelling of hydrocarbon migration is performed along cross section III. A subeconomic gas deposit was detected in the Ludersdorf-Wollsdorf region. Li, Litzelsdorf 1; *S*, Stegersbach 1; *Wa*, Waltersdorf 1; *Bl*, Blumau 1, 1a; *F*, Fürstenfeld 1; *U*, Übersbach 1; *B*, Binderberg 1; *J*, Jennersdorf 1; *W*, Walkerdorf 1; *Wo*, Wollsdorf 1; *L*, Ludersdorf 2; *M*, Mitterlabill 1; *N*, St. Nikolai 1, 2; *P*, Perbersdorf 1; *Pi*, Pichla 1; *Mu*, Mureck 1

shallow northern part the section contains middle Badenian algal reefs, which host the only significant gas accumulation in the Styrian Basin known today. The basin fill reaches a thickness of 4000 m at the southern end of the transect. The geological evolution of the cross section was defined by a total of 19 gridpoints including two shallow wells in its northern part and 28 geological events (Fig. 7.46):

- *Ottnangian* (events 3–4): Non-marine (vitrinite- and alginite-rich) source rocks were deposited along the southern part of the transect.

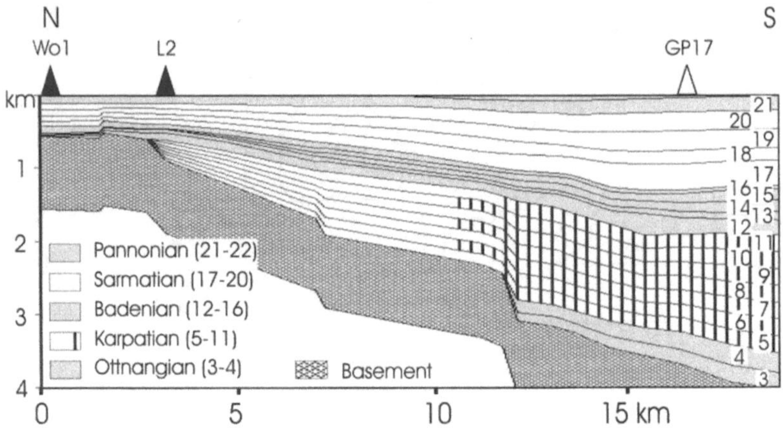

Fig. 7.46. Stratigraphy along cross section III. Potential source rocks occur in Karpatian (basinal facies; vertical hatching) and in Ottnangian levels; potential carrier rocks occur in post-Karpatian sediments. The only (subeconomic) gas deposit in the Styrian Basin was found in middle Badenian algal reefs in the Ludersdorf area. For well names see Fig. 7.45; *GP17*, gridpoint 17

- *Karpatian* (events 5–11): Marine sediments were deposited in a fault controlled setting. A deltaic facies with coarse-grained sediments along the fault-bound northern basin margin grades southward into a basinal facies with fine-grained sediments. Only the latter are considered as potential source rocks. Measured source rock parameters and parameters used for modelling are summarised in Tables 7.9 and 7.10 for both Ottnangian and Karpatian sediments.
- *Badenian* (events 12–16): The Badenian sedimentary cycle starts with the deposition of conglomerates and sandstones. During middle Badenian times algal reefs up to 50 m thick, interfingering to the south with shaly sedi-

Table 7.9. Key parameters of potential source rocks in the Styrian Basin (the generative potential is described using the nomenclature of Peters 1986)

			Generative potential	
	TOC (%)	HI (mg HC/g TOC)	Oil	Gas
Non-marine Ottnangian and Karpatian				
Coaly samples	0.35–6.5	30–160		Fair
Alginite-rich samples	6.0–8.0	325–450	Very good	
Marine Karpatian				
Basinal facies	0.5–1.3	350–400[a]	Fair	
Deltaic facies	0.2–1.5	30–80		Poor-Fair

[a]HI measured on kerogen concentrates

Table 7.10. Source rock and petroleum parameters used for modelling of hydrocarbon generation and migration in the Styrian Basin

	TOC (%)	HI (mg HC/g TOC)	Kinetic parameters
Non-marine Ottnangian and Karpatian	1.5[a]	214	Brent coal[b] (Espitalié et al. 1988)
Marine Karpatian			
Basinal facies	1.0	375	Karpatian basinal facies
Deltaic facies	–	–	–

Migrating petroleum type: "high viscous gas". (Reason: gas was found in Ludersdorf and condensates occur in several other wells). Primary migration: saturation model. Boundary conditions: open basin sides.
[a]To be conservative, a relatively low TOC content is assumed for non-marine samples
[b]Kinetic parameters of Brent coal are used, because most samples contain type III kerogen and the kinetic parameters are similar to those for alginite-rich samples

ments, developed along the northern part of the cross section. These sediments were buried by shaly and marly sediments in late Badenian times. Early Badenian conglomerates and sandstones and middle Badenian algal reef limestones are potential reservoir rocks.
- *Sarmatian* (events 17–20): Sarmatian sediments were deposited in a sea with reduced salinity. Normal fault tectonics along the northern margin caused the formation of the horst-structure in the Ludersdorf area (L2 in Fig. 7.46) during Sarmatian times.
- *Pannonian* and *Pontian* (events 21–23): Marine conditions were replaced by limnic and fluvial conditions with deposition of intercalations of shaly and sandy sediments.
- *Pliocene* and *Quaternary* (events 24–28): Uplift started in Pliocene times and resulted in erosion of 260–340 m of sediment.

Heat Flow Model

There are only few calibration data along the studied cross section. However, according to the well-calibrated heat flow model presented in Fig. 7.47 the following heat flow model is considered as the most probable one (heat flow model "middle" in Fig. 7.47).

The influence of high Miocene heat flows around the volcanic centres was restricted to a relatively narrow halo. It is therefore thought that the area of cross section III, which is located more than 10 km from the next volcanic centre, was not influenced by extremely raised heat flows during Karpatian and Badenian times. An Ottnangian to middle Badenian heat flow of 120 mW/m^2 seems reasonable. Probably heat flows decreased afterwards until the end of the middle Miocene, when they reached their present values. Present heat flow along the northern end of the transect is about 75 mW/m^2 and decreases toward the central and southern part to 63 mW/m^2.

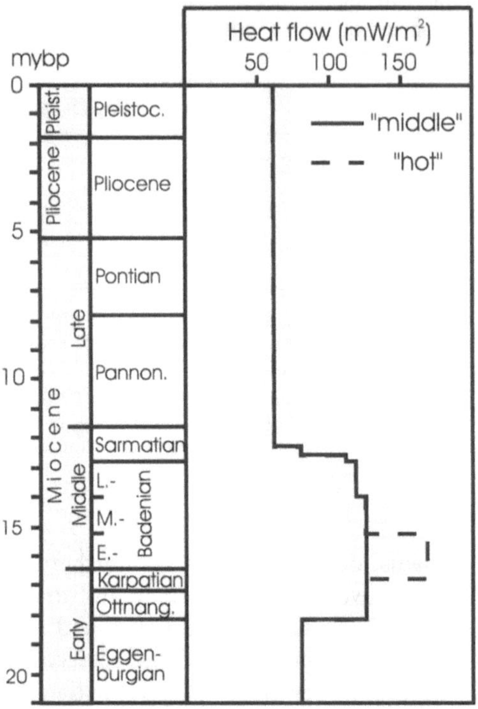

Fig. 7.47. Heat flow model applied to the central and southern part of cross section III. *Thick line* (*"middle"*), the original scenario; *dotted line*, a "hot" scenario; *shaded area*, range of geological meaningful heat flow histories. Heat flow along the shallow northern part of the cross section is 75 mW/m^2 since late Miocene (Sarmatian) times in all scenarios

Hydrocarbon Generation and Migration

The following hydrocarbon generation and migration histories were calculated using the heat flow model "middle". Results on the evolution of temperature, vitrinite reflectance and oil and gas potential of three stratigraphic levels versus time are displayed in Fig. 7.48a. All calculations are for gridpoint 17, which is situated about 2.5 km north of the southern end of the transect (Fig. 7.45).

At this location main oil generation from Ottnangian sediments (layer 3) occurred during early to middle Badenian times, main gas generation occurred during late Badenian to early Sarmatian times. Hydrocarbon generation from Karpatian levels started at the same time, but continued into late Badenian (layer 5) or even early Sarmatian times (layer 8). Gas generation is limited to the lowermost Karpatian layers. Only small amounts of hydrocarbons were generated since Sarmatian times.

According to the calculations, migration started contemporaneously in early Badenian times (Fig. 7.49). Due to the lower density the petroleum migrated vertically upward. In addition, migration was accelerated by excess pressure in Ottnangian sediments at the southern margin of the profile (about 0.2 MPa). This excess pressure was even higher (about 1 Mpa) during late Karpatian times. A great part of these early generated and vertically moving hydrocarbons was lost to the surface. From middle Badenian times on, hydrocarbons were trapped in lower Badenian sediments. The change of flow direction is a consequence of the difference between low capillary pressures of early Bade-

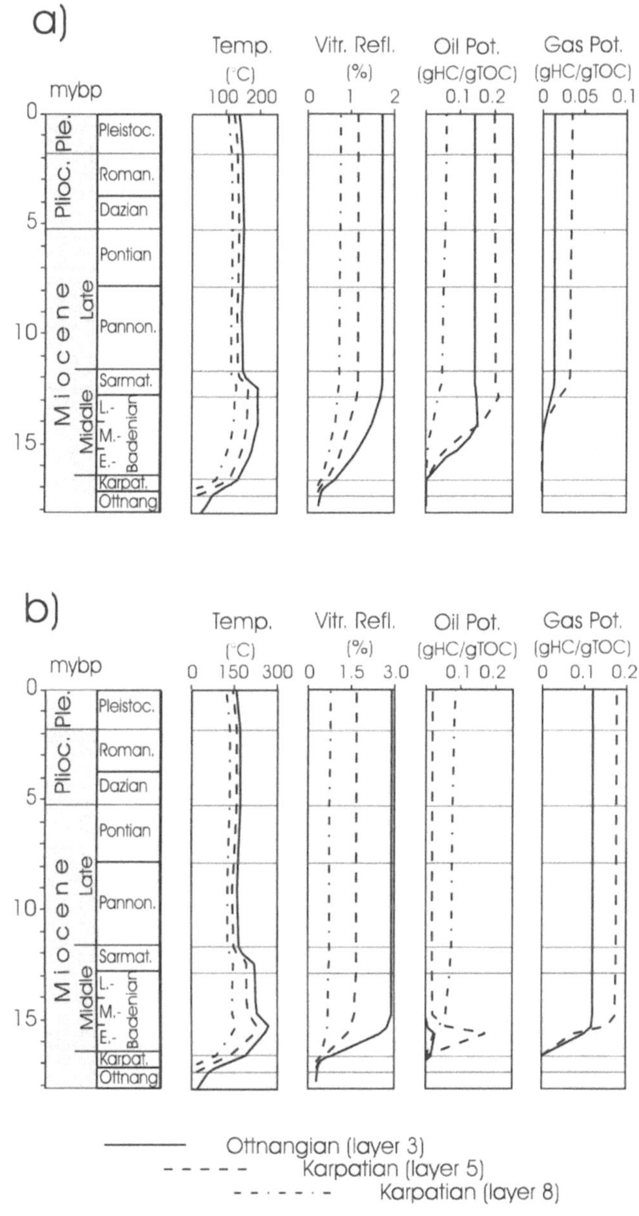

Fig. 7.48a,b. Calculated temperature, vitrinite reflectance, oil and gas generation histories for three lower Miocene levels of gridpoint 17 (*GP17*), cross section III. **a** Heat flow scenario "middle". **b** Heat flow scenario "hot" (petroleum potential of the Ottnangian layer was calculated using kinetic data of type III kerogen (Espitalié et al. 1988), petroleum potential of Karpatian layers was calculated using kinetic data of the Karpatian basinal facies

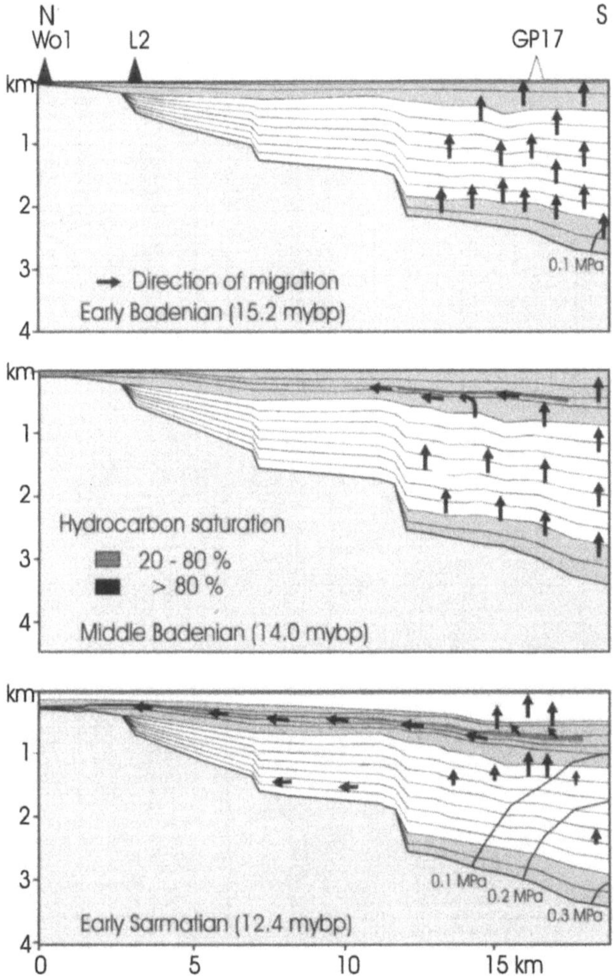

Fig. 7.49. Petroleum saturation and migration along cross section III for middle Miocene and present times. Isolines show excess pressure (MPa)

nian reservoir rocks (sand- to siltstones) and high capillary pressures of middle Badenian seal rocks (sandy calcareous shales). In Sarmatian times extensional tectonics caused an increase in sedimentation rates along the southern part of the cross section and the development of a horst structure in the Ludersdorf area. This situation led to the development of regions with excess pressure in the south (up to 0.4 Mpa) and the northward migration of hydrocarbons to the Ludersdorf area. At the same time the overpressure caused by the accumulated hydrocarbons in the southern part of the profile initiated a breakthrough of hydrocarbons through the middle Badenian seal. These hydrocarbons were partly lost to the surface and partly trapped in

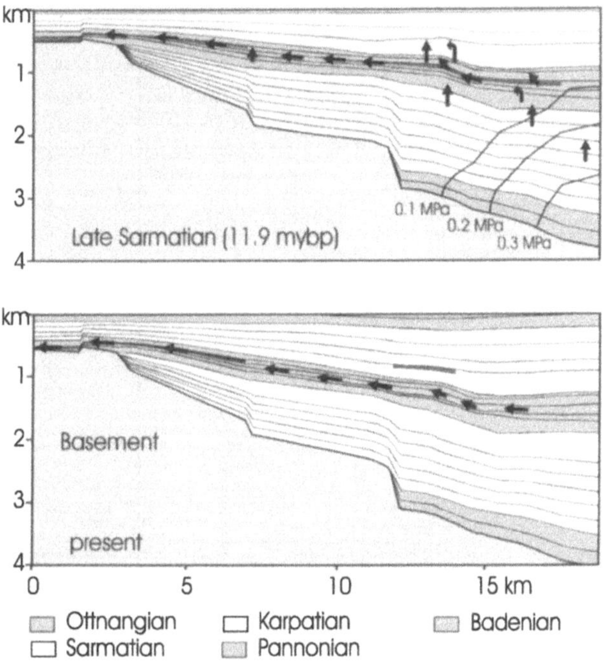

Fig. 7.49 (*Contd.*)

Sarmatian reservoirs. The northward migration of hydrocarbons continued until the present. Some hydrocarbons were lost along the northern margin of the cross section.

It is noteworthy that the simulation predicts the drilled hydrocarbon deposit in the Ludersdorf area in a correct way. However, the deposit contains gas in reality whereas oil and gas (mainly oil!) are predicted by the model. Possible explanations for this discrepancy include: (a) Karpatian and Badenian heat flows were higher than assumed in the model (see sensitivity analysis), (b) additional gas migrated from deeper parts of the basin into the Ludersdorf structure, (c) generated hydrocarbons were expelled out of sediments of the Karpatian basinal facies only after oil to gas cracking because of the low expulsion efficiency of these sediments (Sachsenhofer, 1994), and (d) the kerogen type of Karpatian sediments along cross section III is different from the kerogen type of the studied Karpatian samples in the Pichla area.

A quantification of hydrocarbon generation, accumulation in carrier beds and hydrocarbon losses to the surface through time is presented in Fig. 7.50. According to the calculations, a total of about 40 wt% of the initial hydrocarbon potential of the Ottnangian and Karpatian source rocks has been generated up to now (Fig. 7.50). The generation lasted from early Badenian to Sarmatian times. About 3% of the initial hydrocarbon potential was lost to the surface during early Badenian time. More hydrocarbons (about 10%) were lost during early Sarmatian time. Until late Badenian time about 20% of the initial

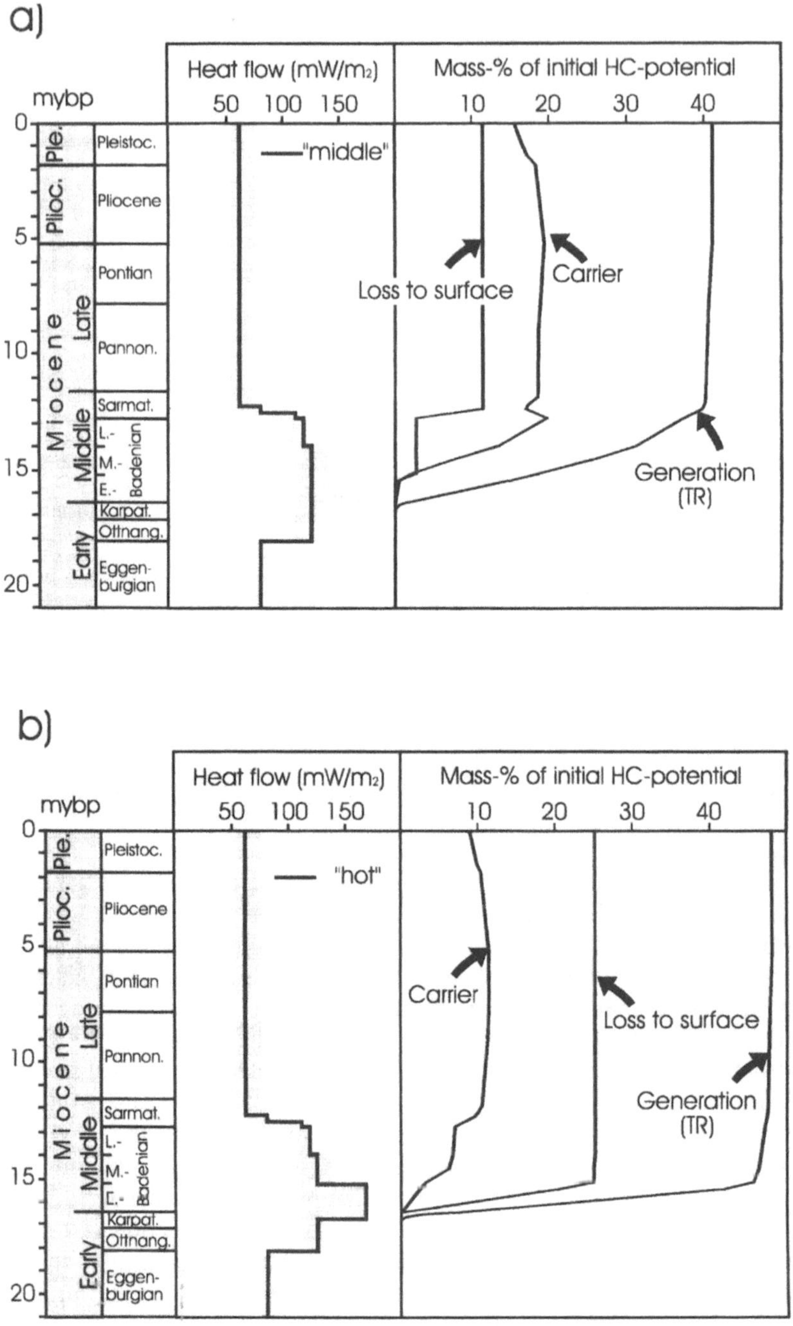

Fig. 7.50a,b. Hydrocarbon generation, accumulation in carrier beds (post-Karpatian sediments) and hydrocarbon losses to the surface through time. **a** Quantification based on the original (and most probable) heat flow model "middle". **b** Quantification based on the alternative heat flow model "hot" from Fig.7.48b. (Note: The sum of accumulated and lost hydrocarbons is not equal to the generated hydrocarbons. This is because some hydrocarbons are held back in the source rocks, and some hydrocarbons are lost to the basin sides. Moreover a mass loss during the oil-gas conversion must be considered)

hydrocarbon potential accumulated in carrier rocks (these are all rocks younger than Karpatian). Afterwards a part of these hydrocarbons was lost to the surface. This loss was partly replaced by newly generated and expelled hydrocarbons. Northward migration and losses along the northern margin of the cross section resulted in a decrease in accumulated hydrocarbons from Pliocene times on.

Sensitivity Analysis

In the framework of a sensitivity analysis the heat flow history along the transect was modified, accounting for a "hot" scenario (Fig. 7.48), which considers the highest geological meaningful Miocene heat flow. This value is derived from the Walkersdorf 1 well; "W" in Fig. 7.45 where heat flow did not exceed 170 mW/m^2, although the well is situated close to a volcanic centre.

The calculated temperature and maturation histories at gridpoint 17 differ significantly from the original scenario. Higher Karpatian and early Badenian temperatures result in an early generation phase, which ended already during the early middle Badenian and in a higher gas/oil ratio (Fig. 7.48). Only small amounts of hydrocarbons were generated after the early Badenian.

Although the general migration pattern is similar to the original scenario, the quantification of hydrocarbon generation and accumulation shows significant differences (Fig. 7.50). Because of higher late Karpatian and early Badenian temperatures more hydrocarbons were generated (about 48% of the primary potential). However, most hydrocarbons were lost to the surface in early Badenian time. Because of the weak post-early Badenian generation only about 10% of the primary hydrocarbon potential was accumulated in carrier rocks. The pressure build-up by the relatively small amount of accumulated hydrocarbons did not result in a Sarmatian break-through through the middle Badenian seal. Therefore, and in contrast to the original scenario, no hydrocarbons were lost to the surface during the Sarmatian. Again some hydrocarbons were lost since the Pliocene along the northern basin side.

Conclusions

The heat flow evolution and the time-relation between hydrocarbon generation and trap formation have an important effect on the hydrocarbon potential of the Styrian Basin. The elevated heat flow in close vicinity of Miocene volcanoes resulted in an early (Karpatian to early Badenian) petroleum generation phase near the volcanoes. Because no traps were formed at that time, the generated hydrocarbons will have been lost. Hydrocarbons generated along cross section III had a chance to become trapped and form accumulations. The calculated amount of accumulated hydrocarbons depends on the applied heat flow history. Although more hydrocarbons are generated in the hot scenario, only small deposits evolve and most hydrocarbons are lost to the surface. In the case of the scenario "middle", which is considered as the most probable one, more hydrocarbons are trapped. This is above all a consequence of the relatively late generation phase. Because of higher early Badenian temperatures and the loss of early generated oil during early Badenian times, a more realistic higher gas/oil ratio is calculated for the Ludersdorf deposit in the "hot" scenario.

Gas migration in the Northwest German Basin

Geological Setting

The geological evolution of the Northwest German Basin, the burial history of Carboniferous source rocks and the heat flow history within the study area is summarised according to the following five principal stages.

- Stage 1: Deposition of thick coal-bearing strata occurred during the late Namurian and early Westphalian. These units are the source rocks for most of the gas in the Northwest German Basin. Also, some commercial gas fields exist in Carboniferous sandstones.
- Stage 2: After a short period of erosion, deposition of volcanoclastics (0–600 m) occurred during the Lower Permian (Autunian). This time of enhanced heat flows represents also the first gas generation from the Carboniferous strata. It is followed by fluvio-deltaic sedimentation during the Saxonian. This upper part of the Lower Permian "Rotliegend" series contains the most prolific reservoir sandstones of the area.
- Stage 3: Marine evaporite sequences characterise the following Upper Permian ("Zechstein") which serves as the major regional seal.
- Stage 4: During the Triassic and Jurassic, the depositional environment fluctuates between shallow marine and continental. This period represents the second phase of gas generation in the Carboniferous and is also the time of the beginning doming of Zechstein salts, which caused the formation of many important gas traps. Reservoir rocks occur in the Lower Triassic, whereas the overlying formations are of limited significance with respect to gas accumulation.
- Stage 5: After some erosion during Early Cretaceous times, sedimentation continued and maximum depths were reached by all formations in the Cenozoic. The Cenozoic is the third period of gas formation from the Carboniferous source rocks.

The gases in the Lower Permian as well as in the Triassic and Carboniferous reservoir sandstones are characterised by variable, commonly high, nitrogen concentrations (<5 to >90%). The nitrogen province in which Lower Permian reservoirs contain less methane than molecular nitrogen is shown in Fig. 7.51. As also shown in this figure, the high nitrogen concentrations occur in an area, where the maturity at the top of the Carboniferous exceeds 3% vitrinite reflectance. Krooss et al. (1995), Krooss and Leythaeuser (1996) and Littke et al. (1995) evaluated the coaly organic matter of the Carboniferous as possible source for the molecular nitrogen in the region and concluded "that nitrogen generation from organic matter is one, probably the most important, source for nitrogen gas, provided that sufficient organic matter at high levels of ma-

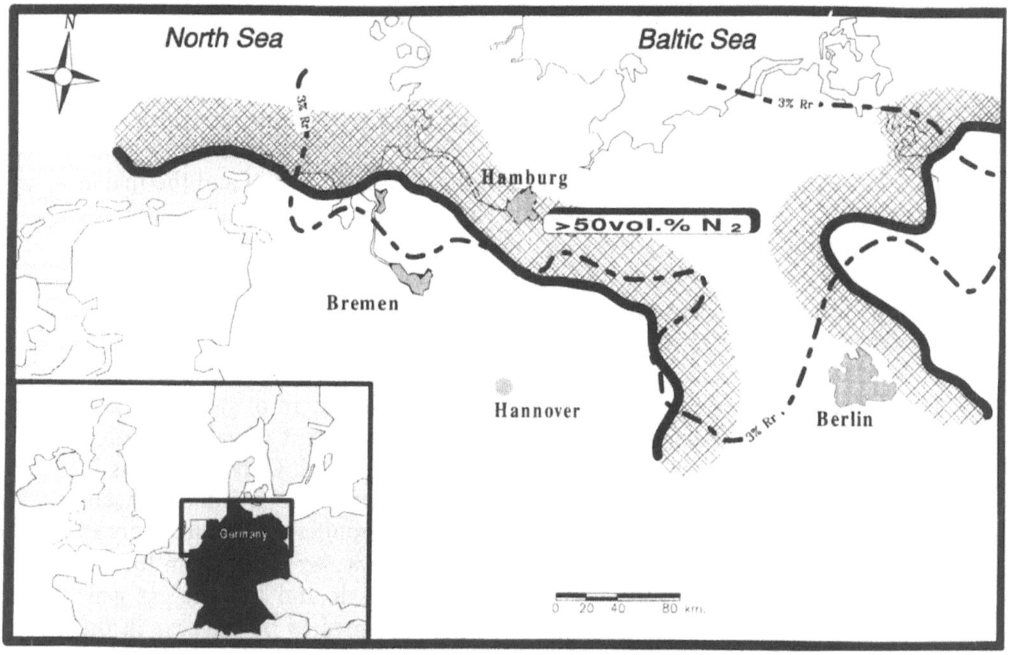

Fig. 7.51. Regional distribution of nitrogen-dominated gases in Rotliegend (Lower Permian) reservoirs. Underlying source rocks for methane are coal-bearing Westphalian strata, except for the southeastern part of the area, where the Westphalian is missing. *Area partly shaded gray*, nitrogen province, where molecular nitrogen percentages in Lower Permian reservoirs generally exceed methane percentages. (Compilation from Boigk and Stahl 1970; Teichmüller et al. 1984; Eiserbeck et al. 1992)

turation is attained" (Littke et al. 1995). In essence, coals generate upon maturation about 20 times more methane than nitrogen, but the ratio of nitrogen over methane increases with increasing maturation, and becomes greater than one at maturity stages corresponding to vitrinite reflectance levels above 3%.

For applications in petroleum exploration the knowledge of this sequence of events is useful only if it is possible to evaluate whether a given gas in a reservoir accumulated during early or late generation phases. A prerequisite for this reconstruction of the timing of gas accumulation is, however, the knowledge of the temperature history of the source rock horizons and of the related gas generation. These aspects are presented below. For the simulation of the migration of both methane and molecular nitrogen the physical parameters described in chapter 2 (section 2.6.6) section xy for dry gas were used, as well as the kinetics of methane and nitrogen generation from coaly organic matter which is presented in the kinetics chapter (see Krooss and Leythaeuser 1996).

Gas Saturation and Migration

Within the study area the first gas generation occurred during the early Permian (Autunian) as discussed in the heat flow chapter. This first gas generation from the Westphalian and Namurian source rocks was due to the volcanism and high heat flow and happened despite the low burial depths of the coal-bearing sequences. However, the present traps in the Saxonian and the major Zechstein seals were not yet deposited. Therefore the model predicts that gas accumulations could not persist and that the early gas generated during this period was completely lost to the atmosphere (Figs. 7.52, 7.53). This first gas was mainly methane, regardless of whether the Namurian or the Westphalian are considered as effective source rock formations.

Much greater burial depths were reached during the Triassic and Jurassic, representing the second phase of gas generation. The extent of gas generation during this period depends to a great extent on the extent of gas generation during the previous, early Permian gas generation period. Only in regions in which the Permian heat flows remained on a "moderate" level will significant gas generation have occurred from the Carboniferous source rocks (Fig. 7.53A,B). Where, however, Autunian heat flows were very high, for example, in the direct vicinity of volcanoes, the Triassic and Jurassic gas generation remained insignificantly small. This is assumed to be the case in the northern part of the study area, where volcanic rocks reach a thickness of 600 m and more (Fig. 7.53C,D). The gas generated during the Triassic and Jurassic was predominantly methane at all locations. Part of the present trap structures were already available at that time, but due to the fact that salt doming only started, another part of the traps was not yet constructed. Furthermore, gas accumulations were possibly destroyed during the following phase of uplift and erosion, which affected the basin in late Jurassic/early Cretaceous times.

Greatest burial depths were reached during the Cenozoic which is accordingly regarded as the third period of gas generation. The composition and total volume of gas generated at that time strongly depends on the previous temperature history of the Carboniferous source rocks: In areas in the south, where the methane generation potential was not yet completed great amounts of methane and only small amounts of nitrogen were generated (Fig. 7.52A,B). In the north, in contrast, the methane generation from the source horizons was already almost finished and mainly molecular nitrogen was generated (Figs. 7.53C,D).

The gas composition of the individual reservoirs in the north and south of the study area depends certainly not only on the gas charge from directly underlying source rocks by purely vertical migration, but also on the lateral migration within the most permeable rock types. According to the simulation results, this lateral migration is generally directed from north to south at present (Fig. 7.52B) with some exceptions. If this lateral migration had been the dominant transport processes for the fluids in the subsurface, molecular nitrogen generated in the kitchen area in the north would be expected to occur in all reservoirs in high percentages. This scenario is, however, not predicted

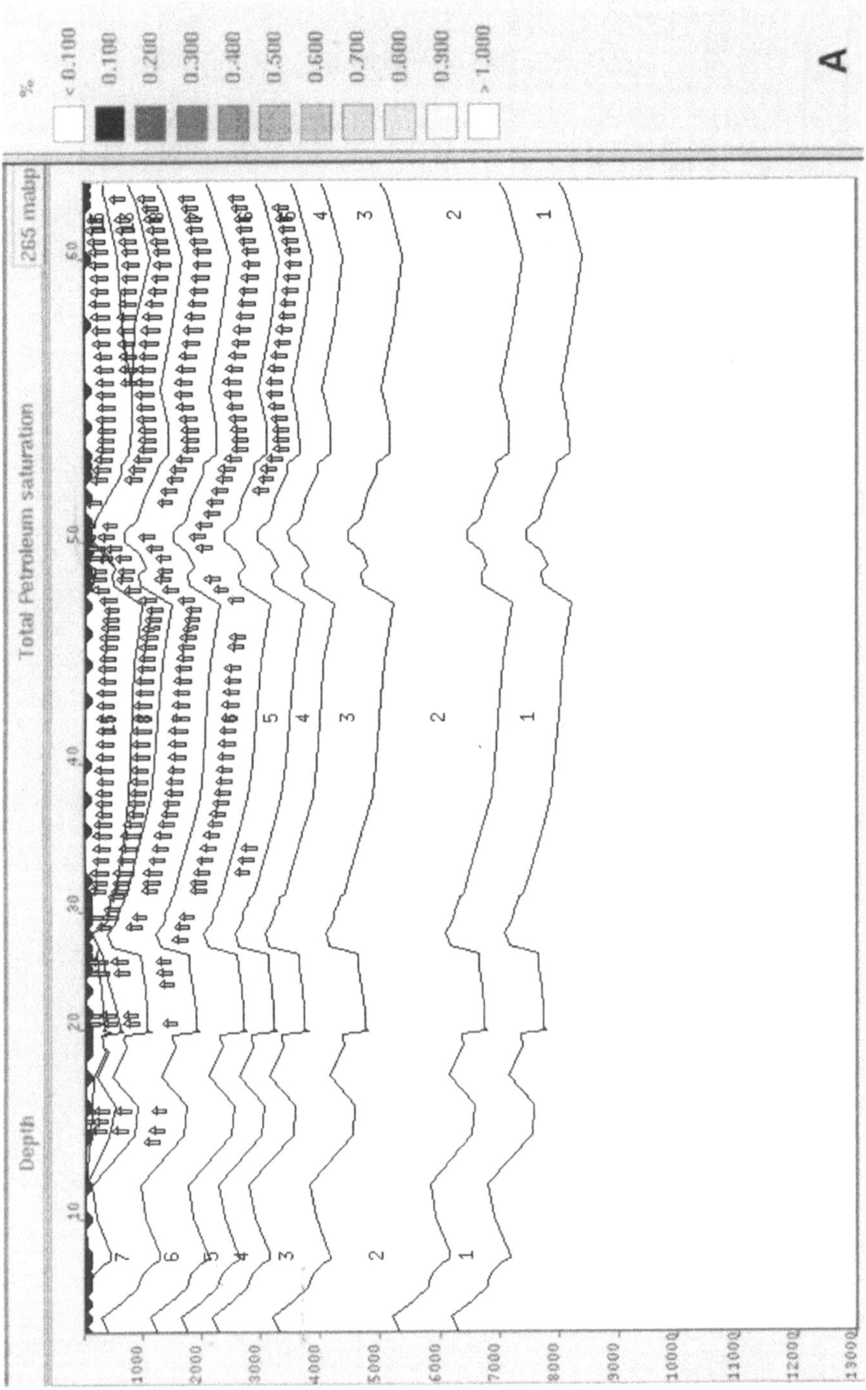

Fig. 7.52A,B. Gas saturation and migration as simulated along a 200-km-long geological transect in northern Germany. **A** At the end of the Lower Permian period which was characterized by high heat flows. **B** At present

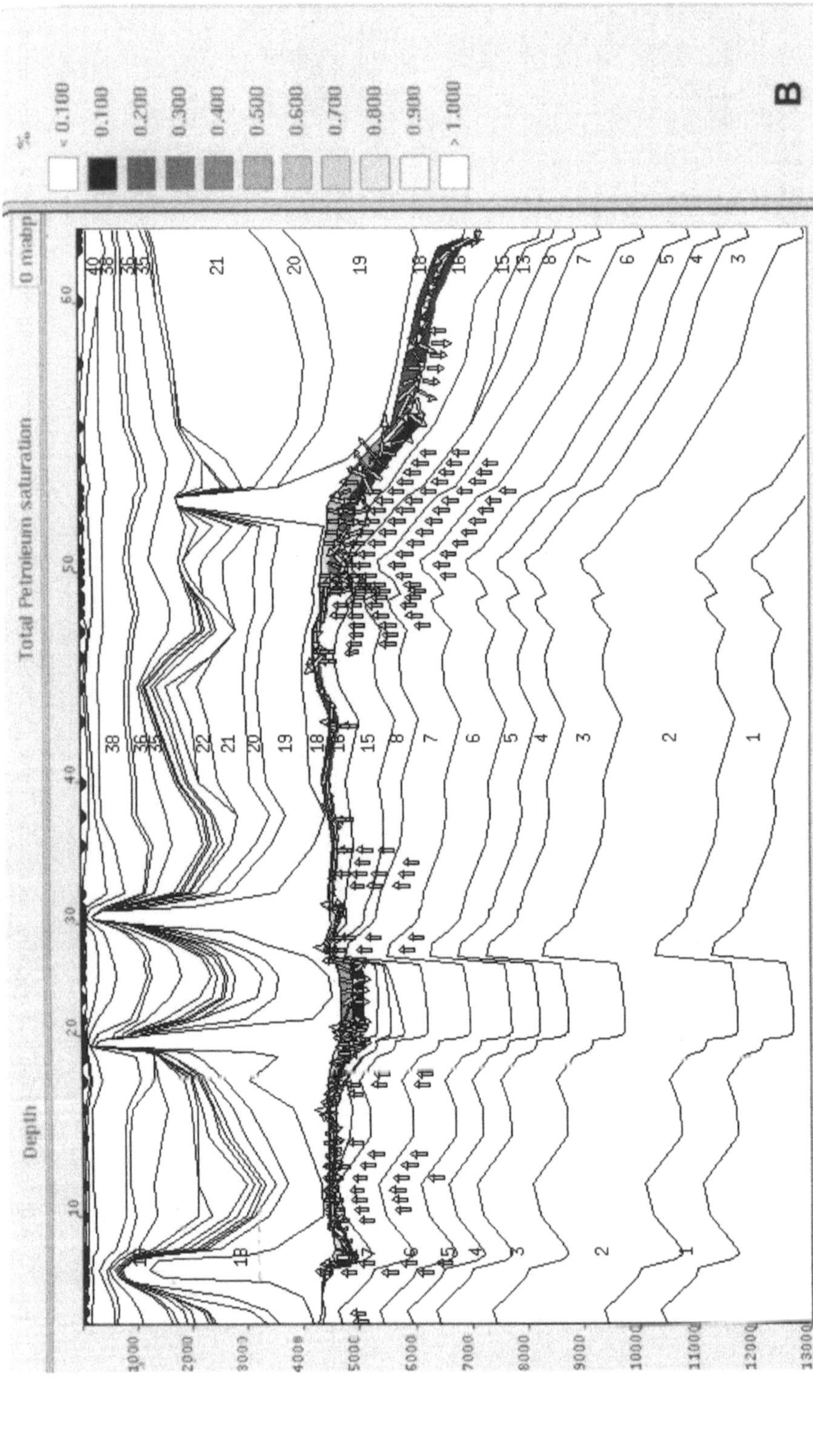
Depth
Total Petroleum saturation
0 mabp
%
< 0.100
0.100
0.200
0.300
0.400
0.500
0.600
0.700
0.800
0.900
> 1.000
B

Fig. 7.52B

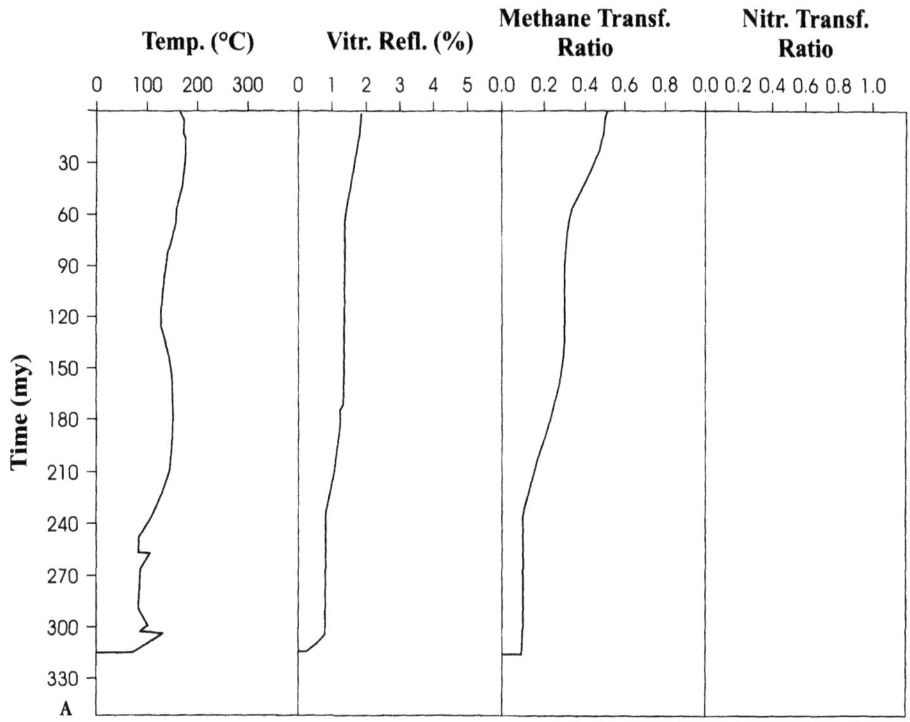

Fig. 7.53A-D. Evolution of temperature, vitrinite reflectance, methane generation and nitrogen generation for the lowermost Westphalian and the Namurian at a site in the south of the study area (**A**, **B**) and for a site further north (**C**, **D**)

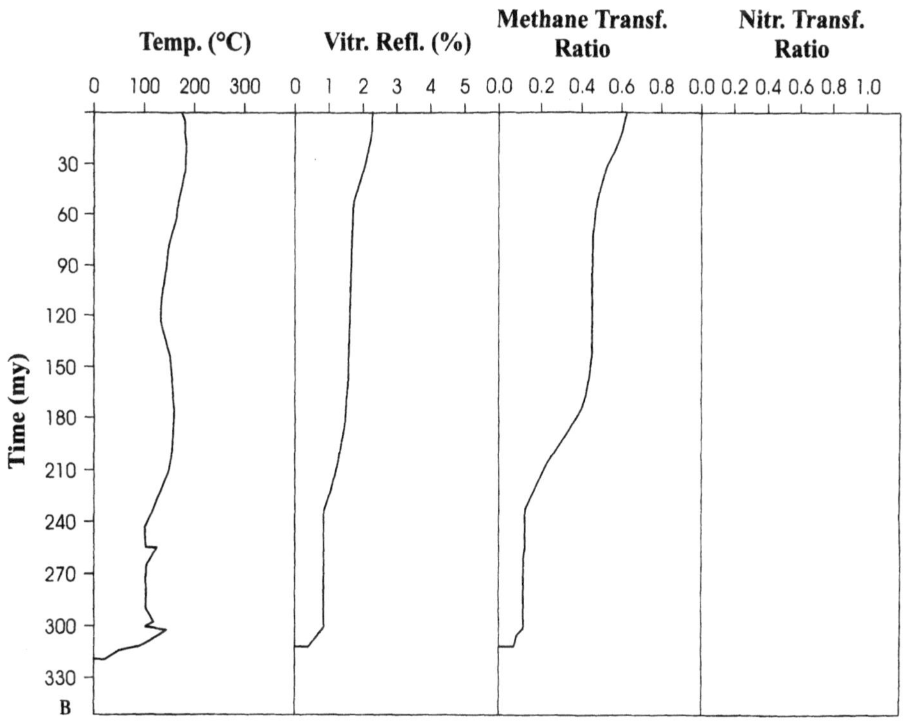

Fig. 7.53B

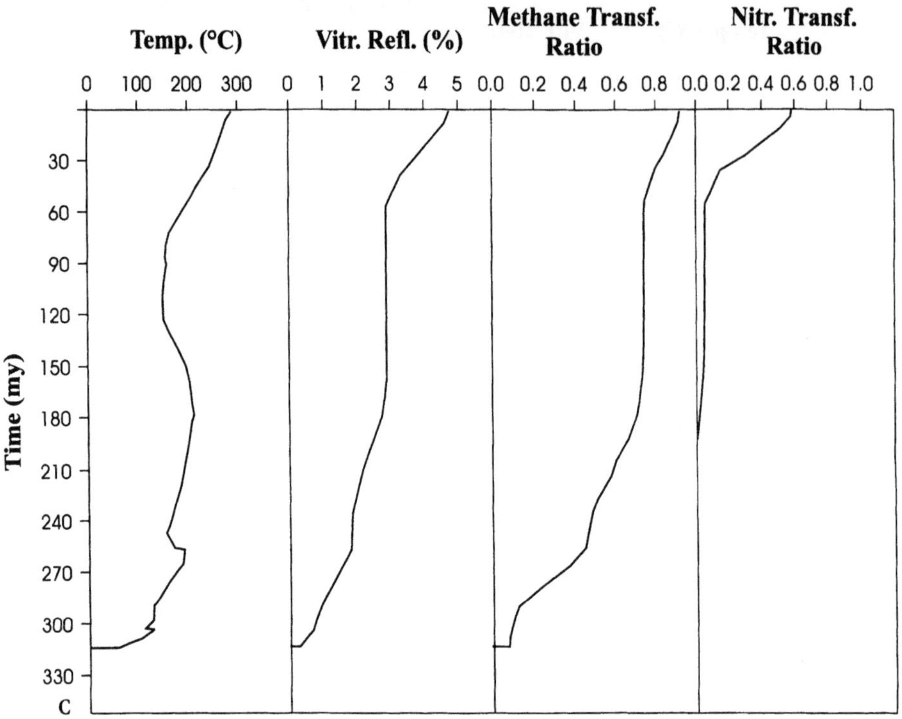

Fig. 7.53C

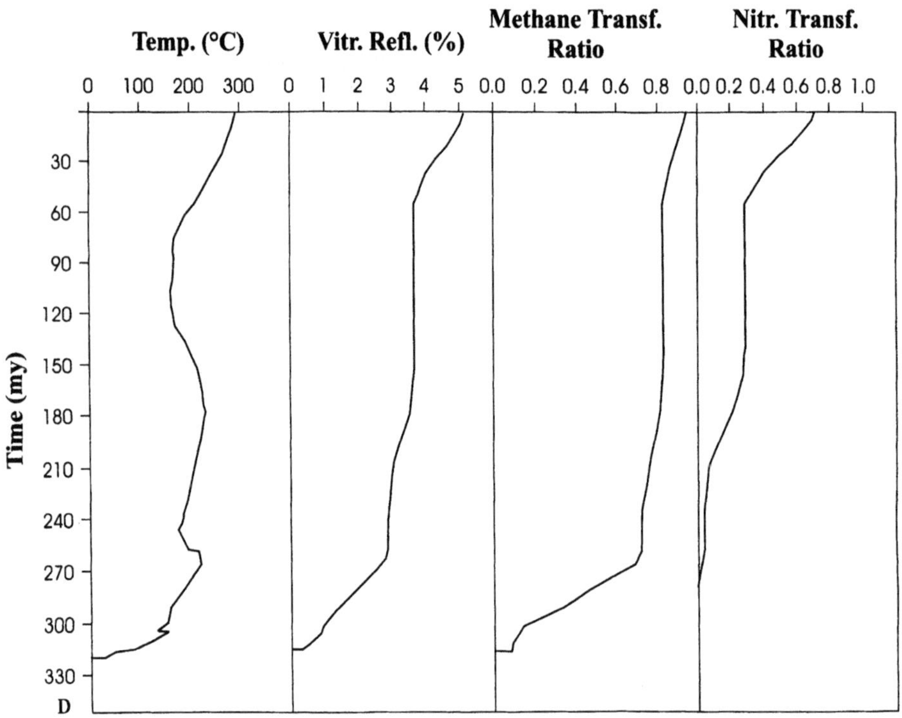

Fig. 7.53D

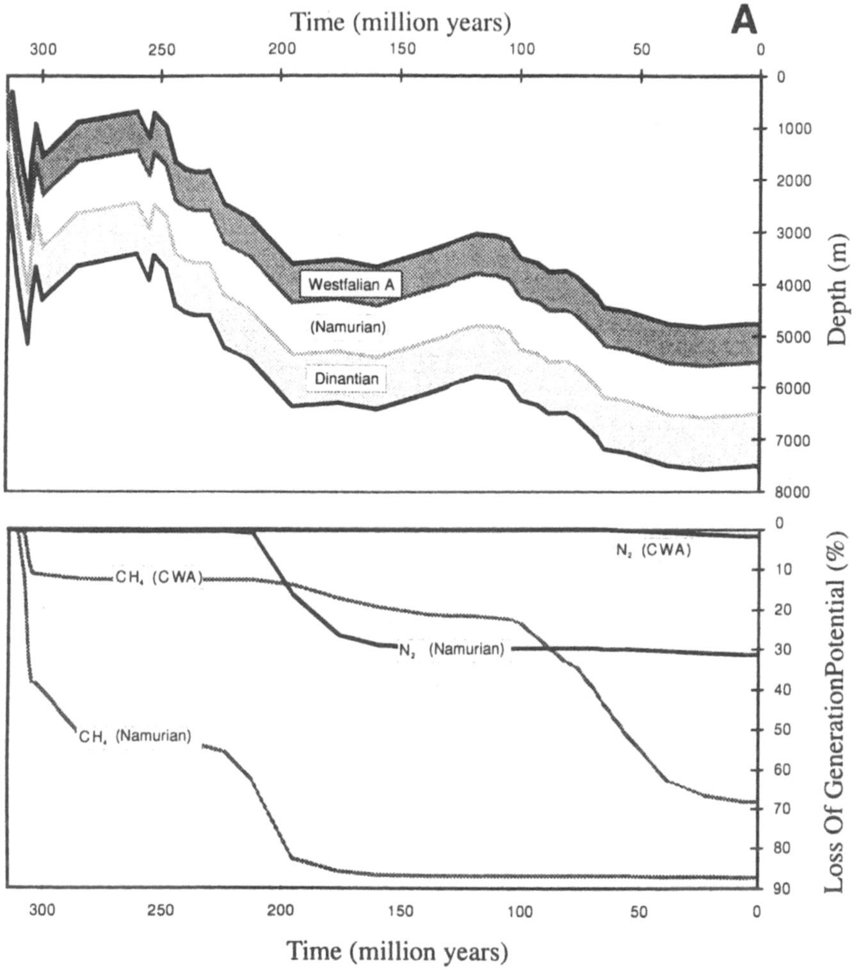

Fig. 7.54. Dynamics of methane and nitrogen generation and expulsion since the Carboniferous for one site situated in the south of the study area (**A**) and one site in the north (**B**). *Above*, for each of the two sites the burial history since the Carboniferous (note the different depth scales); *below*, the loss of initial gas generation potential (methane and molecular nitrogen) for the lowermost Westphalian and the Namurian

by the numerical two-dimensional simulation mainly because of the lack of homogeneous, widespread and highly permeable lithologies such as sandstones. The major lithologies in the uppermost Carboniferous and lowermost Permian are interlayered shales and sandstones, which are of limited lateral continuity and do not allow long-distance lateral migration. Furthermore, faults either act as vertical barriers or conduits for vertical migration.

Therefore, near-vertical migration predominates over long-distance lateral migration according to the simulation.

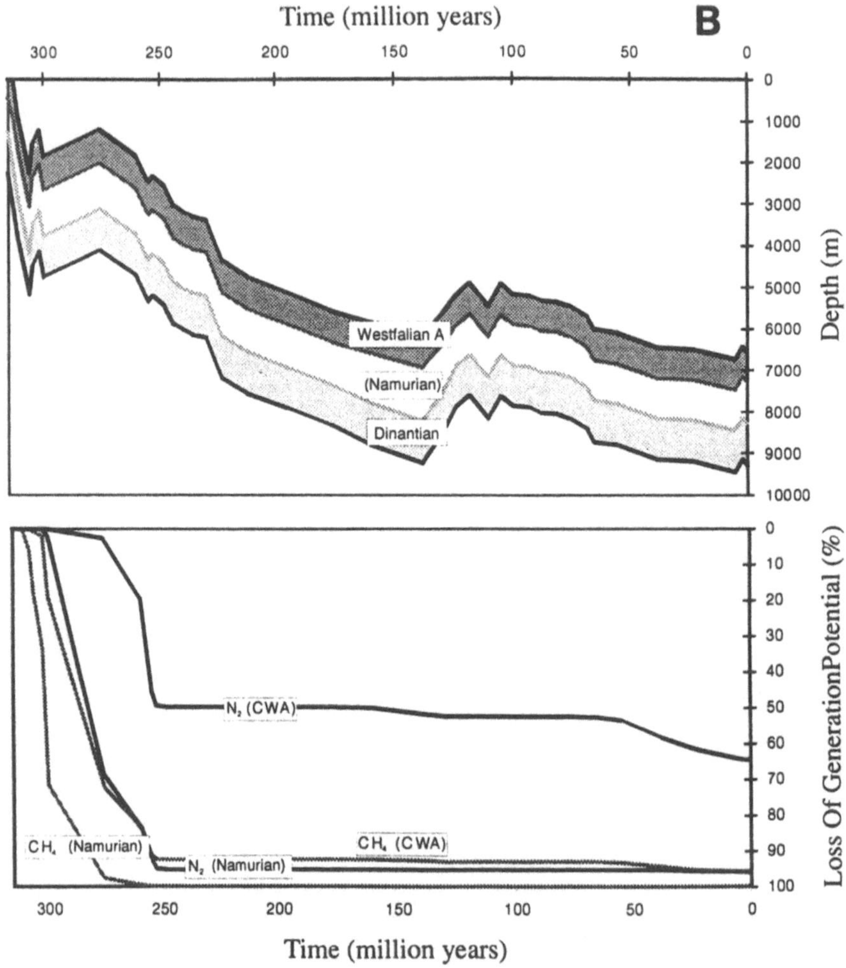

Fig. 7.54B

With respect to the timing of gas accumulation in the Northwest German Basin the numerical simulations provide – together with the kinetics of methane and nitrogen generation from coal – some interesting clues for hydrocarbon exploration. When the methane/nitrogen volume ratios found in the present reservoirs are compared to those calculated by the modelling for the three different periods of gas generation (Fig. 7.54), it is clear that the gas generated during the Permian is overwhelmingly methane. A similar composition is calculated for the gas generated during Triassic/Jurassic times, and both gas compositions contrast dramatically with the present gas composition. Also, a mixture of all gas generated during the three generation periods would result in a methane dominated gas (> 95 vol%) at all locations, which sharply contrasts with reality. A reasonable, if imperfect match between real gas compositions and modelled gas compositions is achieved only if a loss of all pre-Cenozoic gas from the basin (into the atmosphere) is assumed, and if only the latest, Cenozoic gas fills the reservoirs. This scenario results in low predicted percentages of molecular nitrogen in the southern reservoirs (about 5%) which closely fits to data from nearby gas fields, and in high percentages of molecular nitrogen in the north (> 50%), where real percentages of molecular nitrogen approach 90%. A more nearly perfect fit would require a better knowledge of the masses of organic matter in the deep-lying Palaeozoic horizons and a better constrained kinetics for nitrogen and methane generation at very high levels of maturation (corresponding to > 3% vitrinite reflectance). The general pattern of the accumulation and distribution of nitrogen-rich and nitrogen-lean gases in the Northwest German Basin, as shown in Fig. 7.54, is, however, well explained by the above described modelling of gas generation and migration.

References

Amyx JW, Bass DM Jr, Whiting RL (1960) Petroleum reservoir engineering. McGraw-Hill, New York

Anderson RN, Flemings P, Losh S, Austin J, Woodhams R (1994) Gulf of Mexico growth fault drilled, seen as oil, gas migration pathway. Oil Gas J 6: 97–104

Ayres MG, Bilal M, Jones RW, Slentz LW, Tartir M, Wilson AO (1982) Hydrocarbon habitat in main producing areas, Saudi Arabia. AAPG Bull 66: 1–9

Aziz K, Settari A (1985) Petroleum reservoir simulation. Elsevier, London, 476 pp

Bally AW (1975) A geodynamic scenario for hydrocarbon occurences. Proc 9th World Petroleum Congress Tokyo 2: 33–34

Barker C (1972) Aquathermal pressuring – role of temperature in development of abnormal-pressure zones. AAPG Bull 56: 2068–2071

Barker CE, Halley RB (1986) Fluid inclusion, stable isotope and vitrinite reflectance evidence for the thermal history of the Bone Spring Limestone, southern Guadaloupe Mountains, Texas. In: Gautier DL (ed) Roles of organic matter in sediment diagenesis. SEPM Spec Publ 38: 129–203

Baum GR, Vail PR (1988) Sequence stratigraphic concepts applied to paleogene outcrops, Gulf and Atlantic basins. In: Wilgus CK et al. (eds) Sea-level changes: an integrated approach. SEPM Spec Publ 42: 209–327

Berg RR (1975) Capillary pressures in stratigraphic traps. AAPG Bull 59: 939–956

Bjørlykke K (1994) Fluid-flow processes and diagenesis in sedimentary basins. In: Parnell J (ed) Geofluids: origin, migration and evolution of fluids in sedimentary basins. Geological Society 1974, London, Spec Publ 78, pp 127–140

Bjørlykke K, Ramm M, Saigal GC (1986) Sandstone diagenesis and porosity modification during basin evolution. Geol Rundsch 78: 243–268

Boigk H, Stahl W (1970) Zum Problem der Entstehung nordwestdeutscher Erdgaslagerstätten. Erdöl-Kohle-Erdgas Petrochem 23: 325–333

Brace WF (1980) Permeability of crystalline and argillaceous rocks. Int J Mech Min Sci Geomech Abstr 17: 241–251

Bray EE, Foster WRA (1980) Process for primary migration of petroleum. AAPG Bull 64: 107–114

Brederhoeft JD, Norton DL (1990) Mass and energy transport in a deforming earth's crust. In: The role of fluids in crustal processes. Studies in geophysics. National Academy, Washington, pp 27–41

Bruce CH (1984) Smectite dehydration – its relation to structural development and hydrocarbon accumulation in northern Gulf of Mexico. AAPG Bull 68: 673–683

Burnham AK, Sweeney JJ (1992) Influence of compaction models on petroleum expulsion. 203rd Am Chem Soc Natl Meet, 5–10 April 1992, San Francisco, Division of Geochemistry, Pap No 101

Burrus J, Osadetz K, Gaulier JM, Brosse E, Doligez B, Choppin de Janvry G, Barlier J, Visser K (1993) Source rock permeability and petroleum expulsion efficiency: modelling examples from the Mahakam delta, Williston Basin and the Paris Basin. In: Parker JR (ed) Petroleum geology of northwest Europe. Geol Soc Lond, pp 1317–1332

Burruss RC, Cercone KR, Harris PM (1985) Timing of hydrocarbon migration: evidence from fluid inclusions in calcite cements, tectonic and burial history. In: Schneidermann N, Harris PM (eds) Carbonate cements. SEPM Spec Publ 26: 277–289

Burst JR (1969). Diagenesis of Golf coast clayey sediments and its possible relation to petroleum migration. AAPG Bull 53: 73–93

Catalan L, Xiaowen F, Chatzis I, Dullien FAL (1992) An experimental study of secondary oil migration. AAPG Bull 76: 638–650

Chung HM, Claypool HE, Rooney MA, Squires RM (1994) Source characteristics of marine oils as indicated by carbon isotope ratios of volatile hydrocarbons. AAPG Bull 78: 396–408

Clauser C, Neugebauer HJ (1991) Thermisch relevante Tiefenwasserzirkulation in der Oberkruste unter dem Oberrheingraben? Eingrenzungen mit Hilfe hydrothermischer Modellrechnungen. Geol Jahrb E48: 185–217

Connan J, Cassou AM (1980) Properties of gases and petroleum liquids derived from terrestrial kerogen at various maturation levels. Geochim Cosmochim Acta 44: 1–23

Connan J, Bouroullec J, Dessort D, Albrecht P (1986) The microbial input in carbonate-anhydrite facies of a sabkha paleoenvironment from Guatemala: a molecular approach. In: Leythaeuser D, Rullkötter J (eds) Advances in organic geochemistry 1985. Pergamon, Oxford. Org Geochem 10: 29–50

Cooles GP, Mackenzie AS, Quigley TM (1986) Calculation of petroleum masses generated and expelled from source rocks. Org Geochem 10: 235–245

Corbett PWM, Ringrose PS, Jensen JL, Sorbie KS (1992) Laminated clastic reservoirs – the interplay of capillary pressure and sedimentary architecture. Soc Petrol Eng Annu Tech Conf, Washington, DC, 4–7 Oct 1992, Pap SPE 24699

Cornford C (1984) Source rocks and hydrocarbons of the North Sea. In: Glennie KW (ed) Introduction of the petroleum geology of the North Sea. Blackwell, Oxford, pp 171–209

Craig FF (1971) The reservoir engineering aspects of water flooding. Soc Petrol Eng, Monogr Ser 3, Richardson, Texas

Curtis CD (1978) Possible links between sandstone diagenesis and depth-related geochemical reactions occurring in enclosing mudstones. J Geol Soc Lond 135: 107–118

Curtis CD (1983) Geochemistry of porosity enhancement and reduction in clastic sediments. In: Brooks J (ed) Petroleum geochemistry and exploration of Europe. Blackwell, Oxford, pp 113–126

Daines S (1982) Aquathermal pressuring and geopressure evaluation. AAPG Bull 66: 931–939

Demaison G (1984) The generative basin concept. In: Demaison G, Murries RJ (eds) Petroleum geochemistry and basin evaluation. AAPG Mem 35: 1–14

Demaison G, Huizinga BJ (1991) Genetic classification of petroleum systems. AAPG Bull 75: 1626–1643

Dickey PA (1975) Possible primary migration of oil from source rocks in oil phase. AAPG Bull 59: 337–345

Di Primio R (1990) Organisch-geochemische und mikroskopische Untersuchungen zur Kohlenwasserstoffgenese und primären Migration in karbonatischen Muttergesteinen der Trias Italiens. Diplomarbeit, RWTH Aachen

Dow WG (1974) Application of oil-correlation and source-rock data to exploration in Williston basin. AAPG Bull 58: 1253–1262

Düppenbecker SJ (1990) Genese und Expulsion von Kohlenwasserstoffen in zwei Regionen des Niedersächsischen Beckens unter besonderer Berücksichtigung der Aufheizraten. Dissertation, RWTH Aachen, 304 pp

Düppenbecker SJ, Welte DH (1992) Petroleum expulsion from source rocks – insights from geology, geochemistry and computerized numerical modelling. 13th World Petroleum Congr Proc 2: 165–177

Dubinin MM (1975) Progr Surface Membr Sci 9: 1

Durand B (1983) Present trends in organic geochemistry in research on migration of hydrocarbons. In: Bjorøy M et al. (eds) Advances in organic geochemistry. Wiley, Chichester, pp 117–128
Durand B (1988) Understanding of HC migration in sedimentary basins (present state of knowledge). Org Geochem 13: 445–459
Durand B, Oudin JL (1980) Example de migration des hydrocarbures dans une série deltaique: le delta de la Mahakam. 10th World Petroleum Congr Proc 2: 3–11
Durand B, Ungerer P, Chiarelli A, Oudin JL (1984) Modélisation de la migration de l'huile: application de deux exemples de bassins sédimentaires. 11th World Petroleum Congr Proc 2: 3–16
Durham LS (1994) Finding a field of streams - fluid flow concept could boost reserves. AAPG Expl May Issue, pp 1–17
Du Rouchet J (1981) Stress fields, a key to oil migration. AAPG Bull 65: 74–85
Eiserbeck WS, Mehlhorn S, Müller EP, Schimanski W, Schretzenmeyer S (1992) Zum Kohlenwasserstoffpotential des Präperm in den östlichen Bereichen der Bundesrepublik. Dtsch Wissenschaftliche Gesellschaft für Erdöl, Erdgas, Kohle (DGMK), Tagungsbericht 9203, pp 31–54
England WA, Mackenzie AS (1989) Some aspects of the organic geochemistry of petroleum fluids. In. Polechau HS, Mann U (eds) Geologic modelling - aspects of integrated basin analysis and numerical simulation. Geol Rundsch 78(1): 291–303
England WA, Mackenzie AS, Mann DM, Quigley TM (1987) The movement and entrapment of petroleum fluids in the subsurface. J Geol Soc Lond 144: 327–347
Espitalié J, Laporte JL, Madec M, Marquis F, Leplat P, Paulet J, Boutefeu A (1977) Méthode rapide de caractérisation des roches mères, de leur potentiel pétrolier et de leur degré d'évolution. Rev Inst Fr Pet 32: 23–42
Espitalié J, Ungerer P, Irwin I, Marquis F (1988) Primary cracking of kerogens. Experimenting and modeling C_1, C_{2-5}, C_{6-15}, and C_{15+} classes of hydrocarbons formed. Org Geochem 13: 893–899
Foscolos AE, Powell TG (1980) Mineralogical and geochemical transformation of clays during catagenesis and their relation to oil generation. Can Soc Petrol Geol Mem 6: 153–172
Freeze RA, Cherry JA (1979) Groundwater. Prentice-Hall, Englewood Cliffs, 604 pp
Fuex AN (1977) The use of stable carbon isotopes in hydrocarbon exploration. J Geochem Expl 7: 155–188
Gehmann HM (1962) Organic matter in limestones. Geochim Cosmochim Acta 26: 885–894
Gize AP (1993) The analysis of organic matter in ore deposits. In: Parnell J, Kucha H, Landais P (eds) Bitumen in ore deposits. Springer, Berlin Heidelberg New York, pp 28–52
Goff JC (1983) Hydrocarbon generation and migration from Jurassic source rocks in the E. Shetland Basin and Viking Graben of the northern North Sea. J Geol Soc Lond 140: 445–474
Grabowski GJ (1984) Generation and migration of hydrocarbons in Upper Cretaceous Austin Chalk, south-central Texas. In: Palacas JG (ed) Petroleum geochemistry and source rock potential of carbonate rocks. AAPG Stud Geol 18: 97–115
Graham SA, Williams LA (1985) Tectonic, depositional, and diagenetic history of Monterey Formation (Miocene), Central San Joaquin Basin, California. AAPG Bull 69: 385–411
Green AR (1983) Future petroleum province exploration. 11th World Petroleum Congr Proc 2: 167–177
Gussow WC (1954) Differential entrapment of oil and gas: a fundamental principle. AAPG Bull 38: 816–853
Haas J Jr (1987) An empirical equation with tables of smoothed solubilities of methane in water and aqueous sodium chloride solutions up to 25 weight percent, 360 °C and 138 Mpa. US GS Open-file Rep 78–1004
Halbouty MT (ed) (1970) Geology of giant petroleum fields: introduction. AAPG Mem 14, Am Assoc Petrol Geol, Tulsa, pp 1–7
Hanebeck D (1994) Experimentelle Simulation und Untersuchung der Genese und Expulsion von Erdölen aus Muttergesteinen. Dissertation, RWTH Aachen
Hedberg HD (1974) Relation of methane generation to undercompacted shales, shale diapirs and mud volcanoes. AAPG Bull 58: 661–673

Heinen V (1991) Der Einfluß der Lithofazies im Lias epsilon (Unteres Toarcium) auf die Erdölabgabe. Diplomarbeit, RWTH Aachen, 101 pp
Heller-Kallai L, Miloslavski I (1992) Reactions between clay volatiles and calcite reinvestigated. Clays Clay Miner 40: 522–530
Heller-Kallai L, Miloslavski I, Aizenshat Z (1987) Volatile products of clay mineral pyrolysis revealed by their effect on calcite. Clay Miner 22: 339–348
Heller-Kallai L, Miloslavski I, Aizenshat Z, Halicz L (1988) Chemical and mass spectrometric analysis of volatiles derived from clays. Am Mineralogist 73: 376–382
Heller-Kallai L, Miloslavski I, Aizenshat Z (1989) Reactions of clay volatiles with n-alkanes. Clays Clay Miner 37: 446–450
Hinch HH (1980) The nature of shales and the dynamics of hydrocarbon expulsion in the Gulf Coast Tertiary section. In: Roberts WH, Cordell R (eds) Problems of petroleum migratiom. AAPG Stud Geol 10: 1–18
Hirsch LM, Thompson AH (1995) Minimum saturations and buoyancy in secondary oil migration. AAPG Bull 79: 696–710
Hobson GD (1954) Some fundamentals of petroleum geology. Oxford University Press, London
Hobson GD, Tiratsoo EN (1975) Introduction to petroleum geology. Scientific, Beaconsfield 300 pp
Hoffmann CF, Mackenzie AS, Lewis CA, Maxwell JR, Oudin JL, Durand B, Vandenbroucke M (1984) A biological marker study of coals, shales and oils from the Mahakam Delta, Kalimantan, Indonesia. Chem Geol 42: 1–23
Horstad I, Larter SR, Dypvik H, Aagaard P, Börnvik AM, Johansen PE, Eriksen AS (1990) Degradation and maturity controls on oil field petroleum column heterogeneity in the Gulfraks field, Norwegian North Sea. Org Geochem 16: 497–510
Hower J, Eslinger EV, Hower ME, Perry EA (1976) Mechanism of burial metamorphism of argillaceous sediment. 1. Mineralogical and chemical evidence. Am Geol Soc Bull 87: 725–737
Hubbert MK (1953) Entrapment of petroleum under hydrodynamic conditions. AAPG Bull 37: 1954–2026
Huc A, Hunt JM (1980) Generation and migration of hydrocarbons in offshore south Texas gulf coast sediments. Geochim Cosmochim Acta 44: 1981–1989
Hunt JM (1979) Petroleum geochemistry and geology. Freeman, San Francisco 617 pp
Hunt JM (1990) Generation and migration of petroleum from abnormally pressured fluid compartments: reply. AAPG Bull 74: 1–12
Hunt JM (1991) Generation and migration of petroleum from abnormally pressured fluid compartments: reply. AAPG Bull 75: 336–338
Hunt JM, Whelan JK, Cathless LM (1994) Gas generation – a major cause of deep Gulf Coast overpressures. Oil Gas J 18: 59–63
Irwin H, Hurst A (1983) Application of geochemistry to sandstone reservoir studies. In: Brooks J (ed) Petroleum geochemistry and exploration of Europe. Geol Soc Lond, Spec Publ 12: 127–146
Isaacs CM (1984) The Monterey – key to offshore California boom. Oil Gas J 82: 75–81
Isaacs CM, Pisciotto KA, Garrison RE (1983) Facies and diagenesis of the Miocene Monterey formation, California: a summary. In: Iijimaa A, Hein JR, Siever R (eds) Siliceous deposits of the Pacific region. Elsevier, New York, pp 247–282
James KH (1990) The Venezuelan hydrocarbon habitat. In: Brooks J (ed) Classic petroleum provinces. Geological Society Spec Publ, vol 50. Geological Society, London, pp 9–35
Jochum J (1988) Untersuchungen zu Bildung und Karbonatmineralisation von Klüften im Posidonienschiefer (Lias epsilon) der Hilsmulde. Diplomarbeit, RWTH Aachen
Jochum J (1993) Karbonatumverteilung, Mobilisation von Elementen und Migration von Erdöl-Kohlenwasserstoffen im Posidonienschiefer (Hilsmulde, NW Deutschland) in Abhängigkeit von der Paläotemperaturbeanspruchung durch das Massiv von Vlotho. Ber Forschungszentrum Jülich No Jül-2750
Jochum J, Leythaeuser D, Littke R (1991) Oil-bearing fluid inclusions in calcite-filled horizontal fractures from mature Posidonia Shale (Hils Syncline, NW Germany). In: Manning DAC (ed) Organic geochemistry: advances and applications in energy and the natural

environment. Poster Abstr 15th Meet of the European Association of Organic Geochemists, Manchester University Press, Manchester, pp 160–162

Jochum J, Friedrich G, Leythaeuser D, Littke R, Ropertz B (1995) Hydrocarbon-bearing fluid inclusions in calcite-filled horizontal fractures from mature Posidonia Shale (Hils Syncline, NW Germany). Ore Geol Rev 9: 363–367

Jones RW (1984) Comparison of carbonate and shale source rocks. In: Palacas JG (ed) Petroleum geochemistry and source rock potential of carbonate rocks. Studies in Geology. AAPG Bull 18: 163–180

Karlsen DA, Nedkvitne T, Larter S, Bjørlykke K (1993) Hydrocarbon composition of authigenic inclusions: application of petroleum reservoir filling history. Geochim Cosmochim Acta 57: 3641–3659

Karlsen DG, Larter SR (1991) Analysis of petroleum fractions by TLC-FID: applications to petroleum reservoir description. Org Geochem 17: 603–617

Katsube TJ, Mudford BS, Best ME (1991) Petrophysical characteristics of shales from the Scotian shelf. Geophysics 56(10): 1681–1689

Khanin AA (1968) Evaluation of sealing abilities of clayey cap rocks in gas deposits. Geol Nefti i Gaza 8: 17–21(in Russian)

Kingston DR, Dishroon CP, Williams PA (1983) Hydrocarbon plays and global basin classification. AAPG Bull 67: 2194–2198

Krooss B (1988) Experimental investigation of the molecular migration of C1–C6 hydrocarbons: kinetics of hydrocarbon release from source rocks. Org Geochem 13: 513–524

Krooss B, Leythaeuser D (1988) Experimental measurements of the diffusion parameters of light hydrocarbons in water-saturated sedimenrary rocks. II. Results and geochemical significance. Org Geochem 12: 91–108

Krooss B, Leythaeuser D (1996) Molecular diffusion of light hydrocarbons in sedimentary rocks and its role in the migration and dissipation of natural gas. In: Schumacher D, Adams MA (eds) Near surface expression of hydrocarbon migration. AAPG Spec Vol (in press)

Krooss B, Littke R, Müller B, Frielingsdorf J, Schwochau K, Idiz RF (1995) Generation of nitrogen and methane from sedimentary organic matter: implications on the dynamics of natural gas accumulations. Chem Geol 126: 291–318

Krooss B, Brothers L, Engel MH (1991) Geochromatography in petroleum migration: a review. In: England WA, Fleet AJ (eds) Petroleum migration. Geological Society, London, Spec Publ 59: 149–163

Krooss B, Leythaeuser D, Schaefer RG (1992a) The quantification of diffusive hydrocarbon losses through cap rocks of natural gas reservoirs – a reevaluation. AAPG Bull 76: 403–406

Krooss B, Leythaeuser D, Schaefer RG (1992b) The quantification of diffusive hydrocarbon losses through cap rocks of natural gas reservoirs – a reevaluation. Reply. AAPG Bull 76: 1842–1846

Kulander BR, Dean SL, Ward BJ (1990) Fractured core analysis: interpretation, logging, and use of natural and induced fractures in core. Methods in expoloration No 8. AAPG, Tulsa

Kvenvolden KA, Claypool GE (1980) Origin of gasoline-range hydrocarbons and their migration by solution in carbon dioxide in Norton Basin, Alaska. AAPG Bull 64: 1078–1086

Lafarge E, Barker C (1988) Effect of water washing on crude oil compositions. AAPG Bull 72: 263–276

Langford FF, Blanc-Valleron M-M (1990) Interpreting Rock-Eval pyrolysis data using graphs of pyrolizable hydrocarbons vs. total organic carbon. AAPG Bull 74: 779–804

Larter S, Mills N (1991) Phase-controlled molecular fractionations in migrating petroleum charges. In: England WA, Fleet AJ (eds) Petroleum migration. Geological Society, London Spec Publ 59: 137–147

Lee Chung-I (1991) Generation and migration in the Miocene Monterey Formation, southern San Joaquin Basin, California. Ph. D.Thesis. Texas A&M University, Houston, 188 pp

Lehner FK (1991) Pore-pressure induced fracturing of petroleum source rocks. Implications for primary migration. In: Imarisio G, Frias M, Bemtgen JM (eds) The European oil and gas conference, a multidisciplinary approach in exploration and production R&D proceedings. Graham & Trotman, London, pp 142–154

Leverett M (1939) Flow of oil-water mixtures through unconsolidated sands. Trans Am Inst Mech Eng 132: 149–171

Leverett M (1941) Capillary behaviour in porous solids. Trans Am Inst Mech Eng 142: 152–169

Levorsen AI (1967) Geology of petroleum, 2nd edn. Freeman, San Francisco, 724 pp

Leythaeuser D, Poelchau HS (1991) Expulsion of petroleum from type III kerogen source rocks in gaseous solution: modeling of solubility fractionation. In: England W, Fleet AJ (eds) Petroleum migration. Geological Society, London, Spec Publ 59: 33–46

Leythaeuser D, Schaefer RG (1984) Effects of hydrocarbon expulsion from shale source rocks of high maturity in Upper Carboniferous strata of the Ruhr area, Federal Republic of Germany. In: Schenck PA, de Leeuw JW, Lijmbach GWM (eds) Advances in organic geochemistry 1983. Pergamon, Oxford. Org Geochem 6: 671–681

Leythaeuser D, Hagemann HW, Hollerbach A, Schaefer RG (1980) Hydrocarbon generation in source beds as a function of type and maturation of their organic matter: a mass balance approach. 10th World Petroleum Congr Proc 2: 31–41

Leythaeuser D, Schaefer RG, Yükler A (1982). Role of diffusion in primary migration of hydrocarbons. AAPG Bull 66: 408–429

Leythaeuser D, Mackenzie AS, Schaefer RG, Bjoroy M (1984a) A novel approach for recognition and quantification of hydrocarbon migration effects in shale-sandstone sequences. AAPG Bull 68: 196–219

Leythaeuser D, Radke M, Schaefer RG (1984b) Efficiency of petroleum expulsion from shale source rocks. Nature (Lond) 311: 745–748

Leythaeuser D, Schaefer RG, Radke M (1987) On the primary migration of petroleum. 12th World Petroleum Congr Proc 2: 227–236

Leythaeuser D, Schaefer RG, Radke M (1988a) Geochemical effects of primary migration of petroleum in Kimmeridge source rocks from Brae field area, North Sea. I. Gross composition of C_{15+}-soluble organic matter and molecular composition of C_{15+}-saturated hydrocarbons. Geochim Cosmochim Acta 52: 701–713

Leythaeuser D, Radke M, Willsch H (1988b) Geochemical effects of primary migration of petroleum in Kimmeridge source rocks from Brae field area, North Sea. II. Molecular composition of alkylated naphthalenes, phenanthrenes, benzo- and dibenzothiophenes. Geochim Cosmochim Acta 52: 2879–2891

Leythaeuser D, Littke R, Radke M, Schaefer RG (1988c) Geochemical effects of petroleum migration and expulsion from Torcian source rocks in the Hils syncline area, NW-Germany. In: Mattavelli L, Novelli L (eds) Advances in organic geochemistry 1987. Pergamon, Oxford. Org Geochem 13: 489–502

Leythaeuser D, Borromeo O, Mosca F, di Primio R, Radke M, Schaefer RG (1995) Pressure solution in carbonate source rocks and its control on petroleum generation and migration. Mar Petrol Geol 12: 717–733

Lindgreen H (1985) Diagenesis and primary migration in Upper Jurassic claystone source rocks in North Sea. AAPG Bull 69: 525–536

Lindgreen H (1987) Molecular sieving and primary migration in Upper Jurassic and Cambrian claystone source rocks. In: Brooks J, Glennie K (eds) Petroleum geology of north west Europe. Graham & Trotman, London, pp 357–364

Littke R, Rullkötter J (1987) Mikroskopische und makroskopische Unterschiede zwischen Profilen unreifen und reifen Posidonienschiefers aus der Hilsmulde. Facies 17: 171–180

Littke R, Baker DR, Leythaeuser D (1988) Microscopic and sedimentologic evidence for the generation and migration of hydrocarbons in Toarcian source rocks of different maturities. Org Geochem 13: 549–560

Littke R, Krooss B, Idiz E, Frielingsdorf J (1995) Molecular nitrogen in natural gas accumulations: generation from sedimentary organic matter at high temperatures. AAPG Bull 79(3): 410–430

Mackenzie AS, Quigley TM (1988) Principles of geochemical prospect appraisal. AAPG Bull 72: 399–415

Mackenzie AS, Leythaeuser D, Schaefer RG (1983) Expulsion of petroleum hydrocarbons from shale source rocks. Nature (Lond) 301: 506–509

Mackenzie AS, Price I, Leythaeuser D, Müller P, Radke M, Schaefer RG (1987) The expulsion of petroleum from Kimmeridge clay source-rocks in the area of the Brae Oilfield, UK

continental shelf. In: Brooks J, Glennie K (eds) Petroleum geology of north west Europe. Graham & Trotman, London, pp 865–877
Mackenzie AS, Leythaeuser D, Müller P, Quigley TM, Radke M (1988) The movement of hydrocarbons in shales. Nature (Lond) 331: 63–65
Magoon LB (1987) The petroleum system – a classification scheme for research, resource assessment and exploration. AAPG Bull 71: 587
Magoon LB (1988) The petroleum system – a classification scheme for research, exploration and resource assessment. In: Magoon LB (ed) Petroleum systems of the United States. US Geol Surv Bull 1870: 2–15
Magoon LB, Dow W (1994) The petroleum system – from source to trap. AAPG Mem 60: 655 pp
Mann U (1985) Early diagenesis of biogenic siliceous constituents in silty clays and claystones, Japan Trench. N Jahrb Miner Abh 153: 33–57
Mann U (1987) Veränderung von Mineralmatrix und Porosität eines Erdölmuttergesteins durch einen Intrusivkörper (Lias epsilon 2–3, Hilsmulde, NW-Deutschland). Facies 17: 181–188
Mann U (1989) Revealing hydrocarbon migration pathways. In: Poelchau HS, Mann U (eds) Geologic modeling – Aspects of integrated basin analysis and numerical simulation. Geol Rundsch Spec Vol 78/1: 337–348
Mann U (1990) Sedimentological and petrophysical aspects of primary petroleum migration pathways. In: Heling D, Rothe P, Förstner U, Stoffers P (eds) Sediments and environmental geochemistry. Springer, Berlin Heidelberg New York, pp 152–178
Mann U (1991) A multidisciplinary approach to primary migration pathways of petroleum – the Lower Toarcian of NW-Germany as a model source rock. In: Imarisio G, Frias M, Bemtgen JM (eds) The European oil and gas conference, a multidisciplinary approach in exploration and production R&D. Graham & Trotman, London, pp 142–154
Mann U (1993) Multiple source tracing – a potential new calibration tool for petroleum migration during basin modeling. KFA – industry cooperation, internal status report, Kassel, Jan 1993
Mann U (1994) An integrated approach to the study of primary petroleum migration. In: Parnell J (ed) Geofluids: origin, migration and evolution of fluids in sedimentary basins. Geological Society London, Spec Publ 78, pp 233–260
Mann U (1996) Zur Erdölmigration in karbonatischen Muttergesteinen (Petroleum migration and expulsion from carbonate source rocks). Zentralbl Geol Paläont Teil I 1994, H 11/12: 1311–1330
Mann D, Mackenzie AS (1990) Prediction of pore fluid pressures in sedimentary basins. Mar Petrol Geol 7: 55–65
Mann U, Jochum J, di Primio R, Ropertz B (1990) Expulsion pathways from petroleum source rocks. 3rd Conf European Association of Petroleum Geoscientists, Florence, preprint
Mann U, Düppenbecker S, Langen A, Ropertz B, Welte DH (1991) DH Pore network evolution of the Lower Toarcian Posidonia Shale during petroleum generation and expulsion – a multidisciplinary approach. Zentralbl Geol Paläontol Teil I 8: 1051–1071
Mann U, Klaus S, Neisel JD (1994a) Kryo-REM-Petrographie: Porenflüssigkeiten in Bausteinen und Erdölreservoirgesteinen. Jahrestagung Deutsche Geologische Gesellschaft, Heidelberg
Mann U, Neisel JD, Burchard W-G, Heinen V, Welte DH (1994b) Fluid-rock interfaces as revealed by cryo-scanning electron microscopy. First Break 12: 131–136
McAuliffe CD (1966) Solubility in water of paraffin, cycloparaffin, olefin, acetylene, cycloolefin and aromatic hydrocarbons. J Phys Chem 70: 1267–1275
McAuliffe CD (1979) Oil and gas migration – chemical and physical constraints. AAPG Bull 63: 761–781
McDougall SR, Sorbie KS (1992) Network simulations of flow processes in strongly wetting and mixed-wet porous media. 3rd Conference on the Mathematics of oil recovery, 17–19 June 1992, Delft University Press, Netherlands, pp 169–181
Meissner F (1978) Petroleum geology of the Bakken Formation, Williston Basin, North Dakota and Montana. In: Williston Basin Proceedings 1978, Montana Geological Society, Billings, pp 207–227

Miloslavski I, Heller-Kallai L, Aizenshat Z (1991) Reactions of clay condensates with n-alkanes: comparison between clay volatiles and clay condensates. Chem Geol 91: 287–296
Momper JA (1978) Oil migration limitations suggested by geological and geochemical considerations. In:. Roberts WH, Cordell R (eds) Physical and chemical constraints on petroleum migration. AAPG Geol Course Note 8 B1–B60, AAPG, Tulsa
Müller EP (1985) The genesis of crude oil in carbonate rocks: an example from the Upper Permian of the GDR. In: Al-Hashami WS (ed) 2nd Geological Congress on the Middle East. Arab Geol Assoc, Baghdad, pp 113–123
Murris RJ (1980) Middle East: stratigraphic evolution and oil habitat. AAPG Bull 64: 597–618
Okui A, Waples DW (1993) Relative permeabilities and hydrocarbon expulsion from source rocks. In: Doré AG et al. (eds) Basin modelling: advances and applications. Elsevier, Amsterdam, pp 293–302
Özkaya IA (1988) A simple analysis of oil-induced fracturing in sedimentary rocks. Mar Petrol Geol 5: 293–297
Palacas JG (1984) Carbonate rocks as sources of petroleum: geological and chemical characteristics and oil-source correlations. 11th World Petroleum Congr Proc 2: 31–43
Palacas JG, Anders DE, King JD (1984) South Florida Basin - prime example of carbonate source rocks of petroleum. In: Palacas JG (ed) Petroleum geochemistry and source rock potential of carbonate rocks. AAPG Stud Geol 18: 71–96
Palciauskas VV, Domenico PA (1980) Microfracture development in compacting sediments: relation to hydrocarbon-maturation kinetics. AAPG Bull 64: 927–937
Parnell J (1991) Timing of hydrocarbon-metal interactions during basin evolution. In: Pagel M, Leroy J (eds) Source, transport and deposition of metals. Balkema, Rotterdam, pp 573–576
Parnell J (1994) Hydrocarbons and other fluids: paragenesis, interactions and exploration potential inferred from petrographic studies. In: Parnell J (ed) Geofluids: origin, migration and evolution of fluids in sedimentary basins. The Geological Society, London, 78, pp 275–292
Poelchau HS, Mann U (1989) Geologic modeling - aspects of integrated basin analysis and numerical simulation. Geol Rundsch 78: 440 pp
Perrodon A (1980) Geodynamique petrolière: genese et repartition des gisments d'hydrocarbures. Masson-Elf-Aquitaine, Paris, 381 pp
Perry EA Jr, Hower J (1977) Late-stage dehydration in deeply buried pelitic sediments. AAPG Bull 56: 2013–2021
Peters KE (1986) Guidelines for evaluating petroleum source rock using programmed pyrolysis. AAPG Bull 70: 318–329
Peters KE, Moldowan JM (1993) The biomarker guide. Prentice Hall, Englewood Cliffs, 363 pp
Philp RP, Gilbert TD (1982) Biomarker distributions in oils predominantly derived from terrigenous source material. In: Leythaeuser D, Rullkötter J (eds) Advances in organic geochemistry 1985. Pergamon, Oxford, pp 73–84
Posamentier HW, Vail PR (1988) Eustatic controls on clastic deposition II-sequence and system tract models. In Wilgus CK et al. (eds) Sea-level changes: an integrated approach. SEPM Spec Publ 42: 109–124
Poulet M (1968) Problèmes posés par la migration secondaire du pétrole et sa mise en place dans les gisements. Rev Inst Fr Pét 23: 159–173
Powell TG (1984) Some aspects of the hydrocarbon geochemistry of a Middle Devonian barrier reef complex, western Canada. In: Palacas JG (ed) Petroleum geochemistry and source rock potential of carbonate rocks. AAPG Stud Geol 18: 45–62
Pratsch JC (1982) Focused gas migration and concentration of deep-gas accumulations. Erdö l Kohle Erdgas Petrochem 35: 59–65
Price LC (1976) The solubility of petroleum as applied to its origin and primary migration. AAPG Bull 60: 213–244
Price LC (1989) Primary petroleum migration from shales with oxygen-rich organic matter. J Petrol Geol 12: 289–324
Price LC, Clayton JL (1992) Extraction of whole versus ground source rocks: fundamental petroleum geochemical implications including oil-source rock correlation. Geochim Cosmochim Acta 56: 1213–1222

Radke M (1987) Organic geochemistry of aromatic hydrocarbons. In: Brooks J, Welte DH (eds) Advances in petroleum geochemistry, vol. 2. Academic, London, pp 141-207

Rasmussen B, Glover JE, Alexander R (1989) Hydrocarbon rims on monazite in Permian-Triassic arenites, northern Perth Basin, Western Australia: pointers to the former presence of oil. Geology 17: 115-118

Redwine L (1981) Hypothesis combining dilation, natural hydraulic fracturing, and dolomitization to explain petroleum reservoirs in Monterey Shale, Santa Maria area, California. In: Garrison RE, Pisciotto KE, Isaacs CM, Ingle JC (eds) The Monterey formation and related siliceous rocks of California. SEPM Spec Publ, pp 221-248

Reitsema RH (1983) Geochemistry of the north and south Brae areas, North Sea. In: Brooks J (ed) Petroleum geochemistry and exploration of Europe. Geological Society Spec Publ 12: 203-212

Ringrose PS, Corbett PWM (1994) Controls on two -phase fluid flow in heterogeneous sandstones. In: Parnell J (ed) Geofluids: origin, migration and evolution of fluids in sedimentary basins. Geological Society London, Spec Publ 78, pp 141-150

Roberts WH, Cordell RJ (1980) Problems of petroleum migration: introduction.In: Roberts WH, Cordell RJ (eds) AAPG Series in Geology 10. AAPG, Tulsa, pp 6-8

Robin M, Rosenberg E, Fassi-Fihri O (1995) Wettability studies at the pore level: a new approach by use of cryo-SEM. SPE Formation Evaluation, March, pp 11-19

Roedder E (1984) Fluid inclusions. Rev Mineral, Miner Soc Am 12: 644 pp

Ropertz B (1989) Petrophysikalische, organisch-chemische und organisch-petrographische Untersuchungen zur Verteilung des organischen Materials in Ton- und Mergelsteinen in zwei ausgewählten Katagenesestadien (Unteres Toarc, Hilsmulde, NW-Deutschland). RWTH Aachen

Ropertz B (1994) Wege der primären Erdölmigration: Eine Untersuchung über Porennetze, Klüfte und Kerogennetzwerke als Leiterbahnen für den Kohlenwasserstofftransport. Dissertation RWTH Aachen, Ber. Forschungszent. Jülich No Jül 2875: 299

Rullkötter J, Leythaeuser D, Horsfield B, Littke R, Mann U, Müller PJ, Radke M, Schaefer RG, Schenk HJ, Schwochau K, Witte EG, Welte DH (1988) Organic matter maturation under the influence of a deep intrusive heat source: a natural experiment for quantitation of hydrocarbon generation and expulsion from a petroleum source rock (Toarcian shale, northern Germany). In: Mattavelli L, Novelli L (eds) Advances in organic geochemistry 1987. Pergamon, Oxford. Org Geochem 13: 847-856

Sachsenhofer RF (1994) Petroleum generation and migration in the Styrian Basin (Pannonian Basin system, Austria): an integrated organic geochemical and numeric modelling study. Mar Petrol Geol 11: 684-701

Sajgo C, Maxwell JR, Mackenzie AS (1983) Evaluation of fractionation effects during the early stages of primary migration. Org Geochem 5: 65-73

Sandvik EI, Mercer JN (1990) Primary migration by bulk hydrocarbon flow. Org Geochem 16: 83-89

Sandvik EI, Young WA, Curry DJ (1992) Expulsion from hydrocarbon sources: the role of organic absorption. Org Geochem 19: 77-87

Sassen R (1988) Geochemical and carbon isotopic studies of crude oil destruction, bitumen precipitation, and sulfate reduction in the deep Smackover Formation. Org Geochem 12: 351-361

Sassen R (1990) Geochemistry of carbonate source rocks and crude oils in Jurassic salt basins of the Gulf Coast. In: Brooks J (ed) Classic petroleum provinces. Geological Society Spec Publ 50, Geological Society, London, pp 265-277

Sassen R, Moore CH (1988) Framework of hydrocarbon generation and destruction in Eastern Smackover Trend. AAPG Bull 72: 649-663

Sassen R, Moore CH, Meendsen FC (1987a) Distribution of hydrocarbon source potential in the Jurassic Smackover Formation. Org Geochem 11: 379-383

Sassen R, Moore CH, Nunn JA, Meendsen FC, Hexdari E (1987b) Geochemical studies of crude oil generation, migration, and destruction in the Mississippi Salt Basin. Gulf Coast Assoc Geol Soc Trans 37: 217-224

Schaefer RG (1992) Zur Geochemie niedrigmolekularer Kohlenwasserstoffe im Posidonienschiefer der Hilsmulde. Erdöl-Kohle-Erdgas Petrochem 45: 73-78

Schaefer RG, Littke R (1988) Maturity-related compositional changes in the low-molecular-weight hydrocarbon fraction of Toarcian shales. In: Mattavelli L, Novelli L (eds) Advances in organic geochemistry 1987. Pergamon, Oxford. Org Geochem 13: 887–892

Schidlowski M (1981) Uraniferous constituents of the Witwatersrand conglomerates: ore microscopic observations and implications for Witwatersrand metallogeny. US Geol Surv Prof Pap, 1161 pp

Schoell M (1984a) Stable isotopes in petroleum research. Adv Petrol Geochem 1: 215–245

Schoell M (1984b) Recent advances in petroleum isotope geochemistry. Org Geochem 6: 645–663

Schowalter TT (1979) Mechanics of secondary hydrocarbon migration and entrapment. AAPG Bull 63: 723–760

Schwarzacher W (1993) Cyclostratigraphy and the Milankovitch theory. Developments in sedimentology, vol 52. Elsevier, Amsterdam, 225 pp

Schwarzkopf TA (1990) Relationship between petroleum generation, migration and sandstone diagenesis, Middle Jurassic, Gifhorn Trough, N. Germany. Mar Petrol Geol 7: 153–168

Seifert WK, Moldowan JM (1981) Paleoreconstruction by biological markers. Geochim Cosmochim Acta 45: 783–794

Seifert WK, Moldowan JM (1986) Use of biological markers in the sedimentary record. In: Johns RB (ed) Methods in geochemistry and geophysics 24. Elsevier, Amsterdam, pp 261–290

Selle OM, Jensen JI, Sylta Ø, Anderson T, Nyland B, Broks TM (1993) Experimental verification of low dip, low rate two-phase (secondary) migration by means of γ-ray absorption (extended abstract). Geofluids '93 Conf, 4–7 May 1993, Torquay, UK, pp 72–75

Setzer MJ (1977) Einfluß des Wassergehalts auf die Eigenschaften des erhärteten Betons. In: Deutscher Ausschuß für Stahlbeton. Ernst, Berlin

Shi Jixi Li Benchao, Fu Jiamo, Liu Dehan, Peng Pingan (1988) Organic inclusions and their relation to oil and gas orrurrences. In: Mattavelli L, Novelli L (eds) Advances in organic geochemistry 1987. Part II. Analytical geochemistry. Pergamon Press, Oxford, pp 1101–1108

Sibson RH (1994) Crustal stress, faulting and fluid flow. In: Parnell J (ed) Geofluids: origin, migration and evolution of fluids in sedimentary basins. Geological Society London, Spec Publ 78, pp 69–84

Sibson RH, Moore JM, Rankin AH (1975) Seismic pumping: a hydrothermal fluid transport mechanism. J Geol Soc Lond 131: 653–659

Silvermann SR (1967) Carbon isotope evidence for the role of lipids in petroleum formation. J Am Oil Chem Soc 44: 691–695

Silvermann SR (1971) Influence of petroleum origin and transformation on its distribution and redistribution in sedimentary rocks. 8th World Petroleum Congr Proc 2: 47–54

Simoneit BRT (1994) Organic matter alteration and fluid migration in hydrothermal systems. In: Parnell J (ed) Geofluids: origin, migration and evolution of fluids in sedimentary basins. Geological Society London, Spec Publ 78, pp 261–274

Singh KSW, Everett DH, Haul RAW, Moscou L, Pierotti RA, Rouquerol J, Siemieniewsky T (1985) Pure Appl Chem 57: 603–605

Sloss LL (1950) Paleozoic stratigraphy in the Montana area. AAPG Bull 34: 423–451

Sluijk D, Nederlof MH (1984) Worldwide geological experience as a systematic basis for prospect appraisal. AAPG Mem 35: 15–26

Snarsky AN (1962) Die primäre Migration des Erdö ls. Freiberger Forschungshefte C123: 63–73

Soedder DJ (1990) Application of fluorescence microscopy to study of pores in tight rocks. AAPG Bull 74: 30–40

Sokolov VA (1964) Essays on the genesis of petroleum. Gostoptekhizdat 447 (1948)

Sokolov VA, Zhause TP, Vassoyevich NB, Antonoc PO, Grigoriyev GG, Kozlov VP (1964) Migration processes of gas and oil, their intensity and directionality. Proc of the 6th World Petroleum Congress, Frankfurt 1963, pp 493–505

Stahl W (1977) Carbon and nitrogen isotopes in hydrocarbon research and exploration. Chem Geol 20: 121–149

Stainforth JG, Reinders JEA (1990) Primary migration of hydrocarbons by diffusion through organic matter networks, and its effect on oil and gas generation. Org Geochem 16: 61–74
Stonely R (1981) Fibrous calcite veins, overpressures and primary migration. AAPG Bull 67: 1427–1428
Talukdar S, Gallango O, Vallejos C, Ruggiero A (1987) Observations on the primary migration of oil in the La Luna source rocks of the Maracaibo Basin, Venezuela: In: Doligez B (ed) Migration of hydrocarbons in sedimentary basins. Technip, Paris, pp 59–78
Teichmüller M, Teichmüller R, Bartenstein H (1984) Inkohlung und Erdgas - eine neue Inkohlungskarte der Karbonoberfläche in Nordwestdeutschland. Fortschr Geol Rheinl Westf 32: 11–34
Thomas MM, Clouse JA (1990a) Primary migration by diffusion through kerogen. I. Model experiments with organic-coated rocks. Geochim Cosmochim Acta 54: 2775–2779
Thomas MM, Clouse JA (1990b) Primary migration by diffusion through kerogen. II. Hydrocarbon diffusivities through kerogen. Geochim Cosmochim Acta 54: 2781–2792
Thomas MM, Clouse JA (1990c) Primary migration by diffusion through kerogen. III. Calculation of geologic fluxes. Geochim Cosmochim Acta 54: 2793–2797
Thomas MM, Clouse JA (1995) Scaled physical model of secondary oil migration. AAPG Bull 79: 19–29
Tissot BP, Pelet R (1971) Nouvelles données sur les mèchanismes de genèse et de migration du pétrole, simulation mathématique et application á la prospection. 8th World Petroleum Congress Proc 4. Wiley, Chichester, pp 35–46
Tissot BP, Welte DH (1984) Petroleum formation and occurrence, 2nd edn. Springer, Berlin Heidelberg New York, 699 pp
Ulmishek G (1986) Stratigraphic aspects of petroleum resource assessment: In: Rice DD (ed) Oil and gas assessment - methods and applications. AAPG Stud Geol 21: 59–68
Ungerer P (1990) State of the art of research in kinetic modelling of oil formation and expulsion. Org Geochem 16: 1–26
Ungerer P, Behar E, Discamps D (1983) Tentative calculation of the overall volume expansion of organic matter during hydrocarbon genesis from geochemistry data. Implications for primary migration. In: Bjorøy M et al. (eds) Advances in organic geochemistry 1981. Wiley, Chichester, pp 129–135
Vail PR (1988) Sequence stratigraphic workbook. Fundamentals of sequence stratigraphy. AAPG Annu Conv Short Course Notes, Tulsa
Vandenbroucke M (1972) Étude de la migration primaire: variation de composition des extraits de roche à un passage roche-mere/réservoir. In: von Gaertner HR, Wehner H (eds) Advances in organic geochemistry 1971. Pergamon, Oxford, pp 547–565
Vandenbrouke M, Durand B, Oudin JL (1983) Detecting migration phenomena in a geological series by means of C_1–C_{35} hydrocarbon amounts and distributions. In: Bjorøy M et al. (eds) Advances in organic geochemistry 1981. Wiley, Chichester, pp 147–155
Van Wagoner JC, Mitchum RM, Campion KM, Rahmanian VD (1990) Siliciclastic sequence stratigraphy in well logs, cores, and outcrops. AAPG Methods in Exploration Series 7, Tulsa, p 55
Verweij JM (1993) Hydrocarbon migration system analysis. Developments in petroleum science, vol 35. Elsevier, Amsterdam, 275 pp
Waples D, Suizo M, Kamata H (1992) The art of maturity modeling. Part 1. Finding a satisfactory geologic model. AAPG Bull 76: 31–46
Wardlaw NC (1979) Pore systems in carbonate rocks and their influences on hydrocarbon recovery efficiency. AAPG Continuing Education Course Note Series No 11. AAPG, Tulsa, E1–E24
Watts H (1963) The possible role of absorption and diffusion in the accumulation of crude oil deposits; a hypothesis. Geochim Cosmochim Acta 27: 925–928
Welte DH (1987) Migration of hydrocarbons. Facts and theory. In: Doligez B (ed) Migration of hydrocarbons in sedimentary basins. Technip, Paris, pp 393–413
Welte DH, Yükler A (1980) Evolution of sedimentary basins from the standpoint of petroleum origin and accumulation - an approach for a quantitative basin study. Org Geochem 2: 1–8
Welte DH, Yükler A (1981) Petroleum origin and accumulation in basin evolution - a quantitative model. AAPG Bull 65: 1387–1396

Wilhelms A, Larter S (1994a) Origin of tar mats in petroleum reservoirs. I. introduction and case studies. Mar Petrol Geol 11: 418–441

Wilhelms A, Larter S. (1994b) Origin of tar mats in petroleum reservoirs. II. formation mechanisms for tar mats. Mar Petrol Geol 11: 442–456

Willingham TO, Nagy B, Nagy LA, Krinsley DH, Mossman DJ (1985) Uranium-bearing stratiform organic matter in paleoplacers of the lower Huronian Supergroup, Elliot Lake-Blind River region, Canada. Can J Earth Sci 22: 1930–1944

Witherspoon PA, Saraf DN (1965) Diffusion of methane, ethane, propane and n-butane in water from 25 to 43° C. J Phys Chem 69: 3752–3755

Wygrala BP (1989) Integrated study of an oil field in the southern Po basin, northern Italy. Dissertation, Köln University, Jül-Rep 2313, Research Centre Jülich, Jülich, ISSN 0366-0885

Yükler A, Dow WG (1990) Temperature, pressure and hydrocarbon generation histories in San Marcos arch area, de Witt County, Texas. In: Schumacher D, Perkins BF (eds) Gulf Coast oils and gases: their characteristics, origin, distribution, and exploration and production significance. Proc 9th Annu Res Conf SEPM, pp 99–104

Zumberge JE (1984) Source rocks of the La Luna Formation (Upper Cretaceous) in the Middle Magdalena Valley, Columbia. In: Palacas JG (ed) Petroleum geochemistry and source rock potential of carbonate rocks. AAPG Stud Geol 18: 127–134

Outlook

The subject of petroleum and basin evolution continues to be a scientific challenge in the geosciences. It extends beyond aspects of hydrocarbon exploration and production and now belongs to the mainstream of modern geoscientific research - which can loosely be described as the understanding and quantification of geological processes.

The understanding of petroleum must be approached on the molecular level. Research in petroleum migration must include investigation of the microscopic world of the pores of rocks and conduits offered by permeable rock strata or fault systems. All of this makes up the framework of sedimentary basins, which are, geologically speaking, continuously changing through their course of evolution. In comprehending petroleum and basin evolution we therefore need to combine the very small scale of the molecular level with the large regional scale of sedimentary basins.

The processes of petroleum generation, migration, and accumulation and of basin evolution are highly complex. In approaching this web of interdependent factors, we need to simplify the complex without distorting and falsifying the main chains of events and their courses. Basin modeling is the only methodology that offers a better chance for understanding and quantifying geological processes. However, a thoughtful reflection upon the material presented in this volume shows that we still have quite a way to go before reaching this goal.

The holistic approach as suggested in this book confronts us with a problem of scale, with respect not only to resolution in space but also to that in time. In basin evolution the aspects of sediment deposition, subsidence, tectonics, and temperature regimes are more of a regional scale. By contrast, processes associated with chemical reactions, either in mineral or rock diagenesis or in petroleum generation, and processes connected to transport phenomena of fluids are determined more by small-scale factors. Hence in basin modeling we have a dilemma with respect to gridding. Regional aspects would tolerate a wider mesh of grid points, whereas small-scale local phenomena need to be addressed with a very dense grid system.

In considering the question of dense gridding versus wider grid point distances we are confronted with the problem of the availability and equal distribution of data points for ensuring a fairly homogeneous data matrix. Clearly the optimal data source - boreholes and the drill cores taken from them - is rare and has a highly irregular distribution on a basin-wide scale. The best compromise for basin modeling as a data source is seismic, combined with well data. However, what kind of data do we obtain from seismic interpretation? Although seismic interpretation has made enormous progress over the past decade, the important insights to satisfy basin modeling are still hard to come by, especially with respect to lithological interpretation and porosity – permeability information. Subdividing seismic sections appropriately into meaningful stratigraphic entities to define events in the conceptual model may also cause problems. Nevertheless, it should be recognized that the establishment of a direct link between seismics and basin modeling constitutes a great advance in improving the data matrix.

Due to the nature of seismic waves the best possible resolution throughout the stratigraphic rock sequence is in the order of several tens of meters. Resolution is better at shallower depth levels than at deeper levels. It is usually difficult to interpret basin floors, or depth zones where potential source rocks tend to have reached an overmature stage at levels of 4000–8000 m. Typical source layers or carrier rocks at greater depth can no longer be identified. Thin source rocks, important for petroleum generation, and carrier rocks and fault systems, important for migration in the range of 2–3 m, often escape seismic interpretation.

Classical seismic interpretation focused on the recognition of structural features and, more recently, on lithological characterization. Although seismic- and sequence stratigraphy introduced the time element and hence aspects of the dynamics of basin evolution, this dynamic approach must be expanded and improved. An interactive link between seismic interpretation, synthetic seismic, and basin modeling through the use of an interactive computer environment is probably the best method for reaching this goal (Kevin Donihoo, IES, personal communication). One way out of the dilemma of poor resolution in space, due to the blindness of seismic waves over short distances, is a combination with other geophysical tools and data sources which can provide information in the range of meters. Such opportunities arise when the first wells have been drilled in a basin, and when the stage of field development has been reached. Then the integration of well-logging results is of great help. Again, an interactive computer environment linking log and seismic interpretation and basin modeling is indispensable for such a task. In the end, our goal must be to expand and homogenize the geological data base, if possible on a regional, basin-wide scale.

Our problems with resolution in time become obvious in attempting to extrapolate kinetic findings derived from laboratory experiments to chemical reactions occurring at a geological time scale. Even with the most powerful computers it is meaningless to reduce ongoing chemical reactions, which have taken place over millions of years, into time steps of hours or days – not to speak of the complexity of the thousands of reactions leading to thousands of compounds in petroleum generation.

Experience so far has shown that an acceptable compromise for time resolution are time steps of about 0.1–1 million years when performing usual heat flow simulations and 1000–50000 years for fluid flow problems. For some special situations, such as for example sudden heat pulses and magmatic systems, finer time loops are used during the simulation. Average basin modeling runs have shown that computer time is not overstreched when applying such time loops.

Nevertheless, with the ambitious goals of expanding and homogenizing the data base and refining our understanding of the dynamics of processes, the trend is toward more grid points, shorter time steps, and increased numbers of variables. Therefore, ensuring acceptable computing times will require ways to limit the computations involved. This may mean a search for better modular architecture of the software, the introduction of parallel computing, and/or a better choice of algorithms.

In reviewing briefly the subjects treated in this book we would like to make the following comments.

Constructing the conceptual basin model is the foundation for the entire integrated approach, culminating in the numerical simulation of a given basin. This conceptual model cannot be better than the available input data. At this stage the importance of integrating seismic data via a direct link between basin modeling and seismic interpretation cannot be overemphasized. Furthermore, testing and adjusting the conceptual model requires not only supporting data for calibration but also a very broad geological familiarity with the region under scrutiny as well as sound scientific judgment.

The most important single parameter as a driving force for chemical reactions is temperature. However, because temperature history is determined by many factors, the simulation of temperature is correspondingly complex. Some of the variables are mutually dependent. This means that temperature "pathways" that are elaborated with too little calibration are often insufficiently constrained. In such cases alternative pathways may lead to the same "present-day" results. In other words, a whole sequence of calibration points along the time axis and, if possible, with independent calibration parameters (e.g., organic parameters and fluid inclusions) is necessary to constrain the model. In this way an acceptable control of a temperature history can be established, and reliable data for intensity and timing of petroleum generation etc. can be expected.

The reconstruction of temperature histories has formed the basis for a meaningful application of kinetics. Up to now, however, only the generation aspect has been quantified by the use of kinetic models. These models correctly reflect the principal structural features of various kerogens and allow extrapolation from experimental to natural heating conditions. In view of the magnitude of this extrapolation and in view of the increasing use of laboratory-derived kinetic parameters within the framework of basin modeling programs it will become indispensable to develop methods that allow researchers to assess and to enhance the reliability of kinetic timing predictions in each individual case. This requires systematic investigations of maturation sequences of various types of source rocks regarding not only hydrocarbon generating reactions but also the accompanying and competing structural reorganization reactions within the residual solid organic matter; it is hardly conceivable that the former could take place independently of the latter.

Considerable progress has been made in predicting petroleum products since experimental pyrolysis was introduced, and means for recognizing source rock facies were established. However, in terms of applicability in basin modeling, the relationship between basin fill models and "classical" basin modeling could considerably improve product predictability.

The present understanding of petroleum migration and accumulation certainly extends beyond mere qualitative knowledge and has reached a stage where we see initial successes in the attempt to quantify migration processes. However, we still need to improve our knowledge about the efficiency of migration or, on the other hand, about losses during migration. Likewise, many aspects of timing and duration of migration must be studied in greater detail in the future.

A very important subject for improvement is the recognition and definition of migration pathways. This of course is related to the problem of providing a better geological data matrix. This has been discussed above; however, in addition to that, we definitely need to address at least two other problem areas. One is restoring and analyzing the kinematics of tectonically deformed rock sections, and the other is a more thorough investigation of the role of faults as migration avenues. Concerning the former, research is under way to investigate and hopefully to combine in the near future (two- and three-dimensional) palinspastic reconstruction models with basin modeling.

A task for the future is to close the gap between basin modeling and reservoir simulation. This is desirable for several reasons: firstly, for the more thorough assessment of petroleum drainage areas and the better delineation of targets during the stage of field development drilling, and, secondly, because seismic and seismic interpretation cannot really identify small geological features (in the range of 2–3 m).

Basin modeling works with big time steps (millions of years), changing geometries, and continuously changing temperature and pressure conditions in a changing rock matrix. Reservoir simulation works with short times (days, months, years), fixed geometries, and a more or less rigid rock matrix. Hence a link between the two modeling techniques is not easy to establish. However, as a first step in linking the two modeling procedures "reservoir characterization" could be established as a bridging technique for the two. Then, as an initial step in bridging, geometries and simulation data can be exchanged and used in both modeling methodologies for mutual benefit.

From the standpoint of basin modeling the individual disciplines and fields of expertise on basin evolution, source rocks and organic matter, generation of hydrocarbons, sedimentology and petrophysics, seismic data acquisition and seismic interpretation, production geology, and reservoir engineering are quite well developed as individual activities. However, considering the need for more synergy in the future, we certainly need to focus attention on the interfaces between these disciplines and fields of expertise. Only then can we make further progress in basin modeling.

This integration of disciplines demands the existence of research teams, or centers of excellence, with a range of scientific skills including geology, geophysics, geochemistry, physics, chemistry, and mathematics. This means that for the future of the geosciences, as we understand it, research teams consisting of a critical minimum number of scientists and disciplines are needed. Such efforts should lead to a new geoscience research culture.

August 1996 D.H. Welte

Subject Index